1997

THE EXCITEMENT AND
FASCINATION OF SCIENCE:
Reflections by Eminent Scientists

Volume 3, Part 2

Dedicated
with admiration and affection to
J. MURRAY LUCK
founder of Annual Reviews Inc.

THE EXCITEMENT AND FASCINATION OF SCIENCE:
Reflections by Eminent Scientists

VOLUME 3, PART 2 (1990)

Compiled by JOSHUA LEDERBERG
The Rockefeller University

Reprinted from the following Annual Reviews:

Part 1

Anthropology
Astronomy and Astrophysics
Biochemistry
Biophysics and Biophysical Chemistry
Earth and Planetary Sciences
Fluid Mechanics
Genetics
Immunology
Microbiology

Part 2

Neuroscience
Nuclear and Particle Science
Nutrition
Pharmacology and Toxicology
Physical Chemistry
Physiology
Phytopathology
Plant Physiology and Plant Molecular Biology
Sociology

ANNUAL REVIEWS INC.   4139 EL CAMINO WAY   P.O. BOX 10139   PALO ALTO, CALIFORNIA 94303-0897

ANNUAL REVIEWS INC.
Palo Alto, California, USA

International Standard Book Number: 0–8243–2603-2
Library of Congress Catalog Card Number: 65-29005

Library of Congress Cataloging-in-Publication Data
(Revised for volume 3)

The Excitement and fascination of science.

    Vol. 2, with subtitle Reflections by eminent scientists, compiled by W. C. Gibson.
    "Reprinted from the Annual review of biochemistry, the Annual review of pharmacology, the Annual review of physical chemistry, the Annual review of physiology," etc.
    Vol. 3 has subtitle: Reflections by eminent scientists, compiled by Joshua Lederberg.
    "Reprinted from the Annual review of anthropology," etc.—Vol. 3, t.p.
    Includes bibliographies.
    1. Science. 2. Scientists—Biography. I. Bishop, George Holman, 1889–. II. Gibson, William Carleton. III. Lederberg, Joshua. IV. Annual Reviews, inc.
    Q171.E98   500   65-29005
    ISBN 0-8243-2601-6 (v. 2)

PRINTED AND BOUND IN THE UNITED STATES OF AMERICA

500
2473
V. 3 Part 2

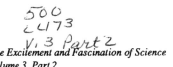
The Excitement and Fascination of Science
Volume 3, Part 2

CONTENTS

(The contents of Part 1 are listed on pages ix-xi)

Note: These chapters have been repaged for this reprint collection. The numbers appearing in the source line on the opening page of each chapter are as they appeared in the original publication.

158,030

CONTENTS OF PART 1

For the convenience of readers, a detachable order form/envelope is bound into the back of this volume.

NEUROSCIENCE

Ann. Rev. Neurosci. 1979. 2:1–15

THE METAMORPHOSIS
OF A PSYCHOBIOLOGIST

Seymour S. Kety

Harvard Medical School, Laboratories for Psychiatric Research,
Mailman Research Center, McLean Hospital, Belmont, Massachusetts 02178

My introduction to the challenge of the brain and human behavior came about, some fifty years ago, at the hands of Edwin Landis, a remarkable teacher of science at the Central High School in Philadelphia. Cognizant of the interest of several of us, he had organized a science and philosophy club for which we found an appropriate dedication from Virgil: "Felix, qui potuit rerum cognoscere causas." Through him I learned that to understand the causes of things was an extremely remote goal, particularly in the case of the brain, but that there was excitement enough in the process of trying to understand and in the individual small steps along the way. I also learned that there were many approaches to the understanding of a complex phenomenon and that the ultimate understanding was likely to involve most of them. Although fortune or reasoned choice might set us on one of these paths by virtue of idiosyncratic aptitudes, training and experience, that did not make it the best or the only path. The unparalleled complexity of the brain and of behavior, both individual and collective, calls for great humility. Yet, in its stead, we have often seen complacency and conviction, or the advocacy of one approach and the disparagement of others.

Neuroscience, whose very name represents a juxtaposition of disparate roots, is a discipline that grew out of the manifold approaches to the study of the nervous system. It may be unnecessary to remind ourselves that for the most part neuroscience deals with one important aspect of behavior—the machinery that mediates it. After the networks of the brain have been delineated to their uttermost detail, as well as the mechanisms for processing, storing, and acting upon information, and the physical and chemical mechanisms that modulate these processes, we will still need to define and

1301

0147-006X/79/0315-0001$01.00

understand the informational content and its significance in terms of human experience. It is unlikely that either aspect can be reduced to the other, and neither alone will constitute a sufficient explanation of the diversity and versatility of human behavior.

In particular types or syndromes of behavior, however, it is possible that these two spheres of influence may have different weights, account for larger or lesser fractions of the variance, or be susceptible to more proximate or remote elucidation. Thus, there are few neurobiologists, indeed, who would feel that their techniques and competence would usefully be directed at uncovering differences between the brain of a Democrat and that of a Republican (Kety 1961), yet that proposition is no more absurd than the opposite presumption that the biological disciplines have little to contribute to an understanding of the major mental disorders, on the unproven assumption that these are essentially social and psychological problems.

Modern neurochemistry had its origins in the less restrictive and more constructive hypothesis that many forms of insanity were biochemical in nature. That was the basis on which, nearly one hundred years ago, Johann Thudichum received a grant from the Privy Council of England. It is particularly significant that, armed with that hypothesis, Thudichum did not then go to a mental hospital for the urine or blood of patients; instead, he went to the slaughterhouses for cattle brain and spent the next ten years isolating and characterizing the major components. I have recounted that glimpse of scientific history many times because it illustrates in a compelling way, the motivation of most scientists to contribute to the understanding and solution of a human problem, the value of enlightened public support that leaves the strategy of the effort to the scientist himself, and the wisdom of laying firm foundations of basic knowledge on which others may build, rather than directly but prematurely attacking the problem with inadequate tools, concepts, and information.

Thudichum did not discover the biochemical disturbances that occur in psychosis, but he developed one of the foundations on which modern neurochemistry was built (Thudichum 1884). It may turn out that in the balance and relationships among the minute quantities of dopamine, noradrenalin, serotonin, acetylcholine, GABA, or in neurotransmitters yet to be discovered, or the growing list of polypeptides that occur in the brain, lie some of the crucial disturbances of insanity. What would have been the chance that Thudichum would have discovered this at the time when those substances were still undreamed of? He would have frittered away the public funds and ten years of his career in futile research, which a more naive judgment would have called "relevant."

Forty years after Thudichum's death, renewed research on the chemical structure of the brain was undertaken by Jordi Folch-Pi, among whose major contributions were the characterization of three distinct phospholipids from Thudichum's "cephalin" and the discovery of the proteolipids, a new class of chemical compounds and the major components of myelin.

After the discovery that cells and tissues respire in vitro, brain slices were found to be the seat of active respiration. It was Ralph Gerard, the late Honorary President of the Society for Neuroscience, who coupled metabolism with neural function. As a post-doctoral fellow with A. V. Hill he measured the heat of axonal conduction, and with Otto Meyerhoff, the importance of oxidative metabolism in that process. Seven years later, Gerard and Hartline were successful in measuring the micro-quantities of oxygen utilized by optic nerve fibers during physiological conduction. At the same time physiologists were laying another foundation of neurochemistry. Dale, Loewi, Cannon, and, in time, many other scientists were elucidating the chemical processes involved in transmission at peripheral synapses.

Oxygen utilization and energy metabolism of the brain rapidly became the main focus of neurochemistry. Studies on the respiration and intermediary metabolism of brain slices and homogenates by Quastel, Elliott, and many others produced a wealth of information on the rates of oxygen and substrate utilization by different regions of the brain and in various species. The preference of brain for glucose as substrate was demonstrated as was the ability of neural tissue to engage in anerobic glycolysis. McIlwain produced a model of neuronal activation by electrical stimulation or with potassium ion, and Larrabee, in the isolated sympathetic ganglion, examined metabolism in relation to transsynaptic neuronal activation. An important conclusion that followed from his observations was the primary action of anesthetic agents on synaptic transmission rather than neuronal metabolism, the latter falling as a result of decreased bombardment. This was one of several early hints of the importance of synaptic transmission as a crucial site of action for a variety of chemical substances that act on the brain.

Valuable as in vitro techniques were in their ability to control most of the variables in the neuronal environment, they provided only elegant but highly simplified models of what the living brain was like and there was a need to study it directly.

STUDIES ON THE CIRCULATION AND ENERGY METABOLISM OF THE BRAIN

Nature has protected the brain, however, not only against gravitational and traumatic stresses, but also against the prying eyes and hands of scientists. The obvious way to measure the amounts of oxygen or other substances used or produced by an organ is by assaying their concentration difference

between blood entering and leaving the organ. The amount of blood entering or leaving per minute, multiplied by the arteriovenous difference of the
substance yields the amount utilized per minute. It was not, however, easy
to measure the cerebral blood flow in animals under reasonably physiological conditions. In most species the brain receives its arterial blood not only
from the internal carotid and vertebral arteries, but also from the external
carotid by way of a network of anastomoses—the *rete mirabile* of Galen.
It requires rather drastic surgery to restrict the arterial inflow to one or two
channels that can be monitored. In primates, however, the cerebral arterial
supply is much more discrete. Recognizing this, Dumke & Schmidt (1943)
made the first reliable quantitative measurements of blood flow in the
mammalian brain, using the rhesus monkey and inserting an ingenious
bubble flow meter into the arterial supply. I was in Boston at the time,
working on traumatic shock with Joseph Aub and his associates at the
Massachusetts General Hospital. Alfred Pope and I, the two young postdoctoral fellows in the laboratory, had just written a review on the circulatory system in shock and had become impressed, as others had before us,
with the wealth of homeostatic mechanisms that had been evolved to pre-
serve the circulation of the brain.

I was so impressed with the paper by Dumke and Schmidt that I decided
to return to the University of Pennsylvania to work with Carl Schmidt on
the cerebral circulation. Shortly after I arrived, he invited me and Harry
Pennes, another newcomer to the Department of Pharmacology, to assist
in the studies he was about to undertake, of the oxygen consumption in the
brain of the monkey. Besides obtaining values for that function in the
mammalian brain under light anesthesia, that study also demonstrated a
striking increase in metabolism, exceeding the increase in cerebral blood
flow, during convulsions induced by metrazol or picrotoxin.

But just as the brain is unique among organs for its complexity, so is the
human brain unique in its capacity, its versatility and plasticity, its ability
to conceptualize and create, to experience ecstasy and deep grief, and to
describe to outside observers the results of its inner processes. It is also the
human brain that falls prey to serious disorders of these functions, for which
no comparable animal models exist. To study the metabolism of the human
brain while it was engaged in these functions and experiences might teach
us something about these processes, and its study in disease might be of
benefit to those suffering from neurological or mental disorders.

While in Boston, I had heard an inspiring lecture by Andre Cournand,
describing his early work on the output of the human heart in health and
disease, and was impressed with the possibility that clinical studies could
be more physiological, relevant, and fundamental, under certain circumstances, than studies in animals. Cournand had used the Fick principle,

calculating pulmonary blood flow, equivalent to cardiac output, from the oxygen taken up by the lung and the oxygen content in blood entering and leaving, all of which he could measure independently.

Much earlier, in 1927, the Boston psychiatrist Meyerson had described a simple technique for obtaining cerebral venous blood in man from the superior bulb of the internal jugular, making it possible to measure the arteriovenous difference across the brain for substances utilized or produced in significant amounts by that organ. Lennox and Gibbs then used that approach to infer cerebral circulation from the arteriovenous oxygen difference, assuming that cerebral oxygen consumption, which they could not measure, was unchanged. About the same time, Himwich used the arteriovenous oxygen difference to infer metabolic rate for the brain with the assumption that blood flow was constant. Interestingly enough, many of the findings of both of these groups were later found to be valid by virtue of a fortunate choice of problem to which each approach was applied. Lennox and Gibbs were able correctly to infer that carbon dioxide increased cerebral blood flow because metabolism is not, in fact, appreciably altered at the concentrations of carbon dioxide they used, and both they and Himwich discovered that the respiratory quotient for the human brain was close to unity since that calculation does not require a measurement of cerebral blood flow. However, they sometimes made inferences that were incorrect, i.e. that cerebral blood flow is markedly increased in coma (because of the narrowed arteriovenous difference, which is the result of diminished metabolism). What had limited the acceptance of that approach was that the arteriovenous oxygen difference, being a function of both blood flow and oxygen consumption, was not a valid measure of either. The oxygen consumption of the brain could not be measured independently and certainly it could not be assumed to be constant under a wide variety of functional states, since it would be expected to vary with the state of activity or disease that was the object of investigation.

By 1943 the Fick principle had been applied by Homer Smith to the kidney, and two years later by Stanley Bradley to the measurement of hepatic blood flow; in each case they took advantage of the ability of these organs specifically to excrete a foreign substance at a rate that could be independently measured. Unlike the kidney and liver, however, the brain was not known to remove selectively and specifically a foreign substance from the blood and excrete it for accurate measurement.

The brain does, however, absorb by physical solution an inert gas, which reaches it by way of the arterial blood. The accumulation of such a gas in the brain should be independent of the state of mental activity and determined instead by relatively simple physical principles such as diffusion and solubility, which I felt should be quite constant in the brain whether asleep or awake, working out a complex mathematical problem, or suffering from schizophrenia. In the brain of a monkey I found that an arteriovenous difference did exist for nitrous oxide as it was breathed, and that it narrowed as the brain approached equilibrium with the arterial partial pressure. The variable difference could be integrated to give the denominator of a Fick equation. The next problem was how to determine the numerator, representing the amount or concentration of the inert gas accumulated in the brain. One could, of course, do this by using a radioactive inert gas and external counting, but a simpler approach presented itself. The arteriovenous difference became progressively narrow with time because the brain was coming to equilibrium. If that was the case, there would be a time at which the venous blood draining the brain would be in equilibrium with the brain itself and could be used as a measure of the partial pressure of gas in the brain, which, multiplied by the ratio of solubilities between blood and brain, would yield the concentration in the brain.

In a series of determinations on the monkey, carried out with Carl Schmidt in 1944 and 1945, it appeared that the major assumptions were valid and that the indirect method was capable of yielding values that correlated well with direct measurement using the bubble flow meter. Then began a number of efforts to examine more rigorously these and a number of other assumptions on which the accuracy of the method depended.

The solubility of nitrous oxide in blood and brain was determined in vitro in collaboration with Harmel, Broomell, and Rhode. No significant difference was found in the solubility for blood, normal brain of animals and man, nor in the brains of patients dying with a variety of disorders. Moreover, the equilibration between blood and brain could be studied in animals and was found to be sufficiently complete at the end of ten minutes. Studies with radioactive krypton indicated that in man as well, ten minutes was sufficient for nearly complete equilibrium. Two other series of studies indicated that the venous blood was well mixed by the time it emerged and that contamination from the distribution of the external carotid was minor.

The first systematic study in man was carried out on 14 healthy young men who volunteered to serve as subjects (Kety & Schmidt 1948). The values for blood flow and oxygen consumption were in the same range as those we had previously found in the monkey, when both were reduced to unit weight of brain. I am still impressed, however, with how large a share of the body's economy is used in supporting the brain—about a fifth of the cardiac output and of the oxygen consumption at rest—but how small the utilization of energy by the human brain—a mere 20 watts—in comparison with what man-made computers require. The number of problems to which we wanted to apply the new technique were legion but we tried to select those that might contribute to fundamental knowledge about the brain and its physiological functions, or in the case of disease, where there was reason to think that an alteration in circulation or metabolism would be crucially involved. The first application was to the effects of altered concentrations of oxygen and carbon dioxide. Here we confirmed previous studies in animals and the remarkable insight of Roy and Sherrington, half a century earlier, that, by dilating cerebral vessels, the products of metabolism (diminished oxygen, increased carbon dioxide and H^+) played an important role in the homeostatic linkage of blood flow to metabolic requirements locally. We were not surprised to find that in deep anesthesia or coma, the cerebral oxygen consumption was reduced by 50%.

The study on sleep, however, produced some surprising results (Mangold et al 1955). Although it was not unusual for some of our subjects to fall asleep and have to be awakened during our previous studies, it was not easy to get them to sleep when we wished. I was a subject in a sleep study and although I was quite comfortable, the effort of trying to sleep kept me awake. In the course of more numerous unsuccessful trials we did have six subjects sleep for the ten minutes necessary to make a measurement. The results were unexpected. The utilization of oxygen by the brain was quite normal, in spite of the prevailing belief, which stemmed from Pavlov and Sherrington, that sleep was characterized by a suppression of neuronal activity. It was not until Evarts succeeded in recording from individual cortical neurons of sleeping animals, and the confirmation provided by our later studies of regional circulation that the observations in man showing sleep to be an active process became credible.

In disease states, the finding that cerebral circulation was normal in essential hypertension despite a perfusing pressure that was twice normal, was interesting. This occurred without intervention by the known sympathetic supply, suggesting a humoral vasoconstriction or a homeostatic autoregulation, both of which are now known to occur. We were unprepared for Sokoloff's finding that the brain did not share in the generalized increase in metabolism that occurs in hyperthyroidism. This led him to surmise, and eventually to demonstrate, that the important action of thyroxin was on protein rather than carbohydrate metabolism.

The studies on schizophrenia were begun because it had been proposed that a deficit in oxygen utilization occurred in the brain, but there was an equally cogent reason. The development of the nitrous oxide technique was supported in part by a small grant from the Scottish Rite Schizophrenia Program. Although the directors had never asked about the relevance of cerebral circulation to schizophrenia, when an application to that problem was possible we were eager to make it. The results were not illuminating: the brain of the schizophrenic patient has the same blood flow and utilizes oxygen at the same rate as that of normal individuals (Kety et al 1948).

From this we concluded that if there was a biochemical derangement in schizophrenia it was in processes more subtle and more specific than oxygen utilization. But as recently as thirty years ago we had no idea what those processes might be. We also suggested that alterations in oxygen utilization could occur in particular regions of the brain so small as not to be reflected in the overall measurements to which the nitrous oxide technique was limited. In the intervening time much more has been learned about both possibilities.

In 1948 I joined Julius Comroe in the Graduate School of Medicine at Penn where he had developed an exciting new approach to the teaching of the basic medical sciences to physicians. With his encouragement I decided to tackle the theoretical considerations involved in the uptake of inert gases by the lungs and their distribution in the tissues. Although the nitrous oxide technique was based upon such theory, its assumptions had been examined only empirically. Moreover from the arterial concentration curves for nitrous oxide obtained in a variety of clinical conditions it was obvious that the shape of that curve was altered by changes in ventilation or cardiac output. My rudimentary knowledge of mathematics did not go very far beyond linear differential equations, but using these it was possible to derive expressions that described the uptake and distribution of inert gases (Kety 1951), or for that matter, any diffusible and nonmetabolized substance. That description was a first approximation and has since been further elaborated by others, using more sophisticated approaches. The expression for the uptake at the lungs in terms of ventilation, cardiac output, and solubility was of some value to anesthesiology, and equations derived for capillary-tissue exchange facilitated the development of various techniques for the measurement of capillary permeability or circulation depending on the choice of tracer.

THE INTRAMURAL BASIC RESEARCH PROGRAM OF THE NIMH

In 1951 I had an unexpected visit from Robert Felix, the director of the newly established National Institute of Mental Health in Bethesda, who invited me to join him as its first scientific director. Although I was very happy with academic life, Dr. Felix had stimulated a receptive area. The studies on schizophrenia, which were what brought me to his attention, had served to impress on me the magnitude of the problem of mental illness and the depths of our ignorance about it. For one who was motivated to contribute to an understanding of the problem, yet did not feel that a quick breakthrough was likely, what better opportunity existed than to plan and develop a program of basic research that might help to provide the basis of an eventual understanding. After thinking about it for two months, visiting Bethesda, seeing the 200 laboratories being constructed for the new institute, meeting its small but dedicated staff, conferring with James Shannon and Harry Eagle, the scientific directors of the Heart and Cancer Institutes, I accepted the challenge and spent 16 years there.

Dr. Felix was the ideal director for the National Institute of Mental Health. He appreciated the need for substantially increased research and defended it valiantly. He did not presume to know in what directions our research program should go, or if he did, he did not permit that to influence me. For my part, these mysterious illnesses that had baffled the human race for centuries had not revealed any of their secrets to me. I could think of no better investment of these new and unprecedented resources than using them to establish a broad program of fundamental research, representing all of the disciplines concerned with the brain and behavior. Perhaps because my background was more deficient in the social sciences than in any of the others, I decided to establish the Laboratory of Socio-Environmental Studies first, with John Clausen as its chief. Shortly thereafter Wade Marshall became head of Neurophysiology. Alexander Rich was appointed to represent neurochemistry, and Giulio Cantoni, soon joined by Seymour Kaufman, to represent biochemistry. When the Clinical Center was completed Robert Cohen was asked to be Director of Clinical Research and together, we recruited David Shakow to direct a large laboratory of psychology, representing a wide spectrum of experimental, developmental, and clinical psychology.

I was also charged with organizing the Basic Research Program of the National Institute of Neurological Diseases and recruited several additional neuroscientists: William Windle, soon joined by Sanford Palay in neuroanatomy, Roscoe Brady in neurochemistry, Kenneth Cole in biophysics, Ichiji Tasaki in neurobiology, Karl Frank in neurophysiology, and, for a short time, Roger Sperry in developmental neurobiology, until he was wooed away by Cal Tech. One concern that some had expressed was rapidly put to rest. A government institution with the proper philosophy could attract a faculty as distinguished as that of any university. Of the initial group that joined me, six eventually became members of the National Academy of Science. The two programs were organized simultaneously and merged into a single basic research program as they should have been, since the neurosciences have as much pertinence to mental illness as to neurological disease. It is unfortunate that they were separated some years later for parochial reasons. Much more serious, however, has been the segregation of disciplines that has prevailed in the extramural research programs of the two institutes over the past decade or more, as the result of a tacit agreement that the Neurological Institute would support fundamental research in the neurobiological disciplines while the Mental Health Institute recognized the social and psychological sciences as its domain.

LOCAL CEREBRAL CIRCULATION
AND FUNCTIONAL ACTIVITY

I had accepted the position in Bethesda with the understanding that I would be able to continue some personal research, and this was made possible by the excellent administrative support that the National Institutes of Health provided. Most important, however, were the scientific collaborators with whom I became associated. Louis Sokoloff joined me as soon as a modest laboratory was available for our studies in cerebral metabolism and we were joined from time to time by a substantial number of post-doctoral fellows such as William Landau, Lewis Rowland, and Martin Reivich, and visiting scientists, including Niels Lassen from Denmark and Cesare Fieschi from Italy, all of whom have achieved well-deserved recognition in neurology, neurochemistry, and physiology.

The nitrous oxide technique that yielded average values for the whole brain was useful for the study of physiologic or pathologic changes that were generalized or affected a significant fraction of the brain. But that organ is remarkable for its complexity and heterogeneity, and for this, another approach was necessary. In the theoretical analysis of inert gas exchange at the tissues, I had developed an expression relating local tissue concentration of a freely diffusible, nonmetabolized substance, to the history of its concentration in arterial blood and the rate of blood flow through the region. Since the arterial and tissue concentrations of a radioactive tracer could be evaluated, it should be possible to measure regional blood flow. The theory was tested using trifluor-iodo-methane labeled with iodine-131, and measuring tissue concentrations by means of autoradiograms of frozen tissue sections (Landau et al 1955). Later, antipyrine-C^{14} was substituted in a simpler technique, which gave better resolution (Reivich et al 1968). Lassen and Ingvar applied the theory to measurement of regional blood flow in man, with techniques that have become widely used.

In the autoradiograms from animals or the visual display obtained in clinical studies, many of the structures of the brain were differentiated by means of their blood flow. The magnitude of the blood flow and the changes that could be induced by activation, suggested a linkage between perfusion, metabolism, and functional activity, although metabolism could not be measured at that time.

Recently, Sokoloff (1977) has succeeded in visualizing and measuring metabolic rate throughout the brain in terms of glucose uptake, using a nonmetabolized, radioactive congener (carbon-14 labeled deoxyglucose). So close is the coupling between glucose utilization and neuronal activity that the autoradiograms visualize, with a high degree of fidelity and resolution, small regions of the brain in terms of their functional activity. With the use of positron emission tomography and a suitably labeled deoxyglucose, application of the technique to man is feasible, so that one can look forward to visualizing levels of functional activity throughout the conscious human brain, in addition to the presently recognized value of the technique in fundamental neurophysiological, anatomical, and pharmacological research.

THE LABORATORY OF CLINICAL SCIENCE

By 1956 several investigators had joined the Intramural Research Program of the Mental Health Institute whose interests were in the interface between the basic neurobiological sciences and clinical psychiatric problems, and a new grouping was established designated the Laboratory of Clinical Science. By that time the Basic Research Program was fully staffed, I was eager to involve myself more in research and less in administration, and the implications of the new neurobiological knowledge to psychiatry attracted me. I asked to be allowed to step down from the scientific directorship to join the new laboratory as its nominal chief.

The initial group consisted of Edward Evarts, Julius Axelrod, Louis Sokoloff, Marian Kies, Roger McDonald who was succeeded by Irwin Kopin, Philippe Cardon, and Seymour Perlin who was succeeded by William Pollin. Although quite diverse in their scientific interests, they shared a commonality of motivation and a mutuality of spirit, which made the 11 years I spent in that laboratory one of the most exciting and rewarding periods of my life.

What were these new and promising implications that attracted many of us? It was not the numerous enthusiastic claims that were being made regarding abnormal proteins, metabolites, or toxic factors in the blood or urine of schizophrenic patients, which lacked plausibility and did not survive replication. Rather, it was a number of less spectacular but more credible observations with more remote relevance—observations that suggested that the synapses of the brain, like those in the periphery, were chemically mediated switches rather than electrical junctions. Acetylcholine by that time had achieved the status of a putative neurotransmitter in the brain, but there were other substances like serotonin, noradrenalin, and dopamine, which could conceivably serve in such a role, and which had only recently been identified in the brain. Lysergic acid diethylamide, a drug which had attracted wide attention because of its hallucinogenic properties, had also been found to block some of the pharmacological actions of serotonin. Three other psychotomimetic drugs, dimethyltryptamine, mescaline, and amphetamine, were substituted forms of serotonin or dopamine. Only a few years before, chlorpromazine had been found to be remarkably effective in the alleviation of psychotic behavior and reserpine was being used extensively as a major tranquilizer. Although it was to take ten years for the action of chlorpromazine on dopamine synapses to be discovered and substantiated, knowledge of the remarkable ability of reserpine to deplete the brain of serotonin was literally around the corner—in Bernard Brodie's laboratory at the Heart Institute.

If the synapses involved in the mental states and behaviors produced or ameliorated by such drugs were chemically mediated, that would offer a plausible site at which these drugs could act. Moreover, if central synapses in general were chemical switches, then a biochemistry of behavior was conceivable, and at the synapse, not only drugs, but genetic factors, dietary constituents, hormones, metabolic, immune, and infectious processes, could all be seen to act, altering the patterns of transsynaptic interaction and affecting behavior and mental processes. For the first time, plausible and heuristic approaches could now be opened and explored, that might some-day explain the biological disturbances of mental illness and the symptoms that depend on them.

The most productive way of exploring these new approaches was not by way of a crash program. The gap between the knowledge we had and the clinical problems was still too wide to be spanned all at once by any concerted effort. What was needed was to narrow the gap by an increase in knowledge on both sides, which is best done by relying on the creativity and judgment of individual scientists who know better than anyone else what their next step should be. The members of the laboratory pursued their own research goals, some studying the clinical problems in greater detail and in the light of new knowledge, most expanding the base of fundamental knowledge in areas that they perceived to be relevant. Where appropriate, collaborative efforts developed within the laboratory, and quite as often, outside of it. I believe that subsequent events have justified this approach.

Among the claims that were being made at that time was one postulating the formation of a toxic, hallucinogenic metabolite of circulating epineph-rine in schizophrenia, which was sufficiently provocative that some of us decided to examine it further. The difficulty was that in 1956 we knew little enough about the normal metabolism of epinephrine, let alone its metabo-lism in disease. One strategy would be to administer labeled epinephrine in pharmacologically insignificant amounts and compare the urinary chromatographic profiles of radioactivity. The carbon-14 labeled material that was available would not provide sufficient specific activity. It was possible that a tritium-labeled epinephrine could be prepared with the requisite stability and activity, and arrangements were made to have 7-H^3-epinephrine of high specific activity synthesized. By the time the labeled compound arrived, however, that strategy was no longer necessary. In the year that had elapsed, Axelrod, taking off from a brief report in the literature, had demonstrated the enzymatic O-methylation of catecholamines in vitro, characterized the enzyme responsible, predicted the major catecholamine metabolites, and then went on to extract and identify them in the urine of animals (Axelrod 1959). When the radioactive epinephrine became available, it was a simple matter to examine its metabolism in normal subjects (LaBrosse et al 1961) and in schizophrenics. No evidence was found for an abnormal metabolism of circulating epinephrine in that disorder.

There is not the space nor the necessity of indicating Axelrod's contributions to our present knowledge of catecholamine metabolism and inactivation at the synapse. They have not solved a major psychiatric problem as yet, but when the final chapter to our understanding of mental illness is written, his work will occupy a prominent place in it.

Studies by Axelrod, Glowinski, and Kopin, on the metabolism of norepinephrine in the brain and the effects of psychoactive drugs, further supported the role of that amine as a neurotransmitter in central synapses. It required evidence from many quarters before there was general agreement that the biogenic amines were important as neurotransmitters in the brain. Electron microscopy and fluorescence histology were the most direct and compelling, but these were reinforced by microinjection and electrophysiological studies, and most recently, by the demonstration in vitro of specific receptors for several of the transmitters.

The list of neurotransmitters and synaptic modulators has gotten longer, extending from the biogenic amines to include amino acids and polypeptides. The involvement of specific members of the list in the pharmacological action of most of the psychoative drugs has been reasonably well established. It is also clear that neurotransmitters and modulators play important roles in the mediation of certain mental states and types of behavior, although their precise action and interactions remain to be elucidated. It is not possible to state, at present, how they are involved in mediating the symptoms of mental illness or whether they play an etiological role. Those who choose to point out that no morphological or biochemical lesion has been identified in the major psychoses are still correct, although the inference they draw, that therefore none exists, is clearly a non sequitur.

GENETIC FACTORS IN SCHIZOPHRENIA

When I reviewed the field in 1959, I found no compelling evidence for a biochemical disturbance in schizophrenia, except for the observations on families and twins, which was compatible with the existence of forms of the illness with a genetic component. The inability to control or randomize environmental variables, however, made that conclusion less than rigorous, and suggested the desirability of studying a national sample of adopted individuals whose genetic endowment and environmental influences could be studied separately through their biological or adoptive relatives.

In 1962, David Rosenthal, Paul Wender, and I learned of our respective interests in such a strategy and began a collaboration with Fini Schulsinger in Copenhagen where it has been possible to compile such a national sample, identify adoptees or biological parents with schizophrenia or affective disorder, and to study their relatives or offspring, reared in another environment. In each of our studies and in those of others these mental illnesses continue to run in families, but in the biological and not in the adoptive families. The evidence permits the conclusion that genetic factors operate significantly in the transmission of these disorders, or at least in some of their major subgroups (Kinney & Matthysse 1978). Since the genes can express themselves only through biochemical mechanisms, this should constitute rather compelling justification for the relevance of biological research to mental illness.

The directions that such research can take are legion, limited only by what fraction of the staggering costs of mental illness we invest in its ultimate understanding and prevention. There is the development of a new neuropathology (Matthysse & Pope 1975), representing the application to the brain of powerful new morphological, histochemical, and molecular techniques. Studies, such as those by Nauta and his associates, of the interconnections of the mesolimbic system are now possible by autoradiographic and electron microscopic techniques (Cowan 1975). There is the further study of the manifold neurotransmitters—the systems in which they are involved, their interconnections, the properties of their receptors. It is possible to elucidate the neural basis of psychological functions, as Hubel & Wiesel, Mountcastle, Evarts, or Kandel have done in areas of perception, attention, voluntary behavior, or adaptation to experience. One could go on, through all of the neurosciences. The opportunities for significant research are many, and there is a new generation of eager and competent scientists to explore them. The prospects have never been more promising.

Those prospects have been recognized not only by neuroscientists, but by academic medicine, and many departments of psychiatry have been moving in the direction of their realization. On that basis I accepted the chair of psychiatry at Johns Hopkins in 1961, which would have been a most gratifying and important role if I had been a psychiatrist. When I saw the clinical responsibilities outstripping my involvement in research, I reluctantly resigned. In 1967 I moved to Harvard, not as a department chairman, but as a professor of psychiatry with my major responsibilities in research. At our laboratories at the McLean Hospital I have the good fortune to be associated with a number of scientists who represent a wide range of basic disciplines that sustain psychiatry—from Walle Nauta in neuroanatomy to Philip Holzman in psychology. They and their counterparts in genetics, pharmacology, molecular biology, and biochemistry exemplify the mature and multidisciplinary field that psychiatric research has become. The chemical nature of synaptic transmission, the efficacy of the new psychotherapeutic drugs, and the recognition of the significant genetic contribution to the etiology of the major psychoses, make clear the pertinence of the neurosciences, as well as the psychological and social sciences, to modern psychiatry and to further progress in the understanding, treatment, and prevention of mental illness.

Literature Cited

Axelrod, J. 1959. Metabolism of epinephrine and other sympathomimetic amines. *Physiol. Rev.* 39:751–76

Cowan, M. 1975. Recent advances in neuroanatomical methodology. In *The Nervous System,* ed. D. Tower, R. O. Brady, 1:59–70. New York: Raven. 685 pp.

Dumke, P. R., Schmidt, C. F. 1943. Quantitative measurements of cerebral blood flow in the macacque monkey. *Am. J. Physiol.* 138:421–28

Kety, S. S. 1951. The theory and applications of the exchange of inert gas at the lungs and tissues. *Pharmacol. Rev.* 3:1–41

Kety, S. S. 1961. A biologist examines the mind and behavior. Many disciplines contribute to understanding human behavior, each with peculiar virtues and limitations. *Science* 132:1861–70

Kety, S. S., Schmidt, C. F. 1948. Nitrous oxide method for the quantitative determination of cerebral blood flow in man: Theory, procedure, and normal values. *J. Clin. Invest.* 27:476–83

Kety, S. S., Woodford, R. B., Harmel, M. H., Freyhan, F. A., Appel, K. E., Schmidt, C. F. 1948. Cerebral blood flow and metabolism in schizophrenia. The effects of barbiturate semi-narcosis, insulin coma and electroshock. *Am. J. Psychiatry* 104:765–70

Kinney, D. K., Matthysse, S. 1978. Genetic transmission of schizophrenia. *Ann. Rev. Med.* 29:459–73

LaBrosse, E. H., Axelrod, J., Kopin, I. J., Kety, S. S. 1961. Metabolism of 7-H^3-epinephrine-d-bitartrate in normal young men. *J. Clin. Invest.* 40:253–60

Landau, W. M., Freygang, W. H., Rowland, L. P., Sokoloff, L., Kety, S. S. 1955. The local circulation of the living brain; values in the unanesthetized and anesthetized cat. *Trans. Am. Neurol. Assoc.* 80:125–29

Mangold, R., Sokoloff, L., Conner, E., Kleinerman, J., Therman, P. G., Kety, S. S. 1955. The effects of sleep and lack of sleep on the cerebral circulation and metabolism of normal young men. *J. Clin. Invest.* 34:1092–1100

Matthysse, S., Pope, A. 1975. The approach to schizophrenia through molecular pathology. In *Molecular Pathology,* ed. R. A. Good, S. B. Day, G. Yunis, pp. 744–68. Springfield, Ill.: Thomas

Reivich, M., Jehle, J., Sokoloff, L., Kety, S. S. 1968. The effect of slow wave sleep and REM sleep on regional cerebral blood flow in cats. *J. Neurochem.* 15:301–6

Sokoloff, L. 1977. Relation between physiological function and energy metabolism in the central nervous system. *J. Neurochem.* 29:13–26

Thudichum, J. W. L. 1884. "A Treatise on the Chemical Constitution of the Brain." London: Balliere, Tindall & Cox

Berta Scharrer

Ann. Rev. Neurosci. 1987. 10 : 1–17

NEUROSECRETION: Beginnings and New Directions in Neuropeptide Research

Berta Scharrer

Department of Anatomy and Structural Biology and Department of Neuroscience, Albert Einstein College of Medicine, Bronx, New York 10461

The Early Years

During the past decade, spectacular advances in our knowledge of the far-reaching significance of neuropeptides have ushered in a new era in neurobiology. It may be of interest, therefore, in this prefatory chapter to present some personal recollections of the gradual evolvement of these new insights, which, at the outset, were entirely unforeseen and difficult to accept.[1] In 1928, when Ernst Scharrer reported his discovery of gland-like nerve cells in the hypothalamus of a teleost fish, *Phoxinus laevis*, neurobiologists considered neuronal signaling largely as an electrophysiological process. What made this small group of unusual hypothalamic neurons so striking, therefore, was their content of impressive amounts of a secretory material comparable to that found in typical protein-secreting gland cells, such as those of the pancreas. For this reason, they were called "neurosecretory neurons." E. Scharrer made the bold proposal that their role may be endocrine, and at the outset suggested a functional relationship with the pituitary gland. He was the first to point out that the neurosecretory neuron, because of its dual (neural and glandular) capacities, seems to be ideally suited for the very special task of

---

[1] This account is by necessity selective. Much of the information not documented in detail will be found in the following comprehensive publications: E. Scharrer & B. Scharrer 1963, B. Scharrer 1970, Meites et al 1975, 1978, Gainer 1977, Barker & Smith 1980, Bloom 1980, Burgen et al 1980, Costa & Trabucchi 1980, Krieger et al 1983, Bloom et al 1985, Kobayashi et al 1985, White et al 1985.

1317

0147–006X/87/0301–0001$02.00

conveying neural directives to the endocrine system in its own "language" (E. Scharrer 1952). These pioneering ideas foreshadowed the growing realization of the close interdependence of the two systems of integration, and the central role played by the neuroendocrine axis. In the early 1960s, when neuroendocrinology, i.e. the study of the interactions between the nervous and the endocrine systems, became established as a discipline in its own right (see E. Scharrer & B. Scharrer 1963), the examination of mutually "understandable" signals received due attention.

However, initially the idea that neurons may be capable of dispatching neurohormonal, i.e. blood-borne, signals, an activity heretofore attributed only to endocrine cells proper, met with powerful resistance. It did not conform with the tenets of the neuron doctrine, and the fact that it was proposed on the basis of cytological evidence was considered preposterous.

The theory of the chemical transmission of the nervous impulse proposed by Loewi in 1921 provided no tangible support for the concept of neurosecretion. The type of neurohormonal messenger substances envisioned to function in these special hypothalamic cells appeared to differ significantly from "neurohumors" (classical neurotransmitters) with respect to their proteinaceous character, the amount of material produced, and the extracellular pathway from site of release to site of action.

When I joined forces with Ernst Scharrer in the search for the functional significance of the phenomenon of neurosecretion, which became a lifetime endeavor, we had the territory practically to ourselves for quite a long time. Eventually, disbelief gave way to mild interest, as demonstrated by a memorable meeting on the hypothalamus of the Association for Research in Nervous and Mental Disease in 1939, at which we were invited to present our views (E. Scharrer & B. Scharrer 1940). Nevertheless, for a number of reasons, among them a lack of facilities and appropriate techniques, progress was slow.

A search of the literature yielded very few reports that appeared to be relevant, and it was not until much later that such reports found their proper place in the emerging saga (see below). One was the description of "gland-cells of internal secretion" in the caudal spinal cord of skates (Speidel 1919). Another called attention to the presence of a protein material, called "Substance P," extractable from mammalian gut as well as nervous tissue (von Euler & Gaddum 1931). A third proposed, on the basis of extirpation experiments in a lepidopteran insect, *Lymantria*, that pupation is controlled by a brain hormone (Kopeć 1917, 1922).

What is so fascinating, in retrospect, is that each milestone reached in the elucidation of neurosecretory activities is marked by the breakdown of once powerful conceptual barriers and their replacement by major new and seminal insights. Within the space of 50 years, there occurred not only the emergence of the new discipline of neuroendocrinology but, as an outgrowth of these endeavors, the discovery of new modes of interneuronal communication that reach far beyond the confines of neuroendocrine phenomena.

To return to the early phase of our search, as a first step a broadly based comparative approach was undertaken in which E. Scharrer focused on vertebrates and I on invertebrates (see E. Scharrer & B. Scharrer 1937, 1945, 1954). It revealed the virtually universal occurrence of distinctive peptide-producing neurosecretory cell groups throughout the animal kingdom, up to mammals, including man. This laid to rest recurrent criticisms suggesting their pathological nature. It also showed that neurosecretory neurons are not restricted to a small magnocellular component of the hypothalamus of vertebrates.

The first invertebrate ganglia in which impressive signs of secretory activity were detected were those of the opisthobranch snail *Aplysia* (B. Scharrer 1935), an animal that has since proved to be an excellent model for experimentation in this area of research (e.g. Scheller 1984, Strumwasser 1985). What turned out to be of even greater heuristic value was the study of the brain-corpus cardiacum-corpus allatum system of insects, because of its remarkable analogy with the hypothalamic-hypophysial system of vertebrates (B. Scharrer & E. Scharrer 1944). A number of new insights gained from the comparison of the structure–function relationships observed in these two neuroendocrine organ complexes could not have been obtained from mammalian material alone.

Extensive cytophysiological studies, undertaken by E. Scharrer and his collaborators in vertebrates and by myself and others in invertebrates, were focused on structural differences in the neuroendocrine systems of animals examined under different physiological or experimentally altered conditions. These efforts provided indirect, though persuasive, evidence in support of a functional role of neurosecretory products in the control of developmental, reproductive, and metabolic functions. The search for structural parameters of known functional states remained an absolute requirement even after increasingly sophisticated methods permitted a more direct approach to the elucidation of neuroglandular function. In fact, it is difficult to estimate where this field of research would stand today without the continued essential contributions made by morphologists.

The advent of electron microscopy made possible a precise characterization of the sites of synthesis and release of peptidergic neurosecretory products. As shown in early contributions by Palay, Bargmann, Bodian and others, as well as our own, this material consists of strikingly electron-dense, membrane-bounded granules that are easily identified and localized. The demonstration of their formation in packaged form by the Golgi apparatus (E. Scharrer & S. Brown 1961), as well as their axonal release by exocytosis, provided further evidence in support of a close similarity with the products of other protein-secreting gland cells. In combination with sensitive and selective immunocytochemical procedures, electron microscopy now permits us to determine not only the distribution but also the chemical nature of neurosecretory messenger substances in various parts of the organism. In this program, the pioneering contributions by Sternberger (1986), especially his peroxidase-antiperoxidase method, have proved to be most productive in the hands of many investigators, including those searching for analogous phenomena in invertebrates (B. Scharrer 1981).

The Hypothalamic-Neurohypophysial System

The first documentation of a neurohormonal activity attributable to peptides released by neurosecretory neurons did not pertain to the postulated role of the hypothalamus in the control of adenohypophysial function. Instead it solved an enigma of long standing by demonstrating the hypothalamic origin of the so-called posterio-lobe hormones, vasopressin and oxytocin. Here again, a morphological observation led the way. A new histological staining procedure, developed by Gomori for the demonstration of pancreatic beta-cells, the chrome-alum hematoxylin phloxin technique, turned out to be selective for neurosecretory products, and for the first time it made possible the visualization of their presence throughout the entire neuron. This convenient marker enabled Bargmann (1949) to trace the neurosecretory fiber tract connecting the hypothalamic cell bodies, where the material is synthesized, with the posterior pituitary, where it is released into the circulation. These cells constitute the magnocellular component of the nuclei supraopticus and paraventricularis of the mammalian hypothalamus, which in fishes and amphibians are represented by the singular nucleus praeopticus. The correct interpretation of this topographic relationship (Bargmann & E. Scharrer 1951) was subsequently substantiated by appropriate experimental procedures. Evidence supporting the concept of an intraneuronal transport of such neuron-derived hormonal messengers to their release sites in analogous neurohemal organs located outside of the brain (posterior lobe and corpus cardiacum, respectively) was obtained by surgical interruption of the neurosecretory pathway. In mammals (Hild 1951) as well as insects (B. Scharrer 1952) the active material accumulated proximal and became depleted distal to the level of severance. Hild & Zetler (1953) further showed a clear correlation between the functional potency of tissue extracts, determined by pharmacological tests, and the amount of selectively stainable neurosecretory material present in the tissues under investigation.

This breakthrough, i.e. the establishment of valid criteria for the existence of a new class of neurochemical mediators, for which the term "neurohormone" is appropriate, posed a conceptual problem, namely the rationale for such one-step neurohormonal activities. The question of why neurons should deviate so profoundly from the norm to function as gland cells of internal secretion in organisms endowed with endocrine glands proper was eventually resolved by the examination of the evolutionary history of bilogically active neuropeptides (see below).

The Hypothalamic-Adenohypophysial System

The search for a solution of the central tenet, i.e. the operation of neural mechanisms controlling adenohypophysial function, was encouraged once a case was established for the existence of neurohormonal regulators. Not only were neurohormones now considered to be the most likely candidates for neuroendocrine mediation at the level of the hypothalamic-adenohypophysial axis, but the first active principles suspected to serve in this capacity were the "posterior-lobe hormones" themselves. Such a role has, in fact, been demonstrated quite recently for vasopressin, which participates in the control of corticotropin release.

However, the majority of stimulatory and inhibitory neuropeptides directed to the anterior lobe of the pituitary gland was found to originate in hypothalamic centers other than those known to provide vasopressin and oxytocin. The operation of a "semiprivate" vascular pathway for the conveyance of these hypophysiotropic messenger substances, i.e. the hypophysial portal system, was established by the pioneering work of Harris, in collaboration with Green and Jacobsohn (Harris 1955). They demonstrated that reproductive function can be interrupted by section of the hypophysial stalk and restored after the regeneration of the portal vessels. The median eminence was recognized as the neurohemal release organ for hypophysiotropic neurohormones, and an explanation had thus been found for the virtual absence of a nerve supply to the adenohypophysial parenchyma. The rich contributions by Everett, Gorski, Martini, McCann, Reichlin, Saffran, Sawyer, and others brought further advances in the clarification of the neuroendocrine control of reproductive phenomena.

These investigations made use of various techniques, among them organ culture, electrical stimulation, or lesions in discrete hypothalamic centers (e.g. Szentágothai et al 1962), and transplantation experiments (e.g. Nikitovitch-Winer & Everett 1958), and showed a comparable dependency on hypothalamic directives for other adenohypophysial, including thyrotropic, adrenocorticotropic and metabolic functions. Further contributions demonstrating neuroendocrine control systems in nonmammalian vertebrates were made by, for example, Benoit and Assen-macher (1959) and Oksche and Farner (Oksche et al 1964) in birds and by Etkin (1963) and his collaborators in amphibians.

Parallel studies established that two-step neuroendocrine control systems are not restricted to vertebrates. In insects, cerebral neuropeptides, such as the pupation hormone discovered by Kopeć (1917, 1922), have been shown to act via the regulation of non-neural glands of internal secretion, i.e. the corpus allatum and the prothoracic gland. Evidence for an inhibitory effect of the brain on the corpus allatum, an analogue of the adenohyphysis (B. Scharrer 1952), parallels that for the inhibitory neural control of the pars intermedia of tadpoles reported by Etkin (1962).

The isolation, chemical characterization, and de novo synthesis of several of these hypothalamic regulatory principles, referred to as "releasing" or "regulating" factors (RFs) or hormones (RHs), had to await the development of an appropriate biochemical technology. Achieved largely by Guillemin (1978) and Schally (1978) and their collaborators, this momentous accomplishment placed neurosecretion and neuroendocrinology on firm ground. An intensive multidisciplinary effort, aimed at the investigation of the functional capacities and modes of operation of these active principles, became a major concern of many laboratories.

At this juncture, the widely held view was that, in addition to a vast majority of conventional, synaptically transmitting neurons, a small and quite separate class of peptide-producing neurosecretory cells, which dispatch their signals exclusively via circulatory channels, had to be accommodated in the spectrum of neuronal elements. Their deviation from the norm was then generally understood as being related to their unique role in the transduction of cues to the endocrine system.

Nonendocrine Functions of Neural Peptides

Another quite unexpected turn of events necessitated a shift in our interpretation of neurosecretory phenomena. A variant by which neuro-secretory signals may address endocrine cells at close range was discovered by means of electron microscopy. The observation of synapse-like ("synap-toid") terminals on cells of the mammalian pars intermedia led Bargmann et al (1967) to introduce the term "peptidergic neuron," which today extends far beyond its original conception. The same localization of neuro-secretory release sites in close spatial relationship to effector cells was found in an analogous endocrine organ of insects, the corpus allatum (B. Scharrer 1972), and subsequently in several nonendocrine tissues of vertebrates as well as invertebrates.

What had become increasingly clear was that peptidergic neurons have available several alternate, i.e. non-neurohormonal, possibilities for neuro-chemical communication. Some of them come very close to the pattern of classical synaptic intervention, others are neither strictly "private" (in loco) nor neurohormonal. The latter case, in which a relatively narrow zone of extracellular stroma intervenes between nerve terminal and effector cell, represents an intermediate mode of action, both in space and time. The recognition of this interstitium has given rise to the concept of "synapse à distance."

The most intriguing outcome of these studies was the realization that not only somatic cells but even neurons can be addressed by neuropeptides in a nonconventional manner. One of the possible pathways is the cere-brospinal fluid. The dispatch of a neuroregulatory peptide into an extra-cellular compartment wider than the synaptic gap may operate in a special, nonsynaptic form of neurocommunication known as "peptide-mediated neuromodulation" (see below). Other variants are synapse-like contacts in which either the presynaptic partner of a pair of neurons or else two (or more) neurons in close contact are of peptidergic nature. Some junctional complexes of the latter type (B. Scharrer 1974) give the impression of providing for a reciprocal exchange of signals. It has been suggested that peptidergic synaptic transmission closely resembles that involving nonpeptidergic transmitters, the difference being a slower time course, which is necessitated by the different mechanism of inactivation of a peptidic neuroregulator.

The Spectrum of Neuropeptides

The new direction in neurobiological research discussed in the preceding section was greatly stimulated by yet another spectacular, almost explosive, step forward, i.e. the discovery, in rapid succession, of a multitude of "new neuropeptides." A current estimate of their number is about 100, and the search still continues. By the same token, the number of known peptide-producing cell types has steeply risen.

The family of peptidergic neurons now encompasses, in addition to the classical or archetypal neurosecretory cells, a long list of others known to synthesize and make use of biologically active peptides in one form or another. Sites of neuropeptide production are widely distributed in the central and peripheral nervous systems. Maps prepared by means of immu-nocytochemical visualization, at both the light and electron microscopic levels, show a distinctive pattern in the distribution of cell bodies and their axonal projections containing specific neuropeptides. Noteworthy examples are the precise laminar organization of several different neuro-peptides in the optic tectum (Kuljis & Karten 1982) and their localization in distinct retinal cell populations, primarily amacrine cells (Brecha & Karten 1983).

Moreover, the same or very closely related chemical principles are now known to be manufactured by various non-neuronal cells, e.g. those of the "diffuse endocrine cells" of the digestive apparatus (Pearse & Takor Takor 1979). An example of the commonality of such peptides is somatostatin, first known as a hypophysiotropic neurohormone, now identified, and functionally characterized, also in the endocrine pancreas and in other non-neural tissues (Patel & Reichlin 1978). Conversely, substances heretofore better known as digestive hormones (e.g. gastrin and cholecystokinin) or as adenohypophysial products (e.g. ACTH, α-MSH, and β-lipotropin) have now been identified also in the brain.

Much has been learned about the chemical identities of numerous neuropeptides, as well as the biosynthetic process that gives rise to them, by the use of strategies developed by molecular genetics. Virtually all of these bioactive peptides (and other proteins manufactured by the body) are known or postulated to be derived from high-molecular-weight precursor substances. These inactive pro-proteins (or pre-pro-proteins) are synthesized under the direction of mRNA templates on ribosomes bound to the endoplasmic reticulum (rough surfaced ER) of the perikaryon. Examples of such precursor molecules are pro-opiomelanocortin (containing the amino acid sequences of ACTH, α-MSH, β-endorphin, and β-lipotropin), pro-enkephalin (containing multiple copies of met-enkaphalin plus leu-enkephalin), and pro-pressophysins (precursors of vasopressin and oxytocin plus their specific neurophysins, respectively).

Posttranslational processing of pro-neuropeptides includes sequential proteolytic cleavage of the precursor as well as additional steps of conversion resulting in active substances, among them glycosylation, phosphorylation, acetylation, and amidation. Synthesis and processing of pro-opiomelanocortin in the arcuate nucleus of the hypothalamus are similar, though not identical, chemical steps as compared with those occurring in the adenohypophysis.

All of these processes occur within membrane-bounded compartments (organelles) of the cytoplasm, presumably during the axoplasmic transport of these products to their sites of release, primarily at the axon terminal.

The degradation of the active principles is considered to be brought about by the action of specific peptidases.

Evolutionary History of Neuropeptides

Immunobiological tests, i.e. both radioimmunoassays and immunocytochemical probes, have shown that neuropeptides are widely distributed in the animal kingdom, including unicellular organisms (Roth &

Le Roith 1984). The use of antisera raised against a number of mammalian-type neuropeptides has yielded distinctly localized reaction products in the nervous systems of lower phyla, including a variety of invertebrates (see, for example, Rémy & Dubois 1981, Hansen et al 1982). Conversely, such positive responses have been elicited in vertebrates with antisera raised against a few physiologically and chemically identified invertebrate neuropeptides, among them a molluscan cardioexcitatory factor, the tetrapeptide FMRFamide, which resembles gastrin/cholecystokinin. Perhaps the most striking parallelism demonstrated is the occurrence of an undecapeptide, identical with the growth-promoting "head activator" of the coelenterate *Hydra*, in the human hypothalamus (Bodenmüller & Schaller 1981).

This similarity, or even identity, in molecular configuration indicates that biologically active neuropeptides are stable elements with a long evolutionary history. Their origins seem to reach back to when the first primitive nervous systems began to form and even farther. Apparently, when no endocrine cells proper were as yet in existence, the nervous system was in charge of all required integrative functions. In fact, in primitive nerve cells endocrine functions must have taken precedence over synaptic activities, which seem to have taken over at a later date.

It is tempting to speculate that the bioactive neuropeptides of today are derived from ancestral protein precursors (see B. Scharrer 1978) which, step by step, have developed the capacity of splitting off active principles in a manner illustrated by the present mechanism of peptide biosynthesis from pro-proteins. As the demand for increasingly complex signaling devices arose, the assignment of specific physiological functions to selected amino-acid sequences had to evolve in concert with that of corresponding receptor molecules. Some of the neuropeptides identified in various tissues may be functionless, being mere signposts of a process of molecular evolution.

It makes sense to propose that, after the appearance in more advanced organisms of an endocrine system proper, the pluripotential neurosecretory neurons have taken over a new and highly significant role, that of presiding over the endocrine system. However, their capacity for direct communication with terminal effector cells has not been lost. One-step neurohormonal activities, exemplified by the antidiuretic effect of vasopressin, make sense when viewed as carryovers from a time in the distant past when neurons had to perform all integrative functions directly. In this evolutionary framework, the neurosecretory neuron, far from being a newcomer and rare exception, can be considered as ancestral. It seems to have more in common with the nerve cell precursor than has the "conventional" neuron with its specialization for synaptic transmission.

158,030

The Neurosecretory Neuron in a New Light

Quite obviously, the original definition of the neurosecretory neuron, based on the hypothalamic prototype dispatching its signal via the general circulation, no longer suffices. The fact that this classical type, which was the first discovered, represents the most unorthodox known to date accounts for the difficulty with which the scientific community accepted it. This difficulty has been largely overcome by the subsequent demonstration of a spectrum of transitional peptidergic neuronal types bridging the gap between the classical neurosecretory and the conventional nonpeptidergic neurons. Now the multiplicity of neurons known or presumed to be engaged in peptidergic activity poses a quite different problem, that of where to draw a sharp line of demarcation between them and neurons synaptically transmitting by means of nonpeptidergic conventional neurohumors.

It is a problem that has become accentuated by one of the most recent and quite challenging developments in neuropeptide research. There is rising evidence for the coexistence in one and the same neuron of two or more potentially neuroactive mediators (Hökfelt et al 1984, Chan-Palay et al 1978, Chan-Palay & Palay 1984). This principle applies to the colocalization, not only of different neuropeptides, but also of peptides and classical neurotransmitters. The presence of, for example, ACTH, α-MSH, and β-endorphin in neurons (even within the same secretory vesicles) of the arcuate nucleus can be explained in biosynthetic terms, i.e. by their origin from the common precursor, pro-opiomelanocortin. That the same or a very similar combination of neuropeptides also occurs in endocrine cells of the adenohypophysis is an example of the commonality of these chemical mediators. What is still unclear, however, is how these multiple peptides operate after being released.

Even less can be said about the function of neuropeptides found to be colocalized with conventional neurotransmitters. The number of neurons shown to contain combinations such as catecholamine and enkephalin, serotonin and Substance P, acetylcholine and vasoactive intestinal peptide (VIP) has risen sharply within recent years. There is some speculation that in this situation conventional neuroregulators may function as primary neurotransmitters, and peptides present in the same neuron may function in an auxiliary capacity as "cotransmitters." However, as long as this role remains unclear, the question of whether or not a "peptidergic activity" should be attributed to these neurons remains in abeyance. Be this as it may, these cases of colocalization contribute much to the general conclusion that neuropeptides are now a solid majority in the armamentarium of neurochemical messenger substances.

By the same token, neurons engaged in internal secretion can no longer be considered to be the only nerve cells releasing functionally important glandular products. With the exception of those engaged in electrotonic transmission, not involving neurochemical mediators, the vast majority of neurons must be accorded at least a small measure of secretory capacity. How large a proportion of these may operate entirely without the participation of neuropeptides remains yet to be determined. Therefore, the terms "neurosecretory" and "peptidergic" should be understood to refer to the degree to which they apply rather than to the characterization of strictly separate cell types.

Even neurons as far removed from the conventional end of the spectrum as classical hormone-producing neurosecretory cells show a number of structural and functional features characteristic of neurons in general. They are capable of generating action potentials. They are known to respond to afferent stimulatory as well as inhibitory neuronal signals. In the case of neurons dispatching hypophysiotropic signals, afferent directives are provided by various intero- and exteroceptive factors, which, in integrated form, result in the "final common pathway" to the endocrine apparatus (E. Scharrer 1966). Major examples of exteroceptive influences on hypophysiotropic activity are olfactory and photoperiodic stimuli. Afferent interoceptive directives of importance are contributed by circulating hormones that become part of the three-step feedback mechanism by which the dispatch from their sites of production is controlled.

The release of neuropeptides at axon terminals, like that of "regular" neurotransmitters such as acetylcholine, occurs by the process of exocytosis. This process is triggered by an influx of Ca^{2+} through voltage sensitive channels following depolarization of the axon terminal, and presumably involves the interaction of Ca^{2+} with calmodulin and specific protein kinases (Thorn et al 1985).

In both situations, the operation of stereospecific receptors at the respective sites of action of the neuroregulators has been demonstrated. This concept, based on the recognition between conformations of two reacting molecules, is instrumental in the determination of putative sites of action of neuropeptides reaching them via circulatory pathways. The available information refers primarily to receptors for opioid peptides, e.g. enkephalin, and for somatostatin, which have been identified in mammals and several nonmammalian vertebrates (Way 1980). Among invertebrates, high-affinity binding sites that suggest the operation of such receptors in a mollusc and an insect have been demonstrated with the use of a radio-labeled enkephalin analogue, selected as an exogenous ligand because of its stability (see Stefano & Scharrer 1981). These binding sites, demonstrated in the insect brain, may be involved in a modulatory role played by "enkephalinergic" neurons identified by Rémy & Dubois (1981) immunocytochemically in this organ.

The wide variety of available neuropeptides, outnumbering all other, previously recognized neuroregulators combined, goes hand in hand with the greater versatility in their mode of operation. Depending on the extracellular pathway used (ranging from general circulation to intersynaptic gap), the amount of neuropeptide released varies, and so does the type of signal. It may closely resemble "standard" synaptic transmission, as is the case with Substance P or gastrin.

A different form of close-range activity appears to be that of neuromodulation. In this new and intriguing mode of neurochemical intervention, neuropeptides may either augment or depress synaptic signals passed on between two synaptically joined "conventional" neurons. This role, as well as the still enigmatic function of neuropeptides colocalized with other neuroregulators, has been referred to above.

As to the neural directives aimed at the adenohypophysis, the type of operation is flexible. A given signal to one of the specific "target cells" may be accomplished by the collaboration of several hypophysiotropins. Conversely, one hypophysiotropic factor may contribute to the regulation of more than one type of pituitary cell.

The principle that one and the same neuropeptide can function in different capacities, depending on where it is put to use, is best illustrated by vasopressin. This neuroregulator, primarily known for its antidiuretic neurohormonal role, also participates in hypophysiotropic signaling, specifically in collaboration with the corticotropin-releasing factor, CRF (e.g. Blume et al 1978). A third role of vasopressin is its presumed effect on extrahypothalamic neurons that are contacted by terminals containing this nonapeptide. The relationship between this ultrastructural and immunocytochemical evidence (Buijs et al 1978, Dogterom et al 1978, Sofroniew & Weindl 1978) and the demonstration by De Wied and his collabrators (De Wied 1978) of vasopressin effects on certain memory and learning functions remains to be clarified.

Reference should be made here also to interactions between peptidergic neural centers and two regulatory systems that have recently attracted much attention.

One of these links is with the immune system (see Goetzl 1985), which shares with neuroendocrine structures a set of regulatory substances, including ACTH, TRH, and endorphins, as well as the corresponding receptors (Blalock et al 1985a,b). Neuroimmunologic communication is bidirectional.

Numerous studies indicate that immune functions are subject to neuroendocrine regulation, as shown by their susceptibility to stress. Such external regulatory signals received by the immune system are now considered to operate in addition to its well-documented autoregulatory mechanism (Besedovsky et al 1985). Peptide signals dispatched by neuronal or endocrine cells can reach lymphoid tissues in two ways: (a) These centers have access to blood-borne messengers, such as vasopressin and oxytocin, that are known to act as "helper signals" in the regulation of lymphokine production (Johnson & Torres 1985); (b) some lymphoid structures, e.g. thymus and spleen, have been shown to receive peptidergic innervation, in particular by fibers showing vasoactive intestinal peptide (VIP)-like, neuropeptide-Y-like, met-enkephalin-like, cholecystokinin (CCK)-like, and neurotensin-like immunoreactivity (Felten et al 1985).

Conversely, there is increasing evidence that signals dispatched by immunomodulator agents reach the nervous system and thus provide part of the information recorded and processed by the neuroendocrine regulatory apparatus. For example, microiontophoretic application of α-interferon to various parts of the rat brain, combined with simultaneous single neuron recordings, revealed alterations in cellular activity in all brain structures examined. Moreover, systemic administration of α-interferon resulted in certain behavioral responses (Dafny et al 1985).

A second relationship involves the control over fluid and electrolyte balance (and blood pressure) by vasopressin and by the atrial natriuretic factors (ANFs, atriopeptins) produced by the heart. These recently discovered polypeptide hormones are synthesized by atrial cardiocytes that exhibit ultrastructural features of protein-secreting cells (De Bold 1985). Their potent diuretic and hypotensive effects counteract those of vasopressin. The existence of a feedback mechanism between these antagonistic regulatory substances is demonstrated by the fact that the administration of arginine vasopressin stimulates the release of atriopeptins from the cardiac endocrine system (Manning et al 1985). Moreover, the presence of atriopeptin-immunoreactive neurons in the hypothalamus has been revealed by tests with antisera to atriopeptin III and to the precursor molecule, atriopeptigen (Saper et al 1985).

In light of our present insights, the three early, seemingly disparate, contributions mentioned at the beginning of this chapter fall into line. The study of the gland-like neurons in the spinal cord of fishes discovered by Speidel (1919) was followed up by several groups of investigators (Bern et al 1985, Lederis et al 1985). These cells are now recognized to be part of a neurosecretory system whose neurohemal release site is the urophysis. The chemical and physiological properties of the neurohormones produced by this system, neurotensin I and neurotensin II, have much in common with those of the products of the hypothalamic-hypophysial system of higher vertebrates.

Substance P, discovered by von Euler & Gaddum (1931), has been chemically identified by Leeman (1980) and found to be a multifunctional neuropeptide. The initially puzzling fact that it is extractable from the gut as well as the brain now places it in line with an array of other neuropeptides.

The report on the "pupation hormone" produced by the insect brain (Kopeć 1917, 1922), initially viewed with much skepticism, is now fully supported by multiple studies on the endocrine role as well as the structural properties of the neurosecretory cells of this group of invertebrates.

Concluding Remarks

The elucidation of the phenomenon of neurosecretion has become an enormously productive area of research. New insights gained with accelerating speed from multidisciplinary studies focused on peptide-producing neurons have wrought major changes in neurobiological concepts long considered to be firmly established. Once interpreted as a small, possibly aberrant minority among a well established group of neuroregulators, neuropeptides have now moved to center stage. Their importance in the control over a variety of fundamental biological processes can hardly be overestimated. Their versatility seems to account for much of the complexity and subtlety in neurochemical signaling that could not be accomplished by classical synaptic transmission alone.

The wide distribution of neuropeptides within and outside of the central nervous system, including their coexistence with nonpeptidergic neurotransmitters in many nerve cells, leaves much less room for "conventional neurons" than that assigned to them in the past. The borderline between neurons and other cellular species engaged in chemical communication, especially endocrine cells, has become somewhat indistinct. The focus of attention on features shared by neurons and non-neurons producing bioactive peptides has resulted in two proposals for classification: "APUD cells" (Pearse & Takor Takor 1979) and "paraneurons" (Fujita 1985).

The time has come to make use of this new knowledge in clinical neurology and neuroendocrinology, especially in areas dealing with stress, pain, biological rhythms, memory loss, and depression. The future pace of advances in this field promises to be lively.

Literature Cited

Bargmann, W. 1949. Über die neurosekretorische Verknüpfung von Hypothalamus und Neurohypophyse. *Z. Zellforsch.* 34: 610–34

Bargmann, W., Lindner, E., Andres, K. H. 1967. Über Synapsen an endokrinen Epithelzellen und die Definition sekretorischer Neurone. Untersuchungen am Zwischenlappen der Katzenhypophyse. *Z. Zellforsch.* 77: 282–98

Bargmann, W., Scharrer, E. 1951. The site of origin of the hormones of the posterior pituitary. *Am. Scientist* 39 : 255–59

Barker, J. L., Smith, T. G. Jr., eds. 1980. *The Role of Peptides in Neuronal Function.* New York : Dekker

Benoit, J., Assenmacher, I. 1959. The control by visible radiations of the gonadotropic activity of the duck hypophysis. *Recent Progr. Hormone Res.* 15 : 143–64

Bern, H. A., Pearson, D., Larson, B. A., Nishioka, R. S. 1985. Neurohormones from fish tails : The caudal neurosecretory system. I. "Urophysiology" and the caudal neurosecretory system of fishes. *Recent Progr. Hormone Res.* 41 : 533–52

Besedovsky, H. O., Del Rey, A. E., Sorkin, E. 1985. Immune-neuroendocrine interactions. *J. Immunol. Suppl.* 135 : 750s–54s

Blalock, J. E., Bost, K. L., Smith, E. M. 1985a. Neuroendocrine peptide hormones and their receptors in the immune system. Production, processing and action. *J. Neuroimmunol.* 10 : 31–40

Blalock, J. E., Harbour-McMenamin, D., Smith, E. M. 1985b. Peptide hormones shared by the neuroendocrine and immunologic systems. *J. Immunol. Suppl.* 135 : 858s–61s

Bloom, F. E., ed. 1980. *Peptides : Integrators of Cell and Tissue Function.* Soc. General Physiol. Ser. Vol. 35. New York : Raven

Bloom, F. E., Battenberg, E., Ferron, A., Mancillas, J. R., Milner, R. J., Siggins, G., Sutcliffe, J. G. 1985. Neuropeptides : Interactions and diversities. *Recent Progr. Hormone Res.* 41 : 339–67

Blume, H. W., Pittman, Q. J., Renaud, L. P. 1978. Electrophysiological indications of a 'vasopressinergic' innervation of the median eminence. *Brain Res.* 155 : 153–58

Bodenmüller, H., Schaller, H. C. 1981. Conserved amino acid sequence of a neuropeptide, the head activator, from coelenterates to humans. *Nature* 293 : 579–80

Brecha, N. C., Karten, H. J. 1983. Identification and localization of neuropeptides in the vertebrate retina. In *Brain Peptides*, ed. D. T. Krieger, M. J. Brownstein, J. B. Martin, pp. 437–62. New York : Wiley

Buijs, R. M., Swaab, D. F., Dogterom, J., van Leeuwen, F. W. 1978. Intra- and extrahypothalamic vasopressin and oxytocin pathways in the rat. *Cell Tiss. Res.* 186 : 423–33

Burgen, A., Kosterlitz, H. W., Iversen, L. L., eds. 1980. Neuroactive Peptides. *Proc. R. Soc. London Ser. B* 210 : 1–195

Chan-Palay, V., Jonsson, G., Palay, S. L. 1978. Serotonin and substance P coexist in neurons of the rat's central nervous system. *Proc. Natl. Acad. Sci.* 75 : 1582 86

Chan-Palay, V., Palay, S. L. 1984. Cerebellar Purkinje cells have GAD, motilin and CSAD immunoreactivity. Existence and coexistence of GABA, motilin and taurine. In *Coexistence of Neuroactive Substances*, ed. V. Chan-Palay, S. L. Palay, pp. 1–22. New York : Wiley

Costa, E., Trabucchi, M., eds. 1980. *Neural Peptides and Neuronal Communication.* New York : Raven

Dafny, N., Prieto-Gomez, B., Reyes-Vazquez, C. 1985. Does the immune system communicate with the central nervous system? Interferon modifies central nervous activity. *J. Neuroimmunol.* 9 : 1–12

De Bold, A. J. 1985. Atrial natriuretic factor : A hormone produced by the heart. *Science* 230 : 767–70

De Wied, D. 1978. Pituitary peptides and adaptive behavior. See Meites et al 1978, pp. 383–400

Dogterom, J., Snijdewint, F. G. M., Buijs, R. M. 1978. The distribution of vasopressin and oxytocin in the rat brain. *Neurosci. Lett.* 9 : 341–46

Etkin, W. 1962. Hypothalamic inhibition of pars intermedia activity in the frog. *Gen. Comp. Endocrinol. Suppl.* 1 : 148–59

Etkin, W. 1963. Metamorphosis. In *Physiology of the Amphibia*, ed. J. Moore, Chapt. 8. New York : Academic

Felten, D. L., Felten, S. Y., Carlson, S. L., Olschowka, J. A., Livnat, S. 1985. Noradrenergic and peptidergic innervation of lymphoid tissue. *J. Immunol. Suppl.* 135 : 755s–65s

Fujita, T. 1985. Neurosecretion and new aspects of neuroendocronology. See Kobayashi et al, eds., 1985, pp. 21–28

Gainer, H., ed. 1977. *Peptides in Neurobiology.* New York/London : Plenum

Goetzl, E. J., ed. 1985. Neuromodulation of immunity and hypersensitivity. *J. Immnol. Suppl.* 135 : 739s–863s

Guillemin, R. 1978. Peptides in the brain : The new endocrinology of the neuron. *Science* 202 : 390–402

Hansen, B. L., Hansen, G. N., Scharrer, B. 1982. Immunoreactive material resembling vertebrate neuropeptides in the corpus cardiacum and corpus allatum of the insect *Leucophaea maderae. Cell Tissue Res.* 255 : 319–29

Harris, G. W. 1955. *Neural Control of the Pituitary Gland.* London : Arnold

Hild, W. 1951. Experimentell-morphologische Untersuchungen über das Verhalten der "Neurosekretorischen Bahn" nach Hypophysenstieldurchtrennungen, Eingriffen in den Wasserhaushalt und Belastung der Osmoregulation. *Virchows Arch.* 319 : 526–46

1332 SCHARRER

Hild, W., Zetler, G. 1953. Experimenteller Beweis für die Entstehung der sog. Hypophysenhinterlappenwirkstoffe im Hypothalamus. *Pflügers Arch.* 257 : 169–201

Hökfelt, T., Johansson, O., Goldstein, M. 1984. Chemical anatomy of the brain. *Science* 225 : 1326–34

Johnson, H. M., Torres, B. A. 1985. Regulation of lymphokine production by arginine vasopressin and oxytocin : Modulation of lymphocyte function by neurohypophyseal hormones. *J. Immunol. Suppl.* 135 : 773s–75s

Kobayashi, H., Bern, H. A., Urano, A., eds. 1985. *Neurosecretion and the Biology of Neuropeptides. Proc. 9th Int. Symp. on Neurosecretion.* Tokyo : Japan Sci. Soc. Press, and Berlin/Heidelberg/New York/ Tokyo : Springer-Verlag

Kopeć, S. 1917. *Bull. Acad. Sci. Cracovie, Classe Sci. Math. Nat. Sér. B,* pp. 57–60

Kopeć, S. 1922. Studies on the necessity of the brain for the inception of insect metamorphosis. *Biol. Bull.* 42 : 323–42

Krieger, D. T., Brownstein, M. J., Martin, J. B., eds. 1983. *Brain Peptides.* New York : Wiley

Kuljis, R. O., Karten, H. J. 1982. Laminar organization of peptide-like immunoreactivity in the anuran optic tectum. *J. Comp. Neurol.* 212 : 188–201

Lederis, K., Fryer, J., Rivier, J., MacCannell, K. L., Kobayashi, Y., Woo, N., Wong, K. L. 1985. Neurohormones from fish tails. II : Actions of urotensin I in mammals and fishes. *Recent Progr. Hormone Res.* 41 : 553–76

Leeman, S. E. 1980. Substance P and neurotensin : Discovery, isolation, chemical characterization and physiological studies. *J. Exp. Biol.* 89 : 193–200

Manning, P. T., Schwartz, D., Katsube, N. C., Holmberg, S. W., Needleman, P. 1985. Vasopressin-stimulated release of atriopeptin : Endocrine antagonists in fluid homeostasis. *Science* 229 : 395–97

Meites, J., Donovan, B. T., McCann, S. M., eds. 1975. *Pioneers in Neuroendocrinology,* Vol. 1. New York/London : Plenum

Meites, J., Donovan, B. T., McCann, S. M., eds. 1978. *Pioneers in Neuroendocrinology,* Vol. 2. New York/London : Plenum

Nikitovitch-Winer, M., Everett, J. W. 1958. Functional restitution of pituitary grafts retransplanted from kidney to median eminence. *Endocrinology* 63 : 916–30

Oksche, A., Wilson, W. O., Farner, D. S. 1964. The hypothalamic neurosecretory system of *Coturnix coturnix japonica. Z. Zellforsch.* 61 : 688–709

Patel, Y. C., Reichlin, S. 1978. Somatostatin in hypothalamus, extrahypothalamic brain, and peripheral tissues of the rat. *Endocrinology* 102 : 523–30

Pearse, A. G. E., Takor Takor, T. 1979. Embryology of the diffuse neuroendocrine system and its relationship to the common peptides. *Federation Proc.* 38 : 2288–94

Rémy, C., Dubois, M. P. 1981. Immunohistological evidence of methionine enkephalin-like material in the brain of the migratory locust. *Cell Tissue Res.* 218 : 271–78

Roth, J., Le Roith, D. 1984. Intercellular communication : The evolution of scientific concepts and of messenger molecules. In *Medicine, Science, and Society.* Symposia celebrating the Harvard Medical School Bicentennial, ed. K. J. Isselbacher, pp. 425–47. New York : Wiley

Saper, C. B., Standaert, D. G., Currie, M. G., Schwartz, D., Geller, D. M., Needleman, P. 1985. Atriopeptin-immunoreactive neurons in the brain : Presence in cardiovascular regulatory areas. *Science* 227 : 1047–49

Schally, A. V. 1978. Aspects of hypothalamic regulation of the pituitary gland. Its implications for the control of reproductive processes. *Science* 202 : 18–28

Scharrer, B. 1935. Über das Hanströmsche Organ X bei Opisthobranchiern. *Pubbl. Staz. Zool. Napoli* 15 : 132–42

Scharrer, B. 1952. Neurosecretion. XI. The effects of nerve section on the intercerebralis-cardiacum-allatum system of the insect *Leucophaea maderae. Biol. Bull.* 102 : 261–72

Scharrer, B. 1970. General principles of neuroendocrine communication. In *The Neurosciences : Second Study Program,* ed. F. O. Schmitt, pp. 519–29. New York : Rockefeller Univ. Press

Scharrer, B. 1972. Neuroendocrine communication (neurohormonal, neurohumoral, and intermediate). *Progr. Brain Res.* 38 : 7–18

Scharrer, B. 1974. New trends in invertebrate neurosecretion. In *The Final Neuroendocrine Pathway. Proc. 6th Int. Symp. on Neurosecretion,* pp. 285–87. Berlin/Heidelberg/New York : Springer-Verlag

Scharrer, B. 1978. *An evolutionary interpretation of the phenomenon of neurosecretion,* pp. 1–17. New York : Am. Museum Natural History

Scharrer, B. 1981. Neuroendocrinology and histochemistry. In *Histochemistry : The Widening Horizons,* ed. P. J. Stoward, J. M. Polak, pp. 11–20. Chichester, UK : Wiley

Scharrer, B., Scharrer, E. 1944. Neurosecretion. VI. A comparison between the intercerebralis-cardiacum-allatum system of the insects and the hypothalamo-hypophyseal system of the vertebrates. *Biol. Bull.* 87: 243–51

Scharrer, E. 1928. Die Lichtempfindlichkeit blinder Elritzen (Untersuchungen über das Zwischenhirn der Fische). *Z. Vergl. Physiol.* 7: 1–38

Scharrer, E. 1952. The general significance of the neurosecretory cell. *Scientia* 46: 177–83

Scharrer, E. 1966. Principles of neuroendocrine integration. In *Endocrines and the Central Nervous System. Res. Public. Assoc. Nerv. Ment. Dis.* 43: 1–35

Scharrer, E., Brown, S. 1961. Neurosecretion. XII. The formation of neurosecretory granules in the earthworm, *Lumbricus terrestris* L. *Z. Zellforsch.* 54: 530–40

Scharrer, E., Scharrer, B. 1937. Über Drüsen-Nervenzellen und neurosekretorische Organe bei Wirbellosen und Wirbeltieren. *Biol. Rev.* 12: 185–216

Scharrer, E., Scharrer, B. 1940. Secretory cells within the hypothalamus. *Res. Public. Assoc. Res. Nerv. Ment. Dis.* 20: 170–94

Scharrer, E., Scharrer, B. 1945. Neurosecretion. *Physiol. Rev.* 25: 171–81

Scharrer, E., Scharrer, B. 1954. Hormones produced by neurosecretory cells. *Recent Progr. Hormone Res.* 10: 183–240

Scharrer, E., Scharrer, B. 1963. *Neuroendocrinology.* New York: Columbia Univ. Press

Scheller, R. H., Kaldany, R.-R., Kreiner, T., Mahon, A. C., Nambu, J. R., Schaefer, M., Taussig, R. 1984. Neuropeptides: Mediators of behavior in *Aplysia. Science* 225: 1300–8

Sofroniew, M. V., Weindl, A. 1978. Extra-hypothalamic neurophysin-containing perikarya, fiber pathways and fiber clusters in the rat brain. *Endocrinology* 102: 334–37

Speidel, C. C. 1919. Gland-cells of internal secretion in the spinal cord of the skates. *Carnegie Inst. Washington Publ. No.* 13: 1–31

Stefano, G. B., Scharrer, B. 1981. High affinity binding of an enkephalin analog in the cerebral ganglion of the insect *Leucophaea maderae* (Blattaria). *Brain Res.* 225: 107–14

Sternberger, L. A. 1986. *Immunocytochemistry.* New York: Wiley. 3rd ed.

Strumwasser, F. 1985. The structure of the neuroendocrine commands for egg-laying behavior in *Aplysia.* In *Comparative Neurobiology,* ed. M: J. Cohen, F. Strumwasser, pp. 169–79. New York: Wiley

Szentágothai, J., Flerkó, B., Mess, B., Halász, B. 1962. *Hypothalamic Control of the Anterior Pituitary. An Experimental Morphological Study.* Budapest: Akadémiai Kiadó

Thorn, N. A., Chenoufi, H.-L., Tiefenthal, M. 1985. The calcium-calmodulin-proteinkinase system and the mechanism of release of neurohypophysial hormones. *Acta Physiol. Scand. Suppl. 542* 124: 300

Von Euler, U. S., Gaddum, J. H. 1931. An unidentified depressor substance in certain tissue extracts. *J. Physiol.* 72: 74–87

Way, E. L., ed. 1980. *Endogenous and Exogenous Opiate Agonists and Antagonists.* New York/Oxford/Toronto/Sydney/Frankfurt/Paris: Pergamon

White, J. D., Stewart, K. D., Krause, J. E., McKelvy, J. F. 1985. Biochemistry of peptide-secreting neurons. *Physiol. Rev.* 65: 553–97

Ann. Rev. Neurosci. 1987. 10 : 19–40

PERSPECTIVES ON THE DISCOVERY OF CENTRAL MONOAMINERGIC NEUROTRANSMISSION

Arvid Carlsson

Department of Pharmacology, University of Göteborg,
S-400 33 Göteborg, Sweden

The concept of chemical neurotransmission goes back to the beginning of this century but was confined to the peripheral nervous system for a long time. A number of parallel events in the 1950s triggered its penetration through the blood-brain barrier : these included the discovery of the cate-cholamines and 5-hydroxytryptamine (5-HT, serotonin) in the central nervous system, the introduction of the modern psychotropic drugs, and the development of sensitive and specific biochemical and histochemical methods for the detection of the monoamines and their precursors and metabolites in tissues and body fluids.

In the 1950s very little was known about information transfer between nerve cells in the central nervous system. Electrical transmission seemed to be a likely mechanism in many instances, given the generally close synaptic contacts and sometimes short synaptic delays in the central nervous system. In his monograph, *Synaptic Transmission*, McLennan (1963) concludes : "In the vertebrate central nervous system there is only one synapse identified whose operation can, with assurance, be ascribed to acetylcholine." He was referring to the synapses where the Renshaw cells in the spinal cord receive innervation from axon collaterals of the moto-neurons. Since these neurons are undoubtedly cholinergic, the conclusion was reasonable. However, no generalizations from this special case could of course be made. As to the monoamines and other putative neuro-transmitters, McLennan did not find a single case in which any of them

1335

0147–006X/87/0301–0019$02.00

could be ascribed a role as a neurotransmitter in the vertebrate central nervous system.

"Sympathin," i.e. a mixture of noradrenaline and adrenaline, was shown to be a normal constituent of brain tissue by Marthe Vogt in 1954. Simultaneously, 5-HT was discovered in the brain (Twarog & Page 1953, Amin et al 1954).

Impact of Psychopharmacology and New Analytical Techniques

The ability of the hallucinogenic agent LSD to block peripheral 5-HT receptors led to speculations about a role for 5-HT in maintaining sanity (Gaddum 1954, Woolley & Shaw 1954). These speculations prompted Brodie and Shore to study the interactions among 5-HT, LSD, and the newly discovered antipsychotic agent, reserpine. From observations on sleeping times in mice they concluded that LSD could antagonize both 5-HT and reserpine (Shore et al 1955).

At this time an important methodological innovation was being developed in Dr. Brodie's laboratory at the National Heart Institute in Bethesda, Maryland: Dr. Bowman, in collaboration with Drs. Brodie and Udenfriend, was constructing the first prototype of a spectrophotofluorometer. This instrument proved to be extremely useful especially in biochemical research, e.g. for the analysis of monoamines and their precursors and metabolites in tissues and body fluids. It was soon to replace the previous bioassay techniques. Using the new instrument, Brodie and Shore were able to demonstrate the virtually complete disappearance of 5-HT from tissues, including the brain, and the simultaneous increase in the urinary excretion of the 5-HT metabolite, 5-hydroxyindoleacetic acid, following treatment with reserpine (Shore et al 1955, Pletscher et al 1955).

The first demonstration of the effect of a psychotropic agent on an endogenous agonist in the central nervous system had an enormous impact on the research field. Already the first attempts to explain the actions of reserpine on 5-HT were based on the tacit assumption that a neurotransmitter mechanism was involved. Brodie and his colleagues put forward the hypothesis that reserpine, by blocking the storage mechanism without causing any inhibition of synthesis, induced a continuous, uninhibited release of 5-HT onto receptor sites. This interpretation seemed logical in view of the interaction experiments carried out by these workers, mentioned above. However, their interpretation of these experiments was questionable in view of the poor penetration of systemically administered 5-HT into the brain.

I had the great privilege to spend six months in 1955–1956 in Dr. Brodie's laboratory, while on sabbatical leave from my assistant professorship at the University of Lund, Sweden. My contact with Drs. Brodie and Udenfriend was established through Dr. Sune Bergström, who was then Chairman of the Department of Physiological Chemistry at the University of Lund. I am indebted to Dr. Bergström for a lot of support and encouragement during this early part of my research career. Coming from an entirely different field (bone-mineral metabolism), I was introduced by Drs. Brodie and Shore to the fascinating field of biogenic amines and to the use of the spectrophotofluorometer. I was given the opportunity to demonstrate, in collaboration with Brodie and Shore, that reserpine, added in low concentration to platelets in vitro, was capable of releasing 5-HT from the platelets by a stereospecific mechanism (Carlsson et al 1957c).

I suggested to Drs. Brodie and Shore that it might be worthwhile to study the action of reserpine on the catecholamines as well, but they did not consider such an approach very promising, given the lack of sympathomimetic actions of this agent. While still in Bethesda, however, I had read with great interest a paper from my home University by Hillarp et al (1953) that showed the existence of specific storage organelles, called granules or vesicles, in the adrenal medulla, and a subsequent paper (Hillarp et al 1955) demonstrating the co-existence of catecholamines and ATP in these organelles. In preliminary experiments, I found large amounts of ATP also in platelets, as was shortly afterwards reported by Born (1956). Thus a link between the storage of 5-HT and catecholamines was suggested.

Immediately after my return to Lund, Hillarp and I started to collaborate. We very soon discovered the depletion of adrenal catecholamine stores by reserpine (Carlsson & Hillarp 1956). Shortly afterwards, together with my graduate students Bertler and Rosengren, I demonstrated the depletion of noradrenaline stores in the heart and brain by this agent (Bertler et al 1956, Carlsson et al 1957a). Moreover, we showed that the sympathetic nerves ceased to respond to stimulation following depletion of catecholamines by reserpine. Similar observations were made by Holzbauer & Vogt (1956) and by Muscholl & Vogt (1958).

Thus, my colleagues and I had to disagree with my highly esteemed mentors and friends, Drs. Brodie and Shore, on two essential points. We concluded (*a*) that not only 5-HT but also the catecholamines had to be considered in attempts to explain the mode of action of reserpine, and (*b*) that rather than the proposed continuous agonist release onto receptors, one should consider lack of neurotransmitter as the functionally crucial result of reserpine's action. In fact, the release of the adrenergic transmitter by reserpine did not seem to occur into the synaptic cleft, because it was not accompanied by any sympathomimetic actions. It was thus reasonable to assume that the release occurred from the storage organelles into the cytoplasm and was followed by intracellular deamination by monoamine oxidase (MAO). In support of this proposal, reserpine had already been shown to cause a sympathomimetic response as well as central stimulation following pretreatment with a MAO inhibitor, thus suggesting that amine release onto receptors did indeed take place when deamination was prevented.

In order to test the amine-deficiency hypothesis we tried to replenish the stores of reserpine-treated animals by the systemic administration of precursors, which in contrast to the amines themselves are capable of penetrating from the blood into the brain. We thus discovered the dramatic antireserpine action of the catecholamine precursor, DOPA, as well as its centrally stimulating action in nonpretreated animals; the 5-HT precursor, 5-hydroxytryptophan, was not an effective reserpine antagonist, however, thus suggesting that lack of catecholamines rather than 5-HT was responsible for the gross behavioral actions of reserpine (Carlsson et al 1957b).

Dopamine : A Central Agonist in Its Own Right

Since the action of DOPA was strongly potentiated by the MAO inhibitor, iproniazide, we were convinced that the catecholamine(s) formed from DOPA mediated the effect. We were hoping that noradrenaline, which was then recognized as the major central catecholamine, would prove to be responsible for the effect. Much to our regret, however, we found that despite the virtual elimination of the behavioral action of reserpine after DOPA treatment, the noradrenaline levels in the brain remained at or below the limit of detection by the methods then available. We then turned our attention to dopamine, which at that time was recognized as a poor sympathomimetic agonist and believed to be just a precursor of noradrenaline and adrenaline. We had first to develop a spectrophotofluorometric method for detection and quantitative measurement of dopamine (Carlsson & Waldeck 1958). This method was used after the purification of tissue extracts on Dowex 50 columns. The procedure could thus be adapted to our previous method for the determination of adrenaline and noradrenaline in tissues (Bertler et al 1958). These methods soon became widely used in catecholamine research. We were pleased to find that dopamine did accumulate in the brains of animals treated with DOPA following reserpine pretreatment and that this accumulation coincided with the antireserpine response (Carlsson et al 1958). Moreover, we found that dopamine occurs normally in the brain, as shortly before suggested by Montagu (1957). Since the amounts were comparable to those of noradrenaline and since dopamine, too, was made to disappear almost completely by reserpine, we suggested that dopamine besides being an intermediate in catecholamine synthesis, is an agonist in its own right in the central nervous system.

To investigate the regional distribution of dopamine seemed to be an appropriate part of the thesis work conducted by Bertler & Rosengren. They (1959) thus discovered the unique accumulation of dopamine in the basal ganglia. This region had long been recognized as an important component of the extrapyramidal system. Moreover, since the extra-pyramidal, Parkinson-like side effects of reserpine were known, it did not seem farfetched to propose a role for dopamine in the control of extrapyramidal motor functions, a deficiency of which would lead to Parkinsonism, while excessive function would give rise to chorea. These ideas were first proposed at the First International Catecholamine Symposium in Bethesda in 1958 (Carlsson 1959).

My colleagues and I were quite excited by these observations. We felt that the opposite behavioral actions of the monoamine depletor, reserpine, and the catecholamine precursor, DOPA, made a strong case for the catecholamines as agonists in the central nervous system, controlling important functions such as motor activity and alertness. We were very disappointed by the scepticism with which our interpretations were received. Thus the Ciba Foundation Symposium in London in 1960 (Vane et al 1960) revealed considerable disagreement on various points, as evident from the recorded discussions. Marthe Vogt in particular was reluctant to accept the available evidence for a role of catecholamines (or 5-HT) in behavior. Sir Henry Dale was surprised to hear that an amino acid such as DOPA could be a "poison." Unpublished data were quoted, indicating that the catecholamines in the brain were located in glia cells. Sir John Gaddum, in his final remarks, concluded: "The meeting was in a critical mood, and no-one ventured to speculate on the relation between catechol amines and the function of the brain." This was indeed a puzzling comment, since I for one had speculated quite a lot on precisely this issue.

When rereading the proceedings of this symposium today I still find it difficult to understand the resistance to our interpretations of the reserpine-DOPA data. There was little or no disagreement about the actual observations. In fact, others had confirmed or independently demonstrated the most salient findings. As mentioned, Marthe Vogt and her colleagues had also observed the depletion of noradrenaline by reserpine and the resultant loss of adrenergic nerve function. At the meeting Drs. Blaschko and Chrusciel described the antireserpine action of DOPA. As it turned out, our disagreement with Brodie was minor in comparison with these other controversies, which raised doubts about any role for the monoamines as agonists in the brain.

Visualization of the Monoamines in the Fluorescence Microscope

The acceptance of the monoamines as neurotransmitters in the central nervous system was a slow process. Certainly the important discovery of low dopamine levels in the basal ganglia of Parkinson patients (Ehringer & Hornykiewicz 1960) and the demonstration of a therapeutic effect of L-dopa in Parkinson patients (Birkmayer & Hornykiewicz 1962, Barbeau et al 1962) contributed to bringing the central monoamines into focus. However, the real breakthrough came with the histochemical techniques visualizing the cellular localization of the monoamines by means of the fluorescence microscope. This fascinating story, which has recently been reviewed in some detail (Dahlström & Carlsson 1986), is briefly summarized below.

I had the privilege to collaborate for many years with the late Dr. Nils-Åke Hillarp, a highly talented histologist who died at the age of 48 in 1965. Hillarp had already in the 1940s and 1950s acquired a considerable research experience in the area of the autonomic nervous system and made several important discoveries. Our collaboration was not confined to the mechanism of storage of catecholamines; we also engaged in attempts to visualize the catecholamines under the microscope. At that time such visualization was possible only in the chromaffin cells of, for example, the adrenal medulla, by using the so-called chromaffin reaction or, as shown by Eränko (1955), by the green fluorescence occurring in sections of the adrenal medulla following formalin fixation. The high concentrations of dopamine in the basal ganglia seemed encouraging, however, and Falck & Hillarp (1959) had actually tried the chromaffin reaction to visualize the catecholamines in the central nervous system, though without success. Evidently, more sensitive methods had to be developed. It was logical to turn our attention to fluorescence, given the tremendous success of this principle in the biochemical methods recently developed for the analysis of the monoamines.

When I was appointed to the chair of pharmacology at the University of Göteborg in 1959, I was delighted to learn that Hillarp wanted to join me. This was made possible through a grant from the Swedish Medical Research Council, which enabled Hillarp to take leave from his associate professorship at the University of Lund. This was probably one of the most profitable investments of the Council to date. In 1960 we moved into the newly built department of pharmacology in Göteborg. The University had allowed fairly generous funding for the equipment of the new institution, and thus Hillarp and I could start our research in Göteborg without too much delay.

Our first attempt to visualize catecholamines in the fluorescence microscope utilized the principle of the so-called trihydroxyindole method for quantitative analysis of catecholamines, i.e. the principle used, for example, by Bertler et al (1958). Tissue sections were first exposed to iodine to oxidize the catecholamines to adrenochromes, and then exposed to ammonia to rearrange the adrenochromes to fluorescent adrenolutines. Our efforts were very successful insofar as the adrenal medulla was concerned: a highly intensive fluorescence developed; it was reduced, though still clearly visible, after removing more than 95% of the catecholamines by reserpine treatment (Carlsson et al 1961). However, for some unknown reason, perhaps diffusion of the amines out of the tissue, the adrenergic transmitter could not be visualized in nerve terminals or cell bodies. Nevertheless, the results were considered very encouraging, and Hillarp was convinced that the project would ultimately prove successful.

Hillarp decided to try another principle for the development of fluorescence from monoamines, based on the analytical method of Hess & Udenfriend (1959) for measurement of tryptamine. In this method tryptamine is condensed with formaldehyde to form a highly fluorescent product. Together with his skillful Research Engineer, Georg Thieme, Hillarp started to investigate systematically this reaction in histochemical model experiments. Various amines were dissolved in solutions of serum albumin, sucrose, gelatin, or gliadin, spotted on glass slides and air-dried. Upon treatment with formaldehyde vapor, generated from a 35% solution of formaldehyde, a very strong fluorescence developed in spots of noradrenaline or dopamine. Protein catalyzed the reaction. The fluorescence products were identified as tetrahydroisoquinolines, as verified by Corrodi & Hillarp (1964).

The report of these fundamental model experiments, all of which were performed by Hillarp and Thieme in the Department of Pharmacology at the University of Göteborg, was authored by Falck, Hillarp, Thieme & Torp (1962), and was one of the 100 most cited publications in 1961–1982. When Hillarp moved to Göteborg, Falck remained at the Department of Histology, University of Lund. After the model experiments were completed, a considerable part of the experiments on various tissue specimens were performed in collaboration with Falck in Lund, since a histology department was of course better equipped for this purpose. Various attempts were made, though without any clearcut success, until one day in late August 1961, when Hillarp visited Falck and his old department, Hillarp proposed that they try air-dried stretch preparations of thin tissue specimens, such as rat iris and mesentery. Hillarp was very familiar with these preparations from his previous work (1946, 1959), now considered classical treatises on the functional organization of the autonomic ground plexus. Such preparations were exposed to dry formaldehyde gas generated from paraformaldehyde powder. The outcome was dramatic: in the fluorescence microscope Hillarp and Falck saw the same nerve-plexus pattern as previously observed by Hillarp following staining with methylene blue. But this time it was the adrenergic transmitter, which showed up as green fluorescence as a consequence of treatment with formaldehyde. In addition, yellow fluorescence derived from mast-cell 5-HT could be seen in the mesenterium preparations. In principle, a great discovery had thus been made, but needless to say, a lot of work remained. For example, the technique had to be adapted to embedded tissue specimens. This work was first performed in Lund and later, after Hillarp's move to take over the chair in histology at the Karolinska Institute in Stockholm in 1962, also by a rapidly growing group of young, enthusiastic students in Hillarp's new working place. For a detailed account of the further development of the histofluorescence techniques, see Dahlström & Carlsson (1986) and Björklund & Hökfelt (1984). The new technique was a powerful tool that helped to solve a large number of important problems in the monoamine field.

Cellular Localization and Mapping of the Monoamines

A major, initial step in cellular localization and mapping of the monoamines was the demonstration of the monoamines in nerve cell bodies and nerve terminals of the central nervous system (Carlsson et al 1962). The neuronal localization of the central monoamines was confirmed by lesion experiments. In fact, the first lesion experiments, which demonstrated the virtually complete disappearance of monoamines from the spinal cord below a transection, utilized biochemical techniques only (Magnusson & Rosengren 1963, Carlsson et al 1963b), shortly before the demonstration of the descendent bulbospinal monoaminergic pathways by means of the histofluorescence technique (Carlsson et al 1964). The first lesion experiments to confirm the neuronal localization of a monoamine in the brain demonstrated the disappearance of dopamine, measured biochemically, and of green-fluorescent nerve terminals from the rat neostriatum following a lesion of the substantia nigra (Andén et al 1964a). Moreover, removal of the striatum was followed by an accumulation of fluorescent material in the cell bodies of the substantia nigra and of axons proximal to the lesion, thus demonstrating the existence of a nigrostriatal dopamine pathway. Independent work in Lund by Bertler et al (1964) reported similar findings.

But these were only beginnings. An enormous amount of work remained to map out all the monoaminergic pathways in the central nervous system. A large number of workers became engaged in this important mapping.

The first systematic studies were performed by Dahlström & Fuxe (1964) and Fuxe (1965), followed by Ungerstedt (1971) and many others (see Lindvall 1974, Björklund & Hökfelt 1984). Needless to say, the detailed knowledge of the central monoaminergic pathways has been of fundamental importance for the further development of research in this field. For example, it enabled Andén and his colleagues (see Andén et al 1969) to develop simple and useful functional models for the individual monoamines, and Aghajanian and his colleagues (see e.g. Aghajanian & Bunney 1974, 1977, Bloom 1984) to embark on their pioneer studies of the electrophysiological activity of monoaminergic neurons.

Fitting Together the Pieces of the Synapse Puzzle

The impact of the histochemical visualization of the monoamines of course extends far beyond the mapping of monoamine-carrying neuronal pathways. After the introduction of the new technique, previous doubts expressed about the transmitter function of the monoamines gradually changed into a debate on the complex issue of neurotransmitter versus neuromodulator function. Today nobody appears to question the neurotransmitter function of the monoamines in the central nervous system, at least in a broad sense of this term. In fact, the monoamines have become "spearheads" in neurotransmission research, especially in the central nervous system. In particular, the monoaminergic synapse has become a very useful model.

After the localization of the central monoamines to special neurons had been established, it remained to demonstrate their subcellular distribution. In the fluorescence microscope the accumulation of monoamines in the so-called varicosities of nerve terminals was obvious. This corresponded to the distribution of synaptic vesicles, as observed in the electron microscope. In fact, Hökfelt (1968) was able to demonstrate the localization of central as well as peripheral monoamines to synaptic vesicles in the electron microscope.

In 1960, Axelrod et al (see Axelrod 1964) discovered the uptake of circulating labeled catecholamines by adrenergic nerves. This uptake proved to be an important inactivation mechanism. According to Axelrod it could be blocked by a large number of drugs, for example, reserpine, chlorpromazine, cocaine, and imipramine, thus leading to supersensitivity to catecholamines and acceleration of their metabolism. This interpretation was not entirely in line with ours, especially insofar as reserpine was concerned; we considered the action of reserpine on the adrenergic transmitter to be a strictly intraneuronal event. The discrepancy could be resolved by combined biochemical (Carlsson et al 1963a, Kirshner 1962) and histochemical studies (Malmfors 1965). Two different amine-concentrating mechanisms were detected : uptake at the level of the cell membrane, sensitive for example to cocaine and imipramine ; and uptake by the storage granules or synaptic vesicles, sensitive for example to reserpine. Blockade of the former but not the latter mechanism leads to catecholamine supersensitivity (although secondary receptor supersensitivity may develop after blockade of the latter). In the presence of reserpine, extracellular amine is still pumped into the cytoplasm with unabated efficiency. However, since it cannot be stored, it is deaminated by intraneuronal MAO (see also Carlsson 1966).

Another issue emerged after the discovery by Axelrod and his colleagues of catechol-O-methyl-transferase (COMT). Axelrod (1960) proposed that COMT was mainly responsible for the metabolism of catecholamines, whereas MAO was thought to be primarily involved in the metabolism of the O-methylated metabolites of the catecholamines. Axelrod referred to the intracellular geography : since COMT occurs in the cell sap, newly released catecholamines will be primarily exposed to this enzyme and only secondarily to the mitochondrially located MAO. However, our reserpine data (Carlsson et al 1957a), as well as our observations on the accumulation of catecholamines and their metabolites (Carlsson et al 1960), suggested to us that intraneuronally released catecholamines would be primarily metabolized by MAO ; only after release into the extracellular space would they be exposed to COMT (Carlsson 1960, Carlsson & Hillarp 1962). This concept was later generally accepted (see e.g. Axelrod 1964, Jonason & Rutledge 1968, Westerink & Spaan 1982).

In the mid-1960s several controversies still existed in the area of monoaminergic synaptology. This is evident from the recorded discussions of the symposium, "Mechanisms of Release of Biogenic Amines," held in Stockholm in February 1965 (von Euler et al 1966). One of the major issues dealt with the release mechanism in relation to the subcellular distribution of the transmitter. According to Drs. Axelrod and von Euler, a considerable part of the transmitter was located outside the granules (vesicles), mainly in a bound form. This fraction was proposed to be more important than the granular fraction, since it was more readily available for release. Quoting a conversation with Udenfriend, von Euler underlined this by facetiously referring to the granules as "garbage cans." Our group had arrived at a different view, based on a variety of biochemical, histochemical, and pharmacological data. We felt that the granules were essential in transmission, in that the transmitter had to be taken up by them in order to become available for release by the nerve impulse. In favor of this contention was our finding, mentioned above, that reserpine's site of action is the amine uptake mechanism of the granules. The failure of adrenergic transmission as well as the behavioral actions of reserpine were closely correlated to the blockade of granular uptake induced by the drug (Lundborg 1963). Moreover, extragranular noradrenaline (accumulated in adrenergic nerves by pretreatment with reserpine, followed by an inhibitor of MAO and systematically administered noradrenaline) was found to be unavailable for release by the nerve impulse, as observed histochemically (Malmfors 1965). We proposed that under normal conditions the extragranular fraction of monoaminergic transmitters was very small, owing to the presence of MAO intracellularly. Subsequently work in numerous laboratories has lent support to these views. Already, at the Symposium, Douglas had presented evidence suggesting a Ca^{2+}-triggered fusion between the granule and cell membranes, preceding the release. The release is now generally assumed to take place as "exocytosis," even though the complete extrusion of the granule content may still be debatable.

Another, much debated area was the regulation of monoamine synthesis. Shortly after the discovery of DOPA decarboxylase (Holtz 1939), Blaschko (1939) formulated the main pathway for catecholamine synthesis: tyrosine, DOPA, dopamine, noradrenaline, adrenaline. The above-mentioned rapid conversion of administered DOPA to dopamine indicated that the first step in the pathway was rate limiting. Perhaps the first evidence for a regulation of catecholamine synthesis came from the observation that the accumulation of catecholamines and their O-methylated basic metabolites in mouse brain following inhibition of MAO leveled off within a few hours, suggesting that the synthesis was brought to a standstill when the stores had been filled (Carlsson et al 1960). It should be noted that the highly polar catecholamines and their basic O-methylated metabolites do not seem to escape easily through the blood-brain barrier. Subsequent, more sophisticated work in a large number of laboratories has revealed that catecholamine synthesis is regulated by several independent mechanisms. After the isolation of tyrosine hydroxylase, Udenfriend and his colleagues discovered an inhibitory action of catechols on this enzyme (Nagatsu et al 1964), whose affinity for the tetrahydropteridine co-enzyme was reduced. End-product inhibition was thus demonstrated, and evidence that this mechanism operates under physiological conditions was later presented (see Carlsson et al 1976). For several years, end-product inhibition was believed to be the only mechanism of short-term control of catecholamine synthesis. For long-term control, enzyme induction was shown to be responsible (Thoenen et al 1973). However, short-term activation of tyrosine hydroxylase can take place in dopamine neurons in vivo despite an increase in dopamine levels, thus indicating the existence of an additional control mechanism (Carlsson et al 1972). Phosphorylation of the enzyme seems to be involved in this reglation (Lovenberg et al 1975).

Several in-vitro and in-vivo approaches have been used to demonstrate and measure the release of amine transmitters and their dependence on the nerve impulses. Release in vivo has been shown by using push-pull cannulas (Cheramy et al 1981), semi-permeable tubes (Ungerstedt et al 1983), and voltammetry (Adams & Marsden 1982). Indirect but non-traumatic biochemical approaches have also proven useful, such as measurements of the rate of transmitter disappearance following inhibition of its synthesis (see Andén et al 1969) and the rate of accumulation of O-methylated basic metabolites following inhibition of MAO (Kehr 1976). The data thus obtained demonstrate that dopamine, noradrenaline, and serotonin are released from nerve terminals by nerve impulses. However, in the somatodendritic region of the nigral dopamine neurons, a rapid release and turnover of dopamine appears to occur more or less independently of nerve impulses, beyond the reach of receptor-mediated control (see Cheramy et al 1981, Nissbrandt et al 1985). The physiological significance of this observation is not known, but the possible implications are intriguing.

The False-Transmitter Concept

The "false-transmitter" concept is another spin-off of the early attempts to elucidate the mode of action of antipsychotic agents. Udenfriend and his colleagues (Hess et al 1961) had discovered that alpha-methyl-meta-tyrosine is capable of depleting central noradrenaline stores, while leaving the 5-HT stores intact. Costa et al (1962) pointed out that no sedation occurred after treatment with this agent despite the virtually complete depletion of noradrenaline stores; thus the central action of reserpine was unrelated to catecholamine deficiency (dopamine was ignored in this context) but rather was induced by 5-HT release. At the First International Catecholamine Symposium in Stockholm in August 1961, we challenged this interpretation (Carlsson 1962). We had found that alpha-methyl-meta-tyrosine and alpha-methyl-DOPA yield decarboxylation products that displace the catecholamines from their storage sites stoichiometrically. To explain the lack of sedation by alpha-methyl-meta-tyrosine and the fact that only very mild sedation was induced by alpha-methyl-DOPA, we pointed out that, unlike reserpine, these agents do not appear to block the storage mechanism. We proposed that the amines formed from these amino acids are able to take over the functions of the displaced endogenous amines (Carlsson & Lindqvist 1962). The "false-transmitter" concept thus formulated was then thoroughly investigated in numerous laboratories (for review, see Kopin 1968). Not only the aforementioned decarboxylation products but several other amines were found to be taken up in monoamine stores, thereby displacing the endogenous transmitters, and to be released by nerve impulses, thus causing postsynaptic effects.

The false-transmitter concept, as it is nowadays often understood, assumes that the false transmitter is less active than the endogenous transmitter. Thus, the displacement of the latter by the former amine should lead to a deficient transmission mechanism. To what extent this actually occurs, however, seems doubtful. To account for the lack of sedative action of alpha-methyl-metatyrosine, the possibility should be considered that despite the pronounced displacement of the endogenous catecholamines by the alpha-methylated decarboxylation products, newly synthesized endogenous transmitter may still be available for release in sufficient amounts to keep the function intact. The reduction of blood pressure by alpha-methyl-DOPA appears to be induced by a complex mechanism. It is central in origin: the effect persists after pretreatment with a peripheral decarboxylase inhibitor but is prevented by a centrally acting inhibitor (Henning 1969). The hypotensive action may be analogous to that of clonidine, i.e. it may be due to preferential activation of alpha-2 receptors by alpha-methyl-noradrenaline (see Andén 1979). Differences in pharmacological profile between the endogenous and the false transmitter should thus be taken into account.

Receptor Studies

For a long time the presynaptic events attracted most attention in this area of research. During the last 10 to 15 years, however, the monoaminergic receptors have attracted an ever increasing interest. These studies have utilized, for example, the biochemical changes induced by receptor manipulation in vivo, the electrophysiological changes caused by such manipulations in both pre- and postsynaptic neurons, and in-vitro binding studies to characterize and subclassify receptors. Much progress has also been made in elucidating the events occurring beyond the receptors (see Nestler & Greengard 1984).

Not unexpectedly, the discovery of the dopamine-receptor blocking action of the major neuroleptic (antipsychotic) agents has attracted a lot of interest. Our first study in this area (Carlsson & Lindqvist 1963) was undertaken in the hope that our recently improved fluorimetric methods for the determination of basic catecholamine metabolites would enable us to solve the riddle as to why the major antipsychotic agents, such as chlorpromazine and haloperidol, had a reserpine-like pharmacological and clinical profile and yet lacked the monoamine-depleting properties of the latter drug. Earlier, an inhibitory action of chlorpromazine on the turnover of monoamines in the brains of small rodents had been reported. However, when the chlorpromazine-induced hypothermia was prevented, the effect was no longer detectable. We found that chlorpromazine and haloperidol actually enhanced the turnover of dopamine and noradrenaline in the brain: they accelerated the formation of the dopamine metabolite, 3-methoxytyramine, and of the noradrenaline metabolite, normetanephrine, while leaving the neurotransmitter levels unchanged. In support of the specificity, promethazine, a sedative phenothiazine lacking antipsychotic and neuroleptic properties, did not change the turnover of the catecholamines. It did not seem farfetched, then, to propose that rather than reducing the availability of monoamines, as does reserpine, the major antipsychotic drugs block the receptors mediating dopamine and noradrenaline neurotransmission. This would explain their reserpine-like pharmacological profile. To account for the enhanced catecholamine turnover we proposed that neurons can increase their physiological activity in response to receptor blockade. This, I believe, was the first time that a receptor-mediated feedback control of neuronal activity was proposed. These findings and interpretations have been amply confirmed and extended by numerous workers, using a variety of techniques. In the following year, three of my students discovered the neuroleptic-induced increase in the concentrations of deaminated dopamine metabolites (Andén et al 1964b). Despite confirmatory work by others, our findings did not receive much attention until several years later. A possible explanation for this was that in the 1960s most workers in this field were focusing on other aspects of neurotransmission. Since the early 1970s, however, receptors have attracted an ever increasing interest. Moreover, the ability of catecholaminergic, especially dopaminergic agonists and antagonists, to induce and alleviate, respectively, psychotic symptoms, led to the much debated "dopamine hypothesis of schizophrenia." For a review of the historical background and recent developments, see Carlsson (1983).

The further analysis of receptor-mediated feedback control of neuronal activity, which, incidentally, soon proved to occur also in noradrenergic and serotonergic systems (see Andén et al 1969) and in other systems as well, revealed that this control was largely, if not entirely, mediated by a special population of receptors, apparently located on the monoaminergic neuron itself. These receptors have been called *presynaptic receptors* or, perhaps preferably, *autoreceptors*, since they are characterized by being sensitive to the neuron's own neurotransmitter. The first suggestion of the existence of such receptors came from studies on brain tissue slices (Farnebo & Hamberger 1971) that demonstrated inhibition and stimulation of nerve-impulse-induced dopamine release by dopamine agonists and antagonists, respectively. Subsequent in vivo studies demonstrated inhibition of striatal dopamine synthesis by the dopamine-receptor agonist, apomorphine, and blockade of this action by the neuroleptic agent, haloperidol; moreover, this effect persisted after cutting the dopaminergic axons, thus demonstrating that this feedback control was not loop-mediated but was restricted to the nerve-terminal area (Kehr et al 1972). Finally, Aghajanian & Bunney (1974, 1977) demonstrated a similar control in the somatodendritic part of dopamine neurons that causes decreased firing by dopamine-receptor agonists and a blockade of this action by dopamine-receptor antagonists. Further work along this line has led to the discovery of selective dopamine-autoreceptor agonists and antagonists with very interesting pharmacological properties and with potential clinical utility (see Clark et al 1985a,b, Svensson et al 1986a,b).

Speculations on the Molecular Requirements of a Neurotransmitter

In this prefatory chapter I may be permitted to indulge in some speculations on the molecular properties required of a neurotransmitter. A general requirement for any chemical to serve as a messenger must be that it can be readily identified with some degree of accuracy, for example, by binding to a specific receptor molecule. More than one recognition site would of course be an advantage. The monoamines contain a 2-carbon aliphatic chain with an amino group and an aromatic component—benzene or indole—with a phenolic hydroxyl group in a critical position, i.e. they apparently possess at least two binding sites. In noradrenaline and adrenaline, a hydroxyl group is added on the side chain, which confers chirality and an additional binding site to the molecules, and in adrenaline a further methyl group is added on the nitrogen, thus increasing the affinity especially to the beta-adrenoceptor. Why take the trouble to go through these cumbersome synthetic steps, starting out from sometimes scarcely available essential amino acids, when other, apparently satisfactory solutions of the problem are so readily within reach? Cells have to synthesize proteins anyway, for example, and proteins or peptides would obviously serve this purpose. Indeed, dozens of proteins and peptides are used as chemical messengers, especially as hormones.

Whereas proteins and peptides play a prominent role as hormones, the biogenic amines occupy a similar position among the neurotransmitters. A possible rationale for this apparent specialization could be that proteins and peptides represent a phylogenetically older group of chemical messengers that suffered from a serious drawback at the evolutionary stage, however, when some cells started to evolve toward becoming neurons: the messenger molecules could no longer be manufactured close to the site of release but had to be transported from the cell bodies through the axons to reach this site. Thus, the availability might have become inadequate. Smaller molecules would offer the advantage of being producible in the nerve terminals. Nonessential amino acids would fulfill these requirements.

It seems logical that not a single essenial amino acid appears to serve as a neurotransmitter.

To pursue the speculation one step further, we may assume that the amino acids, though undoubtedly useful, do not satisfy all needs for neurotransmitters. The amino acids may be just insufficient in number, or the biogenic amines may possess some useful property not shared by the amino acids. One conspicuous difference between the two groups of transmitters pertains to the storage mechanism: the amines are fairly strong bases that can be stored in high concentration, together with an acid such as ATP. This may provide the basis for an efficient storage-release mechanism. That the amines would lend themselves to a more rapid inactivation seems less likely.

Another entirely speculative possibility to consider in this context would be that the interaction between an agonist and its receptor does not always merely involve a conformational change of the receptor; in addition, the agonist and the receptor might interact like a substrate with an enzyme, hence leading to a chemical conversion of both molecules. Since such a mechanism would probably involve but a small number of transmitter molecules, it might prove fruitful to search for minor metabolic pathways of neurotransmitters. The phenolic hydroxyl groups of the monoamine neurotransmitters might be involved in such a mechanism. Our laboratory has actually engaged in investigating a newly discovered metabolic pathway of catecholamines, occurring in the mammalian brain and apparently involving autoxidation to quinones followed by coupling to glutathione (Rosengren et al 1985, Fornstedt et al 1986). Whether this pathway is at all related to the transmitter-receptor interaction is, however, entirely unknown at present.

The hypothesis that proteins and peptides represent a phylogenetically older type of chemical messengers than the "classical," small neurotransmitter molecules, receives some indirect support from the fact that certain neuropeptides occur abundantly in embryonic and neonatal brain regions but disappear at a later stage (for references, see Bloom 1984). As a corollary, some neuropeptides coexisting with classical neurotransmitters may be rudimentary phenomena of limited functional significance. Shortage of adequate research tools may thus not afford the only explanation of the many failures thus far to demonstrate beyond doubt a transmitter function of neuropeptides.

Functional and Clinical Implications

The functions of the central monoaminergic neurons and their role in neurological and psychiatric disorders are more poorly understood than the morphology and synaptology of these systems. Perhaps the functional and pathophysiological aspects of dopamine are least obscure. That the nigrostriatal dopamine system is involved in extrapyramidal functions and disorders, especially Parkinson's disease, is clearly established. The role of the tuberoinfundibular dopamine pathway for the control of prolactin secretion is also obvious, and even though the pathogenetic role of dopamine in clinical hyperprolactinemia is doubtful, the use of dopamine agonists in such cases is a significant advance. That dopamine plays a role in important mental functions is evident from the therapeutic actions of dopamine antagonists, as well as from the psychotomimetic effects of dopamine agonists. However, no disturbance in dopamine function has as yet been demonstrated beyond doubt in schizophrenia or any other psychotic condition (see Carlsson 1983).

Likewise, our knowledge of the physiology and pathology of central noradrenaline, not to speak of adrenaline, is very fragmentary. Like dopamine, noradrenaline seems to be somehow involved in arousal, and the locus coeruleus seems to respond very actively to incoming stimuli, especially when they signify novelty. Stimulation of noradrenaline neurons, for example by electrical stimulation of the locus coeruleus or by yohimbine-induced blockade of alpha-2-adrenergic autoreceptors, seems to cause strong arousal accompanied by severe anxiety. Conversely, stimulation of alpha-2-adrenergic receptors by clonidine causes, in addition to a decrease in blood pressure, sedation and an anxiolytic response. Bilateral destruction of the locus coeruleus has a similar effect. Both dopamine and noradrenaline appear to be somehow involved in the positive reinforcing action of dependence-producing drugs, and clonidine appears to be capable of alleviating certain withdrawal reactions (for review, see Elam 1985, Engel & Carlsson 1977).

The physiology and pathogenetic significance of 5-HT are also poorly comprehended. Animal data support the contention, however, that serotonergic systems exert an inhibitory function on aggressive behavior (Valzelli 1974). The serotonergic system appears to be more strongly developed in female than in male rats (Carlsson et al 1985), a finding that may at least partly account for the well-known sex difference in aggressive behavior among vertebrates. The serotonergic systems seem to control the mating behavior of both sexes, but here the situation may be more complex, with notable species differences. That 5-HT also plays a part in the control of motor functions is suggested by the "5-HT motor syndrome," but the physiological implications of this phenomenon are obscure. Similarly, the role of 5-HT and the other monoamines in various aspects of sleep seems to be established, although the available data are contradictory in certain respects.

Evidence for a role of 5-HT, noradrenaline, and dopamine in the control of mood and psychomotor activity and in affective disorders and anxiety has also been presented, and some impressive observations relating 5-HT to suicidal behavior have been published (Träskman et al 1981). Moreover, there seems to be no doubt that all the three major monoamines play a role in various neuroendocrinological and autonomic nervous functions.

Enormous efforts are obviously required to obtain a reasonably complete picture of the biological and clinical significance of the monoaminergic systems. More sophisticated animal models are needed, and human studies using, for example, modern imaging techniques will no doubt prove helpful. Although pharmacological tools have contributed much, better ones are needed. Especially, more selective monoaminergic agonists and antagonists, applicable also to humans, would prove useful.

Literature Cited

Adams, R. N., Marsden, C. A. 1982. Electrochemical detection methods for monoamine measurements *in vitro* and *in vivo*. In *Handb. Psychopharmacol.* 15: 1–74

Aghajanian, G. K., Bunney, B. S. 1974. Pre- and postsynaptic feedback mechanisms in central dopaminergic neurons. In *Frontiers of Neurology and Neuroscience Research*, ed. P. Seeman, G. M. Brown, pp. 4–11. Toronto: Univ. Toronto Press

Aghajanian, G. K., Bunney, B. S. 1977. Dopamine autoreceptors: Pharmacological characterization by microiontophoretic single cell recording studies. *Naunyn-Schmiedeberg's Arch. Pharmacol.* 297: 1–8

Amin, A. H., Crawford, T. B. B., Gaddum, J. H. 1954. The distribution of substance P and 5-hydroxytryptamine in the central nervous system of the dog. *J. Physiol.* 126: 596–618

Andén, N.-E. 1979. Selective stimulation of central alpha-autoreceptors following treatment with alpha-methyldopa and FLA 136. *Naunyn-Schmiedeberg's Arch. Pharmacol.* 306: 263–66

Andén, N.-E., Carlsson, A., Dahlström, A., Fuxe, K., Hillarp, N.-Å., Larsson, K. 1964a. Demonstration and mapping out of nigro-neostriatal dopamine neurons. *Life Sci.* 3: 523–30

Andén, N.-E., Roos, B.-E., Werdinius, B. 1964b. Effects of chlorpromazine, haloperidol and resperine on the levels of phenolic acids in rabbit corpus striatum. *Life Sci.* 3: 149–58

Andén, N.-E., Carlsson, A., Häggendal, J. 1969. Adrenergic mechanisms. *Ann. Rev. Pharmacol.* 9: 119–34

Axelrod, J. 1960. Discussion remarks. See Vane et al 1960, pp. 558–59

Axelrod, J. 1964. The uptake and release of catecholamines and the effect of drugs. In *Progress Brain Res.* 8: 81–89

Barbeau, A., Sourkes, T. L., Murphy, G. F. 1962. Les catecholamines de la maladie de Parkinson. In *Monoamines et Systeme Nerveux Central*, ed. J. de Ajuriaguerra, pp. 247–62. Geneve/Paris: Georg/Masson

Bertler, Å., Rosengren, E. 1959. Occurrence and distribution of dopamine in brain and other tissues. *Experientia* 15: 10

Bertler, Å., Carlsson, A., Rosengren, E. 1956. Release by reserpine of catecholamines from rabbits' hearts. *Naturwissenschaften* 22: 521

Bertler, Å., Carlsson, A., Rosengren, E. 1958. A method for the fluorimetric determination of adrenaline and noradrenaline in tissues. *Acta Physiol. Scand.* 44: 273–92

Bertler, Å., Falck, B., Gottfries, C. G., Ljunggren, L., Rosengren, E. 1964. Some observations on adrenergic connections between mesencephalon and cerebral hemispheres. *Acta Pharmacol. (Kbh.)* 21: 283–89

Birkmayer, W., Hornykiewicz, O. 1962. Der L-Dioxyphenylalanin (=L-DOPA)-Effekt beim Parkinson-Syndrom des Menschen: Zur Pathogenese und Behandlung der Parkinson-Akinese. *Arch. Psychiat. Nervenkr.* 203: 560–74

Björklund, A., Hökfelt, T., eds. 1984. *Handbook of Chemical Neuroanatomy*, Vol. 2, *Classical Transmitters in the CNS*, Pt. 1, pp. 1–463. Amsterdam/New York/Oxford: Elsevier

Blaschko, H. 1939. The specific action of L-dopa decarboxylase. *J. Physiol.* 96: 50P–51P

Bloom, F. E. 1984. General features of chemically identifiable neurons. See Björklund & Hökfelt, pp. 1–22

Born, G. V. R. 1956. Adenosinetriphosphate (ATP) in blood platelets. *Biochem. J.* 62: 33P

Carlsson, A. 1959. The occurrence, distribution and physiological role of catecholamines in the nervous system. *Pharmacol. Rev.* 11: 490–93

Carlsson, A. 1960. Discussion remark. See Vane et al 1960, pp. 558–59

Carlsson, A. 1962. Discussion. See Costa et al 1962, pp. 71–74

Carlsson, A. 1966. Physiological and pharmacological release of monoamines in the central nervous system. See Euler et al 1966, pp. 331–46

Carlsson, A. 1983. Antipsychotic agents: Elucidation of their mode of action. In *Discoveries in Pharmacology*, Vol. 1: *Psycho- and Neuro-pharmacology*, ed. M. J. Parnham, J. Bruinvels, pp. 197–206. Amsterdam/New York/Oxford: Elsevier

Carlsson, A., Hillarp, N.-Å. 1956. Release of adrenaline from the adrenal medulla of rabbits produced by reserpine. *Kgl. Fysiogr. Sällsk. Förhandl.* 26(8)

Carlsson, A., Hillarp, N.-Å. 1962. Formation of phenolic acids in brain after administration of 3,4-dihydroxyphenylalanine. *Acta Physiol. Scand.* 55: 95–100

Carlsson, A., Lindqvist, M. 1962. In-vivo decarboxylation of alpha-methyldopa and alpha-methyl metatyrosine. *Acta Physiol. Scand.* 54: 87–94

Carlsson, A., Lindqvist, M. 1963. Effect of chlorpromazine and haloperidol on the formation of 3-methoxytyramine in mouse brain. *Acta Pharmacol.* 20: 140–44

Carlsson, A., Waldeck, B. 1958. A fluorimetric method for the determination of dopamine (3-hydroxytyramine). *Acta Physiol. Scand.* 44: 293–98

Carlsson, A., Rosengren, E., Bertler, Å., Nilsson, J. 1957a. Effect of reserpine on the metabolism of catecholamines. In *Psychotropic Drugs*, ed. S. Garattini, V. Ghetti, pp. 363–72. Amsterdam: Elsevier

Carlsson, A., Lindqvist, M., Magnusson, T. 1957b. 3,4-Dihydroxyphenylalanine and 5-hydroxytryptophan as reserpine antagonists. *Nature* 180: 1200

Carlsson, A., Shore, P. A., Brodie, B. B. 1957c. Release of serotonin from blood platelets by reserpine *in vitro*. *J. Pharmacol. Exp. Ther.* 120: 334–39

Carlsson, A., Lindqvist, M., Magnusson, T., Waldeck, B. 1958. On the presence of 3-hydroxytyramine in brain. *Science* 127: 471

Carlsson, A., Lindqvist, M., Magnusson, T. 1960. On the biochemistry and possible functions of dopamine and noradrenaline in brain. See Vane et al 1960, pp. 432–39

Carlsson, A., Falck, B., Hillarp, N.-Å., Thieme, G., Torp, A. 1961. A new histochemical method for visualization of tissue catechol amines. *Med. Exp.* 4: 123–25

Carlsson, A., Falck, B., Hillarp, N.-Å. 1962. Cellular localization of brain monoamines. *Acta Physiol. Scand.* 56(Suppl. 196): 1–27

Carlsson, A., Hillarp, N.-Å., Waldeck, B. 1963a. Analysis of the Mg^{++}-ATP dependent storage mechanism in the amine granules of the adrenal medulla. *Acta Physiol. Scand.* 59(Suppl. 215): 1–38

Carlsson, A., Magnusson, T., Rosengren, E. 1963b. 5-Hydroxytryptamine of the spinal cord normally and after transection. *Experientia* 19: 359

Carlsson, A., Falck, B., Fuxe, K., Hillarp, N.-Å. 1964. Cellular localization of monoamines in the spinal cord. *Acta Physiol. Scand.* 60: 112–19

Carlsson, A., Kehr, W., Lindqvist, M., Magnusson, T., Atack, C. V. 1972. Regulation of monoamine metabolism in the central nervous system. *Pharm. Rev.* 24: 371–84

Carlsson, A., Kehr, W., Lindqvist, M. 1976. The role of intraneuronal amine levels in the feedback control of dopamine, noradrenaline and 5-hydroxytryptamine in rat brain. *J. Neural Transm.* 39: 1–19

Carlsson, M., Svensson, K., Eriksson, Carlsson, A. 1985. Rat brain serotonin: Biochemical and functional evidence for a sex difference. *J. Neural Transm.* 63: 297–313

Cheramy, A., Leviel, V., Glowinski, J. 1981. Dendritic release of dopamine in the substantia nigra. *Nature* 289: 537–42

Clark, D., Carlsson, A., Hjorth, S. 1985a. Dopamine receptor agonists: Mechanisms underlying autoreceptor selectivity. I. Review of the evidence. *J. Neural Transm.* 62: 1–52

Clark, D., Hjorth, S., Carlsson, A. 1985b. Dopamine receptor agonists: Mechanisms underlying autoreceptor selectivity. II. Theoretical considerations. *J. Neural Transm.* 62: 171–207

Corrodi, H., Hillarp, N.-Å. 1964. Fluoreszenzmethoden zur histochemischen Sichtbarmachung von Monoaminen. 2. Identifizierung des fluorisierendes Productes aus Dopamin und Formaldehyd. *Helv. Chim. Acta* 47: 911–18

Costa, E., Gessa, G. L., Kuntzman, R., Brodie, B. B. 1962. In *Pharmacological Analysis of Central Nervous Action*, ed. W. D. M. Paton, P. Lindgren, pp. 43–71. Oxford: Pergamon

Dahlström, A., Carlsson, A. 1986. Making

1354   CARLSSON

visible the invisible. Recollections of the first experiences with the histochemical fluorescence method for visualization of tissue monoamines. In *Discoveries in Pharmacology*, Vol. 3: *Chemical Pharmacology and Chemotherapy*, ed. M. J. Parnham, J. Bruinvels, pp. 97–125. Amsterdam/New York/Oxford: Elsevier

Dahlström, A., Fuxe, K. 1964. Existence of monoamine-containing neurons in the central nervous system. I. Demonstration of monoamines in the cell bodies of brain stem neurons. *Acta Physiol. Scand.* 62(Suppl. 232): 1–55

Ehringer, H., Hornykiewicz, O. 1960. Verteilung von Noradrenalin und Dopamin (3-Hydroxytyramin) im Gehirn des Menschen und ihr Verhalten bei Erkrankungen des extrapyramidalen Systems. *Klin. Wschr.* 38: 1236–39

Elam, M. 1985. In *On the physiological regulation of brain norepinephrine neurons in rat locus ceruleus*, pp. 1–43. Thesis, Univ. Göteborg, Sweden

Engel, J., Carlsson, A. 1977. Catecholamines and behavior. *Curr. Dev. Psychopharmacol.* 4: 3–32

Eränkö, O. 1955. Distribution of fluorescent islets, adrenaline and noradrenaline in the adrenal medulla of the cat. *Acta Endocrinol.* 18: 180–88

Falck, B., Hillarp, N.-Å. 1959. On the cellular localization of catecholamines in the brain. *Acta Anatomica* 38: 277–79

Falck, B., Hillarp, N.-Å., Thieme, G., Torp, A. 1962. Fluorescence of catecholamines and related compounds condensed with formaldehyde. *J. Histochem. Cytochem.* 10: 348–54

Farnebo, L.-O., Hamberger, B. 1971. Drug-induced changes in the release of ^{3}H-monoamines from field stimulated rat brain slices. *Acta Physiol. Scand. Suppl.* 371: 35–44

Fornstedt, B., Rosengren, E., Carlsson, A. 1986. Occurrence and distribution of 5-S-cysteinyl derivatives of dopamine, dopa and dopac in the brains of eight mammalian species. *Neuropharmacology* 25: 451–54

Fuxe, K. 1965. Evidence for the existence of monoamine neurons in the central nervous system. IV. Distribution of monoamine nerve terminals in the central nervous system. *Acta Physiol. Scand.* 64(Suppl. 247): 37–84

Gaddum, J. H. 1954. Drugs antagonistic to 5-hydroxytryptamine. In *Ciba Found. Symp. on Hypertension. Humoral and Neurogenic Factors*, ed. J. H. Gaddum, pp. 75–77. Boston: Little, Brown

Henning, M. 1969. Studies on the mode of action of alpha-methyldopa. *Acta Physiol.*

Scand. 76(Suppl. 322): 1–37

Hess, S. M., Udenfriend, S. 1959. A fluorimetric procedure for the measurement of tryptamine in tissues. *J. Pharmacol. Exp. Ther.* 127: 175–77

Hess, S. M., Connamacher, R. H., Ozaki, M., Udenfriend, S. 1961. The effects of alpha-methyldopa and alpha-methyl-m-tyrosine on the metabolism of norepinephrine and serotonin *in vivo. J. Pharmacol. Exp. Ther.* 134: 129–38

Hillarp, N.-Å. 1946. Structure of the synapse and the peripheral innervation apparatus of the autonomic nervous system. *Acta Anatomica Suppl.* 4: 1–153

Hillarp, N.-Å. 1959. The construction and functional organization of the autonomic innervation apparatus. *Acta Physiol. Scand.* 46(Suppl. 157): 1–38

Hillarp, N.-Å., Lagerstedt, S., Nilsson, B. 1953. The isolation of a granular fraction from the suprarenal medulla containing the sympathomimetic catecholamines. *Acta Physiol. Scand.* 28: 251–63

Hillarp, N.-Å., Högberg, B., Nilsson, B. 1955. Adenosine triphosphate in the adrenal medulla of the cow. *Nature* 176: 1032–33

Holtz, P. 1939. Dopadecarboxylase. *Naturwissenschaften* 27: 724–25

Holzbauer, M., Vogt, M. 1956. Depression by reserpine of the noradrenaline concentration in the hypothalamus of the cat. *J. Neurochem.* 1: 8–11

Hökfelt, T. 1968. *Electron microscopic studies on peripheral and central monoamine neurons*, pp. 1–30. Thesis, Karolinska Inst., Stockholm

Jonason, J., Rutledge, C. O. 1968. Metabolism of dopamine and noradrenaline in rabbit caudate nucleus *in vitro. Acta Physiol. Scand.* 73: 411–17

Kehr, W. 1976. 3-Methoxytyramine as an indicator of impulse-induced release in rat brain in vivo. *Naunyn-Schmiedeberg's Arch. Pharmacol.* 293: 209–15

Kehr, W., Carlsson, A., Lindqvist, M., Magnusson, T., Atack, C. 1972. Evidence for a receptor-mediated feedback control of striatal tyrosine hydroxylase. *J. Pharm. Pharmacol.* 24: 744–47

Kirshner, N. 1962. Uptake of catecholamines by a particular fraction of the adrenal medulla. *J. Biol. Chem.* 237: 2311–17

Kopin, I. J. 1968. False adrenergic transmitters. *Ann. Rev. Pharmacol.* 8: 377–94

Lindvall, O. 1974. *The glyoxylic acid fluorescence histochemical method for monoamines. Chemistry, methodology and neuroanatomical application.* Thesis, Dept. Histology, Univ. Lund, Lund

Lovenberg, W., Bruckwick, E. A., Hanbauer, I. 1975. ATP, cyclic AMP, and magnesium increase the affinity of rat striatal tyrosine hydroxylase for its cofactor. *Proc. Natl. Acad. Sci. USA* 72: 2955–58

Lundborg, P. 1963. Storage function and amine levels of the adrenal medullary granules at various intervals after reserpine treatment. *Experientia* 19: 479

Magnusson, T., Rosengren, E. 1963. Catecholamines of the spinal cord normally and after transection. *Experientia* 19: 229

Malmfors, T. 1965. Studies on adrenergic nerves. The use of rat and mouse iris for direct observations on their physiology and pharmacology at cellular and subcellular levels. *Acta Physiol. Scand.* 64(Suppl. 248): 1–93

McLennan, H. 1963. In *Synaptic Transmission*, pp. 1–134. Philadelphia/London: Saunders

Montagu, K. A. 1957. Catechol compounds in rat tissues and in brains of different animals. *Nature* 180: 244–45

Muscholl, E., Vogt, M. 1958. The action of reserpine on the peripheral sympathetic system. *J. Physiol.* 141: 132–55

Nagatsu, T., Levitt, M., Udenfriend, S. 1964. Tyrosine hydroxylase: The initial step in norepinephrine biosynthesis. *J. Biol. Chem.* 238: 2910–17

Nestler, E. J., Greengard, P. 1984. Protein phosphorylation in nervous tissue. In *Catecholamines, Pt. A: Basic and Peripheral Mechanisms*, ed. E. Usdin, A. Carlsson, A. Dahlström, J. Engel, pp. 9–22. New York: Liss

Nissbrandt, H., Pileblad, E., Carlsson, A. 1985. Evidence for dopamine release and metabolism beyond the control of nerve impulses and dopamine receptors in rat substantia nigra. *J. Pharm. Pharmacol.* 37: 884–89

Pletscher, A., Shore, P. A., Brodie, B. B. 1955. Serotonin release as a possible mechanism of reserpine action. *Science* 122: 374–75

Rosengren, E., Linder-Eliasson, E., Carlsson, A. 1985. Detection of 5-S-cysteinyldopamine in human brain. *J. Neural Transm.* 63: 247–53

Shore, P. A., Silver, S. L., Brodie, B. B. 1955. Interaction of reserpine, serotonin, and lysergic acid diethylamide in brain. *Science* 122: 284–85

Svensson, K., Hjorth, S., Clark, D., Carlsson, A., Wikström, H., Andersson, B., Sanchez, D., Johansson, A. M., Arvidsson, L.-E., Hacksell, U., Nilsson, J. L. G. 1986a. (+)-UH 232 and (+)-UH 242: Novel stereoselective DA receptor antagonists with preferential action on autoreceptors. *J. Neural Transm.* 65: 1–27

Svensson, K., Carlsson, A., Johansson, A. M., Arvidsson, L.-E., Nilsson, J. L. G. 1986b. A homologous series of N-alkylated cis-(+)-(1S,2R)-5-methoxy-1-methyl-2-aminotetralins: Central DA receptor antagonists showing profiles ranging from classical antagonism to selectivity for autoreceptors. *J. Neural Transm.* 65: 29–38

Thoenen, H., Otten, U., Oesch, F. 1973. Trans-synaptic regulation of tyrosine hydroxylase. In *Frontiers in Catecholamine Research*, ed. E. Usdin, S. Snyder, pp. 179–85. Oxford: Pergamon

Träskman, L., Åsberg, M., Bertilsson, L., Sjöstrand, L. 1981. Monoamine metabolites in CSF and suicidal behavior. *Arch. Gen. Psychiat.* 38: 631–36

Twarog, B. M., Page, I. H. 1953. Serotonin content of some mammalian tissues and urine and a method for its determination. *Am. J. Physiol.* 175: 157–61

Ungerstedt, U. 1971. *On the anatomy, pharmacology and function of the nigrostriatal dopamine system.* Thesis, Karolinska Inst., Stockholm

Ungerstedt, U., Herrera-Marschitz, M., Ståhle, L., Tossman, U., Zetterström, T. 1983. Dopamine receptor mechanisms correlating transmitter release and behavior. In *Dopamine Receptor Agonists*, ed. A. Carlsson, J. L. G. Nilsson, 1: 165–81. Stockholm: Swedish Pharmaceutical Press

Valzelli, L. 1974. 5-Hydroxytryptamine in aggressiveness. *Adv. Biochem. Psychopharmacol.* 11: 255–64

Vane, J. R., Wolstenholme, G. E. W., O'Connor, M., eds. 1960. *Ciba Found. Symp. on Adrenergic Mechanisms*, pp. 1–632. London: Churchill

Vogt, M. 1954. The concentration of sympathin in different parts of the central nervous system under normal conditions and after the administration of drugs. *J. Physiol.* 123: 451–81

Von Euler, U. S., Rosell, S., Uvnäs, B., eds. 1966. *Mechanisms of Release of Biogenic Amines*, pp. 469–77. Oxford: Pergamon

Westerink, B. H. C., Spaan, S. J. 1982. On the significance of endogenous 3-methoxytyramine for the effects of centrally acting drugs on dopamine release in the rat brain. *J. Neurochem.* 38: 680–86

Woolley, D. W., Shaw, E. 1954. A biochemical and pharmacological suggestion about certain mental disorders. *Science* 119: 587–88

Emilio Segrè

NUCLEAR AND PARTICLE SCIENCE

Above: Emilio Segrè around 1953.
Below: P. Zeeman, E. Segrè, and S. A. Goudsmit in Amsterdam 1931.

Ann. Rev. Nucl. Part. Sci. 1981. 31:1–18

FIFTY YEARS UP AND DOWN A STRENUOUS AND SCENIC TRAIL

Emilio Segrè

Department of Physics, University of California, Berkeley California 94720

In response to the kind invitation to write a prefatory chapter to the 31st volume of the *Annual Review of Nuclear and Particle Science*, I have decided to give a personal account of my own participation in the unfolding story of nuclear and particle physics. I am fully aware of the one-sidedness of the account and I emphasize that this is part of the story as I saw it, not the Story.

I started physics research in 1928, when quantum mechanics was becoming established and the interest previously centered on the atom was veering to the nucleus. Twenty years later I witnessed a similar shift of emphasis from the nucleus to particles. At the beginning of my career I could still be helped by P. Zeeman and I at least saw H. A. Lorentz, M. Planck, and J. J. Thomson. Later I learned either directly from, or from the papers of, A. Sommerfeld, W. Heisenberg, W. Pauli, P. A. M. Dirac, E. Schroedinger, O. Stern, and E. Rutherford. Enrico Fermi was a close friend and had the greatest influence on my scientific life. F. Rasetti, E. Amaldi, E. Majorana, F. Bloch, H. Bethe, C. J. Bakker, G. Placzek, G. Racah, E. Teller, G. C. Wick, and O. Frisch were among my friends and colleagues in the Italian period. I learned from them as we studied and worked together.

When I arrived in the United States I was already a mature physicist and my friendships were of a different nature. I. I. Rabi, J. R. Dunning, D. P. Mitchell, R. B. Brode, F. A. Jenkins, L. W. Alvarez, J. Kennedy, G. T. Seaborg, E. McMillan, D. Cooksey, E. O. Lawrence, R. I. Thornton, M. Kamen, R. Serber, and J. R. Oppenheimer were my contemporaries. With some I had close personal friendships, but we had not shared our formative years and that made a difference.

1359

0163-8998/81/1201-0001 $01.00

My students in Italy were few, some in Rome and a couple in Palermo. M. Ageno was my lone listener when I first taught as "Libero Docente" in Rome. I remember having suggested subjects for their work to U. Fano, E. Fubini, and R. Einaudi.

In the United States I had many students: C. S. Wu was the first. O. Chamberlain, C. Wiegand, G. Farwell, J. Jungerman, G. A. Linenberger, and chemists M. Kahn and J. Miskel came with me as students at Los Alamos. After the war I formed a permanent group at Berkeley. Wiegand was its first and most permanent member, soon joined by Chamberlain, who had completed his studies at Chicago with Fermi. The group changed with time. Some of our students, R. Tripp, T. Ypsilantis, and H. Steiner, joined the Berkeley faculty. G. Goldhaber, W. Chinowsky, G. Shapiro, and R. Stiening, coming from other universities, worked in my group and on the faculty for different periods. The official record of the PhDs conferred in Berkeley assigns students to specific professors; it does not reflect the vital contribution of mutual instruction and collaboration nor exchanges of sponsorship dictated by contingencies. Among those I find in my list are: H. York, J. Jungerman, S. C. Wright, S. N. Ghoshal, E. L. Kelly, C. E. Wiegand, R. L. Mather, F. N. Spiess, T. J. Thompson, W. John, T. J. Ypsilantis, H. M. Steiner, J. E. Simmons, D. V. Keller, R. C. Weingart, L. E. Agnew, T. Elioff, R. R. Larsen, W. Lee, E. H. Rogers, R. J. Esterling, R. E. Hill, D. A. Jenkins, and A. R. Kunselman; also D. Cutts, Min Chen, and P. Kijewski, who worked primarily with Stiening. My memory, an uncertain instrument, would add some names to these: M. Jakobson, A. Bloom, M. O. Stern, G. Pettengill, J. Button, C. Schultz, G. C. Tobias, G. Temmer, and G. Lum; I had more than passing exchanges with all of them.

Many post doctoral fellows from the US and from abroad worked in my group and learned in comparable amounts from each other, from staff members, and from the professors. I was not the easiest professor to work with and I recognize that patience is a two-way virtue. I hope, however, that students benefited from me not less than I from them.

I was born into a well-to-do Italian family in 1905, the year of the three immortal Einstein papers. The time of my birth made me too young to participate in the First World War and too old to participate as a soldier in the Second. My father was a successful industrialist. His brothers, who were very close to him, were a geologist and a professor of Law, both distinguished in their professions. My mother, A. Treves, was the daughter of a well-known architect from Florence. I was born in Tivoli, where my father's mills were located; it was then a small town of about 16,000 inhabitants. In 1917 we moved to Rome. I went to school in Tivoli and Rome, and, in 1922, the day Mussolini's black shirts were entering Rome,

I registered at the University as an engineering major. I had been interested in physics from childhood; I still have a notebook on experiments performed when I was seven. Family reasons and the state of physics in Rome dictated my choice of engineering. Moreover, physics was not too far removed from engineering and the first two years of study were common. Among my professors at the University were the mathematicians Severi, Castelnuovo, Levi-Civita, and, later, Volterra.

In 1927 when I had finished my fourth year of study, through lucky circumstances helped by fairly shrewd insight, I met Rasetti and Fermi who had just come to the University of Rome on the Physics faculty. I went with them to the Como International Physics Conference in September, where I could see with my own eyes the most famous physicists, including Lorentz, Planck, Rutherford, Bohr, Compton, Born, Heisenberg, Pauli, Sommerfeld, Stern, Langmuir, and Wood. I transferred shortly thereafter to physics and soon I persuaded my schoolmate and friend Ettore Majorana to do the same. He was a mathematical genius and engineering was not his vocation.

In 1928 I took my doctorate with a thesis on the anomalous dispersion of lithium vapor. During the year, I had been learning quantum mechanics with Fermi. He later said I learned all the physics I ever knew from him in that year. The claim is slightly exaggerated in a literal sense, but symbolically rather true. His teaching was completely informal. Amaldi, Rasetti, and I, plus occasional visitors such as E. Persico, B. Rossi, O. M. Corbino, and others, met in the afternoon in his study, about twice a week. We brought only paper and pencils. Fermi would start on a subject, often prompted by our questions, and improvise a perfect lecture. We saved the papers on which he developed his arguments and wrote them down later. We also devised problems of our own to verify that we had understood. Next time the subject could be completely different.

Our research work followed closely what Fermi was doing. My father once told me, "You are living on Fermi's crumbs." The phrase struck me. However, Fermi was more than glad to let us invent our own research subjects. When this happened he helped wholeheartedly and, of course, most efficiently.

My first original work was the discovery that the forbidden transitions in alkali metal spectra were due to electric quadrupole radiation (1). I still remember the deep impression I received when, in observing visually the Zeeman effect on a potassium line, I saw the unmistakable quadrupole Zeeman pattern. I later investigated several types of forbidden lines (2), including x-ray spectra (3a, 3b). Until 1934 I worked mainly in atomic spectroscopy and spent two fruitful and instructive periods abroad in Zeeman's laboratory in Amsterdam.

In the late 1920s Fermi and his group started to feel that atomic physics was reaching maturity and that the next big advance would be the exploration of the nucleus. We were all in our twenties and we did not want to fall behind. Consequently, we made a deliberate effort to learn whatever was then known on the nucleus, in theory and experiment. Coming from spectroscopy, the first subject that was natural for us to tackle was hyperfine structure. The problem was to determine the magnetic moments of nuclei and, equally important, to make sure that spin and magnetic moments were sufficient to explain all the hyperfine structures with no other interactions except the electromagnetic. Fermi worked on this for a while and, at a later date, invited me to help him. It was an act of kindness to soothe me from some academic disappointment I had suffered. That was the origin of the Fermi-Segrè formula (4).

In 1931, as part of our preparation for nuclear studies, Fermi persuaded the Accademia d'Italia, of which he was the only physics member, to convene a conference in Rome. There were about 30 participants, among them M. Curie, R. A. Millikan, N. Mott, N. Bohr, W. Bothe, and G. Marconi, who acted as host because he was the President of the Italian Academy.

At the end of 1930 I obtained a Rockefeller Fellowship through the recommendations of V. Volterra, O. M. Corbino, and P. Zeeman, in whose laboratory I had worked for a brief period. At Fermi's suggestion I went to Hamburg to learn vacuum techniques and molecular beams from O. Stern. Stern at that time was working with O. Frisch on the magnetic moment of the proton and on helium diffraction on crystalline surfaces. I inherited a problem on the dynamics of space quantization, which I finished with some help from Frisch (5).

This work could be carried to completion only with the inspiration of no one less than James Clerk Maxwell. In perusing his treatise on electricity, I saw a beautiful figure that told me at once how to solve a technical problem that had consumed the fellowship of my predecessor and would have also consumed mine. A few months later Rabi wrote to me asking for some details on the experiment. He showed that unwittingly we had crudely measured the nuclear spin and magnetic moment of potassium.

At the end of my Rockefeller fellowship I returned to Rome as first assistant to Corbino, and worked further on spectroscopic subjects. I was startled when I realized how large an atom in a high quantum state may become. I called them swollen atoms and soon I demonstrated for them the quadratic Zeeman effect (6). To prevent distillation from heated absorption tubes, I added some inert gas to the vapor, but I was surprised that the absorption lines remained sharp and were not broadened by collision. Then, Amaldi and I started to study such effects systematically, and Fermi also became interested in them (7a, 7b, 7c). We found experimentally a shift of the spectral lines. Fermi first told us that it was clearly an effect of the dielectric constant of the added gas. But calculations of that effect resulted, in some cases, in the wrong sign for that explanation. Fermi then had to think for a few days on the subject and finally developed the theory of what are now called Fermi pseudopotentials. He introduced the idea of a scattering length connecting our shifts to the Ramsauer effect. Little did we know that these notions would soon become important in nuclear physics.

Soon after the discovery of the neutron, Majorana formulated a nuclear model based on neutrons and protons. I remember having asked Fermi repeatedly why all nuclei with half-integral spin seemed to be fermions and those with integral spin bosons. He answered (about 1933) that he believed it was a law of Nature, but could not derive it. It was a deep mystery awaiting Pauli for its clarification.

Fermi attended the important Solvay Council of October 1933 and returned to Rome pondering beta decay and Pauli's neutrino hypothesis. (We had given the Italian diminutive name to the particle and it stuck.) During the Christmas vacation I went skiing in the Alps. I wanted to combine sports with a discreet visit to Switzerland, unobserved by Italian authorities, and I crossed the Alps with Giulio Racah and other friends, in bad weather and not very easily. I returned to Italy by train and stopped in the Val Gardena to meet Amaldi, Fermi, and other friends. They noted that I was not walking normally and I had to show them the reason. In crossing the high passes of the Alps, frequent falls on the icy snow had left my back dark blue. Sitting on a bed in the crowded small room, Fermi explained to us his new beta-ray theory. The toughest part was the use of creation and destruction operators, new to us and only recently mastered by him.

Soon thereafter we read the stunning announcement of the discovery of artificial radioactivity by I. Curie and F. Joliot. This was great physics and easy to understand. Fermi saw at once the advantage of using neutrons as projectiles and recognized that here was the awaited occasion for our entry into experimental nuclear physics.

Fermi started exploring systematically all elements in order of increasing atomic number Z. His first success was with fluorine ($Z = 9$). He then generously asked Amaldi and me to help in the investigation. Rasetti, who was temporarily in Morocco, soon returned, and a little later we were joined by O. D'Agostino, a chemist. We investigated all the elements we could lay our hands on. At that time, in my innocence, I even tried to secure "masurium," the name given then to element $Z = 43$, but the experienced supplier with whom I was dealing told me that he had never seen it.

We showed by chemical methods that the atomic numbers of the substances we obtained were either the same as the target or were one or two units less. The reactions were thus presumably (n, γ), $(n, 2n)$, (n, p), and (n, α). We could not decide at first whether we had $(n, 2n)$ or (n, γ).

Rasetti wanted particularly to test uranium, but it was difficult to investigate because of its natural activity. We soon found it gave complicated results and we concentrated on demonstrating that the activities we found did not have atomic numbers between 82 and 92. We also found a case of nuclear isomerism in bromine.

During the summer vacation of 1934, Amaldi and I went to the Cavendish Laboratory. On our arrival in London I met for the first time L. Szilard and F. Paneth. It was the day of the first unsuccessful attempt by Hitler to conquer Austria, and Dollfuss had just been murdered. We spoke more of politics than physics. At Cambridge, Rutherford was very kind to us and we met many of the younger people in the laboratory, among them Chadwick, Oliphant, Cockcroft, Wynn Williams, Maurice Goldhaber, and Kapitza. We were particularly interested in the work of T. Bjerge and H. C. Westcott, who had bombarded various substances with neutrons. By combining their results with those obtained at Rome we could establish that in some cases we surely had an (n, γ) reaction.

When Amaldi and I returned to Rome, we pursued this work further and found the ^{27}Al (n, γ) reaction. This was communicated to Fermi, who was returning from a South American trip and was in London for an International Conference on Nuclear Physics. He reported our results but on his arrival in Rome I had to admit that I could not reproduce the experiment. I remember the rather imperturbable Fermi angry with me only twice; this was one occasion. Meanwhile, measurements on the activation of Ag by Amaldi, Pontecorvo, and others gave amazing yet highly unreproducible results. On October 24, 1934, around noon we discovered that the filtration of neutrons through paraffin greatly increased their efficiency for activating certain substances. By 3 pm, after lunch and siesta, Fermi told us that he believed he had found the reason. Neutrons were slowed down by collisions with hydrogen atoms and became *more* efficient for the (n, γ) reaction. The scales fell from our eyes and the evening of the same day we sent a note to the *Ricerca Scientifica* announcing the discovery (8). I rushed to repeat my (n, γ) reaction on Al by irradiating it with slow neutrons and my reputation with Fermi was reestablished.

Intense work followed for many weeks. Two papers in the *Proceedings of the Royal Society of London* (9a, 9b) summarize this work. Our chemist D'Agostino was convinced that nuclei had a periodic system like atoms. We disagreed and somewhat ridiculed him; he proved to be more correct than we expected!

In pursuing our work on uranium, we proved that the radioactive substances produced did not lie between lead and uranium, and we concluded that they were transuranic. The chemist Ida Noddack complained in a paper that we had not proved that our products were not relatively light elements in the middle of the periodic system. Maybe uranium fragments into two big chunks. We did not take her seriously, to our loss.

During the summer of 1935 I went to Ann Arbor, Michigan, with Fermi and later to Columbia University in New York where I found Rasetti. We were all convinced that Europe was on the verge of a catastrophe and we wanted to establish relations with American scientists and institutions to facilitate emigration if it became necessary.

Otto Stern had spoken to us in glowing terms of Lawrence's cyclotron and Fermi was much interested in the machine, although it seemed to us, correctly, beyond our technical and financial capabilities. While in Ann Arbor, Fermi received a letter from Lawrence in which he spoke of having obtained one millicurie of ^{24}Na. Fermi thought that Lawrence had somehow slipped and that he meant one microcurie. He tried tactfully to make the correction in his answer, but a few days later a reply came from Lawrence. Actually enclosed with the letter was one millicurie of ^{24}Na, as we verified with an electroscope! Later, when I knew the people at the Radiation Laboratory, I could imagine the effect produced by Fermi's letter, although I never asked about it.

At Columbia, Dunning, Mitchell, Pegram, Rasetti, and I performed some good experiments with a mechanical velocity selector for neutrons, including a direct proof of the $1/v$ law for neutron absorption (10a, 10b).

At the end of 1935 I won a competition for a Physics chair in Italy and left Rome for Palermo, Sicily. It was an important promotion, but leaving Rome was a sacrifice. In Palermo I found that the Physics Institute had been neglected and semiabandoned for years, but there was a new building, some money, and much goodwill on the part of the University. I tried to do my best to reestablish the place, concentrating at first on the teaching aspects.

In February of 1936 I married Elfriede Spiro, a young German woman who had fled Breslau to escape the Nazis. I kept visiting Rome during vacations; only Fermi and Amaldi were there. My other physicist friends had disbanded. In looking for a research program for Palermo, I had to consider the local possibilities and my own capabilities. I decided to work in radiochemistry; it was cheap and technically simple *if* one had suitable radioactive substances. Elfriede and I decided to spend the summer of 1936 in the US. She was expecting a child and we thought that this might be our last opportunity, for some time, of visiting the United States. We went first to Columbia University, but later the heat and my desire to see the cyclotron pushed us to California. We explored the state following Rasetti's instructions. In Berkeley I saw the cyclotron and talked with Lawrence and his collaborators McMillan, Kurie, Abelson, Cooksey, Livingood, and others. I was most impressed by the 37-inch cyclotron and by the huge amounts of radioactivity it produced. I obtained from Lawrence several scraps of molybdenum deflectors that were quite radioactive. I planned to study them on my return to Palermo treating them as though they were an ore. I had equipped a small radiochemistry laboratory; samples were just what I needed. To improve my chemical prowess, I asked the help of the professor of mineralogy, Carlo Perrier, who had his Institute in the same building and with whom I had struck a close friendship on my arrival at Palermo.

The cyclotron scrap proved to be a fertile mine of radioactivity. We found in it the missing element 43 (11a, 11b), the masurium I had in vain tried to buy from my chemical merchant, and which much later we named technetium to commemorate the fact that it was the first artificial element. The scrap also contained a substantial amount of ^{32}P, ^{60}Co, and other radioisotopes. I was able to arouse the interest of the physiologist Camillo Artom (later of Wake Forest, North Carolina, Medical School) and together we investigated phospholipid absorption by various organs, (12a, 12b). Of course, he did the physiology and biochemistry. I worked on the radioactive measurements and on a mathematical model of rat metabolism. That year, 1937, I had the great satisfaction of being praised by Fermi who came to visit me. His praise was almost as sparse as his wrath and both counted heavily. Niels Bohr invited me to Copenhagen for one of his restricted conferences. For a debut from the remote island of Sicily, I could not complain.

At that time I also arranged for a Chair of theoretical physics at Palermo, with the intention of inviting G. C. Wick. By a complicated and totally unpredictable chain of events, connected with the establishment of the Palermo chair, E. Majorana, who in the meantime had become a recluse, was appointed professor at Naples. After a few months he vanished from the Naples-Palermo boat, most probably a suicide. Wick came to Palermo in 1937.

Although for the time being things were going well at Palermo, the future looked bleak. The political situation was deteriorating and Hitler's visit to Italy in the spring of 1938 certainly boded ill.

My close friend Count Emo Capodilista, who lived in Berkeley and worked at the Radiation Laboratory, sent me some more cyclotron scrap. I had arranged to put near the cyclotron target test tubes and bottles containing ammonium nitrate, uranium oxide, and sundry other substances to be irradiated with neutrons. From the nitrogen of ammonium nitrate

I hoped to obtain ^{14}C; from uranium I wanted to see whether any new long-lived activity would appear. The projects were sensible but came to naught. In the summer of 1938 I decided to go to Berkeley to study short-lived isotopes of technetium that could not stand the duration of a sea trip from California to Italy. I had a return ticket and my wife and child remained in Italy.

In New York I met Szilard. He inquired about my plans, and commented that I would be unable to return to Italy. By the time "The Challenger," the transcontinental train, had brought me to Berkeley I had read in the newspaper that Mussolini had embraced anti-Semitism and was firing all state employees who did not satisfy his newly discovered criteria of racial purity. It was clear to me, even without an explicit letter from the Italian authorities, that I had lost my job and that there was no future for me in Italy, barring an upheaval and a change of government.

Fortunately I found temporary support at the Radiation Laboratory, which gave me a breathing spell. I proceeded with the investigation of short-lived technetium isotopes and became acquainted with a new PhD in the chemistry department, Glenn T. Seaborg. Together we found the isomer ^{99m}Tc (13). Little did we anticipate that it was to become one of the mainstays of nuclear medicine.

In Berkeley I found a very different physics from the one I was used to. Theory, represented by Oppenheimer, was much more highbrow, at least in form, if not in substance. Experimentalists at the Radiation Laboratory, represented by Lawrence, were primarily occupied with the development of the cyclotron, a spectacular and unique tool that opened undreamed of possibilities. The relations between theory and experiment were looser than in Rome or in Hamburg and detectors for radioactive work were rather primitive. The interest definitely centered on the machine rather than on the results obtainable with it. In the Physics department, spectroscopic work proceeded in a manner more familiar to me. The magnet of the 60-inch cyclotron, then under construction, gave me a chance to resume spectroscopic work together with F. A. Jenkins (14), who had by then become a dear friend and a mentor in my Americanization.

The Radiation Laboratory and the Physics department cooperated harmoniously. There were occasional difficulties, but they were much less important than the true fundamental concord prevailing. R. T. Birge ruled in Physics, E. O. Lawrence in the Radiation Laboratory. The two, in spite of very different personalities, understood each other and understood that "Viribus unitis" had to be their motto.

My own research, apart from the spectroscopic interlude, was devoted mainly to nuclear problems using chemistry as a tool. I worked on the cyclotron, too, because that was part of my duties, but my contributions to machine building or operating were never substantial. In the nuclear work I collaborated with several scientists, who were somewhat younger than I—either advanced students or new PhDs who gravitated around the cyclotron. A. Langsdorf, C. S. Wu, J. W. Kennedy, G. T. Seaborg, D. R. Corson, and K. MacKenzie were among them.

We worked on nuclear isomerism (15a, 15b, 15c) and on new elements. I devised a method for chemically separating nuclear isomers (16) and we studied the conversion electrons accompanying the isomeric transitions, making advances in the classification of their radiation. Astatine was a new element, an upper homolog of iodine, that Corson, MacKenzie and I prepared by bombarding bismuth with alpha particles as soon as the new 60-inch cyclotron became operative (17).

The discovery of fission by Hahn and Strassmann at the end of 1938 was a major event. It cleared up the mysteries of the activities observed in uranium bombarded with neutrons. Not a few of the physicists and chemists who had previously worked on this subject were red-faced. I worked on fission products for some time (18a, 18b, 18c). During a visit by Fermi to Berkeley we produced fission by alpha-particle bombardment; this was the last paper we coauthored (19). When Joliot and, independently, McMillan showed that there was a "nonrecoiling" activity in uranium bombarded with neutrons, I tried to determine its chemical properties. I found that it behaved like a rare earth, but I did not recognize that it was element 93, neptunium, as Abelson & McMillan did some time later (20).

The Second World War was by then imminent. Scientists, even in neutralist and occasionally isolationist America, were turning to radar and similar war research. I was an Italian citizen and, after Pearl Harbor, an enemy alien. Thus I could not join the official war effort. But I was well aware that nuclear fission opened up unfathomable possibilities of energy release. As did many others, I made my private calculations and speculations on the subject. Nuclear energy seemed likely and nuclear explosions possible. My primitive calculations tried to estimate a critical mass, taking into account the discovery that only ^{235}U was fissionable by slow neutrons.

With many professors absent from Berkeley, I was asked to teach several physics courses. This gave me an occasion to refresh or learn much physics and also to meet in classes American graduate students, among them Owen Chamberlain.

During 1940 I was eager to find a more permanent job than the one offered at Berkeley, and I visited Purdue University at the invitation of Karl Lark Horovitz. I also visited an oil exploration company, Wells Surveys in Tulsa, Oklahoma. I consulted on well logging by neutrons, but did not take the job I was offered. It fell to Bruno Pontecorvo, who had just fled a Paris invaded by the Nazis. In December 1940 I went to visit Fermi at Leona in New Jersey. We had a long walk along the Hudson, in freezing weather, during which we spoke of the possibility that the isotope of mass 239 of element 94, called later ^{239}Pu, might be a slow neutron fissioner. If this proved to be true, ^{239}Pu could substitute for ^{235}U as a nuclear explosive. Furthermore a nuclear reactor fueled with ordinary uranium would produce ^{239}Pu. This gave an entirely new perspective on the making of nuclear explosives, eliminating the need to separate uranium isotopes, at that time a truly scary problem.

An experimental investigation of the then-unknown ^{239}Pu was thus decisively important. We took advantage of the presence of Lawrence in New York to interest him in the problem. Fermi, Lawrence, Pegram, and I met in Dean Pegram's Office at Columbia University and developed plans for a cyclotron irradiation that could produce a sufficient amount of ^{239}Pu. In the first months of 1941 the project was carried out by J. W. Kennedy, G. T. Seaborg, A. C. Wahl, and me. By the spring we had good measurements of the fission cross section of ^{239}Pu for slow neutrons and we found a number close to the presently accepted value, comparable to the cross section of ^{235}U (21). The results of this work brought me to the center of the uranium project that was to become the Manhattan Project. In 1941, however, the work did not yet have governmental sponsorship and we conducted it as private citizens.

Pearl Harbor changed many things for physics at Berkeley. Lawrence mounted an all-out effort to separate uranium isotopes by building giant mass spectrographs for which he used the cyclotron magnets. I felt strongly that it was necessary to develop a reliable method for isotopic analysis of small uranium samples to check the performance of the spectrographs. By then I had a number of graduate students, including Clyde Wiegand, and we set up a method of analysis which in essence detected ^{234}U by its alpha activity, ^{235}U by its slow neutron fission, and ^{238}U by its mass. This method was used for a while to check the progress of the performance of the mass spectrographs (22).

During 1942 it became clear that serious planning of a bomb required many nuclear data, especially cross sections for intermediate energy neutrons. The pertinent experiments were coordinated by John Manley and used by a theoretical group that convened at Berkeley under the leadership of Oppenheimer. Bethe, Teller, Serber, Van Vleck, and Konopinski were among its members. The possibility of spontaneous fission was also considered and we started an investigation to find its frequency. Unknown to us, Flerov and Petrjak in the Soviet Union had discovered the phenomenon and were performing similar investigations.

Kennedy, Wiegand, other students, and I worked on spontaneous fission in Berkeley. As time went on, the necessity of a special laboratory devoted to building the bomb became clear and led to the creation of the Los Alamos Laboratory in New Mexico.

Its newly appointed director, Oppenheimer, was trying to assemble the needed personnel. He had clearly laid his eyes on me as a possible recruit. I was invited, quite unusually, to dinner at his home and presented with a lemon plant for my garden. When Los Alamos materialized, I was among the very first to go there, with my family and the group of students that had been working with me at Berkeley.

At Los Alamos our primary task was to study spontaneous fission of uranium and plutonium isotopes. We found that ^{240}Pu had a high rate of spontaneous fission, high enough to influence by its presence the performance of ^{239}Pu as a nuclear explosive (23a, 23b). Because the two would be formed together in a reactor, there would be no practical possibility of avoiding the presence of ^{240}Pu. Weeks of high drama followed this finding. Initially the observation was limited to 1 spontaneous fission, then 2, and then 3 within a few days. As luck would have it, the number we obtained, turned out to be correct in spite of the large statistical error. Spontaneous fission forced a significant change in the plans for making a bomb. Other measurements we performed concerned cross sections, number of neutrons emitted per fission, and similar quantities. They had their importance but did not affect the project as deeply as spontaneous fission did.

In July 1945 I participated in the test of the first bomb at Alamogordo. Our group measured the prompt gamma rays emitted by the explosion and one of our young helpers smuggled a camera at the test. With it he took pictures of the explosion. They were ready before the official ones and found their way, I believe, directly to President Truman. Fermi and I saw the explosion from the same observation point. It was an awesome sight, comparable to great natural phenomena, and it had a sobering effect on the beholders.

The end of the Los Alamos period posed serious problems for me. I did not have a home university to which to return because my Berkeley appointments had always been only temporary. I wanted to return to scientific work. Although I was enthralled by the natural beauty of Los Alamos and its surroundings, I did not want to devote the rest of my life to its work, whatever it might be.

In the Fall of 1945 I had several offers from universities, among them Chicago where Fermi's presence was a very strong attraction. Ultimately Berkeley made an acceptable offer and I decided to return there. New accelerators were in the making and would offer in the near future unprecedented experimental possibilities.

I arrived at Berkeley in January 1946 as a Professor of Physics. For the first time since leaving Italy I had a permanent appointment and I could make long-term plans. First I wanted to finish some work started long ago and interrupted because of the war. In 1940 I had thought that it might be possible to alter slightly the decay constant of a nuclear K capture via chemical means by changing the electron density at the nucleus. ^{7}Be seemed to be a suitable candidate and C. S. Wu and I had started manipulating several grams of beryllium compounds without taking any special precaution, because the toxicity of beryllium was not then known. The war interrupted the work, but in 1947 Leininger, Wiegand, and I completed the work and demonstrated the effect (24a, 24b, 24c). The chemistry of astatine was also unfinished business, but it proved very difficult (25). Other nuclear problems kept us busy for a while; the direction of research was, however, changing.

As the cyclotron energy increased, new problems could be addressed. Theoreticians compiled a list of them, which proved to be very modest in imagination and foresight. However, the nucleon-nucleon interaction seemed central to nuclear physics and there were hopes to repeat Rutherford's feat and derive it from scattering experiments. Even such modest advances as the observation of the p-wave in scattering was considered by theoreticians as extremely important. We undertook a large-scale investigation of np and pp scattering to obtain reliable absolute data. It was a painstaking enterprise that lasted several years. A number of graduate students contributed greatly to it, and for years our data remained the best on the subject. The analysis of the experimental data to obtain the phase shifts was conducted by Metropolis, Stapp, and Ypsilantis at Los Alamos using the then-new computing facilities of that laboratory (26a, 26b).

Tom Ypsilantis, then a graduate student, succeeded in polarizing the beam of the 184-inch cyclotron and added a new dimension to our work. Soon we were measuring several of the polarization parameters (A, R, etc in modern notation), adding significance to our data (27a, 27b, 27c, 27d).

At the end of the war I was acutely aware of the great technical progress made by nuclear physics and at the same time of the lack of a treatise in the style of the old German treatises on which one could rely for complete information. It was clearly unfeasible for one author to write such a book, but by enlisting the cooperation of willing collaborators, I thought it would be possible to complete the enterprise in a reasonable time. I thus edited a treatise *Experimental Nuclear Physics*, published by John Wiley & Sons (28), which for a period was useful to the scientific community. During the same period the *Annual Review of Nuclear Science* was founded. In 1952 I became its editor. The work was instructive, if not lucrative, and I felt it was my duty to contribute to the welfare of the profession. I continued as editor of the series until 1977. I was thus forced to keep abreast of developments in our science, a task that became progressively more difficult because of the increasing amount and technicality of the literature. I take it that the invitation to write this article is a recognition of good behavior and devotion to the cause of science.

In November of 1953 Fermi died. He knew from September that his days were numbered. I visited him repeatedly during his last illness and talked at length with him just a few hours before he died. The event shook me deeply. Corbino had died in 1937; my parents had died during the war while I was in Los Alamos; now Fermi was gone. One of the great satisfactions of any scientific success for me had been to tell it to these persons to whom I owed so much and for whom I had filial feelings. It may be suitably interpreted by psychologists, but approval from a few selected persons was to me very important, much more than popular success.

I have often been asked about my relations with Fermi. I always recognized his obvious superiority and never felt any envy or jealousy of him. The gap was too great. On the other hand, I never had the feeling of being simply his tool. I was fortunate in my early days to have enough independent ideas to preclude too strong a sense of dependency, although his scientific influence on me was deep. Others were his daily collaborators for long periods. I worked continuously with him only in the exciting neutron work of 1934–1935. For all our close and sincere friendship, I knew very well his aloofness and reserve even when masked by an apparent lighthearted cordiality and I tried not to intrude. My labor on his collected papers, as Chairman of the Editorial Committee, was a heartfelt tribute of gratitude and admiration, and one, I thought, he might have liked.

By 1954 the bevatron, designed chiefly by W. Brobeck, had reached its planned energy of approximately 6 GeV, a value chosen to give protons above the threshold for the production of antiprotons in p-p collisions. It was thus clear that the experimental proof of the existence of antiprotons was within our grasp. The chief difficulty of the experiment was the extraction of a reliable antiproton signal out of a huge noise produced by the many reactions occurring in the target. In fact, only about one in 50,000 of the negatively charged particles emerging from the target was an antiproton.

Two main lines of attack were possible: a determination of the e/m ratio for the particles produced, or an observation of the terminal event that was sufficiently detailed to identify the annihilation process. Both approaches were tried. Chamberlain, Wiegand, Ypsilantis, and I built a mass spectrograph that could detect antiprotons in a heavy background of charged particles. We determined the momentum and the sign of the charge of single particles by deflection in a magnetic field, assuming that the magnitude of the charge was equal to that of the proton. On the same particle, we measured the velocity with a Cerenkov counter as well as by measuring the time of flight between two detectors. From momentum and velocity values we derived the mass of the particle as equal to that of the proton to within a few percentage points. This investigation, conducted in October of 1955, gave the first conclusive proof of the existence of the antiproton (29).

The other line of approach was an exposure of nuclear photographic emulsions to a beam enriched in antiprotons. A study of the terminal events of the tracks would show the annihilations. For this method we needed a large number of scanners, and Amaldi brought an Italian contingent to the task, while Gerson Goldhaber, who had worked for some years in my group at Berkeley, headed the American side of the effort. The Italians found a beautiful star with all the connotations of an antiproton annihilation; a little later in Berkeley we found a star produced by a particle at rest and showing a visible energy release larger than $m_p c^2$ and thus necessarily due to an antiproton (30a, 30b).

Other groups were working on the antiproton in Berkeley. After our success they demonstrated the antineutron, obtaining it by charge exchange. In cooperation with the Wilson Powell group we, too, obtained evidence in a propane bubble chamber for the formation of antineutrons by charge exchange (31a, 31b).

After the initial antiproton work, I tried to pursue it as far as possible, but we did not have a strong enough group structure to take full advantage of our headstart. Furthermore the bevatron was soon superseded by more powerful accelerators that produced many more antiprotons; and even at Berkeley the upcoming hydrogen bubble chamber preempted the field.

My way of doing experiments had always been to keep apparatus as simple and inexpensive as possible, to have a solid theoretical foundation, and to try to keep the ratio between results and expense as high as possible. This attitude derived in part from my nature and in part from my upbringing in Rome. It was appreciated wherever I worked and I never had serious financial strictures nor difficulties in obtaining time on the accelerators, at least after the war. In my early days in Berkeley I carried on most of my work using an ionization chamber modeled on one we used in Rome. I was an expert on its use and it proved invaluable not only for me, but also for colleagues such as Kamen, McMillan, and others. There was no question of technical or secretarial help, but some of the graduate students, for instance Clyde Wiegand, had outstanding electronic skills, acquired independently of their study. After the war the financial situation changed radically and then, although I kept a tight ship, we had all we needed.

After the antiproton work, larger and larger collaborations became necessary. Our group did not fit too well into this scheme of things. Chamberlain for a while formed a separate group and devoted himself to the development of polarized targets. Later he rejoined our group. Clyde Wiegand with Ypsilantis and some students undertook several important experiments to observe the pion-pion interaction. They built an apparatus (fly eye) that at the time seemed large and complicated. It would be considered naive and too small now. This apparatus incorporated several features that later became standard and was coupled to a small computer. They found indications of the ρ and ω resonances, but the whole effort was too small for the problem.

Later they studied the beta decay of the positive pion; that is, the rare branching $\pi^+ \rightarrow \pi^0 + e^+ + \nu_e$. This was an important and accurate piece of work. The results were possibly the best obtained on the subject at the time. Ray Stiening came to our group from MIT. He started a series of investigations with Wiegand and Min Chen on the polarization in $K_{\mu3}$ decay. He also introduced us to flying. I was a passenger on several trips with him. Wiegand bought a plane and emulated him.

Ypsilantis left Berkeley for personal reasons and Stiening was lured to Fermilab, where he did distinguished work, before moving to the Stanford Linear Accelerator Center (SLAC). Wiegand entered the field of mesic atoms, which had interested me for a long time. He detected K mesic and Σ mesic atoms and for several years conducted a series of successful investigations on their intensity, on mass determinations of the K meson, and on related subjects. I did not participate directly in any of this experimental work.

In 1964 there appeared *Nuclei and Particles* (32), a text that was translated into several languages. It enjoyed success and represents fairly my own scientific generation. Nuclei and particles were not yet separated, and indeed a good fraction of the particle physicists came from the ranks of nuclear physicists or cosmic-ray physicists. I prepared a second edition of this book in 1977. It will certainly be the last. The attempt to keep nuclear and particle physics together was justified in 1964 especially as a countermeasure to the increasing specialization of the students. Now, however, the fields have grown so much, and so much apart, that it is futile to try to keep them together, at least in text books. The *Annual Review of Nuclear and Particle Science* attempts to encompass both fields at the research level.

Since my retirement in 1972 I have devoted time to historical studies. I am learning the history of physics in part by reading, in part from colleagues who specialize in the history of science, and much by reflection on my own experiences.

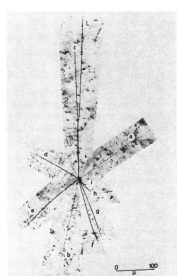

Near right: Ionization chamber of a model used extensively in Rome, Palermo, and Berkeley for nuclear work. *Far right*: Antiproton annihilation star in a photographic emulsion (from Reference 30a). *Below*: Antiproton group—C. E. Wiegand, O. Chamberlain, E. Segrè, and T. J. Ypsilantis.

Literature Cited

1. Segrè, E. 1930. *Nature* 126:882
2. Segrè, E., Bakker, C. J. 1931. *Nature* 128:1076
3a. Segrè, E. 1931. *Accad. Lincei, Atti* 14:501–5
3b. Segrè, E. 1932. *Accad. Lincei, Atti* 16:442
4. Fermi, E., Segrè, E. 1933. *Z. Phys.* 82:729–49

5. Frisch, R., Segrè, E. 1933. *Nuovo Cimento* 10:78–91
6. Segrè, E. 1934. *Nuovo Cimento* 11: 304–8
7a. Amaldi, E., Segrè, E. 1934. *Nature* 133:141
7b. Segrè, E. 1934. *Accad. Lincei, Atti* 19:595–99
7c. Amaldi, E., Segrè, E. 1935. In *Pieter Zeeman, Verhandelingen*, pp. 8–17. The Hague: Martinus Nijhoff
8. Fermi, E., Amaldi, E., Pontecorvo, B., Rasetti, F., Segrè, E. 1934. *Ricerca Sci.* 5(2):282–83
9a. Fermi, E., Amaldi, E., D'Agostino, O., Rasetti, F., Segrè, E. 1934. *Proc. R. Soc. A* 146:483–500
9b. Amaldi, E., D'Agostino, O., Fermi, E., Pontecorvo, B., Rasetti, F., Segrè, E. 1935. *Proc. R. Soc. A* 149:522–58
10a. Dunning, J. R., Fink, G., Pegram, G. B., Segrè, E. 1936. *Phys. Rev.* 49:198(A)
10b. Rasetti, F., Segrè, E., Fink, G., Dunning, J. R., Pegram, G. B. 1936. *Phys. Rev.* 49:104(L)
11a. Perrier, C., Segrè, E. 1937. *Nature* 140:193
11b. Perrier, C., Segrè, E. 1937. *J. Chem. Phys.* 5:712–16; 1939. 7:155–56
12a. Artom, C., Perrier, C., Santangelo, M., Sarzana, G., Segrè, E. 1937. *Nature* 139:1105–6
12b. Artom, C., Sarzana, G., Segrè, E. 1938. *Arch. Int. Physiol.* 47:245–76
13. Seaborg, G. T., Segrè, E. 1939. *Phys. Rev.* 55:808–14
14. Jenkins, F. A., Segrè, E. 1939. *Phys. Rev.* 55:52–58; 545–48
15a. Kennedy, J. W., Seaborg, G. T., Segrè, E. 1939. *Phys. Rev.* 56: 1095–97
15b. Langsdorf, A. Jr., Segrè, E. 1940. *Phys. Rev.* 57:105–10
15c. Segrè, E., Helmholz, A. C. 1949. *Rev. Mod. Phys.* 21:271–304
16. Segrè, E., Halford, R. S., Seaborg, G. T. 1939. *Phys. Rev.* 55:321(L)
17. Corson, D. R., MacKenzie, K. R., Segrè, E. 1940. *Phys. Rev.* 58:672–78
18a. Segrè, E., Wu, C. S. 1940. *Phys. Rev.* 57:552(L)
18b. Segrè, E., Seaborg, G. T. 1941. *Phys. Rev.* 59:212(L)
18c. Wu, C. S., Segrè, E. 1945. *Phys. Rev.* 67:142–49
19. Fermi, E., Segrè, E. 1941. *Phys. Rev.* 59:680(L)
20. McMillan, E., Abelson, P. H. 1940. *Phys. Rev.* 57:1185
21. Kennedy, J. W., Seaborg, G. T., Segrè, E., Wahl, A. C. 1946. *Phys. Rev.* 70:555–56

22. Kennedy, J. W., Segrè, E. 1943. *Atomic Energy Commission Report* (Manhattan District Declassified Document) 973, 26 pp.
23a. Segrè, E. 1952. *Phys. Rev.* 86:21–28
23b. Chamberlain, O., Farwell, G. W., Segrè, E. 1954. *Phys. Rev.* 94:156
24a. Segrè, E. 1947. *Phys. Rev.* 71:274(A)
24b. Leininger, R. F., Segrè, E., Wiegand, C. 1949. *Phys. Rev.* 76:897–98; 81: 280(E)
24c. Segrè, E., Wiegand, C. E. 1949. *Phys. Rev.* 75:39–43; 81:284(E)
25. Johnson, J. L., Leininger, R. F., Segrè, E. 1949. *J. Chem. Phys.* 17:1–10
26a. Hadley, J., Kelly, E. L., Leith, C. E., Segrè, E., Wiegand, C., York, H. F. 1949. *Phys. Rev.* 75:351–63
26b. Chamberlain, O., Segrè, E., Wiegand, C. 1951. *Phys. Rev.* 81:284–85(L)
27a. Chamberlain, O., Segrè, E., Tripp, R., Wiegand, C., Ypsilantis, T. 1954. *Phys. Rev.* 93:1430–31(L)
27b. Chamberlain, O., Donaldson, R., Segrè, E., Tripp, R., Wiegand, C., Ypsilantis, T. 1954. *Phys. Rev.* 95: 850–51(L)
27c. Chamberlain, O., Segrè, E., Tripp, R., Wiegand, C., Ypsilantis, T. 1954. *Phys. Rev.* 95:1104–5(L)
27d. Ypsilantis, T., Wiegand, C., Tripp, R., Segrè, E., Chamberlain, O. 1955. *Phys. Rev.* 98:840–42(L)
28. Segrè, E., ed. 1953–1959. *Experimental Nuclear Physics*, 3 vols. New York: Wiley
29. Chamberlain, O., Segrè, E., Wiegand, C., Ypsilantis, T. 1955. *Phys. Rev.* 100: 947–50(L)
30a. Chamberlain, O., Chupp, W. W., Goldhaber, G., Segrè, E., Wiegand, C., Amaldi, E., Baroni, G., Castagnoli, C., Franzinetti, C., Manfredini, A. 1956. *Phys. Rev.* 101:909–10(L)
30b. Chamberlain, O., Chupp, W. W., Ekspong, A. G., Goldhaber, G., Goldhaber, S., Lofgren, E. J., Segrè, E., Wiegand, C., Amaldi, E., Baroni, G., Castagnoli, C., Franzinetti, C., Manfredini, A. 1956. *Phys. Rev.* 102: 921–23
31a. Button, J., Elioff, T., Segrè, E., Steiner, H. M. Weingart, R., Wiegand, C., Ypsilantis, T. 1957. *Phys. Rev.* 108: 1557–61
31b. Agnew, L., Elioff, T., Fowler, W. B., Gilly, L., Lander, R., Oswald, L., Powell, W., Segrè, E., Steiner, H., White, H., Wiegand, C., Ypsilantis, T. 1958. *Phys. Rev.* 110:994–95(L)
32. Segrè, E. 1964. *Nuclei and Particles*. New York: Benjamin (1977, Reading, Mass. 2nd ed.)

NUTRITION

Ann. Rev. Nutr. 1982. 2:1-20
Copyright © 1982 by Annual Reviews Inc. All rights reserved

PERSONAL REFLECTIONS ON CLINICAL INVESTIGATIONS

William B. Bean[1]

Department of Internal Medicine, University of Iowa, Iowa City, IA 52240

Introduction

I was fortunate to participate in one of the explosive phases of clinical nutrition more than 40 years ago, just before World War II began. At that time water-soluble vitamin B was becoming the vitamin-B complex, budding off many separate new vitamins. These were being discovered by students of animal nutrition and biochemistry, and by those interested in the nutritional growth requirements of such lowly organisms as bacteria, yeasts, and fungi.

As an undergraduate medical student at the University of Virginia, I saw pellagra. We were told about malnutrition and Goldberger's work. I saw one pellagra patient on the Osler Service at Johns Hopkins in 1935–1936. The next year, when I went to the Thorndike Laboratory, Soma Weiss and Robert Wilkins were in the midst of their exciting studies demonstrating that vitamin B_1, thiamin, quieted the alarming and precarious hyperdynamic state of the circulation in beri-beri, then so prevalent among the alcoholic gentry who crowded into the Boston City Hospital.

In 1937 I went to work in Marion Blankenhorn's Department of Medicine in Cincinnati. It was an exciting place to be. Gene Stead, Gene Ferris, Johnson McGuire, Leon Schiff, Lee Foshay, and others there were stimulating young investigators and teachers. Tom Spies had done excellent clinical studies on pellagra at the Lakeside Hospital in Cleveland and the Cincinnati General Hospital. Near the end of my year as a senior medical resident an

[1]William B. Bean, medical scholar—physician, clinical investigator, editor, teacher, writer, and medical philosopher—was intimately involved in clinical nutrition research during the exciting era of discoveries of the 1930s–1950s. His reflections on clinical research in and knowledge of nutrition during that period of more direct, relatively uncomplicated, less restrictive investigations provide much wisdom for consideration by investigative-minded physicians today.

1379

0199-9885/82/0715-0001$.02.00

opportunity arose to work as a Fellow in nutrition. I had no special competence in the field of human nutrition but had helped Soma Weiss and Bob Wilkins in some of their clinical experiments. A good friend of Blankenhorn, Dr. James McLester of Birmingham, Alabama, still skeptical of the ideas introduced by Goldberger, welcomed Tom Spies to Birmingham to see if he could get the same good results in treating desperately ill pellagrins there that he had obtained in Ohio. The mobility of the staff, which shifted to Birmingham for the spring and summer months and then returned to Cincinnati for the fall and winter months, was remarkable. Richard and Sue Vilter and I had the main clinical and laboratory responsibilities. In the considerable outpouring of papers from the nutrition clinic, Spies's name led all the rest. Charles Aring, Cincinnati neurologist, essayist, and scholar, made prominent contributions to the thiamin and beri-beri studies. Many physicians, biochemists, nurses, dentists, and others from various medical schools in the country were eager participants. In addition to A. B. Chinn and Blankenhorn, the following co-workers are listed alphabetically: W. F. Ashe, A. E. Axelrod, W. Beckh, C. E. Bills, Hugh Butt, C. Cogswell, Clark Cooper, Zola K. Cooper, G. Delfs, R. Eakin, Conrad Elvehjem, Joe P. Evans, E. Gross, Morton Hamburger, Harold E. Himwich, M. P. Hudson, T. H. Jukes, Walter F. Lever, J. B. McLester, A. W. Mann, V. Minnick, Carl V. Moore, Robert A. Moore, Gordon R. Morey, Milton Rosenbaum, J. M. Ruegsegger, E. E. Snell, S. R. Stanbury, R. E. Stone, E. P. Swain, Emory D. Warner, R. J. Williams. After I had left Carl Vilter and Wally Frommeyer were important contributors.

A Note on the History of Pellagra

Three recently published books (25a, 27a, 40a) afford detailed accounts of the history of pellagra. These and the earlier compilation of selected reprints of Joseph Goldberger and an evaluation of his classical studies of the disease, *Goldberger on Pellagra* by Milton Terris (62a), make it redundant to detail here an account of the evolution of understanding of this fascinating deficiency syndrome. Rather, I recall here but a few especially poignant personal recollections pertinent to my own early involvement with clinical evidence concerning pellagra.

For more than two hundred years the nature of pellagra had been a subject of great debate and confusion (32). Many drugs had been tested on severely ill pellagrins brought into a hospital, put to bed, and given fluids and good nursing care. For a brief time many experienced clinical improvement. Spies had shown that this might occur even while they were eating a pellagra-producing diet or given nothing but salt solution and dextrose. Such a program, however, could not be continued long or the patients would suddenly get sicker and might die if the elements missing from the diet were not restored in ample quantities.

Perhaps the first controlled study in human nutrition after Lind's observation on scurvy was Cerri's study of pellagra carried out in Milan, Italy in 1795 and 1796 (32). Cerri was convinced that a diet made up largely of corn and corn products somehow explained the very high incidence of pellagra among the country folk. He selected ten of them, who for two years moved to town and ate the diet common to those in the city. On this regimen no signs of pellagra developed, while in the peasant controls in the country the incidence remained high. The following year the subjects were allowed to revert to their polenta (corn meal) diet, and pellagra returned as usual. One, however, was so much impressed that he got work in the city and never had pellagra again. These observations have been largely neglected by those who have written on pellagra.

A good example of the pre-Goldberger confusion is found in a 1912 book entitled *Report of the Pellagra Commission of the State of Illinois* (40). It describes the distressing outbreak of pellagra in Illinois mental hospitals from 1909–1911. Of the 258 patients with pellagra in Peoria, 128 died. In Kankakee, with fewer patients, 40% died, and about a third died at the Elgin State Hospital [subsequently the site of a long series of remarkable studies by Horwitt et al, in which deficiencies of niacin, thiamine, riboflavin, or tocopherol were induced by feeding diets deficient in the nutrients. These studies were made in collaboration between the Food and Nutrition Board of the National Research Council and the Elgin State Hospital over a 23-year period (33a)]. At that time all the Commission could think of were infections. Study of fecal bacteria occupied 105 pages of the special report, which then proceeded to complement fixation tests, cutaneous tests with corn extracts in pellagrins, and a learned discourse on black flies and buffalo gnats, including beautiful drawings of the various larval phases of the Simulium in Illinois. An essay on the protozoan infections was followed by dietary studies from the hospital concerned only with the food as issued, not the food as eaten. A corn-free diet was compared with a mainly-corn diet without definite conclusions. The experiment is a good example of how not to do a study comparing one food with another. In the summary the authors wrote that "the lack of definite information regarding the food requirements and metabolism in the class of subjects experimented upon has made it difficult to interpret the results obtained in these studies." A rather wistful comment added that "the experimental work here reported is in itself brief and is not extensive enough to allow any broad interpretation."

Joseph Goldberger Solves the Problem

The spectacular prevalence of pellagra in the Southern states seemed related to poverty. Of course, infection was known to follow poverty fairly closely. In 1914, Joseph Goldberger of the US Public Health Service, the old Marine Hospital Service, was sent to Georgia to see if the peculiar epidemiology of pellagra indicated an infection. The seasonal peak of pellagra occurred in the late spring and early summer, some two or three months before the peak intensity of sunshine and heat. In cotton towns the disease clustered among the homes of workers who lived in poor quarters, often along the banks of streams or rivers subjected to flooding, and largely without plumbing. Goldberger's work in solving certain epidemics of insect-borne skin disorders made him a logical person to find some kind of infecting organism. In the South, Goldberger was under the triple jeopardy of being a Yankee, a Federal Government worker, and a Jew. His assignment seemed to add Northern insult to Southern injury, though eventually some southerners accepted the surprising results of his heroic experiments.

Goldberger not only tormented himself as an experimental subject, he also used his wife as a subject in hair-raising ingestions or injections of horrible samples of the effluvia of very ill victims of pellagra. While this ranks high as an example of heroic autoexperimentation and of a wife's devotion, it failed to reveal any infecting organism. Indeed, Goldberger was well aware of the fact that nurses, attendants, and physicians almost never got pellagra in asylums, hospitals, or prisons. Thus having found no infection, he turned to study of diet. After he learned what the people who developed pellagra customarily ate, he obtained convict volunteers as subjects, promised them early reprieve, and fed them the same food as customarily eaten by those developing pellagra. After many months the prisoners developed pellagra, which cleared up on a good diet. Goldberger found, in particular, that inexpensive brewer's yeast was a good biological food for treatment or prevention. He suspected that an amino acid deficiency, particularly that of tryptophane, was important—a suggestion made by Sandwith as early as 1913 (29–31, 41a).

Nicotinic Acid: The New Era

In 1937, Elvehjem, Woolley, and their associates (27) found that canine black tongue, which Goldberger had identified as a good experimental model of human pellagra, was relieved dramatically by nicotinic acid. It was most interesting to see the spectacular and sometimes astonishing efforts to establish priority of publication in nailing nicotinic acid to the clinical mast. In New Orleans, Durham, Augusta, Birmingham, and Cincinnati there was a great scurrying around to find out what nicotinic acid might do for sick pellagrins.

The name nicotinic acid for this derivative of nicotine was rather frightening (48–52), and the term niacin was coined to avoid any implication of a toxic material. A dramatic and alarming, but fortunately innocuous, event occurred when Spies and his associates each took a quantity of nicotinic acid by mouth. Soon, to their astonishment, they developed a blotchy erythema of the face and upper body. Some experienced a pounding headache and a little nausea. No doubt anxiety was a part of the reaction, which fortunately soon passed without damage.

Later on I studied whether nicotinic acid and related compounds that did not produce flushing were effective in the prevention and treatment of pellagra. Several, indeed, had this property. In order to study the great variety of compounds to determine what specifically caused the flushing, it was necessary to measure skin temperature in fasting subjects. Since many compounds had to be prepared and were available in small quantities, the tests involved the intravenous injection of 20 mg of the experimental compounds, each of which I had previously tested on myself. Subjects lay flat on their backs, lightly clad, in a constant-temperature room at 20°C. This produced a steady state of vasoconstriction so that vasodilation could be measured readily. Skin temperatures were read off a galvanometer. The specific molecules that produced flushing were those of nicotinic acid and its various salts. Pellagra could be treated effectively with nicotinamide, which did not produce the unpleasant flush (3).

The Self as Subject

As illustrated by this experience, autoexperimentation was usually the beginning of clinical research in nutrition. I have done hundreds of such experiments with no apparent ill effects. A near miss occurred, however, when I was a busy nutrition Fellow with many clinical and teaching responsibilities. For nearly two weeks I took a pellagra-producing diet and lost ten pounds in ten days, mostly from an inability to eat enough of the cornbread, hominy, fat back, and molasses to supply calories. I was exhausted and unable to do my work, so I stopped the test, which was not expected to produce results for months. This experience made me acutely aware of the other side of the patient-physician interaction. Therefore, in testing various compounds for effects on malnourished persons we ate or injected them into ourselves before giving them to patients. Later, extensive experiments of many kinds were done in the Army with the medical officers undergoing practically all of the test routines before or while our soldier subjects were tested.

The Inertia and Momentum of Clinical Signs of Malnutrition

Food provides material for constructing the fabric of the body and for the energy for the multitude of functions that constitute cell life. After growth is complete, repair and renewal require energy, as do the daily metabolic functions of the body.

Malnutrition begins when supplies are low. For water-soluble vitamins this occurs more rapidly than with fat-soluble vitamins, some of which are retained with great tenacity. A deficiency at the biochemical level begins to interfere with various functions, causing pathophysiological changes. The organic lesions of niacin deficiency are at first microscopic but become grossly visible as clinical hallmarks of disease of gut, skin, and nervous system. Stomatitis and diarrhea occur. The initial changes in the skin of the malnourished look like a second degree burn, a mild scald, or a severe sunburn. The skin becomes pigmented, desquamates, and sometimes regenerates with the cracking of crazy-pavement epithelium. In localization, regional forces are important. A skin lesion may be progressing in one place while it is healing and regenerating in another. Individual variation in susceptibility is also significant.

It was logical to suppose that nicotinic acid played its role through its function in respiratory enzyme systems. For this reason it was important to find chemical methods to obtain information about the blood level of such enzymes. In 1939, Richard and Sue Vilter published with Tom Spies a microbiological method for determining the level of codehydrogenase (cozymase) in the blood and urine of pellagrins and normal persons (64, 65, 67). The assay of the enzymes was a logical approach to classification and measurement, but later clinical study provided most of the key information on nicotinic acid deficiency. Levels of these pyridine compounds were low in people with pellagra, but also in many persons with diabetes, leukemia, Roentgen sickness, and pneumococcal pneumonia (9, 64, 65). Sometimes the levels were decreased in other infections, in various fevers, and after extreme physical exercise.

It was a puzzling problem to Spies and others that while eating the pellagra-producing diets, pellagrins who are hospitalized might have their lesions improve or even disappear. It began to be obvious that the upset balance between demand and supply could be rectified temporarily by reducing the need as well as by normalizing or increasing the intake. Any agent given at this time that had no specific effect might be credited for the therapeutic improvement that followed.

Biosynthesis, which is important in ruminants, is much less significant in human beings since the colon, where most bacterial action occurs, is too far down the alimentary canal for very effective absorption.

Indirect effects of test substances must also be remembered in therapeutic testing. For instance, if nicotinic acid is given to someone with pellagra and beri-beri while the individual remains on a poor diet, digestion and absorption may improve enough so that thiamin is absorbed in quantities adequate to correct the thiamin deficiency and thus relieves the beri-beri (16).

Deficiencies may take months or years to produce clinical changes. It is not surprising that restoring the missing vitamins and providing a balanced diet do not always produce rapid effects. This is particularly true in lesions involving the nervous system. When a nerve cell is killed, no treatment can repair the damage. However, in many patients with the peripheral neuropathies of beri-beri, slow improvement occurs over a period of months of carefully continued treatment (16).

Because of the diversified metabolic missions of various cells, specific deficiencies may interfere with certain functions of cells before others. The study of nutrition is full of surprises. Krehl, Elvehjem, and colleagues (34) found that the essential amino acid, tryptophane, can be converted into niacin and thus can prevent or cure pellagra. This was a unique example of the conversion of an essential amino acid into an essential vitamin. On the other hand, an abundance of niacin does not take the place of tryptophane in human nutrition, so it is a one way street. It is of note that Goldberger suggested tryptophane deficiency as a contributory mechanism in the development of pellagra (31).

Vitamins undoubtedly may serve as placebos. The confidence engendered by coming to a famous nutrition clinic well-known for benefiting many persons may assist in generating great subjective improvement. Where most manifestations are subjective, the interpretation of improvement must be made in the light of this significant point. Hospitals have lost some of their terror and famous clinics have achieved famous cures. Faith has helped ailing humanity, as at Lourdes. Psychosomatic factors must be investigated, controlled, and evaluated, particularly where manifestations contain so much that is subjective (16).

My observations on pellagra led me to believe that a major source of the arguments and disagreements among observers resulted from significant variation in the quality of corn in different southern regions—i.e. from the amount of nicotinic acid and perhaps of tryptophane in the diet. Certainly when we learned about tryptophane a number of mysteries were cleared up. Diets with more nicotinic acid were sometimes associated with more pellagra, rather than the other way around. Later studies of bound niacin further clarified concepts.

It is still a matter of conjecture why mechanical defects such as intestinal shunts may be followed now by pellagra, now by beri-beri, now by macrocytic or microcytic anemia, or by apparently reasonable health in what seem to be precisely comparable circumstances (4, 6). (See the review by Young & Blass in this volume.)

Unilateral and Experimental Lesions Produced Without Sunlight

Goldberger, who made so many capital observations on pellagra, observed that atypical lesions may readily occur.

> It may be stated as the rule, that if the back of one hand, or one foot, one elbow, one knee, one side of the neck, one cheek, or the lid of one eye is affected then the corresponding part of the other side of the body is assumed to become similarly affected and affected

to almost exactly the same degree. This rule, however, is not without many exceptions. It must not be hastily assumed, therefore, that the possibility of pellagra is necessarily excluded because the back of the head or of one foot or of one side of the neck alone seems to be involved or is involved to so slight an extent to be almost nothing in comparison with the other side.

Stannus, in his classic review of the theories of the cause of pellagra (62), observed that "the facts in regard to the distribution of the exanthem in pellagra may be stated in reality quite simply, though they appear to have escaped the observations of most pellagrologists. The exanthem tends to appear in those areas of the skin which in any particular individual have *undergone certain changes* as the result of the action *in the past* of traumata of various kinds including solar radiation, exposure to cold, friction, pressure, irritants, etc." This is also true of skin lesions in many other diseases.

I made a careful study in 1940 and 1941 of a series of patients who had asymmetric or unilateral lesions as well as bilaterally symmetrical ones (8). Five cases of asymmetric lesions were observed in persons with varicose veins that were more severe on one side than the other. Trauma, pressure, and irritation were found to be the cause in five other cases. Infection was responsible in two, paralysis was found in one, and no cause could be found in two. Asymmetrical lesions on the elbow occurred only in patients confined to bed who, if right-handed, rested on the left elbow. A one-sided lesion was associated with asymmetrical varicose veins in patients who were ambulatory. In none of the patients we observed was sunlight a provoking cause, although Plunkett (39) noticed that a one-sided exposure to sunlight or wind damaged the skin and gave rise to asymmetrical localization.

In the middle 1930s Spies had demonstrated that glossitis was a much more sensitive gauge of the clinical stage of pellagra than the skin; skin lesions might clear up while patients were eating a pellagra-producing diet. Such a diet might ultimately lead to skin lesions in persons not exposed to sun. Dermatitis in the pudendal region had been known for a long time. There was thus much argument about the specificity of sunlight in the disease. For this reason, early in 1941 I undertook to determine whether by reducing the blood flow to the skin and increasing its metabolism by the use of a heating pad the localization of skin lesions could be encouraged. This had to be done during the prodromal stage of the disease, perhaps months and certainly weeks before the clinical manifestations were expected. For one or two hours a day for a period of more than two weeks, I put a cuff on my arm, set the pressure well over the level of systolic blood pressure, and wrapped an electric heating pad around. After several days' testing the skin developed the so-called fire stains, or *erythema ab igne,* the reticulated, mottled, pigmented network seen in those who have used a hot water bottle or an electric heating pad for a long time or have sat long in front of an open fire or a hot water radiator. No other manifestation occurred, and after a few weeks my arm resumed its normal appearance. I concluded that the method was probably safe to use in persons whose history led us to believe they would develop an annual relapse in pellagra. We used the tourniquet method of producing ischemia in several persons before encountering an example of neuropathy. This mishap occurred in a malnourished man in his early 30s who had had annual attacks of pellagra for several years. I used the method of heat and ischemia that I had used on myself, but only for 15 minutes. To my surprise and consternation, this produced a neuropathy that included both paralysis of the muscles and a loss of the sensations of touch and pain. Fortunately, the condition cleared up in a few hours. This led me to use simple sandbags to weight the electric heat pad. The weight of the sandbags in place 30 minutes was sufficient without heat to cause reactive hyperemia, indicating effective oxygen deprivation. No further neuropathy occurred, and we used the method in a study of thiamin metabolism and beri-beri. An effective variation of the method was to have the patient lie supine on a firm examining table. The weight of the leg induced ischemia on the back of the calf. In two of five persons with latent pellagra, the induction of ischemia plus heat produced changes in the skin that resulted in a unilateral lesion during a clinical relapse of the disorder in which bilateral dermatitis developed in other parts.

Secondary Pellagra

My studies of diarrheal diseases that give rise to pellagra got me interested in going over the records of all patients with pellagra in the modern period at Lakeside Hospital in Cleveland and at the Cincinnati General Hospital (6). This review was supplemented by an intensive literature search which I had been conducting for years on the history and current understanding of pellagra. Strictly secondary pellagra can be established only when a person on a fixed diet develops pellagra following the development of a disease and gets rid of the pellagra when the disease is reduced or cured. To be sure, the diet is likely to be marginal, but not in itself sufficient to produce the disease. The current state of medicine and science has generally determined current thoughts about pellagra. When first described, its cause was thought to be a toxin or poison on corn, analogous to ergot on rye (32a). In the last third of the 19th century infection dominated thoughts of etiology and convinced many that pellagra was an infection. Sambon incriminated the buffalo gnat. For sixty years the belief that pellagra was an infection prevented progress.

Clearly any disease can dispose to the development of pellagra by disorganizing a person's relationship to his environment. Disorders of the body politic give rise to war and famine, upheavals in personality lead to food fads, dietary cults, and addiction to alcohol; other disabling upsets, including insanity, are important in the background upon which pellagra develops. Even epilepsy, hemiplegia, and parkinsonism may have pellagra engrafted upon them.

It is not easy to separate symptoms of pellagra from those of the diseases of the alimentary canal upon which it may be engrafted. We must consider intestinal hypermotility, decreased enzymatic digestion, inadequate absorption, abnormal bacterial flora, destruction or utilization of vitamins by bacteria, inactivation or binding by various constituents, liver damage and such abnormalities of function as nausea, vomiting, loss of appetite, infection and fever. Thus the alimentary canal from one end to the other can influence nutrition—e.g. damaged teeth or their absence, disease in the gullet, trouble in the stomach including cancer, ulcer, inflammation, syphilis, and a host of diseases of the remaining alimentary canal: parasitism, obstruction, granulomas, tuberculosis, ulcerative colitis, Crohn's disease, shunts, operations, strictures, and functional disorders of the alimentary canal. All such conditions cause conditioned malnutrition. Liver disease, operations, anesthesia, infection, and fever are all important. Women need more nutrients during pregnancy, lactation, and childbirth in order to produce the fetus and placenta and to supply milk later. In addition, a mother is likely to sacrifice her own food in order to feed her child. Hypertension, congestive heart failure, and radiation sickness have had meager systematic study but may have special mechanisms of reducing the pyridine coenzymes. Drugs, chemicals, hemorrhage, and a vast miscellany of nonspecific disorders may also condition occasional instances. It has not been possible to study in isolation a genetic liability to pellagra, mainly because people inherit not only their genes but also, to a surprising degree, their environments. During the period of the Birmingham studies among a large Negro population, blacks provided only 9% of all pellagra patients whereas in Ohio pellagra was two or three times as common in blacks as in whites. It has been my impression that in the old South blacks usually had white friends or a white family who were genuinely concerned with their well-being and saw to it that they got food.

Pellagra in Ohio Hospitals in 1941 and 1949

I did two studies (5, 15) of pellagra in Ohio Hospitals, in Cleveland and Cincinnati (1942) and the other in Cincinnati (1949), where pellagra had been under intensive study for nearly 15 years and underestimation of its frequency, therefore, was unlikely when we studied the effectiveness of new remedies. During the depression in Cleveland and Cincinnati the disease was prevalent, constituting one or two percent of the admissions to the medical services. The figure doubled if readmissions were counted. Despite emphasis on early signs of deficiency, the 1949 study disclosed a spectacular decrease in the frequency beginning between 1939 and 1940. In 1939, 44 pellagrins were admitted to the Cincinnati General Hospital. In 1940 there were three, and by 1946–1947 none. Changes in the vitamin content of white flour began in 1942, two years after pellagra experienced its sharp decline. Vitamin tablets were not used by the Cincinnati patients. The improvement of the economy with retooling for World War II must have improved the diet of many persons who during the depression rarely had food adequate in quality or quantity.

The Birmingham Scene

The social milieu of Birmingham, its southern culture and graces, made it a pleasant place to work. The many visitors to the Spies clinic were entertained as guests at the Mountainbrook Country Club—not only scientists and nutritionists but also publicists, those in pharmaceutical and manufacturing firms, and a number of private persons who were generous and sometimes lavish contributors to the work of Tom Spies and the nutrition clinic. It was my good fortune to more or less fall into the responsibility of looking out for visitors, meeting them at trains and planes and otherwise spending time with them. In this way I had the uncommon and pleasant experience of getting to know Paul DeKruif (26), John Steinbeck, and others who spent weeks or months there gathering information. De Kruif, the gifted author of *Hunger Fighters,* wrote many popular articles about the colorful Dr. Spies and his nutrition clinic. (I do not think Steinbeck ever used his experience in a book.) Joseph Goldberger's widow, who belonged to a prominent New Orleans family, was a charming guest enormously interested in anything having to do with pellagra, nutrition in the South and, naturally, her husband. From the back numbers of the *Southern Medical Journal* I compiled into one volume all the pellagra articles by Goldberger I could find; this volume, supplemented with his papers in the US Public Health Reports, made a nearly complete file of Goldberger's published studies. Mrs. Goldberger was astonished that we knew so much about her husband, that our admiration of him was so great, and that his contributions had been every bit as valuable as she believed. Despite a steady inflow of bourbon whiskey, DeKruif was a vivid and entertaining conversationalist; and once one had penetrated the protective cover of the very shy Steinbeck, whose face was disfigured with acne scars, he, too, was delightful, perceptive, and extremely warm and gentle as well as profoundly intelligent. Just as knowing history gives us a clearer view of reality, so knowing individuals gives us special insights into their creative works and achievements.

In the spring of 1940, Tom Spies told me that I was to present a paper at the New York meeting of the American Medical Association (AMA) and discuss a hundred patients with diarrheal diseases who had developed pellagra (4). I had been in the clinic only a couple of months and could not find any satisfactory records. It was clear to me that we needed to devise accurate, quantitative clinic records and that I would have to find some way of straightening out the strange fascination with round numbers that seemed to exist in the clinic. The only records available had been those kept by the excellent nurses and social workers in the clinic, so I designed a record that became the standard nutritional form in the clinic and is displayed in Spies's paper in the 1943 handbook (38).

I was interested in a variety of clinical observations by that time on fingernails, skin mottling, the triple response, and vascular spiders. These, therefore, were included on the form. The nutrition clinic in Birmingham was the focus of such enormous interest that visitors from all over the country, indeed from many parts of the world, were coming and going. Each was given a copy of the nutrition form, and I became inadvertently responsible for a rather casual assumption that anything found in a pellagra clinic must have something to do with pellagra. Some of the visitors even wrote learned articles about vascular spiders and such things as signs of pellagra. It took a long time to get this heresy removed.

Urinary Pigments in Pellagra

One of the interesting misinterpretations of the early studies was in the Beckh, Ellinger & Spies (BES) tests for pigments in the urine of pellagrins (25). In Spies's article, "The Principles of Diet and the Treatment of Disease" in the 1943 American Medical Association Handbook of Nutrition, he stated that the BES Test for porphyrins in the urine was of "considerable clinical use in detecting small quantities of abnormal pigments" in persons with subclinical and clinical deficiency states. While it is perfectly true that the BES Test will be positive in people who excrete much corproporphyrins and other porphyrins, Cecil Watson and his colleagues (69) in Minneapolis demonstrated that in most malnourished persons the main pigment was indole acetic acid or its chromogen, again a tryptophane-related compound.

In 1940 I presented a paper on the subject before the American Society for Clinical Investigation, but by this time it was recognized that the test was not in any sense diagnostic of porphyrin. My paper may have increased interest in the matter, since I discussed the notion of photosensitivity and pellagrous dermatitis. At this time it was not clear that exposure to sunlight was unnecessary to the production of the pellagraderm.

Adenylic Acid

Many studies were done with adenylic acid from yeast and muscle in malnourished patients (55). We reported improvement of ulcers in the mouths of several patients following daily intravenous administration of 50 mg of adenylic acid. Because the adenylic acid produced disagreeable symptoms we did not recommend its use clinically. Such lesions generally healed promptly in pellagrins after administration of nicotinic acid and an improved diet. Often ulcers were swarming with Vincent's organisms and this condition also cleared up. Neither diet nor adenylic acid administration was effective in treating the ordinary aphthous ulcer or canker sore, a lesion that produces discomfort out of proportion to its usual small size.

Roentgen Sickness

In 1944 Bean, Spies & R. W. Vilter (9) published a series of studies on irradiation sickness—the nausea, vomiting, headache, cramps, diarrhea, and general feeling of illness that may follow therapeutic irradiation. We found that persons on a diet very poor in B-complex vitamins developed Roentgen sickness that could be prevented or ameliorated by giving supplements of nicotinic acid or thiamin a few days before irradiation. Well-fed persons had no untoward reaction to 400 roentgen units administered from a distance of 27 cm.

Glossitis

In the effort to compare the speed with which various compounds, diets, or procedures brought about the healing of glossitis, Vance and I published a table in the *Journal of Clinical Nutrition* in 1953 (18) indicating the rapidity with which glossitis underwent characteristic improvement (Table 1).

Ration Testing For Military Use

My introduction to ration testing in the Army in 1942 was to follow Ancel Keys in the California desert and study the acceptibility of the K-ration. He had written a favorable report. We found that one could track a light tank batallion on maneuvers by following a trail of discarded K-rations. In

Table 1 Speed with which various compounds, diets, or procedures bring about the healing of glossitis

Agent	Time for effect
Cozymase	few minutes to several hours
Nicotinic acid	few hours to days
Tryptophane	few hours to days
ACTH + diet	one to three days
Rest in bed	one to three days
Crude liver extract	one to five days
Yeast	three to fifteen days
Diet alone	three to thirty days

the winter of 1942–1943 it became obvious to everyone that the US battle rations were not good. In North Africa our military debut was anything but a success. In order to get into each day's ration and preferably each item a proportion of the daily vitamin requirement while staying within the space and weight allowances it was necessary to make the ration biscuits with yeast, liver extract, and soybean flour. They had a very high satiety value, soon became rancid, and thus were not eaten. Then serious difficulties arose. The Surgeon General ordered me to write a critique of Army Emergency Rations. I emphasized two disregarded facts (7): (*a*) A ration is no good if it is not eaten. All the vitamin king's horses and all the procuring king's men cannot feed an Army if the soldier will not eat the food provided. Soldiers are people and eat things they are familiar with. (*b*) Variety encourages eating. I emphasized that a ration had to be palatable; it must "fill the belly," and be usable in extremes of heat and cold. The fare should consist of common foods, bland and not heavily seasoned. I was directed to design a ration and, using a thousand soldiers, test it against the various rations already procured (11). I obtained from the large food stores a list of the 20 most popular meal items purchased by the American housewife. From the list I selected foods that could be procured readily and developed a ration in which there was no repetition of major item for nine meals. Such meals as chicken stew and ham with lima beans were most popular.

I was then put in charge of a ration test in the Pike National Forest. Our subjects were the 201st Infantry Regiment, whose officers had seen combat in a 20-month stay in the Aleutians and were serious about discipline and training. Expert consultants included John B. Youmans, Virgil P. Sydenstricker, Frederick J. Stare, W. Henry Sebrell, Julian Ruffin, R. H. Kampmire, W. F. Freedman, and M. Corlette. We organized a complicated experiment involving 10 physical fitness tests, 30 biochemical tests, and a complete nutritional examination, recording 200 items on every soldier. By means of a questionnaire each ration item was rated at each meal. The regiment's six companies, separated from each other so there was no opportunity to exchange food, underwent a rigorous training program. Daily records of ration evaluation, biochemical tests, the fitness scores and evaluations by the officers were transferred to IBM cards and sent to the Corps Area Headquarters in Omaha, Nebraska, from which we got reports twice a week. We compared rations as issued, one ration per day of varying calorie content (some being 20% greater than others), with rations issued so that each man received the same number of calories per day. Rapid data processing enabled us to complete a book nearly 250 pages long for distribution 3 months after the test. We found no vitamin deficiencies. The sun and the semiarid weather chaffed the soldiers' lips, and trauma to their gums from chewing K-ration biscuits at first caused concern. The biochemical tests revealed no abnormalities; physical fitness improved; muscle mass increased. The thin soldiers gained weight, those of average weight added muscle and lost fat, and the heavy ones lost a great deal more fat than they gained in muscle mass. Every tenth soldier, the total coming to about 100, was examined independently on four occasions by at least three different clinicians. After careful instructions and agreement on the criteria for diagnosis, we published a detailed account of the subjective factors in these clinical examinations in nutrition which had accounted for some of the polemics of the past. These are important in all clinical judgments. Since then, diagnostic variability has been demonstrated in measurement of blood pressure, interpretation of X-rays, and numerous other clinical matters. Extremes present no problem of observer differences, but borderline manifestations—potentially so important in finding trouble early—are difficult. On our Army ration evaluation team, each observer had a pattern of diagnostic habits. The less experienced tended to overdiagnose and the more experienced found no significant lesions.

The surprising outcome of this study was that military authorities in Washington procured the new and improved C-ration, shipped millions of dollars worth to the Pacific, and sent me with three colleagues to test various old and new rations in ten different places in the Pacific, including observations made on Iwo Jima and Luzon on soldiers immediately out of combat (12–14). The most spectacular results we saw were in the 38th Infantry Division in Luzon, who had been fighting the Japanese for four and a half months, subsisting on the improved C-ration. The soldiers, though low on praise, were still eating the food. They had the best physical fitness record of any group we tested and exhibited no clinical or biochemical abnormalities.

The rations had been designed without much sophisticated metabolic theory but with practical common sense. Here was a unique example of the recognition of a clinical Army problem, a diagnosis, a proposal for change in the form of preventive therapy, a test under conditions of training but not combat, a procurement of the ration, and a final study right out of combat. The food did indeed work in the circumstances for which it was designed.

The development of large-scale rations tests and the improvement of Army emergency rations were a natural outgrowth of the work I had done in the field of nutrition in association with Spies, the Vilters, and many others. Several important lessons came from these studies. The most important, but little recognized, was that the acceptability of a ration is essential for its use. It is better to have an imperfect ration that is eaten than a theoretically perfect one that is not eaten or is eaten only in unbalanced fractions.

Tropical Nutrition

Our ration study group (R. E. Johnson, C. R. Henderson, L. M. Richardson, and myself) in the US Army and that of R. M. Kark and associates in the Canadian Army used identical reagents and methods to demonstrate that there was no form of deterioration peculiar to the tropics (33). Isolation, boredom, homesickness, and abuse of alcohol were just as prevalent in various parts of the United States as they were in tropical regions. Such deterioration as occurred was not related to any peculiar tropical requirement for food or supplements.

The Canadian study of Kark and associates in Southeast Asia demonstrated that animal protein, riboflavin, and ascorbic acid levels were definitely lower in East Indian natives and Japanese prisoners than in US troops. Racial and religious dietary preferences resulted in lower values for hemoglobin, serum proteins, serum and urine ascorbic acid, and urinary riboflavin in Indian soldiers than in Americans; but interestingly enough, no classical disease of nutritional deficiency was found. Japanese prisoners were seriously malnourished, and many had vitamin deficiency diseases. A regiment of Gurkhas with the highest fitness record and almost no physical signs of deterioation or malfunction had much lower values of serum protein and urinary riboflavin than US troops. Food habits, customs, and adaptation must explain this.

Establishment of a Metabolism Unit in Iowa

After extensive clinical research on nutrition and human work and climate physiology in World War II, I established a metabolism unit at Iowa City, where P. C. Jeans had done such splendid work in pediatric nutrition. Kate Daum's section dealt with the biochemistry of nutrition, to which was added my own interest in B-complex vitamin deficiencies and the use of vitamin antagonists (16–22, 24, 35, 63). We published extensively on pantothenic acid deficiency and later on combinations of pantothenic and pyridoxine deficiency. We were trying hard to see whether vitamin antagonists would produce vitamin deficiencies that might be important in reducing antigen-antibody reactions and, thus, play a useful role in organ transplant. We did, indeed, find that many antibody reactions were diminished or prevented by the combination of pantothenic acid and pyridoxine deficiency. The catch was that this deficiency also prevented normal wound healing, perhaps partly owing to local infections. Thus the idea had to be discarded because of interfering complications.

Moral and Ethical Problems in Human Experimentation

I discussed Walter Reed and human experiments in the Garrison Lecture (24). Nearly thirty years before that I had presented my concern with moral responsibilities in clinical research in the Presidential Address of the Central Society for Clinical Research in November, 1951 (17a). This has recently been updated in the December 1981 issue of the *Journal of Laboratory and Clinical Medicine* (17b). Though I saw no examples of disregard for the patients' safety and well-being in studies in Birmingham or Cincinnati, this owed more to the backgrounds and training of those involved than to any formal evaluation or obtaining of written consent. During World War II we had done all manner of experiments on physical fitness, adaptation to extremes of heat and cold, and water and salt requirements; we had done extensive ration tests involving a great variety of clinical, biochemical, and performance evaluations without formally obtaining consent, though the procedures had been explained to the subjects in detail. Though an occasional soldier departed without leave, in general the morale was splendid and the rate of AWOL was much smaller in our experimental subjects than in the units from which they were drawn. We agreed with the English investigator, L. J. Witts, who believed that if one does any investigation on oneself first then it is permissible to do it on others after genuine efforts to explain it to the subjects and after obtaining informed consent. No set of formal and codified rules or laws will be satisfactory in all cases, particularly if the experimenters do not have high ethical standards.

Despite the fact that we tested on ourselves the drugs we used and the compounds that had been effective in animal experiments, we made no systematic efforts to get what we now call informed consent, nor was there any document for the patients to sign. The only signed releases I saw were those that physicians and patients signed for photographs or moving pictures for education and promotion. The subject usually was paid, as token payment, a dollar and signed without any question. There was exhibited a further very important, but scarcely studied influence of faith and belief at the nutrition clinic in Birmingham, Alabama, where the famous Dr. Spies had created an effect resembling, but by no means as evident as that at Lourdes. This similarly exists in some centers today, centers of legitimate as well as pseudoscientific nutrition fame.

From the moral and ethical point of view, the fact that we had tested new compounds or diets on ourselves, orally or by injection, made us feel, in those informal days of confidence and trust, that we had done what was required to safeguard our subjects. I believe that most if not all of our studies would have been approved by the array of committees set up today to monitor moral and ethical guides in experiments on volunteers. I am unaware of any harm that resulted from our studies. A great deal of good resulted.

1396 BEAN

Literature Cited

1. Aring, C. D., Bean, W. B., Roseman, E., Rosenbaum, M., Spies, T. D. 1941. The peripheral nerves in cases of nutritional deficiency. *Arch. Neurol. Psychol.* 48:772–87
2. Bean, W. B., Vilter, R. W., Spies, T. D. 1939. The effect of Roentgen-ray on the blood codehydrogenase I and II. *Ann. Intern. Med.* 13:783–86
3. Bean, W. B., Spies, T. D. 1940. A study of the effects of nicotinic acid and related pyridine and pyrazine compounds on the temperature of the skin of human beings. *Am. Heart J.* 20(1):62–78
4. Bean, W. B., Spies, T. D. 1940. Vitamin deficiencies in diarrheal states. *J. Am. Med. Assoc.* 115:1078–81
5. Bean, W. B., Spies, T. D., Blankenhorn, M. A. 1942. The incidence of pellagra in Ohio hospitals, *J. Am. Med. Assoc.* 118:1176–79
6. Bean, W. B., Spies, T. D., Blankenhorn, M. A. 1944. Secondary pellagra. *Medicine* 23:1–77
7. Bean, W. B. 1944. A critique of army rations: acceptability and dietary requirements. *Armored Med. Res. Lab. Rep.*
8. Bean, W. B., Spies, T. D., Vilter, R. W. 1944. Asymmetric cutaneous lesions in pellagra. *Arch. Derm. Syph.* 49:335–45
9. Bean, W. B., Spies, T. D., Vilter, R. W. 1944. A note on irradiation sickness. *Am. J. Med. Sci.* 208:46–54
10. Bean, W. B. 1944. Remarks on the incidence, manifestations and treatment of nutritional deficiency diseases. *Nebr. State Med. J.* 29 (8):241–44
11. Bean, W. B., Youmans, J. B., Nelson, N., Bell, D. M., Richardson, L. M. Jr., French, C. E., Henderson, C. R., Johnson, R. E. 1944. Final report on tests of the acceptability and adequacy of U.S. Army C, K, 10-in-1 and Canadian Army mess tin ration. *Armored Med. Res. Lab. Rep.*
12. Bean, W. B., Johnson, R. E., Henderson, C. R., Richardson, L. M. 1946. Nutrition survey in Pacific theater of operations. *Bull. U. S. Army Med. Dept.* V:697–705
13. Bean, W. B. 1948. Field testing of army rations. *J. Appl. Physiol.* 1:448–57
14. Bean, W. B. 1948. An analysis of subjectivity in the clinical examination in nutrition. *J. Appl. Physiol.* 1:458–68
15. Bean, W. B., Vilter, R. W., Blankenhorn, M. A. 1949. Incidence of pellagra. *J. Am. Med. Assoc.* 140:872–73
16. Bean, W. B. 1950. Control in research in human nutrition. *Nutr. Rev.* 8:97–99

17. Bean, W. B., Franklin, M., Daum, K. 1951. A note on trytophane and pellagrous glossitis. *J. Lab. Clin. Med.* 38:167–72
17a. Bean, W. B. 1952. A testament of duty: some strictures on moral responsibility in clinical research. *J. Lab. Clin. Med.* 39:3–9
17b. Bean, W. B. 1981. "A testament of duty" revisited. *J. Lab. Clin. Med.* 98(6):795–99
18. Bean, W. B., Vance, M. 1953. Some aspects of the tongue in pellagrous glossitis. *J. Clin. Nutr.* 1:267–74
19. Bean, W. B., Hodges, R. E., 1954. Pantothenic acid deficiency induced in human subjects. *Proc. Soc. Exp. Biol. Med.* 86:693–98
20. Bean, W. B., Hodges, R. E., Daum, K. 1955. Pantothenic acid deficiency induced in human subjects. *J. Clin. Invest.* 34:1073–84
21. Bean, W. B. 1955. Research: prelude and first movement. *Circ. Res.* 3:317–19
22. Bean, W. B. 1955. Vitaminia, polypharmacy and witchcraft. *Am. Med. Assoc. Arch. Intern. Med.* 96:137–41
23. Bean, W. B. 1963. Presidential address: the clinician interrogates nutrition. *Am. J. Clin. Nutr.* 13:263–74
24. Bean, W. B. 1977. Walter Reed and the ordeal of human experiments. *Bull. Hist. Med.* 51:75–92
25. Beckh, W., Ellinger, P., Spies, T. D. 1937. Porphyrinuria in pellagra. *Q. J. Med. New Ser.* VI:305–19
25a. Carpenter, K. J. 1981. *Pellagra.* Stroudsburg, PA: Hutchinson Ross. 391 pp.
26. de Kruif, P. 1940. Famine fighters. *Readers Digest* Dec., 1940, pp. 11–16
27. Elvehjem, C. A., Madden, R. J., Strong, F. M., Woolley, D. W. 1939. Relation of nicotinic acid and nicotinic acid amide to canine black tongue. *J. Am. Chem. Soc.* 59:1767
27a. Etheridge, E. W. 1972. *The Butterfly Caste: A Social History of Pellagra in the South.* Westport, CT: Greenwood. 278 pp.
28. Frommeyer, W. B. Jr., Spies, T. D., Vilter, C. F., English, A. 1946. Further observations on the antianemic properties of 5-methyl uracil. *J. Lab. Clin. Med.* 31:643–49
29. Goldberger, J. 1916. The transmissibility of pellagra. *Pub. Health Rep.* 31:3159–73
30. Goldberger, J., Wheeler, G. A. 1920. Experimental pellagra in white male convicts. *Arch. Int. Med.* 25:451–71

31. Goldberger, J., Tanner, W. F. 1922. Amino-acid deficiency probably the primary etiological factor in pellagra. *Publ. Health Rep.* 37:462–86

32. Harris, H. F. 1919. *Pellagra.* NY: MacMillan

32a. Hirsch, A. 1855. *Handbook of Geographical and Historical Pathology.* Tranl. from 2nd German ed. by C. Creighton, M. D., vol. 2, Ch. 6. London: New Sydenham Soc.

33. Kark, R. M., Aiton, H. F., Pease, E. D., Bean, W. B., Henderson, C. R., Johnson, R. E., Richardson, L. M. 1947. Tropical deterioration and nutrition. *Medicine.* 26:1–40

33a. King, C. G. 1976. *A Good Idea: The History of The Nutrition Foundation.* NY: Nutr. Fnd. pp. 210–41

34. Krehl, W. A., Tepley, L. J., Savma, P. S., Elvehjem, C. A. 1945. Growth retarding effect of corn in nicotinic acid-low rations and its counteraction by tryptophane. *Science* 101:489

35. Lubin, R., Daum, K. A., Bean, W. B. 1956. Studies of pantothenic acid metabolism. *Am. J. Clin. Nutr.* 4:420–33

36. Moore, C. V., Vilter, R., Minnich, V., Spies, T. D. 1944. Nutritional macrocytic anemia in patients with pellagra or deficiency of the vitamin B complex. *J. Lab. Clin. Med.* 29:1226–55

37. Moore, R. A., Spies, T. D., Copper, Z. K. 1942. Histopathology of the skin in pellagra. *Arch. Derm. Syph.* 46:100–11

38. Handbook of Nutrition. A symposium prepared under the auspices of the Council of Foods and Nutrition of the American Medical Association. 1943. Chicago: American Medical Association. 586 pp.

39. Plumbett, O. R. L. L. 1939. Observations and clinical notes on some cases of pellagra seen in cypress. *J. R. Army M. Corps.* 72:317

40. Report of the Pellagra Commission of the State of Illinois. November, 1911. Published in Springfield, Illinois, Illinois State Journal Co., State Printers, in 1912. 250 pages.

40a. Roe, D. A. 1973. *A Plague of Corn: The Social History of Pellagra.* Ithaca, NY: Cornell Univ. Press. 217 pp.

41. Ruegsegger, J. M., Hamburger, M. Jr., Turk, A. S., Spies, T. D., Blankenhorn, M. A. 1941. The use of 2-sulfanilamidopyrazine in pneumococcal pneumonia. A preliminary report. *Am. J. Med. Sci.* 202:432–35

41a. Sandwith, F. M. 1913. Is pellagra a disease due to a deficiency of nutrition? *Trans. Soc. Trop. Med. Hyg.* 6:143

42. Spies, T. D. 1932. Pellagra: etiology, response to a deficient diet. *South. Med. Surg.* 44:128–36

43. Spies, T. D. 1932. Pellagra: improvement while taking so-called "pellagra-producing" diet. *Am. J. Med. Sci.* 184:837–46

44. Spies, T. D. 1933. Skin lesions of pellagra. An experimental study. *Arch. Intern. Med.* 52:945–47

45. Spies, T. D. 1935. Relationship of pellagrous dermatitis to sunlight. *Arch. Intern. Med.* 56:920–26

46. Spies, T. D. 1935. The treatment of pellagra. *J. Am. Med. Assoc.* 104:1377–80

47. Spies, T. D., Chinn, B., McLester, J. B. 1937. Treatment of endemic pellagra. *South. Med. J.* 30:18–22

48. Spies, T. D. 1938. The response of pellagrins to nicotinic acid. *The Lancet* 1:252–59

49. Spies, T. D., Aring, C. D. 1938. Effect of vitamin B_1 on the peripheral neuritis of pellagra. *J. Am. Med. Assoc.* 110:1081–1084

50. Spies, T. D., Aring, C. D., Gelperin, J., Bean, W. B. 1938. The mental symptoms of pellagra: their relief with nicotinic acid. *Am. J. Med. Sci.* 196:461–75

51. Spies, T. D., Bean, W. B., Stone, R. E. 1938. The treatment of subclinical and classic pellagra; use of nicotinic acid, nicotinic acid amide and sodium nicotinate, with special reference to the vasodilator action and the effect on mental symptoms. *J. Am. Med. Assoc.* 111:584–90

52. Spies, T. D., Cooper, C., Blankenhorn, M. A. 1938. The use of nicotinic acid in the treatment of pellagra. *J. Am. Med. Assoc.* 110:622–27

53. Spies, T. D., Bean, W. B., Ashe, W. F. 1939. A note on the use of vitamin B_6 in human nutrition. *J. Am. Med. Assoc.* 112:2414–15

54. Spies, T. D., Bean, W. B., Vilter, R. W., Huff, N. E. 1940. Endemic riboflavin deficiency in infants and children. *Am. J. Med. Sci.* 220:697–701

55. Spies, T. D., Bean, W. B., Vilter, R. W. 1940. Adenylic acid in human nutrition. *Ann. Intern. Med.* 13:1616–18

56. Spies, T. D., Ladisch, R. K., Bean, W. B. 1940. Vitamin B_6 (pyridoxin) deficiency in human beings. Further studies, with special emphasis on the urinary excretion of pyridoxin. *J. Am. Med. Assoc.* 115:839–40

57. Spies, T. D., Butt, H. R. 1942. Vitamins and avitaminoses. In *Diseases of Metabolism,* ed. G. G. Duncan, pp. 366–502. Philadelphia: Saunders

58. Spies, T. D., Bradley, J., Rosenbaum, M., Knott, J. R. 1943. Emotional disturbances in persons with pellagra, beriberi and associated deficiency states. *Res. Publ. Assn. Nerv. Ment. Dis.* 22: 122–40

59. Spies, T. D., Cogswell, R. C., Vilter, C. 1944. Detection and treatment of severe atypical deficiency disease. *J. Am. Med. Assoc.* 126:752–58

60. Spies, T. D., Vilter, C. F., Koch, M. B., Caldwell, M. H. 1945. Observations on the anti-anemic properties of synthetic folic acid. *South. Med. J.* 38:707–9

61. Spies, T. D., Frommeyer, W. B. Jr., Vilter, C. F., English, A. 1946. Antianemic properties of thymine. *Blood* 1:185–88

62. Stannus, H. S. 1937. Pellagra, theories of causation. *Trop. Dis. Bull.* 34:183

62a. Terris, M. 1964. *Goldberger on Pellagra.* Baton Rouge, LA: Louisiana State Univ. Press. 395 pp.

63. Thornton, G. H. M., Bean, W. B., Hodges, R. E. 1955. The effect of pantothenic acid deficiency on gastric secretion and motility. *J. Clin. Invest.* 34:1085–91

64. Vilter, R. W., Vilter, S. P., Spies, T. D. 1939. Relationship between nicotinic acid and a codehydrogenase (cozymase) in blood of pellagrins and normal persons. *J. Am. Med. Assoc.* 112:420–22

65. Vilter, R. W., Bean, W. B., Ruegsegger, J. M., Spies, T. D. 1940. The role of coenzymes I and II in blood of persons with pneumococcal pneumonia. *J. Lab. Clin. Med.* 25:897–99

66. Vilter, R. W., Mueller, J. F., Bean, W. B. 1949. The therapeutic effect of tryptophane in human pellagra. *J. Lab. Clin. Med.* 34:409–13

67. Vilter, S. P., Spies, T. D., Mathews, A. P. 1938. A method for the determination of nicotinic acid, nicotinamide, and possibly other pyridine-line substances in human urine. *J. Biol. Chem.* 125: 85–98

68. Warner, E. D., Spies, T. D., Owen, C. A. 1941. Hypoprothrombinemia and vitamin K in nutritional deficiency states. *South. Med. J.* 34:161–63

69. Watson, C. J. 1939. Further observations on red pigments of pellagra urines. *Proc. Soc. Exp. Biol. Med.* 41:591–95

William J. Darby

Ann. Rev. Nutr. 1985. 5:1–24

SOME PERSONAL REFLECTIONS ON A HALF CENTURY OF NUTRITION SCIENCE: 1930s–1980s

William J. Darby

Department of Biochemistry, Vanderbilt University, Nashville, Tennessee 37232

CONTENTS

PROLOGUE: THE FORTUITOUS SETTING OF A CAREER DIRECTION

In 1927 the flood waters of the Arkansas River filled the streets of North Little Rock, Arkansas, rising to within a few inches of our small home; the hit tune was "My Blue Heaven"; pellagra reached an all time peak in the South; Joseph

1401

Goldberger, Surgeon, USPHS, was touring the Mississippi Valley and promoting the distribution of brewer's yeast by the Red Cross and other measures to combat pellagra. Students at the University of Arkansas Medical School were sent into the field to assist with the health emergencies (infectious diseases, typhoid, diarrheal diseases, pellagra, and emergency feeding) created by the flood. My future mentor, Paul L. Day, received his doctoral degree at Columbia University; his thesis concerned the relative stability of vitamin A activity from plant and animal sources. His major professor was Henry C. Sherman. Dr. Day promptly was appointed Head of the Department of Physiological Chemistry in the University of Arkansas Medical School, where he established a beachhead of research (20), the spirit of which successfully penetrated the bastions of complacency toward research that had reigned in the school during the previous fifty years (2).

In 1930 it was my good fortune to fall under the influence of Paul Day as a consequence of the Great Depression that followed the stock market crash of 1929. The economic depression forced individuals and institutions to survive at subsistence levels. Jobs were virtually nonexistent for my 1930 high school graduating class (my first post–high school job was selling Fuller brushes) and in many families finances were destroyed by bank failures and unemployment. My high school chemistry teacher, Miss Ora Park, encountering me in the school hall in the autumn of 1930 exclaimed, "I thought you were at the University of Arkansas entering Chemical Engineering!" She understood immediately, however, why I was seeking a job instead and told me that she had just had a telephone call from a Dr. Paul Day at the medical school who was seeking a high school graduate to help with animal work and cleaning in the laboratory. This fortuitous coincidence effectually launched my life-time career in nutrition science.

THE MILIEU OF THE 1930s

The academic milieu was considerably different in the 1930s. The professor personally gave all (at least a vast majority) of the series of 55–60 lectures and participated actively in the 120 hours of laboratory sessions that comprised "Physiological Chemistry." Dr. Day personally supervised the laboratory sessions and conducted at least one quiz session weekly. There were no NIH or NSF grants. Hence, an investigator personally conducted such mundane operations as preparation of experimental diets, the weighing and feeding of supplements, and daily examinations of experimental animals. At Arkansas he even constructed his own cages, routinely distilled water and laboratory solvents used in his research, and personally carried out all chemical analyses. Writing of research reports was limited to preparation of manuscripts for publication or preparation for the one (not more than two) national scientific meeting(s) attended annually, attendance at which was largely or entirely the personal expense of the scientist. No doubt this contributed to the relatively small size of these meetings, where it was possible for several scientific societies to meet simultaneously in concurrent sections of the different societies (not concurrent sections of a single society). This resulted in good interchange at FASEB meetings between the physiological chemists, nutritionists, experimental pathologists, and physiologists. Discussion was critical and strictly scientific.

The close association between faculty, participating students, and the laboratory assistants fostered by this milieu was exemplified by the encouragement that Drs. Day and W. C. Langston gave me to pursue a premedical curriculum part-time in a local college and then enter medical school, which I did, completing my M.D. degree in 1937. During these years I was meaningfully involved alongside of Dr. Day in all of the nutrition research underway in the Department of Physiological Chemistry. He generously made me a co-investigator and introduced me to scientists at annual FASEB meetings. He instilled in all who worked with him the necessity of disciplined study of the scientific literature including the critical examination and discussion of original reports and the reading and abstracting of earliest landmark papers. The superb lectures and graduate seminars of Howard B. Lewis, my major professor during my subsequent graduate studies at the University of Michigan, Ann Arbor, were punctuated also with assigned reading of landmark reports of nutrition research. These reports were to be abstracted by the student and reviewed by Dr. Lewis. The substance of these reports and personal anecdotes pertaining to investigators introduced by Dr. Lewis and others at Michigan (particularly Dr. Harry Newburgh) provided an appreciation of our rich scientific heritage and sharpened the student's insight into the scientific method. The then small universe of biological scientists is strikingly illustrated by the fact that Dr. Lewis operated, from his departmental office in Ann Arbor, the placement service of the FASEB, and personally met each applicant seeking employment in any one of the disciplines.

NEW ESSENTIAL NUTRIENTS

The excitement of new discovery was intense in the 1930s and 1940s, created not only by that which occurred in one's own laboratory, but by rapidly growing evidence and identification of new essential nutrients and related metabolic phenomena (water-soluble vitamins, fat-soluble vitamins, essential amino acids, trace elements, and emerging metabolic cycles and biologic phenomena relating food and nutrition to human disease). The *experimental* nature of the evidence was such as to make for rapid resolution of discrepancies or controversies in contrast to the inconclusive nature of so much of the evidence offered today. I refer to evidence based upon association of phenomena observed in epidemiologic studies, the interpretation of which is compounded by complexities of time, uncontrollable events, inappropriate dietary methodology, and confusion between "statistical significance" and physiologic importance. Table 1 presents a chronology of selected developments in nutrition during this half century.

NUTRITION SCIENCE IN THE UNIVERSITY

In the early 1930s the loci of nutrition science in universities were schools of agriculture and the associated experiment stations, departments of home economics and dietetics, schools of medicine and of public health, and a few university departments of chemistry (17). The orientation of and focus on nutrition by these groups were influenced by the perspective of the faculty, but all academic groups viewed nutrition science as the chemistry of foods and their use in metabolism. Emphasis was placed upon material "susceptible of scientific proof" (14), which material constitutes "the scientific substratum upon which rests present day knowledge of nutrition both in health and in disease." Indeed, Graham Lusk introduced the first chapter of his book *The Science of Nutrition* with the quotation

> Blessed is he that maketh due proofe . . .,
> With due proofe and with discreet assaye
> Wise men may learn new things everyday.

Thomas Norton (b. 1493)
in *Ordinall of Alkimy*

NUTRITION IN THE MEDICAL CURRICULUM

In medical schools the scientific substratum of nutrition was a major portion of the course in physiological chemistry (now biochemistry) and in physiology. Perusal of textbooks and journals of the 1930 era reveals the remarkable degree to which nutrition pervaded these subjects. Turning to the clinical years of medical school, the scientific substratum of pediatrics treated the nutrient needs of infants and children for growth, maintenance of health, and disease; deficiency diseases of the young child (scurvy, rickets, infantile beriberi, xerophthalmia); it reviewed, extended, and interpreted nutrition information for practical application in pediatrics. Pediatricians devoted interest and time to nutritional aspects of their subject, not only in didactic teaching, but at the bedside, in rounds, and in the out-patient clinics.

Similarly taught in internal medicine were the nutritional metabolic considerations pertaining to dietary deficiency diseases (e.g. scurvy, pellagra,

Table 1 Illustrative chronology of selected developments in nutrition: 1928–1980[a]

1928	*Journal of Nutrition* began publication
1929	Essential fatty acid deficiency described
	Role of extrinsic and intrinsic factors in pernicious anemia elucidated
1930	Conversion in vivo of carotene to vitamin A demonstrated
1931	Essentiality for the rat of Mn and Mg demonstrated
	Mottled enamel identified with high fluoride content of water
1932	Vitamin C isolated from lemon juice
	Crystalline vitamin D prepared
	Flavoprotein discovered
1933	Pantothenic acid and riboflavin identified as members of B_2 complex
	AIN becomes a national society
1934	Vitamin K discovered
	Zinc found to be essential in rats
1935	Threonine discovered as an essential amino acid
	Curative effect of Co in sheep demonstrated
	First coenzyme identified as NAD
1936	Synthesis of thiamine published
	Vitamin E isolated from wheat germ oil and named tocopherol
1937	Nicotinic acid identified as anti–black tongue factor
1938	Vitamin B_6 crystallized
	Riboflavin deficiency in man described
	Vitamin M (folic acid) demonstrated as essential for the monkey
1939	Choline shown to be lipotropic factor
	Structure of vitamin B_6 determined
	Microbiologic estimation of vitamins introduced
1940	AIN admitted to membership in FASEB
	Sequence of clinical events developing in experimental scurvy described in man
1941	Folic acid proposed as name for growth factor for bacteria
	British Nutrition Society inaugurated
	First RDAs adopted; first standards adopted for enrichment of flour in US
	The Nutrition Foundation incorporated
1942	Biotin synthesized
	Nutrition Reviews began publication
1945	Pteroylglutamic acid (PGA) synthesized
	Response of cases of sprue and of nutritional macrocytic anemia to treatment with PGA (Vitamin M) reported
1946	National Vitamin Foundation established
1947	Vitamin B_{12} identified
	British Journal of Nutrition began publication
1948–49	Isolation of crystalline vitamin B_{12} and identification of Co in the molecule
1949	The Framingham Study initiated
1953	Essentiality of pyridoxin (vitamin B_6) for infants established
1955	Zinc deficiency in swine described
	Structure of vitamin B_{12} determined
	ICNND organized

Table 1 *(continued)*

1957	Se reported to be essential in experimental animal studies
1960	American Society for Clinical Nutrition incorporated
1963	Zinc deficiency established as a cause for syndrome of dwarfism and sexual infantilism in adolescent boys
1964	Pyridoxine deficiency in adults described
1966–74	Role of dietary "fiber" in so-called saccharine disease postulated
1968	Ten-State Nutrition Survey launched
	Report of successful maintenance of young child by total parenteral nutrition (TPN)
	25-OH cholicalciferol identified as an active metabolic form of vitamin D_3
1969	First case treated with home TPN
1970	Nutrition Canada National Survey launched
1971	1,25-dihydroxycholecalciferol identified as an active hormonal form of vitamin D_3
1973	Deficiency of carnitine associated with syndrome of lipid storage myopathy
1979	Se in prevention of Keshan disease reported
1980	Publication of first volume of *Annual Reviews of Nutrition,* marking the 50th anniversary year of Annual Reviews, Inc.

[a]Based in part on E. Neige Todhunter 1976. Chronology of some events in the development and application of nutrition sciences. *Nutr. Rev.* 34:353–65 (Ref. 19).

beriberi, starvation, iodine-deficiency or endemic goiter); secondary or conditioned deficiencies (e.g. iron-deficiency anemia, pernicious anemia, sprue, adult osteomalacia, hypoproteinemia, or hypoalbuminemia); dietary management of metabolic disorders (especially diabetes, gout), as well as the nutritional aspects of acute and chronic diseases of various organs and systems, and of infectious diseases.

These aspects of nutrition were taught in medicine, but maintenance of nutriture and rehabilitation of the surgical patient also was taught in surgery. Vividly I recall one surgeon's impressive underscoring of the importance of maintaining the protein nutriture of patients. Early in my junior year in medical school Dr. George V. Lewis, Professor of Surgery, abruptly asked, "Darby, you are in biochemistry. Now tell me about protein deficiency." After I stuttered out in response a standard explanation of protein-deficiency hypoalbuminemia he commented abruptly, "Don't you ever read the *JAMA?* You had better read the paper by Dr. Youmans." Knowing that I definitely had better read it by the next session with him, immediately I sought out and digested the paper (22) by a Vanderbilt University Professor of Medicine, Dr. John B. Youmans entitled, "Endemic Edema." Not only did I learn from it a great deal about clinical protein deficiency and was introduced to the concept of chronic moderate deficiency syndromes (so-called subclinical), but I also became aware of one of the remarkable men who subsequently enormously influenced my career and became my warm friend and demanding mentor.

Teaching in preventive medicine and public health incorporated the epidemiology of nutritional conditions such as endemic goiter; cretinism; endemic deficiency diseases, especially pellagra; and public health controls, standards and regulatory responsibilities concerning the quality and safety of foods and problems of food-borne diseases.

The curriculum of few medical schools included a separate course labeled "Nutrition" during the 1930s, but the subject obviously received appreciable attention. At Vanderbilt University a course in Nutrition was initiated in the 1920s by Drs. John B. Youmans, Professor of Medicine, and C. S. Robinson, Professor of Biochemistry. It was a "required elective" during the second semester of the second-year for Vanderbilt medical students throughout the 1930s and through the 1950s, later as an elective. Curricular emphasis on nutrition in teaching institutions depended then as now upon the presence of strong leadership with a keen interest in the subject and established collaboration between members of preclinical and clinical departments.

SCIENTIFIC ORGANIZATIONS WITH PRIMARY FOCUS ON NUTRITION

Out of national emergencies have come many significant developments in medicine and in nutrition. Faced with the demands of hospitals and of war needs at home and abroad in 1917, 98 persons attended a dietitians' conference in Cleveland and there organized the American Dietetic Association (3). Thirty-nine charter members organized the first annual meeting of the association in conjunction with that of the American Hospital Association. Professor E. V. McCollum, Ph.D., from the Johns Hopkins University, reported at this first meeting recent developments with the then new vitamins A, B, and C. Shortly thereafter the rapidly growing membership of the ADA embraced such outstanding nutrition authorities as Lafayette B. Mendel, E. V. McCollum, and Mary Swartz Rose.

It was a decade later, 1928, before the initial organizational phase of the American Institute of Nutrition (AIN) occurred (11) around the *Journal of Nutrition*. The eleven founders (see Table 2) represented diverse disciplines, from which emerged the nutrition sciences. I recall the modest surprise that Paul Day expressed at being invited in 1933 to become a charter member of the American Institute of Nutrition and his obvious enthusiasm upon attending in 1934 the first annual meeting at Cornell University Medical College. In retrospect it is easy for me to appreciate his feelings because of similar ones I experienced when I was involved in the founding of the American Society for Clinical Nutrition. The colleagues involved included Bob Goodhart, Ted Van Itallie, Norm Jolliffe, Maury Shils, Michael Wohl, Bob Olson, Bob Hodges, Bill Krehl, Bill Bean, S. O. Waife (who edited the *American Journal of*

Clinical Nutrition), Grace Goldsmith, and D. W. Woolley representing the AIN. Establishment of the new society was not without travail. Many were seriously troubled that the new organization threatened the AIN. Some were especially critical of the sometimes inferior quality of papers that had appeared in early issues of the *American Journal of Clinical Nutrition,* and some feared the potential of an undue influence of the pharmaceutical industry. Despite the *sturm und drang* the two societies now are much alive and their respective journals are internationally prestigious. These societies and their publications, along with similar national groups (e.g. the British Nutrition Society, the Canadian Nutrition Society, and the Latin American Society of Nutrition, SLAN), exemplify the world-wide scientific interest that now constitutes the "nutritional community" which has emerged during the last half century.

Consolidation of the world-wide interest was symbolized in the establishment and positioning of the International Union of Nutritional Sciences, the formation of which was succinctly summarized by Dr. Charles Glen King (13):

> The International Union of Nutritional Sciences (IUNS) had grown substantially in terms of international congresses, held at three-year intervals since the formal organization was established, as suggested at the London meeting in 1946, followed by formal action at a London meeting in 1948. However, there had been very little organizational activity during the periods between congresses, except through the offices of the President, Dr. E. J.

Table 2 The founders of The American Institute of Nutrition

Eugene Floyd DuBois, M.D., Pathological physiology.
 June 4, 1882–February 12, 1959.
Herbert McLean Evans, M.D., Anatomy.
 September 23, 1882–March 6, 1971.
Ernest Browning Forbes, Ph.D., Animal nutrition.
 November 3, 1876–September 8, 1966.
Graham Lusk, Ph.D., Physiology.
 February 15, 1866–July 18, 1932.
Elmer Verner McCollum, Ph.D., Nutrition.
 March 3, 1879–November 15, 1967.
Lafayette Benedict Mendel, Ph.D., Physiological chemistry.
 February 5, 1872–December 9, 1935.
Harold Hanson Mitchell, Ph.D., Biochemistry, Nutrition.
 January 22, 1886–March 28, 1966.
John Raymond Murlin, Ph.D., Physiology, Nutrition.
 April 30, 1874–March 17, 1960.
Mary Swartz Rose (Mrs. Anton R.). Ph.D., Biochemistry.
 October 31, 1874–February 1, 1941.
Henry Clapp Sherman, Ph.D., Chemistry.
 October 16, 1875–October 7, 1955.
Harry Steenbock, Ph.D., Agricultural chemistry.
 August 16, 1886–December 25, 1967.

Bigwood in Belgium and the Secretary, Dr. L. J. Harris at Cambridge University. Many of the active scientists recognized this situation and urged action to meet the need for a more vigorous and continuous program.

As a member of the National Academy of Sciences and the Society of Biological Chemists, I was aware of the very constructive work being done by the international unions in chemistry, biology, physics, biochemistry, physiology and other sciences. Soon after Sir David Cuthbertson, Director of the Rowett Institute in Scotland, was elected President of the IUNS at the general assembly meeting in Paris (1957) he asked me to serve as a committee-of-one to prepare recommendations for a revision of the By-Laws along with recommendations for changes in the structure and activities of the organization.

Edmund Rowan in the Foreign Secretary's office at the National Academy of Sciences was most helpful and generous of his time and advice in preparing a report that would reflect the experience and best practices among other international unions. The recommendations were adopted practically *in toto* at the next General Assembly of the IUNS (1960) in Hamburg, and the traditional fate of a committee chairman followed—my election as President of the IUNS, with an implied: "Now make it work!"

Grants from The Nutrition Foundation and the Research Corporation made it possible for me as a Lecturer at the Institute of Nutritional Sciences, School of Public Health, Columbia University, to have essential secretarial and office facilities.

A Council and appropriate officers were proposed within the IUNS to provide broader and more efficient administration. Increased and regular dues were proposed on a scale related to resources available within the national societies.

Working commissions and committees were proposed to study and report on: ". . . a variety of subjects including nomenclature, procedures and standards; operational programs; human nutrition with special reference to the preschool child; genetic patterns of special nutritional importance; nutritional education and training; nutrition of agricultural animals and fish.

Developments of this nature in addition to the congresses made it possible for the IUNS to be of greater service and to accelerate both pure and applied research in the science of nutrition.

The national societies that constitute the adhering bodies of the IUNS are not the only important scientific and professional organizations that have emerged to further nutrition research, education, and application. Especially important in the United States has been the more recent establishment of the American Society for Parenteral and Enteral Nutrition (ASPEN) with its journal and the medical-practice-oriented American College of Nutrition, as well as a variety of more narrowly focussed organizations devoted to particular nutritional conditions like obesity.

COUNCILS, COMMITTEES, AND BOARDS

The period of the 1930s to 1960s was marked by the establishment in the United States of a remarkably productive series of councils, committees, and boards concerned with assessing scientific knowledge in nutrition and utilizing that knowledge for the improvement of health and welfare of mankind. These organizations served to identify gaps in our knowledge and understanding of nutrition science and to promote research to fill these gaps. They generally interfaced well with each other and maintained liaison with federal agencies that have responsibilities for nutrition programs as well as with responsible segments of the food and food-related industries. They were comprised primarily of recognized leading scientists from the academic community. They maintained an independence that assured objective assessments and recommendations free of self-serving bias. The members served without compensation. The public attitude was receptive to constructive, objectively drawn, scientifically based, apolitical guidance. In that period these committees were not subject to attacks by organized, politically motivated activists operating under the guise of "public interest groups." Some examples of these councils, committees, and boards and their accomplishments are worthy of note.

The Council on Foods and Nutrition of the American Medical Association

Initially organized as a subcommittee on foods of the Council on Pharmacy and Therapeutics of the American Medical Association, this council was comprised of leading scientists (biochemists, home economists, physicians, physiologists, nutritionists), provided with an outstanding executive secretary (Dr. Franklin Bing), and was funded by the American Medical Association. It was responsible for initiating and monitoring the vitamin D fortification of milk, a major factor in the disappearance of infantile rickets; it was a moving, active force in promotion of iodinization of salt, of enrichment of flour, fortification of margarine with vitamin A and similar programs devised to improve the nutritive value of foodstuffs in the United States. This included establishing principles and guidelines for honest educational advertising of foods and continuous control of advertising content through award of the "AMA seal of approval," which conveyed to the public and professionals assurance that the scientific statements pertaining to a given product in advertising were in fact sound and in keeping with current knowledge (4).

Particularly noteworthy was its work with the infant food industry, which resulted in developing and marketing nutritionally sound and tested products and the appropriate promotion of these, i.e. the formulas to the profession and informative educational information concerning products to the public. The Council, for example, set levels of vitamin C content to be retained in juices sold for infant foods and sequential analytical values were required in reports to the Council to assure compliance. The Council obtained authoritative review articles and statements that were periodically published in the *Journal of the American Medical Association* and also compiled books that became widely used classic reference works such as the *Handbook of Nutrition* (5, 6) published by the AMA at a low price so that it received wide distribution. The Council served further as an instrument for convening workshops to assess questions or developments and help provide scientific efforts to obtain needed information. It initiated the Western Hemisphere Nutrition Congresses and thereby extended its influence not only throughout this hemisphere but throughout the world. Many of these important and highly beneficial activities of the Council were reduced when the American Medical Association abandoned the seal of approval program because growing legalities encroached upon the ability of individuals and organizations to state scientific positions and assessments. Other activities were reduced or ceased when in 1974 the American Medical Association discontinued its several scientific councils.

The Food and Nutrition Board

In view of the imminence of the entry of the US into the Second World War, the government turned to the National Academy of Sciences for advice on "all problems related to the food and nutrition of the people" (15). An obvious initial consideration was the nutrient needs of the armed forces and civilians. The Academy responded by setting up a Committee on Nutrition, soon restructured as the Food and Nutrition Board (FNB) under the chairmanship of Dr. Russell M. Wilder of the Mayo Clinic. The initial Committee responded with the first draft of a table of recommended dietary allowances (RDAs) that were presented to and adopted by the National Nutrition Conference called by President Franklin D. Roosevelt, May 1941. These RDAs served as useful guidelines for the planning of feeding programs during World War II. Initially derived by the reasoned judgment of highly competent scientists knowledgeable concerning the soundest scientific evidence then available, the need for periodic revision was recognized to accommodate the rapidly expanding body of knowledge of new essential dietary factors, of nutritional biochemistry and physiology, of food science, and of increasingly quantitative evidence pertaining to requirements. Each of the approximately quinquennial revisions of the RDAs has been the product of some five years of subsequent committees' continuous assessment of new information and reappraisal of older data. The composition of membership on the Committee on Dietary Allowances has differed during each of the revisions. The ninth revision was published in 1980; currently a new committee is preparing the tenth revision. All controversy aside, and despite many evident misuses or misinterpretation of the RDAs, these allowances have served well to advance the use of nutrition science in public policy, in planning the feeding of groups and individuals, and even in assessing national levels of food sufficiency. They have catalyzed and broadly influenced the generation by other nations, as well as by international agencies, of standards or recommended levels of nutrient needs.

Recent unjustified politically motivated attacks upon the validity of the RDAs and other authoritative statements by reputable nutrition scientists reflect the changed political milieu that has emerged in the United States. Politicians and self-appointed, self-serving "public advocates" with their creed of antiscience, antiestablishment, anti-industry have found nutrition and food to be sensitive public issues. Unfortunately even some nutritionists have blurred the boundaries between science and advocacy, thereby creating additional confusion in the public mind. Elsewhere (7, 10) I have dealt with this and related problems arising from actions of those whom Maurice Arthus (1) termed "Theoreticians . . . (who) elevate themselves above the materiality of science and float in the sphere of ideas." He considered "the Theoreticians" extremely dangerous individuals who, in performing an act of faith become ". . .like the attorney who defends a client in spite of the evidence of his crime, like the politician who exalts his party even for its mistakes and vile actions which he proclaims to be acts of virtue and of courage . . . and has recourse to all means, honorable or not, in order to defend it. . . ." By "honorable or not" methods, the activists or Theoreticians stage "an irresponsible effort to impugn the reputation of public servants, scientists, scholarly institutions, and societies through innuendoes, through implication that does not exist among one or another group of scientists." They deliberately make misleading and false statements in Congressional hearings and other situations where a "good press" may be obtained. This behavior has aptly been termed "scientific McCarthyism."

There must evolve some effective means of exposing those members of the scientific community who so behave, those who fail to maintain a responsible level of interpretation of scientific evidence, or who, for self-aggrandizement unjustifiably make public claims that unduly alarm or mislead the public. Similarly, there is urgent need for protection of the individual, responsible, competent scientist who suffers these defamatory attacks from politicians and from self-appointed activists under their guise of representing or protecting the public, using as a shield some respectable-sounding institutional name. The efforts of politicians and some members of the scientific community to impugn the integrity of individual members of the Committee on Dietary Allowances, and even the scientific basis of the allowances, is to be deprecated. The necessity of contending with public defamation deters many responsible, knowledgeable, and conscientious scientists from serving on important national advisory committees.

The Food Protection Committee

Following a recommendation of the National Health Assembly, May 1948, the FNB appointed an *ad hoc* committee to examine the extent to which pesticides or other toxic substances occurred in foods and to develop a plan by which the safety of chemicals used in foods could be assessed (15). The *ad hoc* committee recommended establishing a permanent committee to serve in all aspects of the problems. This followed a National Conference in December 1949 encompassing representation from government, public health, agriculture, manufacturing chemists, food manufacturers, and academic and private research organizations. The board acted in May 1950 (eleven years prior to Rachel Carson's *Silent Spring* of 1961, which nowhere acknowledged these responsible actions of government, scientists, and industry in the public interest) to establish the Food Protection Committee (FPC), on which I served as a member from 1950 until 1971, and as chairman from 1954 to 1971.

The reports of the FPC were evolved by critical study of all available scientific evidence, objectively assessing it, then following with recommendations of general or specific nature; the sole objective was to attain maximal public benefit with minimal risks. The Committee early recognized the principle that the concept of absolute safety is invalid and set down guiding principles concerning the use or exclusion of food ingredients or additives. These principles have repeatedly been revalidated by numerous national and international bodies, including Expert Committees of the World Health Organization. In the early 1950s there existed no complete list of substances employed as food additives. Compilation and publication of such a list was promptly undertaken by the FPC, encountering considerable opposition on the part of certain interests. I remember the strong support for this effort given by a number of the leaders in the food industry on the part of the FPC. At one meeting a particularly forceful vice president of the Quaker Oats Company, Dr. F. C. Peters, vigorously pointed to industrial colleagues and stated: "Every collaboration must be given to this effort by the FPC to prepare and make public a listing of additives in foods. If any food company is putting into foods anything they are ashamed of, they damn well better stop. The public must know about it!" It was with such responsible support that the Food Protection Committee was able to give not only national but international leadership over three decades, elaborating general principles, responding to needs to evaluate the safety of specific categories of substances such as surfactants and sweeteners, publishing two classical monographs on naturally occurring toxic materials inextricably present in foodstuffs, and producing the *Food Chemicals Codex,* which not only is the first official codex of food chemicals in the United States but has been adopted by many of the major countries of the world. This codex is as basic to the field of foods as are pharmacopoeias to medicine.

Despite years of study and assessment of potential carcinogenic properties of long-used substances, even some essential nutrients, there arose irrational demands for exclusion of traditional intentional and incidental additives generally regarded as safe (GRAS). These demands were accompanied by unrealistic clamor for mandatory testing of thousands of ingredients and additives. This testing would have created an impossible demand for personnel and resources, which was totally unjustified in view of rational priorities for scientific study of materials or environmental chemicals. Accordingly the FPC prepared and the National Academy of Science published guidelines for judging when evidence or use was inconsequential and, hence, further extensive study deemed unnecessary.

This report was misconstrued by certain members of the press and seemingly deliberately by a limited number of members of the scientific community who vigorously attacked both the individual members of the Food Protection Committee and the National Academy of Science and its president. The ugly side of the politics of science emerged. At the request of the president of the Academy, Dr. Philip Handler, members of the FPC and the adversaries met in the Academy building to resolve the differences of understanding and interpretation of the report. One of the adversaries promptly wrote a statement that resolved this difference. It was unanimously adopted by those present and transmitted verbatim to the president of the Academy. In a subsequent budgetary hearing before the Congress concerning funding for oncology, one of the involved adversaries stated that he had never seen the statement and he disagreed with it! This disavowal occurred despite unanimous concurrence with the statement prepared by his colleague and agreed to by him at the earlier meeting. Fortunately I had filed the original copy in the handwriting of this adversary. It was given to Dr. Handler and was identical with the previously transmitted typed statement summarizing our meeting. No news report nor public statement from our adversary ever corrected the untrue "testimony" in the hearing.

I cite this instance now over a decade old as an example of the continuous emotionally charged wrangling pertaining to nutrition and cancer and the lack of integrity of those who subjugate science to politics in order to take a public position, even at sacrifice of veracity. It was precisely such actions that led to Maurice Arthus' warning that Theoreticians would take ". . . recourse to all means, honorable or not . . ." to defend their positions because "the daily newspapers, magazines of the most different kind entertain their readers with them in glowing language. Lecturers talk about them to their audiences in high tones. Government authorities give formal approval. . . . The Theoreticians find themselves in the limelight of publicity. How could they change their opinion without losing face?" (1).

The Interdepartmental Committee on Nutrition for National Defense (Development)

Personally rewarding, uniquely scientifically productive, and of remarkable humanitarian benefit appropriately describes almost two decades of work and association with the Interdepartmental Committee on Nutrition for National Defense (Development) (ICNND) (12). This novel committee resulted from an astute observation by Ambassador Lodge concerning night blindness in a

Korean servant, which aroused his suspicion that endemic malnutrition might be reflected in the Korean armed forces. These events led President Dwight Eisenhower to dispatch in 1953 two officers, Harold R. Sandstead and C. J. Koehn, to assess and report on the nutritional status and the messing in the armed forces of Korea. Their findings promptly led the President to establish as an interagency committee the ICNND with Dr. Harold Sandstead of USPHS as Executive Director. The ICNND, organized in 1955, had as its primary objective assisting developing countries to assess their nutritional status, defining problems of malnutrition, and identifying means for solving these problems by taking advantage of the country's own resources. Immediately prior to assuming the executive directorship of the ICNND, Dr. Harold ("Sandy") Sandstead had been assigned to the Vanderbilt Nutrition Program by Dr. W. H. Sebrell, Jr. and was investigating the potential nutritional significance of certain chronic oral lesions and the effect of nutrient supplementation. These studies were planned in cooperation with Dr. Russell Wilder of the Mayo Clinic.

Dr. Sandstead promptly organized the headquarters of the new committee in Stonehouse at the NIH with the highly efficient and loyal support of two remarkable associates, Mrs. Harriet Martin, Administrator, and Dr. Arnold E. Schaefer, Deputy Executive Director. Initial participation and support of the Committee came from the Armed Forces, the US Public Health Service, the US Department of Agriculture, the State Department, and cooperating universities. Support soon was extended by USAID, the Atomic Energy Commission, the Department of Interior, the Food and Drug Administration, and other agencies. Subsequently liaison members were added from Food for Peace, the Pan American Health Organization, United Nations Children's Fund, Food and Agriculture Organization, and World Health Organization of the UN.

The experience during World War II of nutrition officers and of clinical nutrition teams both in military and civilian areas plus the laboratory expertise centered at Vanderbilt University and in the US Army Nutrition Laboratory made it possible rapidly to design the broad survey of food production, processing, and use and the clinical and biochemical assessment of nutriture that was to provide data for guiding national nutrition improvements in the many host countries around the world. Shortly before the departure of the first US team to the host countries, Iran and Pakistan, Dr. Harold Sandstead died tragically as a passenger in the first bombed civilian plane; the plane was over the Grand Canyon en route from Denver. Dr. Arnold E. Schaefer was appointed to fill Dr. Sandstead's post. Dr. John B. Youmans, who during the Second World War was in charge of the Army Nutrition Program and who had but recently been appointed Dean of the Vanderbilt School of Medicine, responded to the emergency and directed the surveys in these two initial countries. A continuing panel of consultants from governmental agencies and the academic community counseled on technical aspects of methodology and survey findings.

Over a 15-year period, surveys were completed in 32 nations, in several of which repeated or continuous evaluations were made. Most of these latter were conducted entirely or in major part by national personnel who had worked with the US teams during the initial survey and not infrequently some of those personnel subsequently had pursued advanced scientific studies in US universities under the tutelage of faculty members who had served on the respective nation's initial survey team. Through follow-up visits by academic members of the teams as well as by government scientists there was invariably maintained a close interpersonal and scientific collaboration between participants. This helped catalyze action on recommendations developed jointly by host country and US scientists and it provided continuing educational opportunities and material resources. The resulting enormous widening of nutrition understanding by faculties and students in the participating US institutions created a reservoir of knowledgeable nutritionists that likely has been unparalleled in any country and a richness of scientific research themes that if continuously explored would have immeasureably advanced the science and application of nutrition throughout the world. The scientist-to-scientist, university-to-university, and university-to-agency relationships so established were unique. It is regrettable that administrative and political leadership has failed to see the opportunities and rich national rewards of this quite modestly costing program with its very high yield/cost ratio not only in health benefits to mankind, but also in great scientific and cultural returns.

In retrospect it appears to this observer that a major force in the demise of the ICNND was the use for political reasons of (or misuse of) findings and claims concerning domestic nutritional problems. A Congressional Committee instructed the Executive Director of the ICNND to conduct a similar survey in the United States and to report back to that Committee. This was at the height of much political attention and speculation concerning the extent and severity of malnutrition in the US. Adequate funding for a domestic study was not forthcoming, which further compromised the efforts to carry out the instruction of the Committee. Despite these difficulties a remarkable series of nutrition assessments were made in ten states (the so-called Ten-State Nutrition Survey) albeit the sampling was limited by socioeconomic factors, a limitation that favored misinterpretation and misuse of data. The database so acquired continues, however, to serve for meaningful scientific analyses. The visible efforts of the survey likely influenced the more orderly Nutrition Survey of Canada. The residue of these programs in the US is the National Health and Nutrition Examination Survey (NHANES) (18), which, although continuing, unfortunately has not provided the consolidated descriptions needed for national guidance of the food, food use, nutriture, and health of populations by socioeconomic, ethnic, and regional samples. By contrast with the ICNND program, the organization of the NHANES program fails to involve meaningfully the academic nutrition community and students in the manner that was accomplished by the ICNND and during the Ten-State Survey.

The Vanderbilt-NAMRU-3 Nutrition Study (Cairo)

One spin-off of the ICNND activities was 25 years of nutrition research in the Middle East based on personnel and resources at Vanderbilt University and NAMRU-3 (Naval Medical Research Unit Number 3) in Cairo. This development served as the site of an unpredictably varied and productive series of researches, and 20 years ago attracted a group of young investigators who today are senior leaders in their respective areas of nutrition science throughout the United States and in the Middle East. It evolved in this manner: As early as 1949 I had served a tour in Egypt as a consultant on pellagra for the World Health Organization. This experience led to subsequent investigations there of pellagra and anemia participated in by Dr. Richard Vilter and members of his laboratory. During these studies we established contact with personnel at NAMRU-3 as well as nutrition scientists in Cairo and Alexandria. As plans were formalized for an ICNND study in Ethiopia it was logical to enlist the more sophisticated resources and some personnel from the Cairo-based NAMRU-3. Scheduled studies in Ethiopia coincided with WHO-sponsored nutrition events in Egypt and I was invited to participate in these while serving as Director of the ICNND nutrition survey in Ethiopia. Dr. John Seale, Commanding Officer of NAMRU-3, suggested a continuing collaboration between that unit and the nutrition program at Vanderbilt University to enhance the capabilities of NAMRU-3 for addressing malnutrition, a major problem of the region.

The resulting cooperative program with its access to ambulatory patients, the wards at NAMRU-3, field investigations, laboratories both at NAMRU-3 and Vanderbilt University, and a remarkable degree of cooperation with institutions in Egypt and elsewhere in the Middle East afforded a nearly unparalleled spectrum of nutritional challenges. Initial studies of severe endemic iron deficiency anemia and the availability of iron from foods led to recognition and elucidation of the clinical syndrome of dwarfism and sexual infantilism due to zinc deficiency. This exciting finding served to trigger not only diverse studies of zinc deficiency and metabolism of humans but also interest in the role of other trace elements in man, especially selenium deficiency in malnourished infants and copper and chromium deficiencies.

The panmalnutrition of kwashiorkor was documented and cooperative regional studies clearly demonstrated that in protein calorie malnutrition there were associated differing types of anemias that resulted from deficiencies of different hemopoietic factors—iron, folic acid, tocopherol, and possibly other essentials. The dominant deficiency varied from one geographic area to another within the Middle East (from Turkey to Lebanon, Jordan, and Egypt). Active collaboration was readily established with competent scientists throughout the region, in large measure because of the ICNND studies and followup that had taken place in country after country in the region plus the high regard in which the NAMRU-3 organization was held. The presence on the Vanderbilt University project of the distinguished nutrition scientist Dr. V. N. Patwardhan, then recently retired as chief of the nutrition program for the World Health Organization, further enhanced the prestige of this activity and facilitated cooperative research.

In order to assure successful execution of field studies in villages, we recruited a young anthropologist, Louis Grivetti, who now holds a joint appointment in the Departments of Nutrition and of Cultural Geography in the University of California. His presence not only greatly facilitated the studies in villages but immeasurably strengthened bonds with the Egyptian Ministry of Antiquities and the historical section of the Ministry of Agriculture of Egypt. A tangible result of his effectiveness was the unlimited access permitted us to monuments and museums as we delved into the rich heritage of the ancient Egyptian civilization in preparation of the two volume monograph *Food: The Gift of Osiris* by Paul Ghalioungui, an Egyptian authority on Pharonic medicine, Louis Grivetti, and me (9). The accumulation of the material summarized in this work, the travels incident to it, and the personal acquisition and study of literally hundreds of beautiful, romantic volumes on history, travels, religion, folklore, sciences (medicine, botany, agriculture, archaeology, Egyptology), military expeditions and other aspects of culture opened unimagined vistas that I never could have viewed except through the guidance of incredibly generous colleagues and scholars—Egyptian, American, Swedish, Greek, French, Lebanese, Jordanian, Indian, German, English, and so forth.

The 15 years during which these volumes were being assembled permanently fixed my long-existing interest in history of food, food science, and nutrition that had been initiated by teachings of Paul Day and Howard B. Lewis. During these years I became an avid bibliophile of such compulsiveness that the syndrome is best described as that of a "bookaholic." Another especially gratifying facet of this period was the stimulating opportunities for learning from young associates whom I continue to watch with pride and satisfaction as their contributions extend the frontiers of nutrition science and its applications.

Institute of Nutrition of Central America and Panama

Some of the successful strategies employed in those international cooperative ventures in nutrition with which I have been associated were adopted from the experiences of the Institute of Nutrition for Central America and Panama (INCAP). Founded in 1948, its initial role was seen basically as that of a food laboratory, but this organization matured to become a major world resource in food and nutrition science. The insistence of its first director, Dr. Nevin Scrimshaw, upon developing a staff of highly competent regional scientists with laboratories, library, and other scholarly resources conducive to creative scientific research was a strong factor in modernization of not only nutrition but medicine and university education throughout Central and South America. INCAP became a leading influence in international nutrition and a base for research and training of personnel from all over the world. This is attributable in part to the reciprocal exchange and effective communication established with leading institutions and organizations world-wide. Its personnel and programs have served as examples for other nutrition centers that have emerged during the past three to four decades in other areas of the world, e.g. Mexico, Chile, Brazil, the Philippines, India, Thailand, Indonesia.

Other productive or highly promising centers have lost their momentum and in some instances have fallen into oblivion. Such has occurred in Uganda, Ethiopia, Iran, and in the former Belgian Congo. The demise of these centers and of others that could be so identified is attributable to the disruptive force of political upheavals and instability. It is evident that a politically stable environment is essential not only for nurture of the population but also for the nourishment of science. The health and well-being of the science of nutrition in both industrialized and developing countries seems inextricably linked to and influenced by the political milieu. Despite this fact of survival it is fatal to permit the essence of science in any manner to be twisted by political considerations or for scientists to prostitute objectivity, integrity, or veracity for political favor.

THE ROLE OF FOUNDATIONS

Those organized philanthropies, foundations, devoted to public good possess an autonomy of organization that permits flexibility of action without the constraints inherent in the university or governmental agencies. They are primarily a product of the present century. Their beneficial influences on scientific development, medical knowledge and health care, humanities, arts and human relations, agriculture and food production, education, and improvement of the quality of life are evident nationally and internationally. Farsighted support by foundations has contributed to every significant nutrition advance during the five decades described in this essay and to much before.

In the late 1930s the International Health Division of The Rockefeller Foundation sponsored the development of methodologies and studies of nutrition of populations and initiated a series of Special Fellowships to develop medical personnel for needed leadership in nutrition in public health and medicine as then envisioned by the Foundation. This program was launched in the very early 1940s; I had the good fortune of being selected as one to benefit from it. Dr. John Youmans of Vanderbilt was sent to France by The Rockefeller Foundation to assess the food and nutrition needs there in the early phases of World War II. It also was support from The Rockefeller Foundation that established the important Typhus Commission in Cairo during this period. Upon completing its mission of successfully developing control methods for this crucially important disease in the Allied military, The Rockefeller Foundation turned over its laboratory resources to the US Navy and thereby established NAMRU-3. Another foundation, the W. K. Kellogg Foundation, was intrigued by a concept put forward by Dr. Robert Harris of MIT, a concept that involved the need to determine the nutritive value of numerous foodstuffs available in the Central American region and thereby better plan the utilization of these for the benefit of nutritional health. INCAP was founded with Dr. Harris' assistance, the organizational and financial aid of the Kellogg Foundation, and participation of the governments of Central America and Panama and the then Pan American Sanitary Bureau.

Three foundations—the Milbank Fund, the Williams-Waterman Fund, and The Nutrition Foundation—made long-term grants to the Food and Nutrition Board to maintain "a core budget from independent sources instead of being dependent on a single government agency or industry" for the first three decades of the Board's existence. The Rockefeller Foundation generously supported a far-reaching international research program on the suitability of unconventional sources of protein for infant and child feeding, which program was administered by the Food and Nutrition Board.

For more than four decades the Williams-Waterman Fund (21) made grants to support a wide variety of basic nutrition research in universities and countries throughout the world. These supported studies of analytical methodology, the biosynthesis of vitamins, fats, enzyme metabolism, physiology, proteins and amino acids, human requirements, education and training, nutrition surveys, clinical nutrition, and cereal enrichment. Its leadership in the latter area was an outstanding commitment to the stated purpose of the Fund "for the combat of dietary diseases." Since this fund administered by the Research Corporation is no longer active it is particularly important to record that it was established through an agreement in 1935 between the three inventors of the process for synthesis of thiamin, R. R. Williams, R. E. Waterman, and E. R. Buchman, and the Research Corporation; and it was funded from the proceeds of the inventors' patent. By agreement, 25% of all net proceeds went to the Research Corporation to support research in that corporation's field of choice, 50% of all net proceeds were placed by the corporation in a special fund, "Williams-Waterman Fund for Combat of Dietary Diseases." It is difficult to perceive of a more beneficial, unselfish use of patent proceeds than this! A similar use of proceeds from the patent on vitamin D served to establish the Wisconsin Alumni Research Foundation. The direct recipient of this latter funding, however, was research and related programs at the University of Wisconsin.

Unparallelled properly describes the ferment at work during the ten years from 1935 to 1945 to promote advances in the science and application of nutrition. A conversation during lunch in 1938 between the president of General Foods Corporation, Mr. Clarence Francis, and the General Council of the Associated Grocery Manufacturers of America (GMA), Charles Wesley Dunn, initiated planning by leaders of the food industry for the establishment of The Nutrition Foundation, officially incorporated on Christmas Eve of 1941 (13). Its first scientific director and full-time president, Dr. Charles Glenn King, had isolated from lemon juice and identified vitamin C in 1932. Dr. King's remarkable leadership marked by impeccable integrity, broad appreciation of the science and application of nutrition, and keen sensitivity to the needs of mankind set the pattern of the general policies of the Foundation. He was fully supported by unselfish, philanthropic members of the board of trustees, who were chief executive officers of supporting companies. The board of trustees was presided over by the president of MIT, Dr. Karl T. Compton. The Scientific Advisory Committee was a true "Who's Who" of nutrition scientists from universities, government, and industry. The history of The Nutrition Foundation (13), prepared by Dr. King, has been described as "a veritable record of the progress of the science and application of nutrition during the . . . three and a half decades" from 1941 to 1976. These decades encompass an era of rapid and widespread nutritional changes that permanently affected human health throughout the world.

Subsequent developments supported by The Nutrition Foundation in the decade of my presidency (1971–1982) were summarized in a special 40th anniversary issue of *Nutrition Reviews* (16), the influential, critical scientific review journal founded by the Foundation and initially edited by Dr. Frederick J. Stare. Support of investigators through research grants was an especially noteworthy feature of the program of The Nutrition Foundation because at that time there were no project-type federal grant programs available for support of research in biochemistry, nutrition, or other areas. The first of the categorical research institutes of the NIH, the National Cancer Institute, came into being in 1937. The meager funds it provided were primarily for instruction and education concerning cancer. It was not until 1948 and onward that project-type research grants in an increasing number of fields became available, and not until 1950 was the National Science Foundation authorized. Recognizing this, one well can appreciate the key importance of those grants from The Nutrition Foundation made early in the career of then young scientists including Carl and Gerty Cori, George W. Beadle, Edward C. Tatum, Fritz Lipmann, Vincent Du

Vigneaud, and Konrad Bloch, all of whom subsequently became Nobel Laureates.

An important force in promoting good clinical research in nutrition during the exciting years of the 1940s and 1950s was the National Vitamin Foundation under the directorship of Robert S. Goodhart, M.D. Established in 1946 by producers and distributors of vitamins and related products, its stated objective was promoting and supporting research on vitamins and nutrition in order to improve the health and welfare of mankind. Its grants, symposia, fellowships, and other contributions greatly advanced clinical investigations on a broad front. Dr. Goodhart's encouragement and support was especially helpful in establishing the *American Journal of Clinical Nutrition* and organizing the American Society of Clinical Nutrition. Some of the members of the nonmedical nutrition community were unjustly apprehensive concerning the consolidation of a new, strong, clinical force in the field that received support from the pharmaceutical sources. The integrity and scientific quality of advisors, grantees, and of the Foundation's leadership by Bob Goodhart minimized such undue concerns. In the dissolution of the National Vitamin Foundation in the early 1960s an important nutrition resource was lost.

The outburst of support for nutrition in this period was a matter of timeliness. There was still an impressive awareness of the scourge of the endemic deficiency diseases, goiter, scurvy, pellagra, and other B-vitamin deficiencies, in the United States; the excitement of and fascination with newly discovered, dramatically effective, essential nutrients was everywhere evident; World War II was creating a public concern for food shortages; medical science was exhiliratingly successful in developing chemotherapeutic agents and ushering in the promising new antibiotics. Two decades later the success of these and other scientific developments served greatly to alter the priority that had existed for nutrition education in the medical curriculum in most institutions. In recognition of the inadequacy of attention given to the subject the Council on Foods and Nutrition of the American Medical Association, with assistance from The Nutrition Foundation, convened a national conference of representatives of medical schools throughout the US to address the needs for nutrition teaching in 1962. Despite forceful arguments concerning the importance of and need for intensified teaching on nutrition and despite the evident identification of its placement in medicine, there followed during the next ten years little apparent improvement in the national situation.

The same two organizations cooperated again in a similar conference at Williamsburg in 1972, since which there has been a virtual rennaissance of interest in nutrition (8) and its role in medicine in a majority of the medical schools in this country and in many abroad. This development reflects several influences: increasing awareness of the gravity of malnutrition and hunger in the developing world, an enhanced concern of students and young physicians for societal problems, availability of laboratory diagnostic tests for objectively assessing nutriture, and professional recognition that many new therapeutic advances have but limited application unless accompanied by sustained nutritional support. Despite early efforts of the surgeon Robert Elman in the late 1930s to utilize parenteral protein hydrolysates and/or amino acid mixtures, and despite intensive investigations of the feasibility of parenteral lipids during and after the Second World War, it was the Philadelphia group's success in 1968–1969 with the technique of administering hypertonic dextrose and amino acids solutions through intravenous catheters in the superior vena cava that sharply focused attention of surgeons and other physicians on meeting short-term (and now very long-term) nutrient requirements. Simultaneously, the clinical importance of trace elements was being recognized. Greater understanding of and interest in the seemingly endless variety of more recently recognized inborn errors of metabolism and the potential of nutritional manipulation for alleviating harmful effects of these genetic disorders have underscored anew the importance of nutrition in medical practice and research. Since the Williamsburg meeting there have been countless national, international, regional, state, and institutional symposia and conferences on nutrition education in medicine and health sciences. It is safe to predict that this concern will be effectively expressed in the reassessment of medical education and curricular content that now is expected to follow the 1984 report on the medical school curriculum from the Association of American Medical Colleges.

EPILOGUE

These semi-autobiographical reflections reveal some portion of philosophic interpretations of past events that may or may not foreshadow the future. They are incomplete both as to personal experiences and historical happenings. I feel more secure in interpreting causative influences such as those that created The Nutrition Foundation or those that modified interest in nutrition in the medical curriculum than I do in identifying the basic reasons for demise of institutions or movements. One can hope that like the legendary phoenix some of the important species now reduced to ashes may rise again in future youthful vigor—perhaps that is the phenomenon we are observing in regard to medical nutrition. I am less optimistic, however, for several of the organizations with which I have had rewardingly productive associations. Despite my concern for some, I am confident that this *Annual Review of Nutrition,* which I helped to initiate five years ago, will increasingly serve as a force in identifying the scope of nutrition science and critiquing its progress. Bob Olson, my successor, and I long have held common values and shared scientific experiences. We both appreciate the firmness of the scientific foundations of nutrition science and deprecate the efforts of those "theoreticians," politicians, and activists who would shake these foundations.

Further Readings

1. Arthus, M. 1943. *Maurice Arthus's Philosophy of Scientific Investigation.* Transl. Henry E. Sigerist. Baltimore: Johns Hopkins Press
2. Baird, W. D. 1979. *Medical Education in Arkansas, 1879–1978.* Memphis State Univ. Press. 505 pp.
3. Barber, M. I. 1959. *History of the American Dietetic Association: 1917–1959.* Philadelphia: Lippincott. 328 pp.
4. Council on Foods of the American Medical Association. 1939. *Accepted Foods and Their Nutritional Significance.* Chicago: American Medical Association. 492 pp.
5. Council on Foods and Nutrition of the American Medical Association. 1943. *Handbook of Nutrition.* Chicago: American Medical Association. 986 pp.
6. Council on Foods and Nutrition of the American Medical Association. 1951. *Handbook of Nutrition.* Chicago: American Medical Association. 717 pp. 2nd ed.
7. Darby, W. J. 1973. Acceptable risk and practical safety: philosophy in the decision-making process. *J. Am. Med. Assoc.* 224:1165–68
8. Darby, W. J. 1977. The renaissance of nutrition education. *Nutr. Rev.* 35:33–38
9. Darby, W. J., Ghalioungui, P., Grivetti, L. 1977. *Food: The Gift of Osiris,* Vols. I, II. London: Academic. 877 pp.
10. Darby, W. J. 1980. Science, scientist, and society; the 1980's. *Fed. Proc.* 39:2943–48
11. Hill, F. W., ed. 1978. *50th Anniversary of the Journal of Nutrition: A History of the American Institute of Nutrition.* Bethesda, Md: American Institute of Nutrition. 198 pp.
12. Interdepartmental Committee on Nutrition for National Defense. 1963. *Manual for Nutrition Surveys.* Bethesda, Md: National Institutes of Health. 327 pp. 2nd ed.
13. King, C. G. 1976. *A Good Idea: The History of the Nutrition Foundation.* New York, Washington: The Nutrition Foundation. 241 pp.
14. Lusk, G. 1928. *The Elements of the Science of Nutrition,* Philadelphia: Saunders. 844 pp. 4th ed. Reprinted with biographical notes by E. Neige Todhunter as a Nutrition Foundations' Reprint, 1976, Johnson Reprint Corporation, New York
15. National Academy of Sciences–National Research Council. 1965. *The Food and Nutrition Board, 1940–1965: 25 Years in Retrospect.* Washington: NAS-NRC. 38 pp. plus addenda v
16. Olson, R. E., ed. 1982. 40th Anniversary issue of Nutrition Reviews. *Nutr. Rev.* 40:321–52
17. Rose, M. S. 1935. University teaching of nutrition and dietetics in the United States. *Nutr. Abstr. Rev.* 4:439–46
18. Swan, P. B. 1983. Food consumption by individuals in the United States: two major surveys. *Ann. Rev. Nutr.* 3:413–32.
19. Todhunter, E. N. 1976. Chronology of some events in the development and application of the science of nutrition. *Nutr. Rev.* 34:353–65
20. Trotter, J. R., Darby, W. J. 1984. Paul Louis Day (1899–1980): A biographical sketch. *J. Nutr.* 114:241–46
21. Williams, R. R. 1956. *Williams-Watermann Fund for the Combat of Dietary Diseases. A History of the Period 1935–1955.* New York: Research Corporation. 125 pp.
22. Youmans, J. B. 1932. Endemic edema. *J. Am. Med. Assoc.* 99:883–87

NOTE ADDED IN PROOF

Three valuable complementary accounts of these years are

Navia, J. M. 1984. Robert Samuel Harris (1904–1983). *Vitam. Horm.* 41:xv–xxix
Sebrell, W. H. 1985. Recollections of a career in nutrition. *J. Nutr.* 115:23–28
Kampmeier, R. H., Darby, W. J. 1985. John B. Youmans (1894–1979). *J. Nutr.* In press

PHARMACOLOGY AND TOXICOLOGY

H.W. Kosterlitz

Ann. Rev. Pharmacol. Toxicol. 1979. 19:1–12
Copyright © 1979 by Annual Reviews Inc. All rights reserved

THE BEST LAID SCHEMES
O' MICE AN' MEN
GANG AFT AGLEY

Hans W. Kosterlitz

Unit for Research on Addictive Drugs, University of Aberdeen,
Aberdeen AB9 1AS, Scotland

Some 25 years ago I attended a symposium with about 30 participants. On the evening before the first working day, the organizers had what appeared to me to be a very original idea. At first, they dined and wined us very well; when we continued a lively conversation in comfortable surroundings they suggested that each of the main speakers give the reasons why he or she had taken up research as an important part of their career, and detail the circumstances leading to the work they were going to present at the symposium. The atmosphere was so congenial that nobody found it difficult to reveal thoughts that in other circumstances might have been considered to be rather personal. I found many of the points made of great importance to me since I have always been interested in knowing what had moved scientists whom I admired to choose a problem and approach its solution in a manner which was so often admirable. Originally, it had been planned to include these presentations in the proceedings of the symposium but this idea was eventually abandoned for a number of reasons.

When I was asked to write this prefatory chapter for the *Annual Review of Pharmacology and Toxicology* I was somewhat surprised that this honor should come to me because my pharmacological activities are of relatively recent origin. Then I remembered the events at the symposium and thought of relating the reasons for my rather unorthodox career, at least as far as I understand them.

My school and university education took place in Germany. My father was a medical practitioner and he tried to dissuade me from taking up medicine and suggested the legal profession. However, after six months during which I was fascinated by the intricacies of the legal system of the ancient Romans, I finally decided to study medicine. In those days, it was

1427

0362-1642/79/0415-0001$01.00

still possible for medical students to arrange the course in such a way that they could follow their inclinations without neglecting the formal requirements too much. Thus, I laid the basis for my future interest in basic research by spending a great deal of time in the departments of histology and physiological chemistry and learning the basic experimental techniques of these disciplines. However, the most important stimulus was the experience I had in the laboratories of Rona who had been a pupil of L. Michaelis. In my third year of medicine, I attended lectures and clinics from 8 to 11 A.M. and spent the remainder of the day, often until late in the evening, learning biochemical microanalysis and the fundamentals, both theoretical and practical, of physical chemistry in relation to biology and medicine. Rona was a hard taskmaster who would not tolerate untidiness or sloppy techniques; I have always been grateful for what I learned in his laboratories.

After I had obtained my medical degree and qualification in 1928, I worked in the first medical department of the University of Berlin, whose head was W. His of auriculoventricular bundle's fame. In those days, one was lucky to be allowed to work on the wards and laboratories of a university department and, as was usual, I did not receive any remuneration for two or three years. Although I had been involved in research during my student days, I now developed the first major research interest of my own, in the area of carbohydrate metabolism and its relation to liver disease and diabetes mellitus. I tried to imitate some of the early clinicians who combined clinical work with experiments in the laboratory designed to elucidate and analyze problems they encountered in patients. O. Minkowski was my hero at that time; in the late nineteenth century he had established that removal of the pancreas caused diabetes in dogs. I was particularly interested in the clinical observations that diabetic patients had a better tolerance for fructose and sorbitol than for glucose. I was able to confirm Minkowski's finding that fructose could form liver glycogen in diabetic dogs in the absence of insulin and extended this observation to sorbitol. I also showed that diabetics utilize galactose, which is converted to glucose in the liver, better than glucose, even to the extent that it had antiketogenic properties in severe diabetes. This finding was the beginning of my quest for the mechanism that leads to the Walden inversion of galactose to glucose.

So great was the pressure for posts in medical university departments at that time that when I obtained my first paid appointment, it was as an assistant diagnostic radiologist. The clinical and laboratory work in which I was interested had to be done in my spare time; in this effort I had the full and sympathetic support of W. His, the head of the department. I should point out that I liked the radiological work although it took up a large part of the working day; at that time, modern radiology of the gastrointestinal tract was in its early stages of development and, when in doubt, gastroscopy was used to verify the diagnosis.

During the five years after I had obtained my medical degree, I had acquired a basis for my later work. The two people who had the most influence on me were Rona, whom I have already mentioned, and one of my senior clinical colleagues, Petow, who supported me through all the ups and downs of the early stages of a scientific career. In particular, I had the kind of experience which is probably shared by many young people; manuscripts, which I thought I had prepared carefully and in an intelligible way, were returned to me again and again with the remarks that they were too long, that my experimental procedure would be impossible to repeat from the description, and that the discussion was too long, too speculative, and to a considerable extent irrelevant. This was a painful lesson which stood me in good stead in later days.

When I left Germany in 1934, I went to Aberdeen where J. J. R. Macleod was professor of physiology. His long experience in carbohydrate metabolism and the actions of insulin in diabetes attracted me to work in his laboratory. This was many years after the controversies which surrounded the discovery of insulin in Toronto in 1922 where Macleod was then chairman of the Department of Physiology. He made a great impression on me, both as a scientist and a person who unfortunately was already suffering from the illness that caused his death a year later. I shall be forever grateful for the encouragement he gave me in this phase of my career. There were several people in the department with a lasting influence on me. Among them were J. M. Peterson, who was my mentor through many a difficult year, and D. J. Bell, from whom I learned a great deal of synthetic carbohydrate chemistry. I took my British medical qualification and my PhD; I had also to improve my knowledge of biochemistry and physiology which I was soon asked to teach to medical students.

As far as research was concerned, I decided to attempt to resolve the problem of the conversion of galactose to glucose. By using two different strains of yeast, both of which fermented glucose but only one could ferment galactose, I could differentiate between the two monosaccharides and their phosphate ester and determine them quantitatively. It was much more time consuming than modern chromatography. By differential fractionation of the barium salts of the organic phosphates extracted from the liver, it was possible to demonstrate that after feeding of galactose to rabbits a phosphoric ester of galactose accumulated in their livers. This was shown to be galactose-1-phosphate, identical with synthetic α-galactose-1-phosphoric acid, and circumstantial evidence was adduced for the view that phosphorylation at C_1 was the first step in the conversion of galactose to glucose, as was later documented step-by-step by the brilliant investigations of Leloir.

In the early 1940s, the first chapter of my research career was brought to an end. Throughout that time, I had the encouraging support of the Coris and of F. G. Young, and I received the degree of DSc from the University of Aberdeen. However, it seemed to me that this type of work, although intellectually very stimulating, was incongruous with the efforts required by a very bitter war. One of these was the production of more medical graduates and therefore the teaching of medical students was made much more intense. As a research project I selected one that dealt with the more fundamental aspects of nutrition, namely the influence of the quality and quantity of dietary protein intake on the composition of the liver.

In collaboration with Rosa Campbell, the earlier findings of Addis and his colleagues were confirmed, that changes in protein intake cause rapid alterations in the protein content of the liver. What was surprising to us was the observation that these findings could not be explained by increases or decreases in stored protein since there were proportionate changes in the contents of phospholipids and RNA; the only constant constituent was DNA. These responses of the liver served as a basis for the bioassay of the nutritional quality of dietary protein. Of considerable interest were changes found in pregnant rats; the constituents of liver cytoplasm increased with RNA taking a prominent part. This phenomenon was only partly due to the presence of the fetuses since it still occurred, although to a lesser extent, when the fetuses were removed as long as the placentas remained intact.

During the course of this research I was elected a Fellow of the Royal Society of Edinburgh which gave me a great deal of encouragement. However, this chapter came to an end in the early fifties, partly by intent and partly by circumstances. I transferred my attention to the physiology and pharmacology of the peripheral autonomic nervous system, a subject with which I am still concerned. This interest was kindled by two factors: Ian R. Innes and I were intrigued by a paper published by W. B. Cannon in 1922 reporting that liberation of sympathin on stimulation of the hepatic nerves and its action on the denervated heart was greater in cats fed on milk and meat than in those given mainly fat and carbohydrates. In those days the nature of sympathin was unknown and Cannon could not match the chronotropic responses of the denervated heart against noradrenaline and thus exclude variations in sensitivity of the heart. In the early 1950s this had become possible mainly due to the work of U.S. von Euler and, on repeating Cannon's experiments, we found to our disappointment that diet had no influence on the amount of noradrenaline released after stimulation of hepatic nerves. If it had been true, a great deal of very interesting work could have been done.

The second factor was an invitation by O. Krayer to come to Harvard and spend a few months in the laboratories of the pharmacology department. This invitation was accepted with much pleasure because I had known Dr. Krayer ever since his Berlin days and admired him for the strength and integrity of his personality. The five months I spent in Boston were hectic: Apart from some teaching, I accepted the challenge to analyze some of the actions of veratramine. To our great surprise this compound caused a periodic sinus rhythm in the cat heart. We thought we had made a fundamental discovery until we learned that Luciani working in Ludwig's laboratory had published a paper in 1873 in which he described a similar phenomenon when the frog heart was tied above the auriculoventricular groove and the ventricle filled with serum. We further learned that these Luciani periods attracted a great deal of attention during the last two decades of the nineteenth century. Is this proof for the saying that one of the ways of making discoveries is to search the literature of more than 50 years ago for unresolved problems? This period at Harvard had a very formative influence on me and I was very pleased and flattered when I was invited to Boston to give the Otto Krayer Lecture for 1977.

Back in Aberdeen, I started to become interested in the physiological role of the myenteric plexus. Having learned my lesson from the Luciani periods, I extended my reading more and more into the past. I came across the paper in which P. Trendelenburg showed in 1917 that morphine in very low concentrations inhibits the peristaltic reflex in the guinea pig isolated ileum mounted in an organ bath. Since the myenteric plexus is a relatively simple nervous system, we spent a great deal of work designed to elucidate the physiological mechanism of the peristaltic reflex. It was hoped that such an understanding would help in the analysis of the mode of action of morphine; it was also planned to develop a method for the assay of the pharmacological potency of morphine and its congeners. By about 1960, the work had reached a stage when it was thought to be possible to use this reflex and the inhibition of the response of the nictitating membrane of the cat to stimulation of its postganglionic fibres for this purpose. I had become more and more convinced that it would be difficult and perhaps impossible to initiate such an investigation in the central nervous system in view of its complexity and the absence of a simple pharmacological model. The action of morphine is highly selective although widespread, and therefore morphine-sensitive and morphine-insensitive neurones will occur side by side in varying proportions.

In 1962 I paid a brief visit to the United States to become familiar with the intense research efforts in the narcotic analgesics field. My first call was at the laboratory of Julius Axelrod. After a most interesting morning, he took me for lunch at the Cosmos Club in Washington to meet Dr. Nathan B. Eddy, the undisputed Nestor in the field of narcotic analgesics. During the hour or two we had together he asked me about my plans in his inimitable searching manner, familiar to everyone who knew him. It was the beginning of a friendship between the master and a newcomer to the field who was grateful for advice and encouragement unstintingly given until Dr. Eddy's death. It gave me therefore the greatest pleasure when the Committee on Problems of Drug Dependence selected me as the winner of the Nathan B. Eddy award for 1978.

My next port of call was the laboratory of Dr. M. H. Seevers at Michigan University in Ann Arbor where I was particularly interested in the colony of morphine-dependent monkeys. I was asked to give a seminar on the work of my group on the use of the peristaltic reflex of the guinea pig ileum and of the cat nictitating membrane in the assay of narcotic analgesics. I gave my talk and at the end Dr. Seevers said in his frank manner which I learned to cherish later, "Dr. Kosterlitz, this has been very interesting but I am afraid I cannot believe that such simple models are of real use." I then asked him what I could do to convince him. "Well, if you are agreeable, I shall send you six coded samples for you to test. If you can assess the majority correctly, I shall accept that there is something in what you told us today." I was somewhat taken aback but agreed to his proposal. After my return to Aberdeen, a packet arrived and we started to assay the samples with our models. Five of them did not cause us too much trouble. One of them was likely to be morphine since it agreed in behavior and potency with authentic morphine but there was another compound which with the experience we had had up to that time, also seemed to be morphine. We sent Dr. Seevers our results and to our surprised satisfaction we were right in all but one of our findings. The exception was the second morphine-like compound which turned out to be nalorphine; until then, we had not studied compounds with dual agonist and antagonist actions. Anyhow, we had gained Dr. Seevers' confidence and he remained our friend until his recent death.

In the course of further work it was found that the action of the opiate alkaloids was difficult to express in quantitative terms. Some years earlier W. D. M. Paton of Oxford had devised a simple and ingenious method of exciting the fibers innervating the longitudinal muscle of the guinea pig ileum either by stimulating an intestinal segment coaxially or by applying field stimulation to strips consisting of myenteric plexus and longitudinal muscle. It was found that these preparations could be used as a model for the quantitative assessment of the agonist and antagonist actions of opiate alkaloids and the study of their kinetics. Over the years, many compounds were investigated, pure agonists, compounds with dual agonist and antagonist actions, and pure antagonists. From investigations on the release of the transmitter, acetylcholine, it appeared that the effect of the opiates was presynaptic and that these receptors were distinct and separate from inhibition by presynaptic α-adrenoceptors. The predictive power of these tests was considerable and research laboratories of the pharmaceutical industry often asked us to assay newly synthesized compounds with agonist and antagonist action which frequently had potent analgesic effects in animals and man without necessarily leading to compulsive drug abuse.

Until about this time I was a member of the physiology department of Aberdeen University. In the earlier years, the teaching was mainly to medical students. Courses to science students started to develop in the late fifties, and these were attended also by the brighter medical students who wanted to become acquainted with the scientific basis of medicine. Some of these students joined me or other members of staff in ongoing research projects. One of the problems in the teaching of physiology, which has been and still is a challenge to many teachers of physiology, is the transition from the preclinical to the clinical years and the difficulty of convincing students of the clinical relevance of the basic aspects of physiology. Robert Aitken, who was professor of medicine during and for some years after the war, and I both felt that an attempt should be made to resolve this problem. So we started to teach the medical students in their last preclinical term in a joint course in which we tried to illustrate the significance of physiology on patients with classical signs of disturbance of the cardiovascular or respiratory systems. We were personal friends but this did not prevent us from attacking each other's preclinical or clinical stances without much inhibition. The more heated the discussion became, the greater was the interest and enjoyment of the students. They obviously learned something to take away because they turned out in force at 9 A.M. on a Saturday morning!

In 1968 I was asked to become chairman of the new Department of Pharmacology. For historical reasons, the Department of Materia Medica had so far taught Pharmacology; this was so in Aberdeen as in all Scottish Universities but Edinburgh. Thus, my main task during the next five years was to establish a new department and to develop new teaching courses, particularly for science students. In this effort I had the never failing support of G. M. Lees, J. C. Gilbert, and J. Hughes. One of the problems that was successfully resolved was the collaboration between the new department and the Department of Therapeutics and Clinical Pharmacology whose chairman, A. G. Macgregor, had been the driving force in the creation of the new department. In spite of additional preoccupations of an administrative nature in the university, I was able to continue with our research successfully, mainly because a number of enthusiastic young people had joined the department.

In 1972, when I still had one year before retirement from the chair, my scientific friends, particularly in Britain and the United States, suggested to me in quite forceful terms that I should continue with the research after my retirement. Applications for funds were successful and the University of Aberdeen set aside sufficient laboratory and office space for a team of 6 to 8 academic members to establish a new Unit for Research on Addictive Drugs.

One of the problems that was to be tackled was the further assessment of new compounds and particularly the search for new antagonists. In 1972, a new model had been developed in the Department of Pharmacology. In the white mouse, transmission from the stimulated hypogastric nerve to the vas deferens was depressed by morphine due to a reduction in the release of noradrenaline from the nerve terminals. Again, pure agonists, components with dual agonist and antagonist actions and pure antagonists, and the kinetics of the interaction between agonists and antagonists were studied. In principle, the pharmacological responses of the mouse vas deferens to opiates were similar to those of the guinea pig ileum. There were, however, important differences in detail and more were to be found later. At that stage, it was shown that compounds classified by W. B. Martin of Lexington as agonists exciting κ-receptors, represented by the ketazocines, were distinct from the agonists exciting μ-receptors, represented by morphine. These differences between the two types of agonists were also present in our models. The mouse vas deferens was throughout less sensitive to κ-agonists than the guinea pig ileum, and naloxone did not antagonize the κ-agonists as readily as the μ-agonists.

The most important task of the Unit was based on possibilities which we, and certainly also other workers in the field, had been considering for a number of years. The receptors in our models, the guinea pig ileum, the mouse vas deferens, and some other tissues, e.g. the cat nictitating membrane, were so well developed that we became more and more convinced that they were not designed only to interact with the alien alkaloids of the opium poppy. This view was strongly supported by the almost simultaneous findings of E. J. Simon in New York, S. H. Snyder in Baltimore, and L. Terenius in Uppsala, that membrane fragments of brain tissue had high and specific affinities for both agonist and antagonist narcotic analgesic drugs. Therefore, it was decided to search for compounds in the central nervous system that interacted specifically with the receptors in the peripheral models and the binding sites of brain membranes.

I discussed this problem with John Hughes who was then one of my lecturers in the Department of Pharmacology. He was enthusiastic about the scientific possibilities; after the university authorities made it possible for him to retain his official status, he joined the Unit at the outset in October 1973.

In an effort of this kind, three aspects had to be considered with great care. First, it had to be as certain as possible that the concept of the presence of endogenous opioid ligands was correct. Second, the methods of extraction had to take into account the possible destruction of the compounds. There was no preconceived idea of the possible structure of opioid ligands: The assumption was that they were probably of relatively small molecular weight with some similarities to morphine and other opiates; for example, they would have a positively charged N-atom. If they were neurotransmitters or neuromodulators, they would probably be liable to destruction by enzymes present in the central nervous system. Third, a good monitoring system was essential to ensure an efficient control of the process of purification. In this latter respect, our situation was probably particularly favorable. In the two models, we had receptor systems in which the responsive tissue, the smooth muscle, acted as a minicomputer and gave the answer in about one minute; the specificity could be controlled by specific antagonists, some of which were quaternary and therefore particularly rapid in their action. We had also pairs of isomeric synthetic antagonists and thus could test for stereospecificity of the antagonism.

John Hughes extracted large quantities of pig brain which he collected at an unearthly early hour at the local slaughter house. Within a few weeks he obtained a crude extract in which our bioassay systems indicated the presence of opioid compounds beyond any doubt. At that stage, when the nature of the endogenous opioid material was unknown, one of the great worries was contamination by one of the many opiates investigated in the laboratory. For this reason, the room set aside for the isolation of the endogenous opioids had never been used for work on opiates. It was a relief when the solubility characteristics of the endogenous compounds indicated that we did not deal with basic alkaloid-like substances. The isolation of sufficient material for analysis and the identification of the structure took about two years. It was demonstrated quite early that the compounds in question were peptides, as had also been shown by Lars Terenius who discussed this problem with us during a two-month visit to our laboratories in the spring of 1974.

The sequencing of the suspected peptide-like structure was complicated because both thin-layer chromatography and electrophoresis falsely indicated that we dealt with a single compound. The Edman-degradation gave the sequence Tyr-Gly-Gly-Phe for the first four amino acids but Linda Fothergill of the biochemistry department of Aberdeen University found it difficult to allocate positions for the methionine and leucine which were shown to be present by amino acid analysis. At this stage, we were very fortunate in obtaining the collaboration of Howard Morris of the Imperial College of Science and Technology in London, who established by mass spectrometry the presence of the two pentapeptides, methionine-enkephalin and leucine-enkephalin.

One of the most surprising coincidences occurred shortly after Howard Morris had established the structure of the enkephalins. Derek G. Smyth of the National Institute for Medical Research in London had been invited to Imperial College to give a talk on the prohormone β-lipotropin originally isolated by C. H. Li of the Hormone Research Laboratory at San Francisco who also had established its structure. Derek Smyth showed the structure of several fragments of β-lipotropin. Among those was C-fragment, now more commonly known as β-endorphin. It is an understatement to say that Howard Morris was greatly surprised when he realized that the first five amino acids of this fragment had the sequence of methionine-enkephalin. Thereafter, Derek Smyth quickly established in collaboration with his colleagues at the National Institute for Medical Research the so far unknown fact that β-lipotropin$_{61-91}$ had a high affinity for the opiate receptor. We showed that it depresses the electrically evoked contractions of the guinea pig ileum and mouse vas deferens and that this phenomenon is reversed by naltrexone.

Our paper revealing the structure of the two pentapeptides was published in *Nature* on 18 December 1975. At the beginning of December we had posted copies of the printer's proofs to our friends, competitors in the field. Some time later I received a letter from Dr. C. H. Li, in which he described some of the events that had occurred. On December 11, 1975, Dr. Avram Goldstein of Stanford wrote a letter to Dr. Li asking him for peptide samples related to the pentapeptides derived from β-lipotropin. On December 16, Dr. Li showed this letter to Dr. R. Guillemin of the Salk Institute who visited Dr. Li on that day. I learned from a recent discussion with Dr. Guillemin that he commenced the isolation of α-endorphin (lipotropin$_{61-76}$) in October 1975, and succeeded in isolating the pure peptide. He had started the sequencing when we published the structure of the enkephalins.

Thus, within a few days of our publication, intensive work on the long-chain opioid peptides present in the pituitary and, as we know now, in certain regions of the brain, began in London, San Francisco, and La Jolla. As was to be expected, many pharmaceutical laboratories also synthesized the enkephalins and began to design many analogues, for which numerous patents have been registered.

Thus started the story of the opioid peptides. Many laboratories the world over are engaged in analyzing their distribution by various methods of assay and by immunohistochemical techniques, but we still know relatively little about their physiological functions. In this respect, the best guide is an analysis of the side effects of morphine, which would suggest physiological roles in the control of respiration, motor activity, mood, endocrine control, and probably many other functions. While morphine and the opioid peptides or their analogues can be used as analgesics, it would appear that analgesia is a pharmacological effect rather than a physiological function. It is of considerable interest that we are not dependent on the opioid peptides as shown by the lack of a withdrawal syndrome after the injection of the antagonist naloxone; on the other hand, they will produce such dependence after prolonged administration to animals. It will take many years before we shall have a real understanding of the significance of these and other peptidergic neurones. Much information will be gained from a study of their biosynthesis, metabolic inactivation, and the receptors involved. Moreover, these systems are sufficiently complex that disturbances may well lead to pathological disorders.

Before leaving the subject of the opioid peptides, I would like to put on record that this work would have been impossible without the enthusiastic and outstanding collaboration of my young colleagues and friends. John Hughes is well known for his excellent achievement when he undertook the hard work of isolating the peptides. Others, who are or have been members of the Unit, are Maureen Gillan, Graeme Henderson, Michael Hutchinson, Frances Leslie, John Lord, Alexander McKnight, Alan North, Stewart Paterson, Linda Robson, Terence Smith, Roberto Sosa, and, last but not least, Angela Waterfield. My recent election as a Fellow of the Royal Society of London may be taken as a recognition of the successful work of the team. It also gave me great pleasure when John Hughes and I were given Pace-Setter Awards by the National Institute on Drug Abuse and I received the Schmiedeberg Plakette of the German Pharmacological Society.

At this point, it may be of interest to relate a conversation I had with A. Gilman about a year ago. He thought that, apart from the scientific outcome, the work of the Unit has shown that collaboration between two people at opposite ends of their careers can be very productive, provided their personal characteristics are complementary. I fully agree with this sentiment. We have found it important that all hierarchical gradings usually associated with differences in age have to be suppressed uncompromisingly.

Our success would have been impossible without the generous support of the UK Medical Research Council, the US National Institute on Drug Abuse, the US Committee on Problems of Drug Dependence, and again, last but not least, the authorities of the University of Aberdeen.

It is generally accepted that there are mainly two possibilities regarding the selection of one's research interests. The first leads to a high degree of expertise in a well-circumscribed area in contrast to the second where the research area is changed at more or less irregular intervals. Temperamentally, the second variant suited me better, probably because such a change creates a challenge to become familiar with new concepts and then try to compete as a newcomer. This may lead to complications which can be worrying at the time but, I believe, are helpful in postponing the inevitable losses in flexibility and adaptability.

Looking back on my own professional career, I am aware that such decisions were often made for me by factors over which I had no or only very little control. I had planned to become an academic physician who would be active in clinical research. Circumstances prevented me from pursuing this goal. Then I became interested in the basic aspects of galactose metabolism and thought of following a biochemical career. It was probably lack of courage to live for a number of years on grant money which made this impossible for me. Thus I became a physiologist and was active as such for twenty-five years until about ten years ago when I moved to the new Chair of Pharmacology, from which I retired after five years. Finally, the exciting new developments induced me to continue with my research on a scale sufficiently large to make the effort viable and productive. When all these circumstances are considered it is not surprising that, when looking for a title for this prefatory chapter, I recalled the quotation from a poem with which Robert Burns addressed a mouse: "The best laid schemes o' mice an' men gang aft agley."

Hermann Blaschko

Ann. Rev. Pharmacol. Toxicol. 1980. 20:1–14

MY PATH TO PHARMACOLOGY

H. K. F. Blaschko

University Department of Pharmacology, South Parks Road,
Oxford OX1 3QT, England

When I first came to Oxford I was occasionally surprised when I heard myself addressed as a pharmacologist, for when I began my research career it was in a very different field. Of course, nothing has really changed in the forty-five years since I began my work. What has probably changed is pharmacology: It has moved closer to my own interests.

Here I want to tell how it all started. First of all, I was most fortunate in my home background. I come from a family of doctors. When I began to demonstrate in the histology class at Cambridge, after my arrival there in 1934, I was surprised to find that the students immediately accepted me as an expert. Eventually I discovered that their text, Schafer's *Essentials of Histology* (1), contained a picture with the legend: "Section of Skin of Heel (Blaschko)." I cannot remember now if I ever revealed that this was a reference to Alfred Blaschko (1858–1922), my father who, as a young man almost fifty years earlier, had published a paper on the anatomy of the human skin from Waldeyer's laboratory in Berlin (2). A picture of my father has recently appeared in the *British Journal of Dermatology* (3) in connection with a paper on the "lines of Blaschko" (4).

By the time I was old enough to remember, my father's chief interests had shifted to problems of social medicine, especially the control of leprosy and of venereal diseases. He was a well-known figure in Berlin, especially in the working-class districts where he had practiced all his life and where he had been a pioneer in the field of social insurance; however, he always retained his interest in natural science and its modern developments, and we learned much from him. He was a keen naturalist and he knew the flora of his native countryside, the area around Berlin, intimately. We were very inquisitive, and he was always conscientious in finding answers to our innumerable questions.

1441

0362-1642/80/0415-0001$01.00

Many of his friends were well-known doctors and scientists. I never knew Paul Ehrlich, who gave him salvarsan for trials before it was released for general practice, but I met Albert Neisser (1855–1916), the discoverer of the gonococcus. Most important for me was Arnold Berliner (1862–1942), the founder of *Naturwissenschaften,* a magazine responsible for much of my early introduction to science. It was through Berliner that we met Max Born (1882–1970), the physicist, who in his recently published autobiography has lovingly described the influence my father exerted on him (5). Much later Max Born's son Gustav, now professor of pharmacology at King's College London, worked in my laboratory here at Oxford.

Through Max Born I have met many of the great physicists of our time; including Albert Einstein, James Franck, Enrico Fermi, and Fritz London.

My own introduction to laboratory work came early. During the First World War we were allowed to take our school-leaving examination early provided that we intended to do work considered essential to the war effort. I was not keen to go into a munitions factory; instead my father found a place for me in the spring of 1917 with Nathan Zuntz (1847–1920), a pupil of Pflüger, who was then director of the Institute of Animal Physiology at the Agricultural Academy in Berlin. Years later, at Cambridge, Joseph Barcroft told me that he had gained his first experience of high altitude physiology with Zuntz. That had been in 1910, on an expedition to the Pic of Tenerife (6). Later, in 1924, I accompanied Zuntz's principal co-worker, Adolph Loewy, on his studies at Davos, in the Engadin and on the Jung-fraujoch (7).

As a laboratory assistant in 1917, I learned simple techniques like weighing and titrating. Zuntz's task was to report on the nutritional value of innumerable ersatz foodstuffs. I had to carry out many Kjeldahl determinations and Soxhlet extractions. I did not mind the repetitive nature of this part of my job, and I enjoyed the freedom of the laboratory atmosphere. There were occasional distractions. For instance, I had to collect the frogs for the physiology class which were easily caught in the flood meadows near the Berlin rivers. On one occasion, based on the suggestion that the seeds of shepherd's purse (*Capsella bursae-pastoris*) might serve as a useful source of oil, I was given the task of collecting seeds in the cobbled streets near our home, a back-breaking job that took several hot and sunny days in 1917. I then did the determinations in the laboratory and found that the seeds were indeed rich in fat; however, it was concluded that the task of collecting them was too onerous.

I enjoyed my medical studies which I started, at first part-time, in the autumn of 1917. During my time as a medical student I formed friendships that have continued throughout my life. The medical course at Berlin was old-fashioned, still based firmly on anatomy. For two winters we were kept busy dissecting the human body, an occupation that I found tedious at first but that began to interest me as I became more skillful. Unfortunately the lectures were boring. Waldeyer had just retired at the age of eighty—I still remember listening to him once in the University, where he gave a course on anatomy for artists, and I can see the old man drawing on the blackboard a picture of a neuron, the structure to which he had given that name many years before. The new professor was R. Fick, a son of the famous physiologist and a very poor lecturer. Our professor of histology was Oskar Hertwig, who had first described fertilization of a sea urchin's egg. He was old and ailing and often absent; his deputy at that time was Poll, to whom we owe the term *phaeochromocytoma*. I most enjoyed the lectures of the two botanists: Haberlandt was the plant physiologist, and Correns, the geneticist, one of the rediscoverers of Mendel's laws. I met Correns again as the director of the Dahlem Institute, which I joined in 1925.

I am afraid I was not a model student. I used to go to the lectures that interested me but would skip those that I found unprofitable. German students enjoyed privileges that we very much cherished. There was much freedom in the choice of lectures, a privilege I made much use of. For instance, I went to a course of popular lectures by Albert Einstein, and I took part in a seminar led by the psychologist Wertheimer, one of the founders of Gestalt theory. I became quite friendly with him. I had first met both Wertheimer and Einstein through Max Born.

German students also had freedom of movement. At the end of each semester we were free to move from one university to another, and so I divided my time between my native city, Berlin, and the university of Freiburg im Breisgau, where I spent four of my ten semesters.

In Freiburg I found some of the professors more interesting than in Berlin. In contrast to Berlin, Freiburg had a Department of Biochemistry, under F. Knoop, whose lectures and practical classes I found useful. I was most attracted by the physiologist, Johannes von Kries (1853–1928), a pupil of Ludwig and Helmholtz. At the end of my preclinical period, I did most of the experiments for my MD thesis in his laboratory; I must have been his last student.

I stayed on for a while in Freiburg after my preclinical period ended. I attended the pharmacology lectures given by W. Straub. He was a little uneven as a lecturer, but when he was in good form he was worth listening to, and I did profit from what I learned from him. He took much care over the demonstrations that accompanied his lectures. By far the most stimulating academic teacher that I have encountered was Ludwig Aschoff, the pathologist. Not only was he a very good morbid anatomist, but he had assimilated modern cytological and biochemical ideas, and was outstanding at conveying his own thoughts to the students.

Because I found the system of clinical instruction boring, my attendance at the clinics was rather poor. In German medical education, clinical work, like the preclinical course, was based on the university terms and on lectures, and lecture demonstrations. The professor demonstrated a case, had a proforma dialogue with a few selected students, and after a cursory examination of the patient lectured for the remainder of the hour on the subject of the disease he had just shown. There were some good academic lecturers among the professors, e.g. the ophthalmologist Axenfeld in Freiburg and one or two of the internists in Berlin, but on the whole they were in the minority, and altogether this method of teaching was very wasteful. Most of what we learned of clinical medicine was picked up during the long vacations between the semesters, when we hired ourselves out as assistants, chiefly in the municipal hospitals that were often understaffed and where there were greater opportunities for work involving responsibility. In Freiburg the clinical students, dissatisfied with the curriculum, appointed a committee for curriculum reform. I was briefly a member of this committee until my father's illness necessitated my return to Berlin. In March 1922, shortly before his death, my father had a visit from Abraham Flexner. I told Flexner of my interest in curriculum reform, and he told me of his own studies of medical education in Europe and the United States. I still own the two volumes, published under the auspices of the Carnegie Foundation, that he sent me after his return to the United States (8, 9).

At the end of my preclinical studies I was not sure whether I should continue my medical course or switch to biology. I am glad I followed my father's advice to remain in the course and get my medical qualification. Much of my later life has been spent teaching medical students, and my own work has gained by an awareness of its relevance to medicine.

After my final examinations in 1922, I had to spend a year in a hospital as an intern to gain my qualification. On the advice of Max Born, I decided to go to Göttingen, where I spent about 18 months in the Medical Clinic of the University Hospital, under Erich Meyer (1874–1928). Eric Meyer was an outstanding representative of the great German tradition of clinical medicine and was the best clinical teacher I ever had. His own outlook and his teaching were based upon his understanding of modern physiology, biochemistry, and pharmacology, and he was able to use this understanding as the basis for his clinical lectures and presentations, which were models of clarity. I owe him much, and he gave me the opportunity for responsible work, unusual in Germany at that time. So, before I left Göttingen I was put in charge of one of the medical wards for the periods when my immediate superior was on holiday. Also, Erich Meyer sent me to the small town of Alfeld an der Leine, halfway between Göttingen and Hannover, where a severe typhoid epidemic had broken out, in order to assist one of the two local practitioners there. The few weeks that I spent there in the winter of 1923–1924 were full of experiences that have inprinted themselves firmly in my memory, maybe because this was almost my last contact with the practice of medicine.

Göttingen at that time was one of the world's great centers of mathematics and physics, and through the Borns I met many of the people whose names are now familiar to all students of modern science.

A coincidence determined my subsequent history. One of Born's friends was Richard Courant (1888–1972), professor of mathematics with whom I became friendly (see 10). One day he asked me, "What are your plans for the future?" Haltingly I tried to describe to him that I would like to do research, and if possible in the borderline field between physiology and biochemistry. After a little thought he asked me, "Have you ever heard of Otto Meyerhof?" I told him I had read an article of his in *Naturwissenschaften*; that was all. He told me, "He might be the right man for you; I know him because we worked for a while in the same office during the war. If you would like me to, I could write to him."

Shortly afterwards a colleague introduced me to a friend of his who was on a visit to Wolfgang Heubner, our professor of pharmacology. This friend was Rolf Meier (1897–1966), on leave of absence from Heubner's institute in order to work at Kiel with Otto Meyerhof. I still remember Rolf Meier's tales of the laboratory at Kiel; he was convinced that this would be the right place for me.

Rolf Meier eventually became a director of CIBA at Basel, and he is remembered by many pharmacologists as the founder of the firm's modern pharmacological laboratory.

The coincidence of these two experiences was much in my mind when I left Göttingen. I first went to Switzerland, where I finished my work with Loewy on the Jungfraujoch, and I then went to a meeting at Innsbruck. I have recently described my first experience there of Otto Loewi (11). When I returned to Berlin I found the search for a place to work a discouraging experience. So I remembered Courant's promise and I asked him to get in touch with Meyerhof on my behalf. He promised to do so.

The outcome was that on January 1, 1925, I started work at Dahlem in Meyerhof's laboratory in Berlin. Meyerhof had received the Nobel Prize in 1923 jointly with A. V. Hill, and he had moved from R. Höber's institute at Kiel to the Kaiser Wilhelm Institute for Biology a few months before my arrival.

Much has been written in recent years about the Dahlem of the 1920s (12, 13). As for myself, I could not have arrived at a better place, and most of my remaining time in Germany was spent in Meyerhof's laboratory, first at Dahlem and later at Heidelberg. My active period of work with Meyerhof was twice interrupted by longish periods when I was laid up with pulmonary tuberculosis, with deleterious effects upon my output.

I have always considered my period of apprenticeship with Meyerhof as one of the great formative experiences of my life. The work was determined by Meyerhof's research interests, which centered around the energetics of muscle. There were usually only about four or five people in the department, most of these visitors from abroad. At first I found Meyerhof a little unapproachable; he spent most of his working day at the bench, and one did not like to interrupt him too often, but I remember a number of occasions when we discussed the general implications of the work I was doing. Once his interest was aroused and he began to "think aloud," it was possible to have a long and fruitful conversation that left one with a feeling of satisfaction and eager to carry on with the program of work.

Much of the stimulus of the laboratory came from my contemporaries. Meyerhof's laboratory has given me many good friends.

When I came to Dahlem, Karl Lohmann (1898–1978) was already in the laboratory. He is best remembered as the discoverer of ATP, but he was a great biochemist, who made many important discoveries, all of them in the stimulating environment of Meyerhof's laboratory. Lohmann was a modest man. I owe him much; in a quiet and unobtrusive way he helped me along and taught me chemical technique. And later there came David Nachmansohn, Ralph Gerard, Frank O. Schmitt, Severo Ochoa, Fritz Lipman, Karl Meyer, and many others. Under the same roof, with Otto Warburg, there was Hans Krebs whom I had already known when we were medical students at Freiburg in 1919. We had a good time together, and the continuous contact with bright and lively contemporaries is one of the gains from my period with Meyerhof.

Dahlem was an exciting place. The Institute of Biology had several other departments, in addition to Meyerhof's and Warburg's. The director was Correns and there was another geneticist, Rudolf Goldschmidt. Viktor Hamburger, later professor of zoology at Washington University, had the laboratory next to mine. Close by there were the Institutes of Chemistry, with Hahn and Meitner; Physics with Haber, Freundlich, and Polanyi; and Biochemistry with Neuberg. We had a seminar in our own institute; I particularly remember an exciting talk by Correns on non-Mendelian inheritance, and another one by Jollos on relative sexuality. Then there was the renowned seminar presided over by Haber which we attended regularly. This was a great occasion, and we admired Haber's grasp of a great variety of subjects and his ability to force the speakers to speak clearly and understandably. Most of the topics discussed I have forgotten. One of Haber's aims at the time was to extract gold from sea water, in the hope of paying the reparations that had been imposed on Germany at the end of the war. One occasion that I still remember is Freundlich's talk in which he described and named the phenomenon of *thixotropy.*

On the other hand, we had no contacts with the university departments; we were separated from these by a considerable distance.

Meyerhof, although older than Warburg, had begun research in Warburg's laboratory; that had been before the war. Meyerhof came late to science, having first studied philosophy and only later taking up medicine, I believe, in order to become a psychiatrist. When he met Warburg in Heidelberg, the latter persuaded him to join his laboratory. When I went to Meyerhof in 1925 he still very occasionally did some work on topics that were essentially in Warburg's sphere of interests. My paper must have been the very last in this line (14). I studied the reversibility of the cyanide inhibition of a number of so-called autoxidation reactions, reactions that, as Warburg had shown, were in reality metal-catalyzed. This modest piece of work later assumed great significance for me when I chose my own line of study.

Of course, all the work of the department was determined by Meyerhof. I remember that while we were still in Dahlem, I asked him if I could do some work on the effect of caffeine on muscle. He was not enthusiastic about this, for he thought this might become too much like pharmacology! In view of the work done on the effect of caffeine in recent years, it still seems to me that we might have discovered something worthwhile. Also, I might have had my introduction into pharmacology at an earlier stage.

In Dahlem Meyerhof could offer only one salaried position; this was held by Lohmann. When he accepted me in 1925 he warned me that he would never be able to pay me a salary. I told him I had not expected a paid job; at that time it was quite usual to have no pay during one's apprenticeship. After a few months he told me that he would be able to give me a modest grant. However, when in 1928 I was asked by E. von Skramlik, who was going to Jena as Professor of Physiology, whether I was interested in the University assistantship vacant in his Institute I thought I should accept his offer. I had met von Skramlik in Freiburg where he had been a lecturer in the Institute of Physiology.

The chair in Jena had become vacant through an unusual accident. It had been held for many years by Biedermann, who had done good work in electrophysiology as a young man. One of the assistants, Professor Noll, wanted to write a congratulatory article on the occasion of Biedermann's seventieth birthday, but there was some doubt about the exact date of his birth. The University authority gave a date in summer, but Biedermann seemed to remember that his mother had told him his birthday was in

January. Noll thereupon wrote to the Burgomaster of Bilin in Bohemia, Biedermann's birthplace, to find out. The Burgomaster wrote back saying he was pleased to learn that a son of his town had done well, and that Professor Biedermann was quite right: his birthday was in January. However, he also had to tell Professor Noll that the birthday that Biedermann was going to celebrate was not his seventieth but his seventy-second! The congratulatory article was never written, and Biedermann, when he was told, was delighted and retired immediately.

For me, the time at Jena might have been a complete loss had I not been given much of the responsibility for preparing the demonstrations for the Professor's lectures and also for the work in the practical classes. This experience I found useful when I came to Cambridge in 1934, where I had to demonstrate in the practical classes in physiology.

The gloomy prospects for my future at Jena were relieved when Meyerhof offered me a position in the new Kaiser Wilhelm Institute for Physiology at Heidelberg, where he moved in 1929. I gladly accepted, but I was first sent for one year to University College London, to learn methods in the laboratory of A. V. Hill, with whom Meyerhof had shared the Nobel Prize in 1923.

My first piece of good luck was to have been raised in the right family environment, the second was to have been accepted by Otto Meyerhof. The third stroke of luck was to work with A. V. Hill. I had been happy enough in Meyerhof's laboratory, but the spirit of freedom and friendship that I experienced during my year with A. V. Hill, 1929–1930, was exceptional. The close and friendly contact with Hill was very different from what I had experienced before. Every day, when he came to the lab, he used to come to my room, and we discussed everything that seemed to matter, from the leading article in the *Times* to the experiment to be done. One was not only entirely in the picture about the general implications of what one was doing but also aware of everything important that was happening in the world. My relationship with A. V. Hill was an experience that I still treasure and that was reinforced in the years to come. Also, I owe to Hill the first independent piece of research that I have done. During his visit to the International Physiological Congress in Boston I worked in Plymouth at the Marine Biological Laboratory on the mechanism of facilitation in the adductor muscle of the crustacean claw (15). I returned to Plymouth again in the summer of 1930. In 1929 I worked there with J. L. Kahn, a visitor to A. V.'s laboratory from Moscow, and in 1930 with McKeen Cattell, later Professor of Pharmacology at Cornell Medical School. The piece of work that Kahn and I had started in 1929 owes much to Cattell's great knowledge and skill.

I visited Meyerhof once during this period, a few days before Christmas 1929, just after he had moved to his still unfinished Institute in Heidelberg. He told me of a letter that he had from Denmark, from Einar Lundsgaard (1899–1968), who reported his observations on the alactacid contraction of the muscle poisoned with iodoacetic acid (16). The school led by Gustav Embden had maintained for some time that the lactic acid formation was not coincident with the muscle's contraction, but followed it. Meyerhof had always stoutly repudiated that claim, chiefly because the evidence brought forward was considered inadequate. But Meyerhof could not see any flaw in Lundsgaard's experiments, and I was impressed with his readiness to give up his long cherished ideas. He told me that Lundsgaard was expected to arrive in Heidelberg in the spring of 1930.

When I came to Heidelberg, in May 1930, Einar was already there. He told me that when the first experiment did not come off, Meyerhof again became doubtful, but that the second experiment worked! As a matter of fact, I was able to clear up that point after my return: The critical effective concentration of iodoacetate in winter frogs and in summer frogs differed by a factor of about ten.

Lundsgaard was the first of many visitors to Meyerhof's Heidelberg laboratory. He was one of the most impressive of all my contemporaries. We soon became friends, and his friendship lasted until his untimely death in 1968. Although his great discovery of 1929 is probably his best-known work, he made other contributions of great importance. For instance, I remember the paper he read to the International Physiological Congress at Rome, in 1932, in which he showed that phosphocreatine breakdown could also be made to lag behind muscular contraction, provided the experiments were carried out at low temperature. It was this discovery that pointed to ATP breakdown as the ultimate source of energy for contraction, although Lundsgaard was too careful to say so. In 1930 I was fortunate to have Lundsgaard as a traveling companion to Plymouth. There he studied the effect of iodoacetate on the crustacean muscle, where phosphoarginine replaces phosphocreatine (17).

In the much bigger laboratory at Heidelberg, Meyerhof was able to accommodate more visitors, and the list of those who worked with him there covers many names well known in biochemistry: Severo Ochoa, Alex von Muralt, Andreé Lwoff, and George Wald. The other assistant, Hans Laser, is still at work at Babraham (Cambridge) at the time of writing.

My own output at Heidelberg was again affected by illness. Little of the work that I did there was published; much of it appears in the form of footnotes in other people's papers. The advent of the Nazi regime in 1933 found me again in hospital, this time at Freiburg im Breisgau, where Hans

Krebs was one of my physicians, a particularly apt coincidence, since Freiburg was the place where we had first met in 1919. While convalescing I improved my skill in manometry in his laboratory.

While still in hospital, a letter from A. V. Hill arrived, inviting me to return to University College London. I did not hesitate, and so in May 1933 I was again installed at University College. I could afford to do this, as my English relatives were willing to offer me hospitality.

So here too was another bit of good luck. I had been very happy in my first year at University College and had made many friends in England. Thus my emigration was unlike that of many of my contemporaries: It felt more like a homecoming. Also, in the last year or two I felt that I had been with Meyerhof long enough, and I was ready for the day when I could start finding my own field of research.

That day, however, was yet to come. I was still convalescing, and most of that year in London I was busy assisting A. V. Hill with his work for the Academic Assistance Council (now the Society for the Protection of Science and Learning). This body was founded by Archbishop Temple, Sir William Beveridge, and A. V., as an organization to help refugee scholars. This organization still serves a much needed function in a world that is still a far from perfect place.

In 1933 it was necessary to interview new arrivals, mainly from Germany, to help them with their initial language difficulties and to help A. V., who interviewed many of them for the purpose of directing them to suitable places. This was work that at the time took precedence over my own research. Also, it taught me a new aspect of A. V.'s conception of the scientist's duty, an experience that has stayed with me throughout my life.

By 1934 I was fit to start doing some work of my own. A. V. Hill and Hans Krebs decided that I should leave London, and so I was glad to accept an invitation from Joseph Barcroft (1872–1947), then Professor of Physiology at Cambridge; at the same time I was given a grant from the Academic Assistance Council. I held this grant for two years, and although I did not get a permanent position until 1946, I managed from then on to maintain myself by teaching, both in the Physiological Laboratory and through giving tutorials.

Coming to Cambridge was another of those fortunate events that happened to me at the right moment. The German centers of learning were already under the threat of extinction, and what Dahlem and Göttingen had been in the 1920s, I found again at Cambridge in the 1930s. I did not have much contact with the physicists except that I had the use of Rutherford's glassblower, the latter also a German expatriate, but the Physiology Department with Barcroft, Adrian, Roughton, Matthews, Rushton, and Willmer was the Mecca for physiologists. Biochemistry under Hopkins had equally become the leading center in the world. I became particularly friendly with David Keilin (1887–1963), who at the Molteno Institute had built up a fine center of enzyme research. I had constant support from him, particularly when I was despondent about my future.

I began with some work on catalase and catalase inhibitors (18, 19). I had bought, with my last German money, a manometer bath and some Warburg manometers, and it was this precious possession that for quite a while dictated my choice of topics. Barcroft, who, jointly with J. S. Haldane (20), had been one of the initiators of manometry, was interested in what I was doing. I much enjoyed my close contacts with that interesting and lovable man. My laboratory adjoined his, and at the end of the day he often showed me the results of his experiments. One day he asked me, "How is adrenaline destroyed?" He had seen some differences in the time course of the pressor response to adrenaline in mother and fetus. I knew nothing about adrenaline, but I promised to look it up in the library. After all, adrenaline was a substance that had been around for a long time, and I felt certain that the answer to Barcroft's question could readily be looked up in the library. However, when I went to the Biochemistry Library, I soon discovered to my surprise that the fate of adrenaline was not known. The textbooks said adrenaline was a readily autoxidizable substance, and it was obvious that its actions were evanescent.

Autoxidation was what I had been concerned with in Dahlem in 1925, and I thought that my experience might possibly be useful in helping to answer Barcroft's question. Moreover, I was sharing a laboratory with Hans Schlossmann, until recently a lecturer in Pharmacology at the Medical Academy of Düsseldorf, who had just written a long article on methods of bioassay (21). So I proposed to Schlossmann that we might have a go at this problem.

My hunch that the Dahlem experience from 1925 might prove useful was soon proved to be right. The "autoxidation" of adrenaline was readily inhibited by cyanide. However, when one incubated the adrenaline with tissue homogenates, an inactivation of adrenaline remained that was resistant to cyanide. This cyanide-insensitive inactivation reaction was an oxidation reaction in which half a molecule of oxygen for each molecule of adrenaline was consumed. It could easily be shown that this inactivation had the characteristics of an enzymic reaction.

This is how the action of monoamine oxidase on adrenaline was discovered. We were fortunate: A few weeks after we started we were joined by Derek Richter, who was working in the Dunn School of Biochemistry, under Hopkins. It was Hans Krebs who introduced us to Derek, who had taken his PhD with O. Wieland at Munich, where he had become familiar with oxidation reactions.

We soon established that both noradrenaline and dopamine were oxidized similarly to adrenaline. Derek Richter knew of earlier work, carried out in Hopkins' laboratory by Miss Hare (later Mrs. Bernheim) on "tyramine oxidase" (22, 23). In a few weeks the identity of our enzyme with tyramine oxidase was established (24, 25). Actually, Mrs. Bernheim had tested adrenaline as a possible substrate, but without success. I suspect she too had run into difficulties over the autoxidation, like some other earlier authors.

We had been most fortunate in the timing of our work. The acceptance of the idea of humoral transmission of nervous impulses had prepared the way for biochemical studies in this field, and I had immediate reactions from many physiologists and pharmacologists; these included, as I have recently described (11), Otto Loewi, then still professor of pharmacology at Graz.

The work of many observers was required before the physiological role of monoamine oxidase in the biological inactivation of adrenaline could be fully established. This is not the story I wanted to tell here; my aim has been to describe how I was led to pharmacology.

However, my tale is not yet quite complete. I was still in Cambridge, in the physiological Laboratory. I was still without a job, and from 1937 onward I was working again on my own. I had, most of the time, no laboratory assistant, and I had to do all my own work, including the washing of glassware. At the beginning of each term I had to wait until I knew whether there would be enough teaching for me to earn my living. However, when the War began and the more active and enterprising people left for more exciting work, I even made myself quite useful. In addition to the work in the laboratory, I acted as supervisor in physiology for St. John's College, an occupation I enjoyed because it brought me into closer contact with younger people. When Joseph Barcroft came back from his war work in 1942 at the age of seventy, I even got a part-time research grant from the Agricultural Research Council.

In the autumn of 1943 I had an offer from J. H. Burn to join him in his department at Oxford. He had just returned from the United States, where he had worked as a reporter for the Medical Research Council. There he had witnessed the upsurge of biochemistry and the contributions that biochemical methods had made to pharmacology. These experiences are described in his letters to the Medical Research Council. When I moved to Oxford early in 1944 copies of these letters were precious reading matter for us, because we had been very much cut off from contacts with our colleagues overseas.

Burn had promised me an established position at Oxford and it was this offer that determined my move to Oxford. Also, I had had a very large teaching load at Cambridge during the war, and I was very tired. The job at Oxford was mainly a research appointment.

So, at the age of 44 I was for the first time housed in a Department of Pharmacology. The established post had to wait until 1946, since once again I was laid up and out of action for a period of nine months. That was the last time my work was interrupted by a spell of pulmonary tuberculosis. However, thanks to the help given by Ruth Duthie and Isabel Wajda the work in the laboratory did not entirely come to a standstill while I was out of action. From then onwards I never again worked entirely on my own.

The present story comes to an end with my transfer to Oxford. As far as my research work was concerned this made little difference at first. I had brought two major problems with me from Cambridge, and they were both essentially pharmacological: monoamine oxidase and adrenaline biosynthesis. Both these lines were continued. Also, shortly before I left Cambridge, I had discovered an enzyme, L-cysteic acid decarboxylase (26), and we continued to do some further work on this enzyme (27, 28). However, this line was not continued. The biological significance of taurine is a problem that remains unsolved.

Berlin, Freiburg, Göttingen, Dahlem, London, and Cambridge, these were the stages that brought me to Oxford and to pharmacology. I can trace my interest in pharmacological problems to a very early period, but it took a long time for me to get there. In the meantime I have been very fortunate in having had many close personal encounters with outstanding people, from an early age onward and right through my development. These experiences not only have helped to shape my own outlook but also have enabled me to hand on a tradition to those who are continuing where I left off.

Literature Cited

1. Sharpey-Schafer, E. 1934. *The Essentials of Histology.* ed. H. M. Carleton, p. 281. London, New York & Toronto: Longmans, Green. 13th ed.
2. Blaschko, A. 1887. Beiträge zur Anatomie der Oberhaut. *Arch. Mikroskop. Anat.* 30:495–528
3. Jackson, R. 1977. Correspondence. *Br. J. Dermatol.* 97:341–42
4. Jackson, R. 1976. The lines of Blaschko: A re-evaluation. *Br. J. Dermatol.* 95:349–60
5. Born, M. 1978. *My Life.,* pp. 178–79. London: Taylor & Francis. 308 pp.
6. Franklin, K. J. 1953. *Joseph Barcroft 1872–1947,* pp. 70–74. Oxford. Blackwell. 381 pp.
7. Loewy, A. 1925. Beiträge zur Physiologie des Höhenklimas. *Pflüger's Arch.* 207:632–70

8. Flexner, A. 1910. *Medical Education in the United States. Bull. Carnegie Found. Adv. Teaching,* No. 4, New York
9. Flexner, A. 1912. *Medical Education in Europe. Bull. Carnegie Found. Adv. Teaching,* No. 6, New York
10. Reid, C. 1976. *Courant in Göttingen and New York. The Story of an Improbable Mathematician.* New York, Heidelberg & Berlin: Springer. 314 pp.
11. Blaschko, H. 1978. Early meetings. *Trends Neurosci.* 1:IX–X
12. Nachmansohn, D. 1972. Biochemistry as part of my life. *Ann. Rev. Biochem.* 41:1–28
13. Krebs, H. A., Lipmann, F. 1974. Dahlem in the Nineteen Twenties. In *Lipmann Symposium: Biosynthesis and*

Regulation in Molecular Biology, pp. 7–27. Berlin & New York: de Gruyter.

14. Blaschko, H. 1926. Über den Mechanismus der Bläusaurehemmung von Atmungsmodellen. *Biochem. Z.* 175: 68–78

15. Blaschko, H., Cattell, McK., Kahn, J. L. 1931. On the nature of the response in the neuromuscular system of the crustacean claw. *J. Physiol. London* 73:25–35

16. Lundsgaard, E., 1930. Untersuchungen über Muskelkontraktionen ohne Milchsäurebildung. *Biochem. Z.* 217:162–77

17. Lundsgaard, E. 1931. Über die Bedeutung der Arginin-phosphorsäure für den Tätigkeitsstoffwechsel der Crustazeenmuskeln. *Biochem. Z.* 230:10–18

18. Blaschko, H. 1935. The mechanisms of catalase inhibitions. *Biochem. J.* 29:2303–12

19. Blaschko, H. 1935. Cell respiration and catalase activity. *J. Physiol.* 84:52P

20. Barcroft, J., Haldane, J. S. 1902. A method of estimating the oxygen and carbonic acid in small quantities in blood. *J. Physiol.* 28:232–40

21. Schlossmann, H. 1935. Technik der Pharmakologischen Analyse. In *Handbuch der Biologischen Arbeitsmethoden,* ed. E. Abderhalden, IV, 7B. pp 1695–1780

22. Hare, M. L. C. 1928. Tyramine oxidase. I. A new enzyme system in liver. *Biochem. J.* 22:968–79

23. Bernheim, M. L. C. 1931. Tyramine oxidase. II. The course of the oxidation. *J. Biol. Chem.* 93:299–309

24. Blaschko, H., Richter, D., Schlossmann, H. 1937. Enzymic oxidation of amines. *J. Physiol.* 91:13P

25. Blaschko, H., Richter, D., Schlossmann, H. 1937. The inactivation of adrenaline. *J. Physiol. London* 90:1–17

26. Blaschko, H. 1942. L-Cysteic acid decarboxylase. *Biochem. J.* 36:571–74

27. Sloane-Stanley, G. H. 1949. Aminoacid decarboxylases of rat liver. *Biochem. J.* 45:556–59

28. Hope, D. B. 1955. Pyridoxal phosphate as the coenzyme of the mammalian decarboxylase for L-cysteine sulphinic and L-cysteic acids. *Biochem. J.* 59:497–500

K. K. Chen

Ann. Rev. Pharmacol. Toxicol. 1981. 21:1–6

TWO PHARMACOLOGICAL TRADITIONS: NOTES FROM EXPERIENCE

K. K. Chen

Department of Pharmacology and Toxicology, Indiana University School of Medicine, Indianapolis, IN 46223

I was born at the end of the Manchu Dynasty, a period of national decay. The Empress Dowager had usurped the power of the government and, with money appropriated to build a navy for China, she constructed a luxurious summer palace. Japan and western nations took advantage of Chinese weakness by establishing wide interests for themselves in China. Desirous of expelling foreign influence, the Empress Dowager encouraged a group of civilians to murder foreign diplomats and Christians, later known as the Boxer Rebellion of 1900.

When I was 5 years old, my parents hired a tutor for me to memorize and pronounce individual characters written on sheets of paper about two inches square. It was merely for recognition of the words, not their meaning. After about one hundred characters were committed to memory, I was given a book of surnames (about one hundred in number). This was followed by four books and five classics. Slowly I was taught the meaning of sentences, starting with a collection of Confucius's teachings. I began to learn writing and composition with the hope of passing literary examinations, which unfortunately were abolished in 1908. This led me to enter an organized public school, reading elementary books of history, geography, and arithmetic. In 1916 I graduated from a middle school, and by competitive examination, was accepted to the third year class of Tsing Hua College of Peking. This school had been established by the United States in 1908 with a portion of the "Boxer Indeminity." Most of the faculty members were American scholars. They prepared Chinese boys to continue their

1457

0362-1642/81/0415-0001$01.00

education in American universities. I was thus exposed to the English language for the first time. I graduated in 1918 and joined the junior class at the University of Wisconsin.

It took me a short time to appreciate the generosity and kindness of my teachers and schoolmates. One of my classmates tried to help me improve my English by introducing me to a debating society (Hesperia), which also helped me gain confidence in speaking before an audience. To encourage my interest in Chinese materia medica, Edward Kremers had me steam distill 300 pounds of leaves and 200 pounds of twigs of *Cinnamomum cassia* that were imported from China. He asked me to obtain the cassia oils, but he published the paper entirely under my name (1). After I received a BS in pharmacy, I transferred to the School of Medicine. For personal reasons I completed only the first two years of medicine. Then I earned a PhD after graduate studies with Harold C. Bradley and Walter J. Meek in biochemistry and physiology.

In 1923 I returned to China, and through Bradley's recommendation to Carl F. Schmidt, I became a member of the pharmacology department of Peking Union Medical College (PUMC) and its hospital. This school was financially supported by the China Medical Board of the Rockefeller Foundation. In addition to teaching, I continued my study of Chinese herbal medicine. My uncle Zao-Nan Chou told me of the important properties of Ma Huang as recorded in Chinese dispensatories. Without a literature search, I purchased a sample of Ma Huang from a Chinese drug shop, and applied the phytochemical techniques I learned from Kremers' laboratory at Wisconsin. By use of immiscible solvents, I succeeded in isolating an alkaloid in crystalline form within a short time. Fortunately I found an early reference of Nagajosi Nagai (2, 3) in which he had named the base *ephedrine* in 1887. Stapf (4) of Kew Gardens gave the Linnean name of *Ephedra sinica* to Ma Huang.

Carl F. Schmidt directed the pharmacological investigation of ephedrine with his elegant techniques, and revised the manuscript we were preparing repeatedly, but refused to accept senior authorship (5–7). Our results were analogous to those of sympathomimetic amines of Barger & Dale (8). As compared with epinephrine, ephedrine has the advantage of oral effectiveness and long duration of action. Thus, clinically ephedrine is useful in allergic diseases, bronchial asthma, and hay fever and in spinal anesthesia, where it prevents the fall of blood pressure. Its mechanism of action has been thoroughly reviewed by Aviado (9). It has both α- and β-adrenergic effects. Its protracted pressor action is due to the inhibition of monoamine oxidase, and its tachyphylaxis following repeated injections of small doses may be the result of the depletion of norepinephrine storage.

C. F. Schmidt was my tutor for only one year because he had to return to the University of Pennsylvania to continue teaching and research in the Department of Pharmacology. My contract with PUMC was completed after the second year, when our Chairman, Bernard E. Read, returned from Yale University. At that time, in 1925, Arthur S. Loevenhart invited me to become an associate of his Department of Pharmacology at the University of Wisconsin. This year turned out to be a busy period: I completed credits for the third year of medical school; W. J. Meek assisted me in working on circulatory effects of ephedrine; William S. Middleton allowed me to serve as an intern while doing clinical work on ephedrine; and Loevenhart arranged through John J. Abel for me to finish my fourth year of medicine at Johns Hopkins University School of Medicine. After all the credits were approved in the Dean's Office, I was fortunate to graduate in 1927. Upon Eugene M. K. Geiling's recommendation Abel appointed me as an associate for two years. My teaching duties were light, for lectures were divided among the staff members, and the laboratory periods were assigned to few of us. Our contact with Abel was best at lunch hour in the laboratory because his wide knowledge often led to discussion of various subjects which benefited us all.

In 1912 Abel & Macht (10) obtained two crystalline substances, bufagin and epinephrine, from the skin secretion of a tropical toad *Bufo aqua*. It was my good fortune to discuss with him my intention to study a Chinese toad venom, Ch'an Su, used in Chinese materia medica. With a grant from the American Medical Association, I purchased 1729 kg from a Chinese drugstore in Peking. This commercial preparation of toad venom comes in round, hard black discs, 7 to 8.6 cm in diameter. Hans Jensen in the same laboratory joined me in the chemical investigation. We succeeded in isolating epinephrine, cholesterol, and two digitalis-like products later known as cinobufagin and cinobufotoxin (11).

In 1929 Eli Lilly and Co. wrote to Abel offering me Directorship of Pharmacologic Research. J. K. Lilly, Sr., and Eli Lilly, father and son, told me in an interview in Indianapolis that they would give me complete freedom of research, particularly in Chinese materia medica by buying crude herbs through their Shanghai Branch. Abel and I both realized the disadvantage of accepting the offer, which was that permanent employees of drug firms were not allowed membership in the American Society for Pharmacology and Experimental Therapeutics (ASPET). I had been elected a member of ASPET in 1925, and with a heavy heart, I resigned from membership in 1929. Abel again came to my rescue by nominating me to the American Physiological Society (APS) in 1929. By so doing, I could request transferring my paper to ASPET in its annual meetings. Indeed I presented a paper to ASPET every year. No rejection was exercised by either society in those days. It was not until 1942 that my membership was reinstated.

Upon my arrival in Indianapolis in 1929, I took full advantage of the opportunities at Eli Lilly and Co. to investigate Chinese materia medica. My work on the Chinese toad venom was continued and extended to 16 species of toads (*Bufo*) (12) collected from different parts of the world. By elegant methods of isolation as employed by T. Reichstein of the Institute of Organic Chemistry and K. Meyer of the Pharmaceutical Institute of the University of Basle, Switzerland, a single toad such as *B. gargarizans* can yield 25 derivatives of bufadienolide and cardinolide. We determined their potencies, each in 8–10 etherized cats. Some of the members, arenobufagin and bufotalidin, are even more potent than ouabain. In our work we expressed the "parotid" glands of 13,800 live toads. We found that the cat is the most sensitive laboratory animal to assess the digitalis-like activity of each pure compound. As with mammalin epinephrine, one can detect a trace of norepinephrine, 2–4% (13) from the Chinese toad epinephrine. While each *Bufo* species biosynthesizes its own bufodienolides and occasionally cardinolides, the larva of North American monarch butterfly and a North African grasshopper consume a glycoside, calactin, of a species of milkweed family, Asclepiadaceae (14).

With the collaboration of T. Reichstein of Basle, R. C. Elderfield of Ann Arbor, and several other organic chemists, we extended our investigation to many natural glycosides, aglycones, synthetic esters and glycosides, and Erythrophleum alkaloids—a total of 463 principles. In order to assay their cardiotonic activity we used 5100 cats. We concentrated on structure-activity relationships. Our clinical work on cinobufagin, thevetin, and acetylstrophanthidin was focused on their possible utility. Toward the end of 30 years, Harry Gold became interested in comparing a few of our compounds with digitoxin, ouabain, and digoxin. Economically, the last three glycosides can easily compete with the galenical preparations.

In my 34 years at Eli Lilly and Co., the list of Chinese herbs I selected was generously secured by the administration. γ-Dichroine, having a Q value of 148 (15) against *Plasmodium lophurae* in ducklings, almost became an important antimalarial drug. Unfortunately, it causes vomiting and hydropic degeneration in the liver of experimental animals. It seems very difficult to develop a superior product from Chinese herbs—much like Henry Wagner's recent experience (16), "Screening these herbal medicines is a little like drilling for oil." More successful results may be often achieved by collaborating with organic chemists and other specialists (17). Much aid was received from my wife. For example, she thought of using isopropyl alcohol to increase the yield of crystalline thevetin. No wonder Kahn (18) mentioned her contributions in the laboratory on the eve of our Golden Anniversary.

In 1937 I was invited by Indiana University School of Medicine to attend Friday evening seminars, give a few lectures in Pharmacology, and occupy space for research, particularly after my retirement at Lilly Laboratories. I surely appreciate the privilege of sharing the finest facilities in the large University now covering seven campuses in addition to the Indianapolis and Bloomington Centers.

In 1947 I was elected Treasurer of the ASPET. After five years of service ASPET nominated me President-Elect. Automatically, I became President next year and by rotation I served as the Federation President in 1953–1954. Being stimulated by the confidence of ASPET Council and Membership, I attempted to initiate certain projects for the good of the Society and the Federation.

The first item was the annual J. J. Abel Award offered by Eli Lilly and Co. to a young pharmacologist not older than 36 years of age, entirely administered by ASPET. At the spring meeting of 1980 the thirty-fourth candidate received this award.

At the Federation Meeting I asked Milton O. Lee, the Executive Secretary, for duties other than presiding at board meetings and arrangement of the Tuesday evening program. He lost no time in saying the Federation needed a permanent headquarters. He and I surveyed the Washington DC area and were attracted to a large private property, Hawley Estate, in Bethesda, near the National Institutes of Health and the National Naval Research Institute. The house and land have been thoroughly described by Lee (19). William F. Hamilton and Lee named it Beaumont House, after the US Army Medical Surgeon (20). To accelerate the interest payments, two thirds of the land was sold off. A new building named after Milton O. Lee was completed and dedicated on October 12, 1962.

Another assignment was the publication of the first volume of ASPET History (21). W. DeB MacNider was our first historian, but resigned because of illness in 1951. E. M. K. Geiling immediately succeeded him, but as a perfectionist he could not finish the work in 16 years. Harold C. Hodge, then President of ASPET, appointed me as the third historian in 1967. It was with the assistance of Allan D. Bass, Ellsworth B. Cook, Robert M. Featherstone, Maurice H. Seevers, and access to the minutes of all society secretaries that the volume was published in two years.

In my humble life, I am grateful for the recognition I have received for my contributions. The China Foundation Prize came from the Mainland in 1927. An honorary ScD was conferred on me by Philadelphia College of Pharmacy (1946), University of Wisconsin (1952), and Indiana-Purdue University (1971). The Remington Honor Medal, 1965; honorary membership in Finnish Pharmacological Society, 1975; and honorary presidency of International Union of Pharmacology, 1972, were all received with appreciation. I have been assigned some government responsibilities, such as preparing Panel Report to the President (22) and serving as State Department Delegate to the First General Assembly of International Union of Physiological Sciences at Brussels, 1956.

The past five decades in the pharmaceutical and chemical industries have seen tremendous gains in the development of effective research products. I am happy to have had a chance to contribute to these advances.

ACKNOWLEDGMENTS

I thank Professor Henry R. Besch, Jr. for his editorial assistance and Ms. Betty Rode, secretary, for her careful typing.

Literature Cited

1. Chen, K. K. 1923. Cassia oils from leaves and twigs. *J. Am. Pharm. Assoc.* 12:294–96
2. Nagai, N. 1887. Ephedrin. *Pharm. Ztg.* 32:700
3. Nagai, N. 1892. Über die Untersuchungen einer Base der chinesischen Drugue Ma Huang. *Yakugaku Zasshi* 120:109
4. Stapf, O. 1927. *Ephedra sinica. Kew Bull.* 3:133
5. Chen, K. K., Schmidt, C. F. 1924. The action of ephedrine. *J. Pharmacol. Exp. Ther.* 24:339–57
6. Chen, K. K., Schmidt, C. F. 1926. The action and clinical use of ephedrine. *J. Am. Med. Assoc.* 87:836–41
7. Chen, K. K., Schmidt, C. F. 1930. Ephedrine and related substances. *Medicine* 9:1–117
8. Barger, G., Dale, H. H. 1910. Chemical structure and sympathomimetic action of amines. *J. Physiol. London* 41:19–59
9. Aviado, D.M. 1970. *Ephedrine, Sympathomimetic Drugs,* pp. 95–156. Springfield, Ill: Thomas
10. Abel, J. J., Macht, D. I. 1912. Two crystalline pharmacological agents obtained from the tropical toad *Bufo aqua. J. Pharmacol. Exp. Ther.* 3:319–77
11. Chen, K. K., Jensen, H. 1929. Crystalline principles from Ch'an Su, the dried venom of the chinese toad. *Proc. Soc. Exp. Biol. Med.* 26:378–80
12. Chen, K. K. 1973. Toads. In *Biology Data Book* ed. P. L. Altman, D. A. Dittmer, 2:723–25 Bethesda, Md: Fed. Am. Soc. Exp. Biol., 2nd ed.
13. Lee, H. M., Chen, K. K. 1951. The occurrence of *nor*-epinephrine in the chinese toad venom. *J. Pharmacol. Exp. Ther.* 102:286–90
14. Chen, K. K. 1970. Newer cardiac glycosides and aglycones. *J. Med. Chem.* 13:1029–37
15. Henderson, F. G., Rose, C. L., Harris, P. N., Chen, K. K. 1949 γ-Dichroine, an antimalarial alkaloid of Ch'ang Shan. *J. Pharmacol. Exp. Ther.* 95:191–200
16. Wagner, H. 1980. Taking Tiger Mountain. *Johns Hopkins Mag.* 31:29
17. Chen, K. K. 1974. Half a century of ephedrine. *Am. J. Chinese Med.* 2:359–65
18. Kahn, E. J. Jr. 1875. *All in a Century.* Indianapolis: Eli Lilly & Co. 211 pp.
19. Lee, M. O. 1954. A home for the federation. *Fed. Proc.* 13:821–24
20. Beaumont, W. 1833. *Experiments and Observations on the Gastric Juice and the Physiology of Digestion.* Plattsburg, NY: Allen. 280 pp.
21. Chen, K. K. 1969 *ASPET, The First Sixty Years, 1908–1969.* Bethesda, Md: ASPET. 225 pp.
22. Chen, K. K. 1965. D. E. Wooldridge, *Biomedical Science and Its Administration,* pp. 167–74. Wash. DC: GPO

Thomas Maren

Ann. Rev. Pharmacol. Toxicol. 1982. 22:1–18
Copyright © 1982 by Annual Reviews Inc. All rights reserved

GREAT EXPECTATIONS

Thomas H. Maren

Department of Pharmacology and Therapeutics, University of Florida College of Medicine, Gainesville, Florida 32610 and Mount Desert Island Biological Laboratory, Salsbury Cove, Maine 04672

I never expected to find myself in these pages. I have been reading my predecessors in the *Annual Review of Pharmacology and Toxicology* and the *Annual Review of Physiology;* many are inspiring, illustrious, and have high tales to tell. We do have in common that age has taken us unaware. My own advent into medical science was through a series of near-comic accidents and chance events; yet, there is a theme to share with my readers, and that is my title. I have responded to the expectations, which often seemed blind faith, held out to me freely and lovingly by family, friends, teachers, and colleagues from my childhood to this day.

One spring in 1943 I found myself in a strange new place, a Pharmacology Department. It was at Columbia where there was a meeting of people interested in a disease with an even stranger name—filariasis. I had become interested in this although I soon realized that I was mispronouncing it. I cannot recall how I gained entry to the meeting, but I must have invited myself since it would be an exaggeration to say that I was on the fringes of wartime biomedical research. I was a chemist in a small cosmetics company, Wallace Laboratories, much later to gain renown as the maker of Miltown. But then I was toiling enjoyably in the synthesis of thioglycollic (mercaptoacetic) acid to make depilatories, and as a byproduct, we started home permanent waves. I read that antimony thioglycollates had been used in tropical diseases by John J. Abel (a portent) in his one excursion into chemotherapy. It was easy to synthesize a few of these compounds, and my boss, Dr. John Wallace, thought no harm could come of it. It even might give our raffish operation some respectability. Fortunately, canine filariasis (heartworm), although usually confined to the South, had a pocket in New Brunswick, New Jersey where I worked. I arranged with a veterinarian, Dr.

1465

0362-1642/82/0415-0001$02.00

Morris, to inject monosodium antimony thioglycollate into a few infected dogs. What happiness! The microfilariae disappeared from the blood within a few hours. How nice if one could do this every day. Such an effect was known, but this action seemed fast, effective, and not (another fine new word) toxic. Armed with these data, I came to the meeting in New York where there were not only professors, but also naval commanders, officers of the National Research Council, and representatives from the (real) pharmaceutical industry. I was asked to speak first, which seemed odd at the moment, but this was soon clarified. I gave my report, and was asked to leave. Thus, I was introduced to the amenities of academic medicine, but as I crept away, I did notice a squat, energetic civilian wearing a bow-tie, who eyed me from the back of the room.

The next week this man called, saying that he was Dr. Gilbert Otto, a parasitologist at the Johns Hopkins School of Hygiene and Public Health, and that he would like to drive up to see me. When he came, he revealed his secret; he had a contract from the Office of Naval Research to study the pharmacology of antimony compounds related to filariasis and schistosomiasis (another new word!), but he knew no chemistry. Nor, in wartime, could he find even a barely qualified Ph.D.; he would be willing to settle for me. I told him that I knew less biology than he did chemistry, and the partnership sounded ideal. So I left Wallace; I said that I'd be back after the war. At the same time, I suspended my underground operation of getting a Ph.D. in English at Princeton. I said that I'd be back after the war. But my move to Baltimore closed the one life and began another.

EARLY DAYS AND COLLEGE

I was brought up in Mount Vernon, New York, then a quiet suburb of New York City, which offered country pleasures and good local schools. My father had grown up as a tough kid in the city, a disciple of the local blacksmith, Tommy Burke, and his apprentice, nine-fingered Smitty, who had been prey to one of many accidents which befell city kids who copped free rides on streetcars or the new subways. Indeed, legend had it that my father had been stepped on by a horse. Fortunately, the lesion was localized in the nose which gave him an appearance not unlike the late actor Victor McLaglan. But there was a difference: my father was a scholar! When the blacksmith needed a name for his new horse lotion, my father named it Equisalve—an early advertising triumph! He was also an athlete; somehow he had pulled himself out of Yorkville to DeWitt Clinton High School and Columbia where he played first base. He learned tennis—a game then associated with ladies, England, and Newport—and was occasionally challenged to fights by local bruisers who thought they could take on anyone walking about with a racket and possible pretensions to the middle class. But he was good with his fists, too. Before he was 20, he had a license to teach school—all subjects from Physical Education to Physics. He scorned educational theories, but advanced to Principal in New York City School System through examinations and his record without ever going to Teachers College. He became a natural sort of authority on delinquent children (without, so far as I know, ever writing anything). In about 1940, Mayor LaGuardia asked my father to organize a school to be called Youth House, so that juveniles who had been accused or convicted of crimes could continue with their educations. My father recruited a faculty, many of whom were also athletes and from poor families. The best way to get to "difficult" children was through athletics and humor. He was kind to these boys, and often very funny, and surely could not have appeared as a jailer. His motto was "laissez faire" at school and certainly at home. He never interfered with anything, and rarely gave advice, and seemed to have no neuroses at all. He could not understand the word "headache," since he had never had one. He loved literature in an idle sort of way, about the way he did his family. He was totally supporting and accepting as long as I "gave him a run for his money." Of course, he did not mean money, and there wasn't much anyway, but the concept was clear. In 1981, he would be called "laid back." My mother was an accepting, loving, and attractive lady who taught me to play tennis. My parents lived rather obscurely, had little social life, and never bothered with an automobile.

In high school, I played tennis, and dabbled unsuccessfully in politics. I had a secret problem: unlike my friends, I did not know what I wanted to be "when I grew up." There seemed to be only three choices: medicine, law, or engineering. Medicine was out since my mother belonged to a sect that forbade it, and my father passively went along with the insanity. I never saw a doctor until I was 18 and in college with a threatened appendix rupture, and he was as frightening as I had dreaded he would be. Law seemed alien; I could not see why my close friends were attracted to it. That left engineering, so I dutifully applied to MIT. In those days there were no terrors in considering the "prestige" schools, or if there were, I was far too naive to perceive them. But when the catalog came, it showed no course in literature, art, or history. What of the poems by Scott and Browning that my father told of? So I withdrew the application. A few weeks later my parents made a rare appearance outside the home at my high school. The history teacher, a Yale man, said to them, "Tommy ought to go to Princeton." I'm still trying to figure that out. And, a curious tour de force was at hand. Someone in the Princeton admission office (of which more later) figured that 85% in New York State Regents was equivalent to College Boards, and so these (very difficult for high school kids, particularly if their minds were chiefly in sports) were waived. The calculation was off by at least 10 points, but the chance was clear. None of us had heard of the place, but there was a remarkable man in Mount Vernon, a lawyer named Leroy Mills. He was the prime world authority on the science of kicking a football, and being a Princeton graduate, it was his hobby to coach the backfield in this lovely skill. Mr. Mills drove me down, and as I watched the balls float high in the fall air over Palmer Stadium, I entered a new world. The admissions procedure was an interview with a gentle lady in Nassau Hall, the assistant and only staff to a formidable relic from the 19th century with the awesome name of Radcliffe Heermaance. He was both Dean of Admissions and of Freshmen. His Mrs. Williams seemed to wonder what this little 15-year-old was doing away from his mother and not even at Lawrenceville. Perhaps, I was admitted by some maternal compunction. I still intended to become an engineer, but a few weeks into freshman year, I saw that the curriculum was hopelessly restrictive, and I was destined to live in this great atmosphere of liberal arts, and yet outside of it. I told this to the Dean of Engineering in my finest hour of all academic discussions in the next 47 years. I found that if instead I majored in Chemistry, there was full latitude for English, History, Architecture, and yet in those parlous depression years, training for that remote possibility, a job at graduation.

I am too far afield already, and cannot recount the sociological and emotional and intellectual play of undergraduate years. I was relatively indifferent to the great social and economic gulfs separating many of the students from each other. I was too naive or too self-absorbed to be seriously affected by the cruelties that were, and still are, engendered by this situation, but I was part of the first attempt to establish "non-selective" clubs at Princeton. Academically, I was greatly influenced by a remarkable group of young English professors, notably, Donald Stauffer, Willard Thorp, and Carlos Baker. Their scholarship was enormous, the entire great swell and sweep of English literature; usually they had *two* fields of specialization— a "classic" and a "modern"! Somehow this made a deep impression on me, and I tried, perhaps subliminally, to do the same in what was to be a subject as different as could be imagined. In the summer of 1937, a friend's mother treated us to a trip to Europe. My astonished father could afford the passport and a copy of *Tom Jones* brought to the boat. In those far-off days, such a trip was a remarkable adventure, and brought the languages and peoples of France and Germany closer, beginnings which were to continue and become most meaningful in the scientific future. Of Hitler we heard and saw remarkably little; Germany seemed under cover, and we lacked the training and perception to see beneath.

When graduation drew near, I had no prospects. I had failed to get a fellowship in several universities for graduate work in Chemistry, and my heart was not completely in it. I was not inside the subject the way I was in English, and to a lesser degree, Art and Architecture History. Perhaps, it was the difference in faculties, or that my emotions and perception of the world could not then fit with science. So after my friends were ready for law or medical school, or their Ph.D.'s, or some for business, I was left over. On the evening before graduation in June 1938, a man named Dr. John Wallace came to my room, told me that he was a manufacturing cosmetic chemist with a small laboratory and plant in Jersey City, and that they needed a man to do control work and research. He seemed pleased when he found that I was not going to graduate school, and also that I had not made Phi Beta Kappa. I visited the plant, exchanged a few limericks with his partner, and was hired. The partner, Wilfred Hand, a near-genius, had just invented the antiperspirant, Arrid, and was working on an odorless depilatory and kissproof lipstick. The first fact that I learned was that Arrid Cream contained 21% aluminum sulfate; I had just written in a physical chemistry exam that oil-in-water emulsions were broken by electrolytes. We ordered one of the first pH meters from Beckman, and with great curiosity took the reading with the delicate glass electrode and the lovely little calomel one with the glass sleeve. The pH was 2.7, but in a mysterious way it was buffered by urea (not a buffer) and was non-irritating. It turned out that NH_4^+ was being generated. I checked all batches between stages of the mixing vats and jar filling, but there was also time to begin some chemical syntheses. The chemistry that I had grudgingly learned in college came to life, and I read with excitement the pioneering book of Louis Hammett, *Physical Organic Chemistry.* It seemed that here might be the way out from the cookbook reactions that I was running. I asked the boss if I could go to graduate school at Columbia (I hoped to study with Hammett) part-time, and key it in with my work. The answer, fatefully and surprisingly, was no —I could succeed in the business without any advanced training. Meanwhile my friends were getting on with their lives, and I was isolated in a loft in Jersey City with no discernible future. What to do? I cast about and struck upon what I loved best—English Literature. I had written some stories in college, but seemed to lack the staying power for real creativity. I was greatly attracted to the scholarship, however, and got admitted to the graduate school at Princeton, quit my job (to the surprise of all), "hung loose," hitchhiking, athletics, reading, studying Medieval Latin (a requirement) for some six months, and returned to Princeton in the fall of 1940.

There followed a memorable year, in which I lived in a single room in the tower of the Graduate College, immersed in Old English, Middle English, Etymology (Linguistics had not yet arrived), and American Literature. Other time was spent studying for what would be a comprehensive exam on all of English Literature. But world events were overtaking us, and by the next summer, most of our generation was in the services or involved in "war work." Having failed a military medical exam, and being a chemist, I found myself back at the bench, and by a curious chance, at the same company that I had left for graduate school. By night and weekends, I continued to study for that awesome examination, and took it under these unconventional circumstances with a rather disapproving faculty. To my own and their surprise, I passed the written part, but revealed some fatal weaknesses in the process. After all, what could they expect of a chemist? So at the oral, the truth came out: I had not read Wordsworth's *Excursion.* Faculty opinion seemed divided as to whether I should give up the field and do science or try again in a few years. It was just at this time that I met Gilbert Otto, as recounted above. My schizophrenic life in New Jersey was suddenly over; I left uncertain of the future, but never to return.

JOHNS HOPKINS: THE WAR, TROPICAL DISEASE, AND MEDICAL SCHOOL

I went to work in the Department of Parasitology of the School of Hygiene and Public Health, a classical department which had then trained about one third of the parasitologists in the country. The strangeness of the environment made it fun. I had good facilities and fine cooperation from everyone. It slowly dawned on me that I was seeing legendary figures everyday in the basement dining room: E. V. MacCallum, Lowell Reed, W. W. Cort, Kenneth Maxcy, Mansfield Clark, E. K. Marshall; and significant younger men, Emmanuel Schoenbach and David Bodian. The Office of Naval Research contracted with us for a microanalytical method for antimony in blood and tissues; metallorganic compounds of this type had been used in tropical diseases since the turn of the century with no pharmacological base. Those were the days of colorimetric methods which, if done well and luck held, were accurate in the parts per million range. One of the first and most famous was the Bratton-Marshall method for sulfonamides—the same Marshall that I saw at lunch. Methodology seemed to fail with antimony, which was surprising since there was no trouble with other metals, bismuth, arsenic, or mercury. Antimony did form a beautiful red lipophilic dye with Rhodamine B. But the metal seemed to "disappear" during the hot digestion process, and was assumed to be volatilized. However, this was not the case; a reading of Mellor's great tract, *Inorganic Chemistry,* revealed that antimony on oxidation reached a relatively inert tetravalent form. The problem was solved in a few minutes; addition of a reducing agent made the metal miraculously appear as judged by the development of the dye.

Thus, we were ready to study blood levels and excretion patterns in patients receiving antimony compounds for the treatment of schistosomiasis, leishmaniasis, and our main target, filariasis. Just at that time, we made another discovery and a surprising one. We screened compounds against the organism that caused filaria in dogs and in cotton rats (where the adult worm was in the pleural space rather than the heart). Antimony had no effect at all on the adult organisms, although it had been used in the human disease, Bancroftian filariasis, for some 30 years! The microfilaria were indeed sensitive to antimony, and since they disappeared rapidly from the peripheral blood when drugs such as tar emetic were injected, there was the comfortable illusion that the disease had been cured. I was ready to drop the antimony project, but Otto gave me an early glimpse of wisdom when he refused, saying that the method might yield valuable theoretical information anyway. In the spring of 1945, the Navy sent us to the Virgin Islands to apply the method to patients still under treatment with antimony. I set up a small laboratory in the hospital at Christiansted. Each morning I climbed the hill protected by my pith helmet, feeling Kiplingesque, and curiously dissociated from the world in that fateful time. The town was still sleeping in the 19th century with the historical and social tradition of having been governed by five countries. The Blacks spoke with a beautiful lilting tongue: "Do not vex me, mon" their hardest rebuke. Reward for some 30 determinations each day was swimming in the tempered azure water off the endless empty beaches. Along with Otto and me, was his great friend and colleague, Harold Brown of Columbia, a parasitologist and tropical disease physician of considerable renown. When dinner conversation lagged, he worked on his project to get me to go to medical school, making Otto (who was not a physician) increasingly uncomfortable. A sign of things to come! The bomb was dropped, the world convulsed, the war ended. We celebrated quietly in the island manner with rum drinks flavored by exotic fruits.

Back in Baltimore, we had another year to work, and an exciting lead that organic arsenicals did kill the adult filarial worm. In what I came to see as the Hopkins manner, unselfish help and "Great Expectations," I was aided immeasurably by a great and kindly man, Harry Eagle, who had a unit for the chemotherapy of trypanosomes and spirochetes and who was a modern Ehrlich in his knowledge of the arsenicals and facilities for synthesis. Our screening, some with his compounds, led to the finding that a p-$CONH_2$ group on phenyl arsenoxide gave some specificity for the filariae. It remained to solubilize the compound by condensing it with my old friend, thioglycollic acid. The compound, which we named arsenamide, received a brief trial in human filaria, but was not pursued because of toxicity and the need for intravenous injection. However, based on our work, the veterinarians took it up, and comprehensive studies in Japan showed how it could be used for the prevention of dog heartworm. The drug is still in very wide use.

At lunch in the School of Hygiene, talk was still of wartime medicine. Giant strides had been made, many by these men, in malaria, infectious disease, epidemiology, and the beginnings of the conquest of polio. E. K. Marshall, Professor of Pharmacology, was a leading figure in discussion. His war work was in malaria, following pioneering studies on the sulfonamides, and his earlier work on kidney, in which he had made a major discovery—renal secretion. At 56, he was tall, thin, handsome, rather austere, roughly outspoken, and intractably honest (1). He was at the peak of his intellectual power and influence, and frightened many, but fortunately not me since I had been immunized to these dominant traits by the great Chaucer and Pope scholar at Princeton, Robert K. Root. Marshall did like to hear about my work, and seemed pleased by it.

One day he told me of his plan to train pharmacologists by "taking men from medicine and getting them some chemistry, or taking chemists and seeing them through medical school." Marshall himself had done the latter under the aegis of his predecessor, John J. Abel. Later I was having lunch with Marshall's secretary, a grande dame of Baltimore and also Pharmacology, for she had been Abel's secretary fifteen years before. Miss Wilson said, "Mr. Maren, I think Dr. Marshall is interested in you, and may make it possible for you to go to medical school." I said, "But Miss Wilson I've been out of college eight years, and I'm too old." Said she, "Mr. Maren, you're going to get old anyway." And so, with this dialogue out of Jane Austen, the matter was settled, at least in my mind. Marshall had his plan funded by three drug houses, each contributing $3,000 per year for that many fellowships. So far as I know, it was the first training program in Pharmacology. Over a ten-year period, three of us went to medical school, and ten physicians had the opportunity to study and practice basic science. No Ph.D. degree was involved.

There was the formality of getting admitted, and a brief snag developed when it was discovered that I had never taken a course in Biology. I tried to convince the Dean, Alan Chesney, that this was unnecessary, but that great and kind man (presumably) turned off his hearing device, and gently suggested that I take a summer course in Comparative Anatomy at the Homewood Campus. It turned out to be the hardest course I had ever taken, run by two or three of the most brilliant young biologists in the country. There were surprises for me at every turn. I got a C, and started medical school in the fall of 1946. My mother still "believed" in Christian Science, and I had never been to a doctor. If, as the creed said "sin, disease, and death were unreal," what would happen when I faced that cadaver? Nothing.

Marshall gave me a small and pleasant laboratory office which I had all through medical school. There were three fine Associate Professors:

Kenneth Blanchard, a chemist with universal knowledge of science and its attendant gossip; Thomas Butler, brilliant, incisive, original, becoming one of the best pharmacologists of these generations; and Morris Rosenfeld, a physicist-physician and legacy from Abel. We all had lunch together every day, joined by visitors from all over the world. What we gained and enjoyed from these conversations is incalculable. I was a lab instructor in the Pharmacology course my first year before I had even completed Physiology. I recall not understanding how atropine reversed the hypotensive effect of acetylcholine, and telling the second year students that they must have used the wrong solution. I never did take the pharmacology course, or any other in the subject. The basic sciences were a feast, and quite low keyed under Marion Hines (Neuroanatomy), Philip Bard (Physiology), Mansfield Clark (Physiological Chemistry), and Arnold Rich (Pathology). There seemed ample time for student research. The clinical curriculum offered everything, but economically, and well organized. The major ward clerkships were very rigorous; our examples were interns and residents on 24-hour duty all year. A highlight was the "elective" in dog surgery taken by everyone, an unparalleled opportunity to get the feel of living tissues and palpably understand physiological stress. The heads of specialty departments gave elegant short courses of lectures followed by a week or two in the outpatient clinics. It was a survey of medicine as well suited to the prospective general physician as to the research man. It seemed simple, uncomplicated, and effective. Even the grading system was easy; there were grades, but not given to the students. In light of such experience, what has been accomplished by the curricular convulsions of the past decades?

I finished studies on the distribution and metabolism of the antimony and arsenic compounds, and then began with Marshall on the cinchoninic acids and their relation to pituitary function. He had shown these to be antidiuretic, and gave me the job of finding if they worked through the neurohypophysis or directly on the kidney. He would have been pleased if the latter were the case for such action was (and still is) unknown, and Marshall was ambitious for firsts. Curt Richter, a behavioral biologist in Psychiatry, was one of the few who could perform a clean-cut neurohypophysectomy in the rat (without injuring the anterior lobe), and with his help , I showed that our drugs did act through the neural lobe. Of greater interest was the finding that they also affected the anterior lobe-adrenal system. It was the time of cortisone and ACTH, and Marshall led a clinical experiment at Hopkins using the cinchoninic acids in rheumatoid arthritis. I became interested in the hypothalamus, and another generous and distinguished colleague, David Bodian, supervised the analysis of nuclei in this region in our pituitary ablated rats. We showed for the first time a relation between the anterior lobe and the paraventricular nuclei. In that time was the dawn of neuroendocrinology, and I cherish the memory of a visit from Geoffrey Harris, the great English anatomist who had just discovered the portal circulation in the pituitary.

My seven years at Hopkins drew to a close. There was no way to know it then, but 1951 and the few years to follow were watersheds between the ideas and ideals generated in the 19th century, and the very different future in which we now live. Many of the men that I have mentioned were educated before World War I, and both Princeton and Johns Hopkins had a long unchanging period of stability during the first half of our century. The buildings were unchanged, the traditions, curriculum, and even the songs seemed to stay the same. I cannot evaluate such issues, but in a sentence or two, I would guess that the intellectual and ethical benefit was enormous. However, a heavy price was paid in social and possibly financial oppression. Perhaps, this is worth remembering if we wish to recapture the best of those years, some of which has certainly been lost.

I considered clinical training, as I enjoyed medicine very much. The economics seemed impossible for a man with a family and no private income —interns at Hopkins were still paid $25 per month. More important was the question of whether clinical training was worthwhile if I were to continue in basic research. Marshall thought not, and I recall Harry Eagle saying that his year as an Osler intern (that great plum) was the worst of his life. I was not sure then, but feel now, that I did make the correct decision. I had learned enough in medical school to understand basic principles, and to be part of a medical faculty. Besides, I was 33, and better get on with my life. I had a choice between an assistant professorship at $5,000 per year, or several jobs in the pharmaceutical industry, at twice this amount. There was also the novelty and challenge of industry, at a time when not many men with substantial academic training were taking that route. Not so many years back from then the American Society for Pharmacology and Experimental Therapeutics, Inc. did not admit scientists from industry!

THE CHEMOTHERAPY DIVISION OF AMERICAN CYANAMID: DEVELOPMENT OF THE CARBONIC ANHYDRASE INHIBITORS

In the fall of 1951, I joined the Chemotherapy Division of the American Cyanamid Company in Stamford, Connecticut, housed on the top floor of a converted factory. One had to ride an old freight elevator to get there. The rest of the building contained the main research laboratories of this large company devoted to chemistry, physics, plastics, and agriculture. The drugs developed by our division were marketed by Lederle Laboratories, a wholly-owned subsidiary of Cyanamid. The Chemotherapy Division was remarkable with, as its head, Richard O. Roblin, a talented chemist and true research director. Ten years before, he and his colleagues had discovered sulfadiazine, and from his small chemical team was built his division which now consisted of eight groups: Three were in organic, one in physical chemistry, and these were balanced by the four groups of bacteriology, parasitology, pathology, and pharmacology. The place was strikingly unpretentious; no one had an office except Roblin. Almost no time was spent in administration or meetings; Roblin read the literature all morning and worked in his lab in the afternoon. The eight group leaders were reputable scientists; indeed, a major contribution to theory of drug action had been made here (Bell-Roblin on sulfonamides). They were also pioneers in broad-spectrum and gram-negative chemotherapy (polymixin), and treatment of parasitic diseases in chickens. Roblin believed in a substantial mix of theory and practical development, and he had carte blanche for his programs.

I was hired with the idea of working on the physiology of the hypothalamus and connections with the pituitary. This was related to work that I had started at Hopkins on obesity, using the strain of mice just discovered at the Jackson Laboratory. There also was still some interest in the cincophen derivatives that Marshall had hoped to use in rheumatoid diseases. But one day that winter, Roblin told me that he had had a call from a gastric physiologist named Henry Janowitz to whom he had given a drug called 6063 which reduced acid output from the stomach. He confused me utterly by talking about enzymes and inhibitors of which I scarcely knew the difference. Finally, he asked if I would work on this drug, and of course, I said yes, being anxious to leave and do some reading. Once again I felt the spirit of "Great Expectations" conveyed by Roblin, unspokenly then and in the years ahead.

I have told the story of the development of the carbonic anhydrase inhibitors elsewhere (2, 3), and now I will try to show some of the strands reaching back into physiology and biochemistry, and how these became woven into my own life. The enzyme was a Cambridge product, discovered by Roughton in 1933, with vital work showing its metallic nature and susceptibility to sulfonamides by Mann and Keilin in 1940. In the following decade, carbonic anhydrase became linked with renal acidification and bicarbonate reabsorption through the work of Davenport, Höber, Pitts, and William Schwartz. The inhibitor used was sulfanilamide, but Roblin, sitting in quiet isolation at Stamford, and drawing on his knowledge of sulfonamide chemistry, realized that possibilities existed for more specific and powerful drugs of this type. There were other strands: The role of the enzyme in gastric and pancreatic secretion, in respiration, and the possibility that inhibition might control epilepsy through alteration in acid-base balance. All of these ideas were in the air on that winter afternoon in 1951, and by then, Roblin and his associate, Jim Clapp, had synthesized several dozen new compounds, some of which were 1000 times more active against the enzyme than sulfanilamide. He had supplied a few of these to academic physiologists; the best one seemed to be 6063 (later named Diamox, and the generic, acetazolamide—the idea being that this was so difficult to pronounce that only the trade name would be used) on the basis of high activity, stability in vivo, and sharp effect in alkalinizing the urine. Robert Berliner had just published a paper in which he had reported the use of 6063 in a study of K^+ secretion and the action of mercurials. I did not know until much later, however, that Roblin had been advised by some of the leaders in renal medicine not to pursue the project since they thought (correctly, as it later turned out) that such drugs would not be ideal diuretics. Roblin, through broader vision, a taste for gambling, and a chemist's vague sense for physiology, went ahead.

I was assigned the development of this drug with the idea that it might be the first oral diuretic, a treatment of ulcers, epilepsy, and renal stones. Every few days another disease was suggested. How to develop a drug of an entirely new type, and inhibit an enzyme that was thought essential to life, and whose absence had been predicted to lead to "speedy death"? Only in retrospect do these seem formidable, then it was stimulating and great fun. My training with Marshall made the first step reflex; we must have an analytical method for the drug in body fluids. Attempts to develop a chemical method failed, and I fell back on a simple enzymic procedure which is still used. Acute and chronic toxicity work began; to our amazement, the drug was entirely nontoxic; we could not even kill an animal with grams/kilogram by vein! What was going on? Why didn't the diuretic effect and alkaline deficit continue and prove fatal? How about respiration when the red cell enzyme was inhibited? What would be the effect of long-term inhibition on the morphology of the tissues?

The answers emerged slowly over the years, always it seemed with implications for my own life as well as for the underlying physiology. The first question led to a meeting with the now legendary Homer Smith who had just found that seagoing fish did not have a renal response to 6063. He asked me to come to the Mount Desert Island Biological Laboratory (in Salsbury Cove, Maine) to see if there was carbonic anhydrase in fish kidneys. There was not; yet, the fish could absorb any amount of filtered HCO_3^-. At the same time, I was finding in Stamford that acidotic dogs could reabsorb HCO_3^- when all carbonic anhydrase was blocked. Clearly, there was a second and quantitatively major mechanism for HCO_3^- reabsorption, and this protected the animal or the patient (as it turned out) from further electrolyte loss after the first day of enzyme inhibition. Thus, we had a solution with theoretical and practical implications, and a triumph of the comparative method. I have been working at this laboratory every summer since; it has immeasurably enriched my scientific and personal life. E. K. Marshall was still at Salsbury Cove, where he had discovered renal secretion 30 years before. Our friendship deepened as the years passed. There is a curious parallel in his use of the goosefish (no glomeruli) to prove tubular secretion, and mine of the dogfish (no renal carbonic anhydrase) to illustrate ionic HCO^-_3 reabsorption. Both ideas were fought by the renal establishment for at least ten years (1, 2).

The respiratory problem was not solved until much later, as I shall tell. The critical issue of morphological damage was answered in the negative based on a most meticulous chronic toxicity study carried out at Stamford under the direction of my colleague in pathology, Edmund Mayer. From this remarkable man, I seemed to be learning, in those four years, about as much medicine as I had learned at Hopkins. My luck had held; in those spare and obscure laboratories at Stamford, I was still in the presence of masters.

I had no training in renal or electrolyte physiology, and was working in comparative isolation with a few excellent technicians, a fine beagle colony, and a good library. Ten days spent with the young Bill Schwartz and Arnold (Bud) Relman in Boston were of seminal importance. I watched their dog experiments and we traded ideas: chemistry for physiology. By the end of 1953, I emerged and faced the Establishment for the first time at a seminar in Pitt's department at Cornell Medical School. We had given 6063 to dogs and rats for over a year, studying electrolyte balance, drug distribution, effects on growth, and general toxicity. No one had hurried me, but the drug was ready for clinical trials, and the market in the next year. Just then came a totally unforeseen development; Bernard Becker at Johns Hopkins showed that 6063 was effective in glaucoma. Roblin's gamble on the unexpected in science had paid. I have told elsewhere the remarkable story leading to Becker's discovery, based on the ideas and genius of Jonas Friedenwald, and the fine experiments of Everett Kinsey and Per Wistrand (4). Within a few years, acetazolamide became a fundamental part of practice and research in ophthalmology.

This discovery, together with our data showing restricted penetration of acetazolamide to eye and brain, led to development of a second and more diffusable drug, methazolamide, better suited for glaucoma treatment. Attention was also given to other sulfonamides of different physico-chemical types, and correspondingly different affinity for tissues. From these studies, came a third drug, benzolamide, whose affinity for proximal tubule cells made it a specific renal carbonic anhydrase inhibitor. [Attempts to develop benzolamide commercially have failed, and it now has the interesting status of an orphan drug (5).] The Merck, Sharpe and Dohme group under Karl

Beyer, competing with our work on renal carbonic anhydrase inhibition, had the good fortune and technical excellence to find a new series of sulfonamides with a different mechanism, better suited to the treatment of congestive heart disease: the thiazides, and later furosemide and all their congeners, were born.

The Stamford Laboratories offered a great deal. I was able to complete a study on the obese mice, showing that diet restriction caused a preferential loss of protein over fat. Also, we discovered the long-acting antibacterial sulfonamides when it was found that a methoxy group on the N^4 or heterocyclic ring induced avid renal reabsorption. But our Chemotherapy Division had within it the seeds of destruction. First, it was simply too good— perhaps the best pharmaceutical research group in the country. Second, Roblin had unusual power. And third, the drug development units at Lederle were floundering. So, in a stroke combining jealousy, an attack on the "elite," and an attempt to save Lederle, the division was destroyed by the parent company. The various units were to be dispersed at the Lederle Laboratories in Pearl River. (From the vantage point of 25 years, this was a dreadful corporate decision; small *was* beautiful and successful, and big *was* ugly.) But *deus ex machina,* my phone rang; at the other end was George T. Harrell, who said that he was dean of a new medical school at the University of Florida. I recall this conversation with the same embarrassment that I mentioned in connection with Dean Chesney at Hopkins. Only ten years had passed, and I guess I had not mellowed much. I asked Dr. Harrell about himself and his qualifications (which were pretty good) after which he said that he would like to consider me for Chairman of the Pharmacology Department at the school in Gainesville. So that summer (of 1955), I went down and faced the 100 degree heat, walking on the girders of the fifth floor of the building rising from a sinkhole. My department! The deal was settled in a decaying hotel-restaurant at Cedar Key on the Gulf of Mexico, as we ate stone crabs, hearts of palms salad, and as a summer hurricane blew open the doors and out the lights. There were "Great Expectations" in Harrell's voice, and in that wind.

THE UNIVERSITY OF FLORIDA SCHOOL OF MEDICINE: TEACHING OF PHARMACOLOGY

The original faculty of the Unversity of Florida College of Medicine were four basic science chairmen: the others were James G. Wilson (Anatomy), Frank Putnam (Biochemistry), Joshua Edwards (Pathology). In the next year came Emmanuel Suter (Microbiology) and Arthur Otis (Physiology). All, happily, are still alive, and may share with me these brief reminiscences. At first, it was sort of a pick-up game, and we were referred to simply and conveniently as "the boys," until one day we realized it was a pretty serious business; students were coming in the fall, the building was almost finished, and we were the Executive Committee. We did admissions, curriculum, departmental layouts, and (with the dean) the hiring of the clinical faculty. We were unified by common ideals, a visionary dean, absurdly proud of our little departmental fiefs, but held the school's goals high as well. Classes were small (40–64), there was extensive laboratory work, two free afternoons per week, and in the second year, some 150 hours set aside for independent research for all students. This continued for 13 years, during which nearly 1000 physicians were trained in a system which did encourage independent thinking and inquiry. Changes in the 1970s blew ill winds, and this milieu has vanished without a trace. I mention this with nostalgia, but also in the hope that these notes will encourage my successors in Pharmacology and the other medical sciences to try as hard as possible to reverse the present faceless medical education (in many schools, fortunately not all) in which there are no laboratories, little personal work, limited student-faculty contact, and robot-like examinations. We have turned the clock back a hundred years; the toll is already being counted in the severe decline of M.D.'s in research, an inevitable result of no exposure in the critical years of training. The effects will soon begin to show even more tragically in the decline of clinical skills and good patient care.

I organized the Pharmacology Department, bringing to life ideas that had been germinating for some years. We added "Therapeutics" to the name (*not* "Experimental Therapeutics") to indicate that we had a role in both the theoretical and practical aspects of the subject. The first appointment was a biochemist, Kenneth Leibman, who rapidly and with obvious enjoyment learned physiological as well as chemical pharmacology. Then came Bohdan Nechay, trained in veterinary medicine, David Travis, a clinical physiologist with particular knowledge of respiratory and renal medicine, and Aaron Anton, a real Ph.D. pharmacologist. This fine group assembled a course in pharmacology for second-year medical students with heavy emphasis on laboratory work. The planning and teaching in the lab was a vital part of our own training, particularly important since we were all so new at it. (The Dean tried hard to keep our guilty secret: The Pharmacology Chairman was only four years out of medical school.) By 1962, we were ready for a good experiment in education: We would split our course in half, and give the second part in the fourth year. We were still in the "honeymoon" period, and the Executive Committee, including the clinical heads, were willing to give up November of senior year for Pharmacology. And Roger Palmer, a graduate of our first class, fresh from a medical residency at Hopkins, was back, full of idealism for the new program. It worked! Fears that fourth-year students would begrudge time spent in a basic science were not realized. The split was tried several ways—perhaps the most satisfactory was to emphasize the chemical aspects in the second year and physiological in the fourth. It may make no difference how the topics are arranged: The main advantage was the milieu furnished to the students, and what they brought in the way of attitudes and questions in the two years since we had seen them. This system is still working, and may be unique among American medical schools. I have set down elsewhere some ideas about Pharmacology in medical education (6).

RESEARCH AT FLORIDA: A BACKWARD LOOK, 1981–1958

My work on carbonic anhydrase was done backwards. Although I sought connections between chemical and physiological events, the latter work, as I have recounted, was done first. In Florida, I tried to relate the quantitative data on renal excretion of HCO^-_3 following acetazolamide to catalyzed and uncatalyzed rates of CO_2 hydration, and fractional inhibition of the enzyme. Because the chemical reactions were simple, the same in vitro as in vivo, and the inhibitors completely specific, this did seem possible, and we were able to add a new small dimension to pharmacology. Both the kinetic and the drug data showed, independently, a great "excess" of enzyme so that 99% inhibition was zero on the dose response curve for renal effect. Superficially, it seemed that there was unnecessary enzyme present, but this concealed the principle that a high concentration of enzyme ensures very rapid reactions in a near-equilibrium state, where there is a narrow gradient between substrate and product.

The writing of a comprehensive review on the subject in 1967 (2) showed that we did not understand the role of the enzyme in CSF formation, or in brain fluid. Being already wedded to the Mount Desert Island Biological Laboratory, our initial experiments were done in fish (the small shark, *Squalus acanthias*), a lucky event since an increase in peripheral pCO_2 caused a large and rapid elevation of CSF HCO^-_3, which was blunted when carbonic anhydrase was inhibited. Later experiments with Dr. Betty Vogh confirmed this in the mammal, and showed the linkage of HCO^-_3 formation to Na^+ transport. The excitement of a new and unexpected finding was modified by the difficulty of getting it accepted or even published; it ran counter to prevailing ideas and even seemed unreasonable since bulk CSF is a neutral, not an alkaline, fluid. In relaxed moments, it seemed fun to fall into the great tradition of iconoclasm, and ten years and the work of other groups have now smoothed the concept into the stream of physiological thought.

In about 1970, I decided to finish with carbonic anhydrase. Another wartime malaria program was under way, and I wanted to be a part of it. I was a bit tired of my identification with this enzyme and drug ("Are you still working on Diamox?"), and did people think I was hopelessly stuck? The excursion was fun, and we learned some interesting things about sulfones, but the carbonic anhydrase problem was too central and compelling, so my transgression was brief.

The problem posed by the near-discoverer (Henriques) and discoverer (Roughton) of carbonic anhydrase concerning its quantitative role in respiration had yet to be worked out. It was still not clear how survival was possible in the absence of enzyme activity, or what its role was in exercise. I was fortunate to have a young colleague, Erik Swenson, attack this problem, both as investigator and subject. The answer explained why our high dose animals survived in spite of the dire predictions of the biochemists. In normal respiration, CO_2 is evolved almost entirely from HCO^-_3. When the enzyme is absent, newly established high gradients of CO_2 make its elimination possible by simple molecular diffusion. But this auxiliary non-enzymic route is not adequate in exercise. This remarkable enzyme, with its magically great turnover number, evolved so that HCO^-_3 and metabolic CO_2 are instantly convertible, and that virtually any amount of the gas can be generated from narrow gradients of the ion.

By good fortune, work at the Mount Desert Island Biological Laboratory continues to give hints to solutions of our physiological questions, in addition to those already mentioned. These concern lens, aqueous humor, endolymph, and drug diffusion across the gills. Presently, we are studying the vertebrate phylogeny of carbonic anhydrase. This place is also my window on the world, in science and in spirit.

In the last few years, association with another talented young man, Gautam Sanyal, has led to fundamental studies on the thermodynamics of catalysis, and work on other substrates for carbonic anhydrase and the properties and possible roles of certain of its isoenzymes. So the 30-year Odyssey has taken me back to the homeland of molecular pharmacology: the active enzymic or receptor sites. How different would the story have been if I had started at the beginning? And am I ending here?

BIOPOLITICS: A SMALL DOSE

By instinct or the example of my father, I had no taste for public affairs, or administration, or power, or the judging of others. I found myself a departmental chairman for 22 years, and on NIH study sections for ten years, and served a term as President of the Association for Medical School

Pharmacology (AMSP). I have an uneasy feeling that I "got away with it," but should have been an even more obscure scholar. And I have a residue of warmth from it all—the (too) few fine men I brought into the department, the wide ranging friendships and expansion of view generated by the study section work, the fanning of life to AMSP by organizing "holiday" winter meetings, and precious and supportive exchange of views among men and women in the same profession. Now I have a chance, as a Research Professor, to test my (alleged) affinity for isolation.

I was thrust briefly, and unexpectedly, into Chinese affairs in the time of Mao, when in 1974, I was part of a National Academy of Science sponsored trip to the Mainland to study Herbal Pharmacology. Scientifically we found little, but I returned with two deep convictions: that personal freedom, which I had never before relinquished, was more precious than I had dreamed, and that the potential for accomplishment by the Chinese, once committed, was greater than any we know in the West.

The reader perceives the curious element in this tale: It goes against the grain of good conventional science which is to do one thing, and do it well. What did Mercutio say: " 'Tis not so deep as a well, nor so wide as a barn door, but 'tis enough, 'twill serve." Whether it is my own disposition, or the ubiquitous nature of the problem I was assigned 30 years ago, or a coincidence between them, I do not know, but I have extended myself most unfashionably across these many fields. A modern Dryden might, if I were lucky, notice me as one "Who in the course of one revolving moon/ Was fiddler, statesman, chymist, and buffoon."

I am greatly fortunate to live as part of a community that has no geographic or social boundaries, linked by interests, affection, and even rivalry. My beliefs should be clear from this essay, so I end only in the passionate hope that scholarship and society maintain their equilibrium in the years to come, and that "Great Expectations" are held out to our heirs.

Literature Cited

1. Maren, T. H. 1966. Eli Kennerly Marshall, Jr., 1889–1966. *Pharmacologist* 8:90–95; *Bull. Johns Hopkins Hosp.* 119:247–54
2. Maren, T. H. 1967. Carbonic anhydrase: chemistry, physiology and inhibition. *Physiol. Rev.* 47:595–781
3. Maren, T. H. 1979. Theodore Weicker Oration: An historical account of CO_2 chemistry and the development of carbonic anhydrase inhibitors. *Pharmacologist* 20:303–21
4. Maren, T. H. 1974. HCO_3 formation in aqueous humor: Mechanism and relation to the treatment of glaucoma. *Invest. Ophthalmol.* 13:479–84
5. Maren, T. H. 1982. Benzolamide: A renal carbonic anhydrase inhibitor. In: *Orphan Drugs,* ed. F. E. Karch. New York: Marcel Dekker. In press
6. Maren, T. H. 1968. Pharmacology: Its nature in medicine. *Science* 161:443–44

Ann. Rev. Pharmacol. Toxicol. 1984. 24:1–18

FROM PHYSIOLOGIST TO PHARMACOLOGIST— PROMOTION OR DEGRADATION? FIFTY YEARS IN RETROSPECT

Börje Uvnäs

Department of Pharmacology, Karolinska Institute, Stockholm, Sweden

When I studied medicine in the thirties at the University of Lund, an old provincial university in southern Sweden, pharmacology was a neglected discipline carrying on its activities on the outskirts of physiology. And to be sure, most pharmacologists of the time, with illustrious (notable) exceptions, expended their efforts on rather unimaginative, qualitative studies of the actions of drugs on animals and isolated organs, the classic targets of physiological research. Trained pharmacologists were nonexistent at Lund—they were rare birds in the whole of Scandinavia—and the pharmacology chair was held by a young physiologist with an interest in tissue respiration, which he studied with the Warburg and methylene blue techniques, in those days the highest fashion in metabolic research. To enter an academic institution a student had to serve as an unpaid assistant, an apprenticeship that could last for months or years. Such positions were very popular, with professors because they meant cheap teaching and research assistance and with students because they could mean the beginning of an academic career. Usually students who had done well on their examinations were offered these jobs. Since there were only about 20 students in our annual courses, the clever ones were chosen before they reached the pharmacology course, the last course of the three preclinical years. In my case, however, even after the pharmacology examination no professor offered me a position, so in desperation I took the rather unconventional step of asking for permission to begin in pharmacology.

My mind was not ready for a scientific career, however, and after one or two years of half-hearted contributions—I collaborated in some minor papers—I left science, temporarily as it turned out. Even though my scientific achieve-

1485

ments during these early years in pharmacology were negligible, I made a contact that was of the utmost importance for my future. Into the little university town a young man arrived one day in answer to an invitation. His name was Georg Kahlson—generally called G. K.—and he was to become a very disturbing and at the same time stimulating member of the little medical faculty. G. K. was preceded by his reputation as an outstanding scientist, with ten years of scientific education in Jena and Göttingen. When he arrived in Lund with his very charming and beautiful wife, Louise, and their fox terrier, Grock, he was received with great expectation. Once he was installed in the department of pharmacology, I found him a stimulating teacher, a generous supporter, and a very good friend. He was to remain all these things to me until personal differences separated us several years later.

G. K. influenced my scientific career considerably. In 1938, when he was appointed professor of physiology in Lund, he invited me to join his staff and I accepted. I stayed in his department for almost 15 years and held a chair in physiology from 1949 to 1953 until I returned to pharmacology, this time in Stockholm, as will be described below.

It is interesting to note the absence of distinct boundary lines between physiology and pharmacology in the middle of the century. The three Swedish chairs in pharmacology were held by physiologists by training: Gunnar Ahlgren at Lund, myself in Stockholm, and Ernst Barany at Uppsala. Of the three Swedish professors in physiology two were trained pharmacologists: Ulf von Euler in Stockholm and Georg Kahlson in Lund. Today trained pharmacologists sit in all the pharmacology chairs in Sweden, at present more than a score of them.

During the thirties the German cultural influence was still very strong in Sweden, as it was in all the Scandinavian countries. Most medical text books were in German and the scientific journal, *Skandinavisches Archiv für Physiologie,* was published in German. Practically all visiting lecturers were from German-speaking countries; few professors and students could follow a lecture in English or French. In this way we were isolated from the English-speaking scientific world before the last world war. With the war things changed rapidly.

As was customary at that time, my work with Kahlson included a scientific problem whose investigation might become a dissertation. Neurophysiology was developing rapidly. Erlanger & Gasser had received the Nobel Prize in 1944 in physiology or medicine "for their discoveries relating to the highly differential functions of single nerve fibers." The relationship between the thickness of a nerve fiber and its excitability afforded the possibility of selectively exciting the fibers with electrical stimuli of selected characteristics. At that time little was known about nervous control of the acid- and pepsin-secreting glands of the gastric mucosa. I was to attack this problem by the selective activation of vagal fibers and was sent to Ragnar Granit at Karolinska

Institute in Stockholm for guidance. Granit was considered a coming man in neurophysiology: with Hartline & Wald he became Nobel Laureate in 1967 "for their discoveries concerning the primary physiological and chemical visual processes in the eye." I spent six months with Granit while one of his engineers worked on constructing a stimulator that would deliver all kinds of impulses— triangular, rectangular, circular, for example—of varying duration, frequency, steepness, etc. Time went by, the apparatus became more and more complicated, and its completion was repeatedly delayed. I became impatient and returned to Lund. There while waiting for the stimulator I began some basic experiments for practice. With an induction coil yielding an alternating current of about 40 periods per second and a metronome for the rhythmic alternating stimulation of the cat's vagus nerves with about one impulse volley per second, I was able to obtain a profuse gastric acid secretion that lasted for hours. In fact, my stimulation technique was a direct copy of the one described by Pavlov at the turn of the century.

What to do next to fulfill my professor's expectations? An accidental observation answered my question. Suddenly my experiments failed; vagal stimulation no longer aroused gastric secretion, and I realized that in order to improve the collection of gastric secretion I had ligated the pyloric area and probably interfered with the antrum vascular supply. Further analysis showed that interference with antrum function either by cocainization of the antral mucosa or extirpation of the antrum abolished or strongly reduced the secretory response to vagal stimulation. Pavlov had already observed that cocainization of the antrum of dogs blocked the gastric secretory response to feeding or sham feeding. He assumed the effect to be due to paralysis of antral secreting reflexes. In 1906, however, Edkins presented his "gastrin" theory according to which gastric secretion was under the control of the antral hormone gastrin, similar to the way pancreatic secretion was controlled by secretin, as postulated by Bayliss & Starling a few years earlier. Edkins's gastrin theory was never accepted by his contemporaries and more or less died. The American physiologist Ivy denied the existence of gastrin in numerous articles and considered histamine the sole humoral agent operating to control gastric secretion.

I postulated that my manipulations with the cat antrum area in some way had interfered with the gastrin mechanisms and I tried to prove it in cross-circulation experiments between two cats. The recipient cat had a cocainized antrum and no gastric secretory response to vagal stimulation; the antrum of the donor cat was intact. As I had observed earlier, vagal stimulation in the recipient cat did not evoke gastric secretion, but in the donor cat, to my delight, gastric acid flowed, indicating the presence of a blood-borne secretagogue.

Current American scientific literature was hard to come by at Lund, but about this time, I happened to come across a proceedings abstract by Komarov, a pupil of Pavlov who had emigrated to the U.S.A., in which he described the extraction of a secretory principle from the dog's antral mucosa. Komarov assumed this principle to be identical with Edkins's gastrin. I made similar extracts from the cat's antral mucosa and found that, even though such extracts did not evoke gastric secretion in cocainized cats, the accompanying infusion of such "gastrin" preparations and vagal stimulation induced profuse gastric secretion. These observations were presented in my thesis, "The Part Played by the Pyloric Region in the Cephalic Phase of Gastric Secretion" (1942), in which I postulated that gastrin was released by vagal impulses and that gastrin and vagal impulses in some way potentiated each other's effects. Unfortunately, radioimmunoassay (RIA) had not been invented at that time and the occurrence of gastrin could not be directly proven, but shown only indirectly as a gastric secretory response. It would take more than 30 years for my daughter Kerstin and I, using an RIA worked out by Berson & Yalow in New York, to directly demonstrate the vagal release of gastrin in cats.

My thesis, published as was customary as a monograph, met with strong criticism and was passed by the faculty only after several stormy meetings. The criticism with which the faculty at Lund received my thesis hurt me very much, it is true. But what almost broke me was the negative attitude I met abroad. When his gastrin theory was rejected, Edkins left science, married his laboratory assistant, and dedicated his life to university politics. I felt inclined to follow his example, but I resisted the temptation and fought on.

I have dwelled so long on the problems surrounding my dissertation because they illustrate a situation common to young scientists in those days. Young scientists were at the mercy of one or a handful of professors who were quick to criticize and loathe to praise. Most young scientists need positive criticism and encouragement, and my chief, G. K. gave me these. With his support I decided not to give up and become a practitioner, but to continue my research despite the criticism with which it had been received.

Although Sweden was not drawn into World War II, the war period meant almost total isolation from the outside world. Foreign scientific journals—especially English-written ones, of course—arrived only occasionally. At the end of the war, a number of Danish scientists of Jewish birth fled to Sweden and found refuge at Lund. Among them were well-known physiologists like August Krogh and Fritz Buchtal. Despite the unfortunate circumstances under which they came, they brought fresh air and a fighting spirit to the somewhat musty atmosphere of the little university town.

Once the war was over and the borders opened again, Swedish scientists hurried abroad. Now, however, their destinations were not Germany and Great Britain, both devastated and impoverished by the fighting, but to the undamaged and prosperous United States. Since I had been working on gastric secretion, it was natural that I choose to study with a specialist in gastric physiology. In those days, there were two alternatives: Ivy, whose laboratory was at Northwestern University in Chicago, and Babkin at McGill University in Montreal.

Ivy was a student of a famous Swedish physiologist, Anton Julius Karlsson, who was still alive at that time. Babkin was a Russian refugee, a student of Pavlov. I chose to work with Ivy, unfortunately, as it turned out, inasmuch as I got neither education nor ideas from him. Ivy was a hard worker. One found him at his desk practically day and night, Sundays as well as weekdays. He had many young scientists in his laboratories but rarely appeared there himself. These young guests were all working busily by themselves in some corner on problems Ivy had given them. He took their reports and wrote papers on their findings. My ticket of admission was to demonstrate my technique for the isolation of gastrin and its secretory activity. For that purpose, I was given a little dark closet, 2×2 m, without a window or equipment. Evidently, no one had worked with a cat in Ivy's laboratory before, and it took months to get the necessary equipment, in fact very simple things, not to mention getting cats. Since I had no technical assistance, my wife helped me, very bravely indeed since she had never set foot in a laboratory. When after a few months I had prepared my first gastrin extracts and could demonstrate their secretory effect on the cat's stomach, I called for Ivy. He had lost interest by that time, however, and flatly denied that the secretory effect could be due to gastrin. Instead he insinuated that it was due to contamination with histamine.

My main purpose in visiting Ivy's laboratory was to have an opportunity to work with conscious dogs. Ivy was especially famous for his studies on pouch dogs. In some preliminary experiments in Lund I had observed that the gastric secretory response to sham feeding of dogs was inhibited by antrectomy. I wanted to confirm these preliminary observations. However, the dogs at Ivy's laboratory—stray dogs caught by the Chicago police—were so filled with lice and the animal quarters were so filthy that I could not allow my wife to work there. Without assistance in the operative procedures I had to give up. After some months Ivy left his laboratory to assume a position in physiology at the University of Illinois. I felt forsaken and began looking for other scientific contacts at Northwestern.

One day I passed a laboratory in which a bald, kind-looking man was sitting bent over his desk. His name was Horace W. Magoun and he was well known for his experiments with stereotactic instruments, with which he could stimulate well-defined areas in the brain. I went in and asked if I could work in his laboratory. He was busy writing a book, he said, but I was welcome to practice with his instruments. I did so and, lacking any real idea of what to do, I put the electrode into the hypothalamus of a cat. To my surprise, the stimulation induced blood-pressure fall and bradycardia instead of what I expected, blood-pressure rise and tachycardia. My curiosity was stimulated and I continued for a few weeks just mastering the technique. The small observations I made were presented with the kind help of Magoun in the proceedings of the Society of Experimental Biology and Medicine.

In another corridor I found pharmacologist C. A. Dragstedt, who was well known for his studies on the role of histamine in anaphylactic reactions. He and Magoun were my first important contacts in the American scientific community. With their help I made many American acquaintances who would be of the greatest importance for my career, and a few months later, after a tour around the country, I left the States.

Back in Lund after the year abroad, we started enthusiastically to explore the possibilities of Magoun's stereotactic instruments. I was very lucky to have as my collaborator a very able young man, Björn Folkow, who later became professor of physiology in Gothenburg with an international reputation for his studies on circulation and hypertension. Our first experiment hit an area in the hypothalamus of the cat—later defined as the defense area—from which a specific reaction pattern, including rather selective vasodilator responses in the skeletal muscles, was induced. We had discovered what was later described as a cholinergic vasodilator outflow in the sympathetics. We worked happily for several years on the distribution and function of this cortico-spinal vasodilator outflow to the skeletal muscles of the cat and dog. We had (to us) exciting skirmishes with the Nestor of British pharmacology, J. H. Burn, who for many years had defended the existence of adrenergic vasodilator fibers. His main argument was the well-known vasodilator action of adrenaline on the skeletal muscle blood vessels. After reciprocal laboratory visits the dispute was settled and the existence of cholinergic vasodilator fibers recognized.

Our vasodilator research led successively to extensive studies on central and peripheral nervous vasomotor control. Both in Lund and later in Stockholm many dissertations were published in this field. I would like to mention two students, Percy Lindgren and Sune Rosell, who were my collaborators in Stockholm for many years. We found later that the vasodilator action of adrenaline occurring after intravenous injection is transformed into vasoconstriction upon stimulation of sympathetic nerves depleted of their noradrenaline by reserpinization and then reloaded with adrenaline. Apparently, the adrenaline released by sympathetic nerve stimulation hits vascular receptors different from the receptors stimulated by adrenaline given intravenously.

Concomitant with these scientific events, things were happening on the political front. G. K. was very active in the developments to come. In large part due to his forceful propaganda and his good relations with influential members of the sitting Social Democratic government, a Swedish Medical Research Council was formed patterned after the corresponding British organization. Increased resources were given to natural science and medical research and new chairs were instituted at the national universities. As a result of this new commitment to research, Sweden took the scientific lead in Scandinavia, and in some disciplines Swedish scientists have reached the international fore-front.

I was lucky enough to profit from these favorable developments and was appointed to the new chair in physiology at Lund University. Plans were also developed to build a new physiology department at Lund and, as G. K.'s favored assistant, I became deeply involved in preparatory negotiations with governmental authorities, architects, contractors, and others. Building began, but I was never to move into the new laboratories. The partnership with my old teacher and benefactor came to a sudden end. I was made uncomfortable by this development and looked for a way out. What I found changed my life. The chair in pharmacology at Karolinska Institute had been vacated with the retirement of Göran Liljestrand. I applied for the position and to my relief—and to the disappointment of others—I was invited to Stockholm in the summer of 1953.

Karolinska Institute was a new world to me. The faculty for generations had been composed of eminent and forceful scientists enthusiastically and success-fully engaged in developing the institute into a scientific medical center of international repute. To sit on this faculty was a tremendous opportunity for a young professor in pharmacology. My predecessor had been a very forceful and influential member of the faculty, but he was not to witness the great expansion in the science of pharmacology waiting around the corner. I was the lucky one who entered the scene at the right moment.

In the stimulating atmosphere of Karolinska Institute, with its recognized international position, I was able to recruit a group of enthusiastic young collaborators within a few years. The competition among disciplines for young talent was tough, and pharmacology was the last course in the preclinical curriculum—two months at the end of the third year. But I had a few tricks to help to attract some of the better students. One I had learned from G. K. in Lund. On his initiative there we bought a sailing boat for the use of staff members and their groups. It was a 40-foot sailing yacht especially built for sailing in the Swedish archipelago. Since histamine was one of the main interests in our department, the ship was named the *Histamina* and its dinghy the *Histaminase*. From the masthead flew our ensign, in honor of our favorite experimental animals two red cross-laid cats on a white ground. Many agree-able summer adventures at sea created a feeling of solidarity and a community of interests among us that were of great value during the everyday activities of the rest of the year.

When I took over the pharmacology department in Stockholm, I repeated the trick. With some friends I bought a sailing vessel, this time a rather large one that could accommodate quite a few people. It was a 30-ton revenue cutter that we rebuilt ourselves into a ketch, with three foresails on a long bowsprit. It was both fast and easy to maneuver. Many young pharmacologists got their training in seamanship on board this ship and our adventures were legion. I remember especially a trip to Åbo in Finland in 1966, where the 12th Scandinavian Congress for Physiology was to be held. We left Stockholm in fine weather with 12 Swedish scientists on board. We never reached Åbo, however. Instead we ran into a hurricane and almost went down, but we found a port of refuge and some leading members of Sweden's scientific community survived to continue their work.

The Department of Pharmacology in Stockholm was relatively new. It was built in 1948 when Karolinska Institute moved from its old site close to Stockholm's center to its new grounds at Norrbacka just north of Stockholm. Unfortunately, the original building plans had for economic reasons been reduced by 50% and as a result the animal quarters and the laboratory equipment, among other things, did not fulfill modern requirements, at least not in my opinion. My first step as chairman was to deliver an address to the faculty in which I presented the future of pharmacology as I saw it. I anticipated the development of various fields of pharmacology, such as biochemical pharmacology, neuropsychopharmacology, clinical pharmacology, and toxicology, and recommended the eventual establishment of chairs in these fields. I also asked for the immediate enlargement of the pharmacology department, especially its animal quarters, and for money to modernize the equipment. Many faculty members shook their heads at this presumptuous newcomer, but I must say to their credit that they were very cooperative; two decades later my demands were fulfilled in good measure, including chairs at the institute in all the subdisciplines I had mentioned.

In retrospect, the pharmacology department at the institute seems to have developed rather smoothly, but at the time some of our victories seemed hard won. Especially difficult was convincing the Ministries of Education and Finance to support our growth. I remember particularly the resistance that met our plans to enlarge the department. When the ministeries rejected even our request for a barrack to be used as temporary animal quarters and erected at our own expense, we were discouraged. But the stricture "that any animal quarters, even temporary ones, should be removable without any delay in case the grounds were requested by the building board for other purposes" gave us an idea. We bought circus wagons from the Stockholm amusement park—the original horse-drawn ones—equipped them with cages for rats, cats, and dogs, painted them bright red with white windowframes, and placed them in a semicircle outside the pharmacology department. This anachronistic sight aroused the public and angered the government authorities. They never forced us to remove the wagons, however, and after a few years of publicity in the leading newspapers we were allowed to build not only new animal quarters but new laboratories. The next problem was equipment. Fortunately, generous support from the National Institutes of Health saved us. In this context, I want to emphasize that financial help from the U.S. in the two decades after the war was of inestimable value in expanding the institute's scientific activities during these years. Unfortunately, Swedish political concerns forced a gradual decline in U.S. financial support. In fact, for a time such support practically disappeared.

Another important force in the development of Karolinska's scientific potential has been its participation in the Nobel Prize Awards in physiology or medicine. In the beginning the institute was reluctant to accept the responsibility, but over the years it has been very well managed and put to profitable use. Serving on the Nobel Committee, as I have done for decades, is an inestimable opportunity to watch international scientific developments. Moreover, studying the achievements of eminent colleagues has been both rewarding and stimulating to my own ambitions.

As I mentioned earlier, of primary interest at the Department of Physiology in Lund in the 1940s was histamine. This amine has intrigued physiologists since its synthesis by Windaus & Vogt in 1907, but its functional role was and still is obscure. Histamine was known to be distributed in tissues all over the body but its precise localization or mode of storage was still unknown, although it was believed to be stored in some way chemically linked to protein. The new extraction technique elaborated by Charles Code at the Mayo Clinic raised new expectations among histamine enthusiasts, but it had still to be assayed biologically on the guinea-pig ileum, which had been found to be especially sensitive to it. This biological assay was both time-consuming and tedious and I felt sorry for my young colleagues, who spent their time on these boring assays. I had resisted all invitations to join the histamine gang, but by the time I arrived in Stockholm things were changing. Chemical techniques had been developed for determining histamine, specific histamine releasers like compound 48/80—discovered by MacIntosh & Paton in 1949—had appeared, and histamine was found to be localized in mast cells by Riley & West in 1953.

Ever since my "gastrin period" in Lund, storage and release phenomena had spurred my interest. How could biologically active substances be stored in an inactive form and "reactivated" immediately on release? In our laboratory cat consumption was high but the rewards were considerable. To make better use of the animals, we decided to perfuse cat limbs in histamine-release studies. Cat skin turned out to be very rich in histamine, as well as a rich source of slow-reacting substance (SRS), so-called because it induces a slow contraction of guinea-pig ileum. SRS was known to accompany histamine release in anaphylactic reactions and was assumed to be at least partly responsible for anaphylactic symptoms. During the 1960s and 1970s we studied the chemical and biological characteristics of cat SRS, and quite a few young scholars defended their theses on SRS problems. We had an almost pure SRS preparation, its UV spectrum identified and ready for mass-spectrographic study, when Samuelsson and his group very elegantly solved the problem and developed the now well-known leucotriene hypothesis that won him the Nobel Prize in 1982. Samuelsson isolated leucotrienes from leucocytes, but in my mind there is no doubt that the cat SRS belongs to the leucotriene family.

Parallel with our research on SRS, our study of histamine developed nicely. Since histamine is located in mast cells, a rational approach to problems concerning its storage and release required access to isolated mast cells. Somewhat pure suspensions of rat mast cells had been obtained by gradient centrifugation of mixed cell populations from rat peritoneal washings on sugar. Unfortunately, such cells, although morphologically seemingly intact, were insensitive to compound 48/80 and other histamine-releasing agents. In our search for other density gradients we decided to try dextran, but by mistake the pharmacological company delivered ficoll. This new polymere turned out to be an excellent gradient material and we were able to obtain up to 95% pure mast cell suspensions. Bertil Diamant, now professor of pharmacology in Copenhagen, many other young collaborators, and I experimented with these mast cells.

In the beginning, our interest focused on the mechanism of mast-cell response to "degranulating" agents like compound 48/80 and antigens. The dependence of the response on oxidative and glucolytic energy production and the role of ATP in the release process were among the questions studied. We theorized that lecithinase A (later phospholipase A) played a role in an initial step of mast-cell reaction. Recently, phospholipase A activation has again been the focus of attention, this time as one of the initial processes in release mechanisms.

As years went on, our interests shifted from mast-cell response to the mechanism of histamine storage and release in mast-cell granules. We observed one day that granules isolated directly from mast cells lysed in deionized water still contained histamine but that they immediately lost their histamine when suspended in salt solutions like 0.9% NaCl or serum. This observation indicated to us that the histamine in the granules was stored in weak ionic linkage to the granule matrix and was then released by cation exchange once the granules were exposed to cation-containing media. For the next 20 years we were spellbound by the idea that the storage and release not only of histamine but also of other biogenic amines and other charged substances, for example neuropeptides, might be effected according to the cation-exchange principle.

The idea that histamine is stored in ionic linkage to anions in mast-cell granules was not original. The presence of heparin in the granules led early to the assumption that histamine is stored in ionic linkage with this strongly acid polysaccharide. We soon found that this was a premature conclusion, however. An analysis of the binding properties of isolated granules showed stoichiometric relationships and pH-dependence, for example, indicating histamine's ionic binding not to the sulphated groups in heparin but to the carboxyl groups. Chemical analysis of the granule matrix, as well as quantitative studies of the storage capacity of isolated granules and of "artificial granules" (a heparin-protein complex) supported our idea that histamine is stored in ionic linkage to carboxyls in a protein-heparin complex in which the sulphated groups of heparin are masked by strong ionic binding to the amino groups of the protein. The fact that histamine is bound in weak ionic linkage to protein carboxyls and not to the strong acidic groups of heparin was an essential observation of importance for our later speculations about the cation-exchange process as a general principle in storage and release mechanisms.

Quantitative studies of the storage capacity of mast-cell granules for histamine and sodium, the stoichiometric relationships between sodium uptake and histamine release and other questions led to the conclusion that in vitro mast-cell granules behave as weak cationic exchangers with carboxyls as the ionic binding sites. Electron microscopic studies strengthened our belief that the release of histamine from the "degranulating" cell also occurs as a cation exchange sodium $\rightleftharpoons$ histamine. Later, a closer inspection of our electron microscopic pictures—the result of a collaboration with the eminent Hungarian specialist Pal Röhlich—cast some doubt on the current idea that degranulation of the mast cell is a primary and histamine release from the expelled granules a secondary step in the release response. Through electron microscopic auto-radiography we observed that granules in the periphery of the cell may swell and loosen their histamine in spite of the fact that they have not yet lost their membranes. In other words, the electron microscopic pictures indicated that histamine release—probably by cation exchange across membranes with increased permeability or through newly formed pores—might occur as the primary step, with the expulsion of granules a secondary phenomenon. I mention this observation because I later defended the idea that the release of neurotransmitters might occur not as an exocytosis but as a fractional release of the transmitter by cation exchange initiated by the nerve impulse. I will come back to these recent studies below.

The observation that led to the idea of cation exchange being a general principle of amine storage and release was the observation that mast-cell granules in vitro take up not only sodium and histamine in a competitive way, but also other amines, biogenic as well as synthetic, as long as they contain a charged amino group. All these amines, among them the transmitter amines noradrenaline, adrenaline, dopamine, serotonin, and acetylcholine, seemingly compete for the same ionic sites in the mast-cell granules and are released by sodium and other inorganic cations. Therefore might the matrix of other amine-storing granules also have properties similar to cation-exchanger materials?

So far we have studied chromaffin granules and adrenergic nerve granules from this point of view. We discovered that the matrices of these two types of granules show properties reminiscent of cation-exchanger materials, with carboxyls as the ionic binding sites. Of special interest to us were experiments that demonstrated great similarities in the cation-exchange properties of the synthetic carboxyl resins, e.g. the Amberlite IRC-50 and Sephadex C-50, and the matrices from these granules. When the synthetic resins and isolated granules from chromaffin cells and nerves were exposed to isotonic NaCl by superfusion, the release of catecholamines and noradrenaline showed the same kinetics. Recently, the in vivo release of catecholamines from cat and pig adrenals induced by supramaximal splanchnic nerve stimulation shows the same kinetics observed in vitro. At present, we are looking for further evidence that the storage and release of transmitter amines follow the principles of cation exchange and that the transmitter release and perhaps the release of other amines and neuropeptides occur not by exocytotic emptying of a few granules but by cation exchange as a fractional release from multiple granules.

The modern and roomy animals quarters approved by the authorities in the face of our circus wagons provided the facilities for gastrointestinal research on conscious animals. Gastrin research began to see movement again with the important contributions of C. A. Dragstedt, previously professor of surgery in Chicago, who was very active after his retirement in experimental gastrointestinal surgery in Gainesville, Florida. He manipulated the dog antrum surgically and found that excluding the antrum either as a separated pouch or as a transplant in the colon led to a hypersecretion of HCl in the remaining stomach, frequently followed by penetrating peptic ulcers. This hypersecretion disappeared when antrum mucosa was acidified or extirpated. Dragstedt considered that the changes in HCl secretion he observed reflected disturbances in the gastrin mechanism, since gastrin release from the antrum is inhibited by acid pH.

These findings revived my old interest in gastrin and we decided to confirm and if possible extend the observations I had made in Lund that the acid secretory response to sham feeding disappears after antrectomy. My young collaborator, Lars Olbe, did an excellent job in conducting the experiments. He invented an esophagal cannula of plastic that allowed the dogs to feed themselves and survive in good health for years. Olbe found not only that antrectomy practically terminates the acid secretory response to sham feeding but also that intravenous infusion of subthreshold doses of gastrin *during* sham feeding induces a dose-dependent acid secretory response. In other words, my previous observations on the interaction between gastrin and vagal impulses in anesthetized cats were fully confirmed in conscious dogs.

There still was no technique for determining gastrin in the blood, but it would come soon. Solomon Berson & Rosalyn Yalow reported in 1955 that they had developed antibodies against insulin and had used them to develop a radioim-munoassay. Their report caused a sensation, since it was considered impossible to induce antibody production against such small protein molecules. In fact, a respected scientific journal refused to publish Berson & Yalow's first paper unless they changed the term *antibody* to a more neutral term. Today everyone knows that Berson & Yalow were correct; their radiommunoassay for insulin revolutionized the whole polypeptide area. Unfortunately, Berson died unex-pectedly of a heart attack—he worked himself to death—but Yalow was awarded the Nobel Prize "for the development of radioimmunoassays of peptide hormones" in 1977, a prize meant for both of them.

I met Berson before his death, when he gave the Ihre lecture at a Nobel symposium in Stockholm in 1970. At a dinner at Bengt Ihre's home I asked him to work out a radioimmunoassay for gastrin. He promised to try and a few months later he called me from New York to say that he had a gastrin RIA ready for me, would I come to New York to see it? Although I did not consider it possible to go myself, I sent one of my young pupils, Göran Nilsson. A few months later in Berson's laboratory, Nilsson and Berson were able to show that sham feeding induces a rise in plasma gastrin level. When the gastrin RIA was brought home to Stockholm, my daughter Kerstin, as already mentioned, demonstrated that vagal stimulation in the cat causes a considerable output of gastrin-17 not only into the blood but also into the stomach. Since then, in our laboratory as well as in others, researchers have shown that the vagus nerve contains several polypeptides—cholecystokinin, vasointestinal peptide, sub-stance P, insulin, somatostatin, gastrin, and many others—that appear in the blood and gastrointestinal canal when the vagus nerve is activated. In fact, in addition to sham feeding, suckling leads to a hormonal release pattern with a concomitant occurrence of gastrointestinal and hypophyseal hormones, in-dicating an intimate central coordination between these hormonal delivery sources.

Our gastrin studies stimulated our interest in the role of the antrum-duodenum area in controlling gastric acid secretion. The Pavlov school consid-ered this area a reflex center for the excitatory and inhibitory control of gastric secretion.

We were able to confirm previous observations that acidification of duodenal content leads to a pronounced inhibition of acid secretion in Pavlov and Heidenhain pouches. However, this inhibition does not occur until the pH is suppressed 2–3. Since such a low duodenal pH is never observed under normal conditions, the functional role of the duodenal inhibitory mechanism has been seriously doubted. However, between meals the stomach empties its acid content into the duodenum, and as a result there might be an acidity gradient with low pH close to the pylorus, the pH rising with increasing neutralization during the distal passage of duodenal content. Could the inhibitory mechanism be located in the bulb and, if so, does bulbar pH reach the required low values? We found in fact that acidification of the duodenal bulb to pH 3–2 or lower does activate the inhibitory mechanism. Moreover, with a series of technically advanced surgical procedures it could be shown that the bulbar inhibitory mechanism is humoral. We baptized the unknown inhibitory principle *bulbogastrone*, but our attempts to identify it chemically in duodenal extracts failed.

In the meantime another of our colleagues, Sven Andersson, was sent to Los Angeles, where Morton Grossman of UCLA had perfected a technique with which it was possible to register the pH-gradient in the duodenum. They found that in both man and dog postprandial pH in the duodenal bulb decreases to 2–3. In other words, the bulbogastrone mechanism might play a physiological role in the inhibitory gastric acid secretion. Later my daughter Kerstin observed that acidification of the duodenal bulb leads to increased somatostatin levels in the peripheral blood. These levels are high enough to exert inhibitory actions. Whether bulbogastrone is identical with somatostatin remains to be established.

As is evident from this somewhat rhapsodic narrative, my scientific activities and those of my collaborators were physiological in nature. One is justified to ask what my department has done for the field of pharmacology. In fact, rather much.

The scientific staff in the Department of Pharmacology at Karolinska Institute has reached considerable size. Last year the list included four professors, four senior lecturers, four assistant professors, two adjunct professors (part-time scientists from the pharmaceutical industry), seven research fellows, and twenty-five postgraduate students. More than 150 scientific papers were published. Out of 80 postgraduate students who have defended their theses at the department in the last 30 years, 25 are professors of pharmacology or of related theoretical and clinical disciplines. Our research activities cover such diverse fields as biochemical pharmacology, cancer chemotherapeutics, prostaglandins, the physiology and pharmacology of adrenergic transmitter mechanisms and of purines, neurohumoral peptide physiology, the pharmacology of gastric secretion, dental pharmacology, and neuropsychopharmacology. If the two chairs in clinical pharmacology and one in toxicology are included—they have branched off to form their own departments—the scientific output is very high indeed.

I suppose such a large and diverse department reflects favorable developments in the pharmacological field, but at the same time the growth of huge departments with many independent research groups has its risks. The diversification of scientific interests can lead to in-fighting among special interests. Harmony and communication among individuals can be jeopardized. Without common goals, the strength and in the long run the future of the department are endangered. I have no solution to the problem, but I can recommend one rather effective measure to prevent or at least retard such an undesirable development. This is the organization of what we have called a manuscript committee, whose members are the "grown up" scientists in the department. This committee reads and edits all manuscripts emanating from the department before they are published. Such a procedure is a great help to young researchers and effectively serves to uphold high scientific standards for the publications, since nobody—especially not the older professors—wants to present bad papers. What is more important, however, is that everyone gets to know what work is being done throughout the department, an effective measure against the isolationism and self-sufficiency that often occur in large organizations.

Teaching pharmacology to medical students, one was confronted with how little doctors, clinicians as well as practioners, knew about drugs. Refresher courses in pharmacology for doctors were instituted to rectify this situation, and such courses led directly to the development of clinical pharmacology, today a strong discipline in Sweden, with chairs and head positions at every teaching hospital in the country. In the beginning, clinical pharmacology met with scepticism, especially from internists, but resistance disappeared when clinical pharmacologists were relieved of their hospital duties to serve as teachers and advisers in drug therapy and drug research. It is still that way in Sweden, and the collaboration between clinicians and clinical pharmacologists is intimate and profitable.

The growth of the medicinal industry and the introduction of chemical and biochemical techniques after World War II opened up new ways to study the pharmacodynamic properties of drugs. Increased insight in the quantitative aspects of drug action on animals and man paved the way for quantitative drug therapy and the development of clinical pharmacology.

These new dimensions within pharmacology weakened the traditional ties between pharmacologists and physiologists. In fact, the new generation of biochemical pharmacologists and other scientists felt no loyalty to physiology at all. Physiologists and the older generation of pharmacologists, who had received their scientific training in physiology, did not seem to realize the explosive power of these new trends.

Except in the leading scientific communities in France, Germany, Great Britain, and the U.S., national pharmacological societies were formed slowly. Nationally and internationally, pharmacologists were represented by physiological societies and there was growing dissatisfaction among pharmacologists with their inability to influence the scientific program at physiology congresses.

In the 1950s, beginning at the 18th congress of the International Union of Physiological Sciences (IUPS) in Copenhagen in 1950, a special pharmacology day was arranged at the end of the meetings. At these gatherings, the question of an independent international organization of pharmacologists was repeatedly put on the agenda. An international committee was set up with Corneille Heymans of Ghent as chairman and Carl F. Schmidt of Philadelphia as secretary. After nearly ten years of discussions and negotiations, IUPS in 1959 agreed to the formation of an independent division for pharmacologists within the organization, the Section on Experimental Pharmacology (SEPHAR). The agreement was a typical compromise. The constitution of IUPS was revised to authorize SEPHAR to "organize international conferences, symposia and congresses and to carry on other activities provided that they do not conflict with the aims and principles of IUPS." This objection led to an agreement that SEPHAR "would do its best not to compete with or otherwise weaken the triennial congresses of IUPS whenever it arranges separate international pharmacological programs." Carl Schmidt was elected the first president of SEPHAR, with Daniel Bovet as secretary. The SEPHAR council had its first meeting in Stockholm in 1961. Council loyalty to IUPS was demonstrated by its appointment of a liaison officer, E. J. Ariëns, as a member of the local organizing committee for the next physiology congress in Leyden in 1962. The council also decided to continue pharmacology day at future physiology congresses.

As years passed and increasing contacts with foreign colleagues widened my horizons, I could not avoid recognizing the increasing tension between physiologists and pharmacologists, as the repeated proposals for an idependent international pharmacological organization showed. The notion of holding an international meeting of pharmacologists in Stockholm was born over a drink in my home one fall evening in 1958. My friend Tom Maren, professor of pharmacology in Gainesville, Florida, and I were discussing the unsatisfactory international position of pharmacology. True, negotiations had by then begun to form a pharmacology division within IUPS. Even so, dissatisfaction was widespread, especially among biochemically oriented pharmacologists, who felt no community with physiologists and who wanted to break loose completely to organize their own scientific programs.

Tom Maren belonged to the group of young non-traditional biochemical pharmacologists who pleaded for the independence of pharmacology. Tom and I decided to make inquiries on the question among prominent friends and colleagues. The correspondence in my files reveals little enthusiasm and encouragement for the formation of an independent pharmacological association. Most responses were ambiguous or passive. Some were directly negative. Most felt that to break with the physiologists was a mistake; many predicted difficulty in raising the necessary funds.

In the meantime SEPHAR was officially established—not without opposition within the parent organization—at the IUPS congress in Buenos Aires in 1959. As mentioned above, according to the revised IUPS statutes, the new division was authorized to organize its own international meetings.

The formation of SEPHAR paved the way for international pharmacological activities. But when we in Stockholm undertook to arrange the first international pharmacological meeting, we were well aware of the divergences of opinion, not only about what form such a meeting should assume but even about whether it should be held at all. Some pharmacologists were enthusiastic advocates of an international pharmacological congress of the conventional kind. Others favored the organization of symposia. Lastly, there were those who rejected the whole idea. In particular I understood the apprehensions of those who felt that an international congress of pharmacologists would weaken the valuable communication between physiologists and pharmacologists long fostered by the international physiological conventions. As a physiologist by training I could well appreciate that point of view. However, like many others, I felt that the danger of a schism among the pharmacologists themselves was so great that some form of an international gathering should be arranged. A feasible compromise solution—at least for the time being—was to organize pharmacological meetings that would provide satisfactory interchange between older and younger generations and offer opportunities for contact not only with physiologists but with biochemists and representatives of other allied disciplines.

At a preliminary informal meeting in Washington early in the winter of 1959, I declared my willingness to arrange a pharmacological program in Stockholm. SEPHAR's agreement not to compete with the 1962 IUPS congress precluded emphasis on physiological presentations. The recent outstanding developments in biochemical pharmacology spoke in favor of a biochemical approach. We therefore decided to put together a series of symposia on the topic, "Modes of Actions of Drugs." One of the most active spokesmen for this program was Bernard B. Brodie, at that time head of the Department of Chemical Pharmacology at the National Institutes of Health in Bethesda, who gave us his enthusiastic and invaluable support from the beginning. K. K. Chen from the Lilly Company in Indianapolis was another indefatigable and influential supporter. C. Heymans of Ghent and C. Schmidt of Philadelphia, who joined when we were well embarked on the adventure, also gave full and unflinching assistance. Even IUPS contributed 2,500 U.S. dollars for preliminary expenses.

In spite of the efforts of the organizing committee to give the program an international character, scientifically as well as geographically, Americans clearly dominated the meetings. Five out of eight organizers and 478 out of 1483 attendees were Americans. In angry letters the Russians, French, Belgians, and others accused us of favoring or giving in to the Americans. The fact is, however, that the program committee's decision to emphasize biochemical pharmacology was more or less forced upon us by our agreement with the IUPS not to focus on physiology and the Americans, and to a certain extent the Germans, were the leaders in biochemical pharmacology at that time. This new branch of pharmacology was still rather undeveloped in most European countries, especially in those where the complainers came from.

The Stockholm meeting was to become not only the first in a series of successful international pharmacological congresses and a strong impetus to the development of the field. It also led to the formation of the International Union of Pharmacologists (IUPHAR). IUPHAR was officially inaugurated by the General Assembly of the IUPS in Tokyo on September 2, 1965. The first ordinary meeting of the IUPHAR council was held in Sao Paulo on July 28, 1966. Since then seven IUPHAR congresses have been held, in Stockholm in 1961, headed by myself; in Prague in 1963, headed by H.·Rašková; in Sao Paulo in 1966, headed by M. Rocha e Silva; in Basel in 1969, headed by K. Bucher; in San Francisco in 1972, headed by R. Featherstone; in Helsinki in 1975, headed by K. Paasonen; in Paris in 1978, headed by P. Lechat; and in Tokyo in 1981, headed by S. Ebashi.

Within IUPHAR, the divisions of clinical pharmacology and of toxicology demonstrate not only the growth of pharmacology as a field, but also IUPHAR's commitment to represent all aspects of the discipline, allowing new branches independence within the framework of the parent organization. With its membership in ICSU, WHO, CIOMS, and various other international scientific organizations, IUPHAR belongs to a global network of government, academic, industrial, and other organizations through which it exerts worldwide influence on all aspects of pharmacological research and teaching as well as on drug development and pharmacotherapy.

My years as secretary of SEPHAR and then as president of IUPHAR were a very challenging and profitable time, both scientifically and personally. My understanding has widened; my circle of friends includes people from all over the world. The steady growth of IUPHAR and the success of its congresses, begun so modestly in Stockholm over 20 years ago, has given me great personal satisfaction.

To be a retired professor in Sweden has its advantages. The law provides an emeritus professor a laboratory and office space for research and teaching activities, that is, if the available resources allow. I am a very lucky man to have retired from a department with such resources, so I spend my time more or less as before, in my office and in my labs, aided by kind, loyal, and experienced assistants and coworkers. My three hunting dogs accompany me to my office and home as well as hunting, and I can work undisturbed by committee meetings and the other official duties that previously were a heavy burden. I consider myself lucky to have entered pharmacology at the beginning of its rise to an independent discipline and to have witnessed its enormous national and international growth from a branch of physiology to an important discipline in the forefront of medical research. I have never regretted my desertion of physiology for pharmacology.

Arnold D. Welch

Ann. Rev. Pharmacol. Toxicol. 1985. 25:1–26

REMINISCENCES IN PHARMACOLOGY: AULD ACQUAINTANCE NE'ER FORGOT[1]

Arnold D. Welch

Division of Cancer Treatment, National Cancer Institute, National Institutes of Health, Bethesda, Maryland 20205

INTRODUCTION

Avoidance of personal pronouns in reminiscences is difficult; nevertheless, the main theme of this memoir will relate insofar as possible to its subtitle (with apologies to Robert Burns). Of necessity, this peripatetic writer has interacted with a great many colleagues during a period of over half a century. Marcia Davenport, in her autobiography (1), wrote beautifully, "I look back across the years and know of course that the real substance of my life, as of all lives, is the men and women with and through whom I have lived." And so it has been with me.

ORIGINS

My life began (7 November 1908) in a small New Hampshire town to which my parents had moved, after illnesses and business reverses, from Lynn, Massachusetts (where we were to return when I was eight). Of my ancestors, many of whom had settled in these areas in the seventeenth century, only the Reverend Stephen Bachiler seems particularly memorable (he even possessed a coat of arms!): he founded the first church in Lynn and then left hurriedly to

---

[1]The US Government has the right to retain a nonexclusive royalty-free license in and to any copyright covering this paper.

launch the town of Hampton, New Hampshire. The Reverend Stephen left a great many descendants, both in the colonies and in England (where he returned in his nineties); one of these, my maternal grandfather, was Colonel Joseph DeMerritt Batchelder (a colonel of what I have not discovered). Although at least doubly descended from Stephen, I had from 500 to 2000 lineal ancestors in the seventeenth century, of whom he was one; of the other 99.9% but little is known: par for the course in the USA. The original Welch in the colonies also derived from England, but the term *Welch* originally meant foreigner (a term appropriately applied to the vexatious Celts of Wales); hence, the surname Welch does not certify a Welshman. I especially regret that I cannot safely claim descent from the illustrious William Henry (Popsy) Welch, one of the founding fathers of The Johns Hopkins Medical School: Popsy died a bachelor. My sometimes mildly annoying middle name, DeMerritt, presumably was once de Mérite; this name, shared with my grandfather, was derived from his mother's family and now belongs to one of my grandsons (he is not yet old enough to regret its other connotations).

EDUCATION (THROUGH 1931)

My education began in the home, where my stern father's demands engendered little mutual affection. When I was seven he began, rather unsuccessfully, to expose me to French and Spanish; nevertheless, I am now grateful for his teaching, because my interests in linguistics, etymology, and grammar resulted (which perhaps contributed to my alleged editorial talents and to my being dubbed a *which*-hunter). The death of my mother when I was thirteen devastated me and adversely affected my scholastic performance. Fortunately, two years later my father retired to Florida, where I had a wonderful science teacher who encouraged my scholarship and devotion to chemistry. [This led to my eventual membership in the *23 Club* (in the Federated Societies 2 stands for biochemistry and *3* for pharmacology); for the older pharmacologists, however, such membership was a kiss of death. Times indeed have changed; otherwise, an invitation to write this prefatory chapter surely could not have been extended.]

My entrance into the University of Florida was helped by my successful competition for a scholarship, which so surprised my father and a bachelor uncle that I received more financial and moral support than I had expected; nevertheless, I always had at least one job while at the university. In spite of such distractions, I attained certain honors, such as Phi Beta Kappa, Phi Kappa Phi, and even Blue Key. In my fifth year, I began my studies in pharmacology; my first paper, with B. V. Christensen, appeared in 1932 in the *Journal of Pharmacology and Experimental Therapeutics*. My initial textbooks were those of Sollmann (whom I was to succeed at Western Reserve University in 1944) and of Meyer & Gottlieb as translated by Velyien E. Henderson (who was to guide my research to a PhD in 1934 at the University of Toronto). I was pleased in 1973 to receive the DSc (hc) from Florida thanks to my good friend Tom Maren.

Before moving to Toronto, I spent a pre-fellowship summer session in physiological chemistry at the University of Minnesota. The then chairman was not inspiring, but the professor of pharmacology at Minnesota, Arthur Hirschfelder, gave me friendship, lab space, and an introduction to toxicology. A belated offer from Toronto's Henderson afforded me, then a lowly MS, the monetary equivalent of a post-doctoral teaching fellowship (then $1500). In those terrible depression years, such a fellowship was manna from heaven. My release from the anticipated fellowship in P-chem made it available to a friend from the University of Florida, Earle Arnow. A few years after finishing his PhD and MD, Arnow joined Sharp & Dohme as director of biochemical research. He succeeded me as director of research in 1944 and went on to an outstanding career at Merck, Sharp & Dohme and later at Warner-Lambert.

Here I must note the almost incredible effects on human lives that the apparently inconsequential act of a single individual can have. Henderson's belated offer to me, which had resulted only from the last-moment defection of a young MD unknown to me, had great impact on my life, on that of Arnow, on those of several thousand of our medical and graduate students, post-docs, and almost innumerable colleagues.

TORONTO YEARS (1931-1935)

The years in Toronto were eventful and sometimes chilling, both literally and figuratively. My initial course in physiology, presented by Best & Taylor before their classic textbook was first published, was stimulating. In Toronto I formed a deep and lifelong friendship with Tom Jukes (we even shared digs[1] for two years before he went off, as Dr. T. H. Jukes, to Berkeley as a post-doc). Both Charlie Best and Fred Banting befriended us; from them I learned a bit about the real story of insulin. I learned much more from Brock, the chief "diener" in pharmacology, a good friend of mine as well as of Sir Frederick-to-be. How Banting handled another of the four main scientific participants at Toronto when Collip refused to disclose his method for partial purification of the pancreatic principle is not to be found in most sources, but perhaps can be surmised. Henderson supported Banting when McLeod, with whom Banting was forced to share the Nobel Award, was unable to find funds for him. Although Best, a medical student, did not share the honor of the prize, Banting

---

[1]My first experiment in clinical pharmacology resulted from Tom's and my discovery that our lonely bottle of medicinal Scotch was being mysteriously emptied. Suspecting the nephew of our motherly landlady, I carefully calculated the necessary (but non-lethal) amount of emetine and added it to the residual spirit. The culprit identified himself very audibly and further problems were avoided without retaliation.

promptly gave him half of his share of the money, which shamed McLeod into doing likewise for Collip, while McLeod was offered a Scottish chair. Best had earned his MD by 1925; then, shortly after a few years with Dale in London (under whom he earned a DSc), he was appointed chairman of physiology at Toronto at the age of thirty. Thus, when I first met Best in 1931 he was only ten years older than I.

I grew very intolerant of the apparent complacency of many pharmacologists, who had neither the training for nor any interest in probing molecularly into drug actions. Certainly the idea that new drugs would some day be designed was yet to come (2). Classical pharmacologists regarded biochemistry with anathema rather than anticipation. Henderson, although he sometimes took a rather dim view of my biochemical leanings, was very tolerant, all things considered. It was Best, however, who encouraged my initial delvings into structure-activity relationships, especially among analogs of choline (later I continued these studies with Tom Jukes). Best and others had already shown choline to be required by depancreatized dogs maintained with insulin, and he encouraged my further studies of mechanisms with Huntsman.

New approaches to mechanistic studies of drug action had been initiated by the splendid book by A. J. Clark, poorly titled *Applied Pharmacology,* which introduced much biophysical analysis, but it was the work of Otto Loewi in Graz in 1920 that began the biochemical revolution. Vagal neurotransmission was shown to involve a biochemically labile entity [later identified as acetylcholine (ACh)] that was enzymically inactivated. Indeed, Loewi demonstrated that one of the most classical of drugs, eserine or physostigmine, exerts its powerful actions by inhibiting the inactivating enzyme. (Henderson moved promptly into this area and carried out his important studies on the chorda tympani nerve, ACh, and the submaxillary gland; however, it was his paper on the mechanism of erection that prompted many requests for reprints!) Loewi also probed the sympathetic neurotransmitters, which had been anticipated by Elliott and by Langley, although their studies had had little impact. The structure-activity studies of congeners of adrenaline by Dale and his associates Barger and Dudley offered clues not fully recognized at first. Nevertheless, the work of the brilliant Dale, by then Sir Henry (who had noted in 1914 that synthetic ACh mimicked the effects of parasympathetic nerve stimulation and that ACh was quickly inactivated by a tissue extract), led to the Nobel Prize for him and Loewi in 1936.[2]

[2]Had I had the courage of my convictions (a dated notebook outlines my reasoning), as well as the necessary experience and encouragement, I might have identified L-noradrenaline (the synthetic DL-form was then termed arterenol) as the sympathin E postulated by Cannon. Indeed, I obtained arterenol as well as D- and L-acids for its possible resolution, but other pressures prevailed. Hence, it remained for U. S. von Euler over ten years later to establish that L-noradrenaline is indeed the excitatory sympathetic neurotransmitter. As one wag recently stated, "You lose some and you lose some."

During my graduate years the adrenal medulla fascinated me, especially the claim by Kendall (of later cortisone fame) that the remarkable stability of adrenaline in the gland, compared with that of the pure compound, was attributable to its conjugation with lactic acid. With post-doc Don Heard, I perfused fresh bovine glands; we reported that gland-derived ascorbate in the perfusate prevented the oxidative inactivation of the catecholamines. This and another report on the mechanisms of oxidation and stabilization of adrenaline caught the attention of the Coris, who then used ascorbate to protect minute amounts of adrenaline during their studies on glycogenolysis. This fortunate circumstance led to correspondence, a meeting, a job and an MD degree for me.

Martin Roepke and I studied ACh as a cation and showed, with a model system, that it possibly was attached to an anionic receptor in or on cells. This led to my conviction that the quaternary N of ACh might be replaceable (so to speak) by quaternary P or As and that such synthetic compounds might behave qualitatively like ACh. Indeed they did, perfectly, although the P-analog exhibited 10–20% of the activity of ACh, while the As-analog had 1–2% of its activity. We also prepared the planar molecule: $(CH_3)_2S^+$-CH_2-CH_2-O-CO-CH_3 (Cl^-). This too was highly active and, like the other analogs, was inactivated by choline esterases, potentiated by eserine, and blocked by the classical drug atropine.[3]

Despite (or because of) these often exciting days, the Chief finally put his foot down and gave me some new alkaloids to study, saying that it would be good for me to learn some neuropharmacology. Perhaps the *real* reason was that these compounds had come from the Canadian National Research Council and had to be studied and I was the rather unwilling victim. The most interesting of these hydrastine-like alkaloids, bicuculline, was a powerful convulsant. Who could have guessed that nearly forty years later, when bicuculline had become a valuable tool in the study of GABA, I would be introduced by Professor Curtis in Canberra as the father of bicuculline? Indeed, Henderson was the father and I only an illegitimate son. The visit in Australia gave me then, as has occurred many times later, an opportunity to renew my warm friendship with Adrien Albert, the author of a real classic, *Selective Toxicity,* new editions of which continue to be in demand.

Henderson lectured to medical students cloaked in a decrepit baccalaureate gown. He always began by peering over his pince-nez with a somewhat

---

[3]In the synthesis of the S-analog of ACh, we were too impatient to wait for $(CH_3)_2S$ to be delivered and chose to synthesize it; fortunately, this was on a Saturday, because our attention wandered for a few moments and exothermia took over. Stinking $(CH_3)_2S$ shot out of the reflux condenser; the mess was cleaned up and the stench, to our relief, was abated by Monday. Had the disaster occurred on a weekday, the medical building would have been evacuated and our own evacuation could well have been permanent.

sardonic grin. After saying, in a very British manner, "In my laẃwst lect-chaw," he would continue in the speech of Upper Canada. The students loved it, although as individuals they were terrified of him, not without reason. The only other staff member, George Lucas, PhD, who was perhaps appropriately listed on the departmental letterhead as Ass. Prof., did not lecture. Henderson regularly took Tom Jukes, the post-docs, and me for an hour or two on Saturday mornings to drill us in scientific German. He even certified (doubtless with his fingers crossed) that Tom and I were qualified for our language requirements in French and German. May this very kindly scholar, who hid behind a mask of acerbity, rest in peace, as I wrote most sincerely in an obituary after his sudden death.

ST. LOUIS YEARS (1935–1940)

Under Carl Cori in St. Louis I became for a second time a scientific great-grandson of the reputed father of pharmacology, Schmiedeberg (actually a pupil of Bucheim in Dorpat). One of Schmiedeberg's pupils was H. H. Meyer; Henderson, in turn, had worked with Meyer in Marburg, and Cori also had worked with Meyer in Vienna and with Loewi in Graz. Cori's pharmacological credentials were as good as or better than those of certain classical pharmacologists who regarded Cori as an unsuitable occupant of the chair of pharmacology at Washington University, where the Coris had moved from Buffalo; their MD degrees had been earned in 1920 from the German University of Prague. Observed in Vienna by a Dr. Gaylord of the New York State Institute for the Study of Malignant Diseases (later to become the Roswell Park Memorial Institute), Carl and Gerty Cori were recruited in 1922; they were world-famous for their work in carbohydrate metabolism by 1931 (hence, a disgrace to pharmacology!). Carl Cori's thoughtful kindness to me was displayed immediately upon my arrival in St. Louis, albeit he mandated that I must obtain medical qualifications. This he made possible by wangling free tuition (as well as advanced standing) and by raising (somehow) my initial stipend from $600 to $800 (per year, not per month!). My wife, Mary, became an assistant mainly to the Coris; however, we also published two choline papers together. I published only once with Cori, a review on adrenaline; all other Cori-Welch papers are by Mary Welch. Initially, I had expected to be an assistant to the Coris, but Carl suggested that I prepare three research proposals; of these, he might approve one for my independent investigation (one was found approvable); otherwise, I would be his assistant. How many times have I wondered what would have have happened had I worked with the Coris rather than independently? Note what happened to Earl Sutherland, who later worked with Cori and then independently on what proved to be cyclic AMP; he succeeded me at Western Reserve in 1953 and won the Nobel Prize, as the Coris had in 1947.[4]

Cori approved my proposal to study arsenocholine as a labeled form of choline (in those days carbon-14 was not yet available, and the mass spectrometer for work with nitrogen-15 was not dreamed of). I hoped that arsenocholine not only would be nontoxic, but also would serve as a metabolic mimic of choline. The resynthesis of arsenocholine (not a job for one man) was made possible with the help of Sidney Colowick, then a new technician Cori loaned me.[5]

Arsenocholine worked like a charm as a lipotropic agent. It was converted to some extent by the liver to the arsenic analog of betaine $[(CH_3)_3As^+\text{-}CH_2COO^-]$. Clearly, the analog had to be synthesized, because betaine $[(CH_3)_3N^+\text{-}CH_2COO^-]$ had been found to be lipotropically active (transmethylation not yet having been discovered, it was thought that betaine might be reduced to choline); however, synthetic arsenobetaine proved not to be lipotropically active. Surprisingly, my S-analog of choline was too toxic to detect lipotropic activity (instability?); sulfobetaine $[(CH_3)_2S^+\text{-}CH_2\text{-}COO^-]$, however, was very effective as a lipotropic agent. It was likely, therefore, that sulfobetaine and betaine were donating one or more of their methyl groups for the biosynthesis of choline. Before this hypothesis could be tested and transmethylation established, duVigneaud, who knew of my work, renamed sulfobetaine dimethylthetine and hastened to publish without appropriate reference to my studies. Even some gods have feet of clay!

During my first post-MD year (1939–1940), Richard Landau, then a fourth-year medical student, did his BSc (Med) with me (Landau for many years has been a professor of medicine at Chicago and the distinguished editor of *Perspectives in Biology and Medicine*). We found great pleasure in working together then and in the friendship that has continued. The gold salts of choline-fractions derived from lecithin isolated from rats fed arsenocholine analyzed correctly with respect to the ratios of N:As:Au; thus, these fractions

[4]In competition for my attention with heavy teaching and research was an uninspiring course in anatomy. The professor (Terry) cared little for function and even less for holders of the PhD. When the female pelvis was dissected, I was attending the Federation meetings; hence, I was given an incomplete and told to report in June. I did, albeit reluctantly, as the speed and poor quality of my dissection displayed. Terry descended upon Cori (who also disliked anatomy and, I suspect, Terry as well) to complain about my performance. In no uncertain terms, Cori told me to get with it and do a perfect dissection. I did one so well that Terry found no hiatuses in my knowledge, but he barely passed me. I take some pride in the fact that, in spite of the magnanimous Terry, I graduated three years later (cum laude) with membership in Alpha Omega Alpha.

[5]Shortly after he had returned to Cori, Colowick, a chemical engineer then without experience with the tools of biochemistry, allowed the ungreased top of a desiccator to crash on the concrete floor directly in front of Cori's office. Cori emerged in horror (how short funds were in those days!) and said in essence, "This guy has gotta go." I like to believe that without my pleading this first of Cori's PhD students might have been lost to science and to his great career in biochemistry.

represented a mixture of choline and arsenocholine, the latter having functionally replaced much of the choline (3).

The days in St. Louis were exciting: the discovery and synthesis of the "Cori-ester" (glucose-1-phosphate); my close friendships with Carl and Gerty, Helen Graham, Gerhard Schmidt, F. O. Schmitt, my late brother-in-law Gordon H. Scott, Evarts Graham, the great surgeon, who offered me a career in anesthesiology, and many others of my colleagues; even those in my own medical class (dear friends still) who had to suffer my lectures and my grading of their papers. The offer to head a new research department of pharmacology at Sharp & Dohme (before its merger with Merck) came in 1940, at a time when I had no intention of ever leaving Cori. One sour note had been heard, however; the dean, despite Cori's urging, refused to promote me unless I did some "proper pharmacology," i.e. work on new sulfonamides. In my "spare time" I had already written (1937) the very first review of the "sulfa" drugs at the request of the *Journal of Pediatrics* (as I recall, it contained less than 40 references!). The mandate of the dean I found intolerable; thus, despite the belated offer of a promotion, an increase in salary from $2500 to $3500 per year, and a technician, I reluctantly parted from my dear friends, although I almost returned as Cori's successor in pharmacology only six years later. I had been elected to membership in ASPET in 1937 at age 28; however, when I became director of pharmacological research at Sharp & Dohme my membership was canceled, as the bylaws then required. Accordingly, the directory indicates that my membership dates from 1942, when, bylaw reform having occurred, I was reelected.

SHARP & DOHME (S & D) YEARS (1940–1944)

William A. Feirer (MD and DSc from The Johns Hopkins University), who had recruited me as director of pharmacological research, had just become director of research of S & D. We remained very close friends until he died only a few years ago. In view of my rejection of the sulfa drugs at Washington University, it was ironical, to say the least, that important discoveries of new sulfonamides by Jim Sprague and M. L. (Mel) Moore made study in this area mandatory for me while I was building a new department of pharmacology at S & D, the first of four. The small initial group included Paul A. Mattis, DSc, a most helpful colleague who later joined me in Cleveland, Albert Latven, a super technician, the animal man, and myself; however, the group grew rapidly to some twenty-five or thirty. The first of the new sulfonamides to show promise was a sulfamethylpyrimidine, which appeared superior to sulfapyrimidine, i.e. sulfadiazine; indeed, equal dosages gave much higher blood levels of the new drug, termed sulfamerazine. The latter was much better absorbed than sulfadiazine after oral administration, while renal clearance took twice as long; hence, equal blood levels of the chemotherapeutically equivalent drugs could be maintained by half as much sulfamerazine given at eight-hour, rather than four-hour, intervals. It seemed that we had a winner. Feirer proposed that our findings be reviewed with E. K. Marshall at The Johns Hopkins, then the panjandrum of sulfa drugs. The discourse had hardly begun when Marshall stopped me, saying, "Welch, you are a damned fool! Don't you know that sulfamethylthiazole causes peripheral neuropathy and that sulfamethylpyrimidine will do the same? It also causes renal damage; look at the holes in this kidney section!" I asked him why a methyl group should be toxic when substituted on a pyrimidine ring compared with a thiazole ring; it seemed to me a non sequitur. He inquired whether I had heard of CH_3OH! No chemist, Marshall apparently really believed that the methyl group per se is potentially toxic; he suggested that we start over with a sulfaethylpyrimidine! The holes in the kidneys reflected hydronephrosis, of course; they resulted from crystal deposition due to overdosage. I stated that no renal differences would be found with equal blood levels of the two drugs. Marshall remained unconvinced and Feirer was nonplussed, to say the least. To make a very long story short, chicks were used to show that the peripheral neuropathy, readily caused in that species by sulfamethylthiazole, was of minor degree and equal with sulfadiazine and sulfamerazine. Extended studies of many phenomena made clear that the earlier contentions had been correct. With a massive paper ready for publication, we went again to Baltimore. Marshall, really a great man, at last was completely convinced and congratulated us. A very long paper was published in 1943 in the *Journal of Pharmacology and Experimental Therapeutics* without any changes; I suspect that one referee of that manuscript can be identified! Why then did the no-more-costly sulfamerazine not completely displace sulfadiazine, with half the dose given half as often? The reasons were entirely economic. S & D, then a relatively small company, could not afford to out-advertise the developers of sulfadiazine, which by then was so well established that few physicians were interested in altering their memory patterns. Science versus the marketplace!

My career appeared to benefit, at least temporarily, from promotion to assistant director of research (but now with seven departments to supervise), while shortly thereafter Feirer became vice president and I became director of research at age 34. I tried to spend half-time in pharmacology but failed. The then assistant director of the department, Karl Beyer, was given full responsibility for research and went on to a brilliant career. Only my studies that led to folic acid (see below) were continued, then with Lem Wright. At the same time several academic posts became available and in 1944 I decided to move to Western Reserve University, where Torald Sollmann was retiring. Solly had been dean at Western Reserve for fifteen years; the Department of Pharmacology was spatially large but almost without equipment, and with the departure of the last assistant professor, the staff was zero, while my new salary, $7500 per year, now seems inconceivably small.

In addition to my lasting friendships with Bill and Jeanne Feirer, I left S & D with great respect for my many scientific colleagues and for the ethics of the company, then led by John Zinsser. [Incidentally, I began my first major review at S & D; it was the first on the design and use of antimetabolites (4).]

WESTERN RESERVE (WRU) YEARS (1944–1953)

I moved to Cleveland in 1944 during the war. A medical officer in the Army Reserve Corps, I had been rejected for active duty: cardiac hypertrophy, diagnosed erroneously. Two of the men who first moved to the department at Western Reserve were Lawrence (Lawrie) Peters, PhD (now deceased), who had worked with me at S & D for two years (I helped him earn his MD), and Ernest Bueding, MD, who had come from New York University. With army support, we promptly undertook chemotherapeutic studies of filariasis, which the military expected to be a major problem in the South Pacific (it proved not to be). Peters and I were soon chasing vicious wild cotton rats naturally infested with the only practicable source of a filariasis resembling that of man. We had to learn that *Litomosoides carinii* in the pleural cavity of these animals responds differently to drugs than *Wuchereria bancrofti* in the lymphatic system of man. We studied new classes of compounds and found activity among a group of cyanine dyes with resonating systems of bonding. We selected a compound with acceptable toxicity that could kill all adult filarial worms in the cotton rats; however, the microfilaria in the blood-stream (infectious for the vector, a mite) were not directly killed. After extensive studies of toxicity in many animal species, we conducted clinical pharmacological investigations in human volunteers with inoperable neoplastic diseases. We carried out clinical trials in Puerto Rico; these were disappointing. The adult worms in the human lymphatics were not killed, although the *microfilaria* were sensitive; they returned in due course, as might be expected. We developed derivatives, however, for the chemotherapy of other nematodal infestations. In the meantime, the army asked us to study schistosomiasis. Bueding, while studying the mechanism of action of the cyanine dyes, began investigations in mice, using a snail colony to produce the miracidia of *Schistosoma mansoni*. In this and related fields, Bueding is today the world's authority.

Other prominent members of the Department of Pharmacology at Western Reserve included George Bidder, MD, Giulio Cantoni, MD, and Harold Chase, MD, about whom space restrictions preclude discussions.

A major dilemma arose in 1946: I was offered the chairmanship of the Department of Pharmacology at Washington University; Carl Cori had moved to the chair of biological chemistry. Despite my nostalgia for Washington

University, I decided to remain in Cleveland; I wanted to follow up on the two years of hard work I had expended in building something out of almost nothing. In addition, my studies leading to folic acid, begun at S & D, had been reinitiated and were moving forward rapidly.

Lem Wright and I had added poorly absorbed succinylsulfathiazole together with the then-known vitamins to highly purified diets of rats. We hoped to learn whether intestinal microorganisms produce essential nutritional factors that might be disclosed in this way. Elvehjem and his associates had similar ideas using sulfaguanidine and they first identified a factor, not then an entity, termed folic acid. To conserve space, the reader is referred to other articles for the history of this research (5–7).

Much of the work on folic acid, and later on vitamin B_{12}, was done in collaboration with Bob Heinle of the Department of Medicine at Western Reserve. He was a superb colleague who was very even-tempered—always mad. Trained with W. B. (Bill) Castle at the Thorndike in Boston, Heinle was an expert hematologist and internist; we worked effectively together for seven years. After Jack Pritchard, MD, joined us, we fed piglets a purified diet containing succinylsulfathiazole and a crude antagonist of folic acid (supplied by Tom Jukes, by then at Lederle); the piglets became relatively huge swine, so strong that only Jack could successfully wrestle them. Eventually they developed striking bone marrow changes and a macrocytic anemia; their marrow became indistinguishable from that of human pernicious anemia in relapse. This condition responded initially to folic acid, but gradually became insensitive to folate unless we injected purified liver extract, as was used in the treatment of pernicious anemia (later replaced by vitamin B_{12}). Combined system disease did not develop in the deficient animals, but sudden deaths occurred eventually unless liver extract (later vitamin B_{12}) was injected. During subsequent army service, Pritchard became involved in obstetrics and gynecology and has been head of that department at Dallas for many years.

Our studies with swine, coupled with evidence that the purified liver extracts contained no folic acid, led me to suggest that the extrinsic factor, utilized orally by pernicious anemic patients only when administered with Castle's intrinsic factor (i.e. normal human gastric juice), could be identical with the parenterally active antipernicious anemia factor of liver. This was established in 1948 in collaboration with Castle and associates, and was extended to pure vitamin B_{12}, which had just then become available (8). Also in 1948, in collaboration with Stokstad, Jukes, and others (9), we demonstrated the antipernicious anemia efficacy of the microbially produced animal protein factor, given parenterally. This was probably the first reported use of a material derived from a source other than liver for the parenteral treatment of pernicious anemia in relapse.

In 1949, two superb post-docs, C. A. (Chuck) Nichol and W. H. (Bill) Prusoff, joined the group. With the latter, I initiated a program to concentrate the intrinsic factor using desiccated pig stomach as a source, while with Nichol I began studies of the effect of rat liver slices on folic acid. Using *Leuconostoc citrovorum,* which requires for its growth reduced forms of folate, e.g. leuco-vorin, we soon obtained evidence for a folate reductase. From the standpoint of chemotherapy, it was exciting to find that aminopterin, the forerunner of methotrexate, almost irreversibly inhibits the reductase (10–12), because this was the first clue to the mechanism of action of this invaluable antileukemic agent. Warwick Sakami and I showed the involvement of folate derivatives in the biosynthesis of labile methyl groups (of betaine, methionine, etc) (13).

Some comments concerning the rather ghastly years of anguishing debates about teaching are necessary, because these led to the nearly total (and rather famous) revision of the curriculum of WRU Medical School. Initially, I had inveighed against the examination system that prevailed (frequent regurgitation after memorization), but I knew not whereof I wrought. Major funds were raised by the dean, Joe Wearn (a man with whom I had great personal rapport and who nominated me for membership in the Association of American Physicians); Hale Ham came from Harvard to coordinate the program, and eventually teaching by committees resulted. One participation I will proudly acknowledge: the multidisciplinary student laboratories. Otherwise, I became essentially antipathetic. I believed then, as I do now, that good instruction comes from within; it is a product of knowledgeable individuals who love to stimulate and debate with willing recipients. No committee can do this—or legislate it—only earnest and dedicated individuals. My departure for Yale in 1953 introduced me to a very different teaching system: no examinations except by the National Board. It was a joy to have real rapport with medical students, who studied without built-in fear of their instructors (in most schools then, teachers also were potential executioners). By mid-1952, however, I was able to accept an invitation from Professor J. H. (Josh) Burn to recover at Oxford University.

Other dear old friends at WRU included Normand Hoerr, John Dingle, and Carl Wiggers (deceased) and Harland Wood and Lester Krampitz.

OXFORD DAYS (1952–JANUARY 1953)

Oxford was an unforgettable experience for me; it was a Mecca for pharmacol-ogists, with such stars as Burn, Edith Bülbring, Hugh Blaschko, and many, many others. Initially Burn and I worked together, but discussions with Blaschko soon led us to attempt to determine (again without the possibility of using isotopes) whether dopa and dopamine were converted to noradrenaline and adrenaline by the adrenal medulla. A fresh homogenate was cold-dialyzed against a very small volume of buffered saline suspension of mushroom catechol oxidase to pump out the pressor amines, forming insoluble melanins, by creating a concentration gradient while minimizing the loss of possible cofactors; O_2 was bubbled through the external compartment and N_2 through the homogenate. The level of the pressor amines (as assayed on the blood pressure of the pithed cat) fell rapidly; however, an asymptote was soon reached at 75–80% of the initial concentration in the homogenate. What was causing retention? Could the amines be held within particles? Using Florey's high-speed refrigerated centrifuge, we found the pressor amines in vesicles that sedimented with the mitochondria; from these particles the catecholamines could be readily released. A new field was opened (14) and lasting friendships were made.

Progress toward the end of these experiments was slowed by my hospitalization for a lumbar disc (after a previous laminectomy, I had not anticipated this); during my return to the States from England on the Queen Mary, I was encased in plaster. On an earlier visit to Yale University at the invitation of Joe Fruton I had met the dean-to-be, Vernon Lippard (not yet moved from Charlottesville), and eventually I accepted an offer from Yale. At the same time, 1953, Peters became chairman of pharmacology at Tulane and moved to Kansas two years later, and Cantoni moved to head a lab at the National Institutes of Health. In 1954, Bueding became chairman of pharmacology at Louisiana State University and moved to The Johns Hopkins six years later.

YALE YEARS (1953–1967)

After the death of the previous chairman, Salter, the Department of Pharmacology at Yale had deteriorated somewhat. When I arrived with several colleagues from Western Reserve [e.g. Charles (Nick) Carter, an assistant professor of medicine strongly grounded in biochemistry, Chuck Nichol, Bill Prusoff, Bill Holmes, and Sheldon Greenbaum, as well as several post-docs, including Bernard Langley, and even two graduate students], only two of the former Yale staff members (and too many graduate students) remained. Desmond Bonnycastle, an associate professor, departed when his five-year term-appointment expired. Nicholas (Nick) Giarman, an assistant professor, proved to be an exceptional teacher and a very competent neuropharmacologist; we urged him to remain. Indeed, at the invitation of J. H. Gaddum, Giarman exchanged positions with Henry Adam for a year (to keep it simple, the two men exchanged jobs, salaries, cars, and homes; everything except wives!). Giarman returned as associate professor. In 1963 he succeeded me as American editor of

Biochemical Pharmacology. He died in 1968, a great loss for Yale, science, and his many friends.[6]

In 1953–1954, Blaschko spent six months with us at Yale and work on the adrenal granules resumed, in collaboration with Joe Demis, and later with Paul Hagen. We obtained proof of the precursory role of ^{14}C-dopa in the formation of the pressor amines, and continued studies of the extraordinarily high levels of ATP in the vesicles.

One outstanding recruit of the pharmacology department at Yale was John Vane, D Phil, whom I had known as a graduate student at Oxford. Vane came to Yale as an instructor and was soon promoted. Eventually, however, the tug of his home country became too great and he joined the Department of Pharmacology of the Royal College of Surgeons. Later, he became group research and development director of the Wellcome Laboratories. It is pertinent to mention here, although out of context, that prior to Vane's joining Wellcome he was a very great help to us at Squibb as the senior consultant in pharmacology. For his brilliant work on the prostaglandins and prostacyclin, as well as the mechanism of action of aspirin, Vane shared a Nobel Prize in 1982; he was knighted in 1984 and became a foreign member of the National Academy of Science of the United States. In December 1982 my wife and I were greatly honored by an official invitation, through Vane, of course, to attend the Nobel ceremonies in Stockholm, one of the most memorable experiences of our lives.

The great strengths of the Department of Pharmacology at Yale, which gained worldwide recognition, lay in the many outstanding young scientists who came to work there, either as young faculty members or as post-doctoral fellows. We were able to secure the funds to retain many of these men and women, and gave them encouragement and help in full measure to develop their ideas and intellectual growth. Over the years, many of that group received various honors and special appointments, e.g. Career Development Awards, Scholars in Cancer Research awards, Markle and Burroughs-Wellcome Scholar awards, as well as career professorships. In addition, I shared in memberships and chairmanships of study sections of the NIH and advisory committees of the NSF, the NRC, and the American Cancer Society. The department grew rapidly and in addition to research gained a deserved reputation for good

[6]Prior to Giarman's American editorship of *Biochemical Pharmacology,* I (and others) had helped Sir Rudolph Peters in the founding in 1958 of that soon rather prestigious journal. As American editor my initial roles were demanding ones. High standards were set and were maintained by Giarman, while I became a vice chairman of the International Board of Editors. Subsequently, Alan Sartorelli carried the standards of the journal to even greater heights (despite his having much to do with the issue in 1979 that commemorated my seventieth birthday). Subsequent to the death of Sir Rudolph, I became chairman of the editorial board, at the time of a symposium in Oxford (1983) that celebrated the twenty-fifth anniversary of the founding of *Biochemical Pharmacology.*

teaching of medical students (much in small discussion groups), as well as of graduate students, while offering an excellent atmosphere for post-doctoral training. Much of the early financial help came in the form of no-strings grants from either the Squibb Institute for Medical Research or the Upjohn Company, where I was a consultant for about eight years. As research and training grants became more readily available (prior to recent years), funds were relatively easy to obtain. Among the many members of the group were such outstanding men as Bob Handschumacher, a scholar in cancer research and a career professor of the American Cancer Society, who was chairman of the Yale department from 1974 to 1977; Van Canellakis, a career NIH professor; Henry Mautner, chairman of biochemistry and pharmacology at Tufts. Jack Cooper was promoted to professor and remained at Yale, while Julian Jaffe was attracted to the University of Vermont, where he is a professor of pharmacology. Chuck Nichol became head of experimental therapeutics at the Roswell Park Memorial Institute and later director of medicinal biochemistry at the Wellcome Research Laboratories (USA). Another man of distinction who moved to Yale from WRU, Bill Prusoff, became a professor and remained at Yale. Nick Carter returned to WRU as chairman of pharmacology and is now scientific director of NIEHS (NIH). After two years at Cornell, Jack Green became chairman of pharmacology at Mt. Sinai Medical School. Paul Hagen moved to Harvard for three years prior to becoming chairman of biochemistry at Manitoba; he is now dean of graduate studies at the University of Ottawa. Glenn Fischer, after eleven years at Yale, transferred to Brown as a professor of biochemical pharmacology.

Alan C. Sartorelli joined the department as an assistant professor and became a professor in 1967; he was a very distinguished chairman of the department from 1977 to 1984 and is now director of the Yale Comprehensive Cancer Center. Since 1968 the American editor of *Biochemical Pharmacology*, Sartorelli is also the executive editor of *Pharmacology and Therapeutics*. Joe Bertino joined the department (and internal medicine) as assistant professor and later became an American Cancer Society professor. Paul Calabresi left Yale in 1968 for the chairmanship of the Department of Medicine at Brown, but he deserves very special mention because of his various contributions to Yale and to his many co-workers there, including myself; he headed the first section on clinical pharmacology at Yale and was a Burroughs-Wellcome Scholar in that field. Dave Johns, a member of the department for seven years, became chief of two laboratories (medicinal chemistry and biology and chemical pharmacology) at the National Cancer Institute. The two-volume monograph edited by Sartorelli & Johns on antineoplastic and immunosuppressive agents remains a classic in these fields after nearly ten years.

Space does not permit other than brief mention of many other friends and important workers in the department between 1953 and 1967 and subsequently;

these certainly include Pauline Chang, Dave Ludlum, Joe Demis, Bob Levine, Maire Hakala, Zygmunt Zakrzewski, Ming Chu and S.-H. Chu, Arnold Eisenfeld, Richard Schindler, Ron Morris, Norm Gillis, Zoe Canellakis, Bill Creasey, John Perkins, John McCormack, Bill Macmillan, Malcolm Mitchell, Bob Roth, Jack Cramer, Morris Zedeck, Karel Raška, and Ed Coleman, who later joined me at Squibb. A dear friend from Germany, Helmuth Vorherr, with whom I studied the selective embryolethality of 6-azauridine and later, with his wife, Ute, that of N-(phosphonacetyl)-L-aspartate (PALA), is now professor of pharmacology and gynecology at the University of New Mexico. Other dear friends from abroad who worked at Yale included Laszlo Lajtha, until recently director of the MRC unit in the Paterson Labs of the University of Manchester; Ronald Girdwood, professor of therapeutics at the University of Edinburgh; Peter Reichard of the Karolinska Institutet; Tony Mathias; Hamish Keir; Charles Pasternak; Margaret Day; and Margaret and Brian Fox. Warm friends in other departments included Paul Beeson, Joe Fruton and his wife, Sofia Simmonds, Aaron Lerner, Sam Hellman, Dan Freedman, Bob McCollum, Bill Gardner, and Nick Greene. Unintentional omissions of other Yale colleagues I hope will be forgiven.

To discuss the investigations of all these exceptional scientists would require an entire volume. Many of my own studies through early 1966 were summarized in part in my Sollmann oration (15), presented at the time of my Torald Sollmann Award from ASPET; these studies involved many of the fine colleagues referred to above. They and other co-workers enabled me to participate in research and contributed greatly to our efforts to build a strong department that continued to prosper. As I said at the time of the dedication of a newly remodeled wing donated jointly to Yale by the Wellcome Trust and the National Cancer Institute, when I announced my imminent departure for the Squibb Institute for Medical Research, "There is nothing that succeeds like *successors!*" At that time also, my old friend George Hitchings, vice president for research at Burroughs Wellcome, presented me with a most appropriate gift: two large bottles of Empirin Compound, which he correctly predicted I would very soon be needing for my new headaches.

If space permitted, I could present much interesting information concerning the remarkably successful oral therapy of severe psoriasis with azaribine (triacetyl-6-azauridine) initiated by Paul Calabresi, Charlie McDonald, and colleagues. The NDA was approved in 1975 but was withdrawn fifteen months later because of the approximately 4% incidence of thromboses, some intraarterial. Evidence now suggests that in a few susceptible individuals a deficiency of pyridoxal phosphate may be induced, with resultant homocystinemia (in rabbits, these are prevented by pyridoxine). At present, azaribine is an orphan drug and psoriasis is an orphan disease! The remarkable efficacy of azaribine, not only in psoriasis but also in mycosis fungoides, choriocarcinoma, polycythemia vera, and perhaps as an embryolethal agent, may now be studied in other countries but not in the USA. After Squibb's initial production of 6-azauridine for us (~700 g) with the aid of a contract from the National Cancer Institute (NCI), the NCI contracted with Calbiochem to manufacture the substance. Handschumacher and I helped that company to initiate production, using 6-azauracil incubated with *E. coli,* a method devised by our friends in Prague, Jan Škoda et al. Thus, Bob, Jan, and I became close friends of Dr. William (Bill) Drell, then president of Calbiochem. These deep mutual friendships have survived time, distances, and the devastating effect of the loss of azaribine on Calbiochem, now a subsidiary of Hoechst. Handschumacher and I went to Prague to conduct studies with Škoda and Šorm, with each of whom, and with Helena Rašková, warm friendships developed. Toward the end of my second sabbatical, in Frankfurt at the Institut f. Therapeutische Biochemie,[7] where I developed friendships with Helmut Maske, Jürgen Drews, and Bill Pratt, Škoda and I began studies in Prague in March 1965; these were extended in July-August of that year.

This second sabbatical leave (1964–1965) was precipitated by a fall while I was skiing with Tom Jukes in Badger Pass in Yosemite; result: a broken fibula and the man who came to dinner. Later complications of the original break were debilitating, and finally a change in scenery was deemed essential. I must comment that one of the finest personal letters I have ever received had been written by Van Canellakis, who felt I was headed for disaster (presumably attributed to premature aging and overwork [?]); hence, I should slow down and rest on my laurels (i.e. contemplate my navel and admire the wonderful department at Yale that the many great colleagues, he, and I had built). My reactions to these concerns were (*a*) to go to Germany in 1964 to work in the laboratory and write a paper (without coauthors) for the Proceedings of the National Academy of Sciences on the selective inhibition of an enzyme induction; (*b*) to go to Prague in 1965 to work with Škoda; (*c*) to end years of unhappiness through divorce; (*d*) to go to India to study the chemotherapy of smallpox with 5-iododeoxyuridine, (*e*) above all, to be most happily married in Prague in 1966 to Erika (Peter) Martinková; and (*f*) to leave Yale in 1967 to head research and development at Squibb, involving about 1000 scientific

---

[7]While working in Frankfurt, I was invited by my old friends Gustav Born and Sir Alex Haddow to dine with them and Sir Henry Dale, Vane, and several others after a lecture in London. Dale was delighted to learn that I had visited the laboratory of Paul Ehrlich, with whom Dale had worked before I was born. Sir Henry then began to reminisce most fascinatingly about his days with Ehrlich, but in German! Born gently reminded Dale of the German limitations of many of his table companions. Sir Henry commented that if Born, who speaks perfect German, was having difficulty, perhaps he (Dale) should speak English. Within a moment, however, when the old gentleman continued his wonderful anecdotes, he spoke again in German! What I would have given then (as well as on other great occasions) for a tape recorder!

colleagues. Now twenty years later and still working, signs of disintegration have not yet become evident.

5-Iododeoxyuridine (IUdR) was first synthesized by Prusoff and studied by him, others, and me; it was the first specific antiviral drug to be used in man. Its efficacy against corneal keratitis caused by herpes simplex virus had led to studies of other DNA virus–induced diseases. The remarkable efficacy in rabbits of parenteral therapy with IUdR on advanced dermal lesions caused by vaccinia virus led to attempts to treat supposedly terminal human smallpox infections in India. Only three moribund patients became available during an entire year (1965–1966) of observation by David Fedson, then a post-doctoral fellow of Calabresi and mine; by a curious coincidence these patients were admitted during my visit to Madras. Intravenous infusions with IUdR were initiated immediately; result: two of the three patients survived (not reported as a 67% cure rate!). Since the disease probably has now disappeared, it is unlikely that the possible value of IUdR in smallpox therapy will ever be known.

During the last eight of my Yale years, I was a research advisor to Upjohn. There the many good scientists and friends are too numerous to name here, with the exception of Earl Burbidge, one of my closest friends in medical school, Bob Heinle (both died several years ago), and Dave Weisblat, then director of research and development. About Charles G. (Chuck) Smith, Upjohn's then director of biochemical research, much more will be said below.

Only three times prior to 1967 did I even consider offers of positions other than that at Yale: two medical deanships (one also a vice-presidency)[8] and one as vice-president for research and development at a pharmaceutical company other than Squibb. None of these appealed to me, however.

SQUIBB YEARS (1967–1974)

As indicated previously, throughout the years at Yale I had had many contacts with scientists at Squibb, and they had supported Yale's Squibb Fellows for fourteen years.

During the years subsequent to the discovery and development of the fluorocorticoids by Gus Fried and other Squibb scientists, however, Squibb had

---

[8]A memorable incident occurred during an interview for one of these positions: A rather distinguished surgical specialist on the search committee asked, "Dr. Welch, where do you think cardiovascular surgery is going in the next ten years?" My reply, as Paul Calabresi reminded me, was, "That's for you to tell me, and, should I accept this position, for me to help you get there." After a brief pause, I continued, "And if you cannot conceive where it should be, it will soon be time for me to appoint a new Chief of Cardiovascular Surgery." According to Paul, this philosophy, which he had observed during our years at Yale, has become a modus operandi for him and others of my colleagues.

made few contributions to drug discovery. Furthermore, with devastating impact on both morale and productivity, E. R. Squibb & Sons, Inc. had been acquired by the Mathieson Chemical Company, and the latter in turn by the Olin Corporation. In 1967 Squibb research and development was floundering. Then, thanks to the genius of a vice president of Olin then in charge of the E. R. Squibb Division, Richard Furlaud, real vision was introduced to the company. When first approached by Mr. Furlaud, I had no desire to leave Yale; in fact, with an endowed chair (the Eugene Higgins Professorship), it had been agreed at last that after fourteen years I would relinquish the chairmanship and return to my laboratory full-time. Nevertheless, with Mr. Furlaud's assurance that Squibb would shortly be a separate entity again (as indeed it became by a complicated separation from Olin), the challenge to try something very different became irresistible. Accordingly, in 1967 I became vice president for research and development and the director of the Squibb Institute for Medical Research. My first major act was to recruit Chuck Smith from Upjohn, initially as associate director. In 1970 he became a vice president. In 1969, another exceptional man, Dennis Fill, became president of E. R. Squibb & Sons, and Mr. Furlaud took on higher responsibilities. My second major act was to accept the resignation of the then head of pharmacology. John Vane became a senior consultant in pharmacology (three one-week visits annually). In due course, Zola Horovitz was promoted from within to head the department of pharmacology; later he became an associate director and a vice president.

Turning research around was not exactly an easy job for us. Company earnings then did not permit any major new research programs to be launched, except when ongoing programs were terminated. Initially, despite the very complex problems involved in increasing the yield of penicillin, further reducing its cost, and seeking better semi-synthetic derivatives of it, we had few other projects. Great improvements in producing penicillin were accomplished, and a valuable cyclohexadiene derivative of penicillin was developed. With great difficulties we entered the cephalosporin field and developed a new drug, cephradine. Effective new corticosteroids were created to replace those soon to be lost by the expiration of patents. These developments had salutary effects on earnings and therefore on the research and development budget. In addition, we made major efforts in the field of β blockers (β-adrenergic receptor-blocking agents), despite the then antipathy of the Food and Drug Administration toward such valuable drugs. As a result, a major entry into the field was later developed.

At last, however, the long-awaited hot lead appeared and we began research that could genuinely be termed basic. Earlier studies by Professor Rocha e Silva in Brazil had shown that the venom of a poisonous local snake, *Bothrops jararaca,* incubated with plasma led to the formation of a vasodilator polypeptide, bradykinin. Another Brazilian worker, Sergio Ferreira, observed that the venom also contained a factor that potentiated the effects of bradykinin; this was later shown to be the result of inhibition of an enzyme that rapidly inactivates bradykinin. Vane had suspected that this bradykininase might be the same enzyme that activates the polypeptide angiotensin I. (The latter substance, then suspected to be involved in renal hypertension, causes no effect on blood pressure until it is enzymically converted to angiotensin II, the most powerful vasoconstrictor known.) In Vane's lab, Mick Bakhle found that the peptide in the snake venom that inhibited bradykininase was also a powerful inhibitor of the angiotensin-converting enzyme (ACE). ACE catalyzes the removal of two terminal amino acids from both bradykinin and angiotensin I; it was later obtained in a homogeneous state by another good friend, Ervin Erdös, at the University of Texas, Dallas. The inhibitor isolated from the venom by Ferreira and Greene was a pentapeptide, while another potent inhibitory peptide, isolated at Squibb, was a nonapeptide. Management's lack of enthusiasm for research in this area was understandable, when one considers that the peptide inhibitors of course were impracticable as drugs. Indeed, it had not even been firmly established that so-called renal hypertension could be explained entirely by the release of the enzyme renin from hypoxic kidneys, with the final hypertensive state being caused by angiotensin II. To validate this concept required a major gamble. Despite the cost and the difficulties of synthesizing the inhibitory peptide in gram-quantities, I decided to go ahead despite a reluctant management; hence, Dr. Miguel Ondetti, an excellent protein chemist, and Dr. David Cushman, an enzymologist and pharmacologist, continued their now well-known studies. To make a very long story short, their colleagues established that the parenterally administered synthetic nonapeptide could prevent a rise in blood pressure from being caused by the intravenous injection of angiotensin I. Furthermore, studies in patients with malignant hypertension showed that the nonapeptide given intravenously could lead to a normotensive state, without evidence that accumulations of angiotensin I or the plasma precursor or renin in themselves were deleterious. Whereas I had reacted with enthusiasm to the potential for new drug discovery, the then president of the Squibb Corporation (neither Dennis Fill nor Richard Furlaud) could only understand the impracticality of the nonapeptide as a drug. The entire program was regarded as a costly exercise in futility. He ridiculed the idea that the concept of the renin cascade had indeed been validated, and rejected the potential importance, if hypertension was to be attacked rationally, of a key enzyme to inhibit having been identified. Indeed, I was told that the major effort to develop a new and practicable inhibitor of the converting enzyme should be terminated. This, I must state, I refused to do—termination of Welch would have to come first! Fortunately for what was to come, that did not happen. The brilliant work of Ondetti & Cushman continued; they eventually determined the essential inhibitory features and synthesized a modified dipeptide with oral activity! By this time, however, my mandatory retirement at 65 (actually 65.98 years) had occurred. In due course, the then president also left the corporation for reasons other than age and the outstanding Mr. Fill was persuaded by the chairman to become president of the entire Squibb Corporation.

In the meantime, angiotensin II had been shown to be much more than a vasoconstrictor; e.g. it stimulates the secretion of aldosterone and thus affects the retention of both sodium ions and fluid. The ultimate drug, termed captopril and now marketed world-wide, is a modified dipeptide of L-cysteine (free SH) and L-proline. A methyl group adjacent to the imid-linkage renders the compound insusceptible to attack by peptidases; hence, it is orally active. It is used not only in the control of malignant, but also essential, hypertension. In addition, captopril is efficacious (for reasons not yet fully understood) in the therapy of congestive heart failure, while other potential activities (as well as promising new ACE inhibitors), are under investigation, not only by Squibb, but also by other companies.

I like to believe that my approximately 7.5 years as director (or president) of the Squibb Institute for Medical Research did help to turn the company upward. In addition to challenges, headaches, and some successes, Squibb gave me wonderful opportunities for scientific edification as well as for travel. Often my wife was able to accompany me. Frequent trips to the research laboratories of Squibb-Germany (Regensburg) were essential, as were trips to London and Liverpool for pharmaceutical research and to Ireland for production, while for other reasons I traveled to Australia, Austria, Belgium, China, Czechoslovakia, Denmark, Finland, France, Greece, Hawaii, Holland, Hong Kong, Hungary, Italy, Japan, Mexico, Nigeria, Sardinia, Sweden, and Switzerland. These and personal trips to Alaska, the Bahamas, Brazil, East Germany, Egypt, India, Portugal, Puerto Rico, Russia, and Spain bestowed insights that could not have been gained in any other way.

Late in 1974, Chuck Smith, a fine executive and innovative scientist who remains our close friend (as does his wife, Angie), also found the previous president intolerable, and six weeks after Smith's promotion to the presidency of the Squibb Institute for Medical Research, he left the corporation. Chuck is now an executive vice president of Revlon with responsibilities for research and development in such component health-care companies as USV Pharmaceuticals, Armour, and many other subsidiaries. Among my other colleagues at Squibb were also many good friends (some now deceased). I think particularly of Oskar Wintersteiner, Helmut Cords, Naomi Taylor, George Donat, Peter Koerber, Fred Wiselogle, Jack Bernstein, Pat Diassi, Jim Knill, Frank Weisenborn, Bill Brown, Zola Horovitz, Miguel Ondetti, and many others.

Now, ten years later, the Squibb Institute is a very different organization than it was and I know but little about it. A period of my life, which in a great many ways was both exciting and memorable, regretfully came to an end. I am "gone, but not forgotten," or forgotten—I know not which.

ST. JUDE YEARS (1975–1983)

Retirement not being an attractive prospect, I accepted an offer from an outstanding institution, St. Jude Children's Research Hospital in Memphis, on whose board of scientific advisors I had served for three earlier years. My goal was to organize a new Division of Biochemical and Clinical Pharmacology. Many things had changed since my last visit in 1972: a new director, Alvin M. Mauer, MD, had replaced Don Pinkel, of whom I had been fond; a new seven-story building was due to open within three months (the new division was to occupy most of the third floor), while an important change for me was a return to cancer research, none having been possible at Squibb. Indeed, a three-month period in early 1975 offered an excellent opportunity to bore into the two then new superb monographs edited by Alan Sartorelli and David Johns. These massive handbooks, which I had actually commissioned as a senior editor of the handbook series, could not have appeared at a more appropriate time if I had designed it that way. I could afford time for planning as well as for recruiting, particularly in the area of medicinal organic chemistry, which I regarded as a mandatory development for the new division. In this area, I was joined as a full member by an outstanding man, Josef Nemec, whom I lured from the Squibb Institute for Medical Research, where I had known him well. Another chemist joined us, T. L. Chwang, who had worked for several years with Charles Heidelberger, while Bill Beck, who had done his post-doctoral work with Van Canellakis at Yale, came from the University of Southern California. Nahed Ahmed became my research associate; her post-doctoral training with Nicholas Bachur at the Baltimore branch of the NCI had qualified her highly. I began studies of the activity of uridine-cytidine kinase in human colorectal adenocarcinomas. Other studies of this enzyme led to the discovery, with Alan Paterson and colleagues at Alberta, that despite the apparent deletion of this kinase activity from certain 3-deazauridine-resistant mutants of a human B-lymphoblast, their proliferation in culture remained quite sensitive to inhibition by either 6-azauridine or 5-azacytidine (16). We suggested that another nucleoside-phosphorylating enzyme is present in such mutants; my assistants and I have now confirmed and extended this hypothesis (17) using a new method for quantifying acidic nucleosides (18). Judith Belt, a post-doctoral trainee of Efraim Racker, and I found that only the undissociated portion of 6-azauridine (pKa ~6.7), at pH 7.4, is transported by the process of facilitated diffusion (19).

Two post-doctoral fellows from the Chester Beatty, Drs. Janet and Peter Houghton, came to St. Jude after we helped them to immigrate. This was done with almost incredible difficulties, which were finally resolved only with the help of Senator Baker. In addition to those named, four scientists already at St. Jude continued as effective members of the division. These were DeWayne Roberts, a biochemical pharmacologist; Thomas L. Avery, an experienced experimental chemotherapist and colleague with whom I worked very closely and with much friendship; Arnold Fridland, a very competent enzymologist; and Thomas Brent, a specialist in the mechanisms of DNA damage and its enzymic repair.

In 1980, my five-year appointment as member and chairman of the Division of Biochemical and Clinical Pharmacology was extended for a year to enable a search committee and the director to select Raymond L. Blakley to succeed me as chairman. Blakley formerly was professor of biochemistry in the University of Iowa School of Medicine. I became member-emeritus and once again a bench scientist at 72. The subsequent two years were among the happiest of my life, although fifty-hour weeks in the laboratory with two splendid assistants, Jim Panahi, BS, and Glen Germain, MS, gave me little time for my personal life or to write papers.

During my continued studies of resistance to 3-deazauridine in both B- and T-human lymphoblastoid cells, I made the unexpected observation that the cytosolic uridine-cytidine kinase activity is not deleted in CEM(T) cells; indeed, in the parent cells that enzyme has no affinity for the cytotoxic nucleoside. Thus was uncovered a hitherto unrecognized enzyme associated with the nuclei of these (and other) cells that catalyzes the phosphorylation of 3-deazauridine and the activity of which is greatly diminished in the drug-resistant mutants. This enzyme activity can either complement or supplant that of the classical cytosolic enzyme.

At this stage (1983), however, I was approached by three government agencies (the EPA, the FDA, and the NCI), and I decided to accept an offer from the National Cancer Institute. We moved from a lovely home in Memphis to a condominium in Chevy Chase, a suburb of Washington, DC, so that I could start still another career just prior to turning 75.

Before leaving the subject of Memphis and St. Jude, I want to present a few impressions of our eight and a half years there. Memphis provided a rather different atmosphere from those of New Haven and Princeton, particularly for Erika, whose almost unaccented British English, acquired in Prague, had already been transformed to American, although not mid-South American. This new learning period was helped greatly by her linguistic talents (she was formerly a scientific translator). The people of Memphis were warm and friendly, and the city progressed remarkably during our stay there. Today, any individual can and should be able to live happily in Memphis, as we did (with the aid of air-conditioning and a swimming pool!). We have left many dear friends there.

St. Jude Hospital only opened in 1962, and in 1975 it was almost incredible to see what had happened there in only thirteen years. In fact, from our arrival until our departure, the continued growth of the hospital was evident. It is a splendid children's hospital, which provides superb (and completely free) medical care to children with all forms of neoplastic disease. I have only the highest praise for the Board of Governors and Danny Thomas for their dedication and their steadily more efficient fund-raising, especially for the clinical activities at the hospital. St. Jude is now as much the pride of Memphis as is the home of Elvis Presley.

The clinical research accomplishments of St. Jude, e.g. the introduction of cranio-spinal irradiation, which increased the apparent cures in acute lympho-blastic leukemia by tenfold (to now more than 50% of the patients), was a major breakthrough. This occurred during the tenure of Dr. Donald Pinkel, the first medical director, in collaboration with Dr. Joseph Simone. The second director, who departed in 1983, was a dedicated pediatric hematologist and oncologist who made every effort to help St. Jude grow and burnish its image brightly. He had little ability to communicate scientifically, however, or to understand the importance of molecular developments. Thus he could not fully appreciate the reasonable needs for the encouragement and financial support of basic scientists, who are developing many new methods for the treatment of cancer.

St. Jude needs many improvements, and the new director, Joseph Simone, a very experienced oncologist and pediatrician, could institute them with strong advisors in basic areas and with the support of the Board of Governors. The basic science divisions not only need to be more strongly supported financially, they also need to be better oriented toward goals more sensible in terms of modern chemotherapy and immunotherapy (for example, the roles of oncogenes and the great promise of monoclonal antibodies). I am and will continue to be an ardent supporter of the institution, and of the best scientists within it. Finally, I am personally very grateful, despite the very difficult and traumatic struggles we experienced at times, for the many courtesies my colleagues there extended to me. In addition to those in our own division, already mentioned, such splendid colleagues as Charles Pratt, George Cheung, the discoverer of calmodulin, Dave Kingsbury, George Marten, Gaston Rivera, and their respective wives, as well as Dolores Anderson, are among those whom I won't forget as are other wonderful Memphians, especially Dr. Gordon and Nancy Mathes, Dr. Eric and Marie Louise Muirhead, and Dr. Henry and Rosalie Rudner.

One other memorable group in Memphis is the Memphis Medical Seminar, of which I was a member (and once the president). It is largely composed of investigative clinicians, for example, Hall Tacket, our splendid internist; Eric Muirhead, an outstanding investigator and pathologist; and Irv Fleming, a dedicated surgeon. These and all the other members are friends whom I will always remember.

NATIONAL CANCER INSTITUTE (NCI) (1983–   )

In October 1983, I joined the NCI and assumed the rather remarkable title of cancer expert, attached initially to the Office of the Chief of the Drug Evaluation Branch (Dr. John Venditti), Developmental Therapeutics Program, Division of Cancer Treatment. The director of this division, Bruce A. Chabner, MD, is a splendid basic scientist and clinical oncologist as well as a fine administrator. My initial responsibility became the coordination of those National Cooperative Drug Discovery Groups (NCDDG) that were launched in mid-1984. These groups include cooperating academic institutions and in some cases industrial organizations. The admirable and I hope attainable goal of these groups is the discovery and development of essentially new approaches to the treatment of human neoplastic diseases, including new entities and new techniques. For the NCDDG, I function as an intellectual participant and facilitator, not as a director. In the meantime, I have become the acting deputy director of the Division of Cancer Treatment, a relatively huge organization whose total budget, intramural and extramural, is about $318,000,000.

During a long series of careers spanning forty-four years of involvements with the administration of four departments of pharmacology and the research and development activities of two pharmaceutical organizations, inevitably I have gained considerable experience and I hope some wisdom and judgment. If so, my presumably terminal contributions may be of value in the multifaceted areas of research and development of the Division of Cancer Treatment in the fields of biochemical pharmacology, chemotherapy, and drug development, the fields with which I have been deeply concerned for so many years.

Certainly, I am now making many new acquaintances who also will be ne'er forgot as long as the gods see fit to help me, not only to be useful, but also to continue to have a time for remembering.

Literature Cited

1. Davenport, M. 1967. *Too Strong for Fantasy*, p. 6. New York: Scribner
2. Welch, A. D., Bueding, E. 1946. Biochemical aspects of pharmacology. In *Currents in Biochemistry*, ed. D. E. Green, pp. 399–412. New York: Intersc
3. Welch, A. D., Landau, R. L. 1942. The arsenic analogue of choline as a component of lecithin in rats fed arsenocholine chloride. *J. Biol. Chem.* 144:581–88
4. Welch, A. D. 1945. Interference with biological processes through the use of analogs of essential metabolites. *Phys. Rev.* 25:687–715
5. Welch, A. D. 1983. Folic acid: Discovery and the exciting first decade. *Persp. Biol. Med.* 27:64–75
6. Heinle, R. W., Welch, A. D. 1951.

Hematopoietic agents in macrocytic anemias. *Pharmacol. Rev.* 3:345–411
7. Welch, A. D., Nichol, C. A. 1952. Water-soluble vitamins concerned with one- and two-carbon intermediates. *Ann. Rev. Biochem.* 21:633–86
8. Berk, L., Castle, W. B., Welch, A. D., Heinle, R. W., Anker, R., Epstein, M. 1948. Observations on the etiologic relationship of achylia gastrica to pernicious anemia. X. Activity of vitamin B_{12} as food (extrinsic) factor. *N. Engl. J. Med.* 235:911–13
9. Stokstad, E. L. R., Page, A., Franklin, A. L., Jukes, T. H., Heinle, R. W., et al. 1948. Activity of microbial animal protein factor concentrates in pernicious anemia. *J. Lab. Clin. Med.* 33:860–64
10. Nichol, C. A., Welch, A. D. 1950.

Synthesis of citrovorum factor from folic acid by liver slices; augmentation by ascorbic acid. *Proc. Soc. Exp. Biol. Med.* 74:52–55

11. Nichol, C. A., Welch, A. D. 1950. On the mechanism of action of aminopterin. *Proc. Soc. Exp. Biol. Med.* 74:403–11

12. Welch, A. D. 1950. New developments in the study of folic acid. *Trans. Assoc. Am. Physicians.* 63:147–54

13. Sakami, W., Welch, A. D. 1950. Synthesis of labile methyl groups by the rat *in vivo* and *in vitro*. *J. Biol. Chem.* 187:379–87

14. Blaschko, H., Welch, A. D. 1953. Localization of adrenaline in cytoplasmic particles of the bovine adrenal medulla. *Arch. Exper. Pathol. Pharmakol.* 219:17–22

15. Welch, A. D. 1967. The Sollmann oration. *Pharmacologist.* 9:46–52

16. Ahmed, N. K., Germain, G. S., Welch, A. D., Paterson, A. R. P., Paran, J. H., Yang, S. 1980. Phosphorylation of nucleosides catalyzed by a mammalian enzyme other than uridine-cytidine kinase. *Biochem. Biophys. Res. Commun.* 95:440–45

17. Welch, A. D., Panahi, J., Germain, G. S. 1984. A nucleoside-phosphorylating activity, distinguishable from uridine (Urd)-cytidine (Cyd) kinase, in the nuclei of human lymphoblastoid cells. *Proc. Am. Assoc. Can. Res.* 25:70 (Abstr.)

18. Welch, A. D., Nemec, J., Panahi, J. 1984. Quantitative determination of nucleosides and their phosphate esters. 1. The acidic nucleosides: 3-deazauridine and 6-azauridine. *Intl. J. Biochem.* 16:587–91

19. Belt, J. A., Welch, A. D. 1983. Transport of uridine and 6-azauridine in human lymphoblastoid cells. Specificity for the uncharged 6-azauridine molecule. *Mol. Pharmacol.* 23:153–58

William Paton.

Ann. Rev. Pharmacol. and Toxicol. 1986. 26:1–22

ON BECOMING AND BEING A PHARMACOLOGIST

W. D. M. Paton

Oxford University, 13 Staverton Road, Oxford, OX2 6XH, United Kingdom

FINDING THE SUBJECT

Like others of my generation, I found my vocation of pharmacology by chance, having originally intended to go into academic medicine. In the 1930's, medical students received teaching in the actions of drugs, but the days of undergraduate degrees or later course work in pharmacology lay far in the future. So I was essentially self-taught, with all the freedom, but also the ignorances of the autodidact. Indeed for much of my life I have been among those with a similar background, not a few of whom would say that they were not "really" pharmacologists, but physiologists, biochemists, pathologists, chemists, or some other species. It was therefore not surprising that there was a good deal of discussion, on faculty boards and elsewhere, about the nature of the subject. I came to recognize that it is the sign of a newly developing discipline to be mildly obsessed with what it "really" is, and to be concerned with establishing an identity. Toxicology is going through something of the same process today. The resistance of established subjects to the emergence of new ones is an old story—going back at least to the opposition to the formation of the Geological Society by Sir Joseph Banks (president of the Royal Society) in the last century. Physicists and chemists seemed lucky; they did not need to ask such questions, which had been settled long before, but could get on with the work.

Nor was some discussion about the nature of the discipline unjustified. Pharmacology has a taproot stretching back into the remote past of man's attempts at healing; but it also has more modern roots. So pharmacology did not lack for those ready to explain that it was "only" a branch or application of physiology, medicine, chemistry, or biochemistry. Like any other discipline, pharmacology needs constantly to renew its vision. For me, the character of the

0362-1642/86/0415-0001$02.00

subject seemed obvious and most of the formal definitions failed to express what it was that caught my own imagination. Before giving my own definition, let me mention how I was caught.

I suppose it was chance that, at the end of my medical residency, after qualifying in medicine during the London "blitz", I had a fifth attack of pneumonia (following a checkered pulmonary childhood). Medical investigation thereafter pronounced me unfit for military service, and also cast doubt on my capacity to do the residencies required for a career in medicine. The immediate outcome, after I failed to get a job at the Brompton Hospital, was that I took a post as a pathologist in a tuberculosis sanatorium. There was a brief training by a London pathologist, Dr. S. R. Gloyne. (It is somewhat ironic that his main life's work was trying to persuade an unbelieving world that asbestos damaged the lungs.) As was customary at that time, I was then on my own, doing sedimentation rates, blood counts, and culturing sputum or pleural fluid for tubercle bacilli. Occasionally I prepared lung biopsy specimens to assess the presence of cancer cells, and Gloyne would check my conclusions. The work was interesting, but my superintendent was an abrasive character, and I did not get on with him. By chance, a nephew of the then Secretary of the Medical Research Council was also on the staff. It was through him that, for the third time, I was offered a job at the National Institute for Medical Research at Hampstead. I had put previous offers aside, wishing to do clinical work of some kind. But now the offer was a relief and I accepted. On my last night at the sanatorium, I told my superintendent what I thought of him, which I had been too naïve to do before. He beamed all over, for he was one of those that like to challenge others; he then said he hoped to start a tuberculosis research unit at the hospital, and asked me to stay on and start it. I refused, but that was perhaps my first adult lesson in distinguishing between appearance and reality.

So off I went to work under G. L. Brown in Sir Henry Dale's old laboratory, joining in their wartime work on diving and submarine problems. I will say a word about the laboratory later. For the moment it will suffice to say that when peace came, the laboratory turned back to its old activities, and I began to turn into a respiratory physiologist. But then F. C. MacIntosh, the second in command, was asked by another member of the Institute, R. K. Callow, to look at the toxicity of an antibiotic that his group had isolated, named licheniformin. MacIntosh asked me to join him, and we started by testing the hydrochloride for its effect on the blood pressure of a cat under chloralose. It produced a quite characteristic response: no action for around 30 seconds and then an abrupt but fairly transient fall in blood pressure, often marked by a tachycardia during the recovery phase. It was a fascinating picture; for the initial latent period meant that the drug had no direct vasodilator action or cardiodepressant action, yet when the fall came, it was as abrupt as that produced by an equidepressant dose of histamine or acetylcholine. What was going on? In itself the simple tracing pointed to the answer: that the drug released a vasodilator from the periphery, which, on recirculation, produced the fall in pressure. It could hardly be other than histamine if it was to persist long enough in the blood. A sample of blood taken from the animal at the nadir of the blood pressure produced an abrupt fall, of 5 to 10 second latency, due to a substance with histamine's properties. J. A. B. Gray and I later verified the circulation times involved.

So licheniformin turned out to be a histamine liberator (1), and I had had the good fortune to see a new, clean, surprisingly specific pharmacological response, which was in itself rather attractive as a tracing. When we went on to find that the same pattern was displayed by any dibasic compound in which the basic groups (amine, amidine, isothiourea, guanidine) were separated by around 6 or more carbon atoms, I was fascinated. A clean action caused by a specific chemical structure opened a new world. I have sketched elsewhere (2) how the follow-up led to the work with Nora Zaimis on the methonium compounds (3). Phenyl biguanide and Compound 48/80 were other dividends. The point, however, is that the fish was caught. That experience and what it led to is the basis of my own definition of pharmacology:

> If physiology is concerned with the function, anatomy with the structure, and biochemistry with the chemistry of the living body, then pharmacology is concerned with the changes in function, structure, and chemical properties of the body brought about by chemical substances. In the same way, pathology is concerned with the changes brought about by disease. For pharmacology there results a particularly close relationship with chemistry; and the work may lead quite naturally, with no especial stress on practicality, to therapeutic application, or (in the case of adverse actions) to toxicology.

Such a pattern seems to me both to do justice to what catches the imagination of those working in the different fields, as well as showing how they all belong to one family.

Do such definitions matter? Most of the time not. But in the recruitment of new blood, it is useful to have some shorthand description; and when it comes to higher policy, the nonscientific administrator is liable to define pharmacology, just from its etymology, as "the science of drugs", with a resulting difficulty in distinguishing it from pharmacy.

Competition for funds, or simple enthusiasm for a subject, often leads to exaggerated claims for a particular discipline, and to the view that other subjects are inferior or merely applications of it. (Sentences containing "merely" or "only" should always be viewed with deep skepticism.) I like to reduce such claims to an absurdity. Let us start with pharmacology;

> which may be said to be merely applied physiology;
> which is, of course, merely an embodiment of biochemistry;
> which is merely chemistry in one development;
> which is merely the working out of certain physical principles;
> which is merely the body of particular solutions of certain mathematical equations.

But mathematics is only a branch of logic, as Bertrand Russell and his peers established, so that philosophy is the fundamental basis of it all.

Yet we know today that philosophical ideas are culturally determined, not absolute, so that sociology is the central discipline.

But sociological thinking must be merely a function of psychological process.

Such processes are, of course, no more than exercises in the pharmacology of the central nervous system.

The whirligig is complete.

Pharmacology, then, can take its own place. There is no one science central to all the others, as various scientific chauvinists would have us believe.

It is equally untrue to real life to say that "all science is one". That suggests a sort of homogenized intellectual mayonnaise, with no differentiation. One can, of course, trace relationships, as indeed I have just done. But each scientist seems to find something particular that catches his imagination. The disciplines that become separate, for the time being, are those areas that seem to fire enough people, are attractive enough to the young, and are worth meeting for, thus creating a kind of mental center of gravity. One need not be exclusive, but can move between one center of gravity and another.

If one asks, when did pharmacology begin, perhaps one can place the start in the remote past of *materia medica,* or with Magendie's first analysis of the action of a drug, or with the founding of chairs and societies, or with the therapeutic explosion from the 1930s onwards (4). But historically, the Stockholm Congress in 1961 now seems to me one of the watersheds. Börje Uvnäs had asked me to join its organizing committee; and there followed a fascinating discussion of the program to be developed in pharmacology's first independent international meeting. It was a critical moment. That meeting might have fallen apart into its biochemical, physiological, chemical, pathological, toxicological groupings. But it did not; the great and marvelous fact was that despite misgivings it fell triumphantly together. The establishment of pharmacology as an independent discipline, a center of gravity in its own right, was certain from then on.

As for my research on diving, I can see now that it too had a pharmacological component. For instance, there had been operational reports of what was called "shallow water black-out", and one of our concerns was the risk to a diver (and also to those in a sunken submarine) of becoming unconscious through CO_2 accumulation. It had previously been thought that respiratory stimulation would always provide sufficient warning, but we had found that if sufficient oxygen was present to remove any hypoxic drive, as was normally the case in diving, then CO_2 could produce unconsciousness readily (even pleasantly) without gross respiratory stimulation. It fell to me to run the O_2/CO_2 mixture breathing experiments to define the range of individual sensitivity.

All the members of the laboratory "blacked out" in their turn, as did any unwary visitors, such as B. H. C. Matthews and D. G. Evans. Judging by subsequent careers, a few minutes of 10–20% CO_2 has no long-term toxicity. But the experiments led to reading about anesthesia. Then G. L. Brown asked me to join him in advising some civil engineers on compressed air work, which resulted in collaboration with D. N. Walder and others on "bends". So an interest in high pressure biology was implanted that survived even when wartime work was all wound up. That interest produced, with E. B. Smith, a colleague in physical chemistry, the Oxford High Pressure Group; and the research still, in the problems presented by the "rapture of the deeps" and the High Pressure Neurological Syndrome, has its pharmacological flavor.

For those whom it suits, pharmacology is a lovely subject, carrying one quite naturally from the molecular level to the whole man, using any and every skill one possesses, intellectually demanding yet with practical usefulness round the corner, not yet too sophisticated technically, and still young and fresh enough for the simpleminded to be able to contribute.

"F4"

Sir Henry Dale's laboratory, the Division of Physiology and Pharmacology, in the National Institute for Medical Research in Hampstead, was one of the classic laboratories. It deserves study in its own right, and I hope that those who worked there have left some account. A latecomer like myself, though deeply influenced by it, can only give a hint of what it must have been like in its prime.

The fourth laboratory on the first floor, it was, in 1944, a large communal laboratory. If one's own experiment was not going too well, or there was a waiting period, one could look at someone else's. The storage cupboards full of equipment reached up to a high ceiling. That ceiling was stained from the overflows in the organic chemistry laboratory above. When we rolled cylinders of gas on the floor, to make new mixtures to breathe in the diving experiments, protests would come from readers in the rather beautiful Georgian-style library below. A small side room contained a bicycle ergometer used for studies at controlled oxygen consumption. The laboratory was entered from a corridor that ran the length of the building, and there was a small office, a small workshop with a lathe and hand-tools, and a lavatory immediately across the corridor. One of the pleasures was that if one went out into the corridor one was only too likely to bump into someone from one of the other laboratories.

The main laboratory was simply laid out: There was a long window bench looking out over the tennis court and other grounds, the Institute's main workshop, and the shed housing J. S. Haldane's old compression chamber, recently transferred from the Lister Institute; desks, arranged according to the inhabitants and needs of the moment; operating tables and kymographs; an old watercentrifuge with a balance for its cups and a stern notice from the head technician above it: "Near enough is not good enough"; paper-smoking and varnishing cabinets in one corner, adapted during wartime to accommodate a Haldane gas-analysis apparatus; a refrigerator, oven, waterpumps, and so on.

At lunchtime we could join other members of the Institute in the lunch room at the top of the building and then go through to the coffee room, which had a balcony looking south over the rest of London. Towards the end of the War, when the time of the "doodlebugs" came, one could see the defensive barrage balloons on the Downs. Across the road outside the Institute was the pub, the "Hollybush", where one could take a visitor after work in the evening or have some celebration.

Between 1944 and 1949, when the Institute moved to Mill Hill, H. B. Barlow, J. A. B. Gray, B. D. Burns, and W. L. M. Perry were other members of staff; M. Goffart, J. L. Malcolm, A. Sand, M. Vianna Dias, and Nora Zaimis were our honored visiting workers.

The head technician, L. W. Collison, had been Dale's personal technician, and was of kindred quality. Not greatly educated, of rather gloomy appearance, formal, courteous, tough, a stern disciplinarian, yet as determined to stand up for his staff's rights as for his own, he was intrinsically very kind, and a man of the highest standards. These he displayed in his preparation for experiments, in his construction of exquisitely delicate Brodie bellows for sensitive plethysmography, and above all in his preparation and mounting of smoked drum tracings for publication. Many of the classical tracings in the *Journal of Physiology,* establishing the theory of chemical transmission, attest to his artistry.

How does one pin down the intangibles that create a successful laboratory? You were expected to stand on your own feet. Yet there was every kind of support by advice, criticism, personal help, and (most notably) general pleasure at any progress you made. No feasible experiment, if you wanted to try it out, was barred. The range of equipment available and the little workshop across the corridor made it easy to develop a new technique. But perhaps the secret lay in the two at the top, both of whom had shared in the *anni mirabiles* of chemical transmission before the War. G. L. Brown ran the laboratory on a very light rein and gave the laboratory high spirits and a wonderfully deft experimental skill—he was a joy to watch. F. C. MacIntosh exhibited a depth of thought and an unquenchable considerateness in everything he said and did. It is impossible to set down what one owes them.

INTO A WIDER WORLD

In 1949, with the move of the National Institute to Mill Hill, G. L. Brown went to the chair of Physiology at University College, F. C. MacIntosh to that at McGill, and W. Feldberg took over "F4", establishing its ethos in a new environment. We worked together on histamine release from skin; devising a perfused skin preparation and learning how to write papers with him were equal delights. (To describe our consumption of cigarillos would, today, doubtless count as pornography.) There was also collaboration with Burns (on endplate depolarization), with Perry (on ganglia), and with new recruits such as W. W. Douglas (on anticholinesterase effects) and M. Schachter (on the effect of antihistamines on the release of histamine), as well as writing papers with Nora Zaimis.

The Director, Sir Charles Harington, was always slightly irritated at comparison with the Hampstead days; yet W. A. H. Rushton's remark, that "moving to a larger laboratory is like an adiabatic expansion: the particles become more distant from each other, and the temperature falls", seemed to have some truth. One bumped into members of other sections less often. Yet it was, in fact, during a corridor conversation that Albert Neuberger remarked to me that he felt his work was not going very well, and perhaps he ought to move. A year later, a chair at St. Mary's and a burst of new work thereafter showed his wisdom.

I am not quite sure what generates this restlessness, which I was feeling, too. The wish to be one's own master does not seem sufficient cause, when one has so much freedom and such facilities. Perhaps, for an experimentalist, it is more the wish to have a shot at doing a new experiment, to head a department oneself. For me, it led in 1952 to a Readership in Applied Pharmacology at University College Hospital, joined by J. W. Thompson, under two professors, M. L. Rosenheim in Medicine and F. R. Winton in Pharmacology. The laboratory was embedded in the Medical Unit, with an additional small room across the road in University College, and £200 a year running expenses. Our formal responsibility was to help Heinz Schild with his practical class in the College, and to give a course of 20 lectures bridging the basic pharmacology taught by Schild and the therapeutics lectures by physicians (that also fell to me to organize). It was quite a challenge to produce a complete lecture course after life at an institute. Work on histamine release went on, and John and I had an interesting time with A. Goldberg testing porphobilinogen, which had been recently isolated from patients with porphyria by R. G. Westall, to see if it was a possible causal agent of their symptoms. Unfortunately it proved rather inactive.

After two years, however, a serpent appeared in the Eden, namely C. A. Keele, with an invitation from the Royal College of Surgeons to start a new department of pharmacology there, for postgraduate teaching. By then I had had many contacts with anesthetists and physicians over the use of the methonium compounds and it seemed to me self-evident that no undergraduate course could possibly equip a medical student with all the basic knowledge he needed in his career—there was too great an annual flow of new drugs and new procedures involving totally new principles. So the idea of postgraduate education, of teaching those already experienced in medical practice who wanted to be updated, made sense. So did the idea of fuller laboratory resources.

In this academic experiment, John Thompson (now in Newcastle), John Vane, and John Gardiner (now in Hong Kong) joined me. I couldn't have asked for anything better than to start a new laboratory with such colleagues. It is remarkable to me now, when academic tenure is so much discussed, that it seemed to matter little to us that I was losing tenure and they were not receiving it. There was only enough money for the new laboratory for 5 years, and its continuance depended on our success. We were lucky with our research fellows and associated staff, all from the clinical world: J. G. Murray, E. Marley, G. C. Clark, and J. P. Payne, later to occupy chairs of surgery, pharmacology, surgery, and anesthesiology, respectively. Of our visitors, B. B. Gaitonde went on to head the Haffkine Institute, and Emile Savini to chairs in France. It was a very happy time, and it was the best interaction of "basic" and "clinical" science I have ever had. That the department did indeed continue, under Gustav Born, John Vane, and Graham Lewis, has always pleased me.

The Royal College of Surgeons broadened my life, but the effect was nothing compared to that of the move to Oxford in 1959. As an Oxford graduate, one was "imprinted" by the University and one's old college; but the experience of an undergraduate, "the toad beneath the harrow", and of one of the dons, the "harrow" itself, is very different. It is too much to describe here, and indeed, it is still in process.

But there are two things from my time at Oxford that I like to mention. The first arises when aspersions are cast on the academic standing of pharmacology. Then I like to recall that in 1962, out of an established staff of six, four were Fellows of the Royal Society: E. Bulbring, H. Blaschko, H. R. Ing, and myself. That is in part a tribute to my predecessor, J. H. Burn, but it seems to me also to show that quality is not impossible even to a small science. Other departments can make equivalent claims.

Second, I have concluded from 25 years at an undergraduate university, as well as from seeing many other forms of research life, that no research institution keeps its freshness and vigor, unless it is continually rinsed through by exposure to new youthful minds. The undergraduate can be infuriating, lazy, stupid, and rude (and also charming, hardworking, intelligent, and courteous). No doubt he feels the same about his seniors. But there is something uniquely invigorating in being obliged to interest minds quite new to a subject, and to respond to the naïve, sometimes foolish, but essentially unspoiled vision of it that they have. New centers of research, free from the "burden" of teaching and the associated tasks, can flourish for a time. But follow their progress over the years, and you will see the complacent inward look developing—unless some wise person has contrived a youthful revivifying stream.

RATE THEORY REVISITED

I am sometimes asked what I think about rate theory today. The question always drives me back to the mental puzzles that evoked it.

The idea of rate theory germinated around 1955. The immediate stimulus was an invitation to give a paper on the mechanism of neuromuscular block at the World Congress of Anesthesiologists in Scheveningen that year. The two types of agents, depolarizing and nondepolarizing, that Eleanor Zaimis and I had distinguished operationally in some detail, were familiar. But examples were accumulating of block of an intermediate type, such as what she called "dual block". Bovet had contrasted "pachycurares" and "leptocurares" corresponding (for example), to tubocurarine and decamethonium, thus emphasizing the role of molecular shape. But the intermediate type of response, with typically stimulant phenomena followed by apparently simple curare-like block, was found even with a "leptocurare" like tridecamethonium. I included in my paper (5) a crude estimate of polarity of the various agents (the ratio of C atoms to N atoms), and suggested that when the ratio rose above 8 to 10, a transition occurred from depolarizing to intermediate type. It therefore seemed to be progressive hydrophobicity that caused the development of a curare-like component of action. The survey suggested that "drugs with some measure of hydrocarbon loading may (through their lipoid affinity) develop an attachment to the membrane of a kind different from, additional to, and interfering with, the attachment which leads to endplate activation. For instance, it might be essential, for normal activation, that a chemical bond be rapidly made and broken, or that the molecule has to move across or through the membrane; fixation of the molecule by its hydrocarbon content to adjacent lipoid regions of the membrane could well interfere with such a process".

But the problem of dual block was not the only stimulus. The work on the methonium compounds had raised in an especially acute form the question of how it was possible that structures so closely alike as hexamethonium and decamethonium could not merely act selectively on different effector organs, but act in totally different ways. Work with W. L. M. Perry had revealed the action of nicotine on the ganglion also to be of intermediate type, with depolarizing block passing over to a nondepolarized hexamethonium-like state; and on the guinea-pig ileum the self-block and bell-shaped dose-response curve were not well understood. Study of the anaphylactic spasm of sensitized ileum, and the extent to which it could be imitated with histamine, also raised some puzzles. Finally, there was a long-standing wish to describe the time course of recovery from neuromuscular blocking agents, ever since our finding, during potency comparisons of decamethonium with tubocurarine, that after apparently complete recovery of a muscle twitch from tubocurarine there might still be enough tubocurarine left to antagonize the action of decamethonium.

I was, however, at that time almost totally ignorant of receptor theory (what there was of it), and had just started a notebook devoted to the subject. It is hard today to understand the paucity of knowledge about, and the almost antagonistic attitude to, the receptor concept at that time. For a start, my greatest help was Haldane's 1930 book on enzymes, to which I turned after initial study of adsorption isotherms in a textbook of physical chemistry. In due course I found Gaddum's 1926 paper, then Clark's monograph of 1937, and Gaddum's formulation of competitive antagonism in the same year. Then there was Schild's pA notation in 1947 and 1949. But there were also a number of papers incompatible with Gaddum's formulation, and in the background were, for instance, the critical remarks by Dale in 1945 of the whole receptor concept, and a conspicuous lack of any direct evidence for it. In 1954 there began a long series of papers by Ariens and his colleagues, developing for drug action a framework comparable to that available for enzyme kinetics. By hindsight, these could have helped me; but at the time the multiplicity of possible models presented, with a growing number of disposable parameters, left me rather confused. I found myself more at home with R. P. Stephenson's work published in 1956. But although his introduction of the concept of efficacy seemed a real advance, it did nothing to answer my question of why some drugs are stimulant, some antagonistic, and some intermediate. To say that efficacy was high, low, or medium just restated the problem.

I think it is unlikely that I would have gone much further had not a bronchial episode led to a period in bed, around 1956. For something to do, I started working through for myself the equations that a bimolecular type of reaction may lead to, particularly seeking formulations that would be experimentally testable. I looked for patterns that could explain the bell-shaped dose response curve of nicotine on the gut, that would give some physical interpretation of efficacy, that would explain the transition from stimulation to relaxation with intermediate-type drugs, and that would help to analyze the time course of blocking agents in the body. Haldane and Briggs' analysis of the results of a two-point attachment, with a bell-shaped substrate concentration-velocity curve interested me very much, and I was carried far out of my depth in trying to extend it and its kinetics. I made rather little progress. When the thought finally occurred that receptor excitation might be a function of rate of association of drug with receptor, it seemed almost magical how many of the questions it solved. If activation of the receptor required that a high rate of drug turnover continually releases receptors for fresh associations, then the characteristic sequence (stimulation then block) of intermediate drug action followed naturally, drug-receptor dissociation rate could provide the independently measurable index of efficacy, and the slowing of dissociation by hydrophobic bonding explained the shift from stimulation to block with increased hydrophobic loading in homologous series.

It was quite a busy time in other ways. There was the move to the Royal College of Surgeons in 1954. From 1951 to 1957 I was secretary of the Physiological Society. In 1959 I moved to Oxford. Most of my experiments were performed in the evening and it shows. There is not nearly enough replication in the paper that I finally wrote (6). It would have been wiser, too, to have separated more distinctly the various themes in that paper. I felt the phenomena of nonspecific desensitization of smooth muscle were important, that they showed that nonreceptor effects could be great and needed to be controlled, and that they seemed to provide an immediate explanation of the existence of spare receptors (for they meant that receptor occupancy would not always be the rate-limiting step). This last point, however, was really a separate issue.

I would treat the ion exchange model differently today. There is no evidence that it plays any role, although I still wonder about the fate of counter ions to charged groupings in the receptor in the presence of an interacting drug. The model was only suggested, because it seemed to me necessary to give some illustrative example of a possible quantal mechanism. MacIntosh and I had considered an ion-exchange mechanism for the action of histamine liberators, with the bases exchanging with histamine in the tissues. So, in my reading round the subject of neuromuscular block, I had noticed Ing and Wright's suggestion of an ion exchange by alkylammonium salts at the neuromuscular junction. It was a simple concept that had stuck in the memory. Finally I was impressed with the evidence about loss of potassium as a result of agonist action. But again, it was a side issue.

Rate theory had its first exposure to the British Pharmacological Society in July 1959. Then, in 1961, I was asked to present the paper before publication at a meeting of the Royal Society, which I did, loaded up with analgesics. The President, H. W. Florey, asked Gaddum, who was present, if he thought the theory was sound, and Gaddum replied that he thought there might be something to it. Bernard Katz asked what quantal mechanism I envisaged. I could only repeat my working analogy, that the drug-receptor interaction was more akin to the playing of a piano (where a finger that is not rapidly withdrawn interferes with the flow of sound), than it is to the playing of an organ. That weekend I went into Hammersmith Hospital for a bilateral antro-ethmoidostomy. One of my pleasantest memories was at a meeting of the British Pharmacological Society around that time, at which John Vane, describing an experimental set-up, referred to "what I understand we must now call an isolated piano bath".

What does one think of the theory now? I think it helped to promote interest in receptor studies, particularly on the kinetic side. I still find its economy conceptually attractive, requiring as it does only two rate constants and one other constant to couple the receptor to the effector. So far, I see no other independently testable interpretation of efficacy. As a possible model it does not seem to mislead the student, since most of its consequences for the general pattern of drug action are quite sound: agonists do wash out quickly; for equiactive doses, antagonists do usually take longer to act and to pass off, the more potent they are; decline of antagonism can be measured, and changes in receptor occupancy do follow an exponential course. With partial agonists, agonism is prompt and then antagonism follows; hydrophobic loading does generally favor antagonist rather than agonist action.

I was very disappointed that the measurement of the rate constants of drug receptor interaction has proved so difficult, particularly with smooth muscle, where the range of values seems wider (and thus potentially more informative) than with striated muscle. It was ironical that it was our own work at Oxford (7) showing how the relatively large uptake of drug by the receptor created a large reserve of drug, that led to the possibility that the kinetics were diffusion, not receptor, controlled, as Douglas Waud had come to suspect and Thron and Waud then showed (8). No way was found to distinguish the two possibilities. I was back in the old position, unable to find a definitive, objective, independent sign of efficacy.

It is pleasing that agonist action has now turned out to be quantal, and it seems likely to be linked to receptor turnover. But it seems uncertain how the quantum originates, whether from the allosteric behavior of the receptor or from some aspect involving the drug molecule too. It was also interesting that an identical theory was formulated within gustatory physiology, quite independently, but for similar reasons.

My guess, however, is that the deeper understanding of the kinetics of drug action is not going to come through the formulation of drug-receptor equations, but from qualitatively new observations. Patch-clamping and conductance-channel biophysics, molecular engineering, and ultramicroscopy offer fascinating openings. I have suspected that our formulations will prove far too simple, ever since seeing some of D. C. Phillips' X ray crystallographic pictures of the interaction of a number of different substrates with lysozyme. With each substrate, the interacting regions were subtly different. Presumably the same is true with drugs interacting with receptors, and the reactions that we blithely represent by a single equation, $D + R \rightleftharpoons DR$, are in fact all different.

ON DOING WHAT YOU ARE ASKED TO DO, AND THE STIMULATED ILEUM

I have already mentioned two lines of research that followed on an "outside" stimulus—histamine liberation and rate theory. This is an important phenomenon, weakening the concept of the autonomous scientist in his ivory tower; even in research he seems to need stirring up from time to time. A third example was an invitation to talk on the pharmacology of the small intestine at a Gordon Conference, presumably because of my ganglionic work. I knew little about the gut, and found considerable doubt in the literature even as to the transmitters. So I began to search for some way of making a "neuroeffector" preparation out of a strip of intestine, on which one could do analyses comparable to those on striated muscle or ganglionic preparations. N. Ambache had found that a strip of gut would twitch if an electric stimulus was applied to its surface. But one then finds that the point of excitation moves and is undefined. It was then that I recalled some old experiments by Rushton, dealing with Lapicque's chronaxie theory of curare action, and Du Bois Reymond's cosine law: that excitation of a nerve in an electric field is in proportion to the cosine of the angle between nerve and field. Suppose one placed an electrode in the lumen of the gut, another outside, and created a field between them. The field would now be defined, independent of gut movement. But would the nerve networks of Auerbach's plexus be at an angle to be excited? The first experiment, in my office-cum-laboratory at the College of Surgeons, is still vivid. I set up the Trendelenburg preparation used to study peristalsis; my only piece of platinum, soldered to a lead, was threaded into the lumen. An alligator clip, dipping into the outer bath fluid, provided the other electrode. It worked like a bomb, with shocks as short as 50 μsec. I had, in a few moments, both a new neuroeffector test-bed (9), and something to say in New Hampshire. G. L. Brown, when he heard about it, was very amused, and called it the "electric clyster" in a mock eighteenth century poem. He also sponsored it for a demonstration at a Royal Society soirée, at which Fellows were able to do an Otto Loewi-type experiment for themselves: they took fluid from one bath holding an eserinised strip of intestine, with or without stimulation, and tested it for acetylcholine content on a second comparable strip arranged for assay.

The guinea-pig ileum longitudinal muscle is a great gift to the pharmacologist. It has low spontaneous activity (unlike rabbit or rat); nicely graded responses (not too many tight junctions); is highly sensitive to a very wide range of stimulants; is tough, if properly handled, and capable of hours of reproducible behavior. A further bonus appeared later when I did electron microscopy on it, and found that Auerbach's plexus lay on the surface of the longitudinal muscle, not penetrating into it. As a result, after Humphrey Rang had shown how to separate strips of longitudinal muscle, Aboo Zar was able to produce acutely denervated strips simply by pulling them off the plexus (10). I know nothing quite like these preparations in physiological pharmacology. Normally, to denervate one cuts a nerve and then waits for it to degenerate; but during that time profound changes take place in the postsynaptic structure, so that one does not have a perfect control. The longitudinal muscle of guinea-pig ileum, with its diffusive type of transmission, is the only tissue I know of that allows you rigorously to compare innervated and totally denervated muscle. This was particularly important at the time for making absolutely certain that the acetylcholine released came from nervous tissue, not muscle. I had found that morphine inhibited the ileum twitch without affecting the response to acetylcholine and reduced ACh output (11), I was now confident in the conclusion that morphine was acting on nervous tissue and that the ileum could be used as a "paradigm of the brain."

The preparation has proved admirable for teaching. The coaxial method of stimulation was only necessary when stimulator power was limited, and simple field stimulation proved to work admirably. Particularly enjoyable was the collaboration with Sylvester Vizi on the effect of catecholamines on transmitter output, opening up a profitable presynaptic vein (12). So has been its use in identifying and assaying opioid peptides and drugs. Perhaps every scientist should be asked to talk on something about which he is ignorant, every five to ten years or so.

COMMITTEES

In my best, or worst, year, I counted service on 72 committees that met at least once, and sometimes ten to twenty times, in that year. I am sure there are others with a similar experience. It seems too many, yet none of them was trivial. The most severe critics were those who wanted me to serve on a seventy-third. Is there anything to say about it, except to warn?

It does require a decision at some stage in one's life. When the invitation first comes, it is not unflattering to find that somebody actually wants your services. Most people would agree that democracy, with continually changing committees, is in the end better than dictatorship or oligarchy. There is, too, the opportunity both to be more "in the know" and to have some influence. Most people find a small dose of one or the other attractive enough to exchange for a little research time or some leisure. In the event, you do indeed lose time, and it can be trying to spend hours awarding money to others to do experiments you would like to be doing yourself. Some say discouragingly that such work is "all right for those that like it". It is a great simplification when you grasp that the only valid reason for doing anything is that it is the right thing for you to do. There is in the end a genuine satisfaction in seeing work progressing that you helped to forward. Even more important, and a reflection of the fact that shared endeavor is the most potent source of friendship, are the friends you make. So, if advice were wanted whether or not to undertake such work, the heart of it must be the importance of diagnosing one's own capacities and values. It proves to be a balance between one's own work on one side, and, on the other, new friendships while forwarding the work of others. Not a very obvious choice. I recall a famous remark by F. M. R. Walshe, the great neurologist at University College Hospital, a Catholic and vigorous controversialist for whom I served as house officer. After seeing a pair of outpatient twins he said, "I wish one could baptize one twin and use the other as a control". (If we could live a second life, would we go back to do the controls to the choices we have made?)

Nevertheless, committee work can certainly go too far. I suspect that pharmacology may be especially vulnerable. The pool of trained pharmacologists is in any case not very large. The academic part has to teach doctors, physiologists, pharmacologists, biochemists, chemists, and pharmacists. It has to attract students, arm them with an undergraduate training, introduce them to research, and then supply them to industry as new recruits to an extent that the other preclinical sciences are not called upon to do, proportionate to their total numbers. It must also supply the research leaders for industry. In the United Kingdom, I reckon that industry has taken away from academic life men who would have filled about a third of the professoriate. (By and large industry does not repay to academia its training and recruitment debt, despite the great scientific contributions of industrial pharmacologists.) That brain drain cannot strengthen the remaining pool to whom it chiefly falls to provide the general servicing (and defense) of pharmacology in its private and public functions. Finally, the sheer usefulness of pharmacology frequently brings its practitioners into practical affairs and issues that call for expert advice.

It would be better if the work was shared more widely. Committees might more often invite younger scientists to serve, instead of relying largely on older people. This would widen the decision-making base, and keep the decision process nearer to the front line of new knowledge, the true source of freshness and invigoration. Getting the right balance between age and youth is quite an art. It is wrong to burden the young while they are still establishing themselves, and while they are in a particularly fruitful phase. They also may lack patience. The old often have a mass of useful case history to call on, and have seen many "experiments" tried. Yet there is a tendency for old men to fight old battles; and while sometimes those battles are over issues that are still alive, too often the arena has moved elsewhere. I can see no rules, except perhaps to prefer for service those who are a shade reluctant to serve.

Secondly, it would be interesting to try building into the standing orders of any committee a terminating rule, that it must be disbanded after three years unless a really strong positive case for its continuance is made.

Thirdly, when a statutory committee simply must meet, it should be expeditious. My best example is my old "boss", G. L. Brown, whom I saw get through a statutory faculty board meeting at University College in 1.5 minutes flat, having accomplished all the required business.

CHOOSING AND JUDGING

Among the most important of committees are those concerned with making choices over appointments and awarding grants. I must have been involved in making hundreds, if not thousands, of such choices. Some years ago I was suddenly struck with repugnance at the whole procedure; and I wished never to make a judgment on another scientist again, nor have it in my power to influence his career in any way. Unfortunately one cannot responsibly just contract out. I suspect it was a reaction that was shared by others, for a general movement has occurred toward examining the efficiency of the decisions made by grant-giving bodies, and toward testing the validity of peer review.

For some time I have made little lists, as opportunity offers, of two things. One was of the names of those whom I had helped to choose for positions, together with their competitors' names. The other was of unforeseeable events (for example an unexpected death of someone in a key position, or a surprise career decision). Taking the approach of the naive empiricist, as opposed to that of the a priori thinker, I hoped that the first list would allow me to look back and see how wise a choice had been and that the second would enable me to test how far decisions seen in hindsight as incorrect could be understood as partly inevitable.

I cannot pretend to any great enlightenment. One thing that struck me was the mind's resistance to the idea that events indeed cannot be foreseen. My list of unpredictables has now lost all the impact each item had at the time it was entered; the choices now seem historically inevitable. Among the protective mechanisms of our minds there is something that rejects uncertainty and the cautious suspension of belief that an awareness of uncertainty requires. Since then, I have been struck by the rarity with which, for instance, journalists are willing to suspend judgment. In public utterance, the "all-or-none" law operates, with premature polarization and sharp antithesis clouding the possibility of any intermediate ground on public issues. Perhaps that is one reason why the scientist may find public affairs difficult.

A second thought was really a question, and it arises from the teaching and examining of students too. What is the proper time span for judging academic and professional performance? Should the competence of a medical student be judged by performance in an examination, in the first residency, after 10 years practice, or by the mortality rate among his patients throughout his life? Is a professor chosen for what he will do immediately, in five years, or over his whole career? Does one judge by the latest work of an individual, by the integral over his whole career thus far, or by some extrapolation into the future based on previous work?

Discussion of such questions does not prove very useful. One tends perhaps to favor that way of assessment that would put the best light on one's own work. But I do share the general suspicion of publication lists. The enormous growth of multi-author papers, and the breaking up of work into multiple papers, has made bibliographies extremely hard to interpret, and time rarely allows the reading of them. Yet if one sets them aside, what evidence is one to use? Referees are often flatly contradictory of each other and themselves need refereeing. It is only a matter of luck if there is an authority on the topic in one's decision group.

I believe, however, that for career choices at least, there is information that could be better used. Among the signs of peer trust and confidence are: the appointment of an individual to editorial boards, or to some office in a society, or to be a representative of some sort, or to be an examiner, or to be an organizer of some meeting or event. Success with graduate students is another sign. In each of these situations, other scientists are entrusting some part of their own concerns to another individual. It is true that none of these tests directly measure, say, creativity. But it is also true that people do not entrust matters of these kinds to the ignorant, foolish, or unresourceful. If one looks back at office-bearers and editorial boards and the like, it seems to me that many of those whose research I have admired do indeed appear among the names.

I would not wholly abandon the Citation Index type of approach, provided one restricts oneself to well-refereed journals of substance, and corrects for self-citation. But I would also like a Scientific Service Index, which records those activities where a scientist has been chosen by his peers. My only gloomy thought is that, once established, the Law of Indeterminacy (that the act of observing a phenomenon itself changes the phenomenon) would operate, as it has begun to do with the Citation Index.

More important, now that we are so much in one another's hands, is to keep thinking about assessment. It is common enough for bodies to look backward at their successes. It will take more courage to extract the rejections and the alternative policies not adopted, and to examine the failures. A good clinical trial notes adverse reactions and deaths, as well as benefits and survivals. Why should one assess one's own past choices less rigorously?

THE SOCIAL RESPONSIBILITY OF SCIENTISTS

Do scientists have, by virtue of their skills and knowledge, any particular social duty outside their science? It is a sign of the times that the question would, today, almost always be interpreted politically, although it is equally valid at the level of individual action on behalf of other individuals. The answers are immensely varied, and the causes taken up equally so, whether nuclear war, preservation of endangered species, pollution in all its facets, holistic medicine, or civil liberties. There would likewise be no agreement about the range over which the scientist speaks with particular authority, and where with no more than any citizen's. He may like to claim an objectivity conferred on him by training and practice, but the nonscientist does not always find this convincing.

For my own part, I have been pushed out of the purely academic path over two issues. In each case what got under my skin was what seemed to me, to quote a phrase of my grandfather's, "suppressio veri and suggestio falsi". The first was the antivivisection movement. Soon after I had qualified in 1942, I saw an advertisement by an antivivisection society on a hoarding near our flat in London. It claimed to give the number of children dying from diphtheria over a term of years, who had been vaccinated against the disease. None of the other relevant figures, about numbers at risk and the fate of the unvaccinated, were given. It stuck in my mind. Perhaps it prepared me for joint work while secretary of the Physiological Society, later as Chairman of the British Research Defense Society, and later still for writing "Man and Mouse" (13). Such tasks took time away from the laboratory; yet how could anyone who has seen what medicine can do as a result of animal experimentation, who then sees that work traduced, fail to defend it?

There may be a trace of genetic predisposition here—I am very proud of my grandfather. David Macdonald was a Presbyterian minister in Derby in 1901, when Stephen Coleridge, the leading antivivisectionist of his day, came to lecture on animal experimentation. My grandfather, who was interested in science and bought *Nature* every week to circulate round his parish, did not like the style of what he heard. He criticized it vigorously at the meeting, citing Ferrier, Keith, and Spencer Wells, and then wrote to the local paper referring to "suppressio veri" and "suggestio falsi" and saying that no lover of truth could support Mr. Coleridge's society. Coleridge then launched a suit for libel, demanding a public apology, damages and costs. Although David Macdonald was a poor man, he refused these. He had always been friendly with the local doctors, and in due course they came to his rescue, bringing in Lauder Brunton and Victor Horsley. When the case came up, David Macdonald needed to do no more than show that there had been no personal attack; after 30 minutes, the jury asked if they needed to hear any more, returned a not guilty verdict and said the case should never have been brought. It remains an exceptional case to my knowledge, of a member of the church putting himself at risk for the world of science. A gentle, affectionate man, he retired to the Scottish Highlands, and died after overtiring himself in the mountains. One of his sayings was: "You should not criticize people unless you have been in the same position and done better."

The other "outside" activity has to do with drug dependence. Work with morphine led me in 1966 to attend a meeting on adolescent drug dependence, where I heard about cannabis use for the first time. Two points were of interest. One was Dr. Rasor's remark that 95% of his New York heroin users had also used cannabis, but only 30–50% had used other drugs like amphetamine, alcohol, and barbiturates. The other point was the flat contradiction of one psychiatrist by another as to the reasons for cannabis use. So I went away to look up this curious material. What I found did not seem to correspond at all with some of the assertions being made (about harmlessness, lack of aftereffects, nonaddictiveness, power to improve mental function, and the like). As

I then began to see the effects in students, my overwhelming feeling was that a vulnerable generation had been gulled. I felt some duty at least to publicize more of the original information, rather than allowing secondary and tertiary citations appear to be the sum of information available. This led to a good deal of lecturing (14) and some writing, and the discovery of Gabriel Nahas as an ally. E. W. Gill, R. G. Pertwee, and I started some work, which later was extended with mass spectrometry under David Harvey. No one can be happy about the incidence and effects of drug dependence today. But now, three satellite conferences (15) and many symposia and reviews later, there is at last some solid information, and the subject is no longer so briskly brushed under the carpet of sociological bromides.

As always, the work brought new friends. It also proved interesting in new ways, because of the political and media dimensions. It is irksome at first to find oneself classified by the antivivisectionist activists with slave traders and Nazi war criminals or by the "pot lobby" with Commissioner Anslinger. One learns the edge behind the advice "If you can't stand the heat, get out of the kitchen". But it is also fascinating to learn to what extent the dictum that "the medium is the message" is true. There is no doubt that a truly effective publicity campaign can both induce a widespread belief in quite erroneous or misleading ideas, and also cause not so much the rejection but the discounting of evidence. But only for a time. In the end, evidence wins through. Perhaps the tendency of the human mind to question everything is ultimately a very great protection. For it means that even if false ideas become current (e.g. that millions of animal experiments are done in the cosmetics industry, or that one cannot develop tolerance to or dependency on cannabis), once such ideas become in any sense orthodoxy, they become themselves subject to doubt and scrutiny. Then the evidence begins to make its mark. But there is, of course, a price, paid by the continuance of unnecessary ignorance, and the suffering of its victims.

EMPIRICAL AND A PRIORI THINKING

As the trajectory of a pharmacologist's life moves from the laboratory to the committee world and back again, inevitably patterns appear. One is the division between those who have done experiments and those who have not. It correlates closely with the division between those with an empirical habit of mind and those whose approach is a priori. I suspect it is what really underlies C. P. Snow's two cultures. Most of the time we inhabit both worlds, but sometimes one meets each mode of thought in all its purity.

The a priori mode of thought needs some defense, if one is writing for scientists, for it seems so absurd to the scientifically trained not to learn from experience all that one possibly can. But there is more to it than that. It is only too familiar a sight to find the conclusions drawn from some body of empirical evidence being bitterly disputed. The onlooker soon appreciates the adage about "lies, damn lies, and statistics". If someone then says, in effect, that there are a limited number of possibilities, that all the outcomes can be envisaged (using imagination, intelligence, experience, and wide reading), and that simply by reflecting on these one can reach rational decisions, it can be attractive and cogent. The empiricist himself uses such methods in deciding if something is "plausible". (The weakness is that in fact all possible outcomes cannot be envisaged.) If the approach is combined with skill in presentation, a powerful a priori case can be built up. It is here that training in the humanities or in law is so powerful; for it is a training that takes a vast range of human experience as its raw material, and teaches how to think deeply and argue cogently about it. Yet, in the end, one must see it as dealing entirely with the given. The facts of a case simply serve, not to suggest further inquiry, but to identify the category of previous thought and analysis to be applied. It can then be quite a battle to get new evidence sought for–or considered.

The experimenter has ultimately a different approach. He knows that there are still things not known, which are also not deducible, and that they are out there waiting to be discovered, by the method of deliberate experiment. That method is easy to write about, but not so easy to acquire. I reckon that I did not really learn what it was about until I had been at Hampstead for around six months. There seems to be a similar interval with my graduate students, before they move from making the observations they feel they "ought" to make, to trusting what they actually see with their own eyes and having the confidence to act on it. A little later comes the marvelous time when they actually make a discovery, however minute, of something that was not known before. It is then that the feeling for when something is proved or not proved begins to grow.

Philosophically, of course, some argue that nothing is ever proved. So let us put it a little more practically: that the empiricist learns that, with the right procedures, he can reach conclusions on which both he and others seem to be able to build, conclusions that are not dependent on his own prejudice, nor on his own powers of advocacy, but which remain true when the experiment is in the hands of others. At that moment, he joins the community of experimental scientists, a community spread over the world among his contemporaries, and reaching back in time through the centuries and forward into the future. We are not just of the same family as our colleagues of today, but also of Ehrlich and Dale and Cushny and Abel and Stephen Hales and Robert Hooke and Robert Boyle.

The lesson of the experimental method is not pain-free, for it is accompanied by repeated demonstrations of how wrong one's ideas had been, of one's experimental inadequacies, of one's stupidity in framing rational possibilities. But once learned, the reward is great. Today, relativism underlies a great deal of thinking; and it often seems fair and cautious to believe that there is no certain answer to some question. Yet if that leads on to the conclusion that it is never possible to say whether or not this or that "is the case", the outlook is depressing indeed; for then force would be the only way to settle uncertain questions. But the experimenter who finds that he can confirm another person's findings, or has his own unexpected results verified in another laboratory, begins to have some confidence that there is a world where something objective exists, that movement forward in some sense, however modest, is possible.

The empiricist's danger may be that of too great an attachment to the latest evidence. But he should be an optimist; for he knows there is more to find, and in that finding, new opportunities are born.

SOME CONCLUSIONS

By the time one retires, any generalizations one might make are of small use. Looking back, I think it was only rarely that I actually took advice; example was a much commoner source of information. At the same time, there are a few points that, if asked, I like to hint at.

The first is the importance of keeping your nerve. There are always ups and downs; but there are always new and unexpected opportunities. My own career allowed me several opportunities to take the road that clearly suited me; and others have told me the same. So one can make mistaken choices, and yet have other chances. I think, too, that the saying is right that "the main difficulty is almost always muddle not malice". Conspiracy theory seems to me historically inaccurate in almost every case; and it can waste a lot of time and emotion.

A second point is the importance of "self-diagnosis", of discovering where your real interests and skills lie. It is not easy, for at the start one does not know the options, and once you are launched there are plenty of pressures from others to do this or that. One useful measure is to discover what your mind turns to when it is "free-coasting", i.e. when there are no pressures on it, nothing is expected of you, and you are even a little bored. A similar test is to recognize those activities where time seems to disappear. Charles Morgan puts it romantically in "The Judge's Story": "ask yourself in what work, what company, what loyalty your own voice is clear and in what muffled. By that answer rule your life". One's mind has a "grain", and it is better to work along it.

A third point is distinguishing between the various meanings of the superlative case of the adjective "good". One meaning is simply "very good". A second meaning is "the best". This has the interesting feature of being a function of the surroundings. The level is set by the competition. While competition may be a potent stimulus, it may also be virtually absent. The paradox may then arise of the "best" not even being particularly "good". This is the weakness of relying totally on the competitive impulse. Finally comes "the best possible", where both the competition and that internal impulse to do yet better are both drawn upon. Here is where the new ground is broken. Wordsworth describes it, in his passage on the bust of Newton in King's College Chapel, as "the prism and silent face, the marble index of a mind voyaging through strange seas of thought alone." That level is for only a few, but perhaps L. W. Collison's notice in F4, mentioned earlier, "Near enough is not good enough", was in the same spirit.

Lastly, accuracy is a quality that can be most helpful. It need not be restricted to measurement. Sometimes a situation can get complicated, with allowances here, and compensating severity there. If one sets aside all the adjustments, and simply tries, at the start, to make as accurate a description or diagnosis or account as possible, it can release a cramp. Everyone respects accuracy; and once the initial ground is clear, solutions become easier to find. None of these matters are easy; but a strategy that seeks to begin with accuracy is off to a good start.

Literature Cited

1. MacIntosh, F. C., Paton, W. D. M. 1949. The liberation of histamine by certain organic bases. J. Physiol. 109:190–219
2. Paton, W. D. M. 1982. Hexamethonium: contribution to symposium honouring Sir John McMichael. Br. J. Clin. Pharmacol. 13:7–14
3. Paton, W. D. M., Zaimis, E. J. 1952. The Methonium Compounds. Pharmacol. Rev. 4:219–53
4. Paton, W. D. M. 1963. The early days of pharmacology, with special reference to the nineteenth century. In Chemistry in the Service of Medicine, ed. F. N. L. Poynter, pp. 73–88. London: Pitman
5. Paton, W. D. M. 1956. Mode of action of neuromuscular blocking agents. Br. J. Anaesth. 28:470–80
6. Paton, W. D. M. 1961. A theory of drug action based on the rate of drug-receptor association. Proc. R. Soc. London Ser. B. 154:21–69
7. Paton, W. D. M., Rang, H. P. 1965. The uptake of atropine and related drugs by intestinal smooth muscle of the guinea-pig in relation to acetylcholine receptors. Proc. R. Soc. London Ser. B. 163:1–34
8. Waud, D. R. 1968. Pharmacological receptors. Phamacol. Rev. 20:49–88
9. Paton, W. D. M. 1954. The response of the guinea-pig ileum to electrical stimulation by co-axial electrodes. J. Physiol. 127:40–41P
10. Paton, W. D. M., Zar, M. A. 1968. The origin of acetylcholine released from guinea-pig intestine and longitudinal muscle strips. J. Physiol. 194:13–33
11. Paton, W. D. M. 1957. The action of morphine and related substances on contraction and on acetylcholine output of coaxially stimulated guinea-pig ileum. Br. J. Pharmacol. 12:119–27
12. Paton, W. D. M., Vizi, E. S. 1969. The inhibitory action of noradrenaline and adrenaline on acetylcholine output by guinea-pig ileum longitudinal muscle strip. Br. J. Pharmacol. 35:10–28
13. Paton, W. D. M. 1984. Man and mouse: animals in medical research. Oxford/New York: Oxford Univ. Press. 174 pp.
14. Paton, W. D. M. 1968. Drug dependence—a socio-pharmacological assessment. Public lecture to British Association, Dundee, Adv. Sci. 25:200–12
15. Paton, W. D. M., Nahas, G. G., Braude, M., Jardillier, J. C., Harvey, D. J., eds. 1979. Marihuana: Biological Effects: analysis, metabolism, cellular responses, reproduction and brain. Oxford/New York: Pergamon 777 pp.

PHYSICAL CHEMISTRY

Henry Eyring

Ann. Rev. Phys. Chem. 1977. 28: 1–13

MEN, MINES, AND MOLECULES

Henry Eyring

Department of Chemistry, The University of Utah, Salt Lake City, Utah 84112

My ancestors were drawn together from northern Europe by the new Mormon religion. My mother's people, the Romneys and Cottams, migrated from around Preston, England, arriving in Nauvoo, Illinois in 1839. My grandfather, Henry Eyring, came from Coburg in Germany, and Grandmother Eyring came from German Switzerland. My Eyring grandparents met while crossing the plains and arrived in Salt Lake City in the same pioneer company in the fall of 1860. My mother's people had reached Salt Lake City 10 years earlier. During the next three decades, colonization of the intermountain area spread from Salt Lake City to Alberta on the North and to Chihuahua and Sonora on the South, wherever the water could be turned out onto the parched land. As a result of these migrations, all my grandparents ended up in the late 1880s in Colonia, Juarez, in northern Mexico, about 100 miles straight south of Columbus, New Mexico.

This part of Chihuahua lies in the foothills of the Sierra Madre Mountains at 5000–6000 feet of elevation in an excellent region for raising cattle. At that time Don Luis Terrasas owned about a third of the state of Chihuahua and had one of the largest herds of cattle in the world. My father owned the Tenaja ranch of 10,000 fenced acres where he raised high-grade cattle and had about 4000 acres of pasture and farmland nearer town where he kept 50–100 head of horses. The 10,000-acre ranch seven miles from town pastured about 600 head of shorthorn Durham cattle. The bull calves were sold to surrounding ranchers to upgrade the herds of Spanish longhorn cattle.

I was born into this well-to-do family in 1901 in Colonia Juarez. I was the third child and the first son. I have no memory of learning to ride, since I was riding as soon as my legs were long enough to straddle a horse. My earliest memories are of my father coming home from the ranch, putting me on his horse after the horse had been unsaddled, and leading him to the river to drink. Mother told a story of such an occasion when she accompanied us to the river. After drinking, the horse shook himself the way a horse wet with sweat often does, and I tumbled off into the river. After being fished out of the river, my first remark (as reported by my mother) was, "Put me back on the horse." This was at about the age of three.

At four I almost died of typhoid fever. Our prosperous little town had electric lights and telephones, but the clear water from the river was piped directly into our homes with no previous purification. After the first cases of typhoid appeared, the practice was to boil the drinking water, but this was rather a case of locking the barn after the horse was stolen. One of my vivid memories of the spring of 1905 is of the magnolias in our front yard in full bloom while I was recovering from typhoid. At about this time my father gave me two small goats, but they soon became a nuisance by getting into the neighbor's lot. My father's way of getting rid of them while leaving me contented was interesting. One day he came riding into the yard leading a pretty, little sorrel horse and asked me if I would be willing to trade my two goats for the sorrel horse. I was delighted with the proposition. My goats disappeared, and I started riding my sorrel horse, which was named Chivo, the Spanish word for goat. However, Chivo turned out to be too much horse for a small boy to handle, so Father traded me a dun horse, Grullo, for Chivo, and finally a pretty, little, black horse for Grullo. The black horse was called the Spanish equivalent of "black baby" and was the horse that I rode the most until we left Mexico in 1912. We left Mexico because of the unrest accompanying the Mexican revolution, which started in 1910. About 4800 colonists from our area left Mexico at this time. This migration to El Paso, Texas, occurred within about a week in mid-July 1912. Since everyone expected to return within a short time, everything was left behind except the requirements for a few weeks' stay. However, we, like most of the colonists, never returned as a result of the continued unrest. Our large, refugee family was thus suddenly transformed from affluence to humble circumstances. My first job in Calisher's department store paid $2 for a six-day week. At that time five cents would buy a loaf of bread or a quart of milk. After a year in El Paso waiting for the Mexican situation to improve and a second year spent in Safford and Thatcher, Arizona, we moved to nearby Pima in southeastern Arizona, where Father had purchased a farm by making a small down payment. Part of the farm was under cultivation. Brush was cleared from the rest by hitching two teams, one to each end of a steel rail, and dragging it across the land after the mesquite had been dug up with ax, pick, and shovel. This involved a lot of hard work, but we were all healthy and anxious to get on our feet again.

I had finished the fifth grade in Mexico, missed a year's schooling in El Paso, and graduated from the eighth grade in Pima in 1914, having skipped the first and seventh grades.

My high school had an important bearing on my subsequent career. I went to a church academy in Thatcher, Arizona, six miles from Pima. My predilection for mathematics and the sciences showed up early. My science teacher, Alma Sessions, who had been a star basketball player at the University of Arizona, and was much admired, gave me career advice. He said I should study either electrical or mining engineering at the University of Arizona. I chose mining as the less hazardous. A $500 state fellowship, won in the competition in Graham County, launched my mining career at the University of Arizona.

My father's advice upon departure for the university made a strong impression on me. He said the constraints our religion placed on one are "to be dedicated to the truth, wherever one finds it, and to live in such a way as to make one comfortable in the company of good people." This advice has had an appeal for me that has lasted through the years.

Arriving at the university in 1919, the same time as the veterans from the First World War were returning to school, made this an interesting time. The first day at the university, I saw President Rufus von Kleinschmidt driving his carriage, drawn by two well-matched black horses, through the campus grounds. This was still the customary way to travel. Hazing was also still in full swing. One resisted it, but the battle was often lost. The war veterans in the freshman class added a disciplined resistance to hazing which I found refreshing. I found it distasteful that, as a freshman, I was expected to wear a green "beanie." The sophomores would throw freshmen not wearing their beanies into the pool. However, it was against the rules to carry hazing into the dormitories. Since I thought I could outrun my tormentors, I left the dormitory without my beanie, believing they would have to quit the pursuit when I reached the door, or else they would be in trouble themselves. It was a miscalculation. I beat them to the dormitory and then to my room where I slammed the door, but they kept coming. Since my door was now locked, they tried to climb in through the transom, but I blocked their entry by working on their hands with a broom handle. By this time, they were annoyed and threatened to break down my door if I did not come out. I was pretty sure I would be paying for the door if they broke it down, so I came out. There were plenty of sophomores, and I was soon face down in the air with someone holding on to each arm and leg and another fellow enthusiastically swinging a large wooden paddle where it would do the most good. This did not engender love for authority in me, but it did engender respect for it when it is backed up by sufficient "lynch law."

The four years I spent studying mining at the university were very pleasant. I made my own way by assisting in classes and waiting on tables. I was able to send a little money home to help with payments on the farm. I enjoyed my studies and made high grades. I particularly liked mathematics and wrote a senior thesis with Professor Cressy of the mathematics department on the theory of the aerial tramway.

The summer after my junior year I worked as a miner in the Inspiration Copper Company in Miami, Arizona. The work was interesting and paid about twice as much as farming. Since I was a prospective mining engineer, they changed me rapidly from one job to another to give me added experience. After I had been underground for only a few weeks, I was given the job of timberman, with a man assigned as my helper who had been mining for 15 years.

The caving system used in the mine was economical, but extremely hazardous. One half of a square mile of rock would be blown up at one time and drained by gravity down raises (tunnels) into square sets where a man would be stationed to regulate the rate of flow to a "grizzly" 20 feet below. At the grizzly, a second man would break up the larger boulders with a sledgehammer, or with dynamite, until the fragments passed between the six steel rails of the grizzly, spaced about 10 inches apart. The ore then continued its downward course another 20 feet, where it was drained into cars of a train driven by compressed air. The train carried the ores to a tipple which turned the cars upside down, dumping the ore into a bin. The ore was drained out of the bin into elevators that carried it to the surface, where trains carried it to the mill. Fifteen thousand tons of ore and gangue would be taken from the mine in three eight-hour shifts. There were many separate operations paralleling the one just described. On one memorable occasion a shift boss assigned my helper and me to repair one of the square sets, which had been all but destroyed by the tumbling boulders that had passed through it. We climbed up the raise to inspect the square set, and my veteran helper declared that he had no intention of being trapped in this particular hell hole and, if it was to be repaired, it was up to me. He would wait down at the grizzly, and I could call when and if I needed help. I had the usual ax and crowbar with which I could gingerly pry out the loose rocks until I had cleared enough space to put in new timber before it all caved in on me. I went about it as carefully as I could, but it was not long before a rock somewhat bigger than my head fell from the ceiling and hit my boot. I soon had a boot partly filled with blood, and I was taken out of the square set and brought up on top. When I returned to the mine some days later, I was assigned to another part of the mine.

I was once on another shift where there were three separate, fatal accidents. There were two other boys from Pima working in the mine that summer, and one of these had his arm crushed and later amputated. This accident occurred as he was driving his train onto the tipple. The customary indemnity paid to the family of a deceased miner was $6000. This rapid method of mining born of wartime and economic needs has been replaced by the much safer opencut mining of today.

Although as a mining engineer I would not personally have to take these risks much longer, I would still have to send others to take them. Therefore, I reluctantly decided to graduate in mining but planned to change over to metallurgy after receiving my degree. The next summer, 1923, my younger brother, Edward, and I worked together as miners at Sacramento Hill in Bisbee, Arizona, for the Phelps Dodge Corporation. This work was interesting and a much safer type of operation than the caving system since it involved driving drifts into undisturbed terrain. This summer ended pleasantly, terminating my active mining career. In the fall I began my thesis on *The Separation of Heavy Sulfide Ores by Selective Flotation*. I had a Bureau of Mines fellowship and worked under the direction of Thomas Chapman, professor of metallurgy, at the University of Arizona.

After completing my master's degree, I spent the summer of 1924 working in the United Verde smelter at Clarksdale, Arizona. Again I had a favored position as a prospective metallurgist and was rapidly shifted among different phases of the smelting operation. After being there a few weeks I was assigned to take samples from the blast furnaces. The sulfur dioxide smoke was especially strong, and I was holding a handkerchief soaked in baking soda over my face when the smelter superintendent came by, slapped me on the shoulder, and said, "Eyring, I plan to put you in charge of the blast furnaces in a few weeks." The problems were intriguing, but the sulfur smoke made it easy for me to return to the University of Arizona as a chemistry instructor.

In 1925, toward the end of the teaching year, Professor Theophyl Buehrer recommended me to Berkeley, where he had taken his Ph.D. Dr. Lathrop E. Roberts recommended me to the University of Chicago, where he had worked under Harkins. I accepted the invitation to Berkeley, where I went in August 1925 and finished my Ph.D. in June 1927. This was a stimulating experience. The emphasis was on research and, since I was no stranger to work, everything went well. I started on a problem involving the lowering of freezing points with Professor Merle Randall, but because Dr. Vanselow needed the equipment longer than expected to finish his Ph.D., I changed to working on an exciting problem with Professor George E. Gibson. A long vacuum tube filled with hydrogen at low pressure was bombarded with an 11 MV high-frequency discharge from a Tesla coil. The protons were expected to pass through a thin aluminum window at the bottom of the tube and strike a beryllium target, which we hoped would emit interesting radiation. Unfortunately, one got lots of unspecified radiation and a frequently punctured, evacuated tube. Nevertheless, I had gained useful experience from this study. We next turned to the study of the amount of ionization, the stopping power, and the straggling of alpha particles from polonium in various gases. We found that stopping power depends very little on how the atoms are bound together, but the total number of ions formed depends very much on the bonds being broken. This is readily understandable since the primary ionization induced by the alpha particles involves a large energy transfer to the electron being ionized, with the result that the fraction of energy lost by the alpha particle as a result of a difference in the molecular bonding of the atoms is negligible. Conversely, in the ionization by the fast secondary electrons only about half of the energy transferred from the secondary electrons goes into ionization, so that differences in the bonding energy are an appreciable part of the total energy expended in ionization.

In the fall of 1927, as a brand new Berkeley Ph.D., I became an instructor at the University of Wisconsin. During my first year there, I continued my research on ionization, stopping power, and straggling of alpha particles in different gases. The second year I took a full-time experimental research position with Professor Farrington Daniels, studying the decomposition of N_2O_5 in a wide variety of solvents. This was the beginning of my active interest in reaction kinetics, which has continued unabated.

In 1929 I was granted a national research fellowship to work with Professor Bodenstein in the University of Berlin. However, before my expected date of departure, I received word that Professor Bodenstein was going to be at Princeton at the dedication of the new Frick chemical laboratory. Professor Frumkin, who was visiting the University of Wisconsin at the time, suggested that, in view of Bodenstein's absence, I should work with Professor Michael Polanyi at the Kaiser Wilhelm Institute in Berlin. Acting on this advice, my wife, Mildred Bennion, and I sailed to Europe by way of Bergen and Oslo, Norway, through southern Sweden, to Copenhagen and Berlin. Arriving in Berlin, we found Professor Polanyi had likewise gone to the Frick dedication. Just at this time Bonhöffer and Harteck were front page news with their study of the rate of conversion of *para*- to *ortho*-hydrogen. Fritz Haber was directing his laboratory effectively. Workers in Haber's laboratory included Fritz London, Eugene Wigner, the Farkas brothers, and Hubert Alyea, along with many others. We had not been in Berlin long before I was visited by Professor Robbins from the Paris branch office, who was responsible for the National Research fellows. He was greatly disturbed to find that neither Professor Bodenstein nor Polanyi was in Berlin to greet me. Professor Polanyi, however, returned soon thereafter.

My first research was the study of light-emitting reactions. A sodium vapor jet meeting a jet of chlorine precipitated NaCl with the emission of a bright light. Spectroscopic examination of the light told the story behind the mechanism. Although we were intrigued by this subject, Professor Polanyi and I turned our attention to the construction of a potential surface for the reaction $H_2(para) + H = H_2(ortho) + H$. We made use of Fritz London's approximate equation for the potential energy as a function of the distance between the atoms. This involved using the Heitler-London-Sugiura exchange and coulombic integrals for the energy of attraction between atomic pairs. The results were disappointing. We then changed to a spectroscopic estimation of the attraction between pairs of atoms using Morse curves. The theoretical calculations gave us the needed guidance in apportioning the bonding energy between the exchange and coulombic integrals. This way we got an exciting, if only approximate, potential surface and with it gained entrance into a whole new world of chemistry, experiencing all the enthusiasm such a vista inspired. We perceived immediately the role of zero-point energy in reaction kinetics, and our method of using Morse curves made it possible to extend our calculations to all kinds of reactions. I continued this work with enthusiasm whenever opportunity permitted.

I received a disturbing letter from my father early during my stay in Berlin. He informed me that, contrary to the opinion of lawyers whom he had consulted when we returned to the United States, the State Department had ruled that my younger brother, Joseph, was a Mexican citizen. This meant that I, too, was a Mexican citizen. On that basis I held a passport and fellowship to which I was not entitled, and I had voted for Hoover. I spent a very restless night, to say the least. Early the next morning, I consulted the American consul, who ruled that I had acted in good faith and should continue my stay in Berlin. He recommended that when I returned to the United States, I should turn in my passport, at which time I could proceed with naturalization. This I did. Five years later, I took out my naturalization papers with a judge in Trenton, New Jersey, who had just recently naturalized Professor Einstein. I was then on the faculty at Princeton, and the judge was so interested in telling me about his earlier experience with Professor Einstein that I passed the tests without difficulty.

My year with Professor Polanyi was fruitful and altogether delightful. He was a very gracious and gifted human being. Toward the end of our stay in Berlin, Professor Wendel Latimer of the University of California at Berkeley visited me and, upon hearing of our involvement in applications of quantum mechanics to chemistry, suggested to Professor G. N. Lewis that I be invited back to Berkeley for a year as lecturer, to take over some of the duties that Professor Hildebrand's impending absence would leave open. As a result of Latimer's proposal, Mildred and I found ourselves back in Berkeley at the beginning of the fall quarter of 1930.

This was an exciting year. My duties at Berkeley were not heavy, so I was able to develop further the quantum mechanical attack on reaction kinetics. Our oldest son, Edward, was born in Oakland on January 7, 1931. By spring I had a paper to present at the Indianapolis meeting of the American Chemical Society, using potential surfaces to explain why iodine was the only halogen-hydrogen reaction to involve four atoms in the activated complex, while the other reactions all went by three-atom complexes. The paper engendered much excitement, and Professor Hugh Taylor invited me on the spot to go to Princeton and present two lectures on quantum mechanical calculations of reaction rates. This visit led to an invitation to go to Princeton, which was the beginning of an exciting fifteen years to be spent in what turned out to be an ideal scientific environment. Hugh Taylor was an inspiring departmental chairman, incisive in chemical discussions and always generous in his encouragement of others.

It will not be possible to speak of more than a few of my 560 scientific papers and 9 books, but a few highlights may be interesting. Twenty years as editor of the *Annual Review of Physical Chemistry* were useful in keeping me conversant with the field. Also valuable was my coeditorship, with Douglas Henderson and Wilhelm Jost, of the 14 volumes of *Physical Chemistry: An Advanced Treatise*. Douglas Henderson and I are continuing to edit volumes entitled *Theoretical Chemistry: Advances and Perspectives.*

My second paper at Princeton was inspired by Professor Charles P. Smyth's suggestion that it would be useful to calculate the effective resultant dipole for a molecule having various dipoles lying along bonds that rotate with respect to each other. This is also essentially the same problem as calculating the distance between ends of a flexible chain, since lengths also lie along bonds. The procedure adopted was to choose coordinate systems such that the origins were at successive atoms along a chain, and the x axis coincided with the bonds of interest. One could then transform all coordinates to an initial set of coordinates and add vectorially the transformed dipoles, or lengths, in this final coordinate system. The rotations around axes could be appropriately averaged according to the potential energies of rotation. The procedure has since been extensively used by others for high polymers, both in calculating the mean lengths of a chain and in evaluating partition functions in thermodynamic calculations.

The task of developing methods of calculating potential energy surfaces for more than the four electrons treated by Fritz London was straightforward but tedious. In the April 1933 issue of the *Journal of Chemical Physics*, George Kimball and I presented a quick way of getting the secular equation for any number of electrons. This, together with the use of the Morse equation, enabled us to construct approximate potential surfaces for any molecular system of interest. It is interesting that, in the same issue, Pauling published a parallel procedure for getting the secular equation for any number of electrons. At this time George Kimball and I tried using difference equations to send a wave packet over the $H_2 + D = HD + H$ surface. This effort was premature. Joseph Hirschfelder carried out the first classical trajectory calculation in his doctoral thesis.

In 1932 Bethe's group theory treatment of orbitals in fields of different symmetry caught my attention. Arthur Frost, John Turkevich, and I incorporated group theory into our bond eigenfunction solution of the methane problem. From then on, group theory became standard procedure with us and naturally found its way, as a chapter, into the book, *Quantum Chemistry*, by Eyring, Walter, and Kimball. Interestingly enough, a contract for the book was signed in 1933. Publication, however, must always await completion of the manuscript, and the book did not come out until 1944. (Not unseemly haste!) This book has had wide use, there having been over 20 printings of the unrevised first edition and a considerable number of translations into other languages.

Our second son, Henry Bennion, was born in Princeton on May 31, 1933. At about this time I was invited to participate in a symposium on molecular quantum mechanics held by the Physical Society, over which Niels Bohr presided. The three other participants were Mulliken, Pauling, and Slater. The audience was most attentive. I was immersed in all kinds of approximate quantum mechanical calculations that could be carried out at that time, without computers. Papers on the relative rates of isotopic reactions followed naturally from my first paper with Polanyi and from experiments with Professor Hugh Taylor. Very early in our experiments, Taylor and I prepared large amounts of heavy water by electrolysis. We found that 92% heavy water proved fatal to tadpoles of the green frog *Rana clamitans*, the aquarium fish *Lebistes reticulatus*, flatworms (*Planaria maculata*), and the protozoan, *Paramecium caudatum*. This work was reported by Taylor, Swingle, Eyring, and Frost in October 1933. During this period, my associates and I were investigating potential energy surfaces for a wide variety of reactions. In November 1934 I submitted my paper on "The Activated Complex in Chemical Reactions" to the *Journal of Chemical Physics*. I showed that rates could be calculated using quantum mechanics for the potential surface, the theory of small vibrations to calculate the normal modes, and statistical mechanics to calculate the concentration and rate of crossing the potential energy barrier. This procedure provided the detailed picture of the way reactions proceed that still dominates the field.

The activated complex has a fleeting existence of only about 10^{-13} sec and is situated at the point of no return or of almost no return. It is much like any other molecule except that it has an internal translational degree of freedom and is flying apart. This concept describes any elementary reaction involving the crossing of a potential barrier. If the activated state is really a point of no return, there is no perturbation of the forward rate by the backward rate, so that the rate at equilibrium applies unchanged to the rate away from equilibrium. For thin barriers, as in the inversion of ammonia, leakage through the barrier is faster than the rate of surmounting the barrier and so must be taken into account for cases involving light atoms and thin barriers. Of the nine books I have coauthored, all but *Quantum Chemistry* and *The Theory of Optical Activity* with Dennis Caldwell involve reaction rates. They deal with such diverse subjects as rates in plastic deformation, liquids, gases, solids, biology, physics, and engineering.

In 1937, E. U. Condon, W. Altar, and I, working at Princeton, published our paper on one-electron optical activity. This was precipitated by the general belief, current then, that a one-electron transition, even in a dissymmetric field, would not contribute significantly to optical activity. Werner Kuhn had shown earlier that Drude's calculation of the optical activity of an electron moving in a spiral neglected a term that reduced the outcome for his model to zero. The result was that Kuhn, Max Born, and other workers in the field adopted the coupled oscillator model as the sole source of optical activity. This seemed unrealistic to us in view of the success of treating spectroscopy as approximately due to one-electron transitions. Adding the simplest perturbing potential that would give optical activity, $Axyz$, to the potential $k_1x^2 + k_2y^2 + k_3z^2$, a system which could be readily solved, we found one-electron transitions were indeed optically active for reasonable values of A if the k's were unequal. This treatment gave rise to what has later been popularized as the octant rule. Interestingly enough, theory showed that this one-electron optical activity also persists in the classical limit. These considerations were extended in my work with Walter Kauzmann, John Walter, Daniel Miles, Dennis Caldwell, and many others. This interest has evolved into a major concern of ours with absorption spectra, circular dichroism, magnetic circular dichroism, and, not surprisingly, nuclear magnetic resonance. These tools are of course invaluable in establishing molecular structure.

In the autumn of 1937 I went with my wife, Mildred, and two oldest sons for a four-month visit to Manchester, England, where Professor Polanyi headed the chemistry department. This visit was prompted by invitations to address the Faraday Society and the Chemical Society, and included lectures at various English universities. The trip included memorable envelopments in the famous English fog, but was altogether delightful.

Returning to Princeton in December, I continued my active involvement in research and teaching. Our third son, Harden Romney, was born in Princeton on August 20, 1939.

In early 1942, I was talking to Professor Newton Harvey, head of the biology department at Princeton, about problems of shock brought on by broken bones when Professor Frank Johnson came into the room. Harvey immediately began to discuss the problem that Johnson, Brown, and Marsland had encountered. They had been studying bioluminescence of bacteria as affected by temperature and pressure. Bioluminescence was a field in which Harvey was preeminent. Bioluminescence is absent at ice temperature and becomes maximal about halfway to blood temperature. It then drops to a very low value at blood temperature due to inactivation of the enzyme at higher temperatures. The problem that interested Johnson could be stated as follows: Why, when the bacteria in a suitable solution are subjected to 200 atm pressure, is there a marked decrease in luminescence in the low-temperature range with a rise of luminescence in the high-temperature range? Since, as Braun and Le Chatelier pointed out long ago, increased pressure shifts an equilibrated system toward lower volumes, and since the activated complex is in equilibrium with reactants, the luminescent response must follow the same laws governing other equilibria. Hence, the reactants are less voluminous than the activated complex in the low-temperature range, and the reverse is true in the high-temperature range.

Soon after this discussion John Magee and I published a paper that quantitatively explained this behavior. This discussion started a collaborative investigation with Frank Johnson into biological reactions that has continued and eventuated in our writing two books dealing with reactions important in biology and medicine. Study of the pressure, temperature, and narcotic effects on bioluminescence has been a powerful tool in understanding many physiological problems.

Prior to the Second World War, I was able to show that gases that penetrated gas masks more readily than those used in the First World War were not to be expected. This conclusion derived from the study of various freshly prepared and aged smokes with the new electron microscope available in Camden, New Jersey. Fairly frequent conversations with Dr. Irving Langmuir at this time revealed that his was a brilliant, determined mind uncompromising on principle.

During the war I worked with the Navy as a consultant on high explosives. Dr. Stephen Brunauer was in charge of high-explosive research for the Navy and suggested that we consult with Professor Einstein, who also lived in Princeton. After a very pleasant morning in discussion with Professor Einstein at the Institute for Advanced Study, we walked together at noon through what had been a rose garden but was now planted with a field crop. I plucked a sprig and asked Professor Einstein what it was. He did not know. We walked a little farther and encountered the gardener sitting on his wheelbarrow. His reply to the same query was, "It is soybeans." Even for a first-rate mind, what gains attention is not just propinquity but interest. Professor Einstein's mind was too busy with more important things. Einstein's manner was never ostentatious but, indeed, on the very infrequent occasions when I talked to him, always kindly.

I also talked frequently with Professor Hugh Taylor about the nickel barriers to be used for separation of uranium isotopes, and I occasionally discussed the properties of uranium with Professor Wigner. I must confess that I did not expect the atom bomb to materialize soon enough to influence the war's outcome.

During the war the study of detonations took up a lot of my time and eventuated in a 112-page report entitled *The Stability of Detonations*, written with Richard Powell, George Duffey, and Ransom Parlin. The article was published in the August 1946 issue of *Chemical Reviews*. The curved front of the detonation was related to the diameter of the cylindrical explosive and the thickness of the reaction zone behind the wave front. The curved front theory, which established these relationships, is still widely used. The fact that the individual solid particles in an explosive burn from the outside, in layer after layer, accounts for the slow burning rate. The consequent outward burning of layer after layer of the bubbles in liquid explosives gives this same type of delayed reaction. This is a case where reaction rate is slowed down by delay in heat conduction and is to be expected, generally, at very high temperatures.

In 1944 I was asked by Professor Taylor to head up research for the Textile Research Institute, which was coming to Princeton. The position was vacant because Dr. Milton Harris, who had made distinguished contributions in this field, had accepted a position in industry. The position carried with it the responsibility for appointing 15 fellows who would work for their Ph.D. at Princeton University, and I would have Professor Eugene Pacsu, Professor John Whitwell, and Dr. Robert Rundell as my associates. This activity required 40% of my time; the remaining 60% was devoted to my regular duties at Princeton. This was a rewarding experience and drew on my engineering as well as my chemical experience.

The Institute was housed on the shore of Carnegie Lake, with extensive grounds whose graveled roads were never meant to carry the increased traffic. As a result, we would receive an occasional call for help from a stalled motorist, which we would answer with enthusiasm and dispatch, even though the consequences were often detrimental to our shoes and trousers. I enjoyed sharing in these rescues as much as my fellow workers did.

The study of spinning, weaving, dyeing, and the measurement of the physical properties of fibers and fabrics involved us in new problems of physical chemistry. My active interest in deformation kinetics dates from this time. Professor A. S. Krausz, of the University of Ottawa, and I coauthored the book, *Deformation Kinetics*, which was published in 1975. George Halsey made notable contributions to the study of the physical properties of fibers, as did Howard White and many others. I carried on parallel work with Arthur Tobolsky and others in the chemistry department. I became a member of the National Academy of Sciences in 1945.

In the spring of 1946, Ray Olpin, newly installed president of the University of Utah, visited us in Princeton to offer me the job of dean of the School of Mines or of the graduate school. My wife said that I should be the one to decide what we should do. Since at Princeton I had nine graduate students working with me, had been a professor since 1938, and had what was considered a high salary at that time, the choice was difficult. At Utah I would have to start the doctoral program and build up my own research program. As a result I decided not to go and wrote a letter to President Olpin to that effect. The next day Mildred asked what decision I had made. I told her, and she was crushed. We had lived away from her family and her Salt Lake mountains for 19 years. Although Princeton had been most pleasant, she wanted to go home. She said nothing at the time I told her of my decision, but prepared a nice letter that she asked me to read when I got to the university. When I realized her need to return home, I naturally agreed.

I immediately told Professor Taylor of the change in plan. His response was, "I told President Dodds the storm wasn't over. We have more money than the University of Utah. What do you want?"

"Nothing," I replied. "We're going."

He asked, "Do you want the Jones Professorship?"

"No. My wife wants to go, and we are going."

"One can't do business with a crazy man," he exclaimed. "Do you mind if I talk with your wife?"

"Help yourself," I said.

He and Mildred were well acquainted, and they had a pleasant conversation, but nothing was changed. So in August 1946 we started our new adventure at Utah, where I was dean of the graduate school and professor of chemistry.

Although the University of Utah had not granted Ph.D.'s before my arrival as dean of the graduate school in 1946, the transition went smoothly since the university already had a strong master's program. I found that two general administrative policies are possible: One involves strict administration from the top, the other encourages individual initiative as long as it is successful. When appropriate procedures are well understood, most decisions can be made at the departmental level, and routine decisions can be made by capable secretaries in the dean's office. Following such a policy of decentralization left me free to devote the required time to administration without seriously curtailing my research during the 20 years I served as graduate dean.

One of my first graduate students at Utah was Tracy Hall, who, after studying chromium complexes at Utah, went on to solve the problem of making diamonds at General Electric in Schenectady. Bruno Zwolinski, who had commenced his work on the transmission coefficient of reaction kinetics at Princeton, finished up at Utah and got his degree from Princeton. Thirty-five of my collaborators completed their doctorates at Princeton and over a hundred at Utah.

Most of the avenues of research I began at Princeton, as well as new ones, were carried out at Utah. In 1935 it was clear to me that the law of rectilinear diameters was understandable in terms of fluidized vacancies in liquids that mimic the behavior of molecules in the vapor both in concentration and behavior. That this is reasonable follows from the fact that to form a molecular-sized vacancy without vaporizing a molecule requires the same number of broken bonds as volatilizing a molecule without leaving a vacancy. Accordingly, the heat required to form a vacancy is the same as the heat required to vaporize a molecule. Further, molecules falling domino-like into a vacancy move the vacancy much as a gas molecule moves. Thus, the fluidized vacancy converts three molecular vibrations into translations. A molecule behaves like a gas for the fraction of the time, $V - V_s/V$, in which there is no neighbor to prevent its free fall. The remaining fraction of the time, V_s/V, it behaves like a solid. Here, V_s and V are the molal volume of the solid at the melting point and the volume of the liquid, respectively. These considerations led to the development, with Professors T. Ree and N. Hirai, of the significant structure theory of liquids, which has evolved into a quantitative treatment of the thermodynamic and transport properties of many types of liquids. Since a liquid is intermediate between solid and vapor, it seems natural to treat it as a mixture of solid-like clusters intermingled with vapor. A quantitative description of a liquid in terms of solid and vapor properties actually rests on first principles insofar as solids and liquids can be so described. The result is a single partition function for solid, liquid, and vapor.

My wife, Mildred, died of cancer June 25, 1969 after a lingering illness that she endured gracefully with a serene faith in happier times to come. During the five years and four major operations spent fighting cancer, we sought help in every quarter where there seemed any prospect of curbing the disease. The most that could be accomplished was to slow, somewhat, its relentless course.

In 1970 Dr. Betsey Jones Stover came to me with some of the results of radiating beagles with radium and plutonium, which caused them to die of bone cancer. She pointed out that the survival curve reminded her of the curve for Fermi-Dirac statistics. The equation that fitted her data for the fraction of a population surviving, S, plotted against the age, t, is

$$S = [1 + \exp k(t - \tau)]^{-1}. \qquad 1.$$

The death rate, $-dS/dt$, is given by

$$-\frac{dS}{dt} = kS(1 - S) = \frac{d(1 - S)}{dt}. \qquad 2.$$

The fraction not surviving is, of course, $(1 - S)$. From Equation 1 we see τ is the age at which half of the population still survives. From Equation 2 we see also that $-k/4$ is the slope of Equation 1 at age $t = \tau$, where $S = \frac{1}{2}$. Chemists will recognize that the equations for S are those for an autocatalytic reaction. If S is the fraction of healthy cells and $(1 - S)$ the fraction of sick ones, one would expect Equation 2 to represent the rate of spread of the disorder as it likewise should represent the growth of any ecological population.

Dr. Stover and I also pointed out that under certain conditions Equation 2 would also represent the rate of mutation. This would be true if the rate of mutation were proportional to the product $S(1 - S)$, where S is the probability one gene is undamaged and $(1 - S)$ is the probability that a neighboring gene is damaged. By use of absolute reaction rate theory, a meaning was given to the parameters k and τ. The last chapter in *The Theory of Rate Processes in Biology and Medicine*, which I coauthored with Frank Johnson and Betsey Stover, develops these considerations further.

In 1971 I married Winifred Brennan and added her daughters Eleanor, Patricia, Joan, and Bernice to our family. This has been a rewarding experience. At the time of writing, January 1977, I still have my regular professorship at the University of Utah and am enjoying excellent health. My research and teaching are going ahead at an undiminished rate with almost 20 collaborators.

As I look back over my efforts, I would characterize my contributions as being largely in the realm of model building. To test a model it is usually advantageous to cast it in mathematical form so that quantitative predictions can be used to compare calculations with experimental findings. Ideally, agreement should be quantitative and complete. Unfortunately, this never happens in the real world. Even Newtonian mechanics must be amended in the realms of relativity and quantum mechanics, and Maxwell's electromagnetic theory fails to predict stationary electronic orbits. The usual statement that the entropy always increases is not mended very successfully by Boltzmann's proposal of known theories of fluctuations. The observed cosmological departures from equilibrium boggle the imagination when considered in terms of fluctuation theory. A better statement of the second law of thermodynamics would seem to be that living things never exist in an environment where there are large decreases in the total entropy. This statement seems to include all we really know about entropy. We still need to find the gigantic Maxwellian demon that winds up worlds and consequently exists.

In model building it is convenient to start out with the following hypotheses: (*a*) There is always a model that will explain any related set of bonafide experiments. (*b*) Models should start out simple and definite enough that predictions can be made. (*c*) A model is of limited value except as it correlates a substantial body of observable material. (*d*) Models that suggest important new experiments, even if the theory must be modified, can be useful.

To be a Newton or a Maxwell it is very convenient to be stimulated by a Kepler or a Faraday, but if one were gifted enough he could still be a Gibbs with very little outside interaction.

Self-analysis is always hazardous but can be amusing. I perceive myself as rather uninhibited, with a certain mathematical facility and more interest in the broad aspects of a problem than the delicate nuances. I am more interested in discovering what is over the next rise than in assiduously cultivating the beautiful garden close at hand. In any event, the study of chemistry is still both exciting and rewarding to me.

HENRY EYRING, 1901–1982*

Henry Eyring was a great scientist who dared to be innovative. His theory of absolute reaction rates has been used to explain the behavior of a wide range of biological and chemical systems. His interests included all natural phenomena. Indeed, he contributed in major ways to the theory of liquids, optical rotations, molecular biology, mutations, aging, and cancer.

In recognition for his research Eyring received a large number of awards, including the 1949 Bingham Medal, the 1980 Wolf Prize in Chemistry, the 1979 Berzelius Gold Medal of the Royal Swedish Academy of Sciences, the 1975 Priestley Medal of the American Chemical Society, and in 1966 the National Medal of Science, in addition to many other honors.

From 1931 to 1946 Eyring was professor of chemistry at Princeton. Then, because of his devout faith in the Mormon religion and in order to please his wife, he moved to the University of Utah, where, in the period 1946–1982 he became dean of the graduate school, Distinguished Professor of Chemistry, and professor of metallurgy.

According to family genealogists, *Eyring* is reputed to be the name of the pagan God of Light, and the German family from whom Henry descended is said to have adopted that name when they accepted Christianity (1).

Henry was born in the Mormon community of Colonia Juarez 100 miles south of the American border in the state of Chihuahua, Mexico. His father was well-to-do and owned a 10,000-acre fenced ranch where he raised 600 head of high grade shorthorn Durham cattle and 100 head of horses—the bull calves were sold to upgrade the herds of Spanish longhorn cattle. As long as Henry could remember he rode horses. Once when he was three years old (2), he fell off into a river. As soon as he was fished out, he cried, "Put me back on my horse!" Henry was always very independent and self-confident.

Henry was also kind and thoughtful. For example, I remember that he advised me to "be nice to the guys on the way up so that they will be nice to you on the way down." He recalled gratefully the care he received when, at the age of four, he nearly died of typhoid fever: "I learned then how important it is to care about people even if they are small and may not seem very important" (1).

Quite suddenly, when Henry was 11, the Eyrings, together with 4800 colonists from their area, were forced by the Mexican revolution to leave their homes and all of their belongings and flee to the United States. Thus

* Reprinted from the *Year Book* with the permission of the American Philosophical Society.

the Eyring family became poor, and little Henry got a job working in an El Paso department store for $2 for a six-day week. After a year or so the Eyrings moved to Pima, Arizona, where the father purchased a farm with a small down payment. Most of the farm was covered with mesquite, which had to be cleared by hand and with the help of a team of horses. Thus from necessity Henry learned to work hard and long hours.

Henry worked his way through the University of Arizona by working in a copper mine as a timberman. On one occasion a large rock fell on his foot, crushing it and disabling him for a short time (1). On another shift three fatal accidents occurred. After getting a BS in mining engineering, Henry applied for and was awarded a US Bureau of Mines fellowship to do graduate work in metallurgy, which he thought would be safer than mining. However, after getting an MS in metallurgy at the University of Arizona, he spent the summer taking samples from blast furnaces in a smelting operation and got his lungs full of sulfur fumes. He then decided to leave metallurgy and he accepted an invitation of an instructorship in chemistry at Arizona for the year 1924–1925. The following year he received fellowships from both the Universities of California and Chicago to continue his graduate studies in chemistry. Since the Berkeley fellowship paid $750 per year and the Chicago fellowship was for $700, Henry went to Berkeley (3).

Eyring was very happy at Berkeley. He wrote (1): "Graduate students mingled with outstanding scientists who entertained no doubt that intelligent research was the most important activity in the world. This contagion infested everyone. Individual success in research was accompanied by shedding of any undue veneration for the embalmed science of the past!" His one regret was that course work was not stressed at Berkeley (3). During the course of his PhD studies he took seven courses in mathematics (his favorite subject), only two in chemistry, and none in physics. In his first attempt at research he bombarded a vacuum tube containing hydrogen at low pressure with an 11 M.E.V. high frequency Tesla discharge. Result: a broken vacuum tube. In 1927 he earned a PhD under the direction of Professor George E. Gibson. His thesis was on the ionization, stopping power, and straggling of alpha particles from polonium passing through different gases.

On the basis of his Berkeley PhD and G. N. Lewis's recommendation, Eyring secured a lectureship in chemistry at the University of Wisconsin, and he taught a physical chemistry laboratory course. During that year he met a charming girl, Mildred Bennion, who was on leave from the University of Utah, where she was chairman of the women's physical education program. In the spring Henry bought a canoe and frequently took Mildred for rides on Lake Mendota. In the fall they were married, and the canoe, having served its purpose, was sold! The following year (1928–

1929), Eyring worked at the University of Wisconsin as the research associate of Farrington Daniels, experimenting on the rate of decomposition of N_2O_5 in a wide range of solvents. This was the beginning of Henry's active interest in chemical kinetics. During his second year at Wisconsin, Eyring was stimulated by the physics lectures of Professor John H. Van Vleck on the new developments in quantum mechanics.

Then, in 1929, Eyring was granted a National Research Fellowship to work with Professor Polanyi at the Kaiser Wilhelm Institute in Berlin. At this time Bonhöffer and Harteck in Germany were making front-page news with their experimental studies of the rate of conversion of *para* to *ortho* hydrogen, $H_2(para) + H \rightarrow H_2(ortho) + H$. On the basis of the new quantum mechanics, it should be possible to calculate the activation energy for this reaction. Sugiura (4), using the method of Heitler and London, had made numerical calculations of the energy of interaction of two hydrogen atoms; and Fritz London had just presented a paper for the sixtieth birthday of Professor Haber showing how it was possible to approximate the potential energy of a 3- or 4-atom molecular system. Eyring & Polanyi substituted Sugiura's calculations into London's equation for the energy of a 3-atom system, and found that it did not agree with the experimental *ortho-para* reaction results. It was at this point that Henry showed his genius for devising semiempirical approximations. Eyring & Polanyi finally used Morse curves and spectroscopic data to estimate the attractions between pairs of atoms as functions of their separation. The Sugiura calculations were also used as a guide in apportioning the bonding energy between Coulombic and exchange integrals. The result was the Eyring-Polanyi potential energy surfaces, which were used for many years to explain the activation energy of not only the *ortho-para* conversion but a large number of 3- and 4-atom reactions.

Soon after Eyring returned to the United States, he explained in a talk at an American Chemical Society meeting that a mixture of hydrogen and fluorine would not explode because (according to his theoretical estimates) the activation energy of $H_2 + F_2 \rightarrow 2HF$ was too large. Hugh S. Taylor (subsequently Sir Hugh), the chairman of the Princeton chemistry department, who was in the audience, knew (on the basis of little known experiments) that Henry was right. Accordingly he invited Eyring to Princeton as a lecturer. Henry soon scaled the academic ladder to become a full professor. During Eyring's Princeton years (1931–1946) he enjoyed a very stimulating collaboration with Hugh S. Taylor, which resulted in a whole progression of important discoveries.

Princeton was a fabulous place in the 1930s. Many famous physicists, such as Einstein, von Neumann, Wigner, Ladenburg, came there to get away from Hitler. Furthermore, Dirac, Pauli, Schrödinger, Slater, and others spent a great deal of time there. I was fortunate to do the chemistry part of my physics and chemistry double doctorate under the direction of Eyring.

Working with Eyring was a lot of fun. He was full of ideas, and thoroughly enjoyed his work; we worked together every night until I was ready to fall asleep. In those days Eyring got the inspiration for most of his research by assiduously studying the literature. Every morning he would pop some new idea for his students to criticize—it was the job of the graduate students to discover which ideas were sound and which were erroneous. Most of his ideas needed modification and reworking—this gave us a good lesson in how discoveries are made!

In 1934, Eyring made his most important discovery, his theory of absolute reaction rates. It was an epoch-making, Grand Concept! His equation (5),

$$\text{Rate constant} = \kappa(kT/h) \exp(-\Delta G_{act}/RT)$$

$$= k(kT/h) \exp((-\Delta H_{act} - T\Delta S_{act})/RT), \qquad 1.$$

extends the concepts and terminology of thermodynamics and statistical mechanics to all sorts of rate processes. Equation 1 implies that there is a single rate-determining bottleneck which is characterized by a free energy of activation—it is immaterial whether the bottleneck is largely in the potential energy (as in small molecule gas phase chemical reactions) or whether entropy considerations play an important role in characterizing the bottleneck (as in the case of most biological reactions). Of course, it should be clearly recognized that Eq. 1 is only intended to apply to single steps of complex rate processes. It has been argued that Eyring's theory follows logically from previous work of Pelzer & Wigner (6) but that paper and all of the other previous treatments of reaction rates considered particular examples and did not anticipate the possibility of there being a formulation that was universally applicable. The only part of the Eyring theory that is not simple and elegant is the catch-all factor κ, which Eyring included in order to account for those systems which pass through the activated state and turn around and come back. (In 1935, Eyring said that whenever he thought of κ, he thought of women drivers! In 1982, he would not dare to think such a thing, let alone to say it!) However, it is the factor κ which is used to account for the very small transmission coefficients that occur in quantum mechanical tunneling. One of the reasons the theory of absolute reaction rates has been so useful is that it is usually easy to make a good estimate of the entropy of activation corresponding to an assumed reaction mechanism—then use a comparison between the observed and the estimated entropies to determine the correct mechanism! Unlike the Arrhenius reaction rate theory, Eq. 1 can explain either extremely slow or extremely fast reactions.

The Theory of Absolute Reaction Rates was so different from any of the previous collision rate formulations that Harold Urey, the Editor of the *Journal of Chemical Physics*, rejected Eyring's manuscript (5) for publication. Fortunately, Hugh S. Taylor persuaded Urey to change his mind, provided that Eyring would add an explanation of how to use his theory to calculate the rate of collisions between two rigid spherical molecules! Thus Eyring added an appendix with this explanation.

In the 1950s Ilya Prigogine followed in Eyring's footsteps in his theories of the thermodynamics and statistical mechanics of irreversible processes, when he assumed that instantaneous configurations of molecular or macroscopic configurations possess meaningful thermodynamical properties even though they are not at equilibrium. This concept and Prigogine's formulation are leading to a better understanding of biologically important chemical reactions that are periodic in both time and space, the theory of turbulence, etc.

One of Eyring's first applications of the theory of absolute reaction rates was to the viscosity, plasticity, and diffusion in liquids. This led to his free volume and significant structure theories of liquids.

Then in 1942, Henry collaborated with Professor Frank Johnson to explain some puzzling problems associated with bioluminescent bacteria. This was the beginning of Eyring's great interest in biological and medical problems.

Eyring's Princeton days came to an end in 1946, when he was asked to become dean of the Graduate School at the University of Utah and build up the chemistry and metallurgy departments. As much as Henry liked Princeton, his family and church ties to Utah were too strong to reject this offer. He accepted this challenge and did an amazing job thereafter of building up the graduate research program at Utah into one of the strongest in the country.

In 1977 Eyring wrote (2):

> As I look back over my research efforts, I would characterize my contributions as being largely in the realm of model building. To test a model it is usually advantageous to cast it in mathematical form so that quantitative predictions can be used to compare calculations with experimental findings. Ideally, agreement should be quantitative and complete. Unfortunately, this never happens in the real world. Even Newtonian mechanics must be amended in the realms of relativity and quantum mechanics.
>
> In model building it is convenient to start out with the following hypotheses: (a) There is always a model that will explain any related set of bonafide experiments. (b) Models should start out simple and definite enough that predictions can be made. (c) A model is of limited value except as it correlates a substantial body of observable material. (d) Models that suggest important new experiments can be useful, even if the theory must be modified.

The Eyring philosophy that I remember best is: "Always ask yourself how might the phenomena occur. Make a Gedenks-model. It will suggest the proper groupings of variables—and this is usually a big help in semiempiricizing." Typical of Eyring's skill at model making is his explanation of why the coefficient of sliding friction is very nearly equal to one-half for most substance: Suppose that the surface is sinusoidal (like corrugated cardboard), with height $z = \sin(\pi x)$, then in order to slide an object from a trough at $x = -1$ to its next trough at $x = 3$ (or a distance of four units), it is necessary to lift the object two units from the trough of $z = -1$ to the crest of $z = 1$.

Henry Eyring had three sons: Edward Marcus, who has followed in his father's footsteps and is now a professor of chemistry at the University of Utah; Henry Bennion, who earned his PhD in business administration and served as president of Ricks College until 1977, and since then has become the commissioner of Mormon Church education; and Harden Romney, who for many years has been the assistant commissioner of Utah's higher education system.

Henry's first wife, Mildred Bennion Eyring, died in 1969 after a long illness; after two lonely years Eyring married Winifred Brennan Clark, a devout Mormon who shared Henry's philosophy.

In 1976 Eyring explained his religious feelings (3): "I know I am never alone in the world. I think you could not describe my feeling in any other way. I'm convinced that this life is meaningful and that it wouldn't make sense without a continuation after death. For me, the idea of living again is a reality. This feeling of not being alone gives meaning to life. I feel that there will eventually be justice. This belief allows understanding of the gallantry of men who, for instance, stay at their posts in time of disaster, such as when a ship is sinking. Or those who help lift the burdens of other people at great sacrifice. This aspect of life has a meaning that I think transcends every other kind of meaning. I try to live up to my ideals, and even though I fail at times, I tremendously admire those around us who have this sense of duty. In the vastness of the universe we are not alone but are really looking to a Higher Power. Religion is a living, real thing for me. I don't see how it is possible to be happy without it. The idea that one is a brother to one's neighbor and the obligation that we have to lift the burdens of those around us are more important than material things. Happiness is more a function of worthwhileness than the possession of material things."

JOSEPH O. HIRSCHFELDER

Literature Cited

1. Heath, S. H. 1980. *Henry Eyring Mormon scientist.* MA thesis, History of Science, Univ. Utah
2. Eyring, H. 1977. *Ann. Rev. Phys. Chem.* 28:1–13
3. Brasted, R. C. 1976. *J. Chem. Ed.* 53:752

4. Sugiura, Y. 1928. *Z. Phys.* 45:484
5. Eyring, H. 1935. *J. Chem. Phys.* 3:107
6. Pelzer, H., Wigner, E. P. 1932. *Z. Phys. Chem. B* 15:445
7. Hirschfelder, J. O. 1966. *J. Chem. Ed.* 43:457

Joseph E. Mayer

Ann. Rev. Phys. Chem. 1982. 33:1–23

THE WAY IT WAS

Joseph E. Mayer

Department of Chemistry, University of California at San Diego,
La Jolla, California 92093

Introduction

At one of Bill Libby's bacchanalian birthday parties in Chicago in the early 1950s, Bill made a solemn pronouncement, which he often did out of the blue sky: "Of all people who have ever lived, 15% of them are living now." The statement was received solemnly by the assembled graduate students and young postdoctorals and elicited very little comment. Sometimes these pronouncements of Bill's formed the theme of discussion for the rest of the party, with a great deal of conversation and argument as to whether the statement was reasonable or nonsense. This one did not, as I remember. It was only months or even years later after I happened to run into some discussion by an historian of census taking in antiquity that I remembered Bill's statement. I tried to see if I could fit some simple algebraic curve to the data given in the article and to a few other figures that I found or remembered. I succeeded; I do not remember the formula; but not to my surprise, the answer was close to Bill's. I concluded that a maximum of about 15% of all humans were still alive, and almost certainly as many as 7%.

Obviously, whatever date one selects as a starting date for such a calculation, one selects a generation that has ancestors for millions of years back to the first lungfish that climbed out of the primordial ocean. One has to start with some arbitrary date. I felt that one might start with ten thousand years ago, which presumably predated any of the large city states in China and also in the Near East and Egypt. If one asks what fraction of *literate* humans who have ever lived are now alive, the number must be greatly increased.

The same kind of difficulty arises when one tries to set numbers for the fraction of scientists living at any year. What does one have to do to be called a scientist? I decided that anyone who spent on science more than

1577

0066-426X/82/1101-0001$02.00

10% of his waking, thinking time for a period of more than a year would be called a scientist, at least for that year. I suppose most young scientists of the present era believe they spend more, very much more indeed, than 10% of their time thinking about science. But I wonder if they do not find that most of that time is spent filling out forms and applying for more research money. In any case, one has to be able to include a man whose paid occupation was director of the British mint, and also one employed by the Swiss patent office.

If one asks what fraction of published science has appeared since I began proudly calling myself "scientist" after my 1927 PhD, the answer is almost all of it! A walk through the stacks of any research library shows much more than ten-fold linear feet of shelf space devoted to "archive" science journals per year now than then. An increase of more than one order of magntidue in any comparison of world science in 54 years from 1927 to 1981 is a completely quantitive change of the characteristics of the milieu.

Childhood

My father was one of 18 siblings, 16 of whom grew to maturity, born to his parents in the small town of Schruns in the Montafon Valley of Vorahlberg, the most westerly province of Austria. As far as I know he was the only one whose education went to collegiate level. He attended the "Technische Hochschule" in Innsbruck in the Tirol. After graduation he went to Paris to the Sorbonne, where he received the degree of "Docent" of applied mathematics, after which he emigrated to the United States and was employed there as a civil engineer by the Union Bridge Company.

My mother, Catherine Proescia, an American-born New York City school teacher, was twenty years my father's junior in age. When I was less than five we moved to Montreal. My father became assistant chief engineer employed by the Canadian government to be responsible for the design of the new Quebec Bridge to replace the first-designed structure which had collapsed.

I learned and understood that if you made a bridge just like the one that stood up, with a given span between piers, and then doubled the span as well as the linear dimensions of every steel member in the design, the strength of the units would be four-fold greater but the weight eight-fold more. Any five-year-old who could count could understand that, if he were sufficiently interested to pay attention to a clearly illustrated lecture by a patient teacher.

My fascination with mechanical construction was greatly stimulated by a fabulous toy brought to me by a young German engineer from Karlsruh: this was a big "Mechano" set. My father was dissatisfied with the mathematical training of American civil engineers. The young German, Hans Grether, became a constant visitor to our household in Montreal until early August 1914, when he avoided the British blockade by shipping as a crewman on a Greek freighter bound for Hamburg, where he jumped ship and assumed his German military duties.

I think my father always considered himself a scientist, an avid reader of Darwin, Huxley, Herbert Spencer, as well as of Freud and Oswald Veblen. Civil engineering was a practical source of income. His family in Schruns were practicing Roman Catholics, but Dad disliked catholicism. He and my mother regularly attended the Unitarian church on Sundays on Sherbrooke Street and I its sunday school. The minister, Frederick Griffin, lived down the street from us. He was a frequent guest at our house, tolerant but disapproving of my father's professed agnosticism.

I have adopted the same agnostic belief, modified by two events. One, through a scolding by a peer when I was a graduate student in Berkeley that the claim of agnostic was a cowardly dodge: Was I not really a convinced atheist? But second, after two round trips of the world, I have attained a theistic realization that all gods worshipped by man have existed or still exist and have been very important in human history. I have seen many! I have seen Shiva's most important organ three meters in circumference and five meters high worshipped by Nandi the bull from without the temple door. I have seen the enormous benign bronze Buddha at Kamakura and the sleeping Buddha in Bangkok. I have seen the great lions of the gods of the Phoenicians shattered by their fall from the lofty pillars of the greatest temple of antiquity dedicated originally to the mighty Baal, but later defeated by Rome who renamed him Jupiter. Beside the fallen lions of Baalbeck is the more modest, but more durable, still-roofed temple to a god at least surreptitiously worshipped throughout the world since the origin of man, whom the Romans called Bacchus.

I suppose that north of cancer 23°N and west of 70°E, for almost two millennia, worship on the Eurasian continent has been pretty much dominated by the various sects that originate from the Judaic tradition—Judaism, Christian, and Moslem—and for the past half millennium, through conquest this is also more or less true of both Americas.

The gods created by man require a priesthood to codify their worship and see to it that temples are erected to demonstrate their power: the might of the doctrine and of the priests of the church. Many are the bloody wars that have been fought about trivia in the theological phraseology of doctrine. It seems as though the teaching of the gentle Jesus has been used as the cause and excuse for many of the cruelest wars of history.

My Introduction to Chemistry

I had a fortunate experience in Hollywood High School, California, in having an excellent chemistry teacher, Mr. Gray, who interested me in the methods of chemistry and with whom I took a second course with a very few (four or five) other students on quantitative analysis. At that time the worldwide price of sugar had recently more than doubled. I spent two campaigns after high school graduation working in two sugar mills, one in Huntington Beach, California, and one in Hooper, Utah. But the price of sugar plummeted at the end of the second campaign and it was clear that chemists were not going to be employed at the bench level. One could guess how to run a sugar mill without paying slaves to titrate the out-put every half hour.

Cal Tech

One of the young high-school educated chemists, Lee Prentice, and I applied to Cal Tech in the midyear class of 1921, the last year that class was given. The name California Institute of Technology had just been adopted by Troop Institute that January with Mullikan's ascension to the presidency of the institution. The midyear class started in January, 1921, and went until the Friday preceding the normal sophomore year beginning on Monday. Without a vacation, we joined the regular sophomore class of the students who had entered the Troop Institute in fall of 1920.

Cal Tech at that time was, as it always has been since, an excellent school. A. A. Noyes was head of the Chemistry Department. He took a great interest in the students although he did not teach any courses. I was amazed to have him greet me by name once in the hall when I was not quite sure who he was. Physical chemistry at that time was pretty much limited to kinetics and thermodynamics.

Richard Tolman came in 1922 as professor of both physical chemistry and mathematical physics. In the fourth year that I was at Cal Tech, Paul Ehrenfest from Leiden, Netherlands, came and spent a semester. Linus Pauling and Paul Emmett came as graduate students in their first year from Oregon Aggie in Corvallis, as I was in my third year as an undergraduate. In my senior year, one of our courses taught by Tolman was based on Summerfeld's *Atomic Structure and Spectral Lines*. The English edition had just come out. The course was listed as graduate, but a few undergraduates took it. Since the undergraduate chemists had not had a thorough course in mechanics, we were pretty much floored. I failed the final examination completely and received a grade of F, which threatened my graduation. Somehow without getting the grade changed they permitted me to graduate.

I had a job as a slavey assistant in Roscoe Dickinson's laboratory, where Linus Pauling was starting his dissertation work on x-ray crystal structure. My first and most serious occupation was to put up a chicken wire grid to protect Linus. Roscoe Dickinson was quite short. The laboratory was in the cellar of Gates Hall and the ceiling was fairly low. The line from the transformer to the x-ray tube went under the ceiling most of the length of the laboratory. Chicken wire had to go under the line to keep Linus's hair from standing on end and shorting the line.

Between Linus's first and second graduate year, Linus purchased Roscoe Dickinson's old Ford—I think it was a 1911 model, in any case it was somewhat antique—and drove up to Oregon where he had done his undergraduate work at Oregon Agricultural. When he came back, he brought his bride, Ava Helen. I was in the lab when Linus walked in upon his return. Roscoe greeted him with obvious relief and the immediate question: How did the Ford behave? "Fine." "It stood up to Corvallis and back without breaking down!?" "Yes." "No troubles at all?" "Well,...we did tip over once on the way back." "My God, weren't either of you hurt?" "Only a few black and blue bruises, but when we got the Ford upright again she started fine."

I had had a very pleasant relationship of my own as a child in Montreal with Mary Bishop. Mary had since gone to Pratt Institute and graduated in fine arts there. Following graduation she had an assignment to interior decorate a new elegant hotel that was being built at the mouth of the Saginaw River where it flows into the Gulf of St. Lawrence, and

Roscoe used Noyes & Sherrill's *Chemical Principles*, which I think was a remarkable textbook. The book was new; the very beat-up edition that I still have was copyrighted in May 1922, but has references to preliminary editions of parts by Arthur A. Noyes alone, copyrighted in 1917 and 1920. The most used phrase in the text that I have is "an important principle is illustrated by the following problem." The student learns nothing unless he does the problems: there are many, and they are mostly hard. I took the course checking, as I now find in my copy, all but a very, very few of the problems. The next year I corrected the problem homework for the following junior class. Since most of the problems were sophisticated, there were several correct approaches as well as an infinity of wrong methods. I learned the material treated in that text very well.

In my senior year I undertook an experimental research problem under the direction of David F. Smith, a National Research Fellow (1). The work was published in the *Journal of the American Chemical Society* in

1924—my first publication! My most lasting lesson from that research was to distrust labels. The title of the paper was "The Free Energy of Aqueous Sulfuric Acid," which we determined by establishing equilibrium at 80°C for the reaction $H_2SO_4 + 6HI \rightleftharpoons 3I_2 + 4H_2O + S_{rh}$, approaching equilibrium from both sides in 20 or more days. Our first attempts used "chemically pure HI" from the stockroom: a clear colorless liquid that reacted instantaneously with even dilute sulfuric acid, due to some phosphorous-containing compound producing H_2S but no I_2, probably phosphonium iodide at about 0.3 normal concentration, which seemed a mite high concentration of impurity in a bottle labeled "C. P." We made our own HI after that.

One of the very long-standing mysteries of science was the behavior of relatively dilute solutions of salts in water. The laws of perfect solutions were known, and relatively low molecular weight, nonpolar molecules were known to obey the laws well at the two ends of the composition mol-fraction diagram. Molecules of similar size and shape deviated but little from the perfect solution laws, even in the middle of the diagram. A. A. Noyes had recognized the quite different behavior of the highly ionized salts in water solution and even before the turn of the century had published some interesting papers calling attention to peculiarities in ionic solution behavior. G. N. Lewis in 1911 and 1912 introduced the concept of "ionic strength," μ, the sum over all ions (i) in a solution of one half the molar concentration times the charge z_i squared, $\Sigma_i \frac{1}{2} c_i z_i^2$. Thus, for a one-one salt like NaCl, the ionic strength is equal to the molar concentration but four-fold greater for the same molar concentration of magnesium sulfate. Lewis then observed that the activity coefficient of any solution of all salts with given valence type z_+, z_- depended on the ionic strength alone at reasonably low concentrations, and its deviation from unity at very low concentration was proportional to $\sqrt{\mu}$.

There was no theoretical explanation until in 1923 the Debye-Hückel treatment of the thermodynamics of ionic solutions was published in the Physikalische Zeitschrift. A. A. Noyes immediately recognized its importance and gave a special colloquium lecture on it. As far as I remember it was the only scientific lecture I ever heard from Noyes, which is a shame because he was so clear and precise that I left with the illusion that I understood it perfectly.

Interlude

After graduating from Cal Tech, my mother, father, and I took a trip to Hawaii. At that time the Moana Hotel was the only hotel in Waikiki except for some bungalow cottages, located where the Royal Hawaiian is now. We stayed at the seaside cottages but visited some of the other islands. It was the first time I had seen true tropical luxuriance. When we returned I stayed in Berkeley, and my mother and father returned to Pasadena. It was the last I saw of my father before his death, which was just before Christmas of that year.

My mother was terribly shocked at my father's death, and more so because she had been so anxious to go to Europe with him and to meet his brothers and sisters in Schruns, most of whom were still there. I agreed to go with her and we arranged to leave by train and boat at the end of the spring term.

I had had a very pleasant relationship of my own as a child in Montreal with Mary Bishop. Mary had since gone to Pratt Institute and graduated in fine arts there. Following graduation she had an assignment to interior decorate a new elegant hotel that was being built at the mouth of the Saginaw River where it flows into the Gulf of St. Lawrence, and she wished to spend her remuneration for that work on a trip to Europe. My mother suggested that she come with us. This was a very happy thing for me. Mary's knowledge of European art was of course exactly what I needed. Mary used the method on museums that I have since learned to adopt, namely to rush through the complete museum, glancing at everything, and then to go back and repeat, stopping only at the things found interesting. I fell in with Mary's tastes very quickly. It was a real revelation to me how much more one saw in a museum with a really good guide than going through by one's self. The present arrangement in many museums of renting a talking machine that tells you what you are looking at and where to go next of course did not exist then. The guides that took groups through then were something of a nuisance.

The three of us "did" Europe together, enjoying the usual tourist traps, but the high point was the visit to Schruns. There we stayed with Uncle Wilhelm and his wife Victoria. Uncle Wilhelm had the commodious second floor over the one bakery in town. I hadn't ever seen a really modern electric kitchen until Tanta Victoria's kitchen in Schruns. Of course another Mayer, a brother of my father's, owned the Electricitätes Werke that supplied the electrical railroad from Bludenz to Schruns as well as the city lighting. Another uncle owned the one inn in town. Schruns at that time was not a well-known tourist resort. Since then it has hosted the International Winter Olympics.

Wilhelm Mayer (Myer is the German pronunciation of the name) had purchased the flour mill in St. Poltern, which is a city much closer to Vienna. Since it was summer when we were there with Mary, the family moved to their summer cottage, away from the big city of Schruns, into the higher mountain valley of Gargellen. We traveled by horse and buggy four hours, from Schruns to Gargellen, a difference of about two thousand meters in altitude. We spent a few days in Gargellen where the cows were for their summer pasture and the cheese was made. We went home about Christmas time so that I would be in Berkeley to begin the spring semester, which began in January.

The last time I visited Schruns with my present wife, Peg, only a few years ago. There were practically no Mayers; the uncles and aunts of course were all dead.

Berkeley and Gilbert Lewis

My textbooks as an undergraduate often contained an introduction quoting Roger Bacon's thirteenth century admonition that no theory was valid if it contradicted observation, that is, experimental fact. Another caution known as Occam's Razor is an added lemma that was then seldom discussed, but states that no unnecessary verbiage should be added to the bare bones of the theory necessary to account for all pertinent facts. These two principles have dominated all good science for several centuries. In my student days there were rare but occasional polemics in the literature concerning the validity of a particular theory versus an apparently contradictory experiment. It was a hardy theoretician who dared enter such a polemic unless he knew the theory had been misapplied or that the experiment had an obvious error. James Franck often remarked that nothing looked more like an important new discovery than a poorly conducted experiment. The other explanation, that the experimentalist misused the theory, also occurs. In the early 1930s I had an evening's discussion with a Johns Hopkins faculty colleague who insisted that themodynamics did not apply to organic chemical reactions! He had a Harvard chemical PhD too!

In the five decades since my student days there are still disagreements between scientists, a negligible fraction of which concern pure science, but more the wisdom of social action in a technical matter. The few serious disagreements in science itself practically never involve identifiable conflict between pure theory and pure experiment. Experimental procedures have become so involved that when conflicts in interpretation occur, it is usually unclear which party is invoking theory and which simple demonstrable experimental fact. Both parties usually use a mixture of both.

The seven-century-old heretical philosophies of Roger Bacon and William of Occam are now so deeply ingrained in the thinking of all scientists, and even, although far less thoroughly, in the consciousness of all scholars, that their principles are seldom now discussed. The methods of science are now, as they always have been, the methods used by the individual practitioners of the disciplines. Since all human individuals are unique, and no two are identical, there is a wide diversity of approach. Nevertheless, there are some characteristics of the ratiocination of scientists, particularly in the physical sciences, that differ from those most commonly met in scholars of other fields.

One of these is both the cause and the effect of the diversity of fields of research. Science seldom, perhaps one should say never, discusses a vague question and never answers one. The vague question is broken down piecemeal into small well-defined parts, and answers to each is painfully sought. The essential feature is that the individual, smaller question is clearly formulated and defined.

One of my early experiences as a graduate student taught me this from a great master. After graduating with a B.S. in chemistry from Cal Tech, I applied to the University of California for a teaching fellowship, which was awarded to me. The next four years I spent in Berkeley. Gilbert Newton Lewis, known as the "Chief" or just as "G. N.," was head of the School of Chemistry. He soon became and remains to this day one of my greatest idols. He was a great and tolerant person, a very important scientist, and a fabulous teacher. The department was a happy one with a very strong leaning towards physical chemistry, which in those days was almost exclusively the application of thermodynamics. Even the organic chemistry faculty knew and used thermodynamics, which was unusual at that time. There were no regularly scheduled graduate courses. Most graduate students from other colleges were advised to take the senior thermo-course. We were advised to take graduate courses in mathematics and physics, and some took courses in biology.

I can imagine no milieu more beneficial to the development of a graduate student than that department at that time. The atmosphere was that of unravelling the intricacies of nature in one of its important aspects. Pure knowledge of an assortment of unconnected facts was seldom emphasized, but a deep understanding of principles and originality in interpretation were most admired. I was never aware of jealousy or friction between faculty members and in four years I grew to know most of them very well. All of them seemed to admire and love G. N. That the atmosphere was good for students has been evidenced by the relatively large fraction of them elected to the National Academy of Sciences some ten to fifteen or more years after their doctorates.

One of the nearest to a required graduate course at that time in the Berkeley Chemistry Department was the Monday evening seminar which all graduate students were expected to attend, and which I think most of us did gladly. Since travel from the East Coast, or even from the Midwest, was then a five-day train ride, we had few visiting scientists, and the notables who did come generally stayed a semester or more and gave a full series of lectures on a special topic. There were seldom more than one on hand. The Monday seminar was rarely attended by any but chemistry faculty, one or two post-docs, and graduate students. During my stay the routine did not vary. A faculty member or a graduate student presented a summary of a paper in the literature that some faculty member had suggested would be interesting. The faculty sat at a long table, students along the wall. The assigned time was twenty minutes. I do not remember slides ever being used, and certainly there was no vue-graph, but blackboard and eraser were in constant use.

After the presentation there was discussion, often for longer than the time taken by the speaker and sometimes involving heated argumentation. When the discussion began to lag G. N. would look around the room, select one of the attendees, student or faculty, and address him: "Mr. John Doe, tell us about your research," or some other equivalent request. If addressed to a beginning student in his first year, the nuance of the request was to describe the question to be answered, the importance of the problem, and the projected experimental approach. Completed research was delivered elsewhere, in special announced lectures by faculty, or in the public PhD examination of students.

I think that the emphasis, which we soon began to recognize, on problems of general importance, rather than on adding only one more example to many similar worked out and well understood cases, was probably the prime characteristic that we, as students, took away from our experience at Berkeley as our most important legacy. Of course, then as now, the research problem was almost always suggested by a faculty member, but the student was expected to be able to critically and dispassionately evaluate its significance and importance to scientific understanding. The answer, occasionally given by students now, "It is part of Professor X's project," by itself would have been regarded as unsatisfactory without an understanding of its place in the project and the project's place in science.

The answer to G. N.'s request seldom exceeded a quarter hour, and the faculty discussion followed another quarter hour with many helpful suggestions of procedure or experimental technique. I think that most of us enjoyed being called on. As I remember, usually two or three students were called upon each week, so that each of us reported several times on the progress of our incomplete doctoral dissertations. The seminar was open-ended in time and often lasted two hours, but the diversity of subject matter kept it from becoming tiresome.

One experience in that seminar I shall always remember. I was elected to give the twenty minute report on a paper, I think in the *Zeitschrift für*

Physik, on a subject I have now forgotten but which involved high vacuum technique as did my doctoral research. Towards the end of my carefully prepared and, I hoped, very clear presentation, I became quite aware of incipient troubles. Unusually dense clouds of aromatic smoke were being emitted from G. N. and his ever present "Fighting Bob" nickel cigar to which he had become addicted in the Philippines. Worse still, the cigar was being shortened from the chewed end rapidly and G. N.'s scowl looked ferocious. I finished and awaited the explosion. Smoke and scowl continued for some time. Finally the cigar came out and after that a question, long and involved and obviously concerned with the scientific conclusion of the paper. I found the question incomprehensible and after some hesitation said, "I'm sorry Professor Lewis, but I don't understand the question." More smoke, more scowl, and finally, "Damn it Mayer, of course you don't. If I understood it I'd probably know the answer." Laughter in the room.

I have often thought that that episode illustrates a rather frequent stage in the development of science. The most important part of any real advance is to formulate clearly the correct question.

It was typical of G. N.'s eclectic interest in all of science that he became interested in relativity, which is as far from physical chemistry as one can imagine in any physical theory. I remember a fascinating lecture on special relativity that he gave, at which I saw for the first time the now oft repeated time-space diagram.

I don't remember exactly when I actually began work on what became my dissertation. It was a fairly difficult experimental stunt, and I think that actually had we started it after we really understood quantum mechanics, we would have thought it not worth doing. The dissertation was finally published under the title, "The Disproof of the Radiation Theory of Unimolecular Reactions" (2). If chemical reagents require activation energy to react, this activation energy must either come from radiation or from collisions between molecules. If it is due to collisions between molecules, the rate should be proportional to at least the square of the pressure and the reaction would not be unimolecular, but bimolecular. This had been observed, of course, and Jean Perrin postulated that all unimolecular reactions were due to the absorption of radiation. I set up a very high temperature radiation field such that, if activation were by radiation, a beam of molecules passing through the field would react. The reaction we studied was the racemization of pinene which, from its known rate measured at room temperature and extrapolated to our field temperature, should show racemization on passing through a few centimeters if it were indeed activated by radiation. The radiation bath was simply a quartz cylinder about three centimeters long, through which the pinene coming through two holes in platinum foil was caught in liquid air at the other end and later transferred to a capillary tube and the optical rotation measured. The effect was nil. Of course this is no mystery now and I think it was recognized at that time or very soon afterwards that at sufficiently low pressure the racemization might indeed go with the square of the pressure. However, the negative result of the measurement was accepted for publication and I was granted a doctorate degree.

I do not know how it is now, but at that time teaching fellows in the chemistry department had the privilege of belonging to the faculty club. I ate there at lunchtime rather regularly, usually sitting at a table with other chemists, although quite often when the chemistry table was full I would sit with faculty members of other departments. The chemists, along with one mathematician, frequently played cards for a half hour after lunch.

I became pretty good friends with Wendell Latimer. Latimer at that time had lost his first wife and was, I guess, in a really unhappy period of his life. However, he met his second wife, Latha, and they were happily married before I got my degree. Before that time we often went to San Francisco. Several of the graduate students, often with Wendell, had dinner at a favorite second floor Italian restaurant, "Mimi's." We were always well fortified with drinks, particularly with good Italian red wine. Mimi used to regale us with stories of the opera singers he knew. He had been an impresario in Vienna and in Rome. We always took Mimi's tales with a grain of salt, but one day the New York Metropolitan Opera was playing in San Francisco. The two stars showed up after the opera at Mimi's while we were there having dinner. The two stars, if my memory is correct, were Martinelli and Galicurci. Galicurci embraced Mimi effusively as her old teacher. We were greatly impressed and we really believed all of Mimi's stories after that. The two opera singers had a postman's holiday and we enjoyed it thoroughly. They sang loud and lustily for several hours after a tiring opera.

Hildebrand, of course, had the freshman chemistry course at that time. I was employed as a teaching fellow; my memory is that every entering student was a teaching fellow. I was impressed by the system. The younger faculty members were in charge of laboratory sections. I think all laboratory sections were about 25 students. At that time, experienced teaching fellows were assigned with the beginning teaching fellows for their first year in a laboratory section which was officially in charge of a faculty member. The faculty member would not necessarily stay throughout the whole afternoon after the first few weeks, if he knew he could trust the teaching fellow.

There was a careful way of correcting for differences in the grading of papers so that the students were not penalized by getting a very strict teaching fellow doing the grading. At the end of the semester there was of course a big final examination, and the grading was done by the teaching fellows all at once in a big room, usually one teaching fellow taking one question. In following years, I did not have a faculty member in the laboratory section but had it all to myself.

As I mentioned above, there were no graduate courses in chemistry listed in the catalog and none given, except that most of the beginning graduate students were advised to take the undergraduate thermodynamics course, which was based on Lewis & Randall. I was excused from that since it was assumed that Cal Tech would have given me a thorough basis in thermodynamics. My final examination was during the 1927 summer session and was open to the public. At that time there was a certain amount of pressure on the high school teachers to know the subject matter that they were teaching rather than only to have had courses in education. There were quite a number of them attending my examination. I had a pleasant introduction to my committee by having lunch with them at the faculty club and playing cards afterwards until it was time to go across the street to Gilman Hall for the examination itself.

Gilbert Lewis at that time was known to have usually one tricky question and he certainly had one for me: Take two independent tungsten filaments, in high vacuum, one of them at a temperature approximately 2000 degrees or below, and the other one really white hot, and both filaments are on separate lines by which the current and voltage across can be controlled. A very small pressure of chlorine is allowed into the apparatus, something like 10^{-4} or 10^{-5} mm pressure. The observation is that the higher temperature filament's resistance decreases: it gets thicker, whereas that at a lower temperature gets thinner. After a few promptings to consider possible chemical reactions, I saw the light: The answer was that at the lower temperature the chlorine attacks tungsten, forming a volatile tungsten chloride, but at the very high temperature this compound dissociates, depositing tungsten on the hotter filament itself. I was very proud that I got the answer correctly.

After the examination, Lewis asked me if I would stay as a postdoc, that he would like to discuss a few problems with me. I do not remember what I was paid but I did get a little more than the teaching fellowship and I felt very flattered by the invitation. I actually stayed until the autumn of 1929 when I went on a National Research fellowship to Göttingen to work with James Franck.

With Gilbert Lewis I had a very stimulating two-year experience. I had no knowledge of statistical mechanics and Lewis had never worked in the field either. He had become interested in the discovery that had just been made of the difference between quantum mechanical statistical mechanics and the classical, and the Bose-Einstein versus Fermi-Dirac systems.

During the day I tried to learn statistical mechanics using Tolman's two books, the first of which I found clear and interesting, but the second seemed to me to be too talkative. I also tried Fowler & Guggenheim, which I did not really like, probably mostly because I was unacquainted with the mathematical methods used.

Gilbert and I spent the evenings together, usually at about eight o'clock, sometimes until about midnight. That was really an experience. It was most interesting to see how Gilbert Lewis thought. He was not infinitely brilliant, but he would go over and over a problem until he really understood. Eventually he decided that we ought to publish and the result was three papers that appeared in the *Proceedings* of the National Academy (3).

I still like the method that we evolved for deriving thermodynamics from statistical mechanics, that is, from the mechanical laws for the motion of molecules. Of course the black body radiation problem had been solved but hardly explained. I remember one remark of Gilbert Lewis at that time. He said that the first papers on black body radiation by Planck were clear and concise. The later papers got fuzzier. Actually, we were quite conscious of the fact that the quantum mechanical interpretation was so strange that it was scarcely believable.

The Michaelson-Morley experiment uses half-silvered mirrors which split a beam of photons into two paths which then are reflected back by two mirrors at right angles to each other, such that there are two beams at right angles going through the half-silvered mirror at 45 degrees, which come back and then half of both beams are coalesced again and interfere. There is a beam of light, half of it moving in one direction, half of it moving at 90 degrees in another direction; both beams are completely reflected approximately the same distance and one-half of each of the beams goes to a receiver after passing through the half-silvered mirrors. If one of the two totally reflecting mirrors is blacked out, the pattern is a simple one: one quarter of the beams arrive at the receiving station. If the other mirror is blacked out, the pattern looks almost identical to the naked eye, but the peaks in intensity are actually shifted. Of course when both of the totally reflecting mirrors operate, the two beams interfere with each other and give the interference pattern. There is no mystery with intense classical beams of electromagnetic waves. The mystery, of course, is that if you picture beams consisting of single photons which always have the energy $h\nu$, how do the photons know what the other photon does at a large distance, that is, whether or not it is reflected back? This, of course, is the essence of the difficulty that Niels Bohr and Einstein argued over years later and is responsible for Einstein's statement, "Rafiniert ist der lieber Gott aber Boshaft ist er nicht." (God is sophisticated but not mean.)

In 1929 I was the proud possessor of a National Research Fellowship paid by the Rockefeller Institute for postdoctoral work in Europe, and I had arranged to be able to study in Göttingen and work with James Franck. Franck had been in Berkeley. I gave him a letter of introduction from Hans Grether. After World War I, Hans Grether had gone to Peru and worked on the railroad that was to run over the Andes into the Amazon Valley. His brother had a ranch in Somis and also one in Salinas, California. The brother, Karl, had come to California from Germany while we were still in Montreal before World War I, and Hans was still in Montreal. Karl stayed with us there in Montreal for a few days. We made contact with Karl in California and I visited the ranch several times. Hans came up once from Peru to visit his brother, and when I told him that I was hoping to go to Germany to work with Franck he gave me a letter of introduction. He had worked with Franck during the war, actually, I think, on poison gas.

Göttingen at that time was known as the source of quantum mechanical knowledge. Heisenberg, working with Max Born, simultaneously developed, at the time that Schrödinger developed wave mechanics, an equally logical system of quantum mechanics. As you may remember, for quite a while there was some difficulty concerning this, as Schrödinger's wave mechanical approach seemed to do exactly the same things as the matrix approach of Heisenberg and Born. It was Schrödinger who finally reconciled the two systems as being really different mathematical formulations of the same theory. Max Born at the time was in Berkeley, and I heard later from Maria that he was rather annoyed that he was scooped by Schrödinger simply because he had too much to do in Berkeley.

Thorfin Hogness, who was a member of the Berkeley faculty, had been in Göttingen for a year previously. He told me that most Americans in Göttingen stayed at a pension on Nicholausberger Weg, the Kreuznacker House, but if I could possibly get a room in a private dwelling, it might be more pleasant than a pension. He mentioned that his pediatrician, Professor Göppert, who was a professor of pediatrics, had died and he understood that Frau Göppert might have a room.

When I went to see James Franck I asked him where I should stay and he gave me almost the identical advice—namely, that there was a pension on Nicholausberger Weg, the Kreuznacker House, and it was very satisfactory and most of the Americans stayed there. I'd probably be happier if I could possibly get a room somewhere alone, and he suggested that Frau Professor Göppert had rented a room to an American (it turned out that the American was Robert Mulliken) a year ago and I might try there.

The maid who answered the door said the Frau Professor Göppert was ill and wouldn't see anybody. Actually she had a nasty cold. However, the maid told me that she would ask the daughter to come down and speak to me. Well, the daughter came, smiled benignly at my frantic German, and then answered in beautiful Cambridge English, that her mother was sick, that it was just a cold, but she did not want to see anybody, that I should come back in a day or two, which I did. I was staying at a hotel close to the railroad station. I was much impressed with the daughter and particularly by her perfect English, which I later found she had acquired in one semester at Cambridge on a student fellowship from Germany, in Rutherford's laboratory. She had lived in Girton College while she was in Cambridge, which was the only girl's student house at that time.

Well, I was feeling relatively wealthy and I purchased on Opel. The Opel was not a General Motors car at that time; the firm was later purchased by General Motors. The Opel was a wonderful car. It was put together with picture-hanging wire and sealing wax. I think the existence of the Opel changed my future life. It was a beautiful machine and I had the only automobile of any of the students or of any of the young faculty.

Maria was the belle of Göttingen, as I soon found out. She and the two daughters of Marianna and Herr Professor Landau, along with Titi Stein, seemed to make up the acceptable female contingent of every student party.

On Wednesday afternoons the students of Göttingen frequented Maria Springs—nothing to do with my Maria. Maria Springs was a natural amphitheater in the woods about 10–12 miles north of Göttingen on the main road and on the railroad. The German band played there and there was dancing on the floored area at the bottom of the amphitheater. The Corps, that is the fraternities, all came in decorated coach-and-fours and made quite a show of their arrival. The girls of Göttingen were welcomed in the early afternoon. The last train from Maria Springs to Göttingen left at 6 o'clock in the evening, which in summer time, of course, was bright daylight still. Proper girls had to go home by the 6 o'clock train. Even with my car I could not get Maria to stay longer than the time the train left.

The German school system, like the American, is a state affair, that is, it is different in the different provinces, but essentially similar, just as the American states have very similar schooling arrangements. Until the students are 10 years old, schooling is uniform, simply a neighborhood school; after that there is a separation. The Volkschule does not prepare students to go into a university, and in general students who start in at the Volkschule never can get into a university. Now I say "in general" because I know at least one exception, and that was Hans Jensen, who was the son of a poor gardener and could not afford to go to an Oberrealschule or Gymnasium, the two schools that do prepare for the university. His instructor at the Volkschule recognized his ability and managed with great difficulty to obtain permission for him to take the abitur, which he passed perfectly. The Oberrealschule, the purely classical school preparation for the university, had both Latin and Greek as required subjects. The Gymnasium permitted one to take an abitur also but was more technical. I think both required calculus. The abitur was very much feared. It was generally regarded as disastrous to fail it and one could not get into a university without passing it. The American department idea with several professors in the university was nonexistent. On the contrary, each full professor had his own institute and apparently operated completely independently of any other full professor in the same field. For instance Göttingen had three physics institutes. The Erstes, the first institute, was under Herr Professor Pohl and was primarily solid state work. The Zweites was James Franck's institute to which I was assigned. The Drittes institute was that of Max Born and was purely theoretical. The only common feature was a weekly seminar that all members of all three institutes attended. The seminar had other attendees, for instance both Hilbert and Courant quite frequently came; less often the Professor of aerodynamics, Prandtl, came.

I know very little about the chemistry departments. The physical chemist was Professor Gustav Tamman, who was regarded as a fierce man to have on an examination. One student, who had Tamman on his PhD examination, came out trembling. Tamman had asked him: "Na Kerl was ist lambda?" "Lambda ist Wellenlänge." "Falsch! Lambda ist specifische Conductivität!" Many physics students had Tamman on their examination, but I don't know of any student who was ever flunked because of Tamman's questions, although a good many students were not passed on their first examination.

Only a few weeks after I got to Göttingen there was a meeting of the German Physical Society in one of the Hartz mountain towns. There I met quite a few German physicists whose names I had heard, but had not met before. Among them was Polanyi, the Hungarian physicist philosopher. His son is now professor at the University of Toronto and was a long time in Ottawa.

In Göttingen, Professor Franck came around on a certain afternoon each week with his entourage of assistants and visited everybody working in the laboratory. The group would always include Herta Sponer, who was Franck's assistant, and later after Franck's wife died he married Herta. Otto Oldenberg was the oldest of the people who came on these tours. He was Auserordentlich professor, which corresponds pretty much to an associate professor in America and implies tenure. Otto Oldenberg later emigrated and was in Harvard.

One of the important holidays in northern Germany is Pfingsten, which I think is the same as Whitsuntide in England. Dick Badger, who was in Göttingen from Cal Tech, and I took our first Pfingsten vacation together walking from Lyon to Marseille down the Rhone Valley. It was a long walk, rather too long, walking is slow; however, we did see the French countryside and enjoyed the Rhone wines enormously on the way. One afternoon a car stopped and the two Canadian occupants asked us if we wanted a ride. We did. Later in the afternoon they stopped at an inn and decided they were going to stay there for the night. Dick Badger and I also found it a good idea. The two Canadians asked for a room with bath. Dick Badger and I were satisfied with a room without a bath. At supper the Canadians suggested that we ought to come up and look at their room, that it was worthy of a visit. We did after supper and found the room gorgeously furnished—an enormous room with a platform in the middle of it and a big bathtub on the platform! Dick Badger and I parted in Marseille and both of us went back to Göttingen independently.

In Göttingen I found several letters from Johns Hopkins University offering me a position as associate. The first one had evidently come just about as I left on the Pfingsten holiday. I responded affirmatively and felt very happy that I had a position assured when I got back to the United States. In the meantime I was getting more and more interested in trying to induce Maria to come back as my wife to the United States. This was not completely trivial; the German immigration quota was filled for several years in advance. However, I found out from the consulate that my wife could get in on a special visa. Well, that worked out. I remember that Maria's favorite aunt, who was not very much older than Maria, the wife of the youngest brother of her father, said to her: "You are fortunate in going to America. My sons will be caught up in the next war." They were. One of them survived but was badly wounded.

Return to USA

Maria and I married, and Maria finished her exam shortly before we left. We traveled on the Nord Deutscher Loyd ship, Europa, on its maiden voyage. Arriving in New York on April Fools day, we were met by my cousin and his wife. We stayed with them a few days and then went on to

Baltimore where we found accommodations in a pension. However, the summer vacation broke out pretty soon and we found that there was going to be a special summer session for graduate work at Ann Arbor, Michigan. Enrico Fermi and Paul Ehrenfest were to be the two lecturers. We both knew Ehrenfest quite well. He had been a regular visitor in Göttingen and on his invitation we had once driven to Leiden and stayed with him and his wife. He locked Maria in his study and scolded her that she was not working on her dissertation. He let her out only after she produced n pages, and I forget the number n. In the meantime I was assigned the guest room; the guest room was on the third floor of the house or maybe it was the fourth. It was a whitewashed room lined with bookcases which contained paperback detective stories. The detective stories were in all languages. There were many in English, some in Russian, and of course various ones in Dutch and German. And there were signatures on the whitewashed walls of guests who had stayed there, signatures with dates attached. I signed under that of the last guest, Albert Einstein!

That particular Ann Arbor summer session was enormously successful. Both Enrico Fermi and Paul Ehrenfest were extremely good lecturers. Each sat in the front row when the other was lecturing and corrected the other's English, much to the amusement of the audience. But both were extremely clear. The audience included Robert Atkinson, an English astronomer and physicist, Lars Onsager, Serge Korff, Donald Andrews, Charles Squire, and of course Sam Goudsmit and George Uhlenbeck, both professors at Ann Arbor at the time.

We became particularly good friends of the Fermi's. Laura Fermi was always a delight and Enrico was always interesting and informative. He was a lot of fun too.

After the session was over I undertook to show America to Maria. We drove in our very conservative secondhand Buick, which we labeled "Connie" for conservative, and tried to see as much of the West as we could. We included the Black Hills, the Tetons, the Yellowstone and Glacier Parks, and then went further west to Seattle, stopping at Mt. Rainier, which I climbed. In Seattle, we visited Henry Frank who was at a summer resort on one of the islands in Puget Sound at that time. Actually we dug clams, much to the horror of the natives, who were not brought up in New England. On driving up to the village at the bottom of Mt. Rainier, as far as we could go in a car, we passed Nisqually Glacier, which was then down almost to the road. We chilled the clams on the glacier and enjoyed them.

We drove south from Seattle, stopped at Crater Lake, and then took the Redwood Highway from there to the San Francisco Bay area and

Berkeley. At Berkeley we visited old friends of mine, including Robert Oppenheimer and his wife Kitty. In the last years that I was at Berkeley, Oppenheimer had come back from Germany and gave a lecture on quantum mechanics. I don't think I understood anything of it but I was enormously impressed and felt that I was getting quite a bit from hearing his stories. I was amused recently at reading an article in the Cal Tech magazine by Carl Anderson who had listened to Oppenheimer's lectures on quantum mechanics in the same year at Cal Tech. According to Anderson's story, the class kept getting smaller and smaller until he was the only one left. Oppenheimer then came to him and said, "Please don't leave, I can't go on with nobody in the class. Let me have at least one student to the end of the quarter." At Berkeley there were several of us who went through the whole semester, or whatever the length of time that Oppenheimer was scheduled to lecture, and we enjoyed it, but I think we were pretty well snowed. Oppenheimer was not a good lecturer at that time. His great facility for making things clear came only later. I think actually that this is a common failing of very brilliant young fresh PhD's in not being able to talk down to students that are not as able as they are. Or even if the students are actually very good, they still tend to talk over their heads.

From Berkeley we went down the coast to Pasadena where we visited the Paulings. We camped in the Pauling yard. They had a house very close to Cal Tech, and we imposed ourselves on them by setting up our tent in their yard. It was a delightful visit for us. I was very sad to see in the newspaper recently that Ava Helen died. She was a delightful person.

The most exciting thing that happened on the trip back from Pasadena to Baltimore was simply a series of tire failures which depleted the ready cash that we had. At that time it was almost impossible to cash checks, although I think we had some money in the bank in Baltimore. If I remember correctly, we simply drove through the bridge at Harper's Ferry without paying any toll. When we got to Baltimore late at night, we had no key to our house, having left it with Frank and Kitty Rice, who lived very close to us on the second floor of an apartment building. We went there and after a while managed to awaken the Rices, went up to their apartment, where we were given a few drinks and then, with the key to our house, proceeded to our home. Frank and Kitty Rice were our most intimate friends at that time. Frank was professor at Johns Hopkins and Kitty was a student in the Johns Hopkins Hospital Medical School, where she later got her degree, Doctor of Medicine with a specialty in psychiatry. Frank left Johns Hopkins for Catholic University in Washington in 1938 shortly before I was fired and went to Columbia. The administration at Johns Hopkins was making it uncomfortable for people to stay. I believe there was a real financial difficulty.

In summer of 1931 we went to Göttingen. Maria was employed by Max Born to help write an article on crystal dynamics in the *Handbuch der Physik*. I went with her, or course, but in addition, my first real student at Hopkins, Lindsay Helmholz, came with us.

It was impractical to start any experimental work in the summer, at least trying to do the experimental work Lindsay and I were interested in, so I worked on crystal theory, that is, lattice energy theory. I consulted with Max Born, and Lindsay helped me. We published two papers, one of them under the names of Max Born and Joseph E. Mayer (4) and the second one Joseph E. Mayer and Lindsay Helmholz (5). This was really very much a copy of the methods used by Born and Haber years earlier, but went into much more detail, attempting to get really good semiimpirical values for the lattice energies of the alkali halide series using exponential repulsive potential. The results were extremely good, really, and I think that for many years they were the best theoretical calculations of the energies for any chemical reactions. Of course, the lattice energy, which is the energy difference between the vapor ions and the normal crystal, is very large, much larger than any directly observable chemical transformation of the ions. Actually there is very little difference in energy between the crystalline salts and the salts in water solution where the ions are dissociated. However, the differences in the chemically observable solubilities were actually fairly well given by our theoretical values at that time. Certainly rather better than almost any quantum mechanical calculation of the energy of the chemical reaction then current. Lattice energies are of the order of 100 kcal per mole or more and the results showed fairly good values to the order of 1 kcal. Later on we evolved a different method of getting at lattice energies. The only unknown at that time was always the electron affinity of the halogen, that is the reaction chlorine neutral atom as a perfect gas to chlorine minus, also a perfect gas. The new method was simply to observe the ratio of electrons to ions coming off a hot filament in an atmosphere of very low pressure of chlorine or any other halide. The electrons could be deflected by a relatively small magnetic field parallel to the length of the filament; they curled themselves up and did not go to a positively charged plate cylinder of a centimeter diameter. This was a far simpler and more satisfactory method than the awkward treatment of the salt crystals. It worked quite well on the halides but did not give good results when used to measure the electron affinity of oxygen.

In the meantime, a Wiley representative got Maria and I interested in trying to produce a book on statistical mechanics and it actually materialized in 1940, the first edition of Mayer & Mayer's *Statistical Mechanics* (6). The book was quite successful as books on statistical mechanics go—enough so that Wiley even tried to get us to produce a second edition very much later. It was in doing that that I got interested in attempting to improve the method of deriving equations for the virial coefficients of normal molecules. Philip Ackermann was working on an experimental problem of seeing whether we could use low energy electron beams for molecular structure studies. Ackermann had trouble getting a satisfactory position and he stayed and helped me with the statistical mechanics calculations. The first two papers on the theory of condensing systems (which was an unfortunate choice of titles) were done with Ackermann's help (7, 8).

In general, I had a most interesting and excellent group of students while I was at Hopkins. It was an unfortunate time, much worse than at present, to get positions. Science was not a popular subject and well supported as it still is now, in spite of our complaints.

Maria and I tried to build an electron microscope. It was a joke and we took it as a joke; we never even tried to publish it. The apparatus was essentially made of wood with window screening to produce the electric fields. I give this only as an example of the sort of tomfoolery that we often were forced to use. Of course, essentially we did not have the courage to think that an electron microscope would have a real use in the future. There are other scientists who have occasionally thought ahead of their time in trying to develop experimental methods, but given them up, wisely probably, because everything has a time to be successful. The only thing I object to is that in many cases these people think they have been cheated. We knew what we were doing was foolish, but it was fun.

The sad thing of that time was how extremely difficult it was to get good positions for our students when they graduated. Particularly the women. I was very fortunate in having several excellent women students at Hopkins: there were Sally Harrison, Sally Streeter, Irmgaard Holdner Wintner, who was the wife of a mathematician, a professor at Hopkins, Aurel Wintner. I also had Willard Bleick and Louis Roberts as well as J. J. Mitchel. Louis Roberts came with me to Columbia and actually took his PhD degree at Columbia instead of Hopkins. Willard Bleick was one of the more impressive students.

In 1938 I was informed that I would be discontinued at Hopkins after a year and a half. It was very fortunate for me, although I was happy at Hopkins and I would have stayed there had I not been fired. I wrote to various friends, including Harold Urey at Columbia and James Franck, who had left Hopkins to go to the University of Chicago. At both places I was given an offer to come as associate professor, presumably with tenure or at least tenure after one year; I felt very happy to go to Isaiah Bowman, the president of Hopkins, and resign long before I was obliged to leave. I finally decided after visiting both Columbia and the University of Chicago to go to Columbia. I think it was probably a very good choice at the time; in any case I have followed Harold Urey ever since. Maria and I went often in summer to Germany. In 1937 she received a telegram that her mother had had a stroke. We managed to get her on the Europa the same evening, which was quite a feat and managed only because we knew the German consul in Baltimore who was also the representative of Nord Deutscher Loyd. Maria was received at disembarkation in Germany with the news that her mother had died. We managed to get a load of furniture out of her house but never succeeded in being paid for the sale of the house.

Of course, the trips to Germany were always sad after Hitler's arrival as Führer.

I learned many things at the time. Americans are too polite and not nearly as direct as the Germans or, particularly, as the Dutch. Several times Germans told me they had a position in the United States and showed me the letters they had received. Those letters did not offer a position at all. They were usually in the tone, "Of course we would love to have you here, all my colleagues would like it, they would all recommend to the administration that we create a position for you, but the financial situation is so bad that it is almost impossible that anything will happen..." and so on.

Literature Cited

1. Smith, D. F., Mayer, J. E. 1924. *J. Am. Chem. Soc.* 46:75–83
2. Lewis, G. N., Mayer, J. E. 1927. *Proc. Natl. Acad. Sci. USA* 13:623
3. Lewis, G. N., Mayer, J. E. 1928. *Proc. Natl. Acad. Sci. USA* 14:569; 14:575; 15:172; 15:208
4. Born, M., Mayer. J. E. 1932. *Z. Phys.* 75:1
5. Mayer, J. E., Helmholz, L. 1932. *Z. Phys.* 75:18
6. Mayer. J. E., Mayer, M. G. 1940. *Statistical Mechanics*, pp. xi, 495. New York: Wiley
7. Mayer, J. E. 1937. *J. Chem. Phys.* 5:67
8. Mayer, J. E., Ackermann, P. G. 1937. *J. Chem. Phys.* 5:74

Joseph Oakland Hirschfelder

Ann. Rev. Phys. Chem. 1983. 34: 1–29

MY ADVENTURES IN
THEORETICAL CHEMISTRY

Joseph O. Hirschfelder

Theoretical Chemistry Institute, University of Wisconsin,
Madison, Wisconsin 53706, and Departments of Chemistry and Physics,
University of California, Santa Barbara, California 93106

INTRODUCTION

As Joel Hildebrand said, "Chemistry is fun!"

I have thoroughly enjoyed my fifty years in Theoretical Chemistry and I am looking forward to the next twenty years. I was fortunate to get started during the Golden Age of the 1920s and 1930s and now I am hoping to participate in the Laser Age of the 1980s and 1990s.

The 1920s and 1930s were years of great discoveries in quantum mechanics and the nature of chemical binding. Research during this period was very exciting: There were many new techniques to be learned, and whenever these new techniques were applied, they led to new concepts—everything worked! It seemed as though all of the secrets of nature were unraveling. Theoretical physicists and chemists worked together on molecular problems.

The honeymoon came to an end at the close of the 1930s with the realization that although nature might be "simple and elegant," molecular problems were definitely complicated. Furthermore, neither the elegant "first order perturbation approximation" that Dirac, Van Vleck, Serber, Nordheim, and colleagues had derived for valence bond wave functions nor the Heitler-London formalism for organic molecules were sufficiently accurate to agree with precise experimental data. At this point, the theoretical physicists left the chemists to wallow around with their messy molecules while they resumed their search for new fundamental laws of nature.

The scientists in the 1930s who made the great discoveries had a very thorough understanding of the existing theories. In addition, they had both

1601

0066–426X/83/1101–0001$02.00

imagination and guts! Indeed, it took a lot of imagination and guts for Max Planck, Erwin Schrödinger, Werner Heisenberg, and Paul Dirac to propose their quantum mechanical conjectures; for G. N. Lewis and Linus Pauling to propose their concepts of the structure of atoms and molecules; and for Eyring to propose his reaction rate theory. I remember Wigner saying that if Eyring had only realized the many approximations involved, he never would have dared. However, Eyring's theory of absolute reaction rates has been very useful for a long time. Unfortunately, now there are too few theoretical chemists with sufficient vision to take the giant step of exploring completely new techniques. Instead, scientists in the 1980s get so immersed in a maze of computational detail that they lose sight of the simple, elegant theories.

Theory is most useful when it is used to predict critical experiments. Only from such comparisons can we develop a better understanding of the experiments and a knowledge of the deficiencies in the theory. These predictions are important even when the theory is in its preliminary, primitive form—the possibility of learning something important usually outweighs the embarrassment of being wrong. (Lee Allen may disagree with me since he was burned with his poly-water!) As Einstein said, "An expert is a person who has made every possible mistake at least once."

During the 1930s, nobody really knew the deficiencies of the crude theoretical applications. Dirac said in the preface of his famous book that, with the advent of quantum mechanics, all of the laws of physics were known that were needed for an understanding of chemistry. For the last 50 years we have been trying (and for the next 50 years we will continue to try) to determine the quantum mechanical equations whose solutions provide the answers to all of our chemical problems!

In the 1930s there were very few full-time professional theoretical chemists in universities: Linus Pauling (California Institute of Technology), Henry Eyring (Princeton), Bright Wilson (Harvard), and Lars Onsager (Yale). However, all of the major universities employed at least one theoretical physicist. The chief function of a theoretical physicist or chemist was to explain experimental results and to suggest critical experiments. Thus, as a graduate student at Princeton, I was expected to become familiar with many types of experimental procedures and equipment. In the Golden Age, a theoretician who didn't have breadth didn't have bread!

Theoretical chemistry seeks to explain and predict quantitatively the physical and chemical properties of materials, to relate these macroscopic properties to the individual molecules, and to predict the structure and the properties of these individual molecules. Since all of the required fundamental laws of nature appear to be sufficiently well known, we should be able to write down the mathematical relations that determine a particular physical or chemical property. The problem then becomes one of finding the solutions of the mathematical equations, developing simplified models that capture the essence of the phenomena, and devising accurate numerical approximations. I think of theoretical chemistry as theoretical physics applied to chemical problems, whereas other people think of it as the empiricising of observed experimental data. However, when we are given a problem, we use whatever approach seems best for getting a solution.

Theoretical chemistry is a natural focal point for interdisciplinary research because its problem areas overlap with physics, mathematics, astronomy, meteorology, chemical engineering, and mechanical engineering. In relating the macroscopic properties to the individual molecules, we serve as the "middlemen" between the theoretical physicists and the practical engineers and experimental scientists.

Writing this chapter gives me an opportunity to review what I have learned that might be useful. With experience and age is supposed to come perspective. I have had very thorough training in physics, mathematics, and chemistry. As the result of various happenstances, I have worked on an amazing assortment of problems—both basic and practical. Some of the techniques that I have used are not familiar to theoretical chemists, although they are frequently employed by other kinds of scientists or engineers. Other techniques were well known by my generation of theoreticians but have become buried in the literature. In discussing various problems, I want to tell you about how I got started, what I accomplished, and in what ways I was disappointed. Very often, it is our failures that are most instructive—and these almost never get published.

In this discourse, I also want to emphasize that science is neither cold nor logical. As the Russian physicist Frenkel said, "Knowledge advances by illogical steps." The University of Wisconsin organic chemist, Homer Adkins, used to say, "In basic research one shoots an arrow into the sky and where it lands one paints the target!" Indeed, I should not be doing academic research for which I know the outcome! Scientific investigation is a fascinating game in which we make "guestimations" or conjectures based upon the skill and experience of ourselves and others. Our research is a giant pyramid with all of us climbing upon the shoulders of our associates, teachers, and myriads of others. I have been fortunate to climb upon the shoulders of many great men. I want to share with you some of the things I have learned from them.

First, I want to tell you a little bit about my background. In the Appendix, I provide a concise listing of the places I have worked, the problems I have tackled, and the people with whom I have worked. I would have liked to discuss each of these topics in the manuscript, but because of space limitations this was not feasible. So, I have tried to give a detailed consideration to a few subjects associated with my early research, leaving other topics for possible future publications.

My Background and Early Education

I have had every possible opportunity to do scientific research—I can only blame myself for not being more productive. What I have lacked in natural ability, I have tried to make up for by hard work and perseverance.

My paternal great grandparents emigrated from Germany to California in 1843 (before the Gold Rush). They took a ship from New York to the Isthmus of Panama, got a native to paddle them up a river and across a lake to within seven miles of the Pacific Ocean, and waited at a monastery for a ship to take them to California. Both my grandfather and father devoted their lives to medical research. Grandpa was the first child born in Oakland; he graduated in the first class at the University of California, and became the first Professor of Clinical Medicine at Stanford. Dad entered the University of California at the age of 13. After receiving his MD, he joined the medical faculty of Johns Hopkins University, where he was the first doctor in the United States to use an electrocardiogram (1). Later, he became very much interested in the colloid chemistry associated with the physiological effects of drugs and accepted a Professorship in Pharmacology at the University of Minnesota.

Thus, I was born in Baltimore and grew up in Minneapolis. When I was five years old, Dad built a chemistry lab for me in the basement of our home. When I was ten, he took me to an American Chemical Society Meeting in Los Angeles. And when I was 15, I helped Dad determine the size distribution of colloidal particles in a Zsigmundy ultramicroscope—my contribution was to suggest a correction factor for the convection currents produced by passing street cars.

When I was an undergraduate (University of Minnesota, 1927–1929 and Yale, 1929–1931), I soon learned that experimental chemistry was not for me! In my Freshman Chemistry course, the professor excused me from attending lectures and set me to work synthesizing chloro-pentamine-cobalt chloride—I got beautiful purple crystals but my product was supposed to be green. In the final laboratory examination in Qualitative Analysis, I reported three ions where there were sixteen. Then in Organic Characterization, the only unknown that I successfully identified was glycerine—I knew that I was right because it tasted sweet! I admire other people who can do both experimental and theoretical research.

In contrast, I found that theory was my meat! During my freshman year, George Glockler, a young professor who had just come from Germany, introduced me to Arnold Sommerfeld's new book on atomic structure (2). I was fascinated and studied it for many years. I was given a key to the physics library stacks and became acquainted with both old and new treatises. Also, I was fortunate to take calculus from Dunham Jackson who explained the logic and thought processes that go into the development of mathematics. Then in my senior year at Yale, I devoted most of my time to studying Leigh Page's graduate course on the *Introduction to Theoretical Physics*, which provided a thorough foundation in dynamics, hydrodynamics, electricity and magnetism, optics, and the old Bohr quantum theory. Classes met six hours a week for the whole year. I solved every problem in Page's book (3) and got an A +. It was then that I decided to spend my whole life applying theoretical physics to chemical problems.

In the spring of 1931, I had to decide where I should go for my graduate training. It was easy to narrow the choice to either MIT or Princeton. The Chairman of the MIT Physics Department, John Slater, was so youthful looking that he frequently was mistaken for a Freshman! He explained his research on atomic structure in great detail and described the excellent Chemical Physics program that he supervised. Had it not been sleeting that day, and had MIT not looked so much like a factory, I probably would have enrolled there.

There were many reasons for my attraction to Princeton. First of all, it was possible to take a double PhD in theoretical physics and chemistry. This involved taking the same courses and passing the same preliminary examinations as were required of students majoring in Theoretical Physics, and then doing research on chemical applications. Thus, I had the opportunity to get thoroughly grounded in theoretical physics and mathematics before trying to apply the theory to chemical problems. Eugene Wigner was my major professor in the physics part (4, 5), while Henry Eyring and Hugh S. Taylor supervised my chemical research (6–12). After getting my PhD in 1936, I spent another year doing postdoctoral work with John Von Neumann at the Institute of Advanced Studies and finishing my research with Eyring and Taylor. This training was extremely valuable in all of my work.

Princeton in the Early 1930s

Princeton was a fabulous place in the 1930s (13). The Institute for Advanced Studies was just getting started and was housed with the Mathematics and Theoretical Physics Departments in Fine Hall. Many famous physicists, such as Einstein, Von Neumann, Wigner, and Ladenburg, came there to get away from Hitler. Furthermore, Dirac, Pauli, Schrödinger, and Slater spent a great deal of time there. For the first two years of my graduate work (1931–1933), my training was in straight theoretical physics. I remember distinctly Dirac's course in Quantum Mechanics, Johnny Von Neumann's Methods of Theoretical Physics, Gaylord Harnwell's Methods of Experimental Physics, Wigner's and Weyl's courses on Group Theory, and E. P. Adam's Classical Mechanics.

In addition, I was privileged to audit the only course in Relativity that Einstein gave in the United States. Einstein thought slowly and he talked slowly, making very sure that everybody in the room understood and agreed with him before he made his next remark. Contrary to public opinion, Einstein was not a great mathematician (he rated as equivalent to a second year graduate student in math). Instead, he always asked himself under what other conditions a particular equation could arise; then, he would study how other people had treated this equation in the other connections. In contrast to Einstein, Von Neumann was a very fast thinker and always very clear and precise. Von Neumann was the most brilliant scientist I have ever known, and he was brilliant in everything that he did. For example, his paper entitled, "Die Notwendigkeit des Bluffens im Poker" (in English, "the necessity of bluffing in poker") was the origin of his theory of games. Again, in contrast, Paul Dirac was very quiet and pensive—the only things that seemed to arouse Dirac's interest were soccer and chess. (One of Dirac's proudest accomplishments was inventing a modified game of chess played with all of the pawns plus a queen.)

The hardest work that I have ever done was to prepare for the preliminary examination that I had to take in competition with the double Nobel Prize winner, John Bardeen, and the former President of the (USA) National Academy of Sciences and the Rockefeller University, Fred Seitz. Parts of our *Molecular Theory of Gases and Liquids* (14) originated with the notes that I prepared at that time. The exam consisted in two days of written and three days of oral questioning—the proudest moment of my life was when I passed!

After the physics preliminary exams I switched, in accordance with the requirements for a double doctorate, to spend the rest of my graduate training in theoretical chemistry. Working with Henry Eyring was a lot of fun. He was full of ideas, thoroughly enjoyed his work, and we worked together almost every night until I was ready to fall asleep. In those days, Eyring got the inspiration for most of his research by assiduously studying the literature. Every morning he would pop some new idea for his students to criticize; it was the job of the graduate students to discover which ideas were sound and which were erroneous. Most of his ideas needed modification and reworking: this gave us a good lesson in how discoveries are made!

In retrospect, I am amazed at the variety of the research problems that I worked on during my Princeton days—I guess it was a combination of "blood and sweat," or ambition, hard work, and energy. Wigner and Von Neumann shared a Distinguished Professorship at Princeton when they were both 29 years old. At that time, Wigner had already published his famous book on Group Theory. When I asked him how he accomplished so much at such an early age, he said: "I was a bachelor and I was lonesome!" This ends my introduction!

THE POLARIZATION OF MOLECULAR HYDROGEN AND H_2^+

When I first started graduate work in the fall of 1931, Professor Edward Condon suggested that I calculate the parallel and perpendicular polarizabilities, $\alpha_\parallel$ and $\alpha_\perp$, of H_2 and H_2^+ (4). The question to be answered was whether $\alpha_\parallel$ is larger than $\alpha_\perp$. The only existing experimental data was for the depolarization of H_2, which only provides a measure of $(\alpha_\parallel - \alpha_\perp)^2$. According to the classical theory of Steensholt and intuitively, $\alpha_\parallel$ should be the larger. I varied two parameters in accordance with the variational method of Hylleraas and Hassé, as described in pages 942–47 of the *Molecular Theory of Gases and Liquids* (14). All of my calculations were made by hand, with the aid of a large table of logarithms. I found that $(\alpha_\parallel - \alpha_\perp)$ is positive for both H_2 and H_2^+, and my results agreed fairly well with the depolarization measurements. Naturally I was proud of my accomplishments, but my happiness ended abruptly.

Between the time that Condon had given me this problem and the time I had completed the calculations, he had received a letter from Herr Professor Mrowka in Germany who had made a calculation (15) showing that $\alpha_\perp$ is larger than $\alpha_\parallel$. It was clear that Professor Mrowka must be right because he was well established and had an excellent reputation, whereas I was merely a beginning graduate student solving his first research problem. Thus, Condon concluded that I must have made a mistake and he would not permit me to publish my manuscript until three years later when Mrowka wrote Condon explaining that the method that he had used was unsatisfactory.

My reason for giving you this detailed explanation is that many people are still using Mrowka's unsatisfactory procedure to estimate the polarizabilities of atoms and molecules! When a molecule is placed in an electric field of strength E, the energy E_j of its jth state is

$$E_j = E_j^{(0)} - \boldsymbol{\mu}_j^{(0)} \cdot \mathbf{E} - \tfrac{1}{2}\mathbf{E} \cdot \boldsymbol{\alpha} \cdot \mathbf{E} - \cdots \qquad 1.$$

Therefore the polarizability of the jth state in a particular direction of the field is given by the second order Rayleigh-Schrödinger perturbation expression $E_j^{(2)}$ [see Ref. (16)],

$$\alpha_j = \sum_k {}' \frac{2|\mu_{jk}|^2}{E_k^{(0)} - E_j^{(0)}} = -2E_j^{(2)} \qquad 2.$$

where μ_{jk} is the transition dipole moment in the direction of the field. Note that the summation in Eq. 2 is over all of the discrete states (for which the denominator is not zero) and it is also meant to be an integral over all of the continuum states. Mrowka had approximated the polarizability by considering only the transition of the hydrogen molecule from its $^1\Sigma$ ground state to its $^3\Sigma$ first excited state.

The Hylleraas variational principle (17), which I used, states that $E_j^{(2)} \leq \tilde{E}_j^{(2)}$ where

$$\tilde{E}_j^{(2)} = \langle \tilde{\Psi}_j^{(1)} | H^{(0)} - E_j^{(0)} | \tilde{\Psi}_j^{(1)} \rangle + \langle \tilde{\Psi}_j^{(0)} | V - E_j^{(1)} | \tilde{\Psi}_j^{(1)} \rangle$$
$$+ \langle \tilde{\Psi}_j^{(1)} | V - E_j^{(1)} | \tilde{\Psi}_j^{(0)} \rangle. \quad 3.$$

Here $\tilde{\Psi}_j^{(1)}$ is an approximation to the first order wave function.

It is instructive to compare the two methods of calculating the polarizability by considering the ground state of the hydrogen atom. The exact value of its polarizability is $\alpha = 4.5a_0^3$; using the one parameter variational approximation, $\alpha = 4.0a_0^3$; and considering only the $1s$-$2p$ transition (à la Mrowka), we obtain $\alpha = 2.96a_0^3$. Furthermore, if one considers progressively larger numbers of transitions, the value of the polarizability only very slowly approaches the correct value. Indeed, Inokuti (18) has shown that 18.6% of the polarizability of the ground state of the hydrogen atoms is contributed by the continuum states! For most other atoms and molecules the contribution of the continuum is an even larger percentage. A few years ago in a lecture at the University of Wisconsin, David Buckingham used this hydrogen example to persuade people to use the variational method. Somehow or other, the variational treatment includes the contributions of *all* of the states, discrete and continuous!

The other lesson to be learned from my polarizability problem is that when graduate students argue with professors, they are frequently right. Or, as G. N. Lewis said, "The *impertinent* questions that students ask often turn out to be *pertinent!*"

Condon was very brilliant, hard working, amiable, and blunt. Both his lectures and his publications were well organized, expressed clearly, and very easy to understand. In his courses, if he did not feel that he had prepared his lecture so that it would be letter perfect, he wrote on the blackboard, "Lecture postponed!" He attributed the clarity of his manuscripts to his undergraduate training at Berkeley, where he worked his way through college by writing a weekly column for the lovelorn in a local newspaper (using a female pseudonym). He maintained that no manuscript is ready for publication until it has been rewritten eight times (and this is true for my papers). Condon was a mystic—he truly believed that death is a transition from Pauli to Anti-Pauli matter! Condon certainly got a bum rap from Joe McCarthy—no one was ever more loyal or could keep a secret better than Ed. The reason McCarthy chastised him was that Mrs. Condon had organized a group of women who sat in the front row of the balcony and harassed the House Unamerican Activities Committee when they had their hearings.

SEPARATION OF THE ROTATIONAL COORDINATES FROM THE N-PARTICLE SCHRÖDINGER EQUATION

After I completed my hydrogen polarizability calculations in 1932, I started to work with Professor Eugene Wigner on the separation of the rotational coordinates from the wave function for an N-particle system in field-free space (5). By making use of group theoretic arguments, we succeeded in expressing the wave function Ψ_u^L for a system having a total angular momentum quantum number L and a Z-component of angular momentum in the form

$$\Psi_u^L = \sum_v D^L(R)_{uv}^* \chi_v^L. \qquad 4.$$

Here the $D^L(R)_{uv}$ are the coefficients of the Lth irreducible representation of the three dimensional rotation group and the R specifies the orientation of the three coordinate axes, which are determined by the relative position of the N particles in the molecular system. The χ_v^L are functions of the 3N–6 relative coordinates of the particle, whereas the Ψ_u^L are expressed in terms of the particle coordinates in the laboratory reference frame.

Thus, it should be easy to determine the set of $(2L+1)$ coupled partial differential equations by simply substituting Eq. 4 into the original Schrödinger equation. Actually, it took us more than a year to carry out the detailed algebraic manipulations. Thus I learned what Einstein meant when he told me, "Some things are easy, they are just hard to do!"

Wigner expressed the rotations relative to the principal axes of inertia of the system. Then he cleverly avoided the necessity of defining 3N–6 independent internal coordinates explicitly by making use of redundant coordinates. In other words, he considered the χ_v^L to be functions of the 3N cartesian coordinates y_{nk} of the particles in a reference frame parallel to the principal axes of inertia, provided that the y_{nk} are restricted to the quasi-hyperboloidal regions that satisfy the six conditions

$$\sum_{n=1}^{N} m_n y_{nj} = 0, j = 1, 2, 3 \quad \text{and} \qquad 5.$$

$$\sum_{n=1}^{N} m_n y_{nj} y_{nk} = 0, j \neq k. \qquad 6.$$

After we had completed this work, I wrote a manuscript explaining the details of what we had done. Wigner then rewrote my manuscript in a simple and elegant manner to correspond to what we would have done if only we had known the results when we first started! When Professor Carl Eckart visited Princeton, he was impressed with our work and congratulated us. I hope that our paper gave Eckart the inspiration for his famous *Separation Conditions*, which he published a short time later (19).

The lesson I learned from the separation of rotation problem was that the derivation of a theoretical formalism is only the first step in solving a problem—the application of the formalism may be very difficult. Indeed, up to now, no one has succeeded in deriving a completely satisfactory procedure for separating the rotational coordinates from the dynamics of a three particle (atom-diatomic molecule chemically reacting collision) system, although a lot of people have worked on it.

In 1950, I interested my colleagues "Chuck" Curtiss and Felix Adler in separating the rotational coordinates from the N-particle Schrödinger equation (20). We followed the viewpoint and procedures that were used in my work with Wigner (5). However, in place of redundant internal coordinates referred to the principal axes of inertia, we used (3N–6) *explicit* internal coordinates and chose the molecular-fixed reference axes so that particle "1" was on the positive z-axis while particle "2" was in the positive half of the x-z plane. Curtiss has written papers (21a–d) on the solution of the resulting $(2L + 1)$ coupled equations for the χ_v^L for a number of dynamic problems involving atom-diatomic and diatomic-diatomic collisions.

Then, in 1968, Russell Pack and I (22) separated two of the three rotational coordinates of N-electron diatomic molecules. This required completely new techniques. The direction of the internuclear axis was expressed relative to laboratory-fixed coordinates, but the electronic coordinates were oriented with respect to the internuclear axis. This permitted the electronic coordinates to be treated symmetrically and also simplified the determination of the (Born-Oppenheimer approximated) electronic wave functions for particular rotational states. The separation process was carried out directly by making use of the angular momentum raising and lowering operators without involving the group theoretic representations explicitly. The treatment was carried out for the following three choices of the origin of the electron coordinates:

1. SA—each electron is associated with a specific nucleus;
2. CMN—all electronic coordinates are relative to the center of mass of the nuclei;
3. GCN—where the geometric center of the nuclei is the origin for all electron coordinates.

This formalism permits the calculations of corrections to the Born-Oppenheimer approximation and accurate treatment of atom-atom scattering. Russell Pack was only a graduate student when he did this work, but it certainly showed his ingenuity and ability!

Potential users of our Wigner formulation have been frightened by the additional problems resulting from the redundant coordinates and the appearance of the difference of two moments of inertia in the denominator of our equations for χ_v^L. Actually, in chemically reacting collision dynamics, in which the internal coordinates that represent the reactants are not suitable to represent the products, the use of redundant coordinates seems like a desirable option. Furthermore, there is a big advantage in using the principal axes of inertia as the molecular reference frame, as any other coordinate axes wobble rapidly and erratically. Thus I believe that the time has come to apply our equations to chemically reacting collision problems!

The use of redundant variables is not strange to chemists. Indeed, redundant variables play an important role in the kinetic theory of gases. In 1931, Urey & Bradley (23) showed how to use *redundant* interparticle separations, r_{ij}, in molecular potential energy functions. In 1955, Wilson, Decius & Cross had an excellent discussion of *redundancy* in their famous treatise on *Molecular Vibrations* (24). A more detailed discussion of redundant coordinates in problems involving small molecular vibrations is given by Crawford & Overend (25). Whereas chemists have used these coordinates for algebraically simple dynamical problems, physicists use them routinely for even complicated problems such as Kracjik & Foldy's (26a,b) semi-relativistic dynamics of N particles in classical electromagnetic fields.

Recently, Russell Pack told me that the apparent singularity that appears in our equations when two moments of inertia, I_1 and I_2, become equal does not cause any difficulties [it is only necessary to identify uniquely the "1" and "2" axes before and after (I_2-I_1) changes sign]. Russ said that this "singularity" is similar to the $(...)/R^2$ centrifugal energy of a diatomic molecule, in which case the wave function either vanishes or else it varies as a sufficiently large power of R so that no singularity appears in the quantum mechanical equations when R approaches zero.

Wigner was born and raised in Budapest within a few blocks of the homes of Von Karman, Von Neumann, Teller, Szilard, and a dozen other famous Hungarian scientists. Wigner cannot recall their having the same outstanding teacher or any other factor that might explain this coincidence. When I started to work with him in 1932, he had just recently come to Princeton from the Kaiser Wilhelm Institute in Berlin. He was very stiff and formal. Everything he said was very precise and very carefully prepared in advance. Of course, I made the mistake that all young graduate students make—I was afraid of asking questions when I did not understand something that Wigner told me. As a result, I wasted a lot of time and made needless mistakes in my research. Wigner was always very modest and generally he would begin a discussion or lecture by saying, "This I do not understand!" Then he would proceed to explain what he did not understand and it soon became evident that he knew more about the subject than anyone else in the world. Wigner is still considered to be the most polite man in the world: For example, when he was angry at a garage man who was trying to cheat him, he said "Go to Hell, *please!*" Ordinarily, he would never swear in English, but occasionally when I was working with him he would say softly in Hungarian, "Egyen mega fenna!" (meaning, "go to Hell") or "Orderg!" (meaning, "the Devil"): He never supposed that I had learned a few things from a Hungarian nurse whom I had had when I was very young. Although he was very polite, Wigner would not agree with his brother-in-law, Paul Dirac, that when the two of them came to a door they should always pass through in alphabetical order.

In 1934 Wigner left Princeton to become Professor of Physics at the University of Wisconsin. He was delighted by the informal atmosphere and he says that it was in Madison that he became an *American*. The Physics Department at that time was a small, congenial group of faculty and graduate students who ate together, hiked together, and had a good time together. One of the graduate students whom he met was Amelia Frank— they liked each other, became engaged, and got married. However, their romance ended tragically when Amelia developed cancer and died six weeks after they were married. Eugene was heartbroken and returned to Princeton. Then in 1937, when his memories of Amelia were not so fresh, he came back to Wisconsin and greeted me in the fall when I came to Madison as a Wisconsin Alumni Research Fellow in Chemistry. Thus, I had a wonderful opportunity to do some research with Wigner on chemical kinetics—our objective was to assess the errors in the Eyring Theory of Absolute Reaction Rates (27, 28). Wigner would probably have remained at the University of Wisconsin if Gregory Breit had not explained to him that there really was not room for two Theoretical Physicists in such a small department. Eugene took the hint and returned permanently to Princeton.

CALCULATION OF POTENTIAL ENERGY SURFACES IN THE 1930s

After passing my preliminary examinations in theoretical physics, I started to work with Henry Eyring on the calculation of chemical reaction rates. Three steps are involved in the rigorous calculation of reaction rates (29). The first is the determination of the energy of interaction of the atoms as a function of their separations—this function is called the *potential energy surface*. Then, by integrating the equations of motion of the reactants, the reaction cross-sections are obtained as functions of the quantum numbers and relative velocities of the reactants. Finally, the reaction rate constants are determined in terms of the reaction cross-sections by considering a statistical mechanical ensemble of the quantum numbers and relative velocities of the reactants. However, in the 1930s, both the theoretical and the experimental values of the reaction rates were far from exact!

In order to calculate potential energy surfaces, Eyring used the spin-theoretic valence bond treatment explained in Eyring, Walter & Kimball's famous book (30), which is based upon the first-order perturbation vector model developed by Dirac, Van Vleck, London et al, and which neglects all integrals involving the interaction of more than two electrons. In addition, Eyring made the very interesting but unpalatable assumption that the coulombic energy of a diatomic molecule is a certain fraction, n, of its total binding energy and he assumed that $n = 0.14$ for *all* diatomic molecules and for all values of their internuclear separation. With these assumptions, it was easy to estimate the binding energy of all states in terms of the potential energy curves of diatomic molecules, and Eyring represented these diatomic potentials by Morse curves,

$$W_{AB} = D_{AB}[-2 \exp(-a_{AB}(R_{AB} - R_{AB}^0)) + \exp(-2a_{AB}(R_{AB} - R_{AB}^0))] \qquad 7.$$

where D_{AB}, a_{AB}, and R_{AB}^0 are constants which are usually determined from spectroscopic data.

I was mystified by the evidence that led Eyring to make the $n = 0.14$ assumption. Eyring explained that he had gotten this idea by studying a graph that Sugiura (31) had made in connection with his evaluation of the exchange integral that appears in the Heitler-London equation for the energy of H_2.

Figure 1 shows n as a function of the separation between the two hydrogen atoms. At separations of less than equilibrium, n is negative, whereas at larger separations, it increases abruptly until it reaches a maximum value of 0.14 at about 1.2 Å and remains almost constant for even larger separations. Eyring reasoned that only the behavior at large separations is important for chemically reacting collisions! [Actually, the universal use of $n = 0.14$ was questionable, because Rosen & Ikehara (32) in 1933 had made calculations similar to Sugiura's which showed that $n = 0.23, 0.32$, and 0.38 for Li_2, Na_2, and K_2, respectively.]

Thus, I was determined not to make semi-empirical calculations of potential energy surfaces! However, when I got to Wisconsin and did post-doctoral research for Prof. Farrington Daniels, he insisted that I calculate the activation energies for a very large number of chemical reactions. Since I

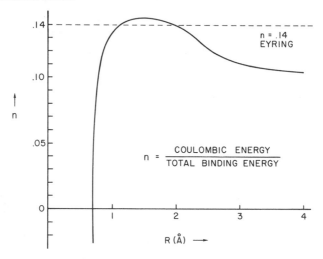

Figure 1 Sugiura's ratio of the coulombic energy to the total binding energy of a hydrogen molecule.

could not avoid the job, I tried to figure out the easiest and fastest way of doing it. So, with the help of Lee Henke, the chemistry department mechanician, I constructed two mechanical gadgets, which are described in my article in Ref. (33). With these devices I could calculate a point (corresponding to a particular set of interatomic separations) on a potential energy surface for either a three atom or a four atom bimolecular reaction in a matter of fifteen minutes. I soon discovered that there are quite a few points on a potential energy surface and this procedure involved a fantastic amount of effort. Thus, instead of using the mechanical contraptions, I made analytical determinations of both the activation energy and the principal curvatures at the activated state for arbitrary three and four atom bimolecular reactions and arbitrary values of n. I soon discovered that for an exothermal three atom reaction AB + C, it was a very good approximation to assume that $W_{AB} = 0$ at the activated state; then, from the definition of the activated state, it followed that at the activated state atoms A and B are at their equilibrium separation, $r_{AB} = r_{AB}^0$, and also

$$W_{BC} = -(D_{AB}/2)(1 - n[3/(1 - 2n)]^{1/2}). \qquad 8.$$

Thus, to this approximation, the activation energy is

$$E_{act} = (D_{AB}/2)[2 - 3n - (3(1 - 2n))^{1/2}]. \qquad 9.$$

Or, for $n = 0.14$, the activation energy is

$$E_{ACT} = 0.055 D_{AB}. \qquad 10.$$

Furthermore, it follows from a similar approximation that the activation energy for an exothermic reaction $AB + CD \rightarrow AC + BD$ is

$$E_{ACT} = 0.285(D_{AB} + D_{CD}).\qquad\qquad 11.$$

In 1941, there was supposed to be accurate experimental data for nine chemical reactions; the experimental activation energy for six of these nine reactions agreed satisfactorily with either Eq. 10 or 11, whereas in the other three cases the agreement was poor. Indeed Eqs. 10 and 11 appeared to be just as accurate as the activation energies obtained by calculations using the Eyring potential energy surfaces without any approximations. Thus, Eqs. 10 and 11 became widely known as Hirschfelder's Rule (although I had no confidence in their validity).

In retrospect, Eyring's semi-empiricism was very useful because it provided a guideline for research in a field in which neither the experiments nor the theory were reliable. I know of only two respects in which the semi-empirical treatment was qualitatively misleading. There exist a number of chemical reactions in which more than two bonds are broken in a single reaction step, whereas the semi-empirical formalism appeared to rule out this possibility. Also, it appeared from semi-empirical considerations that in order for a molecule to become *chemisorbed* on a surface, it must overcome an activation energy barrier—this is true for dirty surfaces (and until quite recently all of the research involved contaminated surfaces); however, for some very clean metal surfaces there is no activation energy required for chemisorbtion by either the theory or the observations.

Figure 2 shows what a hydrogen molecule looks like to a hydrogen atom.

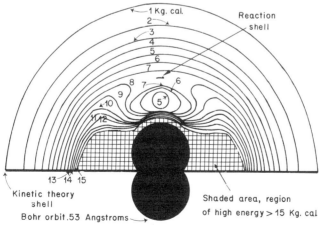

Figure 2   Potential energy contours corresponding to a hydrogen atom approaching a hydrogen molecule (holding the H_2 internuclear separation equal to R_{HH}^0). These contours (34) were calculated using the Eyring semiempirical procedure.

It is easiest for the atom to penetrate deeply into the molecule if it approaches along the internuclear axis. The point marked "Reaction Shell" is essentially the activated state—from the Eyring semi-empirical formulation, $R_{HH} = 1.15\ R_{HH}^0$ at the activated state.

Now, to return to my graduate research at Princeton, I told Henry Eyring that I wanted to make an a priori rigorous calculation of the potential energy surface for the chemical reaction of a hydrogen atom with a hydrogen molecule using 1s-hydrogen orbitals with a screening constant, which could be optimized by minimizing the energy for each nuclear configuration. Eyring was enthusiastic about this project but said that he was much too busy to play an active role in the calculations. It took me almost a year to evaluate the four three-center integrals that occur. My principal motivation was that Wigner said that I would not be able to do it!

I used ellipsoidal coordinates and carried out some complicated analytical manipulations, so that only two dimensions of numerical integrations remained, in order to evaluate each of two of these difficult three-center integrals, whereas each of the other two three-center integrals involved only one dimension of numerical integration. I was very grateful to Nathan Rosen, who was Einstein's Research Associate at that time and who subsequently became the director of a Works Progress Administration computational project at MIT, for teaching me how to organize my calculational procedures so as to reduce the effort required by a factor of ten! My calculations of the energy of H_3 and H_3^+ were as accurate as is possible to obtain by using 1s hydrogen atomic orbitals. Nevertheless, my best calculated activation energy was 13.6 kcal/mol as contrasted to the experimental value of 7 kcal/mol: I realized that it would be extraordinarily difficult to calculate accurate a priori values of chemical reaction rates.

Since the 1930s, there have been tremendous improvements in the calculation of molecular energies and wave functions thanks to high speed computing machines and improved mathematical techniques. Nevertheless, we cannot be completely happy with the state of the art!

Whereas the old timer, Hylleraas, spent years concentrating on the physical significance of each one of the five terms in his helium atom wave function and used coordinates involving interelectron separations, the new generation of theorists uses basis sets composed of 10,000 Gaussian orbitals. Instead of using their own judgment and experience, theorists in the new generation code their giant computing machines with a jumble of physically meaningless input and leave it up to the computer to use a variational principle to unravel the mess and produce a meaningful solution to the molecular problem. It is in this sense that we have regressed—we leave the thinking to the computer and do not put enough stress on the physical significance. Even the Slater orbitals, which were the work horses of the 1930s to 1950s, are considered to be too difficult to use with electronic computers, although one Slater orbital is equivalent to approximately seven Gaussians.

Currently, Herbert Jones and Charles Weatherford at Florida A&M University are developing an algebraic method (similar to the procedure I used for the three center H_3 molecule) to generate the exact formulae for three and four center Slater type orbital integrals that computers can manipulate to insure accurate evaluation over all ranges of parameters. I wish them success. These integrals would be very useful in determining very accurate intermolecular potentials for intermediate to large internuclear separations.

Ultimately, we hope that it will be feasible to use the Löwdin-Shull-Davidson natural orbitals in determining properties of polyatomic molecules.

THE THEORY OF ABSOLUTE REACTION RATES

In the fall of 1934 when I started to work with Henry Eyring, he had just completed the manuscript for his famous paper entitled, *The Activated Complex in Chemical Reactions* (35). In this paper, Eyring thanked Dr. Bryan Topley, who had worked with him during 1933–1934 as his Research Associate, "for valuable discussions, as it was with him that the present calculations of absolute rates were begun." I never met Bryan Topley, as he had returned to England to work with Professor Michael Polanyi. Eyring's paper was published in the February 1935 issue of the *Journal of Chemical Physics*. Shortly thereafter, a very similar paper on the role of the transition state in the theory of absolute reaction rates written by Evans & Polanyi appeared in the *Transactions of the Faraday Society* (36). Eyring got very angry—he surmised that Topley had told Polanyi about the research that Topley and Eyring had worked on together. Eyring accused Evans and Polanyi of stealing his discovery! Wigner sided with Polanyi, as he had great admiration for his honesty and ethics. Furthermore, Wigner could not understand what Eyring and Polanyi were arguing about since Pelzer & Wigner (37) and Wigner (38) had developed the theory of absolute reaction rates three years earlier. Fortunately this dispute did not last long, and in 1937 Wigner & Eyring wrote a paper together (39) on reaction rates. Indeed, it is not unusual for scientific discoveries to be made independently by many people when the state of the art has progressed to the point at which discovery is inevitable: For example, Heisenberg, Schrödinger, and Dirac almost simultaneously derived modern quantum mechanics (in such different representations that it took a long time before the similarities in their theories were recognized).

I was very fortunate to have the opportunity of working with both Eyring and Wigner. With Eyring, I applied the theory of absolute reaction rates to many different kinds of problems; and with Wigner, I studied the quantum mechanical corrections to Eyring's formulation! Let me tell you about some of our work on quantum corrections (27, 28). These papers demonstrate the penetrating logic and simple elegance of Wigner's analysis. Furthermore, the subject matter is just as pertinent now as it was in 1939!

The activated complex or transition state method of calculating the absolute rate of a chemical reaction with an activation energy would be rigorously valid if classical mechanics applied to all of the degrees of freedom. Note that the words "activated complex" and "transition state" are used interchangeably. According to Wigner (27), the idea for this sort of statistical mechanical formulation dates back to 1917, when Marcelin (40) proposed the equation for the rate, k, of reactions with an activation energy,

$$k = (P_t/P_i)(\bar{v}/\delta)\gamma. \qquad 12.$$

Considering that the reacting system is in thermal equilibrium, P_t is the probability of being in the transition state. Here the transition state is considered to be a strip of width δ in configuration space that lies across the deepest saddle on the energy mountain separating the two regions in configuration space that correspond, respectively, to the initial and final state of the reaction; P_i is the probability of the system being in the initial state; $\bar{v}$ is the average velocity with which the configuration points cross the saddle; $\delta/\bar{v}$ is their average time of sojourn in the transition state. Finally, γ expresses the probability that a system that crosses the saddle at complete thermal equilibrium actually originated in the initial state and will proceed to the final state to complete the chemical reaction. The transmission coefficient, γ, is the only quantity appearing in Eq. 12 that cannot be evaluated by well-known methods of statistical mechanics—it can be estimated only from the general shape of the potential mountain. Wigner's contribution (38) was in replacing the classical expressions for P_t and P_i by the corresponding quantum mechanical sums. Eyring (35) then developed formulae of great generality by regarding an activated complex as a molecule in which one of the vibrational degrees of freedom has been replaced by a translational degree of freedom.

Although Eq. 12 is a very general consequence of statistical mechanics, the value of γ and the ease of calculating k depend a great deal on the coordinates that are used to describe the dynamics of the system. In 1930, when Eyring & Polanyi (41) calculated the potential energy surface for reactive collisions of a hydrogen molecule with a hydrogen atom (with the three atoms in a line), Wigner suggested that they take as their coordinates the internuclear distances, $(x_2 - x_1)$ and $(x_3 - x_2)$; however, they should skew these coordinates by 30° so that the collision dynamics would be the same as for a marble rolling on their potential energy surface. Thus, in 1934 when I started to work with Eyring, he had a big laboratory filled with plaster of Paris models so that we could all roll marbles on his surfaces. I remember that one of these was about the size of a ping-pong table. The only difficulty with Eyring's models was that he had misunderstood Wigner in skewing the coordinates and ended up with the angle between $(x_2 - x_1)$ and $(x_3 - x_2)$ being 120° instead of 60°. I thought of the three atoms as billiard balls, and I knew that if the table were flat and the three balls were in a line with "3" and "2" stationary and "1" headed for "2," then when "1" hit "2," "1" would stop and "2" would move toward "3"; and finally when "2" hit "3," "2" would stop and "3" would move away. The kinematics of this collision is shown in Figure 3, in which the angle between the two coordinates is 60°. Such a collision could not be represented by the 120° angle! The surprising thing is that Eyring & Polanyi (41) published a potential energy surface with the 120° angle, and the models had been used for many years without anyone making a simple test to check that they were sensible!

There has been a great deal of research on trying to find the optimum coordinates to use both for the internal coordinates of individual molecules and for describing chemically reacting collisions. Dahler, Jepsen, and I generalized the (corrected) Eyring skewed coordinates to obtain the *mobile* or *Jacobian* coordinates (42a,b); and more recently, Johnson, Yang, and I generalized the mobile coordinates to obtain *GLD* (43a,b), which are the most general linear combinations of the cartesian coordinates that diagonalize the kinetic energy and have the center of mass as one of the coordinates. These can easily be converted into hyperspherical coordinates,

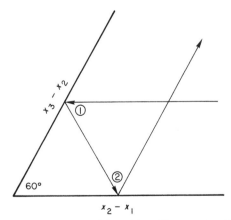

Figure 3   Billiard ball collision of three like atoms in a line.

which have symmetry properties that can be utilized by means of dynamical group theory (44). The principal difficulty in chemically reactive collisions is that a coordinate system that is suitable to describe the reactants is unsuited for the products.

There is one "booby trap" I should mention. The usual Eyring type of activated complex treatment of rate calculations does not apply to trimolecular reactions such as $H + H + H \rightarrow H_2 + H$. In such cases, the molecule in its activated state would have two (instead of one) of its vibrational degrees of freedom replaced by internal translational motion; furthermore, such reactions have no activation energy. It is interesting to note that in 1935, Eyring, Gershinowitz, and Sun (45) recognized that this problem required quite a different sort of treatment. Soon afterwards, Wigner (46) generalized their procedure; later, Huw Pritchard (at York University) refined Wigner's formulation. Also, quite recently Bernstein & Curtiss explained these atom recombination reactions in terms of the quenching of the energy of a metastable diatomic molecule by collisions with an atom. Some experimental data seem to favor the Pritchard mechanism, whereas for other reactants, the Bernstein scheme appears more likely.

In 1938, Wigner (47) showed how the shape of the potential energy surface can play an important role in determining the value of the transmission coefficient. More importantly from the standpoint of present molecular beam research, the shape of the potential surface can determine the vibrational distribution of the product molecules. A good example is the reaction, $H + Cl_2 \rightarrow HCl + Cl$, in which it is apparent from the energy surface [which George Kimball and Henry Eyring determined (48) in 1930] that after the system has passed through the activated state it must hit almost perpendicularly against the repulsive contours of the product channel. Thus, the product molecules are formed in high vibrational states. Truhlar & Dixon (49) have made a thorough study of such steric effects.

Wigner, and to a lesser extent myself, studied most of the quantum mechanical effects that are involved in the calculation of the transmission coefficient (27, 28, 38, 46, 47): the non-adiabatic crossings of energy surfaces, the tunneling through potential barriers, the effect of reaction path curvature, and the effect of changes of the frequencies in the vibrational modes perpendicular to the reaction path.

Recently Truhlar et al (50a,b) made a very thorough and critical study of the transition state method, including its quantum corrections and its applications. They concluded that their version of the absolute reaction rate theory is quite accurate. It is interesting to compare their 1980 treatment with the work in the 1930s. The only qualitative feature that we did not understand was that frequently the most populated path for the tunneling is a short-cut on the potential energy surface that does not go directly under the activated state saddle point. However, because we did not have high speed digital computers, our estimates of the quantum mechanical corrections were crude estimates of the first order effects. In contrast, Truhlar and his associates

1. used the uniform approximation and other more sophisticated procedures for calculating the tunneling through potential barriers (51) and (52);
2. used a correction factor for curvature of the reaction path (53), which is certainly more reliable than using our estimates (28) based on quantum mechanical wave packets expressed in ellipsoidal coordinates;
3. defined generalized transition states—which are narrow zones in configuration space that separate the reactants from the products, but do not necessarily pass through a saddle point—by a variational procedure called the *united statistical theory*, developed by Miller (54) on the basis of Keck's variational principle (55) that the activated state should be the dividing surface across which there is the least flow. (Actually this is not much different from Eyring's definition of the activated state as the region along the reaction path which has the maximum *free energy*.)

In the early days of quantum mechanics, there was a great deal of discussion as to whether the reaction rate in the forward direction is affected if the backwards reaction is taking place at the same time. However, nowadays, Miller (54), Truhlar (56), Anderson (57), and others believe that reactions in the backward direction have *no* effect on the reactions in the forward direction. I like to think of reaction rates in terms of the flow of ensembles in phase space (58)—Hulburt and I made a hydrodynamic analogy with gases in a bottle flowing through a nozzle. If the ratio of the pressure in the bottle to the exit pressure is greater than a critical value, the gas expands adiabatically until it reaches the nozzle (or constriction) where its velocity is the velocity of sound; past the constriction, the velocity is supersonic; thus, no perturbation downstream has any effect on the gas at the constriction. We showed in our paper that the analog of the critical pressure ratio is the condition that the activation energy should exceed kT and the analog of the velocity of sound is the Maxwell mean velocity. Thus, the transition state theory should only apply to reactions having activation energies larger than kT—this is almost certainly true! (Of course, by activation energy, I mean free energy.) Unfortunately, I have never been able to make a rigorous derivation of my analogy.

Although great changes have been made in the theory of reaction rates and it is now possible to apply the theory to many kinds of practical problems, the basic ideas have not changed very much!

In thinking about reaction rates, I am reminded of a question that Norbert Wiener, the famous mathematician, once asked me:

> Thermodynamics determines the *distribution in space* of a system which is in *equilibrium in time*—reaction rate theory explains the *distribution in time* of a system which is *homogeneous in space*. Thus, a relativistic four-dimensional *covariant thermodynamics* should determine the *rate of chemical reactions* since space and time are symmetrical. Why has no one done this?

Clearly, this is a special relativistic problem and hence it should not be difficult! However, in quantum mechanics we come very close to Wiener's conjecture when we express the energy in complex variables by

$$\varepsilon_j = E_j - i\Gamma_j \qquad\qquad 13.$$

where E_j is the real part of the energy and Γ_j determines the *lifetime* of the state. When ε_j is thermally averaged over the states, it resembles the relativistic internal energy and its imaginary part determines a reaction rate.

Stiff Equations and Why the Eyring Absolute Rate Theory Works!

The Eyring Theory of Absolute Rates works because the Principle of Equipartition of Energy does *not* apply to quantum mechanical systems! In Eyring's treatment, the motion along the reaction path is considered to be classical, whereas the vibrational degrees of freedom are supposed to be quantized. At the activated state, the vibrational zero point energies are large compared to kT, which is the mean translational energy along the reaction path. Thus, the vibrational relaxation times are very small compared to the translational relaxation time, $\tau_{vib} \ll \tau_{trans}$. This difference between the orders of magnitude of the vibrational and translational relaxational times makes it possible to separate the vibrational degrees of freedom from the translational.

If, on the other hand, the vibrational motion were classical, then the Equipartition Principle would imply that the energy in each of the vibrational modes would be comparable to kT—and the separation of the vibrational motions from the translational would not be feasible. It is interesting to note that Willard Gibbs tried to use the Principle of Equipartition in his statistical mechanical calculation of the specific heat of N_2. He was puzzled to find that his value of C_v did not agree with experimental results— as a result, he never truly believed statistical mechanics!

The Theory of Absolute Rates is just one of many examples of how simple theories result from suppressing (or separating off) those dynamical modes that have relaxation times inappropriate for the phenomenon under consideration. Some other examples are the following:

1. The Born-Oppenheimer separation.
2. The pseudo-stationary approximation in chemical kinetics (59–62).
3. Heat conductivity in chemically reacting mixtures or polyatomic gases (63–65), which resulted in my generalizing the Eucken correction factor.
4. Bogoliubov's (66) explanation of the Hilbert Paradox, which is the fact that the Navier-Stokes equation of motion of liquids only involves the first three moments of the one-particle distribution function rather than requiring knowledge of the complete N-particle distribution function. The explanation consists in showing that in times of the order of 10^{-12} sec (the duration of a collision), the N-particle distribution function becomes a functional of the one-particle distribution function; and in times of the order of 10^{-9} sec (the time between collisions), the one-particle distribution function becomes a functional of its collisional invariants, which are its first three moments.

The mathematics involved in the separation of variables is closely related to *boundary layer theory*. The theory has developed in three stages: It started with *stiff equations*, which became known as *singular perturbations*, and more recently the singular perturbations have branched out into *bifurcations!*

Stiff or singular perturbation equations have solutions that are either overly stable or unstable, depending upon the direction in which the integration proceeds. Generally they are differential equations in which the coefficient of the highest order differential is very, very small (67–69). When a stiff equation is integrated in its stable direction, regardless of the initial conditions, the solution merges with the *principal solution* after a short interval. The principal solution results from solving a differential equation of lower order, and lacks the constant of integration (or the functions of integration for partial differential equations) that enables particular solutions to satisfy arbitrary boundary conditions. In the language of singular perturbations, the principal solution is called the *inner solution*, whereas the particular solutions or transients are called the *outer solutions*. Because the principal solution is an envelope of the particular solutions, it is almost but not quite a solution of the original equation.

Soon after World War II, Chuck Curtiss and I discovered stiff equations accidentally when we were having difficulties in integrating the equations for the propagation of flames (67). We noticed that the scientists and engineers at the Johns Hopkins Applied Physics Laboratory, who were using analog computers to solve servo-mechanism problems, connected

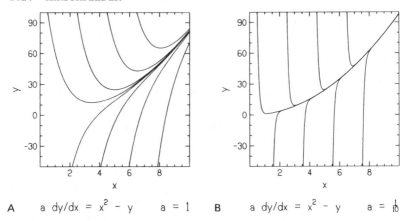

A $a\ dy/dx = x^2 - y$ $a = 1$ B $a\ dy/dx = x^2 - y$ $a = \frac{1}{10}$

Figure 4a   When the coefficient a has the "large" value $a = 1$, the differential equation is "loose" and the solutions are slowly varying.

Figure 4b   When the coefficient a has the "small" value $a = 1/10$, the differential equation is "stiff." Note how rapidly the "transients" approach the "principal solution," which itself is the envelope of the "transients."

their circuits in a strange manner when the coupling between the driver and the slave was stiff. We were curious and studied the mathematics of what they were doing. Figure 4 gives an example showing what happens to the solutions of a differential equation when the coefficient "a" of its highest order is made very small. In Figure 4b, where $a = 1/10$, the distinction between the principal solution and the transients is clear.

The principal solution is often called the *separatrix* or *bifurcatrix* because it separates those transients with very large positive slopes from those with very large negative slopes. In carrying out numerical integrations, the computer has difficulty in deciding whether to follow the transients or the principal solution—away from the boundaries, it is usually the *inner* solution that is desired. Algorithms that use forward interpolation are most successful in remaining attached to the principal solution, and the larger the interval between the calculated points, the more successful it is! Unless one is careful, his calculations can result in hash! The most difficult problems are the simultaneous differential equations (such as those that occur in electronic circuits) in which the solutions may be stiff with respect to some variables in one region and change to being loose in another region. However, new techniques are being developed to cope with all sorts of bifurcation problems.

CONCLUSION

I started to write this chapter with the intention of saying a little bit about each of my adventures and ended up by telling you a lot about a few of my carliest research experiences. Therefore, sometime in the future, I may try to write about intermolecular forces, kinetic theory of gases, perturbation theory, and my attempts to do research on practical problems. Finally, I hope that eventually I will learn enough about laser-molecule interactions to make an interesting story. In any case, I shall try to follow Joel Hildebrand's example as long as I have my marbles!

ACKNOWLEDGMENT

I want to thank Bruce K. Holmer for drawing Figure 4 and my wife for criticizing and proofreading this manuscript. And of course, I am grateful to the National Science Foundation for its financial assistance.

APPENDIX : COMPILATION OF MY RESEARCH ADVENTURES

(Research performed with or suggested by names in parenthesis)

1. *Princeton, 1931–1937*
   (PhD in Physics and Chemistry, 1936; Fellow Inst. Adv. Studies, 1936–1937)
   Parallel and perpendicular polarizability of H_2 and H_2^+ (Condon)
   Separation of rotational coordinates from N-particle Schrödinger eq. (Wigner)
   A priori calculation of energy of H_3, H_3^+, and H_3^- (Eyring & Rosen)
   A priori calculation of rate of $H + H_2$ reaction (Eyring)
   Rates of reactions produced by ionization (Taylor & Eyring)
   Free-volume theory of liquids (Eyring)
   Virial theorem and the scaling of wave functions

2. *University of Wisconsin, Pre-War, 1937–1942*
   (Wis. Alumni Res. Fellow, 1937–1942; Asst. Prof. Chemistry and Physics, 1940–1942)
   Quantum mechanical effects in reaction rates or collisions (Wigner & Hulburt)
   Lennard-Jones $(aR^{-n} - bR^{-m})$ interatomic energies (Lennard-Jones)
   Determination of interatomic energies from Joule-Thomson coefficients (Roebuck)
   Second virial coefficients of simple and complex molecules (McClure)
   Second virial coefficients of organic molecules determined by their liquid vapor pressure (Curtiss)
   The effusion of gases through a pinhole, experimental work interrupted by war (Curtiss)

3. *National Defense Res. Com. (NDRC), Washington, DC ; First Year World War II, 1942–1943*
   (High Velocity Gun Sec., Geophysical Lab and Rocket Secs., NDRC)
   Thermodynamics of powder gases (Curtiss, McClure & Osborne; suggested by Gibson)
   Interior ballistics of guns, rockets, and recoilless guns (Curtiss & Kershner)

4. *Los Alamos and the Atom Bomb, 1943–1946*
   (Group Leader, Ordnance Div. under Adm. Parsons; Group Leader Theor. Div. under Bethe)
   Interior ballistics of the gun bomb, "Little Boy" (Tolman & Parsons)
   Specification of propellant for "Little Boy" (Parsons)
   The dynamics of the early stages of the explosions (Bethe)

The formation of the ball of fire and the shock wave (Bethe & Magee)

The opacity of the shock wave (Magee)

The rise of the ball of fire and the winds it produces (Taylor & Magee)

Prediction of the fall-out (Magee)

Range-energy relations for penetration of high energy protons (Magee; suggested by Bethe)

Multiple Klein-Nishina scattering and penetration of gamma radiation through thick layers (Magee & Mac. Hull; suggested by Von Neumann)

After Nagasaki explosion, worked on setting up Theoretical Physics Group at the Naval Ordnance Test Station at Inyokern in the Mojave Desert (Magee; under direction L. T. E. Thompson)

Summer, 1946, as "Chief Phenomenologist," predicting all of the effects of the atom bomb explosions at Bikini (Magee)

During 1946–1947, editing the AEC treatise, *The Effects of Nuclear Weapons* (as requested by a committee headed by Von Neumann and Teller; assisted by 237 atomic energy experts!)

5. *University of Wisconsin and our Naval Research Laborabory, 1946–1962*
(Prof. of Chemistry and Director of U. W. Naval Res. Lab.)

The theory of flames and detonations (Curtiss)

Stiff or singular equations (Curtiss)

Project "Bumblebee" and the interior ballistics of ram jets (theory, Curtiss; experiment, Olsen & Boyd)

Chemical effects of ultrasonics (theory, Curtiss; experiment, Boyd & Olsen)

Intermolecular forces

Collision cross-sections for atomic collisions (Bird & Spotz)

Kinetic theory of gases (Curtiss)

Numerical calculations of transport properties of gases (Bird, Spotz & Curtiss)

Metastable states of diatomic molecules (Curtiss & Bird)

Van der Waals molecules (Stogryn)

6. *Consulting for Humble Oil Company, 1946–1962*
(Consultant for refining, production, and exploration)

Joe Franklin's concept of industrial research

The other consultants: Von Neumann, Wigner, Debye, G. I. Taylor et al

The Humble Lecture Series

7. *University of Wisconsin Theoretical Chemistry Institute (TCI) and NASA, 1962–1975*
(Homer Adkins Professor of Chemistry and Director of the TCI)

The establishment of TCI as a NASA facility (Townes & Holloway)

The virial theorem and its generalizations

The hypervirial theorem (Epstein & Coulson)

Application of the hypervirial theorem to scattering problems (Robinson & McElroy)

Integration of the first order perturbation equation (Byers Brown)

Iteration-Variational (FOPIM) procedures for quantum mechanical perturbations

Upper and lower bounds and variational methods for Rayleigh-Schrödinger and Brillouin perturbation energies (Sando & Meath)

Long range interaction of two hydrogen atoms expressed in natural orbitals (Löwdin)

Determination of forces between complicated molecules

Relativistic intermolecular forces (Meath & Power)

New type of treatment for exchange perturbations (Silbey)

Explicit determination of the long range $1/R^6$ energy of interaction of two hydrogen atoms by the electrostatic Hellman-Feynman or the Kohn-Hohenberg density functional theorem (Eliason)

Hyperfine splitting of the long range interaction energy of $H + H$, $H + D$, or $D + D$ atoms (Harriman, Larry Curtiss, Milleur & Twerdochlib)

Separation of rotational coordinates from N-electron diatomic molecules (Pack)

Formal Rayleigh-Schrödinger perturbation theory for degenerate states (Certain)

Adiabatic corrections to the Born-Oppenheimer interatomic potentials (Pack)

New partitioning perturbation theory for almost degenerate or electron exchange states (Certain)

Spontaneous ionization of a hydrogen atom in an electric field (Larry Curtiss)

Scattering of a molecular beam from a square two dimensional barrier (Christoph)

General theory of quantized vortices around wave function nodes (Goebel & Bruch)

Quantum mechanical streamlines formed in idealized atom-diatomic molecule reactive collision (Tang)

Quantum mechanical streamlines formed in collision of two spheres which have square well potential wells or barriers (Tang)

Kinetics of nucleation of many component systems

8. *Univ. Wisconsin TCI and Winters at Univ. California, Santa Barbara, 1976–Now* (Became Emeritus Prof. at Univ. Wis. in 1981, but remain Adjunct Prof. at UCSB)

Floquet Modes for sinusoidal perturbation of two state systems (Dion)

Hydrodynamic separation of optical isomers; suggested by Sanibel sea shells (Lightfoot & Howard)

Angular momentum, creation, and significance of quantized vortices

Equivalence of Wigner's delay time and the virial theorem for central field scattering

Virial theorem for inelastic molecular collisions (Pack)

Classical dynamic semirelativistic interaction of molecules with electromagnetic fields (K.-H. Yang)

Generalization of classical Poisson brackets to include spins (K.-H. Yang)

Quantum mechanical semirelativistic dynamics of molecules with moving nuclei interacting with classical electromagnetic fields (K.-H. Yang and Bruce R. Johnson)

Quantum mechanical nonrelativistic dynamics of molecules with moving nuclei interacting with constant homogeneous electric and magnetic fields (Johnson & Yang)

Intermolecular magnetic or current-current force and energy operators either in the presence or absence of external electromagnetic forces (Yang & Johnson)

Literature Cited

1. McGehee Harvey, A. 1978. Arthur D. Hirschfelder—Johns Hopkin's First Full-Time Cardiologist. *Johns Hopkins Med. J.* 143:129–39
2. Sommerfeld, A. 1922. *Atombau und Spektrallinien.* Braunschweig: Friedrich Vieweg, 3rd ed.
3. Page, L. 1935. *Introduction to Theoretical Physics.* New York: Van Nostrand
4. Hirschfelder, J. O. 1935. Polarizability and related properties of molecular hydrogen and H_2^+. *J. Chem. Phys.* 3:555
5. Hirschfelder, J. O., Wigner, E. 1935. Separation of rotational coordinates from the N-particle Schrödinger equation. *Proc. Natl. Acad. Sci. USA* 21:113
6a. Hirschfelder, J. O., Eyring, H., Rosen, N. 1936. Calculation of the energy of H_3 and H_3^+. *J. Chem. Phys.* 4:121, 130
6b. Hirschfelder, J. O., Diamond, H., Eyring, H. 1937. *J. Chem. Phys.* 5:697
7. Hirschfelder, J. O., Eyring, H., Topley, B. 1936. Reactions involving hydrogen molecules and ions. *J. Chem. Phys.* 4:170
8. Eyring, H., Hirschfelder, J. O., Taylor, H. S. 1936. Chemical reactions produced by ionization. *J. Chem. Phys.* 4:479
9. Stevenson, D., Hirschfelder, J. O. 1937. Structure of H_3, H_3^+, and H_3^-. *J. Chem. Phys.* 5:933
10. Hirschfelder, J. O., Kincaid, J. F. 1937. Virial theorem for approximate molecular and metallic eigenfunctions. *Phys. Rev.* 52:658
11. Hirschfelder, J. O., Taylor, H. S. 1938. Alpha particle reactions in CO, O_2, and CO_2 systems. *J. Chem. Phys.* 6:783
12. Hirschfelder, J. O. 1938. The energy of nonlinear H_3 and H_3^+ (and the required integrals). *J. Chem. Phys.* 6:795, 806
13. Hirschfelder, J. O. 1982. My fifty years of theoretical chemistry. I. Chemical kinetics. *Ber. Bunsenges. Phys. Chem.* 86:349
14. Hirschfelder, J. O., Curtiss, C. F., Bird, R. B. 1954. *Molecular Theory of Gases and Liquids.* New York: Wiley
15. Mrowka, B. 1932. *Zeit. Phys.* 76:300
16. Hirschfelder, J. O., Byers Brown, W., Epstein, S. T. 1964. *Adv. Quant. Chem.* 1:255
17. Hirschfelder, J. O. 1966. In *Perturbation Theory and its Applications in Quantum Mechanics,* ed. C. H. Wilcox, pp. 3–33. New York: Wiley
18. Inokuti, M. 1964. *Argonne Natl. Lab. Rep. ANL-6769*
19. Eckart, C. 1935. *Phys. Rev.* 47:552
20. Curtiss, C. F., Hirschfelder, J. O., Adler, F. T. 1950. *J. Chem. Phys.* 18:1638

21a. Curtiss, C. F., Adler, F. T. 1952. *J. Chem. Phys.* 20:249
21b. Curtiss, C. F. 1953. *J. Chem. Phys.* 21:1199
21c. Bodi, L. J., Curtiss, C. F. 1956. *J. Chem. Phys.* 25:1117
21d. Kouri, D. J., Curtiss, C. F. 1966. *J. Chem. Phys.* 44:2120
22. Pack, R. T., Hirschfelder, J. O. 1968. *J. Chem. Phys.* 49:4009
23. Urey, H. C., Bradley, C. A. 1931. *Phys. Rev.* 38:1969
24. Wilson, E. B. Jr., Decius, J. C., Cross, P. C. 1955. *Molecular Vibrations.* New York: McGraw-Hill
25. Crawford, B. Jr., Overend, J. 1964. *J. Mol. Spectrosc.* 12:307
26a. Kracjik, R. A., Foldy, L. L. 1974. *Phys. Rev. D* 10:1777
26b. Kracjik, R. A., Foldy, L. L. 1975. *Phys. Rev. D* 12:1700
27. Hirschfelder, J. O., Wigner, E. 1939. Quantum mechanical considerations in the theory of reactions involving an activation energy. *J. Chem. Phys.* 7:616
28. Hulburt, H. M., Hirschfelder, J. O. 1943. Transmission coefficient in the theory of absolute reaction rates. *J. Chem. Phys.* 11:276
29. Eliason, M. A., Hirschfelder, J. O. 1959. General collision treatment for the rate of bi-molecular reactions. *J. Chem. Phys.* 30:1426
30. Eyring, H., Walter, J., Kimball, G. E. 1944. *Quantum Mechanics.* New York: Wiley
31. Sugiura, Y. 1927. *Zeit Phys.* 45:484
32. Rosen, N., Ikehara, S. 1933. *Phys. Rev.* 43:5
33. Hirschfelder, J. O. 1941. *J. Chem. Phys.* 9:645
34. Hirschfelder, J. O. 1966. *J. Chem. Ed.* 43:457
35. Eyring, H. 1935. The activated complex in chemical reactions. *J. Chem. Phys.* 3:107
36. Evans, M. G., Polanyi, M. 1935. *Trans. Faraday Soc.* 31:876
37. Pelzer, H., Wigner, E. 1932. *Zeit. Phys. Chem. B* 15:445
38. Wigner, E. 1932. *Zeit Phys. Chem. B* 19:203
39. Wigner, E., Eyring, H. 1937. *Sci. Mon.* 44:564
40. Marcelin, A. 1915. *Ann. Chim. Phys.* 3(9):120, 185
41. Eyring, H., Polanyi, M. 1931. *Zeit. Phys. Chem. B* 12:279
42a. Hirschfelder, J. O., Dahler, J. S. 1956. *Proc. Natl. Acad. Sci. USA* 42:363
42b. Jepsen, D. W., Hirschfelder, J. O. 1959.

Proc. Natl. Acad. Sci. USA 45:249; J. Chem. Phys. 30:1032

43a. Yang, K.-H., Hirschfelder, J. O., Johnson, B. R. 1981. J. Chem. Phys. 75:2321

43b. Johnson, B. R., Hirschfelder, J. O., Yang, K.-H. 1983. Rev. Mod. Phys. 55:109

44. Kellman, M. E., Herrick, D. R. 1980. Phys. Rev. A 22:1536; 1982. Phys. Today 35:21

45. Eyring, H., Gershinowitz, H., Sun, C. E. 1935. J. Chem. Phys. 3:786

46. Wigner, E. 1937. J. Chem. Phys. 5:720

47. Wigner, E. 1938. Trans. Faraday Soc. 34:29

48. Kimball, G. E., Eyring, H. 1932. J. Am. Chem. Soc. 54:3876

49. Truhlar, D. G., Dixon, D. A. 1979. In Atom-Molecule Collision Theory, ed. R. B. Bernstein, p. 595 et seq. (see especially pp. 628 ff.). New York: Plenum

50a. Truhlar, D. G., Isaacson, A. D., Skodje, R. T., Garrett, B. C. 1982. J. Phys. Chem. 86:2252

50b. Garrett, B. C., Truhlar, D. G. 1979. J. Phys. Chem. 83:1052; 84:682, 1749

51. Garrett, B. C., Truhlar, D. G. 1979. J. Phys. Chem. 83:2921

52. Skodje, R. T., Truhlar, D. G. 1981. J. Phys. Chem. 85:624

53. Skodje, R. T., Truhlar, D. G., Garrett, B. G. 1981. J. Phys. Chem. 85:3019

54. Miller, W. H. 1976. J. Chem. Phys. 65:2216

55. Keck, J. C. 1967. Adv. Chem. Phys. 13:85

56. Garrett, B. C., Truhlar, D. G. 1982. J. Chem. Phys. 76:1853

57. Anderson, J. B. 1973. J. Chem. Phys. 58:4684

58. Hulburt, H. M., Hirschfelder, J. O. 1949.

J. Chem. Phys. 17:964

59. Hirschfelder, J. O. 1957. Pseudo-stationary state approximation in chemical kinetics. J. Chem. Phys. 26:271

60. Richardson, W., Volk, L., Lau, K. H., Lin, S. H., Eyring, H. 1973. Application of the singular perturbation method to reaction kinetics. Proc. Natl. Acad. Sci. 70:1588

61. Edson, D. 1981. Computer simulation in chemical kinetics. Science 214:981

62. Bowen, J. R., Acrivos, A., Oppenheim, A. K. 1963. Singular perturbation refinement to quasi-steady state approximation in chemical kinetics. Chem. Eng. Sci. 18:177

63a. Hirschfelder, J. O. 1957. Heat transfer in chemically reacting mixtures. J. Chem. Phys. 26:274

63b. See also Secrest, D., Hirschfelder, J. O. 1961. Slowly reacting gas mixture in heat conductivity cell. Phys. Fluids 4:61

64. Hirschfelder, J. O. 1957. Heat conductivity in polyatomic or electronically excited gases. J. Chem. Phys. 26:282

65. Baker, C. E., Brokaw, R. S. 1964. Thermal conductivity of gaseous H_2O–D_2O mixture. J. Chem. Phys. 40:1523; 42:812

66. Bogoliubov, N. N. 1946. J. Phys. USSR 10:265; trans. in Studies in Statistical Mechanics, eds. J. De Boer, G. E. Uhlenbeck. New York: Interscience

67. Curtiss, C. F., Hirschfelder, J. O. 1952. Integration of stiff equations. Proc. Natl. Acad. Sci. USA 38:235

68. Nayfeh, A. H. 1973. Perturbation Methods. New York: Wiley

69. Nayfeh, A. H., Mook, D. T. 1979. Nonlinear oscillations. New York: Wiley

Bruno H. Zimm

Walter H. Stockmayer

Ann. Rev. Phys. Chem. 1984. 35 : 1–21

WHEN POLYMER SCIENCE LOOKED EASY

Walter H. Stockmayer

Department of Chemistry, Dartmouth College, Hanover, New Hampshire 03755

Bruno H. Zimm

Department of Chemistry, University of California at San Diego, La Jolla, California 92093

INTRODUCTION

By the eve of America's entry into World War II, both industrial and academic activity in polymer science had grown to a stage permitting the identification of a number of basic problems in macromolecular physical chemistry and their solution without radical advances in existing experimental or theoretical techniques. Both of us were at Columbia University at that time, loyal to quite other branches of physical chemistry, when to each the "call" came in a different way. In this memoir we hope to convey some notion of the prevailing atmosphere, and to recall and transmit as best we can the way certain subfields looked and how they developed. This is a highly subjective business that lays us open to criticism by historians of science, one of whom in a recent review (1) remarked with some justice, "For scientists, history is not the field upon which they wrestle for truth, but principally their field of celebration and self-congratulation." And again, "Though recollection may add vividness and color, it cannot reliably be used except as embellishment of a picture delineated by written sources from the period." So, even with our bibliography, *caveat lector.*

To set the historical stage slightly, we remind readers that the struggle of Hermann Staudinger and Wallace H. Carothers to establish the macromolecular hypothesis was mainly waged and won in the 1920s and early 1930s (2). The relevant areas of physical chemistry, however, actually go back further than this, since the ultimate constitution and nature of the binding

1633

0066–426X/84/1101–0001$02.00

forces within the particles, be they association colloids or true single macromolecules, are not primary considerations. Thus, for example, Einstein's calculation of the viscosity of a suspension of spheres in a hydrodynamic continuum appeared in 1906 (with a correction five years later), and Svedberg developed the ultracentrifuge in the early 1920s. But in the 1930s the rate of progress increased markedly. Among the most provocative developments were Staudinger's announcement of his (now long abandoned) viscosity-molecular weight rule (3) and the first paper on the molecular theory of rubber elasticity by Eugene Guth and Herman Mark (4). It was also in this decade that Werner Kuhn, Paul J. Flory, G. V. Schulz, and Maurice L. Huggins first appeared on the polymeric scene. Kuhn, a versatile theoretician, was already well known for his work on optical activity and a classical spectral sum rule. His is the classic paper on the conformational statistics of flexible chains (5), in which he foresaw the excluded volume effect; he also considered such distinctly chemical problems as the statistics of random hydrolytic degradation of cellulose (6). Flory at the start of his still highly active career was also much concerned with polymerization reactions, producing seminal works on the statistics and kinetics of polycondensation (7) and free-radical chain polymerization (8). In the latter area we also find the pioneer kinetic investigations of Schulz (9); while Huggins (10) established definitively that the viscosities of polymer solutions showed them to be not the rigid rods envisaged by Staudinger but the random flexible coils of Kuhn, Guth, and Mark.

The Columbia chemistry department in the early 1940s was an exciting place for physical chemists. The chairman was Harold C. Urey, and the staff numbered such luminaries as Louis P. Hammett, George E. Kimball, Victor K. LaMer, and Joseph E. Mayer. The last-named was then at the height of his fame and prowess as a creator of the modern statistical mechanics of fluids (11, 12) and coauthor with his wife of a pioneering text (13). Each of us had gone to Joe Mayer for primary inspiration: W. H. S., one year after taking his degree at the Massachusetts Institute of Technology, had accepted an instructorship in the Columbia Extension Division (requiring evening lecturing in general, analytical and physical chemistry) mainly to be near him; and B. H. Z., after graduating from Columbia College in 1941, chose him as mentor of his doctoral research. Maria Goeppert Mayer was then professor of physics at Sarah Lawrence College, but she managed to spend several days a week at Columbia. She was working on a variety of problems, and had recently published a Thomas-Fermi calculation predicting (14) the start of the actinide series at element 93, but this was some years before her elucidation of the "magic numbers" for nuclear stability which earned her a share of the 1963 Nobel prize in physics. Also, the war already raging in Europe had diverted to the

Columbia neighborhood a number of refugees, of whom Francis Perrin and Ronald W. Gurney frequently dropped by the chemistry department.

Immediately after Pearl Harbor in December, 1941, Urey called the entire Department together to stress that the war would be fought with applied science and that quixotic enlistment in the armed forces would be neither helpful nor wise. In the months that followed, some of the staff and many of the students went off to other locations to take part in war-related activities. By the summer of 1942, Joe Mayer was spending most of each week at Aberdeen Proving Ground, but would return for one or two days of lecturing and counseling. The famed Monday Lunch Club, which he and George Kimball ran with their research groups, now generally met on Tuesdays, and gave us a chance to meet informally many of the now frequent great visitors, including Enrico Fermi, Eugene P. Wigner, Edward Teller, and John G. Kirkwood. Those of us who stayed on at Columbia were of course also drawn into war-related work : B. H. Z. into a project that is described in the later section on Molecular Weights and Light Scattering, and W. H. S. into calculations with George Kimball on deuterium exchange equilibria, commandeered by Urey. The remaining graduate students were permitted to keep their doctoral programs barely alive, and time was somehow stolen to think now and then about other types of research, play a little chamber music or softball, or even to hike in New Jersey or the Catskills.

Although it was Joe Mayer who drew us to the eighth floor of Chandler laboratory, where he and Kimball had their offices and labs, it was another member of that colony, Charles O. Beckmann, who must take most of the blame for our conversion into (not quite full-time) polymer physical chemists. The ways in which this occurred is recounted below. Charlie Beckmann deserved more recognition than he received during his lifetime (1904–1968) for his researches on optical activity and on polymers, particularly starch derivatives. But, like many of us, he didn't enjoy writing papers and preferred to think about new problems. He and his students worked with a home-built ultracentrifuge of which he was justly proud. His informal manner laid him open to tricks from his juniors. One evening when Beckmann addressed the departmental colloquium, we managed to doctor the first slide, a photo of Beckmann at the helm of the centrifuge, by adding a scantily clad bathing beauty—an "MCP" maneuver that neither of us would dare to attempt nowadays!

We left Columbia at different times. W. H. S. returned to MIT in late 1943, shifting his war-related efforts to the Laboratory for Insulation Research under Arthur von Hippel. As LaMer confided to Beckmann, "It's a shame Stockmayer is leaving. His wife is the best dancer in the department!" B. H. Z., on the other hand, stayed on nearly till the end of the war, when he migrated to the Brooklyn Polytechnic Institute to join Paul Doty, Turner Alfrey, and Arthur Tobolsky as a member of Herman Mark's team in polymer science.

The selection of topics for this recollection was perhaps more difficult than the actual writing. The limitations of space, plus the hope of interesting a wider circle of readers than just polymer specialists, were probably the determining factors in our arbitrary choice.

CHEMICAL STATISTICS: TREES, RINGS AND TRANSITIONS (W. H. S.)

In the November, 1941 issue of JACS there appeared a now famous set of three papers by Paul Flory (15) discussing the molecular statistics and gelation of polymerizing systems containing ingredients with more than two functional groups per monomer molecule. It had been known for some time that many such systems (for example, "Glyptal" resins made from equivalent quantities of glycerol and phthalic anhydride) change suddenly, during the course of the polymerization reaction, from liquids of moderate fluidity to infusible, insoluble gels with infinite Newtonian viscosity. Carothers had suggested (16) that the phenomenon was due to the formation of essentially macroscopic molecules—giant trees—and discussed the process in a simple stoichiometric way. For example, a mixture of two moles of glycerol, $CH_2OHCHOHCH_2OH$, with three moles of adipic acid, $HOCO(CH_2)_4COOH$, will form a single gigantic tree (provided cyclic structures are forbidden) at the point at which 5/6 of the carboxyl or hydroxyl groups have been esterified. If some ring structures are permitted, the event will be postponed to higher degrees of reaction. This simple picture is, however, quantitatively inadequate, as the observed gel point always occurs *earlier*; in the above example, it comes at about 75% esterification, corresponding to a number-average molecular weight of only about 10^3 g mol^{-1}. The notion was therefore current that a physical logjam among the relatively small but bristly branched molecules could drive up the viscosity. What Flory had now done was to rescue the Carothers picture by showing that statistically the very polyfunctionality of the reagents produces extremely broad distributions of molecular size, a few structures becoming really huge even though the number-average size remains modest.

Quantitatively, Flory's well-known criterion for the critical degree of reaction at which giant trees first form is

$$\alpha_c = 1/(f-1), \qquad\qquad 1.$$

where f is the number of functional groups on the multifunctional monomer branch units and α is the probability that a chosen functional group on one branch unit is connected, through the necessary sequence of one or more reactions, to another branch unit. The crucial quantity α can be simply evaluated from the reaction stoichiometry; in the above example, again taking equivalent quantities of hydroxyl and carboxyl functions, we have simply $\alpha = p^2$, where p is the fractional extent of esterification, since two ester links are needed to connect a pair of glycerol units. Since $f = 3$ in our example, Eq. 1 predicts $\alpha_c = 1/\sqrt{2} = 0.707$, as compared to the observed 0.75. The difference is attributable (15) partly to a reactivity difference between the primary and secondary hydroxyl groups, and partly to ring formation. In any case, the prediction is startlingly good. Flory has given two excellent expositions (17, 18) of the story we have just sketched.

Early in 1942 Flory's papers were reviewed in Charlie Beckmann's group seminar by a graduate student, Thomas G. Fox, Jr., who went on to a notable career (19) in polymer science, including extensive collaborations with Flory. The effect of Tom Fox's presentation on me was electrical: Here was a kind of phase transition, predictable almost quantitatively from stoichiometry and relatively simple statistics, and not requiring estimates of high-order cluster integrals or intractable lattice combinatory factors! The techniques used by Flory had been atypical, but I got the notion that the sol-gel transition could be treated by methods more imitative of standard statistical thermodynamics. Through the attempt to do this, I fell into polymer science.

To appreciate the impact of Flory's work at that time, it should be realized that knowledge of phase transitions was not then highly developed or organized. Although there was experimental evidence suggesting that the critical region was somehow anomalous (20, 21), most physical chemists did not go beyond the classical treatment of vapor-liquid equilibrium used by Maxwell, Gibbs, and van der Waals: The pressure and the Helmholtz free energy were supposed to be analytic functions of temperature and density everywhere on the primitive thermodynamic surface, and two-phase equilibrium was only to be determined by knocking out the metastable or essentially unstable regions with the requirement of equal chemical potentials. In the later 1930s Joe Mayer had rocked the boat (12) by championing the belief that the canonical partition function itself must contain all the secrets of equilibrium, including the singularities on the binodal locus. It may be conjectured that this view was inspired by the ideal Bose-Einstein condensation and that it was shared by Max Born, earlier a patron of both the Mayers in Göttingen (22). Indeed Born (23), perhaps influenced by the much earlier and cruder notions of Lindemann (24), had proposed that the melting point of a crystalline solid coincided with the vanishing of its shear modulus, a theory that commanded some respect till it was scotched by Joseph Slepian, obviously a man of the world, who remarked (25) that the ice cubes in his highball glass were perceptibly rigid.

Mayer's ideas about vapor-liquid condensation were not so easily disposed of. Developing the thermodynamic functions in cluster series, he and his students observed that singularities might in principle arise from either of two apparently distinct mathematical conditions, which need not be coincident. Thus there emerged the "derby hat" picture of the critical region, which though now long abandoned was an important stepping stone to modern theories of the critical region. For details of this picture the 1940 textbook by the Mayers (13) may be consulted.

The Mayer theory eludes quantitative test because of the necessity of evaluating big cluster integrals, and a similar remark could probably be made about any off-lattice treatment of systems with additive pairwise interactions. As to lattice models, the exact critical temperature for the unmagnetized two-dimensional Ising model had indeed been located by Kramers & Wannier (26), but Onsager's evaluation of the partition function did not appear until 1944. The Flory gelation model, with its inherently tree-like structure ("Bethe lattice"), was much simpler, and its success invited generalization, with the hope of relating it theoretically to other more formidable transitions, as well as of producing a method for finding the molecular species distribution for an arbitrary stoichiometric system.

For ease, I first considered the so-called RA_f model, in which a single kind of f-functional monomer is capable of reacting with itself, and in which case α is just the fractional extent of the bonding reaction. Flory had treated other special systems and developed the statistics mainly through appropriate recursion relationships. With the RA_f model, with exclusion of rings, one can go straight for the goal with the microcanonical ensemble, maximizing the entropy at a fixed degree of bond formation (i.e. constant energy); the details can be read in the original paper (27). The combinatory factor caused a few headaches, but it was finally obtained through conscious mimicry of a similar problem occurring in the Mayer cluster theory, by invoking imaginary Erector-set constructions with frames, bolts, and washers (13, pp. 455–59). Since Joe had just taken on a new graduate student named Harris Mayer, there was a strong temptation to prepare a joint manuscript by Mayer, Mayer, Mayer & Stockmayer, but Joe and Maria never yielded.

The microcanonical treatment can be replaced by a canonical ensemble, if desired (29). In either case, the size distribution for the RA_f model without rings can be derived without any mathematical approximations. It yields the following formula for the weight fraction w_x of x-mers:

$$w_x = K_x(1-\alpha)^2\alpha^{-1}\beta^x, \qquad\qquad 2.$$

with

$$K_x = f(fx-x)!/(fx-2x+2)!(x-1)!$$

and

$$\beta = \alpha(1-\alpha)^{f-2}.$$

The number-average size is

$$\langle x\rangle_n = 1/\Sigma x^{-1}w_x = 1/(1-\tfrac{1}{2}\alpha f) \qquad\qquad 3.$$

while the weight-average is

$$\langle x\rangle_w = \Sigma x w_x = (1+\alpha)/(1+\alpha-f\alpha). \qquad\qquad 4.$$

The breadth of the distribution is evident from the fact that $\langle x\rangle_w$ diverges at Flory's gel point, Eq. 1, while $\langle x\rangle_n$ remains small there.

Before the final polishing, I wrote to Flory (then at Esso Laboratories in New Jersey) and received a cordial invitation to come over and compare ideas. We didn't completely see eye to eye about the theory beyond the gel point, but that didn't prevent our becoming friends.

The Journal of Chemical Physics had gotten thin because of the war. (The minimum was reached in 1944, when Volume 12 had just 531 pages.) Publication was therefore rapid: My manuscript was submitted in late October 1942 and appeared in February. A second paper followed a year later (30). In an Appendix to the earlier work, a simple set of kinetic equations was formulated for the RA_f model, excluding reverse reactions as well as rings, and it was observed that these were satisfied by Eq. 2 if α were properly expressed as a function of time, though the uniqueness of the solution was not established. In recent years such "generalized Smoluchowski equations" have become objects of interest (31, 32).

The old methods of treating tree models have been superseded by the elegant applications of cascade theory due to Manfred Gordon (33), Walther Burchard (34), and their associates. Kinetic schemes of great complexity can be treated, including chain mechanisms (35); radii of gyration of Gaussian molecules can be calculated; and unequal reactivities of functional groups can be readily accommodated (36). Allowance can even be made for ring structures within the cascade approach by the so-called spanning tree approximation (37, 38).

In the last few years theoretical interest in the sol-gel transition has risen to a high level, which unfortunately we do not have space to discuss or document to any significant extent. Three major addenda to the classical tree theories must be addressed: (a) inclusion of excluded volume effects, perhaps not only on the chain dimensions but also on the chemical rates or equilibria themselves; (b) realistic description of ring formation at all stages of the reaction; and (c) full description of the system past the gel point. These problems are by no means mutually independent. A typically modern question is that of the critical exponents in the immediate vicinity of the transition point. For example, according to Eq. 4 the weight-average degree of polymerization $\langle x \rangle_w$ scales as $(\alpha_c - \alpha)^{-1}$ as the gel point is approached. An alternative approach near the transition is that of the *percolation* model, which allows in a prescribed way for excluded volume and ring formation, and which makes $\langle x \rangle_w$ vary as $(\alpha_c - \alpha)^{-\gamma}$ with $\gamma \simeq 1.7$. Existing experimental data have not provided a clear decision between these alternatives. For two contrasting points of view, the reviews of Stauffer (39) and Burchard (34) may be consulted.

It still wasn't clear in the summer of 1942 that polymers would become my main research interest. I was working on a simple quantum-mechanical calculation of the non-pairwise-additive overlap energy of three He atoms,[1] when chance again intervened. I had struck up a friendship with Lester Weil, a graduate student doing organic synthesis in the lab directly across the corridor from my office. His advisor rather suddenly went off to a war job, and Weil opted to begin a new research problem, his old one having proved unfruitful. We soon concocted a program directly inspired by Flory's papers: to measure the effects of dilution (and thus presumably of increasing intramolecular reaction) on gel points, and then to assay the effects of branching on solution and melt viscosities. The green light was soon given, thanks to the liberal policies of Urey and the Columbia department, which imposed no formal segregation of the various branches of chemistry and which allowed mere instructors to be research advisers. Thus suddenly there was a graduate student collaborator, and he was to work on polymers.

The gel point work went very well indeed. Taking pentaerythritol, $C(CH_2OH)_4$ with $f = 4$, and adipic acid, $HOCO(CH_2)_4COOH$, in equivalent amounts and adding varying quantities of the inert polyether diluent, $CH_3(OCH_2CH_2)_4OCH_3$, Weil found (41) the critical degree of esterification p_c [determined by titration, as in Flory's experiments (15)] to increase monotonically with dilution, as would qualitatively be expected if there were increasing amounts of intramolecular ring formation. For the undiluted system Weil measured $p_c = 0.630$, a few percent larger than the theoretical value $1/\sqrt{3} = 0.577$ for a system of pure trees. Quantitatively, we assumed that p_c would increase linearly with the relative $1/c$, where c is

---

[1] This work was abandoned in an almost finished state. A similar calculation was later published by Rosen (40).

the volume concentration of the active ingredients, and argued that no rings would exist at vanishing $1/c$. Extrapolating p_c linearly to zero against $1/c$, Weil obtained the least-squares intercept $p_c(0) = 0.578 \pm 0.005$, in remarkable support of the tree theory. The elation we then felt still seems justified: For the first time (as far as we then knew and now know) a "phase transition" (though admittedly not a physical separation into two distinct phases) had been located in the laboratory, when the conditions of the model were satisfied, at exactly the point predicted *a priori*. The results deserved more complete and rapid publication than they got (42).

In retrospect, it has to be admitted that there is no proof that the extrapolation of p_c against $1/c$ should be linear. Today the same system could of course also be studied by additional techniques, such as equilibrium and quasi-elastic light scattering, to determine weight-average molecular weights, mean square radii of gyration, and hydrodynamic radii, and there are many other chemical systems that would be more suitable for precise studies of this type.

Weil made many measurements on viscosities of branched polyesters in solution and in the melt, but publication was deferred and eventually postponed indefinitely when it became clear that end-group titrations alone (the only method then available to us) could not yield reliable molecular weights, even when corrected for ring formation in an approximate way to force agreement with observed gel points. There was moreover no sound viscosity theory yet at hand with which to discuss the results. Today the influence of branching on solution and melt properties, including viscosity, is again an active research area, in which the ability to synthesize almost monodisperse star polymers (and some other structures) by anionic polymerization has played a key role (43, 44).

Our own interest in the dimensions and properties of branched polymers did not altogether die out with the termination of Weil's efforts. A few years later, when B. H. Z. was at Berkeley and I at MIT, we collaborated at long range in some calculations of mean square radii of gyration for various branched Gaussian structures (45). The extensions of such calculations by the cascade technique came subsequently at the hands of Gordon (33) and Burchard (34). When Marshall Fixman was an MIT graduate student, he and I made some calculations of excluded volume effects and hydrodynamic radii in star polymers (46). These problems are also under active scrutiny today.

In the fall or winter of 1942 at Columbia, Homer Jacobson joined the group as my second graduate student, bringing a variety of talents, including musical aptitude and imagination: When studying music theory for recreation, he satisfied a course requirement by writing classical four-part harmony to a ground bass of "Pistol Packin' Mama," a juke-box pest of 1943 which the instructor failed to recognize. After some preliminary work on gel-point theory for systems polymerizing by chain mechanisms (47) (e.g. vinyl acetate + divinyl adipate, etc), it was decided that Jacobson should pursue both theoretical and experimental studies of ring-chain equilibrium in linear condensation polymers. This seemed like a logical first step before confronting the immeasurably harder problem of ring formation in nonlinear systems—a problem on which we never managed to produce any dents.

Let R_x denote the number of closed-ring x-mers and C_y the number of open-chain y-mers in the system. Then contemplate the reaction

$$C_{x+y} \rightleftarrows C_y + R_x, \qquad\qquad 5.$$

which involves no change in the number of bonds and hence no appreciable isothermal change in energy. The essential component of the equilibrium constant of the above reaction is therefore the conformational entropy change accompanying ring closure. This must depend on the length of the bond to be formed. However, a similar bond is broken when the $(x + y)$-mer chain is cleaved, and so the bond length disappears from the calculation. Assuming Gaussian chain statistics, Jacobson was able to formulate the equilibrium constant (here written only for the simplest case of a self-condensing single monomer species) by following a method taught by Mayer & Mayer (13, pp. 213–17), with the result (49)

$$K \equiv C_y R_x / V C_{x+y} = (3/2\pi\langle r_x^2\rangle)^{3/2}/2x \qquad\qquad 6.$$

where $\langle r_x^2\rangle$ represents the equilibrium mean square end-to-end distance of the x-meric open chain, which for a Gaussian chain is proportioned to x, and V is the volume of the system. The factor $1/2x$ comes from the symmetry number $2x$ of the ring, or can equally well be rationalized kinetically as due to the x different bonds that could be cleaved in the reverse of Reaction 5, followed by its being joined to the y-mer chain at either end.

It should surprise nobody that the physical argument leading to Eq. 6 had been anticipated by Kuhn (50), but this fact was not known to us for many years.

When the above result is applied to the over-all statistics of an equilibrium Gaussian ring-chain system, the number of ring x-mers takes the form

$$R_x = VBx^{-5/2}p^x \qquad\qquad 7.$$

where p is the fractional degree of condensation in·the open-chain molecules, and the symbol B subsumes the details of Eq. 6. The total number of ring molecules and the total number of monomer units contained in them are then given by sums ΣR_x and $\Sigma x R_x$, which are identical to those for the pressure and number density of a perfect Bose-

Einstein gas in terms of fugacity. The latter sums have singularities at unit fugacity, as the polymer sums do at $p = 1$. Thus the Bose-Einstein condensation is exactly reproduced in the polymer problem with the following physical meaning: Below a certain critical density, $\rho_c = B\zeta(3/2)$, where ζ is the zeta function, the system can be driven to 100% condensation to produce a system of rings only.

Readers conversant with double-helix theory (51) will be aware that the "loop factor" in the two-strand partition function is in essentials the same as that of Eq. 6. Indeed, the analogy to the Bose-Einstein condensation was rediscovered (52) in that context. Useful application to formation of large DNA rings has also been made (53).

Experimentally, Jacobson was able to give support to the theory by observing changes in solution viscosity of the predicted magnitude upon dilution and reequilibration (54). In this part of his work he was advised by Charlie Beckmann, for I had by then left Columbia. The proportionality of K to $x^{-5/2}$ for Gaussian chains (theta solvent conditions) has been directly confirmed by Semlyen (55), notably for poly(dimethylsiloxane) but also for a number of other polymers. His results also clearly document the increase of the exponent of $1/x$ above $5/2$ in good solvents where excluded-volume effects come into play.

The theory has been refined for "unperturbed" rings too small to be Gaussian, with regard for bond-angle and internal-rotational angle restrictions in closed rings (56), and improved agreement with experiment is seen. A more challenging theoretical problem concerns the effects of intermolecular excluded volume on polymerization equilibria, including ring-chain equilibria, producing departures from Flory-Huggins thermodynamics (which in their simple form require no "activity coefficient" corrections to Eq. 6). Recent progress is due to Wheeler & Kennedy (57), who have exploited the now well-known correspondence (58) between self-avoiding lattice walks and the $n = 0$ magnetic lattice model. Aside from shifts of detail and exponents, there is at least one striking *qualitative* result: For the ring-chain transition in molten sulfur at 160°C a small hump is predicted on the density-temperature curve (as contrasted to a mere discontinuity of slope in the classical theory), and this is in fact found experimentally (59).

MOLECULAR WEIGHTS AND LIGHT SCATTERING (B. H. Z.)

In the early 1940s the absolute molecular weights of most polymers were either unknown or known only within rough limits, and such data as existed were not considered highly reliable. This situation persisted because of difficulties of measurement. Polymers do not have a gas phase, so that they must be measured in dilute solution, but the ideal, Raoult's-Law, term in the free energy, on which the determination of molecular weight depends, decreases at a given mass fraction as the reciprocal of the molecular weight, while the nonideal terms are roughly independent of the molecular weight. Thus, to suppress the nonideal terms, which are proportional to the second and higher powers of the concentration, it is necessary to measure at lower and lower concentrations as the molecular weight increases, and at some point one runs out of measurement accuracy. With the traditional methods of organic chemistry, such as the depression of the freezing point of a camphor solution, this point comes when the molecular weight is about ten thousand, much less than the molecular weight of almost any significant polymer.

The one thermodynamic method with sufficient sensitivity is osmometry, but here the chemical and mechanical stability, as well as the permeability and selectivity, of the membrane is critical. [There is a good discussion in an article by Wagner & Moore (60).] At that time the membrane usually used with organic solvents was partially denitrated nitrocellulose, which the experimenter prepared himself by treating a film of partially dried collodion, cast on a mercury surface, with ammonium sulfide. The permeability and selectivity of such a membrane depended on the extent to which it had been dried and the solvent to which it had been transferred, while the amount of soluble impurity depended on the thoroughness of the sulfide treatment and the subsequent washing. The spurious osmotic pressure that appeared when a new membrane of this kind was first put into an osmometer with pure solvent on both sides, a pressure presumably arising from soluble components of the membrane, could be spectacular, and frequently decayed only slowly with time. Also, the porosity of such a membrane could easily be great enough to let some macromolecular components pass through. Thus osmotic molecular-weight measurements could, and did, give results that varied considerably from one experiment to another.

The only other known absolute methods also had difficulties. Determination of chain ends by chemical analysis worked with only a few polymers where the end groups were definitely known, such as polyesters, and then only at rather low molecular weights, and in any case the determination was sensitive to small amounts of impurities. Svedberg had invented the ultracentrifuge, but only a few of the instruments were available in the whole world until the Spinco corporation went into commercial production of electrically driven machines at the end of the decade. Moreover, synthetic polymers were polydisperse in molecular weight, and their solutions were highly nonideal, so that molecular-weight measurements on them with the ultracentrifuge were complicated, and data reduction was time-consuming with the laborious methods then in use. [See, for example, papers by Wales and co-workers (61, 62).]

The most widely used measure of molecular size, then as now, was the intrinsic viscosity, but this is a relative, not an absolute, measure. The nature of this relation had been the subject of a vigorous dispute between Staudinger and Mark in the previous decade, and the fundamental theory of it was not satisfactorily developed until the work of Debye & Bueche (63) and of Kirkwood & Riseman in 1948 (64) and of Flory & Fox in 1950 (65). Before that, the intrinsic viscosity was mainly a quantity of empirical significance only.

Thus in the 1940s determination of the values of the absolute molecular weights of synthetic polymers was a subject of fundamental interest, and when P. J. W. Debye (66) introduced a new and very different method, based on measurement of the light scattered from solutions, it was a major event.

Debye's method was actually a new extension of an old theory, one with which the names of Rayleigh and Einstein were primarily associated. After Rayleigh's development in the nineteenth century of the theory of the scattering of light from individual small particles, and following a preliminary discussion by von Smoluchowski in 1908 (67), Einstein in 1910 (68) derived a formula for the scattering from a pure liquid on the basis of a Fourier analysis of density fluctuations. In his 1944 publication, Debye (66) extended this formula to include composition fluctuations in a mixture, and related these to thermodynamics, to the dilute-solution laws, and hence, finally, to the molecular weight of a solute. This remarkable formula is

$$\tau = \frac{32\pi^3 kTn^2}{3\lambda^4} \left[\frac{\rho(\partial n/\partial \rho)^2}{(\partial p/\partial \rho)_{T,c}} + \frac{c(\partial n/\partial c)^2}{(\partial P/\partial c)_{T,p}} \right]. \tag{8.}$$

Here τ is the turbidity, i.e. the fraction of light scattered per unit length of path, kT as usual, n is the refractive index, λ the wavelength in vacuo, ρ the density, p the pressure, c the concentration of solute, and P the osmotic pressure. The first term is Einstein's original term, and represents the scattering from density fluctuations, while the second term represents the scattering from composition fluctuations. The first term is practically independent of concentration and for dilute solutions can be replaced by τ_0, the scattering from pure solvent. If we then introduce van't Hoff's law for P, we get

$$\tau - \tau_0 = \frac{32\pi^3 n^2 (\partial n/\partial c)^2 cM}{3\lambda^4 N_a}, \tag{9.}$$

where c is now the concentration of the solute in mass per unit volume, N_a is Avogadro's number, and M is the molecular weight of the solute. Thus

Debye could say (66, p. 340), referring to a slightly altered form of the above: "Equation 5' can therefore be interpreted as showing how by the combination of two measurements, the first of the turbidity, the second of the difference in refraction of solution and solvent, the molecular weight of the substance in solution can be evaluated, without introducing any kind of empirical constants."

The effective introduction of light scattering as a method for measurement of molecular weight and size of polymers occurred during World War II, and much of the original work was never published in the usual journals, or was published only much later. For that reason it seems best to insert some personal reminiscences.

Light scattering and I (B. H. Z.) had become acquainted during the summer of 1942, at the end of my first year of graduate school, when I worked on a project investigating the optical properties of smokes for possible military use as smoke screens. This project was located in the chemistry department's laboratories and was under the direction of Victor K. LaMer. Paul Doty, also a first-year graduate student, and I were hired for the summer to help with this. The smokes in question consisted of fine spherical particles of dyes with strong absorption bands in the visible, and correspondingly complex dependences of scattering on wavelength. The object was to see whether a smoke could be found that would scatter white light strongly but that would become transparent at a specific wavelength. We were not very successful at finding such a smoke—the features of the scattering-versus-wavelength curve were not pronounced enough—but the project was a good introduction to the optical theory of light scattering. This I remember studying in the best book then available, the original German edition of Born's *Optik* (70), which was lent to me by David Sinclair, a physicist working on the project (and son of the novelist Upton Sinclair). (In addition to learning some excellent physics from this book, I profited from the practice of reading Born's elegant German; nearly everything that I have since had to read in that language has seemed easy.)

Later Doty and I both did research for our Ph.D. theses with Joe Mayer at Columbia. A third research student at the same time with Mayer was William G. McMillan, Jr., who was beginning a study of the vapor pressure of mixtures of triethylamine and water near their lower critical mixing point. His aim was to see whether experimental evidence could be found for the "derby-hat picture" of the critical point that Mayer had proposed on theoretical grounds (71). The derby hat was supposed to be a region adjacent to the critical point within which the vapor-pressure isotherms had zero slope. McMillan was facing the unrewarding task of measuring the vapor pressures so precisely that the almost vanishing slopes of curves through these data could be confidently said to be "zero" over a finite range.

Then one of us, I do not remember who, noticed a section in Fowler's book on statistical mechanics (72) that discussed the Einstein-Smoluchowski theory of light scattering and its relation to fluctuations. This theory showed that the intensity of scattering is inversely proportional to the slope of the isotherm of vapor pressure against concentration, and so was obviously useful for McMillan's problem.

Probably nothing would have come of this, if our conversations had not been overheard by Charlie Beckmann, who occupied the laboratories adjoining Mayer's on the eighth floor of Chandler Hall. Beckmann, as we have mentioned, was interested in the physical chemistry of starch; he had one of the first ultracentrifuges to be built in the United States, and among his other instruments was a turbidimeter from Carl Zeiss of Jena. When Beckmann, who was a friendly man, heard us talking about light scattering, he immediately offered the use of his turbidimeter. This was a simple machine: a visual differential (Pulfrich) photometer, a tungsten-filament bulb with a colored filter, and a cell holder with a water jacket as thermostat. There was also a most important accessory, a turbidity standard in the form of a piece of beautiful smoky glass with the value of its turbidity engraved by the Zeiss firm on the brass holder. [I found the published description of the calibration of this standard (73) for the first time while preparing this account.] To make a measurement, one had to balance the illumination intensities in two halves of the visual field of a telescope eyepiece by adjusting calibrated drums that controlled the aperture stops of two objectives, one aimed at the scattering solution or the standard and the other at a piece of opal glass in the same light beam. Late in 1943, McMillan, Doty, and I made a number of measurements of triethylamine and water in this way, but we did not find any sign of the derby hat. Not knowing what to make of this, we never did anything with the data; I still have them in a file folder. If the modern theory of critical points had been available, we would probably have tried to see what exponents the data exhibited; in fact, we actually plotted them on log-log paper, found straight lines, and noted the slopes, but we had no idea of the significance of the latter. Instead of pursuing the triethylamine work, McMillan developed a thesis on statistical mechanics, which became the well-known McMillan-Mayer theory of multicomponent systems (74).

At the beginning of 1944, Paul Doty finished the work for his thesis on the electron affinity of bromine and took a position with a research project directed by Professor Herman Mark at the Polytechnic Institute of Brooklyn. The project was concerned with various aspects of polymer chemistry and physics, especially their application to processing of plastics for military applications such as covers for guns and packages for supplies. Shortly after joining the project, Doty came back one day with the report that Debye at Cornell had developed an as yet unpublished method for measuring molecular weights of polymers by light scattering. Being familiar with the Einstein-Smoluchowski theory, we understood immediately that combining the theory with Raoult's Law would give a molecular-weight method; since we already had the necessary apparatus, we became interested in actually trying the method out. Doty obtained three samples of polystyrene from Professor Mark, samples that two of Mark's former students, Turner Alfrey and Al Bartovics, had prepared and had measured by osmometry (75). In a few weeks of work we made solutions and measured the scattering and the refractive index increment of these samples in both toluene and methyl ethyl ketone. After some struggles with the units of the constants in the Einstein-Smoluchowski formula, we were pleased to find that the molecular weights of the same sample in the two different solvents not only agreed with each other, but that they also agreed with the osmotic values. Doty, Mark, and I published a short communication in the April, 1944, issue of the *Journal of Chemical Physics* describing these results (76), and we published a longer paper the following year (77). Debye's first paper had appeared shortly before in the *Journal of Applied Physics* (66).

It is obvious that we were extremely lucky in the availability of a calibrated instrument as well as of samples of polystyrene, probably the most suitable of all polymers for light scattering because of its complete solubility and its high refractive index, and in the fact that Alfrey & Bartovics had already measured the samples by osmometry. Later we realized that we had been lucky in some more subtle ways too. The light-scattering molecular weights agreed too well with the osmometric results. Since light scattering gives a weight average and osmometry a number average, and since the samples were only roughly fractionated, the former molecular weights should have been higher than the latter by about 50%. What had probably happened was that Alfrey & Bartovics' osmotic membranes were too permeable, on the one hand, so that their molecular weights were too high, and on the other hand, we had overlooked the necessity of applying corrections to measured luminosities when the light has passed from a medium of one refractive index (organic liquid) to another (air) (78); these corrections would have raised the scattering values. Also we had not taken the angular dependence of the scattering into account; the Zeiss turbidimeter measured at only one angle (135 degrees from the incident beam), and this correction would have raised the results somewhat further.

It took a number of years to sort all these problems out. At first we continued to work with the Zeiss Pulfrich photometer but with two new cell holders, one that allowed measurements at 90 degrees from the incident beam, and one that allowed measurements of the ratio of the scattering intensities at 135 and 45 degrees (the "dissymmetry"). This work was done by Doty and several students at Brooklyn, where I also went late in 1944 after finishing my thesis (far from polymer solutions; it was on the vapor pressures of alkali halides) and working for most of the year making smokes again for LaMer's project. Gradually it became evident that there were difficulties with the calibration of the new light-scattering instrument, for which we no longer had Carl Zeiss to rely on. The Zeiss firm was inaccessible, of course, in an enemy country. The calibrations were different, depending on whether we measured the turbidity of a strongly scattering solution directly by the attenuation of transmission in a spectrophotometer, or whether we attempted to calibrate with the scattering from a highly reflecting magnesium-carbonate surface. The task was not made any easier by the size of the ratio of intensities of the incident beam and of the scattering from liquids at the photometer aperture; this ratio was of the order of one million.

To further confuse the situation, some writers in the older literature proposed that Einstein's basic scattering formula was defective and should have a factor of $[(n^2 + 2)/3]^2$ included, where n is the refractive index, allegedly to take account of the modification of the electric field of the light by the cavity containing the scattering molecule. There was considerable discussion of this in a French book by Cabannes (79). Depending on which calibration method one favored for the instrument, one could easily convince oneself that the experiments verified one form of the theory or the other. That there was no obvious place to include such a factor in Einstein's elegant 1910 derivation tended to be overlooked. Einstein used a phenomenological (optical) dielectric constant in his derivation, and assumed it and its derivatives to be equal to the corresponding macroscopic property; this assumption led to the debate. Of course, differences would be expected at the scale of the molecular dimensions of the liquid. However, the Fourier components of the density fluctuations of interest in light scattering have wavelengths of hundreds of nanometers at the least, much larger than the molecules of ordinary liquids; thus one would expect the bulk dielectric constant to apply. In fact, later careful derivations based on molecular theory lead to the same result; see papers by Fixman (80) and by Zwanzig (81).

In 1946 I went to the University of California at Berkeley where I built a scattering photometer using the newly available multiplier phototube instead of a visual device (82). My student, Clide I. Carr, Jr., elected to do a thesis on the absolute scattering power of pure liquids and solutions and the relation of the scattering to Einstein's theory. In the course of this he rediscovered the effect of refraction at a surface on the apparent brightness of an object, a relation that was well known in optics (78), but which we had overlooked. With this taken into account, everything fell into place; three independent methods of calibrating the photometer agreed, and the measured scattering from pure liquids and from solutions of simple substances was in accordance with theory (83). Using this methodology, Paul Outer, a post-doctoral fellow from Belgium, was able to make an extensive series of measurements on polystyrene solutions and to get all the numbers well pinned down (84).

Confusion about the "high values" and "low values" of the magnitude of the scattering from liquids persisted in the literature for several years until the accumulating experimental evidence came out heavily in favor of the high values. Even in 1953 and 1954 the debate was still going on (85, 86); see some discussion in a review by Stockmayer, Billmeyer & Beasley (87). Later, I found that Debye's group had been aware of the refraction effect at the time of their first work, but mention of it was buried inconspicuously in their writings, and had been missed by everyone else.

In fact, the circumstances surrounding the first publications of the determination of the molecular weight of polymers are curious from a bibliographic point of view. The first work of Debye and his associates at Cornell was published in reports to the Office of Rubber Reserve, Reconstruction Finance Corporation, and were given only limited circulation because of wartime security restrictions. Debye's first paper in the open literature in 1944 (66) omits mention of any previous derivation of his formula for getting the molecular weight. I have often wondered whether he derived it independently, which certainly would not have been difficult for him; much of his previous work had been on other aspects of scattering. Actually the formula, an extension of Einstein's 1910 formula for a pure liquid, had been published long before in extensive discussions by Gans (88) and Raman & Ramanathan (89) in 1923, and these were the references cited by Doty, Mark, and me in our first work (76). Einstein himself had given a short treatment of composition fluctuations, but had unnecessarily limited himself to the case where the vapors of the constituents were ideal gases. Einstein's comment (68, p. 1297) had been: "This formula, which contains only experimentally accessible quantities, completely determines the opalescence properties of binary liquid mixtures, insofar as one may treat their saturated vapors as ideal gases, up to a small region in the immediate neighborhood of the critical point." So the essence of the theory had been available in 1910, if any one had wished to use it. It was not until 1927 that Raman first pointed out the applicability of the formula to colloidal solutions in a paper (91) that we all overlooked; it had appeared in the *Indian Journal of Physics*. In about 1946 several of us unexpectedly received reprints of this paper; they arrived in the mail from India without explanation or comment.

There had also been a few studies in which light scattering was used to compare molecular weights of macromolecules, but in which the full power of the method for determining absolute values was not utilized. Such were papers published by Putzeys & Brosteaux in 1935 (92), by Staudinger & Haenel-Immendörfer in 1943 (93), and by Schulz in 1944 (94); referring to the first two of these, Debye (95) said: "The authors do not yet realize that the [constants in the molecular-weight formula] can be determined experimentally without making assumptions about the particles or their so-called optical constants."

In most of the preceding we have not mentioned the angular dependence of the scattering, which is essentially an interference phenomenon, and which gives information about the linear dimensions of the scattering particles if the particles are sufficiently large. Nor have we said anything about the depolarization of the scattering, which depends on, and gives information about, the anisotropy of the scattering particles. These phenomena complicate the determination of molecular weights, and they are of interest in their own right, but discussion of them would go too far in prolonging what is already a long story. Also we have omitted discussion of the fluctuations of the intensity of the scattering, which have played such a prominent role in "dynamic light scattering" in the last two decades. In the 1940s and 1950s we viewed these fluctuations simply as a nuisance to be averaged over. Even if someone had thought otherwise, it would have been hard to exploit them with the poorly coherent high-intensity light sources that were available before the invention of the laser.

Literature Cited

1. Forman, P. 1983. *Science* 220:824–27
2. Flory, P. J. 1953. *Principles of Polymer Chemistry*, Chapt. 1, pp. 3–28. Ithaca, NY: Cornell Univ. Press. 672 pp.
3. Staudinger, H., Heuer, W. 1980. *Berichte* 63:222–34
4. Guth, E., Mark, H. 1934. *Monatsh.* 63:93–121
5. Kuhn, W. 1934. *Kolloid-Z.* 68:2–15
6. Kuhn, W. 1932. *Z. Phys. Chem. A* 159:363–73
7. Flory, P. J. 1936. *J. Am. Chem. Soc.* 58:1877–85
8. Flory, P. J. 1937. *J. Am. Chem. Soc.* 59:241–53
9. Schulz, G. V., Dinglinger, A., Husemann, E. 1939. *Z. Phys. Chem. B* 43:385–408
10. Huggins, M. L. 1938. *J. Phys. Chem.* 42:911–20
11. Mayer, J. E. 1937. *J. Chem. Phys.* 5:67–73
12. Mayer, J. E., Harrison, S. F. 1938. *J. Chem. Phys.* 6:87–104
13. Mayer, J. E., Mayer, M. G. 1940. *Statistical Mechanics*. New York: Wiley. 495 pp.
14. Mayer, M. G. 1941. *Phys. Rev.* 60:184–87
15. Flory, P. J. 1941. *J. Am. Chem. Soc.* 63:3083–3100
16. Carothers, W. H. 1936. *Faraday Soc. Trans.* 32:39–53
17. Flory, P. J. 1946. *Chem. Rev.* 39:137–97
18. Flory, P. J. 1953. See Ref. 2, Chapter 9, pp. 347–98
19. Casassa, E. F., Mark, H., Markovitz, H., Overberger, C. G., Pearce, E. M., Flory, P. J. 1979. *J. Polym. Sci. Polym. Phys. Ed.* 17:1815–24
20. Lowry, H. H., Erickson, W. R. 1927. *J. Am. Chem. Soc.* 49:2729–34
21. Maass, O. 1938. *Chem. Rev.* 23:17–28
22. Mayer, J. E. 1982. *Ann. Rev. Phys. Chem.* 33:1–23
23. Born, M. 1939. *J. Chem. Phys.* 7:591–603

1652 STOCKMAYER & ZIMM

24. Lindemann, F. A. 1910. *Phys. Z.* 11:609–12
25. Slepian, J., quoted by Siegel, S., Cummerow, R. 1940. *J. Chem. Phys.* 8:847
26. Kramers, H. A., Wannier, G. H. 1941. *Phys. Rev.* 60:252–62
27. Stockmayer, W. H. 1943. *J. Chem. Phys.* 11:43–55
28. Deleted in proof
29. Cohen, R. J., Benedek, G. B. 1982. *J. Phys. Chem.* 86:3696–3714
30. Stockmayer, W. H. 1944. *J. Chem. Phys.* 12:125–31
31. Ziff, R. M., Stell, G. 1980. *J. Chem. Phys.* 73:3492–99
32. Ziff, R. M., Hendriks, E. M., Ernst, M. H. 1982. *Phys. Rev. Lett.* 49:593–95
33. Gordon, M. 1962. *Proc. R. Soc. London Ser. A.* 268:240–59
34. Burchard, W. 1983. *Adv. Polym. Sci.* 48:1–124
35. Whitney, R. S., Burchard, W. 1980. *Makromol. Chem.* 181:869–90
36. Müller, M., Burchard, W. 1978. *Makromol. Chem.* 179:1821–35
37. Gordon, M., Scantlebury, G. R. 1965. *J. Polym. Sci. Pt. C* 16:3933–42
38. Dušek, K. 1979. *Makromol. Chem. Suppl.* 2:35–49
39. Stauffer, D., Coniglio, A., Adam, M. 1982. *Adv. Polym. Sci.* 44:103–58
40. Rosen, P. 1953. *J. Chem. Phys.* 21:1007–12
41. Weil, L. L. 1945. Ph.D. dissertation, Columbia Univ., NY
42. Stockmayer, W. H. 1945. Molecular size distribution in high polymers. In *Advancing Fronts in Chemistry*, ed. S. B. Twiss, 1:61–73. New York: Reinhold. 196 pp.
43. Rempp, P., Decker-Freyss, D. 1965. *J. Polym. Sci. Pt. C* 16:4027–34
44. Bywater, S. 1979. *Adv. Polym. Sci.* 30:89–116
45. Zimm, B. H., Stockmayer, W. H. 1949. *J. Chem. Phys.* 17:1301–14
46. Stockmayer, W. H., Fixman, M. 1953. *Ann. NY Acad. Sci.* 57:335–52
47. Stockmayer, W. H., Jacobson, H. 1943. *J. Chem. Phys.* 11:393
48. Deleted in proof
49. Jacobson, H., Stockmayer, W. H. 1950. *J. Chem. Phys.* 18:1600–6
50. Kuhn, W. 1949. *Helv. Chim. Acta* 32:735–43
51. Poland, D., Scheraga, H. A. 1970. *Theory of Helix-Coil Transitions in Biopolymers.* New York: Academic. 797 pp.
52. Poland, D., Scheraga, H. A. 1966. *J. Chem. Phys.* 45:1464–69
53. Wang, J. C., Davidson, N. 1966. *J. Mol. Biol.* 15:111–23
54. Jacobson, H., Beckmann, C. O., Stockmayer, W. H. 1950. *J. Chem. Phys.* 18:1607–12
55. Semlyen, J. A. 1976. *Adv. Polym. Sci.* 21:41–75
56. Flory, P. J., Semlyen, J. A. 1966. *J. Am. Chem. Soc.* 88:3209–12
57. Kennedy, S. J., Wheeler, J. C. 1983. *J. Chem. Phys.* 78:953–62
58. DeGennes, P. G. 1979. *Scaling Concepts in Polymer Physics*, pp. 265–281. Ithaca, NY: Cornell Univ. Press. 324 pp.
59. Kennedy, S. J., Wheeler, J. C. 1983. *J. Chem. Phys.* 78:1523–27
60. Wagner, R. H., Moore, L. D. Jr. 1959. In *Physical Methods of Organic Chemistry*, ed. A. Weissberger, Vol. 1, pt. 1, pp. 815–94. New York/London: Interscience. 894 pp. 3rd ed.
61. Wales, M. 1948. *J. Phys. Colloid Chem.* 52:235–48
62. Wales, M., Williams, J. W., Thompson, J. O., Ewart, R. H. 1948. *J. Phys. Colloid Chem.* 52:984–98
63. Debye, P., Bueche, A. M. 1948. *J. Chem. Phys.* 16:573–79
64. Kirkwood, J. G., Riseman, J. 1948. *J. Chem. Phys.* 16:565–73
65. Flory, P. J., Fox, T. G. Jr. 1950. *J. Polym. Sci.* 5:745–47
66. Debye, P. J. W. 1944. *J. Appl. Phys.* 15:338–42
67. von Smoluchowski, M. 1908. *Ann. Phys. Leipzig* 25:205–26
68. Einstein, A. 1910. *Ann. Phys. Leipzig* 33:1275–98
69. Deleted in proof
70. Born, M. 1933. *Optik.* Berlin: Springer, 591 pp.
71. Mayer, J. E., Mayer, M. G. 1940. *Statistical Mechanics*, p. 312. New York: Wiley. 495 pp.
72. Fowler, R. H. 1936. *Statistical Mechanics.* Cambridge: Univ. Press. 864 pp.
73. Sauer, H. 1931. *Z. Tech. Phys.* 12:148–62
74. McMillan, W. G. Jr., Mayer, J. E. 1945. *J. Chem. Phys.* 13:276–305
75. Alfrey, T., Bartovics, A., Mark, H. 1943. *J. Am. Chem. Soc.* 65:2319–23
76. Doty, P. M., Zimm, B. H., Mark, H. 1944. *J. Chem. Phys.* 12:144–45
77. Doty, P. M., Zimm, B. H., Mark, H. 1945. *J. Chem. Phys.* 13:159–66
78. Born, M., Wolf, E. 1964. *Principles of Optics*, p. 189. New York: MacMillan. 808 pp.
79. Cabannes, J. 1929. *La Diffusion Moléculaire de la Lumière.* Paris: Presses Universitaires de France
80. Fixman, M. 1955. *J. Chem. Phys.* 23:2074–79
81. Zwanzig, R. 1964. *J. Am. Chem. Soc.* 80:3489–93

82. Zimm, B. H. 1948. *J. Chem. Phys.* 16: 1099–1116
83. Carr, C. I. Jr., Zimm, B. H. 1950. *J. Chem. Phys.* 18: 1616–26
84. Outer, P., Carr, C. I. Jr., Zimm, B. H. 1950. *J. Chem. Phys.* 18: 830–39
85. Rousset, A., Lochet, R. 1953. *J. Polym. Sci.* 10: 319–32
86. Zimm, B. H. 1953. *J. Polym. Sci.* 10: 351–52
87. Stockmayer, W. H., Billmeyer, F. W., Beasley, J. K. 1955. *Ann. Rev. Phys. Chem.* 6: 359–80; p. 370
88. Gans, R. 1923. *Z. Phys.* 17: 353–97

89. Raman, C. V., Ramanathan, R. 1923. *Philos. Mag. Ser.* 6 45: 213–24
90. Deleted in proof
91. Raman, C. V. 1927. *Indian J. Phys.* 2: 1–6
92. Putzeys, P., Brosteaux, J. 1935. *Trans. Faraday Soc.* 31: 1314–25
93. Staudinger, H., Haenel-Immendörfer, I. 1943. *J. Makromol. Chem.* 1: 185–96. (English abstr. in 1945; *Chem. Abstr.* 40: 1719)
94. Schulz, G. V. 1944. *Z. Phys. Chem.* 194: 1–27
95. Debye, P. 1947. *J. Phys. Colloid Chem.* 51: 18–32

Ann. Rev. Phys. Chem. 1985. 36 : 1–30

MOLECULAR SPECTROSCOPY:
A Personal History*

Gerhard Herzberg

Herzberg Institute of Astrophysics, National Research Council, Ottawa, Ontario, Canada K1A 0R6

It was my good fortune to start scientific work in the period between 1926 and 1928, which coincided closely with the high point in the development of the subject of molecular spectroscopy. This fortunate accident made it possible for me to learn the subject while it was being developed. It also gave me the opportunity to meet most of the principal actors in the dramatic development of this field and to get a very personal feeling for their aims and ambitions. In discussing my personal history there will be many occasions of describing the forerunners and contributors to the field, and thus I hope that this article will be useful for the historian of science.

YOUTH

I was born on Christmas Day in 1904 into a middle-class family in Hamburg, Germany. My father, who was co-manager of a small shipping company, came from a small town, Langensalza in Thüringen (central Germany, now in the German Democratic Republic), where his ancestors can be traced for more than 300 years. He had one brother, who studied chemistry and became a resident chemist in a sugar factory in Czechoslovakia (then Austria). I never met him. There is no other evidence of scientific or academic pursuits in the family. My paternal grandfather was a grocer in Langensalza, my maternal grandfather was a master glazier in Hamburg.

All my schooling took place in Hamburg, with a very short interruption in Frankfurt soon after my father's death in 1915. My secondary schooling

---

* The Canadian Government has the right to retain a nonexclusive, royalty-free license in and to any copyright covering this paper.

took place at the Realgymnasium of the Johanneum in Hamburg. There were some very good teachers in this school. I should particularly mention W. Hillers, who was the editor of a well-known textbook of physics originally written by Grimsehl. Hillers first aroused my interest in atomic and molecular physics.

Upon finishing school I had intended originally to become an astronomer but when a vocational guidance officer made enquiries for me about the possibility of doing so, the advice given to him by the Director of the Hamburg Observatory for transmittal to me was that only if I had private funds would it be wise to enter that field. As I did not have private funds and as I was also interested in physics, the adviser suggested that I study technical physics, a subject that had just come into prominence at that time and according to him was particularly developed at the Technical University (at that time Technische Hochschule) in Darmstadt. For the first two years I obtained a private scholarship of the Stinnes Company (a large ship-building and shipping company), and when they went bankrupt in 1926 I was able to obtain a scholarship from the newly-established "Studienstiftung des deutschen Volkes."

STUDENT IN DARMSTADT (1924–1928)

During the last year of my work toward the "Diplom Ingenieur" (equivalent to a Master's degree on this continent) I started research work under Professor Hans Rau, the head of the Physics Department at Darmstadt. He was a pupil of W. Wien (the discoverer of Wien's displacement law) and had made a name for himself by his work at the University of Munich on the Doppler effect of positive rays. He was a first-class physicist and was always familiar with the latest results on research in physics, even though after his call to Darmstadt he did not contribute to these results. He was an exceptionally fine person whom I have venerated all my life.

For my doctor's thesis Rau did not suggest a topic but, rather, wanted me to find my own problem. He asked me to read Sommerfeld's book, *Atomic Structure and Spectral Lines*, which at the time was the main source of information about the status of atomic and molecular physics. After studying Sommerfeld I had the obvious idea of trying to produce a spectrum of Li^{2+}, a one-electron system. Rau liked the idea and a suitable discharge tube was ordered, but I never actually studied the Li^{2+} spectrum because before reaching this goal several other interesting spectra were obtained. I became intrigued by the afterglow of nitrogen and took its spectrum—a task, of course, that many people before me had done. In this work as a byproduct I found a good source for the spectrum of N_2^+. These two spectra formed the subject of my thesis. In studying the spectra of nitrogen I observed for the first time so-called tail bands of N_2^+ and found a way to include them in the main band system of N_2^+. Similar work on the isoelectronic molecule CN was done independently and at the same time by F. A. Jenkins, then at New York University. In the afterglow of nitrogen I found a very intriguing feature connected with CN. Only certain rotational levels occurred, a result that was explained much later by Beutler as being due to perturbations in these particular levels. In larger molecules the same mechanism leads to the phenomenon of internal conversion.

During the beginning of my work on the PhD thesis I was asked by Professor Rau to help another PhD student, M. Blumenthal, with her work on hydrogen. In this work a very clean Balmer spectrum and the continuum extending beyond it was observed. Even now I get occasional requests for a print of the Balmer spectrum. At this time I also had an idea about the continuous spectrum of molecular hydrogen, which turned out to be rather foolish but it did form the beginning of my lifelong interest in the spectrum of molecular hydrogen. Indeed, in my second paper on hydrogen, H_3 is mentioned as a possible source of some of the lines of the many-line spectrum of hydrogen. Of course at the time I did not foresee that 50 years later I would discover the real spectrum of H_3.

When as a beginning graduate student I learned to follow the current literature I saw in a new issue of the *Annalen der Physik* Schrödinger's first paper, "Quantization as an Eigenvalue Problem." It immediately aroused my interest, as I had just taken a course in partial differential equations and had learned a great deal about eigenvalues and eigenfunctions. I was deeply impressed and soon gave a talk about this paper in our Physics Colloquium. Subsequently Rau sent me to Freiburg, where Schrödinger was lecturing, and I had the great experience of meeting Schrödinger and hearing him lecture. He was a charming and modest man; he had a fascinating and lucid style of lecturing, with an engaging Viennese accent.

POSTDOCTORATE FELLOW IN GÖTTINGEN (1928–1929)

I was extremely lucky that the Studienstiftung allowed me to continue my studies for a year at the University of Göttingen. I had visited Göttingen during my PhD work in Darmstadt, presenting a talk on my thesis in their colloquium, and I was much impressed by the scientific atmosphere there.

Formally, during the first six months of my stay in Göttingen, I was assigned to the Department of Theoretical Physics under Max Born and during the second six months to the Second Physical Institute under James Franck. Almost as soon as I got to Göttingen, I started a short collaboration with Walter Heitler, who was then Privatdozent and assistant to Max Born. I told Heitler about my work on N_2^+ and about the result, based on the new Birge-Sponer extrapolation method, that the ground state did not dissociate into normal N^+ and normal N but that the excited state did. Heitler thought that this fitted in very well with the valence theory that he and London had originated, and we seemed to find a number of other examples. Although this work was published, it later turned out to be wrong because we had trusted the extrapolation method much more than we should have done. Even though this in retrospect was a failure, at the time it stimulated my own thinking about electronic structure of molecules, particularly when I learned about the new work of Hund and Mulliken on molecular orbitals.

A second paper with Heitler has stood the test of time. It was based on the rotational Raman spectrum of N_2 published at the time by F. Rasetti. His observation showed quite unambiguously that the even lines were strong in contrast to H_2, where the odd lines are strong. Since in both molecules the ground states are of $^1\Sigma_g^+$ symmetry, it means that the protons and the nitrogen nuclei follow opposite statistics, that is, Fermi and Bose, respectively. At the time of this observation only two elementary particles were known, the electron and the proton, and it was generally assumed that a nucleus consists of A protons and $A-Z$ electrons (A = mass number, Z = atomic number). Because both protons and electrons follow Fermi statistics it would follow that the nitrogen nuclei with an odd number of Fermi particles ($A = 14, Z = 7$) should also follow Fermi statistics and that therefore, just as in H_2, the odd rotational lines should be strong, contrary to observation. This observation that the nitrogen nuclei follow Bose statistics was the first indication that electrons are not present in the atomic nucleus, a result that became immediately clear when after the discovery of the neutron, Heisenberg first postulated that only neutrons and protons occur in the nucleus. Our short paper is not quoted in Heisenberg's paper but the fact that the nitrogen nuclei follow Bose statistics *is* quoted as one of the basic observations.

During the second part of my stay in Göttingen I worked with Professor G. Scheibe (then of the University of Erlangen), who was spending six months in Franck's institute. We observed for the first time the discrete ultraviolet spectra of the methyl halides, a subject that gave me the first femiliarity with electronic spectra of polyatomic molecules.

During this period I was asked by a group of the younger workers in the First Physical Institute (which was under R. W. Pohl) to present an informal introduction to spectroscopy in a series of lectures (I don't remember how many) to which also several people from the other groups came, even some from Born's theoretical group. I was pleased to find that V. Weisskopf, who was a student at the time, mentions this course in his autobiography as having been of value to him. Another listener was Scheibe, who suggested that I should publish these lectures as a book. He helped me in my first approach to a publisher he knew very well, Th. Steinkopff. This was the beginning of the series of books I have written on atomic and molecular spectra, although it was many years before the books were completed.

The year 1928 was a vintage year in the study of molecular spectra and molecular structure. Not only was this the year in which the Raman effect was discovered (by Raman and Krishnan in India and independently but a few weeks later by Landsberg and Mandelstam in the Soviet Union) but was also the year in which Wigner and Witmer published their basic paper on the correlation of diatomic molecular states with those of the separated atoms; the year in which Hund and Mulliken published their fundamental papers on the molecular orbital theory; the year in which Hill and Van Vleck published their paper on the spin splitting in $^2\Pi$ states, and Winans and Stueckelberg their definitive explanation of the continuous spectrum of molecular hydrogen; the year in which Condon made his important contribution to the Franck-Condon principle, and in which Mulliken accounted for the atmospheric oxygen bands (and predicted the infrared atmospheric oxygen bands). During my stay in Göttingen I met many of these people.

Having carefully studied Wigner and Witmer's paper, it became clear to me that the then accepted value of the dissociation energy of O_2 derived from the convergence of the Schumann-Runge bands must be wrong because according to Wigner and Witmer the upper state ($^3\Sigma_u^-$) could not arise from normal (^3P) atoms: the $D(O_2)$ value had to be reduced by the excitation energy of the ^1D state of the oxygen atom (not then known accurately).

The Wigner-Witmer paper as well as the papers by Hund and by Mulliken were also very important for my own small contribution to the molecular orbital theory, which distinguished bonding and antibonding electrons. Although it was partly based on the erroneous paper by Heitler and myself mentioned above, it did lead to a somewhat better understanding of the stabilities of the electronic states of diatomic molecules and was accepted as such in later papers by Hund and also Lennard-Jones (see below). I submitted this paper in the fall of 1929 as my thesis for the Habilitation as Privatdozent at the Technische Hochschule Darmstadt.

During the last six months of my stay in Göttingen, Professor J. E. Lennard-Jones from Bristol came to Göttingen. He was very much interested in the molecular orbital theory, and I had many discussions on this subject with him. Lennard-Jones invited me to spend a second post-doctorate year at the University of Bristol, an invitation that I gladly accepted.

POSTDOCTORATE FELLOW IN BRISTOL (1929–1930)

I arrived in Bristol toward the end of September 1929, just a few days before the start of the General Discussion on Molecular Spectra and Molecular Structure of the Faraday Society. This meeting was organized by Professors Garner and Lennard-Jones of the Departments of Chemistry and Physics, respectively, at the University. It was an exciting meeting, as many of the foremost workers in the field were present and gave talks. The first speaker was O. W. Richardson, who was a terrible lecturer but a great physicist. I did not understand a word of what he said: he mumbled, and I was very concerned about my ability to learn English well enough. However, the second lecturer was C. V. Raman, who gave a very remarkable lecture and whose every word I understood. This was a year after the discovery of the Raman effect and a year before Raman received the Nobel Prize. I met Raman on several occasions during this meeting. He showed considerable interest in the work on nitrogen that Heitler and I had just published.

According to the printed discussion, C. P. Snow was present at the meeting, giving a talk on "Vibration-Rotation Spectra of Diatomic Molecules." I do not remember whether I met him. Other spectroscopists present at the meeting were E. F. Barker, R. T. Birge, G. H. Dieke, O. S. Duffendack, C. F. Goodeve, W. Jevons, V. Henri, J. Cabannes, J. Lecomte, and R. Mecke. R. S. Mulliken contributed a paper on intensity alternation, but was not present at the meeting. I met him for the first time only at the end of my stay in Bristol when he attended the meeting of the British Association.

The head of the Physics Department of the University of Bristol was A. M. Tyndall (as far as I know not related to the Tyndall of the Tyndall effect). He was a very able department head and continued his scientific work on ion mobilities with the help of a young assistant, his former student, C. F. Powell, who 20 years later, in 1950, received the Nobel Prize for his work on the photographic method of studying nuclear processes and his discoveries regarding mesons. The institute that Tyndall headed was very well equipped. It had been built from funds contributed by H. H. Wills, one of the owners of the well-known tobacco company, and was called (and is still called) the H. H. Wills Physical Laboratory. It was located in a very nice botanical garden situated on a hill overlooking the city. Some of the other staff members were W. E. Sucksmith who worked on magnetism, S. H. Piper who worked on X-rays, and L. C. Jackson who worked on solid state physics. In addition, H. W. B. Skinner must be mentioned; he was interested in soft X-ray work on solids but was also sufficiently interested in spectroscopy to design a two-meter vacuum grating spectrograph, which I was to use later. The spectrograph was built in record time. It was only started after I arrived and halfway through my fellowship it was available to me. In the meantime I had at my disposal two Hilger spectrographs, an E1 and an E2.

Naturally, it was difficult to get started in new surroundings with a somewhat unfamiliar language, but gradually I managed to get some work done during that year in Bristol. One interesting study was that of the spectrum of the P_2 molecule, for which I observed a rather striking case of predissociation. This caused me to think more about predissociation and eventually led to a review article in the *Ergebnisse der Exakten Naturwissenschaften*, which was published a year after I returned to Germany. When the grating spectrograph was ready I got some very nice spectrograms of P_2, which were fully evaluated only after my return to Darmstadt. In trying various light sources for the study of P_2 I also came across a spectrum that I showed to be due to CP. For many years it was the only known spectrum of CP.

As a continuation of the work on ultraviolet spectra of polyatomic molecules I tried to study formaldehyde (H_2CO) and acetylene (C_2H_2) in the vacuum ultraviolet. The results were by no means spectacular but further helped my entry into the field of polyatomic spectra.

It was during my stay in Bristol that my paper on molecular orbital theory was submitted for my Habilitation. I went to Darmstadt at the end of 1929 to complete the formalities connected with the Habilitation and shortly thereafter married Luise Oettinger, whom I had met as a student in Göttingen. She returned with me to Bristol and collaborated on some of the problems in which I was engaged.

During the year that I was in Bristol I had, of course, an opportunity to visit various places in England. Particularly interesting for me was my visit to Cambridge, where I had an introduction to P. M. S. Blackett from James Franck. This was my first meeting with Blackett, whom I later met on various occasions, including his visit to Ottawa. Blackett asked a young research worker, Mark Oliphant, to show me around; he introduced me to people like Cockcroft and others who were just on the verge of producing nuclear disintegration by means of artificially accelerated particles. Oliphant, of course, became very well-known later on. He went back to Australia, his home country, and was for many years professor of physics at the National University. Subsequently he became, for a span of five years, Governor of the State of Victoria.

During this time I also met R. K. Asundi, who was spending a year at King's College, London, working with R. C. Johnson (co-discoverer of the Baldet-Johnson bands of CO^+). Asundi introduced me to Indian culture and philosophy in addition to Indian contributions to spectroscopy.

PRIVATDOZENT IN DARMSTADT (1930–1935)

The institution of Privatdozent exists (to my knowledge) only in German-speaking universities. The permission to lecture (*venia legendi*) is given only after a thesis has been submitted and accepted and the candidate has given a test lecture to the faculty—a procedure called Habilitation. The thesis for this purpose is expected to be on a higher level than a normal PhD thesis. Having once passed through this procedure, as I did at the end of 1929, the Privatdozent is permitted to give lectures at the university that are "private," i.e. not prescribed for a course of study—in other words normally they are specialized lectures. The university collects a fee from the students who attend these lectures and that is the only income the Privatdozent gets for his lectures. Unless he has private means he needs another job to support himself. In the natural sciences, such a job is normally an assistantship, i.e. overseeing some of the practical laboratories. This is what I did from 1930 to 1935. It was also my job to guide and advise students proceeding to the Dipl. Ing. (Master's) degree and a very few to the Doctor's degree.

My first concern upon returning to Darmstadt was to get higher resolution for our spectroscopic equipment. A three-meter grating spectrograph was planned and built. We were lucky to obtain an extremely good grating from R. W. Wood at Johns Hopkins University.[1]

Much of the research work during this period was devoted to the study of the spectra of several diatomic molecules. While the grating spectrograph was being constructed I completed the analysis of the P_2 spectra taken in Bristol. There was a breaking-off in two successive vibrational levels on account of predissociation. I soon realized that if such a situation exists, it is possible to establish whether or not the predissociation limit is equal to a dissociation limit and not just an upper bound to it as in several other cases.

Sometime later I found a similar predissociation in N_2 in the upper state of the second positive group and analyzed it in detail together with G. Büttenbender, a PhD student of mine. It was clearly of importance to have a clear-cut dissociation limit in N_2, but because of the uncertainty of the dissociation products, it was only many years later that a definitive value for the dissociation energy of N_2 was established.

Other diatomic spectra studied during this period were those of CP, initiated in Bristol and completed with the help of my wife and H.

---

[1] When shortly after the war I visited R. W. Wood, whom I had never met before, his first question was, "What happened to my grating?" Unfortunately so much destruction had occurred in Darmstadt that there was no way of ever recovering that grating.

Baerwald;[2] PN, studied together with my wife and J. Curry, our first guest from the American continent; and BeO, initiated by my wife in Bristol and completed by her in Darmstadt and submitted as her PhD thesis in 1933 to the University of Frankfurt.

During this period in Darmstadt I became interested in forbidden transitions, initially only in diatomic molecules. For this purpose I set up in the basement of the building an absorption tube 12.5 m long, composed of glass sections, each 2 m long and 6 cm wide, and joined together by ground joints. At that time J. U. White's method of getting a long path by multiple traversals had not yet been invented (see below). Light from a hydrogen discharge went through the absorption tube twice, yielding an absorbing path of 25 m.

One of the main aims was to observe and measure the predicted $^3\Sigma-^1\Sigma$ absorption in N_2 and the $^4\Pi-^2\Pi$ absorption in NO. Neither of these aims was accomplished. However, when the tube was filled with pure oxygen and spectra taken with an ordinary quartz prism spectrograph, a striking progression of new absorption bands was found. Each band showed under the available resolution a single fairly well-resolved Q branch. The bands converged rapidly to a limit that agreed closely with the by then well-known dissociation energy of O_2. Knowing that the ground state of O_2 was a $^3\Sigma_g^-$ state, I immediately concluded that the upper state must be a new $^3\Sigma_u^+$ state because a $^3\Sigma_u^+-^3\Sigma_g^-$ transition, when the spin splitting is unresolved, would consist of Q branches only. When I published a short note about this work I had forgotten that several months earlier on a visit to Berlin, W. Finkelnburg had told me about results by W. Steiner and himself in which in O_2 at high pressure a series of broad bands with three maxima each had been found, and with wavelengths very similar to those of "my" new bands. When Finkelnburg brought this to my attention I felt very embarrassed. Actually, it turned out that the high pressure bands do not belong to the same band system that I had found at low (1 atm) pressure. Indeed at medium pressure both systems appear. Fifteen years later I observed the free molecule analogue of the high pressure bands and showed, as had already been suspected, that they represent a $^3\Delta_u-^3\Sigma_g^-$ transition of O_2. Nevertheless these events taught me to be doubly careful with advance information from scientific friends.

The ultraviolet triplet bands of O_2 appear also in liquid oxygen. In addition, under these conditions, J. W. Ellis and H. O. Kneser observed in

---

[2] Baerwald was the oldest of the Privatdozenten at the Physics Department in Darmstadt. He had not done any previous work in spectroscopy but wanted to get some new experience. He had a tough time during the Nazi period and died shortly before the war in the U.K.

the infrared a small absorption peak at 1.261 μ, which they ascribed to the $^1\Delta$ state predicted by Mulliken. I felt that a confirmation by a study in the gaseous state was necessary and proceeded to install a mirror system that guided sunlight from the roof of the building into our three-meter grating spectrograph. Using the new Agfa infrared plates I succeeded in photographing the solar spectrum to 12900 Å[3] and found the $^1\Delta_g-^3\Sigma_g^-$ transition with all its branches well resolved. The structure of these branches confirmed unambiguously the assignment as a forbidden (magnetic dipole) $^1\Delta_g-^3\Sigma_g^-$ transition and thus the existence of the predicted $^1\Delta_g$ state.

In each of the years 1928–1932, P. Debye had arranged in Leipzig small meetings in which a fairly narrow field was discussed by experts in this field. In 1931 the topic was Molecular Structure. I was fortunate in being invited to lecture on electronic structure of molecules and valence. Here I was able to present molecular orbital theory and my version of its connection with valence, and to discuss bonding and antibonding electrons and related topics. More important for me were the discussions with the participants of this meeting: R. Mecke, who had done so much of the early work on diatomic and polyatomic molecules; F. Rasetti, who had obtained the first high resolution Raman spectra of gases (cf the work on N_2 quoted above); G. Placzek, who had just developed the polarizability theory of Raman spectra, which he described in his lecture; Hertha Sponer, who had contributed so much to the problem of dissociation energies and whom I knew already from my Göttingen period; V. Henri, who had discovered (and named) the phenomenon of predissociation; and R. de L. Kronig, who was one of the first really to understand predissociation.

It was at this meeting also that I first met W. Heisenberg and E. Teller. The latter had just obtained his PhD under Heisenberg. Teller was then very much interested in molecules, especially polyatomic molecules. Our discussions at that meeting later led to our collaboration in the paper on the vibrational structure of electronic transitions in polyatomic molecules, which was written during visits of mine to Göttingen and visits by Teller to Darmstadt. My function was that of a midwife: Teller had the ideas, which I tried to get out of him by describing the experimental results to him and by drafting a tentative form of the paper, which he then corrected. Teller had an extraordinary reservoir of ideas in this field (as well as in other fields) and was always ready to share his knowledge. Working with him was an experience that I shall never forget. Although the ideas came from him, he insisted that on the title page we follow the alphabetical order of the authors. The main result of this work was that in allowed electronic transitions, totally symmetric vibrations are predominantly excited

---

[3] I believe this is still the longest wavelength recorded photographically.

whereas transitions with changes in nontotally symmetric vibrations are generally very weak, and such that $\Delta v_a = 0, \pm 2, \pm 4, \ldots$ while for forbidden electronic transitions $\Delta v_a = \pm 1, \pm 3, \ldots$. Since the publication of this paper 50 years ago, these rules have been confirmed in the analyses of innumerable polyatomic electronic transitions.

One of the best examples to which the new selection rules were applied at that time was H_2CO (formaldehyde). Just before the collaboration with Teller, I had a graduate (Dipl. Ing.) student, K. Franz, whom I asked to try to obtain a fluorescence spectrum of H_2CO. After some effort we were successful in observing the fluorescence and, indeed, with a spectrograph of high f ratio (i.e. $f/0.6$), in obtaining a spectrum. This spectrum showed a simple progression in the C–O stretching vibration, but the first band was not at the place of the exciting radiation, the A band in absorption, but at longer wavelengths coincident with the so-called α band, 1280 cm^{-1} from the A band. However, 1280 cm^{-1} was not a vibrational frequency of the ground state.

To study this problem further my next graduate student, S. Gradstein (who was by far the best student I had had in Darmstadt) obtained a fluorescence spectrum of higher resolution.

While we were working on the interpretation of the new spectra, a political event happened that had a dramatic effect on our lives: the assumption of power in Germany by the Nazis in January 1933. Gradstein, who was Jewish and of Polish nationality, considered it wise (and later events proved that it was) to leave Germany almost immediately. Further work on what was intended to be our joint paper was done by correspondence. By the time it was completed the situation had worsened with regard to myself (see below), and we decided to have the paper under Gradstein's name as the sole author. We did recognize in this paper that the electronic transition of H_2CO in the near ultraviolet was a forbidden transition, i.e. $A_2–A_1$ assuming C_{2v} symmetry, but we did not recognize that the molecule is nonplanar in the excited state, as was done 20 years later by A. D. Walsh and J. C. D. Brand. The wavenumber difference between the A and α band is the sum of the fundamental frequency $v_4 = 1167 \text{ cm}^{-1}$ and the inversion doubling 119 cm^{-1} in the excited electronic state.

In recent years, after the advent of laser spectroscopy, the fluorescence of formaldehyde has been studied under high resolution with excitation of single ro-vibrational levels. It was found that even the apparently sharp levels of the excited state are subject to predissociation, causing the low intensity of the fluorescence. For D_2CO the predissociation is much weaker and therefore the fluorescence much stronger.

In 1932, H. C. Urey, F. C. Brickwedde, and G. M. Murphy discovered heavy hydrogen. Its importance for the study of molecular structure was soon recognized by many investigators. The concentration of heavy water from ordinary water is a long process. At first, no commercial source of heavy water was available, so we set up an apparatus for the concentration of D_2O by electrolysis and obtained, after considerable effort, about 0.5 gm. Soon thereafter, commercial D_2O became available in Norway and we did not continue our own manufacture.

In view of the availability of the Agfa plates I decided to embark on studies of a few molecules containing deuterium in the photographic infrared. To gain experience we first studied hydrochloric acid (HCl), acetylene (C_2H_2), and hydrocyanic acid (HCN). Even though we had only a three-meter grating at our disposal, we were able to improve greatly the data for these molecules, thanks to the Agfa infrared plates.[4] The first deuterated compound we tried was acetylene, that is, C_2HD (the fully deuterated acetylene, C_2D_2, requires much longer optical paths for observation). Several bands were obtained and readily analyzed. The resulting B values combined with those of C_2H_2 give the individual distances r(CH) and r(CC), assuming of course that these distances are the same in C_2H_2 and C_2HD.

During the summer of 1933 I had a letter from a young physical chemist, J. W. T. Spinks, from the University of Saskatchewan in Canada, who asked whether he could work in my laboratory for a year. I had never heard of Saskatchewan then, nor had I heard of this young man. He had obtained his PhD with Professor A. J. Allmand at King's College, London. As I did know Professor Allmand, I wrote to him to inquire about Spinks before committing myself to him. Allmand's reply was very encouraging, so I told Spinks that I would be glad to accept him. He arrived in the fall of 1933, and together we had a very productive year of research in spectroscopy. He took part in the work on HCl, HCN, and C_2H_2. We also had some help from F. Patat of the University of Vienna, whom Spinks had met during a visit to Vienna, and who was more experienced than we were in the preparation of deuterated compounds.

During this time I was also working on the book on *Atomic and Molecular Spectra*. It turned out that the presentation of atomic spectra took as much space as originally assigned for the whole book, and therefore I decided to have a separate book on *Atomic Spectra and Atomic Structure*. The manuscript (in German) was completed just before I left Germany.

During my stay as Privatdozent in Darmstadt I had many opportunities for contacts with physicists and physical chemists at the University of Frankfurt. I especially valued my contacts with K. F. Bonhoeffer. He was one of the most distinguished German physical chemists, and certainly one of the finest people I have known. I was very happy to have the opportunity

---

[4] At that time the resolution of mid- and far-infrared spectra was greatly inferior.

to help him a little with spectroscopic information when he wrote one of the first books on photochemistry. Bonhoeffer had a very difficult time during the Nazi regime, but never compromised with the Nazis. His brother, the theologian Dietrich Bonhoeffer, was executed by the Nazis just before the end of the war.

It soon became clear that I would not be able to stay in Darmstadt indefinitely because of the new laws introduced by the Nazis, one of which was that people with Jewish wives could not teach at universities. Toward the end of 1933 I began to look for a job outside Germany. I approached two refugee organizations, who were very helpful indeed, but jobs at that time, a year after the first exodus, were difficult to find. During his stay with me Dr. Spinks became aware, of course, of my problem and on his return to Canada in the fall of 1934 tried to find out what possibilities existed there. The University of Toronto, where a good deal of spectroscopic work was going on, was unable to find a position for me, but the University of Saskatchewan and its president, Dr. W. C. Murray, became definitely interested in providing a position for me. It turned out that the Carnegie Foundation of New York had made funds available to universities in the British Commonwealth to hire refugees on two-year guest professorships if the university involved was willing to consider the candidate for a regular position after the two years (without making a definite commitment). Even though the University of Saskatchewan was then in a very precarious financial situation, they did offer me this guest professorship starting in the fall of 1935. The former Research Professor of Physics, T. Alty, had just resigned in order to accept a position at the University of Glasgow, and only three months after my arrival in Saskatchewan, Dr. Murray offered me this Research Professorship, which eliminated all concern about my own personal future.

My contract with the University of Darmstadt expired in the fall of 1935 so that events fitted together rather well. I had no difficulty leaving Germany, except that my wife and I were allowed to take with us only ten German marks each, then equivalent to $2.50. We were, however, able to take along our personal belongings, including some spectroscopic equipment that I was able to buy in Germany, thereby helping to overcome one of the principal difficulties in this move, namely the lack of spectroscopic equipment in Saskatoon. Nobody could tell then what the future would bring, and therefore leaving Germany was for us most painful.

RESEARCH PROFESSOR AT THE UNIVERSITY OF SASKATCHEWAN (1935–1945)

We left Germany on August 16, 1935 by boat from Hamburg and arrived in New York on August 22, 1935. However, before proceeding to Saskatoon, I took the opportunity to visit a number of universities in the eastern United States and eastern Canada. Some of the contacts established at this time proved to be very useful later on. In Princeton I met Professor E. U. Condon, who had just taken on the job of general editor of a series of physics texts to be published by Prentice-Hall. I told him about my book, *Atomic Spectra and Atomic Structure*, and soon afterwards I signed a contract with Prentice-Hall for the publication of an English translation of this book. While I was visiting various universities, my wife was visiting her sister in Chattanooga, Tennessee, and we met again in Chicago, from where we travelled by train via Minneapolis, Winnipeg, and Regina and finally to Saskatoon. On travelling between Winnipeg and Saskatoon, we became increasingly worried because the train seemed to stop at places that had only a few houses, and we were wondering what Saskatoon would be like. However, upon our arrival in Saskatoon, we were very pleasantly surprised. Saskatoon was an attractive small city of 40,000 people. The University had a large campus with only a few buildings, but our reception there was most friendly. Everyone went out of their way to help us get settled. There were 1200 students and approximately 100 faculty members, all of whom we soon came to know.

The members of the physics department were E. L. Harrington (head), B. W. Currie, C. A. McKay, and R. N. Haslam. Their interests then were mainly in teaching; they were quite happy that I could take care of some of the graduate students. E. L. Harrington was a former student of R. A. Millikan and had worked with him on the determination of the electronic charge. The head of the chemistry department was T. Thorvaldson, a very active person, who had made a name for himself in the field of cement chemistry. He later became dean of the graduate school.

My appointment had come up through the chemistry department and the administration had almost bypassed E. L. Harrington, knowing that he was the worrying kind. When he found that I was not the demanding sort of person that he had expected, we became good friends. He was an excellent experimenter and an expert glassblower. Haslam, the youngest member of the faculty, had arrived at the same time as I did. He had had an 1851 postdoctoral fellowship, which had enabled him to spend two years in Leipzig with Heisenberg. He helped me a great deal in proofreading my books and in improving my English. Currie was active in research on the upper atmosphere, particularly the aurora and the night sky, and acquired a considerable international reputation in this field.

We arrived in Saskatoon on September 15 and classes were to start about two weeks later. I was to give two lecture courses: one undergraduate course in mechanics, and one graduate course on atomic and molecular spectra. It took me, naturally, some time to get used to the rather different way of teaching on this continent compared to Germany.

Realizing that I would find very little equipment suitable for my research at the University of Saskatchewan, I had worked hard before leaving Darmstadt to collect material for a number of spectroscopic studies. Working up this material to the state at which some results could be published took the first two or three years of my stay in Saskatoon. In addition, the proofreading of the German edition of *Atomic Spectra and Atomic Structure* had to be done. John Spinks volunteered to make a first draft of the translation of this book into English, which he did in record time. By the end of 1936 we were able to submit the manuscript of the English edition to Prentice-Hall. When that was done, I started to work on the book on *Molecular Spectra*, of which I had already prepared a rough draft in German. It soon turned out that the section on diatomic molecules of this proposed book would be fairly voluminous and it seemed impossible to include polyatomic molecules. Thus, I decided to split the book into two volumes, *I. Spectra of Diatomic Molecules, II. Spectra of Polyatomic Molecules.*

I finished work on the German version of the book on *Diatomic Molecules* (Volume I) in the fall of 1938. At nearly the same time John Spinks completed the English translation, which was published in the fall of 1939. Soon thereafter I began to work on what originally was to be the concluding volume (II) on polyatomic molecules (this time directly in English). Again the manuscript became too long for one volume and the book was again split into two: the first one, Volume II of the series, *Molecular Spectra and Molecular Structure*, was to be *Infrared and Raman Spectra of Polyatomic Molecules*, and the second, Volume III of the series, was to be *Electronic Spectra of Polyatomic Molecules.* Somebody jokingly pointed out to me that this history was similar to Richard Wagner's experience when he wrote the Ring Cycle. Originally, he had intended to write only one opera but found that he had to split it first into two and finally into four parts: the first one, *Das Rheingold* was short—the analogue of *Atomic Spectra and Atomic Structure*—and the other three were long, as were the three volumes of my series on molecular spectra that were finally published. Volume II was completed toward the end of 1944 and published in 1945.

When Volume II was completed, Prentice-Hall did not want to take on its publication, even though a contract existed to this effect. The argument was that wartime restrictions on the supply of paper made it necessary for them to reserve this paper for more profitable books. I was therefore faced with the problem of finding another publisher. Contact was made with Van Nostrand through one of the agents who visited the campus of the University of Saskatchewan, and they fortunately agreed to take on the publication of Volume II. The copyright for Volume I had also been returned to me by Prentice-Hall, as they could not undertake to keep it in print by new printings, and so I was faced with the problem of preparing a new edition of Volume I, which Van Nostrand was willing to publish. Volume III was only published in 1966.

In writing these books, I learned a great deal about the subject and not infrequently I found gaps in the literature, which sometimes I was able to fill. A striking example was the discovery in the available spectroscopic data of the phenomenon of *l*-type doubling, resulting in a paper that I contributed to a conference on spectroscopy organized by R. S. Mulliken and held at the University of Chicago in June 1942 (a predecessor of the annual Columbus meetings on molecular structure). I did not actually attend this meeting but the paper was presented, I believe, by R. S. Mulliken. Although everyone seemed to be interested in *l*-type doubling, the short theoretical discussion in my paper caused some controversy. In particular, W. H. Shaffer and H. H. Nielsen objected to it and derived a formula similar to mine but with a factor 2 missing. I was naturally rather pleased when, after considerable and heated discussion by correspondence, it turned out several years later that the factor 2 in my formula was correct.

Going back to the first few years in Saskatchewan, I must mention one result that arose from spectra that I had taken in Germany with the help of F. Patat and H. Verleger, namely spectra of methylacetylene. The C–C single bond distance in this molecule was substantially shorter than C–C single bond distances in molecules such as ethane and others. This fact was so striking that Linus Pauling did not believe the result and asked his colleague, R. M. Badger at the California Institute of Technology, to repeat our experiment. Only when he confirmed our results did Pauling and others accept this shortening of a single bond distance adjacent to a triple bond. Many other examples of this type were found later. Several of the studies after I came to Ottawa were directed to this problem.

Spectroscopic equipment was not plentiful in Saskatoon. However, a medium Hilger spectrograph was available and, together with two bright students, I made good use of it. Later on, a six-meter grating spectrograph was built with the help of funds supplied by the American Philosophical Society. This particular grant arose in the following way. I was able to attend a meeting on molecular structure in the spring of 1937 in Princeton where I had a chance to visit Professor H. N. Russell, the leading American astronomer of his time. I discussed with him two problems in which I knew he would be interested: first, the detection of molecular hydrogen in the outer planets by means of the quadrupole spectrum of H_2, and second, the possibility of making use of the cold climate in Saskatchewan to obtain solar spectra in the red and infrared that were not cluttered by many lines of H_2O. Russell was sufficiently interested in these problems to suggest that he would strongly support an application that I would make to the American

Philosophical Society for a research grant. The total amount of the grant was $1,500, of which $500 was needed to buy a grating and the remainder was needed to construct the spectrograph.

It was thus, four years after I came to Saskatoon, that I had at my disposal a fine spectrograph that was used for a number of interesting studies. I shall mention only three of these. During the years 1940 and 1941 in Saskatoon, I had an exceptionally gifted student, A. E. Douglas, who later joined me in Ottawa. For his Master's thesis I had suggested that he try to find the spectrum of the B_2 molecule. The search for this spectrum was successful and we were able to determine for the first time the structure of the B_2 molecule. One of the reasons for the interest in this molecule was the question of the nuclear spin of ^{10}B and ^{11}B and also the statistics of these nuclei.

The second item was the discovery of a spectrum of CH^+ in the summer of 1941. In the early summer of that year I had attended a meeting at the Yerkes Observatory that was devoted to the problem of the recently discovered sharp interstellar lines. CH and CN had already been identified by means of these lines, but there remained an unexplained progression of three or four sharp lines observed by W. S. Adams at Mt. Wilson Observatory. During the discussion at Yerkes Observatory, Teller and I came to the conclusion that this spectrum was probably due to CH^+. Upon returning from the meeting to Saskatoon, I found that Alex Douglas had a suitable discharge tube available with which we could immediately proceed to the search for CH^+. Within a few days we had a new spectrum, which turned out to be that of CH^+. Fortunately, it was possible for us to look at the fine structure with the help of the new grating spectrograph, and we found that the $R(0)$ lines of these bands agreed exactly with the observed interstellar lines, thus establishing the presence of CH^+ in the interstellar medium.

The third item refers to a feature long known in the spectra of comets. This feature, however, resisted all attempts at identification. It was called to my attention by the Belgian astronomer P. Swings. It was known as the 4050 group and consisted of a number of line-like features. In looking at this spectrum, and remembering what I had learned in writing Volume II of my series, I came to the conclusion that the 4050 group must be due to a molecule consisting of two hydrogen atoms and one heavier atom. The most reasonable assumption for such a system seemed to be CH_2. On the basis of this conjecture, I carried out experiments on a discharge through methane, which I thought might give the spectrum of CH_2. That was very naive. As might have been expected, the spectrum of such a discharge consists of features belonging to H_2, CH, and H, and no 4050 group. I noticed, however, that the initial color of the discharge was slightly different from what it was when it was running continuously. I therefore tried an interrupted discharge in methane and as a result obtained a spectrum in the laboratory that contained, in addition to CH and H_2, precisely the features obtained in comets at 4050 Å.

It is perhaps understandable that I considered the success in finding the 4050 group in the laboratory in methane as confirmation of my conjecture that this spectrum is due to CH_2. I therefore proceeded to try to obtain an absorption spectrum by CH_2 by a method that was suggested by the chemical experiments of F. Paneth, T. G. Pearson, R. H. Purcell, G. S. Saigh, and others, namely to photolyze a compound like ketene that was believed to yield CH_2. I worked hard to prepare ketene, which at that time could not be purchased, and I finally succeeded in preparing it. However, the experiment aimed at finding the absorption spectrum of CH_2 failed miserably. As is explained in the next section, the 4050 group is actually not due to CH_2 but is due to the C_3 radical; however, while in Saskatoon I did not know that.

The success with CH^+ and the 4050 group was of considerable interest to the astronomers, and as a result I obtained an offer from the Director of the Yerkes Observatory, Prof. O. Struve, to join their staff and to set up a spectroscopy laboratory. This offer was made in 1943, but because of the restriction then of the movements of scientific manpower, I was unable to accept it. However, the offer was held open until the end of the war, at which time I felt I should accept the offer. So in the fall of 1945 I left Saskatoon together with my family and joined the Faculty of the University of Chicago, to which the Yerkes Observatory belonged.

Here I must record the births of my two children, Paul and Agnes, during the first three years of our stay in Saskatoon. Paul, although originally tending to physics, eventually became a psychologist, teaching at York University, Toronto, while Agnes became a mathematician, teaching mathematical statistics at Imperial College, London.

An interesting and pleasant interlude occurred just before the war. I was asked by the physics department of the University of Michigan (Professor H. M. Randall) to give a course of lectures on spectroscopy, to substitute for Professor D. M. Dennison, who was on leave. I gladly accepted this invitation, because it gave me a chance to make many contacts with spectroscopists in Ann Arbor and to attend the annual summer school. Here I had an opportunity to meet and attend lectures by E. Fermi and J. A. Wheeler. Fermi was a man of great charm and was an incredibly good lecturer. Wheeler (also a very good lecturer) discussed the work he had done with N. Bohr on fission. L. Szilard was there, urging secrecy on all work on nuclear fission. Many other well-known people participated. S. Goudsmit and R. Sawyer, who were on the faculty at Ann Arbor, were most hospitable to the visitors. I even tried some experiments with the help of Professor Randall's technicians, with the object of obtaining infrared emission spectra of HD^+ and H_3^+. These attempts were unsuccessful. It was only when much more sophisticated methods had been developed 35 years later that these spectra were first observed: HD^+ by W. H. Wing and his collaborators, and H_3^+ by T. Oka.

PROFESSOR OF SPECTROSCOPY AT THE YERKES OBSERVATORY OF THE UNIVERSITY OF CHICAGO (1945–1948)

In contrast to the University of Saskatchewan, which was strongly oriented toward teaching, the Yerkes Observatory was almost exclusively engaged in research, with very little teaching. The Director of the Observatory, who had been responsible for my joining the faculty, was Otto Struve, an extremely able astronomer, known throughout the world for his scientific contributions. He was also a very competent administrator and a hard worker. The other faculty members were G. P. Kuiper, W. A. Hiltner, W. W. Morgan, and S. Chandrasekhar, all of them distinguished astronomers. The association with Chandrasekhar made my stay at Yerkes Observatory especially worthwhile. A man of mathematical genius, he cleared up innumerable theoretical problems in astronomy. We had many interesting and pleasant discussions of his and my work (in very different fields). He had an incredible memory and many illuminating insights on the history of physics and astronomy.

Although I did give a course on spectroscopy, my main job during the first year or two was to set up a spectroscopic laboratory in the basement of the building. A grating spectrograph that was very similar to the one I had in Saskatoon was built in fairly short order. In addition, I was anxious to set up a long absorption tube that would enable me to study various forbidden transitions of interest to astronomy. Fortunately, the Observatory had an optical shop under Fred Pearson, who many years earlier had worked with Michelson and had become an expert optical technician. He produced a set of mirrors for me that proved to be of excellent quality, making it possible to obtain up to 250 traversals of the absorption tube.

I had two students who wanted to take their PhD with me, John Phillips[5] and Narahari Rao.[6] In addition, I had two post-doctorate workers: H. J. Bernstein, a Canadian who had been interned in Denmark during the war and wanted to get back into the field of spectroscopy before taking up his

[5] Now at the University of California at Berkeley.
[6] Now at Ohio State University.

duties at the National Research Council in Ottawa, and S. P. Sinha from Patna University in India, who wanted experience in spectroscopy.

With the new absorption tube, spectra of a number of simple molecules were taken in the photographic infrared and also in the ultraviolet. The work that I was most anxious to do was the observation of the quadrupole spectrum of molecular hydrogen, which was actually accomplished toward the end of my stay at Yerkes Observatory. The resulting spectra were particularly interesting for the study of planetary atmospheres but have later become important also for the interstellar medium and for stellar atmospheres. In addition, photographic infrared spectra of CO_2, N_2O, CO, and CHF_3 were obtained and, of special importance for upper atmosphere physics, the ultraviolet spectrum of O_2, which I had started to study in Darmstadt.

During the time of my stay at Yerkes Observatory the Office of Naval Research was founded and asked for applications for research grants. I made such an application for the new laboratory that I had established, but this application was unsuccessful. It is just possible that this negative result influenced me when I obtained at the end of 1947 an offer from the National Research Council of Canada to set up a new spectroscopic laboratory, with the opportunity of ample funding for equipment and personnel. At any rate, I decided to accept this offer and left Yerkes Observatory in August 1948 to start work in Ottawa. There were, of course, many other personal reasons for making this change. In retrospect, I am certainly very happy that I made it, even though it caused another hiatus in my scientific output, as obviously the establishment of a new laboratory takes time.

DIRECTOR OF (PURE) PHYSICS AT THE NATIONAL RESEARCH COUNCIL OF CANADA (1949–1969)

Originally, my appointment at the National Research Council was as a Principal Research Officer, which is the highest nonadministrative position at the Council. Similar to my coming to Saskatoon, my appointment in Ottawa was engineered by the Director of the Division of Chemistry, Dr. E. W. R. Steacie, later President of the Research Council, but my appointment was in the Division of Physics, nominally under R. W. Boyle, who, however, was just in the process of retiring. The problem of the new director of the Physics Division was discussed with me by Steacie, who was anxious to have a research-oriented person in this position. Steacie strongly urged me to consider accepting this position if it was offered by the President. He pointed out that because I did not know who the new director would be, it would be wise on my part to accept the job myself if it was offered to me. He also made it clear to me that I should not have to spend a lot of time on administration, as there were many qualified people to do that at the Research Council. My main administrative function would be to find first-class people for the positions that were open and let them use their own judgment in selecting their own research projects rather than try to direct them. I found this a very good piece of advice and when in January 1949, C. J. Mackenzie, the President, did offer me the job of Director of the Physics Division, I accepted.

I was extremely lucky when my former student, A. E. Douglas, mentioned above, agreed to join me in starting and running the new spectroscopy laboratory at the National Research Council. He made an immense contribution during a period of over 30 years. Without his constant help to everyone in the laboratory through suggestions and criticism and of course his own scientific work, only a small fraction of what was actually done would have been accomplished.[7] It was Alex Douglas who, by his sense of humour, his ever-ready helpfulness and enthusiasm, his modesty and gentle prodding, succeeded in establishing throughout the years an atmosphere of camaraderie in the laboratory, which still prevails and which was and is an important reason for its success. Another aim that we both shared was a spirit of freedom that we valued and that all members of the laboratory valued, that is, freedom in the choice of problems for anyone who wanted such a choice. Several former members of the laboratory have told me how much they appreciated this freedom and how much they enjoyed the camaraderie.

Two of the contenders for the job of director before it was offered to me were L. E. Howlett and D. C. Rose. Both of them had been with the National Research Council almost from the beginning. During the war, D. C. Rose had been in charge of CARDE (Canadian Armament Research and Development Establishment) but wanted to go back to basic research and had started a laboratory on studies of cosmic rays. He was always most cooperative in any administrative matter that came up. L. E. Howlett was interested more in the applied side of physics, and had already established himself as an effective administrator. He was eventually asked to become co-director of the physics division and was to take care of most of the daily administration. This system worked quite well for a number of years, but in 1955 the decision was made to split pure and applied physics into two separate divisions, of which Howlett took the Applied Physics and I took

[7] See the *Biographical Memoirs of Fellows of the Royal Society*, published by the Royal Society, Vol. 28, p. 91, 1982.

Pure Physics. I did not particularly favor this separation, and after Howlett retired at the beginning of 1969, the division was recombined under A. E. Douglas.

At the time of my coming to the National Research Council, Steacie had just introduced, in the Chemistry Division, postdoctorate fellowships, which in many ways were a substitute for graduate students, whom we did not have. He strongly urged me to institute such a system also in the Physics Division, and that was in fact done. The existence of this system made it possible to prevent stagnation, which easily happens in government organizations when all the positions are filled with permanent appointments. The presence for one or two years of these postdoctorate fellows in all our laboratories helped to keep the group leaders on their toes and had a very favorable effect on the morale of the Division.

Another fortunate development, which, like my acquaintance with John Spinks, I owe to Professor A. J. Allmand of King's College, London, was his recommendation to Steacie of one of his senior technicians who was anxious to come to Canada for family reasons. Jack Shoosmith was all that Allmand had said about him. He was efficient, intelligent, a self-starter, and helped me tremendously in the next 20 years until 1969, when he retired. Without his help, a great deal of my own work would not have been done.

It is, of course, impossible to describe all the work that was done in the years since I joined NRC, and it would certainly be impossible to do that in the exact historical order. I shall restrict myself to a description of my own activities, and among these only the principal points.

Although most of the work of the laboratory was in the field of molecular spectroscopy, I did spend some time on problems in atomic spectroscopy and atomic structure. After having set up a grating spectrograph very similar to the one I had had at Saskatoon and Yerkes Observatory, Douglas and I decided for our first work with this instrument to determine the spin and statistics of the ^{3}He isotope. This was done by way of the intensity alternation in the spectrum of ^{3}He$_2$. The result (spin $\frac{1}{2}$ and Fermi statistics) was in agreement with expectation.

Shortly before I came to Ottawa, the Lamb shift had been discovered in atomic hydrogen. The question that interested me in this connection was: what is the Lamb shift in the ground state of hydrogen? While the Lamb shift of the $2S$ level can be determined by comparison with the nearby $2P$ level, the Lamb shift of the $1S$ level can only be obtained by comparison with the same $2P$ level. For such a measurement, a very accurate determination of the wavelength of the Lyman α-line was necessary. This was not an easy task, but eventually, for the case of deuterium, a Lamb shift of 0.26 cm^{-1} was obtained, which within the rather large uncertainty of the measurement agrees quite well with the value predicted by quantum electrodynamics. It was only many years later than A. L. Schawlow and E. Hänsch obtained by laser techniques a much more precise value of this quantity. Following this result I attempted to determine the Lamb shift in the ground state of helium (a two-electron system). This proved to be more difficult because the resonance line of helium is at 584 Å and has to be accurately measured. Such a measurement was accomplished in this case as well as in the case of the Lyman α-line by the use of the combination principle. These measurements were made with our three-meter vacuum spectrograph. The spectra were taken in the tenth order of this instrument. I was very pleased with the accuracy obtained. The agreement with the predictions of quantum electrodynamics is again of the order of ± 0.05 cm^{-1}. Up to now, to the best of my knowledge, no new measurements of this quantity have been made. The measurement also yielded a precise value of the ionization potential of ^{4}He and ^{3}He that was much more accurate than had been available up to that time.

Several other measurements of Lamb shifts in the He$^+$ and Li$^+$ ions were made, but work on this problem was terminated because other, much more precise ways of checking the results of quantum electrodynamics are now available.

One of the most important activities of the laboratory was in the field of spectra of diatomic molecules and especially in the study of forbidden transitions in these spectra. In particular, the spectra of hydrogen and its isotopes occupied my attention for many years and still do. The study and actual observation of the quadrupole spectrum of H_2 had given me a great deal of satisfaction; its importance for astronomy is discussed further below. The next item concerning hydrogen was the observation of the spectrum of HD. The HD molecule, because of its slight asymmetry, has a very small dipole moment and because of that an ordinary dipole spectrum was expected. One of my first experiments after coming to Ottawa was to set up a glass absorption tube of 5 m in length in which through a White mirror system an absorbing path of up to 1000 m was produced. With this absorbing path two absorption bands (3–0 and 4–0) of the rotation vibration spectrum of HD were actually obtained. Later, one of the postdoctorate fellows (R. A. Durie) and I observed the fundamental (1–0 transition) of HD by means of our two-meter infrared spectrometer.

Another diatomic molecule on which I spent considerable time was the oxygen molecule. I had already found forbidden O_2 bands in the ultraviolet in Darmstadt. Fairly high resolution plates had been obtained at Yerkes Observatory with the long absorption tube established there, but the plates could not be evaluated until I came to Ottawa. It was found that there were actually three forbidden absorption systems in O_2 which fitted in very well with the predicted electronic states of this molecule. In addition, Peter Brix and I studied the ordinary Schumann-Runge bands of O_2 under high resolution with our three-meter vacuum spectrograph and, among other things, obtained what is still the most accurate value for the dissociation energy of the O_2 molecule.

As mentioned above, several of the diatomic spectra that we studied are of astronomical interest, especially for planetary atmospheres and comets, but also for the interstellar medium. Even before the quadrupole lines of H_2 were identified in the spectra of the outer planets, I succeeded in establishing the presence of molecular H_2 in these atmospheres by means of the presssure-induced spectrum of H_2, following the interesting work of H. L. Welsh and his students at the University of Toronto on the fundamental and first overtone. I had a hunch that a diffuse feature first observed in Uranus by G. P. Kuiper might be accounted for by the second overtone of the pressure-induced spectrum. This hunch was actually confirmed by laboratory experiments with H_2 at high pressure and low temperature.

In the interstellar medium there are a number of diffuse lines in the visible region, which even now have not been identified. Much of my spectroscopic work in these 20 years as Director of (Pure) Physics was initiated with the view to solving the problem of the diffuse interstellar lines, but this task has so far been unsuccessful. However, free radicals and molecular ions whose spectra are discussed below are possible candidates.

The spectra of many free diatomic radicals have been known for many years as occurring in electric discharges and flames, for example OH, CH, and others, but the spectra of polyatomic free radicals were not well-known. One of our principal activities was the study of such free radicals, at first almost exclusively by the flash photolysis method. R. G. W. Norrish and G. Porter were the first to develop this method. It was also independently developed in our laboratory by D. A. Ramsay and mainly used for high resolution studies of free radicals. The first result was the observation of the spectrum of the NH_2 radical produced by the photolysis of ammonia. This spectrum, which is rather complicated, was fully analyzed by K. Dressler and D. A. Ramsay. The second free radical was HCO, which presented a number of rather interesting problems.

While still in Saskatoon I had tried to obtain the spectrum of CH_2. My own interest in the flash photolysis technique was motivated by the hope that in this way I would eventually be able to obtain the spectrum of CH_2, which I had failed to get when I decomposed ketene by a mercury lamp. It was ten years after I came to Ottawa and 17 years after the first experiments in Saskatoon that I finally observed the absorption spectrum of CH_2 in the vacuum ultraviolet. Naturally, again ketene was tried as the parent compound but ketene itself absorbs where the absorption of CH_2 occurs.

Only when we turned from ketene to diazomethane (a somewhat dangerous substance) did we obtain almost immediately the first true spectrum of CH_2 at 1415 Å. Before CH_2 was found, the CH_3 radical was observed with much less difficulty. In both cases detailed information about the structure of these radicals was obtained from their spectra. In CH_2 a second absorption spectrum was found in the visible region. This spectrum was due to the singlet states of CH_2, whereas the vacuum ultraviolet spectrum was due to the triplet states. From the relative behavior of these two spectra, it was evident that the lowest triplet state is below the lowest singlet state, but the exact energy difference between the two remained a problem for another 25 years. Only very recently has this energy difference been exactly determined in our laboratory.

Other free radicals in which I became interested were BH_2 (I was really looking for BH_3, but so far nobody has observed a spectrum of BH_3), SiH_2, HNCN, and HSiCl. Many others were studied by other members of our laboratory. The carrier of the 4050 group of comets that I had observed in the laboratory in Saskatoon was identified by Douglas as being the C_3 free radical. The spectrum of this radical appeared in considerable intensity in the flash photolysis of diazomethane. Together with several collaborators, we analyzed this spectrum in detail and in particular found that the bending frequency is extremely low (63 cm^{-1}), which accounts for some of the complications of this spectrum that had made the identification in comets so difficult. Two other free radicals, discussed further in the next section, are H_3 and NH_4. These are radicals whose ground states are unstable, and only their excited Rydberg states are stable; they might be called Rydberg radicals.

My work on the spectra of molecular ions was originally motivated by the possibility that the diffuse interstellar lines might be due to a molecular ion. Therefore, after establishing the flash discharge technique by the observation of the well-known spectrum of N_2^+ in absorption, we immediately turned to the study of CH_4 under similar conditions, because I thought that the CH_4^+ ion might be a good candidate for the diffuse interstellar lines. However, what we did get in this study was the spectrum of the C_2^- ion, which was the first discrete spectrum of a negative ion; this spectrum had many other points of interest, but it did not help in the interpretation of the diffuse interstellar lines. Other ions, whose well-known emission spectra we observed in absorption were CO_2^+ and N_2O^+.

In addition to the study of many diatomic molecules and polyatomic free radicals, I continued my interest in stable polyatomic molecules. Before the development of modern infrared instruments, the study of photographic infrared spectra with a large diffraction grating seemed to be a profitable way of getting information about the rotational and vibrational level of many molecules. Some of this work had already been started while I was still in Darmstadt—in particular, the work on methylacetylene. Now, with the availability of the White mirror system, a number of other stable polyatomic molecules were studied, like CHF_3 and HNCO, as well as some molecules without hydrogen for which the intensities in the photographic infrared were very much smaller; extensive spectra of CO_2 and N_2O were obtained and analyzed with the help of my first wife. Other members of our laboratory studied a large number of molecular structures by these and other techniques, especially B. P. Stoicheff by the use of Raman spectra and C. C. Costain by the use of microwave spectra. Many examples were observed in which the bond distances were found to depend significantly on the nature of the neighboring bonds, as was first established for the C–C single bond in methylacetylene.

At the same time work was done on the ultraviolet absorption spectra of a number of molecules. Keith Innes and I studied the HCN vacuum ultraviolet absorption. Clear-cut evidence was obtained that in the observed excited states the molecule is not linear. The same fact was found in many free radicals and other molecules.

These studies of ultraviolet spectra of free radicals and stable molecules gave me a much better understanding than I had before of electronic transitions in polyatomic molecules, the subject of the third volume of my series on *Molecular Spectra and Molecular Structure*. With all the activities in which I was involved, a special effort was required of me to write this volume. I made sure that all Saturdays and holidays other than Sundays and Christmas Day were fully reserved for work on this book in the years 1960–1965. It was hard work, but the volume was finally published in 1966.

Two years after completing Volume III, I was invited by the chemistry department of Cornell University to present the 1968 *Baker Lectures*. I was asked to prepare these lectures for publication, and a small volume on the *Spectra of Free Radicals* was finally published in 1971. In many ways this book presented a short version of the three volumes on *Molecular Spectra and Molecular Structure*. It was translated into several languages, but on this continent it did not become the best-seller that the publisher had expected it would.

During all this time I had, of course, many invitations for lectures. In one lecture, the *Second Condon Lecture* at the University of Colorado, I discussed the work on molecular hydrogen. Condon, who was in the audience, asked me a question during the discussion about the series limit of the Rydberg series of H_2. I was somewhat embarrassed by his question because I had not thought of it. It caused me to do some more experiments, which produced a slight correction in the value of the ionization potential. Ugo Fano was in the audience of another lecture on the same topic I gave in

Chicago. He got quite excited about the results that I presented, and within two or three weeks after my lecture he sent me a long manuscript on the theory of interacting Rydberg series. This proved to be a very basic paper in this field.

Some of the invitations for lectures took me around the world. On my first trip to India I attended the Indian Science Congress, subsequent to which a lecture tour was organized to a dozen or more universities where I gave a total of 25 lectures in five weeks. I also went to Darjeeling to see the Himalayan mountains. Here I met the Dalai Lama, who at that time was still in power in Tibet and was on a state visit to India. Of course, I saw some of the usual tourist sites of India, like the Taj-Mahal, the Ajanta and Ellora caves, etc. My friend, R. K. Asundi, took great care to make the visit interesting and pleasant. It was, however, quite strenuous.

There were many trips to Europe, including some to the Soviet Union (the meeting of the International Astronomical Union in Moscow in 1957 was the first occasion). In 1960 I spent several months at the Astrophysical Observatory of the University of Liège. Later in the same year (November 17) I presented the *Bakerian Lecture* to The Royal Society of London on my work on CH_3 and CH_2. Three years later I had a lecture tour through Australia. In 1971 I attended a laser meeting in Esfahan in Iran. An excursion to the remarkable remains of Persepolis was especially interesting.

I was fortunate in having several opportunities to visit Japan, including two as a member of an advisory committee of the newly formed Institute of Molecular Science. Japanese hospitality and friendship were often overwhelming and of course very enjoyable. It was a remarkable experience to see the rapid development of science and especially spectroscopy in Japan.

In addition to all these activities, I also had to serve on various international spectroscopy commissions, a somewhat time-consuming activity. Thus the Molecular Spectroscopy Commission of the International Union of Pure and Applied Chemistry requested that I prepare a *Multilingual Dictionary of Spectroscopic Terms* with the help of other members of the Commission. I was also elected President of Commission 14 of the IAU, and had to prepare a lengthy *Presidential Report*.

"DISTINGUISHED RESEARCH SCIENTIST" AT THE NATIONAL RESEARCH COUNCIL (1969–PRESENT)

In the summer of 1969 the Ninth International Symposium on Free Radicals was held in Banff, Alberta and designated in my honor (anticipating my sixty-fifth birthday). At the banquet at the Banff Springs Hotel on August 25, Dr. W. G. Schneider, then president of the National Research Council of Canada, announced my appointment as the first Distinguished Research Scientist, relieving me of all administrative duties and allowing me to devote all my efforts to scientific research, even beyond my then approaching sixty-fifth birthday.

It was at this time that the work on the dissociation energy and the ionization potential of hydrogen was completed as well as the book resulting from the *Baker Lectures*.

In June 1971 my wife died. She had been working on upper atmosphere problems in the Department of Communications in Ottawa. Much of my work in Darmstadt, in Saskatoon, and the earlier work in Ottawa had been greatly aided by her collaboration and independent contributions.

Later in the same year I visited again the Soviet Union and gave some lectures in Moscow and Leningrad. There I was informed by a representative of the Soviet Academy of Sciences that I had been awarded the Nobel Prize. This extraordinary honor meant a great change in my life. The number of invitations for lectures and for functions increased by a considerable factor, and quiet scientific work was impossible for at least a year or two. This becomes very obvious on looking at my list of publications. There is quite a gap at that time except for the publication of the *Nobel Lecture*. It was fortunate that the definitive paper on the ionization potential of H_2 with Christian Jungen was in press then and needed only proof-reading.

In March 1972 I married Monika Tenthoff, whom I had known for a long time as the niece of a close friend from my high school days.

The first really substantial contribution after the prize came only in 1974 and was connected with the appearance of the comet, Kohoutek. Just a year or two earlier Hin Lew in our laboratory had, for the first time, observed a spectrum of H_2O^+. When we saw the first spectra of the tail of the comet, sent to us by astronomers in Italy and Israel, we immediately realized that there was a strong possibility that H_2O^+ might be present, and a more detailed study confirmed this result in detail. Sometime later H_2O^+ was also identified in the spectra of the twilight glow taken by Krassovsky, thus indicating the presence of these ions in the upper atmosphere. However, as yet there is no mass spectrometric evidence of H_2O^+ in the upper atmosphere.

In 1974 an international meeting entitled "Perspectives in Spectroscopy," in honor of my seventieth birthday, took place at Mont Tremblant in Quebec. Dr. Schneider, on behalf of the National Research Council, announced the formation of a new Institute of Astrophysics in which all astronomical and astronomy-related activities of the National Research Council were combined. This Institute was to carry my name and was formally established in 1975. The formation of the Institute did not involve any administrative functions for me but gave me the opportunity to continue scientific work.

During the following years I continued my interest in molecular ions, and together with Izabel Dabrowski, studied HeH^+, $HeNe^+$, and $HeAr^+$. None of these three ions has yet been observed in astronomical sources, but I am firmly convinced that sooner or later the first two will be found in the interstellar medium or in planetary nebulae.

Another molecular ion that I was very anxious to observe in the laboratory was H_3^+. I decided to look for the emission spectrum of this ion. At the same time in our laboratory T. Oka was looking for the absorption spectrum. After four years of hard work Oka finally observed the absorption spectrum of H_3^+, but my own search for the emission spectrum of H_3^+ was unsuccessful. However, in this search I found a new spectrum, which turned out to be the spectrum of neutral H_3. This was a case of sheer serendipity. I was not looking for H_3. It took me one or two months to recognize that the spectrum that we had observed was due to H_3 in Rydberg states, but once I realized this, everything fell into place. The structure of H_3 in these Rydberg states is very similar to the predicted and later experimentally confirmed structure of H_3^+. The three protons in both H_3^+ and H_3 are arranged at the corners of an equilateral triangle with very similar distances. I should emphasize that in the evaluation of the H_3 spectra, I had the enthusiastic cooperation of a number of collaborators, of whom especially Jim Watson and Jon Hougen should be mentioned, who dealt with the intricacies produced by electronic degeneracy in some of the Rydberg states.

Following the work on H_3, it was natural to wonder whether there are other similar systems (Rydberg radicals). The spectrum of one such radical was indeed discovered. It is NH_4, the ammonium radical, part of whose spectrum had been known for 108 years without being recognized as such. The discovery of the other part was due to work by Schüler and his collaborators in Germany in 1955, who did not realize that they were dealing with NH_4. The proof for NH_4 came from a study of mixed isotopes. High resolution spectra of both the Schüler and the Schuster bands have been obtained, and additional bands of these two systems are still being studied. The spectrum of ND_4 particularly lends itself to rotational analysis, because it is sharper than that of NH_4 and also stronger.

The National Research Council of Canada has given me the opportunity to continue my spectroscopic work far beyond the normal retirement age. There seems to be no lack of interesting problems for me to work on. I look forward to my continued association with the younger members of the laboratory that A. E. Douglas and I started 36 years ago, confident that this laboratory will continue to flourish.

J. W. Stout

Ann. Rev. Phys. Chem. 1986. 37 : 1–23

THE JOURNAL OF CHEMICAL PHYSICS: The First 50 Years

J. W. Stout

Department of Chemistry, The University of Chicago, Chicago, Illinois 60637

INTRODUCTION

The Journal of Chemical Physics was founded in 1933 by Harold C. Urey, then an associate professor of chemistry at Columbia University in New York. In an editorial (1) on the first page of Volume 1, Number 1 of the *Journal*, dated January 1933, Urey wrote:

> The Journal of Chemical Physics which makes its appearance with this issue, is a natural result of the recent development of the chemical and physical sciences. At present the boundary between the sciences of physics and chemistry has been completely bridged. Men who must be classified as physicists on the basis of training and of relations to departments or institutes of physics are working on the traditional problems of chemistry; and others who must be regarded as chemists on similar grounds are working in fields which must be regarded as physics. These men, regardless of training and affiliations, have a broad knowledge of both sciences and their work is admired and respected by their co-workers in both sciences. The methods of investigation used are, to a large extent, not those of classical chemistry and the field is not of primary interest to the main body of physicists, nor is it the traditional field of physics. It seems proper that a journal devoted to this borderline field should be available to this group.

The 1920s and 1930s were years of extraordinary ferment in the physical sciences. In nuclear physics great discoveries followed one another in rapid sequence, and the revolution in thinking caused by the newly formulated quantum mechanics had profound implications for chemistry as well as physics. The time was auspicious for the launching of a journal devoted to the new field christened "chemical physics." Urey was an active leader in this field and had written, with Arthur E. Ruark, a book (2) which was the bible of those who, like myself, then a graduate student at Berkeley, wished to learn about the new physics and chemistry. It is a tribute to Harold Urey's extraordinary energy and enthusiasm that he devoted effort

1687

0066–426X/86/1101–0001$02.00

to the founding of the new journal at the time he was actively engaged in the research that led to his receiving the Nobel prize in 1934 for the discovery of deuterium.

According to anecdotal history Urey approached both the American Chemical Society and the American Physical Society in an effort to obtain a society sponsorship for a journal devoted to chemical physics but was turned down by both. The founding institution and publisher of *The Journal of Chemical Physics* was the American Institute of Physics, and the AIP remains the owner and publisher of the *Journal* to this day. The AIP had been organized in 1931 and 1932 by the American Physical Society, the Optical Society of America, the Acoustical Society of America, the Society of Rheology, and the American Association of Physics Teachers. As the first chairman of the governing board of the AIP, Karl Compton, wrote (3):

> In one sense the American Institute of Physics is the child of the five parent national societies which have cooperated in forming it. In another sense, however, it has followed the more usual course of being born of two parents, the one financial distress and the other organizational disintegration.

The Chemical Foundation, a corporation which was given certain German patents after the 1914–1918 World War, had money available that was to be used for the advancement of chemistry and allied sciences. When the American Physical Society approached this foundation seeking financial support for its publications it was told that support could be provided for an association that represented all American physicists. The AIP met this requirement. The AIP undertook the task of the printing and publishing of the journals of its member societies. In addition, beginning in 1933, it assumed the primary responsibility for the *Review of Scientific Instruments* and launched *The Journal of Chemical Physics.*

It is the practice in starting a new scientific journal to assemble a list of distinguished names to adorn the masthead. The *JCP* was no exception. Listed in the first issue as an advisory editorial board were R. T. Birge, A. H. Compton, Irving Langmuir, Gilbert N. Lewis, A. A. Noyes, and John T. Tate. Harold C. Urey was called Managing Editor and the names of 22 well-known physicists and physical chemists of the time appeared as Associate Editors.

From its beginning the *JCP* attracted as authors the leading people in its field. Among the authors of papers in Volume 1 one finds the names of J. D. Bernal, J. B. Conant, P. Debye, Immanuel Estermann, Henry Eyring, R. H. Fowler, William D. Harkins, Herbert S. Harned, G. Herzberg, J. H. Hildebrand, John G. Kirkwood, G. B. Kistiakowsky, Victor K. La Mer, Irving Langmuir, W. M. Latimer, Gilbert N. Lewis, W. F. Libby,

Joseph E. Mayer, Robert S. Mulliken, Linus Pauling, Oscar K. Rice, J. C. Slater, Charles P. Smyth, F. H. Spedding, Hugh S. Taylor, Harold C. Urey, J. H. Van Vleck, E. Bright Wilson, Jr., and W. H. Zachariasen.

GROWTH OF THE JOURNAL

In its first year, 1933, the *JCP* appeared as 12 monthly issues, numbers 1–12 of Volume 1. The total page count was 896. There were 121 articles and 17 letters published in Volume 1, a total of 887 textual pages, and there were nine pages of subject and author indexes. In its fiftieth year, 1982, the *Journal* was published bimonthly, appearing on the first and the fifteenth of each month. There were two volumes, 76 and 77, each with 12 numbers. The total numbered pages was 13,001. In 1982, 1527 articles and 266 letters were published, making a total of 12,797 pages of textual material. Author and subject indexes appeared at the end of each volume, in the issues of 15 June and 15 December, and occupied 204 pages.

The growth of the *JCP* over the years 1933 through 1984 is displayed in Table 1. During its first eight years, 1933–1940, the *Journal* grew slowly, averaging about 3% a year. World War II began in Europe in 1939 and the efforts of many *JCP* authors were gradually diverted from research in pure science to work of military importance. The *Journal* grew thin, reaching in 1944 and 1945 a minimum size less than half that predicted from an extrapolation of the 1933–1940 trend. With the end of the war a rapid growth in scientific publication began. The *JCP* was in the forefront of the pack. In the years 1945–1965 the average yearly increase in the number of textual pages and the number of articles published was 12%. Had this exponential growth continued unabated, the 1985 *JCP* would have had 90,265 textual pages and 13,660 articles. Fortunately exponential growth curves do turn over, and in the years 1966–1984 the growth in material published in the *JCP* fluctuated around a mean of about 1% per year.

In its first five years, 1933–1937, the *JCP* published 707 articles occupying 4291 pages, an average of 6.07 pages per article. In the five years 1978–1982, 7580 articles occupied 59,345 pages, so the mean article length had increased to 7.83 pages. However, like the dollar, a journal page is a nonconstant measure of value. In the early days the printed pages of the *JCP* were beautifully set by monotype operators and were uncrowded with wide margins. Beginning in March 1949 about 12% more material was squeezed into each page, still retaining monotype composition. In 1974 the page size itself was increased in area by about 12%. The 1978–1982 pages were set by typewriter composition, nonjustified on the right. I estimate that a 1978–1982 page contained approximately 25% more

Table 1 Quantity of material published in *The Journal of Chemical Physics*, 1933–1984

Year	Volume	Number of textual pages	Articles	Letters
1933	1	887	121	17
1934	2	891	143	53
1935	3	834	151	42
1936	4	804	140	43
1937	5	994	152	41
1938	6	908	146	53
1939	7	1115	170	60
1940	8	998	158	47
1941	9	880	131	42
1942	10	761	103	39
1943	11	562	81	19
1944	12	522	69	22
1945	13	586	67	25
1946	14	743	97	54
1947	15	886	106	80
1948	16	1176	166	109
1949	17	1358	203	180
1950	18	1687	291	224
1951	19	1615	284	268
1952	20	1983	340	317
1953	21	2247	363	378
1954	22	2099	362	330
1955	23	2469	417	406
1956	24–25	2585	404	419
1957	26–27	3220	538	285
1958	28–29	2695	418	282
1959	30–31	3330	504	307
1960	32–33	3784	602	306
1961	34–35	4482	644	305
1962	36–37	6527	977	350
1963	38–39	6581	936	332
1964	40–41	7763	1111	419
1965	42–43	9132	1305	449
1966	44–45	9412	1323	416
1967	46–47	10,460	1414	448
1968	48–49	11,292	1528	457
1969	50–51	11,164	1451	438
1970	52–53	11,180	1484	442
1971	54–55	11,292	1482	415
1972	56–57	11,895	1595	375
1973	58–59	12,536	1519	346
1974	60–61	10,618	1390	333
1975	62–63	10,472	1366	322
1976	64–65	10,861	1384	318

Table 1 (*continued*)

Year	Volume	Number of textual pages	Articles	Letters
1977	66–67	11,817	1543	304
1978	68–69	11,234	1421	283
1979	70–71	11,315	1413	278
1980	72–73	13,172	1625	311
1981	74–75	12,966	1594	293
1982	76–77	12,797	1527	266
1983	78–79	13,937	1679	262
1984	80–81	12,746	1593	216

characters than a 1933–1937 page. With this correction the mean length per article in 1978–1982 was 1.6 times that in 1933–1937 and the volume of material published was increased by a factor of 17. By what factor has our human ability to read and assimilate increased?

THE EDITING OF THE *JCP*

Harold Urey served as editor of the *JCP* from its inception in 1933 through 1940. In 1939 and 1940 Joseph E. Mayer was listed as assistant editor and in 1941 Mayer succeeded Urey as editor. When Joe Mayer moved from Columbia to The University of Chicago in 1945 he brought the *Journal* with him. Beginning in January 1953, Clyde A. Hutchison Jr. assumed the editorial reins. Hutchison served as editor through 1959, at which time he resigned the editorship to take the chairmanship of the Department of Chemistry at Chicago. In those days the finding of someone to succeed a retiring editor was a very informal procedure, without search committees and with little involvement by the administration of the American Institute of Physics. I remember Joe and Clyde coming into my office early in 1959 and, after a bit of softsoaping listing my sterling qualities, proposed that I become the *JCP* editor. After a few days of debating with myself I agreed to be considered and Clyde wrote Elmer Hutchisson, then director of the AIP, proposing my name. After a formal approval by the executive committee of the Governing Board of the AIP I became the *JCP* editor and served until my retirement at the end of calendar year 1982. The procedure for finding my successor was much more elaborate. The Division of Chemical Physics of the American Physical Society regards the *JCP* as the primary medium for the publication of the scientific work of its members and considers the *Journal* its own. On several occasions representatives of the Division had written me inquiring about plans for the *Journal* after my retirement. In the spring of 1982, H. W. Koch, director of the AIP, appointed a search committee to recommend a successor. The committee was chaired by Bill Klemperer of Harvard University and included the chairman of the Division and the AIP director of publications. The committee unanimously recommended John C. Light as the succeeding editor and, with the proviso that Donald H. Levy be appointed associate editor, Light accepted the editorship. Light and Levy are both at The University of Chicago, so the *Journal* offices and secretarial staff remained unchanged, greatly facilitating the smooth transfer of editorial responsibility.

Editorial Policies

The broad editorial policy of the *JCP* has not changed from the beginning. It is to select and promptly publish the best papers in the field. All papers published must contain new results of original research and must not have been published or submitted for publication elsewhere. No review articles are accepted.

Nearly all papers submitted are sent to referees for review. The only exceptions are those few cases where it is apparent to the editor that a paper is scientific nonsense or well outside the scope of the *Journal*. I always kept in mind that a revolutionary new idea might at first reading seem crackpot to a member of the establishment and would usually send a paper to a referee whom I knew to be fair and open minded even though I was 99% sure that it did not belong in the *Journal*. If, as almost always happened, the referee replied with a strong recommendation against publication I would send this to the author, together with my own evaluation, and decline to accept the paper.

Aside from these few exceptional cases a negative recommendation from a single referee was not sufficient reason for the rejection of a paper by the editor. The author would be sent a form letter enclosing the comments of the referee who had recommended against publication. If the author disagreed he would be asked to write a detailed answer listing his reasons, and additional referees would be consulted if the paper were returned. As a rule second and subsequent referees were sent anonymous copies of earlier referees' comments and the authors' reply. Frequently second referees would say that they had first read the paper themselves and formed an independent opinion before reading the earlier history of the paper. Some authors, returning a paper that had received harsh criticism, would ask that the first referee's comments not be sent to subsequent reviewers. The editor would honor this request if on reviewing the file it appeared to him that the first referee's remarks were unsubstantial and that he should have immediately sent it to someone else. Otherwise it is a question of editorial judgment as to whether it is fairer and wiser to obtain a completely independent evaluation or to make subsequent referees aware of all points in dispute. To avoid the possibility of later referees missing a substantial defect in a paper the second alternative was usually selected.

Some 80% of the articles submitted to the *JCP* are accepted, although many of these undergo substantial revision before reaching final form. The rather low rejection rate reflects, I think, an awareness on the part of authors of the standards of various journals and a preselection by them of the most appropriate journal before submission.

In the *JCP* the editor is the final arbiter of the acceptance or rejection of papers. Unlike some Society journals there is no mechanism for appeal to higher authority. Only once during my tenure as editor did I ever receive a letter from an AIP official asking about a decision I had made, and there was no suggestion that my decision be overruled. An author who feels that his paper has been wrongly rejected by the *JCP* editor could ask one of the associate editors to intervene. I encouraged this. If an associate editor inquired about a particular paper that had been rejected I would send him copies of our complete file, with a warning to keep all referees' names confidential, and welcomed the associate editor's opinion. In all cases that I can recall the associate editors agreed that the original decision had been correct.

The task of referees and editor is to evaluate the suitability of the particular paper under consideration for publication in the *JCP*. Unlike the assessment of a grant proposal we are dealing with the evaluation of finished work, and the past performance or scientific eminence or institutional connection of the authors is irrelevant. Among the authors of rejected papers are the names of several Nobel laureates. Belonging to the faculty of The University of Chicago is not a guarantee that one's papers will obtain a favorable reception by the *JCP*.

Nearly all authors accepted adverse editorial decisions with good grace, but there are always a few who return their rejected papers with strong remarks as to the incompetence and bias of the referees and a demand for reconsideration. Several years may elapse before the folders on these papers reach their final resting place. The associate editors were particularly helpful in such cases, since I asked them to make an independent editorial judgment based on the complete file with all names disclosed and could obtain their opinions as to the suitability of the particular referees that had been previously consulted as well as to comment on the substance of the paper itself.

After two reviewers had recommended rejection of the paper I would reread the complete file and look at the paper itself. The result would usually be rejection but occasionally a paper would strike me as unorthodox but intriguing and worthy of evaluation in the open literature. I would solicit additional opinions and a few such papers were eventually published.

The scope of the *JCP* is loosely stated as being "to bridge a gap between journals of physics and journals of chemistry." A bridge must have substantial foundations at either end and there is inevitably overlap between the area covered by the *JCP* and that belonging to physics or chemistry journals. Perhaps the best definition is one that paraphrases a remark by G. N. Lewis: Chemical Physics is whatever chemical physicists are doing.

An examination of the contents of Volume 1 of the *JCP* reveals that many of the broad areas engaging the attention of chemical physicists in 1933 continue to be active areas of research in the field. There were papers on molecular spectroscopy and molecular structures, both theoretical and experimental, on the quantum mechanical treatment of the electronic structure of molecules and crystals and chemical binding, on understanding the kinetics of chemical reaction from basic physical principles, on the thermodynamic properties of substances and the calculation of these by statistical mechanical methods, on the structure of crystals, and on phenomena at surfaces.

The editors of the *JCP* have avoided setting precise limits on the scope of the *Journal*, relying instead on the opinions of referees and on their own sense of what constituted important chemical physics at a particular time. When Joe Mayer was editor, colloid and surface chemistry was a contentious area with little hard science, and papers in that area were returned to authors without review as lying outside the scope of the *JCP*. By the 1950s modern methods and techniques for investigating surfaces had emerged and as the science hardened the *JCP* publication in surface phenomena increased. In the early days of the *Journal* papers calculating thermodynamic properties of gases from molecular and spectroscopic data were new and important, but as the years went by computation became easy and it was a trivial task to look up in the literature the moments of inertia and vibrational frequencies of some molecule, input the information to a computer program, and produce a paper with impressive tables of thermodynamic properties to add to one's publication list. Shortly after I became editor I noticed several such papers coming to the *Journal* and soon adopted a policy of excluding them unless they formed a small part of a larger paper presenting the primary data. In the beginning there was nothing that could be called biology published in the *JCP*. More recently some chemical physicists have turned their attention to complex problems of biological significance and began to submit their papers to the *JCP*. I tried to draw a line between papers containing new methods or discoveries that would be of interest to the general community of chemical physics and those which used the techniques of chemical physics to obtain results that were of only biological interest.

Letters

The Letters to the Editor section of the *JCP* first appeared in the issue of April 1933. Letters were to be "terse and contain few figures" and were published in a reduced type size at the end of each monthly issue. An explicit length limit of 600 words for Letters was introduced with the April 1939 issue, although an examination of Letters in the 1940s and 1950s shows that this limit was not always rigorously enforced. In the early 1950s the number of Letters began a dramatic increase (see Table 1). When Clyde Hutchison became editor he was concerned with the chaotic growth of the Letters section and proceeded to reorganize it in a systematic fashion. A "Revised Announcement" in the May 1967 *JCP* divided the Letters section into three parts: Communications, Comments and Errata, and Notes. The maximum word count of a Letter was increased to 950 and explicit instructions were included for the count of figures, equations, and tables. The categorization of Letters introduced by Hutchison still remains, with minor modification, in the *JCP*, and a similar categorization, with various names for the categories, has been adopted by other scientific journals. In 1965 I separated the Errata from Comments and in 1974, in response to popular demand, the word limit was increased to 1200.

Communications in the *JCP* are reports of preliminary results of "current and extreme interest to relatively large numbers of workers in the field." It is expected that a fuller description of the work described in a Communication will later be published as a regular article. The only justification for such preliminary publication is that the rapid dissemination of the new results would be of great importance to workers in the field and that the delay in waiting for a complete regular article would substantially impede progress. Speed is of the essence with Communications. They are published in the issue two months following the day that they are forwarded from the editorial offices to the publisher. This tight schedule allows little time for the reading of proof by authors that is normal with regular articles. Publication managers at the AIP have urged that no proof be sent to authors of Communications, and this practice prevailed from time to time. I have always felt that it is essential that an author have an opportunity to read, and correct if necessary, material to be published over his name. Proof is now sent to authors of Communications and they are asked to telephone with any corrections.

Communications were also given expeditous handling in the editorial office. They were sent to experts in the field, usually two in number, with a letter explaining the nature of a Communication and asking for a prompt response. Frequently the experts who had not responded by a Communication date, the first and fifteenth of each month, were telephoned by the editor. During my tenure as editor I read the text of all proposed Communications myself, with an eye as to the reason that this particular manuscript deserved rapid preliminary publication. The criteria for the acceptance of a Communication were both more subjective and more stringent than those for regular articles and some 60% of the papers submitted as a Communication were not accepted as such. The publication of a *JCP* Communication was regarded by some as a mark of unusual scientific distinction, and I have frequently read letters of recommendation of prospective faculty members making this point.

Notes are intended as the final publication of results that can be completely described within the length limitation of a Letter. They are reviewed in the same fashion as regular articles. Errata correct errors in papers published in the *JCP*. If they make sense after a brief scrutiny by the editor they are accepted without review.

Comments are discussion of material previously published in the *JCP*, an essential restriction. A Comment was ordinarily sent to an author of the work commented on who was asked to criticize it and, if he felt it necessary, to prepare a Comment in reply. This often led to an acrimonious exchange of letters through the editor's office and could be a protracted process. A woman scorned hath no fury like a scientist whose work is questioned. Frequently an independent, anonymous referee would be consulted by the editor. If the Comment and Reply contained significant material of scientific importance and were reasonably free of pejorative personal invective, they were eventually published together. Although Comments occupied a very small fraction of the space in the *Journal* they took a considerable amount of the editor's time and judgment.

Sam Goudsmit once said that he suspected that the ratio of readers to authors of *Physical Review* articles was less than unity, since it was evident that some authors had not read their articles before submitting them. I sometimes wondered, as the increasingly thick issues of the *JCP* overflowed the bookshelves in my office, how much of the voluminous material we published was actually read. The Comments we received helped to lay my fears to rest. Two incidents in which a paper in the *JCP* provoked a flood of Comments that, after prolonged correspondence and exchange of views, resulted in the publication of a single clarifying Comment remain vividly in my memory.

In 1965 Ernest Davidson published (4) a brief Note "On Derivations of the Uncertainty Principle," in which he pointed out an apparent paradox in the usual textbook derivation of this principle and proposed a resolution. This is essentially a mathematical question on the foundations of quantum mechanics. Now although most readers of the *JCP* are not professional mathematicians we all love mathematical puzzles. We received six Comments, involving eight authors, on the Davidson Note. I encouraged correspondence among the various authors and attempted, without success, to have them combine their efforts into a single brief Comment that would be the final *JCP* publication on the matter. A total of four independent referees were involved in a correspondence that extended over nine months. The upshot was the publication of a single Comment (5), which in the opinion of reviewers best resolved the question for readers of the *JCP* and which carried a footnote mentioning the other authors.

In 1971 a Note (6) was published proposing an atomic orbital that combined the usual analytical hydrogen atom orbital with a Gaussian. The author had tested this orbital on the hydrogen atom itself, and his computer program, using the variation method, arrived at a ground state energy slightly above the well-known exact value and a ground state function with a nonzero Gaussian admixture. Everyone who has taught an introductory course in quantum mechanics is aware that the variation method applied to a function that contains the exact function as a component will result in the exact function and its corresponding energy value. It was apparent that neither the referee, an eminent expert in molecular orbital calculations, nor the editor had read the Note. At the lunch table in the faculty club I received sarcastic remarks on the decline of the *JCP* and the Comments began to come in. In this case I was successful in persuading the critics to prepare a single joint Comment with five authors, which was published (7) in the 15 January 1972 issue.

The Referees

The selection of reviewers for each paper is the most important function performed by an editor. The referees should be experts in the particular subject matter of a paper, know the relevant literature, and be both highly competent and fair. Nearly all published papers are improved by revisions resulting from a critical reading by a reviewer. Although we hear much of disputes between referees and authors the fact is that such cases, although memorable, are the exception rather than the rule. In most cases authors are very grateful for the advice proffered by referees and would often ask me to convey their thanks to an anonymous referee for pointing out an error or calling attention to an overlooked literature reference.

Unless a referee specifically requests that his name be disclosed to the author his identity is kept confidential by the editor and an anonymous copy of his report is sent to the author. Great care was taken in the *Journal* office to remove identifying marks from copies of reports mailed to authors, but embarrassing mistakes sometimes occurred. We once had a copying machine that made visible the watermark on the paper on which a report had been written; thus we learned to include almost invisible watermarks among the items to delete from material sent to authors. On more than one occasion a paper was mailed for review to a referee whose name appeared among a long list of authors of a paper and we filed the ensuing tongue-in-cheek and glowing report with a red face before sending the paper to an uninvolved reviewer.

The referee file of the *JCP* contained some two thousand names, mostly in Canada or the United States, but with a generous sprinkling of experts residing in other countries. The referees were classified by areas of expertise using a scheme similar to the one employed by the old subject index. The subject file was most useful to me in recovering names that I could not at the moment recall, since much of the information about referees is stored in an editor's head. The referee files of the *JCP* have now been largely transferred to a computer and their subjects of expertise are categorized by key words provided by the referees themselves. The referee files were continuously updated, both to keep track of changing addresses and institutional affiliations and to remove and add names. I have found that the most careful reviews are frequently provided by young people who have not published enough to be well known but who have an expert knowledge of their fields. I often wrote more prominent and busy people asking for such names to add to our referee file and would pick names of likely referees from authors of papers published in the *JCP*. A report that merely says "Publish" makes an editor wonder if the paper has really been read.

A persistent and unsolved problem of refereed journals is the late return of reports of reviewers. If a busy reviewer does not get around to reading a paper in the first week after he receives it the paper may get buried in a pile of unfinished business on his desk and a nudge from the editorial office is required. The *JCP* has a procedure involving a sequence of three letters, of increasing stridency, followed by telephone calls from an editorial assistant, for recovering papers from dilatory reviewers. The date on which a paper was mailed for review is prominently noted on the file of each paper and the tardy referee letters start a month later. The dates when a paper is sent to and received back from a referee are entered in the referee file and the use of chronically tardy referees is soon discontinued. Authors are with justification irritated by delays in reviewing and frequently telephone. The *JCP* has always strived to secure the prompt publication of acceptable papers. I recently looked through the 1 January and 15 January 1982 issues as an example to see how we had done. The dates of receipt and acceptance are listed in the published papers, and I made a histogram of the time between these dates. This time reflects not only the time wasted by tardy referees but includes also the time while a paper is in the author's hands for revision as well as the time lost when the first referee has been sent a paper that for some reason has been promptly returned without review. For the 144 papers published in the two issues sampled, the time from receipt to acceptance ranged from one day to 587 days. The distribution was very non-Gaussian, with a median time of 66 days and a mean of 82 days. Although we usually were able to mail papers out for review on the day of their receipt, one day seemed too short and I checked the file. It turned out that the paper had been sent to a referee whose office is in our building. He had previously read it as a preprint and returned his review to our office the day after he received the paper from us. The minimum review time for papers that had gone through the mails was eight days.

Every year six new associate editors were appointed for three-year terms and the six who had served three years were retired. In choosing associate editors consideration was given to having among the panel of 18 on the editorial board someone expert in each area covered by the *Journal*, but the principal weight was given to finding people who had written excellent papers for the *Journal* and who had in the past served in an exemplary fashion as referees. I continued to use associate editors as referees, taking care not to increase their load, and as mentioned above sometimes asked them for editorial judgments in difficult cases.

Indexes

The *JCP* has always published author and subject indexes at the end of each volume. Beginning in 1956, two volumes a year have been published. In the years 1966–1973 various problems led to delays in the composition of the index and they were mailed to subscribers bound separately, sometimes arriving six months after the scheduled date. Originally the subject index contained an alphabetically ordered list of subject headings that had been chosen by various editors, without systematic analysis, to reflect categories under which a reader might hope to find listed articles published in the *Journal*. There was a typed list numbering these subject index categories and the editor assigned index numbers to each paper when it was forwarded for publication. An index secretary at the AIP transferred this information to typed cards used to prepare the copy for the indexes. A single person at the AIP handled this index preparation. She was intelligent and efficient and the indexes were prepared in timely fashion. However, an automatic sorting operation such as the preparation of indexes seemed well suited for rapid processing by a digital computer, so the preparation of indexes was one of the first operations to be computerized at the AIP. The AIP has created an Information Division, which, in addition to its broader responsibilities involving the general problem of the dissemination of information from the physics literature, assumed the responsibility for the *JCP* indexes. Unlike a good secretary a computer does not recognize when it is outputting garbage and the early computer-produced indexes required extensive correction in proof and were consequently late.

One of the principal tasks of the Information Division was to devise a method of indexing the physics literature by subject, in a logical fashion that would be unambiguous and could be used efficiently by a scientist searching the literature for information on a particular subject. I recall reading lengthy reports by members of this division outlining projected schemes involving mathematical methods such as Boolean algebra to keep the logic straight or statistical techniques to maximize the probability of successful search. The task is difficult since the meaning of words used by physicists to label their work often depends on the context. The upshot of these endeavors was the Physics and Astronomy Classification Scheme (PACS), which to my eye appears a descendant of the scheme formerly used by *Physics Abstracts*. Beginning in 1974 the subject indexes of the *JCP* used the PACS classification. PACS was widely publicized and authors were encouraged to include PACS numbers on their papers. The responsibility for the preparation of indexes was transferred from the editorial office to the AIP indexers. The PACS index is now widely used throughout the international physics literature. The computer programs have been debugged and the subject classification is used for many purposes beyond the preparation of journal subject indexes.

The arrangement of authors' names in alphabetical order for an author index would be a straightforward procedure if there were a one to one mapping of names and identities. When I wrote my first scientific paper, my mentor, W. F. Giauque, strongly advised me to choose a single form for my name and stick to it throughout my scientific career. Not all *JCP* authors followed this advice: one would find separate author index headings for Blow, J. P.; Blow, Joe; Blow, Joseph P.; Blow, Joseph Patrick; and Blow, Joseph Patrick, Jr. even though the same Joe had written all papers. The consolidation of different forms of a name used by a single person and the separation of individuals whose names are the same requires human intervention. For the recent cumulative index covering volumes 72–81 of the *JCP* I was assigned this task. In addition to the indexes at the end of each volume, an author index for each issues was added to the *JCP* beginning with the 1 July 1966 issue. In 1974 the author index appearing in each issue was changed to a cumulative one including all papers in previous issues of the volume. Many readers have remarked that they find the cumulative author index particularly valuable.

Sectioning

As the *Journal* grew in size there were from time to time proposals that it be split into two or more journals, each devoted to a part of the scientific area spanned by the present *JCP*. Virtually all of the readers of and contributors to the *Journal* who expressed opinions were opposed to the splitting of the *Journal*, as was the editor. There is a perceived unity in the papers published, often on apparently disparate subjects, which would be lost if separate journals were established. The Division of Chemical Physics of the American Physical Society has several times strongly recommended against splitting the *JCP*. A committee chaired by Robert G. Parr that reviewed the operations of the *Journal* in 1979 opposed splitting and wrote, "A distinguished broad-scope *Journal of Chemical Physics* should continue indefinitely."

The review committee did, however, urge the editor to try some system of classification of papers by subject matter in the Table of Contents and in the grouping of articles in the body of the *Journal*. Beginning with the 1 July 1980 issue I separated the Table of Contents into five sections: Spectroscopy and Light Scattering; Molecular Interactions and Reactions, Scattering, Photochemistry; Quantum Chemistry, Theoretical Electronic and Molecular Structure; Statistical Mechanics and Thermodynamics; Polymers, Surfaces, and General Chemical Physics. Several people complained that the sectioning was a mistake and in order to find the sentiments among a wider readership I conducted a poll of all present and former associate editors of the *Journal*. The response was amazing, 95 replies in all, almost evenly divided pro and con, with a small majority favoring the continuation of sectioning. Those opposed to sectioning felt that the strength of chemical physics lay in its unity and breadth and that sectioning would weaken this unity and encourage the overspecialization they saw in modern science. All respondents, pro and con, felt that the *Journal* should not be split into separate publications, and all said that sectioning or not, they would continue to read all titles since they often found papers of interest in subfields other than their own. The argument for sectioning was that it makes the Table of Contents page less overwhelming and easier to scan, even when one reads the titles in all sections. As one man remarked, there is a "glazed eye effect." The number of items missed in a long list increases more than linearly with the length of the list.

The assignment of articles to sections is made by the editor, who, of course, welcomes advice from the authors of papers. In assigning a section I often found that a paper could equally well be put in any of several sections and that my choice was very arbitrary. Sectioning should be viewed as a cosmetic device that improves the appearance of the Contents pages and not as a guarantee that related papers will not appear in different sections or as a first step toward the division of the *Journal* into separate publications.

Special Issues

Since 1959 three special issues of the *JCP* have appeared, honoring John G. Kirkwood, Robert S. Mulliken, and Willis H. Flygare, respectively. Other than a picture and a brief introduction or biography, I tried to make sure that papers published in special issues met the usual standards for *JCP* articles and were not reviews of published work.

The November 1960 issue of the *Journal* contained 32 papers honoring Kirkwood. Joe Hirschfelder headed a committee soliciting papers for this issue. All papers were submitted to the *JCP* office and reviewed in the normal fashion. The Kirkwood issue also contained ten regular articles not tied to the Kirkwood memorial and Letters.

The Mulliken issue was published on 15 November 1965 as a supplementary Part 2 of the issue of that date. It was separately bound and paginated. The 50 papers in the supplementary issue had been presented at a symposium at Sanibel Island. Per-Olov Löwdin organized the symposium and served as chairman of a committee that reviewed many of the *JCP* papers. Brief discussions of the papers were included in the supplement.

The Flygare memorial issue was Part 2 of the 15 March 1983 *JCP*. The 116 articles occupying 966 pages of text in the Flygare issue spanned the broad field of chemical physics. George Flynn acted as editor for this issue and succeeded admirably in selecting articles of *JCP*-quality.

The Darker Side

To the casual reader of *Science* it may appear that fraud and deceit in scientific publication is a common result of the pressure to publish. Actually the number of fraudulent papers is very few and the publicity given to them reflects the overwhelming concern of the scientific community that integrity be maintained. During my 23 years as editor there were only three instances of cheating. A perceptive reader wrote me that several sentences in the introductory paragraph of a Note published in the *Journal* were identical to those in an earlier paper by Leo Brewer. When I consulted the file I found that Brewer had been the referee of the Note. In reply to my letter he remarked that he had found the introductory paragraph unusually well written but had not recognized the words as his own. The data and interpretation in the Note were new and had originated with its author. Although a technical plagiarism had occurred I did not think a published correction was necessary and contented myself with a strong admonitory letter to the author.

A more serious breach of scientific ethics was the publication, in a paper primarily concerned with crystal spectra, of magnetic susceptibility data taken from the unpublished thesis of a student at another institution. The original *JCP* paper had presented these data as originating in the laboratory of one of the authors of that paper. An appropriate correction (8) was published. A second Erratum (9) corrected some misleading statements in another paper.

There was one case in which a paper submitted to the *JCP* was an outright copy of published work. A paper on a Monte Carlo study of a polymer chain had been published in Russian in an obscure journal, and the paper that came to the *JCP* was a poor translation of the Soviet publication. Luckily the referee to whom I sent the paper was thoroughly familiar with the Russian literature. The submitted paper was rejected with a harsh letter to the author. The author claimed that, for reasons that security prevented him from disclosing, the paper had been submitted for the purpose of obtaining a referee's opinion and would have been immediately withdrawn had it been accepted. In his last letter he remarked that he had chosen a future path in life other than physics.

There is a file in the *Journal* office labeled "Authors to Watch." Its contents are principally correspondence from former collaborators who have had a falling out and are concerned about the publication of their joint work, or letters concerning authors who have resubmitted papers previously rejected by the *Journal* without alerting the editor, or have submitted a paper simultaneously to two journals.

Potpourri

As the reputation of the *JCP* grew, articles were attracted from authors throughout the world. In the five-year period 1978–1982, 64% of the articles published were by authors whose institutional by-line was in the United States, 6% were from Canada, and the remaining 30% came from a wide assortment of other countries.

The characteristic blue color and austere typographical layout of the cover of the *JCP* have from the beginning uniquely identified it in a pile of journals on a desk. Around 1973 the AIP jazzed up the covers of some of its journals and I was shown a proposed "new look" for the *JCP* cover. I reacted as if it were proposed to paint a moustache on the Mona Lisa and the old-fashioned but well-loved cover remained. Formerly the Table of Contents page was printed on the back cover, running backwards into the Journal if its length required. Several institutions engaged in the business of photocopying contents pages complained of difficulties with the *JCP* and Robert Mulliken once told me at lunch that he would find the contents much easier to read if it were printed on white paper. In 1974 the Table of Contents was moved inside on white pages at the front of each issue.

We have brushed twice with the lunatic fringe. I was once visited by a man, a PhD from the university where I taught, whose paper had been rejected by the *Journal*. After a few minutes of trying to discuss with him the science in his paper I realized that one or the other of us was crazy. Fortunately he had no gun and was not very big. Subsequently I received a stream of telephone calls from this man, and university administrators were bombarded with letters insisting I be fired. It is amusing to note that the paper in question, which still makes no sense to me, was later published in a reputable journal. In 1977 we received a letter signed THETA, PhD Physics, which claimed to represent a colony of disgruntled authors who had gravitated to Salinas, California and had determined to reform scientific publication so that it would allow "creative scientific thought." A "wave weapon" had been developed, which in a test near King City had instantly killed a ten-year old horse a mile away. I happened to mention the letter in a telephone call to the AIP about another matter, and their lawyer got a copy of the letter, which was given to the FBI. We were visited by a delegation of FBI men, whose demeanor reminded me of a television program with that name. A few months later we were assured that the writer of the letter was harmless and we need not fear that the *Journal* personnel would suffer the fate of the horse.

Editors of scientific journals eschew involvement in the struggles for power among nations, at least in times when outright hostilities have not broken out. Nonetheless an occasional wavelet breaks at the door of editorial offices. I was once approached by a gentleman who gave an implausible name and flashed a card identifying him with a national intelligence agency. He proposed to intercept upon arrival and photocopy all manuscripts received by the *JCP* from "behind the iron curtain." I told him of the firm policy of the *JCP* to treat all unpublished papers as confidential information disclosed only to the editor and to reviewers and suggested his agency instead buy a subscription to the *Journal*.

We received in 1974 from a number of sources, all insisting that their identity remain secret, a paper on a generalized Ising model for polymer thermodynamics. At the time the author was writing from a private address in Moscow (USSR not Idaho), and direct correspondence with him involved difficulties. Referees to whom I sent the paper reported that it contained important original ideas and should be published but that problems with English and notation required that it be rewritten. The referee and a colleague, both expert in the subject matter of the paper, performed this task anonymously and it proved possible for the author to see and approve the revised and retyped paper (10) before its publication in May 1975.

PRODUCING THE JOURNAL

A large effort is required to bring the typed papers received from authors into the final printed journal that goes to readers. This function has been performed by the editorial mechanics office of the AIP, originally located in various buildings in downtown Manhattan and now transferred to Woodbury, Long Island.

Copymarking and Printing

Before a paper is sent to the compositor it must be marked to indicate type styles to be used and to put abbreviations in references and for units in standard form. The copymarking should also correct errors in spelling and grammar, which the author would have done had he noted them. Many copymarkers were English majors and in a few cases their impulse to creative rewriting changed the meaning and converted sense into nonsense. In 1963 I wrote some suggestions for copymarkers asking them not to act as rewrite men and to avoid rearranging sentences or altering words. In the main these were followed in *JCP* copymarking.

There is a firm rule that authors must approve proof before a paper is transferred to pages. On the very rare occasions when someone violated this rule and made changes after proof had been approved, the editor was sure to receive an irate letter from the author and an "AIP-paid" Erratum would be published.

When I first became editor the composition of the *JCP* was done by Mono of Maryland using Monotype machines. The resulting pages were beautiful and there were very few errors in the proof sent to authors. For economic reasons a change to typewriter composition, by outside contractors, was made in 1973. The appearance was a bit more ragged but acceptable, and there were few complaints from readers. In the same year the AIP began using a computer for in-house composition of the "heads and tails" of papers: the titles, authors' names, abstract, and references. The early efforts at computerized composition often produced bizarre results in proof mailed to authors. The corrections required resulted in substantial delays in the printing of the *Journal*, and in the latter part of 1973 issues were received by subscribers more than two months after the dates printed on the covers. In Volume 58 a bewildering array of type styles is found for the titles and authors' names. In time the bugs in the computer-controlled composition were ironed out. By 1983 it was possible to compose the entire *Journal*, in-house, by computer with an improvement in both appearance and cost.

Financing the Journal

In some respects scholarly publication resembles the vanity press. The drive for publication comes from authors who submit papers. The authors are not directly compensated for their efforts but are rewarded by the professional recognition and personal satisfaction that result from the publication of their work. When the *JCP* commenced publication most scientific journals were financed by professional societies and, the income from subscriptions being insufficient to cover the costs, were partly supported by dues from society members. Initially the cost of producing the *JCP*, beyond the income from subscriptions, was assumed by the AIP, using resources obtained from a small fraction of the dues paid to its member societies as well as from granting institutions such as the Chemical Foundation.

As the *Journal* grew, costs increased and financial crises caused a continuing examination of the role of the *JCP* and its relation to both the American Physical Society and the American Chemical Society. The problems became particularly acute in the early 1950s, and the editors at that time were much involved. The governing board of the AIP had for some time questioned whether an appropriate function for the Institute was the publication of archival journals such as the *JCP*, and various committees had considered the general question of physics publication. A committee of the APS was appointed to investigate the possibility of joint sponsorship of the *JCP* by the APS and the relatively affluent ACS. This committee found that joint sponsorship was "impractical" and recommended that responsibility for the *JCP* be transferred to the APS. In 1955 another joint AIP-APS committee proposed that the APS assume responsibility for the *JCP*, that the name be changed to the "Journal of Solid State and Chemical Physics," and that the *Journal* become more a physicist's and less a chemist's journal. This last proposal received little support from either the solid state physics or chemical physics communities. In the end the AIP continued as owner and publisher of the *JCP*.

It appears that from the beginning the AIP requested from authors' institutions the payment of page charges to help defray publication costs, although the first public announcement of a "publication charge of $4.00 per page" appeared in the issue of April 1949. The payment of the page charge was not a requirement for the publication of a paper accepted by the editor, and no consideration of whether or not page charges would be honored ever influenced editorial decisions as to the acceptability of papers. With time and inflated costs the dollars per page increased, reaching a maximum of $80 in 1980. The honoring of page charges was accepted by industrial laboratories and after some initial bureaucratic confusion by government laboratories, but the response from academic institutions, particularly departments of chemistry, was poor. As editor, Joe Mayer wrote letters to some department chairmen urging the honoring of page charges, and a later campaign was conducted in 1959 by the Division of Chemical Physics of the APS through the then secretary of the division, Arthur Frost. It is with some nostalgia that one notes that in 1958 money for page charges came from departmental budgets rather than from grants.

In 1969 the AIP instituted a policy that resulted in a delay of publication of some papers on which the page charge had not been honored by an author's institution. At the insistence of the editor the policy was not applied to Communications in the Letters section of the *JCP* but it was applied to other Letters and to all articles. Fifteen percent of the pages published were available for nonhonored papers. If the number of such papers exceeded this quota the papers were delayed until a later issue and appeared, in the order of their originally scheduled publication dates, as nonhonored space became available. During the years 1969 through 1972 the delay for publication of nonhonored papers ranged from zero to a high of 14 months. The handling of publication charges and the imposition of the delays were functions of the publication office at the AIP. Nevertheless the editor became deeply involved both in answering letters from angry authors and in concern for the deleterious effect on the *Journal* caused by the delaying of nonhonored papers. There is no doubt that the delay caused the diversion to other journals of excellent papers that would otherwise have come to the *JCP*. There were cases in which accepted papers were withdrawn by authors after a delay had been imposed. In my annual reports and letters to the publishers I complained *ad nauseum*. The 1979 report of the *JCP* Review Committee listed as its two primary recommendations that page charges be reduced and the delay for nonhonored papers be eliminated. Some of the delayed papers bore acknowledgment of support by the National Science Foundation. In 1980 I wrote to acquaintances at the NSF proposing that publication charges for work supported by the NSF automatically be assumed by that agency at the time a paper is accepted rather than be included as an item in a grant budget. I thought that the costs to the NSF would not be increased much and the delay problem for the *JCP* would be substantially ameliorated. The proposal was seriously considered by the National Science Board but it apparently involved broad questions of government policy and, in the absence of strong support from scientific publishers, nothing happened. The decisions of the AIP governing board in 1983 and 1984 to lower the publication charge and cut the delay time are very welcome. By 1985 the publication charge was reduced to $45 per page and the delay time to two months.

When I first subscribed to the *JCP* in January 1944 the price was $10 per year. In 1948 this was still the price for members of AIP societies but $12 was asked of "others." In the following years the subscription price for people belonging to member societies of the AIP remained relatively low and an increasingly higher subscription rate was asked of nonmembers and institutions. In 1985 a year's subscription cost $100 for members and $760 for nonmembers, with an added mailing charge for those living beyond the borders of the United States. If the 1985 price is adjusted for the inflation of the dollar, members then got more articles or pages for their money than they did in 1944. The price to libraries remains a bargain when compared to that asked by privately published journals.

There were 1619 paid subscriptions to the *JCP* in 1946. The number of subscribers increased as the reputation of the *Journal* grew, reaching a peak of 6488 in 1969. Since then the number of subscriptions has declined slowly. It was 4524 in 1980. Some 69% of the *JCP* subscriptions are by nonmembers, principally libraries. Almost half of the *JCP* subscriptions are mailed to addresses outside of the United States.

CONCLUSION

On 1 January 1986 *The Journal of Chemical Physics* celebrated its fifty-third birthday. It has grown in size and in status and in my admittedly biased judgment is now the leading journal in the world in its field. There is some supporting evidence. The Review Committee in 1979 wrote, "The Journal of Chemical Physics has been the preeminent research journal in the field of chemical physics. It remains so." King & Roderer (11) quote a survey of physicists, which found the *JCP* to be among their five "most important or frequently used" journals. Emsley (12) puts the *JCP* second on a list of journals that are essential to retain in research libraries in times of financial stringency. In the *Journal Citation Reports* of *Science Citations Index* the *JCP* has consistently been in the top six journals in physical and biological sciences, ordered in terms of the total number of literature references to published papers, and in "impact value," a measure of the number of citations per article, the *JCP* ranks high among journals of physics and chemistry.

The community of chemical physicists has grown with the *Journal*. The Division of Chemical Physics is prominent among the divisions of the APS, and the *JCP* is a major publication output for members of that division as well as for those who belong to the Division of Physical

Chemistry of the *American Chemical Society*. With the advent of solid state electronics, digital computers, and lasers the techniques and instrumentation used by chemical physicists have changed tremendously over the years. We no longer find Type K potentiometers with batteries and a galvanometer in the laboratory, and glass blowing is becoming a lost art among graduate students. Elaborate computer-controlled instruments acquire data at a rate beyond human comprehension; one resorts to a computer to reduce the information to something we can understand. The basic problems attacked by chemical physics remain, however, the same, and are solved with ever increasing refinement and accuracy.

From the beginning theoretical papers were prominent in the *JCP*, and the interplay between theory and experiment has always been strong in papers in the *Journal*. The distinction between theoretic ans and experimentalists in chemical physics is less sharp than in some other parts of physics, and we find many chemical physicists who make important contributions to theory related to experiments they also conduct. People perform "computer experiments" with Monte Carlo or Molecular Dynamics methods to solve theoretical problems where the basic theory is understood but the mathematical complexity of analytical solutions is beyond our capabilities.

One reads much of authors' problems with referees and of arbitrary and unfair editorial decisions. One might think the life of an editor is a contentious one, "full of sound and fury, signifying nothing." I am not quite sure about the significance, but in fact an editor presides over an operation whose purpose is to improve the quality of published papers with the cooperation of both authors and reviewers. Instances of acrimonious dispute are atypical, although they stir the blood and remain in the memory. Even rarer are instances of chicanery on the part of authors or referees. The overwhelming impression I retain from 23 years of editing the *JCP* is one of probity and love for the integrity of their science by all protagonists in the publication process.

Literature Cited

1. Urey, H. C. 1933. *J. Chem. Phys.* 1: 1–2
2. Ruark, A. E., Urey, H. C. 1930. *Atoms, Molecules and Quanta.* New York: McGraw-Hill
3. Compton, K. T. 1932. *Rev. Sci. Instrum.* 4: 57–58
4. Davidson, E. R. 1965. *J. Chem. Phys.* 42: 1461–62
5. Yaris, R. 1966. *J. Chem. Phys.* 44: 425–26
6. Rouse, R. A. 1971. *J. Chem. Phys.* 54: 4135–36
7. Goodfriend, P. L., Brownstein, K. R., Katriel, J., Adam, G., Power, J. D. 1972. *J. Chem. Phys.* 56: 1016–17
8. Gruber, J. B., Hecht, H. G. 1975. *J. Chem. Phys.* 62: 311–12
9. Richardson, R. P., Gruber, J. B. 1975. *J. Chem. Phys.* 62: 2926
10. Azbel, M. Ya. 1975. *J. Chem. Phys.* 62: 3635–41
11. King, D. W., Roderer, N. K. 1982. *Phys. Today* 35(Oct.): 43–47
12. Emsley, J. 1982. *New Sci.* 93: 797

Kenneth S. Pitzer

Ann. Rev. Phys. Chem. 1987. 38: 1–25

OF PHYSICAL CHEMISTRY AND OTHER ACTIVITIES

Kenneth S. Pitzer

Department of Chemistry, University of California, Berkeley, California 94720

INTRODUCTION

It is now 55 years since A. A. Noyes invited me, at the end of my freshman year at the California Institute of Technology, to join him in a physical chemical research project. Research in some area that could be called physical chemistry has occupied a majority of my time since then, and I still enjoy very much the challenge of solving chemical and physical problems. But I have also had many interesting experiences in leadership positions in universities and research organizations, with some activities in even more diverse organizations. I welcome the invitation of the Editorial Committee to write about some of these experiences. My primary emphasis is on several areas in basic physical chemistry where my early research had broad impact and opened up fields of study followed by others, but I also comment about my administrative experiences and my recent research. These experiences may be particularly interesting to other active research scientists who are invited to take up major leadership roles.

Early Years

I was born in Pomona, California, in 1914. My father had been trained in the law, but finding few legal clients, devoted his primary efforts to farming. He was a successful grower of oranges and lemons and additionally held part-time leadership roles in the citrus packing and marketing cooperative. Although we lived in town, I had plenty of experience with the orchard operations. But these did not involve milking cows or riding horses, rather they included driving tractors or farm trucks (even when far too young to be licensed for the highway), irrigation operations, and, on cold winter nights, lighting the orchard heaters to prevent loss of the entire crop by freezing.

1711

0066–426X/87/1101–0001$02.00

Pomona was primarily populated by recently transplanted midwestern farmers and small-town people. My father's family was from Iowa and my mother's from Missouri. Both families had high regard for education. My maternal grandmother was an early graduate of Mt. Holyoke and was a member of the School Board at Pomona. My mother had completed an M.A. at Berkeley and then taught high-school mathematics for a while, and she encouraged greatly my interest in mathematics until her premature death in 1927. One uncle was a physician and another was principal of a high school in a town nearby, while an aunt taught geography at Pomona and travelled throughout the world in summer vacations. My schools were good, and I received high grades without difficulty. But family influences were probably more important than teachers in steering me toward a career goal of engineering, which later shifted to physical science.

The California Institute of Technology was recognized as an outstanding school of engineering and science and was only 20 miles away in Pasadena; thus it was an easy choice to seek admission. On the basis of Cal Tech's examinations and an excellent high school record, I received admission with freshman honors, which carried a reduction in tuition. By then, 1931, the great depression was in full force. During the prosperous 1920s my father had paid off the farm loans. Although his income from the citrus fruit was reduced in the 1930s, it still exceeded current operating costs and yielded comfortable living expenses. Thus, while the economic future was a matter of great concern, the immediate costs of my college attendance were met without difficulty.

The educational opportunity for a Cal Tech freshman in 1931 was remarkable. Arnold Beckman gave freshman chemistry lectures, and my laboratory section instructor was graduate student, E. Bright Wilson (who contributed the corresponding prefatory chapter to the *Annual Review of Physical Chemistry* of 1979). In my sophomore year Carl Anderson gave the physics lectures; his discovery of the positron soon made him famous. Although Cal Tech had an active graduate program, the number of graduate students was small, and most of the leaders in research contributed fully to undergraduate teaching.

Cal Tech had been transformed from a local technical school to one of international standing at the urging and with the guidance of the astronomer George Ellery Hale. In 1912 Arthur A. Noyes, the physical chemist, came from MIT to head the chemistry program and to play a major role in the internal leadership, while Robert A. Millikan came soon thereafter to head physics and to become effectively the president, although he preferred the title, Chairman of the Executive Council.

In 1931 Professor Noyes, who was nearing retirement, decided to offer early research opportunities to selected undergraduates, beginning even with the summer after the freshman year. I was both surprised to learn about this program and pleased to be invited to participate. Dr. Noyes was interested in silver in higher oxidation states, Ag^{II} and Ag^{III}. I measured the rates both of the oxidation of Ag^+ by ozone in nitric acid solution and the rate of reduction of the argentic species by the water present. We also measured, by various physical methods, the mean oxidation state of the argentic silver and found it to be close to two. Since the oxidation rate was first order in both Ag^+ and O_3, it had to yield Ag^{III}; hence the rapid following reaction $Ag^{III} + Ag^I = 2Ag^{II}$ was indicated. Others participating in this research, which was reported in three papers in the *Journal of the American Chemical Society* in 1935, were J. L. Hoard, C. L. Dunn, and A. Kossiakoff.

I was given credit for the sophomore analytical course on the basis of the summer research, which extended into the next year, and then completed the rest of the undergraduate chemistry program on an accelerated schedule. This allowed me to take W. V. Houston's course in Introduction to Mathematical Physics, which gave me an excellent and efficient introduction to advanced methods in mechanics, electricity and magnetism, and statistical mechanics. The treatment of each topic was brief, but it was sufficient to allow me to undertake further study individually in later years whenever my research required it. Also I carried out two other research investigations. The first, with Professor Yost, involved helping Fred Stitt make and separate a mixture of SiH_4 and Si_2H_6. We had many explosions when the silicon hydrides were inadvertently exposed to the air, but no injuries resulted. Stitt used the Si_2H_6 for his spectral measurements for his PhD, while I measured the thermal decomposition reaction $SiH_4 = Si + 2H_2$. According to the literature I should have found a measurable equilibrium, but the reaction actually went to completion. We didn't publish anything at the time, but I followed the situation and noted that the thermodynamic properties of SiH_4, as eventually established, corresponded to our results within experimental uncertainty.

My final research at Cal Tech was a crystal structure investigation under Professor Pauling. He had noted that the compound $Cd(NH_3)_4(ReO_4)_2$ had recently been prepared and found to have cubic symmetry. He suggested that I prepare a crystal and measure its structure by X-ray diffraction. Although he had given me relatively detailed guidance, he insisted that the paper carry only my name, and it appeared in *Zeitschrift für Kristallographie* in 1935.

When it came time to consider graduate study, I decided that I wanted to go to a different school. Although I considered Harvard and Princeton, I was inclined from the beginning toward the University of California at

Berkeley. It was obvious that there was great mutual respect between the Pasadena and Berkeley faculties. Linus Pauling regularly gave several lectures at Berkeley each year. On examination of recent journals I found papers from Berkeley that were interesting. Thus, my application went off to Berkeley and an acceptance with an offer of a teaching assistantship arrived surprisingly soon. I didn't bother to complete other applications and accepted the Berkeley offer.

I was married to Jean Elizabeth Mosher in July 1935. We had known one another in the Pomona schools. She had attended Pomona College, and I had taken her to many Cal Tech dances. Thus, in August 1935 we set off for Berkeley.

The University of California differed from Cal Tech in having a large undergraduate student body, but for a chemistry graduate student it was, like Cal Tech, very attractive, with a lively and friendly faculty active in research. It was also similar in that Gilbert N. Lewis had come to Berkeley from MIT in 1912, the same year that A. A. Noyes went to Cal Tech, with the corresponding mission to convert a good but unexceptional department into one of the very highest standing. Although both Noyes and Lewis had made major contributions to chemical thermodynamics, Lewis' interests were more diverse. His ability to offer incisive comments after seminar talks on any chemical or physical subject was remarkable.

Several factors made 1935 a very interesting time to start graduate study in physical chemistry. On the negative side were the economic aspects of the depression and the uncertainty about prospects for professional employment. But the scientific opportunities were great. It had been established that the quantum mechanics of Schrödinger & Heisenberg was probably valid for molecules generally, in view of the successful quantitative treatments of the helium atom by Hylleraas and of the chemical bond in the H_2 molecule by James & Coolidge. The older Bohr quantum theory had been unable to deal with either of these two-electron problems. Also, the simpler but more approximate treatment of H_2 by Heitler & London had offered an approach to the treatment of covalent bonding generally, which Pauling had already developed (the valence bond method). Hund & Mulliken had also treated simple molecules by a different, "molecular orbital," method with considerable success for excited spectroscopic states as well as the ground states.

It appeared that theoretical treatments of a fundamental nature could now for the first time be given for many chemical problems. I lost no time in studying quantum theory from the newly published book of Pauling & Wilson. I would have been glad to take an appropriate course, but Pauling taught it only in alternate years at Cal Tech and not in 1934–1935. I did take a spectroscopy course in physics at Berkeley that included some quantum theory. It was valuable, but the study of the Pauling & Wilson book was more important.

These studies of quantum theory were for future use. As initially discussed with Wendell Latimer, my thesis research was to involve low-temperature heat capacities, together with other measurements and calculations, to yield entropies of aqueous ions. Very significant results of this type were obtained. I was involved in ten papers with Professor Latimer and his other students, O. L. I. Brown, W. V. Smith, L. V. Coulter, and C. M. Slansky, which brought to effective completion Latimer's program of obtaining the entropies of all of the more important aqueous ions. But while this work progressed, other opportunities arose that were more exciting and were possible because of my study of quantum theory.

EARLY RESEARCH

Internal Rotation

In 1936 all of the organic chemical books said that there was "free rotation" about C–C and other single covalent bonds. The evidence was the absence of separated isomers that could interconvert by such rotation. Hence, the evidence really implied only that any barrier to rotation must be small enough to allow rapid interconversion at experimental temperatures. The only theoretical discussions of the possible magnitude of this barrier had been by Henry Eyring and by W. G. Penney, who had offered estimates in the vicinity of 0.3 kcal·mol^{-1} for ethane. The torsional motion in ethane is inactive in both the Raman and the infrared spectra; hence, there was no direct spectral determination.

J. D. Kemp, who had just finished his PhD thesis with Professor Giauque, told me that he had obtained, in a separate non-thesis study, an experimental value for the entropy of ethane and invited me to bring my newly acquired knowledge of quantum theory to the task of interpreting the result with respect to the barrier to internal rotation. The ethane research had been a project of R. K. Witt, who had been a National Research Fellow in 1931–1933, had since left, and had no interest in the theoretical interpretation. Kemp had helped make the measurements, but had then put the results on ethane aside while he finished his thesis research on the NO_2–N_2O_4 system.

This invitation to use the entropy of ethane in a treatment of internal rotation was a major opportunity, out of which many related research investigations developed in the following years. I found that the quantum mechanics of an ethane-like model had been treated by H. H. Nielsen in 1932. The exact solution for the energies of the torsional states is very complex. It is readily handled by modern computers, but in 1936 it was necessary to find approximations that would suffice for statistical calculations of the heat capacity, entropy, etc. These quantum statistics calculations had to be handled numerically, since the energies were just a list of numbers for each assumed barrier height.

Several papers had recently been published that related to the heat capacity of ethane and its dehydrogenation equilibrium with ethylene. In all cases the recommended interpretation called for either completely free internal rotation or a small barrier near 0.3 kcal·mol^{-1}. Yet there were a few discrepancies, and Teller & Topley in 1935 had mentioned that an alternate but unlikely assignment might involve a 3 kcal·mol^{-1} barrier.

I found that the low or zero barrier model yielded an entropy of ethane that was too large by 1.5 cal·K^{-1}·mol^{-1}, with an experimental uncertainty only one tenth this amount. Thus it was with great interest that I carried out the calculation for a barrier near 3 kcal·mol^{-1} and found good agreement with the entropy measured by Witt & Kemp. There was some uncertainty in the vibrational frequency assignment. This uncertainty was trivial for the entropy but substantial for the heat capacity of the gas. Complete agreement could be obtained for the 3 kcal·mol^{-1} barrier model with reasonable frequencies, however.

The entropy value had been obtained by the "third law" method from heat capacity measurements on the solid and liquid together with the entropy of vaporization. This method really determines the entropy increment above the lowest temperature of measurement; therefore, an addition must be made for the entropy at that temperature that is based on an extrapolation of the heat capacity from that temperature to 0 K. A few cases were known in which this extrapolated value was too small for reasons that were also understood. In our view there was no plausible reason that the extrapolated entropy of only 0.24 cal·K^{-1}·mol^{-1} could possibly have been too small by six times that amount. Nevertheless, we knew that this possibility would be argued strongly unless we could counter it. This we were able to do by rearranging the treatment of the equilibrium and enthalpy measurements for the ethylene-hydrogen-ethane system so that the result was a value for the entropy of ethane (since the entropies of H_2 and C_2H_4 were known). Regardless of the model chosen for the heat capacity of gaseous ethane in this calculation, the entropy agreed with Witt & Kemp's value. Thus, we could conclude firmly that there was no extrapolation error for the entropy and that there definitely was a 3 kcal·mol^{-1} barrier to internal rotation in ethane. We published a "letter" in late 1936 and a full paper in 1937. The result for ethane was received with surprise but little controversy. Several confirming results were published within a few years, primarily from Harvard.

With this very interesting investigation on ethane completed, I decided to examine the situation for other simple hydrocarbons. Indeed, L. S. Kassel had published an elaborate and extensive set of calculations all based on zero potential barriers. J. G. Aston and G. H. Messerly reported in 1936 a "third law" entropy for tetramethyl methane (neopentane) that was 8 cal $\cdot$ K^{-1} $\cdot$ mol^{-1} smaller than the statistical value on the free rotation model and had, implausibly I thought, ascribed the difference to a residual entropy, without even mentioning the possibility of restriction of internal rotation.

The quantum mechanical problem for the internal rotation was more complex for the other molecules of interest than for ethane, and different for each molecule. However, in statistical thermodynamics the partition function is a sum of the Boltzmann factors for the various quantum states. If a number of these terms are grouped closely in magnitude, an average value can be used in the sum. Thus, it was possible to present an approximate treatment for any internal rotational motion of a symmetrical group with a sinusoidal potential barrier. There was no closed mathematical form; rather the results had to be presented in a set of tables, one for each thermodynamic quantity, and with two variables, one the ratio V/RT of the barrier height to thermal energy and the other involving the number of potential minima and the reduced moment of inertia. Many of the approximations of this 1937 treatment were reexamined later, and an improved treatment was presented with W. D. Gwinn. Subsequently, more general types of internal rotations were treated in collaboration with John Kilpatrick.

Accompanying the 1937 theory paper was a treatment of the paraffin hydrocarbons through the pentanes and of the olefins through the butenes. The effective barrier for each of the four rotations in neopentane was 4.2 kcal $\cdot$ mol^{-1}, and, from the value 3.0 for ethane, I interpolated the values 3.4 for propane and 3.8 for isobutane. These values were later confirmed but also refined to recognize the small interaction between the various methyl group rotations. In contrast to the simple pattern for these paraffins, the situation for the olefins is more complex. The basic barrier in propylene is lower, and interactions in the dimethylethylenes are sometimes important. In due time the entropies were measured by the third law separately for each of these hydrocarbons, some at Berkeley and some elsewhere, and the barriers to rotation determined by the same method we used for ethane.

In addition to molecules where the only internal rotations involved methyl groups, there were the flexible molecules such as n-butane and n-pentane where there were rotations within the carbon skeleton. The potential barriers for these rotations were, presumably, also in the 3 to 5 kcal $\cdot$ mol^{-1} range, but the threefold symmetry was no longer present. Rather, one could assume rapidly interconvertible isomers and torsional oscillations for each isomer, e.g. *trans* and *gauche* forms of *n*-butane. Since long normal paraffins are made up of varying numbers of identical subunits, a comprehensive treatment seemed possible. I developed a method that combined a classical treatment of the entire carbon skeleton with quantum correction factors and contributions from the hydrogen motions. For low frequency motions, the quantum corrections are small and easily made. The C–C bond stretching vibrations are all near 1000 cm^{-1}. Hence, it was possible to make the correction in this case by the ratio of the quantum to the classical partition function for each C–C stretching vibration in the molecule. The contribution of hydrogen motions was added to give a complete partition function.

I evaluated the classical partition function for the carbon skeleton by integration of the motion of each additional atom with respect to the atoms already included. Thus, after the third atom, there are the same factors for C–C stretching, for C–C–C bending, and for internal rotation for each additional atom. In this method the moments of inertia for overall rotation are not required. These quantities have multiple values for flexible molecules in different conformations; hence it was a major advantage to use this classical approach. Also, it gives a clear picture of the increment for each additional CH$_2$ unit on a long chain.

I still remember this paper on long chain molecules as the only one about which I have had extended controversy with an unknown referee. The objections clearly indicated that the referee had failed to understand this mix of classical and quantum statistics. I rewrote and expanded the explanations several times. I still appreciate the consideration of Joe Mayer as Editor; I suspect that he finally acted as referee himself. It is also true that the first rewriting improved the paper very much, and this experience has made me more agreeable to revisions of misunderstood papers ever since.

In addition to this theoretical paper, a second paper in 1940 gave tables of values of various thermodynamic properties for various paraffins and olefins, and included the increment per additional CH$_2$ group that allowed indefinite extension to higher members of the *n*-paraffin series.

At that time Dr. F. D. Rossini was directing a project at the National Bureau of Standards (NBS) with support from the American Petroleum Institute (API); the objective was a complete set of tables of physical and thermodynamic properties of the lighter hydrocarbons. Fred Rossini invited me to participate by providing the heat capacities, entropies, and other partition-function-related quantities. I agreed and thus began an extended relationship with the API that involved research support, first via the NBS Project, but later directly at Berkeley. In later years George Pimentel was also involved, and we received support for infrared spectral studies as well as for the statistical thermodynamics. The API's Advisory Committee of petroleum industry scientists was very supportive in our work while clearly indicating which information was of greatest value to the industry. This was university-industry cooperation at its best, and in later years when I had broader responsibilities I found it useful to draw from this example the principles to guide policy formation.

The quantum theory of the origin or cause of the 3 kcal·mol^{-1} barrier in ethane remained elusive. I started a calculation in which the terms neglected by Eyring and others would be retained and evaluated. But this calculation would have involved hundreds or thousands of subcalculations by electro-mechanical calculator with personal transcription of interim results and their reintroduction. I knew from experience about how often a mistake arose, usually in transcription but occasionally by calculator failure. I estimated that a single execution of the complete calculation would involve tens of errors and that an enormous number of repetitions and cross-checks would be required to obtain a correct result. I had opportunities more attractive than this.

It was the increased reliability of the electronic computer that made this calculation feasible. Even the early vacuum tube electrical computers made occasional mistakes, but the various improvements gradually reduced the error rate to a very low level. Also the large memory eliminated the human-transcription-error problem.

My son Russell entered graduate study at Harvard in 1959, and Professor Lipscomb proposed to him for his PhD thesis this problem of the electronic quantum mechanics of the barrier in ethane. Russ asked my opinion, and I urged him to go ahead, since I thought the computational equipment was then adequate. Also I sent him the old file of my earlier exploration of the problem. Once the programs were written and the errors were eliminated, which was no small task, his calculation was completed for the simplest realistic molecular orbital wavefunction but with all integrals included without exception. It was with great satisfaction that I learned of the R. Pitzer–W. Lipscomb result of 3.3 kcal·mol^{-1}. This value, published in 1963, was surprisingly accurate in view of the approximate wavefunction used.

The reason for the failure of earlier calculations to find the 3 kcal·mol^{-1} barrier was described by Russell Pitzer in a 1983 review in *Accounts of Chemical Research*. His informal edition is, "In retrospect it is clear that in these calculations the baby was being thrown out with the bath." More formally he writes "The wavefunction itself was sufficiently accurate, but the integrals discarded were not small enough to neglect and should have been retained in evaluating the energies." Barriers to internal rotation can now be calculated theoretically with considerable accuracy and reliability.

Ring Molecules, Pseudorotation

The additional energy (strain energy) of distortion of C–C–C bond angles in ring molecules had long been recognized, but torsional energies related to single bonds had been ignored since free rotation was assumed. Once the torsional barrier was established for ethane and other simple paraffins, it was apparent that this energy term should be added. World War II intervened to delay my attention to this situation. I did find time for an exploratory calculation, which was reported in my lecture in 1943 in acceptance of the ACS Award in Pure Chemistry and in a note to *Science* in 1945.

The situation in cyclohexane is quite straightforward provided the equilibrium orientation in ethane is the one with the hydrogens of one CH_3 group staggered (rather than opposed) to those of the other CH_3. I was confident that this was the correct equilibrium orientation, but absolute proof was not yet available. On this basis there is no torsional strain in cyclohexane in its chair form. The heat-of-combustion values for the normal paraffins and cyclohexane were consistent with this picture.

Cyclopentane was commonly assumed to have a planar C_5 ring on the basis of free rotation about C–C bonds and C–C–C bond angles slightly less than tetrahedral. But this structure places all torsional angles in the opposed orientation. With a 3 kcal $\cdot$ mol^{-1} barrier, this indicates a total strain energy of 15 kcal $\cdot$ mol^{-1}. But the value from heat of combustion data was only about 7 kcal $\cdot$ mol^{-1}; hence, there was a substantial discrepancy. I showed that by allowing a nonplanar C_5 ring, the total strain energy was reduced to about 9 kcal $\cdot$ mol^{-1}. Of this value, 2.6 was an increased bond angle strain, which was more than compensated by a reduction in torsional strain from about 15 to 6.6 kcal $\cdot$ mol^{-1}. J. G. Aston independently concluded from his entropy measurement that cyclopentane had a nonplanar C_5 ring.

After World War II, I decided to examine both the cyclopentane and cyclohexane situations more closely. Ralph Spitzer arrived with a National Research Council Fellowship. He said, with a twinkle, that he wanted to join my group primarily to have papers with (R.) Spitzer and (K.) S. Pitzer authorship. We first measured the heat capacity of gaseous cyclopentane, cyclohexane, and methylcyclohexane and used these results along with recently published entropies for each of these molecules and for all of the dimethyl cyclohexanes. John Kilpatrick also participated in the cyclopentane work and Charles Beckett in that for the cyclohexanes.

For cyclopentane there are two ways of puckering the ring while retaining some symmetry, C_s in one case or C_2 in the other. We calculated the strain energies and found equal values for these two forms. This led to calculations for possible geometries with no symmetry, and then we discovered that the entire problem could be recast in terms of a pseudorotation. The points of maximum distortion from planarity could rotate around the ring with negligible change in energy. We worked out the quantum mechanics and statistical mechanics of the pseudorotation and were able to obtain agreement with the thermodynamic properties. I called this a "pseudorotation" because there is no angular momentum and no rotation of the entire molecule, but it is described by an angular coordinate.

For substituted cyclopentanes or for other five-membered rings, one expects some energy change with pseudorotation. Then the resulting properties are given by the same functions as for restricted internal rotations. Many molecules are now described in terms that include pseudorotations. Indeed a review by Herbert Strauss with the title "Pseudorotation: A Large Amplitude Motion" was published in the *Annual Review of Physical Chemistry* in 1983.

For cyclohexane the treatment of the parent molecule was straightforward. Equilibrium conversion to the "boat" form was included, but its extra energy is so high that the effect is small. More interesting problems arise for methyl and dimethyl cyclohexanes. Hassel had recently noted that the CH bonds in "chair" cyclohexane have either of two geometrical patterns. One H of each CH_2 is in an equatorial location with respect to the ring while the other CH bonds are perpendicular to the mean plane of the puckered ring and alternatively above and below that plane. Hassel used Greek letter labels, but I thought geometrically descriptive labels would be more easily understood so I chose the terms "equatorial" and "polar" by analogy to the Earth. The former was accepted but the latter was subject to confusion, since it might imply an electrical instead of a geometrical characteristic. Thus the term "axial" was agreed upon in 1953 and recommended in a brief note with the multinational authorship of Barton (Great Britain), Hassel (Norway), Pitzer (USA), and Prelog (Switzerland).

Comparison of detailed structures of various simple hydrocarbons indicated that an equatorial methyl group on cyclohexane was essentially strain free, but that there would be a strain energy for an axial methyl of about $1.5 \, \text{kcal} \cdot \text{mol}^{-1}$. The cyclohexane ring can be converted between the two chair forms reasonably rapidly; hence, thermal equilibrium is expected. This converts all axial groups to equatorial and vice versa. The result is a very interesting set of problems for the possible conformational equilibria and the various thermodynamic properties of methyl and the various dimethyl cyclohexanes. H. M. Huffman had measured the entropies of all of these substances, and we found good agreement except that his value for the *cis*-1,3-dimethyl isomer agreed with our value for the *trans*-1,3-

isomer and vice versa. Consequently, we called for a reassignment. This was certainly the first and maybe the only case in which statistical thermodynamics has been the basis for isomeric identification. Later it was called to our attention that Mousseron & Ganger had in 1938 prepared an optically active and therefore *trans*-1,3-dimethylcyclohexane with a boiling point agreeing with our reassignment, but this report had not been recognized by the NBS or other standardizing agencies.

We carried on some further investigations of ring molecules, such as cyclopentene, and I collaborated with W. G. Dauben in the opening chapter on "Conformational Analysis" of a 1956 multi-author book on *Steric Effects*. Subsequently, the physical organic chemists extended these methods to more complex ring molecules, and many important results have arisen.

Corresponding States and the Acentric Factor

The idea of corresponding states for fluid properties was proposed by van der Waals in 1873. The dimensional analysis related to interparticle forces was straightforward, but little was known about intermolecular potential functions until after Schrödinger quantum mechanics was established. Then London showed that the leading term in the attractive potential had inverse sixth power dependence on the distance. The repulsive intermolecular force is much more sudden; thus it became apparent that the interparticle potential for the various rare gases would have essentially the same functional form with separate energy and distance scaling factors. This single functional form with the two scaling factors yields, in classical statistical mechanics, the principle of corresponding states.

Shortly after London's research, several statistical calculations were made for nonideal gas and liquid properties, but the latter involved serious approximations. It seemed to me in 1939 that a somewhat different approach might be fruitful. Since I had now established that argon, krypton, and xenon should follow corresponding states very accurately, I examined the experimental data and found that indeed this was true. Helium deviated substantially and neon slightly due to quantum mechanical effects on the translational motion. De Boer & Michels had noted these theoretical aspects in 1938, but with attention to the gas phase and critical constants only; they ignored the liquid. I thought that one could define the empirical behavior of Ar, Kr, and Xe as that of a "perfect liquid" (or "simple fluid") and usefully discuss the departures therefrom of other fluids. Departures arising from restriction of overall rotation and some other causes were discussed in the 1939 paper, but a more complete consideration was delayed for several years.

In 1955 I reexamined this matter of the deviation of various fluids from the simple fluid behavior of Ar, Kr, and Xe, to which CH_4 could be added since it has effectively a single, carbon-centered electron cloud and free molecular rotation. In the meantime, T. Kihara had treated the second virial coefficient for core molecules. He assumed a Lennard-Jones (L-J) potential for the shortest distance between cores, which could have various shapes including spherical. I investigated alternate models for spherically globular molecules with the L-J potential, which are applicable for increments of core volume or core surface. While these were more realistic for a molecule such as neopentane, the results were essentially the same as Kihara's model, provided the core sizes were adjusted. Then I found that the effects of cores of different shape or of moderate dipole moment were also essentially the same as that for a spherical Kihara core. In all cases the effective potential well is narrowed in comparison with that for the simple fluid. This gave a theoretical indication that the fluids called "normal liquids," with small if any dipole moment but with any shape, should form a single sequence of deviation from simple fluid properties. Metals and hydrogen bonding fluids were excluded on this theoretical basis.

   With this theoretical indication, I persuaded several students, including Robert Curl, to examine the published data for P-V-T and other fluid properties. It seemed best to take the slope of the vapor pressure curve as the most precise and sensitive indicator of the departure from simple fluid behavior. I puzzled about a name for this new parameter and my initial choice was not very good. A friendly referee suggested the term "acentric factor." This general treatment was very successful on a strictly numerical basis. When asked why I didn't use analytical equations initially, I replied that nature gave a better solution to the complex multidimensional integral than any reasonable mathematical expression. With advances in computers, more complex expressions became "reasonable" and the acentric factor method was converted. The first step was taken by J. B. Opfell, B. H. Sage, and myself, but B. I. Lee and M. G. Kessler gave a much better treatment in 1975. This acentric-factor research was recognized by the Institution of Mechanical Engineers of Great Britain by their Clayton Prize awarded to Curl and me in 1958. The acentric factor is now tabulated for appropriate substances along with their critical properties. I was pleased to give an opening paper to the 1977 Symposium on Phase Equilibria and Fluid Properties on the "Origin of the Acentric Factor."

OTHER ACTIVITIES INCLUDING ADMINISTRATION

In 1943 and 1944 and again from 1949 through 1970 I was diverted from teaching and basic research into other activities and held important administrative positions. While I always maintained interest in current research progress, my contributions varied widely in different periods. After commenting briefly on various aspects of these activities, including the major challenges and accomplishments of an administrative character, I return to discussion of some of the research advances in these later years.

As the US involvement in World War II deepened, Professor Latimer undertook research responsibilities related to chemical warfare, and I was asked to investigate the micrometeorology of gas and smoke flow near the surface of the Earth. I studied fluid dynamics in a very fundamental way and provided solutions to certain novel problems. This study of fluid dynamics was of continuing value to me, both for problems such as the design of flow calorimeters and in my hobby of sailing, which has included the design and construction of sails and boats with novel features. In the war research we used a trace of butane to model gas or smoke flow in a variety of conditions.

In the summer of 1943 I was asked by Professor Thorfin Hogness to be his Associate in directing a special laboratory near Washington to be sponsored by the Office of Scientific Research and Development. The objective was to assist the Office of Strategic Services by developing and testing devices for behind-the-lines warfare and intelligence. Thus my wife and I moved to the Washington DC area with our three children Ann, Russell, and John.

Soon after I arrived, Hogness returned to Chicago to help administer a section of the Atomic Bomb project, and I became the Technical Director of what was called the Maryland Research Laboratory and was located at the Congressional Country Club. My office was the bedroom of the Presidential Suite, which had been used occasionally by President Hoover. We assembled a small but excellent scientific staff; most were physical chemists or chemical engineers. One was a Chinese, Lu Jaixi (Chia-si Lu), who had recently received his PhD and whose return to China was then impractical. He was excellent in his technical work and contributed, in addition, knowledge of China, where our devices might be used. Dr. Lu was recently President of the Chinese Academy of Sciences, and it has been a great pleasure to receive his visits in recent years and to visit him in China.

We invented some devices and we tested others that had been developed elsewhere. We often found that a device that worked satisfactorily in the laboratory failed under the range of conditions that might be anticipated in operations. Frequently, we could devise a fix, and this was gratefully accepted in all cases but one. The exception was a water sensitive incendiary developed by an organic chemistry professor at Harvard. He used waxed cardboard packages that failed and caused fires in our tropical humidity test chamber. We suggested a change to a metal-can package, which he resisted. Then a fire in the baggage room of a Boston railway station was traced to a package of these incendiaries. Thereafter I left that problem to higher authority.

I had been back at Berkeley for only five years when I received another invitation to undertake an administrative position. My general view was that a research scientist should be willing to devote a portion of his career to a management position if it were very important to science and to the community generally. This position as Director of Research for the recently formed Atomic Energy Commission certainly met that standard, and I decided to accept even though it would be a serious interruption of my own research. My initial acceptance was on a leave-of-absence basis for the two years 1949–1950, but I was persuaded to continue for half of 1951. So we moved again to Washington.

I had just finished in 1948 building a 28 foot cabin sail boat, which I had designed especially to be towable readily on the highway. This was accomplished by keeping it quite narrow and by using water ballast tanks that could be emptied for trailering and then refilled after launching. I had planned to use the boat on large western lakes and Pacific Coast bays and sounds. Instead we trailered it east and sailed primarily on Chesapeake Bay, with one longer excursion through the inland waterway south into North Carolina. We went east via Oak Ridge and parked the boat next to the Guest House, where it attracted much attention. In a subtle way I think that this extra dimension of my interests helped in my future relations with the Oak Ridge National Laboratory.

The experience at a high level in a new federal agency was very interesting, trying at times, but also yielding satisfaction of accomplishment. My primary responsibilities were to guide the transition of the national laboratories into their peacetime roles with the AEC and to establish a program of AEC support of relevant basic research at universities and other laboratories. The General Advisory Committee, under the chairmanship of Robert Oppenheimer, was both influential and helpful in these activities. The AEC Chairman, David Lilienthal, and the General Manager, Carroll Wilson, favored these research activities and were supportive. Other aspects of their leadership of the AEC became very controversial, however, and I found myself essentially alone in handling Congressional and other external relations for the research program. Senators and Congressmen were generally sympathetic, but I had to learn quickly about their procedures and attitudes. Senator Brian McMahon and Congressmen, later Senators, Albert Gore and Henry Jackson were particularly helpful.

At the AEC I operated with a very small staff of extremely able scientists. Included were Paul McDaniel and Spofford English, who remained at the

AEC for distinguished careers, Joseph Platt, who was later the founding President of Harvey Mudd College, and John Thomas, who has just recently retired as President of Chevron Research.

In this and later administrative positions, it was my policy to delegate specific responsibilities to able associates and to encourage them to proceed with minimal supervision. This allowed me to concentrate on those broad responsibilities which could not be delegated. Also it allowed me to keep up with current scientific advances and to carry out some research. Thus during the AEC period I kept up some work on the hydrocarbon thermodynamics and wrote initial drafts of several chapters of my *Quantum Chemistry* book, which was completed and published in 1953, only two years after I returned to Berkeley. Also there were students finishing their research in Berkeley whose theses were approved and revised into papers for publication. The most able and mature of these students was George Pimentel. Not only was he fully capable of directing his own research, he also ably directed the research of a new postdoctoral associate whom I had accepted just before receiving the invitation to consider the AEC position.

I was elected to the National Academy of Sciences in 1949. It was at my first meeting in 1950 that the membership repudiated the Nominating Committee's selection of James Conant and instead elected Detlev Bronk to be President. The Committee had failed to consult adequately among the members. Thereafter the Academy became much more "national" geographically in participation of its full membership in its central operations. I was later elected to the Academy Council in 1964–1967 and again in 1973–1976 and served on important committees, including, on two occasions, as Chairman of the Nominating Committee for the President.

From 1951 to 1960 I was Dean of the College of Chemistry at Berkeley. This College, committed to both basic and applied chemistry, was established in 1872 soon after the founding of the University of California, and it is the only one of these original academic units that has continued without merger with other units to the present time. I commented above about G. N. Lewis' role in moving the College to the highest level of quality. Immediately after World War II, Wendell Latimer and Joel Hildebrand not only recruited outstanding young faculty in the traditional research areas of the Lewis era but also expanded the College by excellent appointments in synthetic organic chemistry and in chemical engineering. I had been consulted and strongly supported these additions.

One major need in 1951 was modern laboratory space to sustain this expanded program. A strong effort was required to obtain support, first within the University and then with the State Legislature, but we were successful in obtaining two major new buildings. My other special initiative was the reorganization of the College into separate academic Departments of Chemistry and of Chemical Engineering while retaining unified supporting services. This gave the chemists and chemical engineers appropriate relationships with their separate professional organizations and allowed differences in emphasis in evaluation of faculty and in other matters. Also it delegated substantial responsibilities in a manner giving full recognition to those involved.

Any executive in charge of an organization must give careful attention to an array of regular tasks, including the annual budget and operating plan, personnel appointments and promotions, and communications both internal and external. These are important and are initially an interesting challenge. But after repeating the annual cycle several times these tasks become less exciting and more burdensome. Thus, with the building program approved and in progress, with the two separate academic Departments established, and with colleagues fully capable of handling the Deanship, I resigned the administrative position in 1960.

I had been appoined by President Eisenhower to the General Advisory Committee for the AEC and was involved with other important committees and boards. One of these was the Board of Directors of Annual Reviews, Inc. Henry Eyring was also a member. Usually he came first to Berkeley to visit his brother, and the next day I drove him to the meeting in Palo Alto and then to the airport. We had wide-ranging discussions about science and universities, and he told me about his role in the Mormon Church.

Thus I had many interests in addition to my teaching and my immediate research group, which was also very active. Also, I was promoting a unit for interdisciplinary materials research that would be an extension of existing programs within the UC Radiation Laboratory and would have AEC support. Professors Leo Brewer and Earl Parker were key figures in this effort, which became the Inorganic Materials Research Division and was directed by Leo Brewer for many years.

One of my postdoctorals of this period was C. N. R. Rao, who carried out some solid-state investigations; he is now one of India's leading scientists. Another was Enrico Clementi, who made theoretical calculations on polyatomic carbon molecules. In his later work with IBM, he has demonstrated the capacity of computers to deal with physical chemical problems of remarkable complexity.

Brewer and I were also engaged in a major revision of the *Thermodynamics* book of G. N. Lewis and M. Randall. This remarkable book was first published in 1923 and was still in wide use 35 years later. Lewis had declined to revise it during his lifetime; his interests in later years were in other areas. The first part of the book presented general principles with examples in a very clear manner; this had been written by Lewis and was of continuing value. The second part presented best values of the thermodynamic properties for most of the important substances as of 1923. It was work primarily of Randall and was obsolete. The publisher repeatedly asked me to undertake a revision, and I finally agreed to do it with Professor Brewer as a full partner. We made only very limited changes in the first part of the book (the Lewis part) and wrote a totally new second part, which included substantial additions on statistical methods and other general principles as well as examples of their application. Numerical values for various substances were given as tables and not discussed in detail. The book was very well received in 1961 and still enjoys substantial sales 25 years after the revision and 63 years after the original publication.

While I had been approached about university presidencies before 1961, I had declined to be considered. I did not want to give up my research completely and in most universities it would have been impractical for various reasons to continue it. When approached by Rice University in early 1961, however, I was interested. I had had two PhD students come from Rice and both were excellent; also two of my former research students were on the Rice faculty. Rice was small enough that it seemed feasible to continue significant personal research, and the Trustees said that they would welcome it. Further exploration indicated that although Rice was in excellent status in many respects, certain fundamental changes were needed. Specifically, Rice had been founded as a school for whites in an educationally segregated city and state. Also it did not charge tuition. Both of these practices were now inappropriate for a private university seeking the highest standing. The Trustees recognized the need for the changes, although they differed concerning the method and the timing. By then our children were all away from home (the youngest was an undergraduate at UC Riverside), hence our move was not disruptive for them. I accepted, and Jean and I moved to Houston in the second half of 1961, with oscillations back to Berkeley and a month at MIT, where I held an abbreviated Arthur D. Little Professorship.

At Rice University the structural changes were carried out successfully. Legal authorization was obtained in a manner that convinced the great majority of alumni and friends that the action was fully justified. Admission of black students was welcomed on the campus. The introduction of tuition was accompanied by a very generous financial aid program and was accomplished without difficulty. In the meantime a number of excellent faculty appointments were made, including John Margrave and Joe Franklin in chemistry. An ambitious academic development plan was adopted. The introduction of tuition increased income, but even more importantly it confirmed the University's need for increased financial support from all sources. In due time the first general fund campaign in Rice's history was announced and successfully completed. The chairman of the Trustees, Mr. George R. Brown, was a man of great vision who gave strong leadership in these developments.

In these years I was elected to several Boards of Directors or Trustees, appointments that gave me a wide range of contacts and interesting experiences. These included the Rand Corporation, the American Council on Education, the Federal Reserve Bank of Dallas, and Owens Illinois, Inc. The latter two gave insights on finance and business at a very high level. Also I was appointed to the President's Science Advisory Committee (PSAC) for the term 1965–1968. Initially PSAC accomplished much, but toward the end of my term the Vietnam War so dominated the Presidency that the Committee had little influence. In my Priestley Medal address in 1969 I presented my views on a number of issues for which science has importance for public policy.

My plan to continue research activity at Rice was also successful; a few of the accomplishments are discussed below. Although I was approached occasionally about the presidency of larger and more prominent universities, I indicated no interest while the major developments at Rice were in progress. In 1968 when approached by Stanford, however, I was interested and did accept. Although Jean and I had been most cordially received and had enjoyed our life in Texas, we still had deeper roots in California. The repetitive aspect of many administrative functions was having its effect. Also Mr. Brown had retired as Trustee Chairman, and I sensed that the continuing Trustees would not welcome ambitious new ventures.

Although I was aware of the tensions and the tendencies toward violent student protests of the Vietnam War, I underestimated the magnitude and intensity they would take at Stanford. I was able to avoid any long-range damage to the University, but this type of management called for a leader with professional skills and interests different from mine. Thus, when it appeared that this situation was likely to continue for some time, I decided that Stanford would be better served by a different President. I would certainly be happier in teaching and research in physical chemistry than in crisis management. I could have remained as professor at Stanford and received several other invitations, but most happily I accepted the offer to return to Berkeley.

Stanford gave me a one year "sabbatical" and Jean and I spent a most pleasant and interesting fall of 1970 at Indiana University and spring of 1971 at Cambridge, England. At Cambridge we lived for a third time in an official residence, the Masters Lodge of Sidney Sussex College. Jack Linnett, our host, had become Master but hadn't yet reached agreement with the College Fellows on the renovation of the Lodge and so had not moved in himself. Because of Jean's experience on the renovations of the president's houses at Rice and Stanford, her advice was sought by the Linnetts.

Many of my fellow university presidents and chancellors of 1970 soon shifted to less strenuous and tense positions, but I was most unusual in returning successfully to research in my professional field. One must have ideas about important problems and enjoy solving them. It was a very great satisfaction to be honored by my scientific colleagues in these recent years. These honors included the National Medal of Science in 1975, the Willard Gibbs Medal in 1976, and the Robert A. Welch Award in 1984. Since Gerald Ford and I had been two of the "Ten Young Men of the Year" honored by the US Junior Chamber of Commerce in 1950, it was especially interesting to receive the National Medal of Science from President Ford and to have a discussion with him.

I now return to physical chemistry with comments on three areas of recent research.

RECENT RESEARCH

Nuclear Spin Effects in Methane

In the practical use of the third law of thermodynamics it is assumed that the nuclear spins remain randomly oriented at the lowest temperature of measurement (3–14K). Numerous comparisons of statistically calculated entropies with experimental values verify this assumption. One exception known in 1960 was hydrogen with its *ortho* and *para* species, which could be interconverted by appropriate catalysts. With a catalyst present, H_2 converts on cooling to 100% *para* and loses its nuclear spin entropy below 20K. The molecular properties of H_2 related to this effect are the very small moment of inertia and the resulting large spacing of rotational energy levels.

I had been intrigued by the array of λ-type transitions in the range 13–27K of CH_4 and the various deuterated methanes. These must be associated with rotational motions of these nearly spherical molecules of low moment of inertia. The entropies of all of these species checked on the usual third-law basis, assuming also random H–D mixing for the partially deuterated species. These checks were based on measurements down to about 12K. Might there not be interesting effects at lower temperatures?

J. H. Colwell, E. K. Gill, and J. A. Morrison found such an effect in 1962, an anomalously high heat capacity below 10K. T. Nagamiya and later T. Yamamoto, Y. Kataoka, and K. Okada in Japan presented quantum theories for likely models of solid methanes, while other investigators measured neutron scattering effects. This became an active area of investigation for the next two decades. With Robert Curl, my former research student now on the Rice faculty, and postdoctoral Jerry Kasper, I first considered the theory for gaseous CH_4. There are three spin species of symmetries A, T, and E, with net spins 2, 1, and 0, and high-temperature abundances of 5, 9, and 2 parts in 16, respectively. Only the A species can have the $J = 0$ rotational state; hence it is like para H_2 and becomes the stable species as the temperature approaches zero. The $J = 1$ state requires the T spin species. But for the $J = 2$ and higher rotational states of the tetrahedral molecules, some of the states have different symmetries from others of the same J value and the same rotational energy. Thus, in contrast to *ortho* and *para* hydrogen, which never have the same rotational energy, the different spin species of methane have states of equal energies at high values of J. This means that small spin-spin or spin-rotation interactions can rapidly interconvert the spin species of CH_4 at temperatures where there are appreciable populations in $J = 2$, $J = 3$, and higher states. We verified this rapid interconversion experimentally and examined the theory of the conversion processes in some detail. We also made some calculations and matrix isolation spectral studies of partially deuterated methanes that added to the developing picture.

The Morrison group had reported sluggish thermal equilibrium for CH_4 below 10K, which seemed reasonable since I thought that a spin species conversion was involved. A spin conversion catalyst was needed, and I chose O_2 with its unpaired electron spins. Presumably a small impurity of O_2 would condense with the CH_4 and remain in substitutional sites widely distributed. Paul Donoho, on the physics faculty at Rice, and my postdoctoral, Harry Hopkins, combined their skills to test this idea, and we confirmed it. We measured the proton magnetic resonance intensity and showed a very slow rate of change at 4.2K for pure CH_4, whereas the rate was relatively rapid with even 0.01% oxygen present.

My next opportunity for further investigation in this area was delayed until 1971 in Berkeley. Gerald Vogt, a new graduate student, began a thorough study of pure and oxygen-doped CH_4 extending down to 0.3K. Norman Phillips generously loaned us a He-3-cooled cryostat and advised about calorimetry in this range of temperature. Our most exciting result was a large peak in heat capacity of an O_2-containing sample with a maximum at 0.8K. At 0K, CH_4 would be all A species, and we had found the initial conversion to T and possibly also E species. With a small and reasonable adjustment of their parameters, the theoretical model of Kataoka, Okada & Yamamoto yielded good agreement with our results.

Vogt made very complete arrays of measurements on both pure and O_2-doped methane extending through the lambda-type transition at 20.5K and including measurements of the rate of spin-species conversion of the pure CH_4 at certain temperatures. In our 1976 paper we showed that everything fitted together into a coherent picture and confirmed the "eight sublattice antiferrorotational structure." In this structure three fourths of the molecules are in sites of D_{2d} symmetry while the other one fourth are in sites of O_h symmetry. The effect near 0.8K arises purely from the molecules in D_{2d} sites. The spin-species conversion of molecules on O_h sites occurs only at higher temperatures, where other effects also contribute to the heat capacity. In addition, Janice Kim and I made a theoretical study of the O_2 catalysis in solid CH_4. Subsequently many neutron diffraction and other studies on solid methanes have provided further details, but the general picture of 1976 remains valid.

Relativistic Quantum Chemistry

Although Dirac had developed a relativistic quantum mechanics only a few years after Schrödinger's discovery of the nonrelativistic wave mechanics, it was generally assumed that the differences, the relativistic effects, were negligible for most properties of chemical interest. The spin-orbit effect is purely relativistic, but its energy is often small and it can then be appended by perturbation methods to a nonrelativistic treatment. There had been a few relativistic atomic calculations on a self-consistent field (SCF) basis, and these showed substantial relativistic effects for atoms of large atomic number. When I was planning renewed research activities in 1971, I thought that this area might prove to be interesting.

Improved relativistic atomic treatments were given by I. P. Grant in 1970 and calculations for all atoms by J. P. Desclaux in 1973. Pyykko & Desclaux soon made one-center calculations for several hydrides that gave estimates of relativistic effects. My first contributions in this area beginning in 1975 involved arguments relating atomic properties to effects in which the chemistry of very heavy elements deviated from the trends within their groups of the periodic table. All of these deviations were qualitatively explained. Pyykko & Desclaux also considered questions of this type, and they and I presented simultaneous papers in *Accounts of Chemical Research* in 1979.

New methods were needed, however, for quantitative molecular calculations. Yoon Lee, a graduate student, postdoctoral Wally Ermler, and I started to develop a frozen-core, relativistic effective-potential (REP) procedure. We found it possible to formulate the problem such that all relativistic effects were included in the REP. Thus the Schrödinger kinetic operator sufficed for the molecular calculations, which included only valence electrons, but the REP imposed relativistic symmetry on the complete wavefunction and yielded the spin-orbit terms as well as the mass-velocity effect, etc. Phillip Christiansen soon joined this effort, and he contributed substantial improvements in the accuracy of the REP procedures. Various of these individuals together with K. Balasubramanian made calculations for the diatomic molecules of greatest interest, including Au_2, TlH, Tl_2, and Pb_2 and obtained good agreement with experimental spectroscopic properties.

A chapter in the *Annual Review of Physical Chemistry* for 1985 reviews our work in this area, and that of Hay & Wadt at Los Alamos and of others; hence, further detail is omitted here.

Ionic Fluids

For many years physical chemists have investigated dilute aqueous electrolytes at room temperature. The special pattern of their thermodynamic properties was explained by Debye & Hückel in 1923. Joseph Mayer presented a rigorous statistical theory in 1959 that confirmed the limiting law of Debye & Hückel and indicated a direction for further theoretical development, which was pursued by Harold Friedman and others. The application of these theories at finite concentration requires the interionic potential of mean force. Presently, this potential is only beginning to become available from theory, and simplified models have been used. Theories for water are advancing, and we are approaching the time when this interionic potential in water can be calculated accurately, but that remains for the future.

The Mayer theory yields a virial expansion in integral powers of concentration together with the Debye-Hückel term. But the virial coefficients are functions of ionic strength whereas for neutral-molecule systems they are constants. The second virial coefficient had been calculated for simple model potentials and empirical parameters in the potential adjusted to fit experimental solution properties. When I reviewed this field in 1972, I thought that it would be much simpler to empirically adjust the virial coefficients themselves instead of parameters in a model potential. I retained the ionic-strength dependence of the second virial coefficient through a carefully selected simple expression. Scatchard had used a simple series in integral powers of concentration or molality, but many terms were required to obtain good agreement. By allowing for the ionic strength dependence in the binary term, the series converged rapidly and could usually be terminated with the third virial term.

This advantage is especially important for mixed electrolytes. Also, for unsymmetrical mixtures, e.g. Na^+-Al^{+3}, etc, I extended Friedman's theory to yield a theoretical binary interaction term with a strong ionic strength dependence, which, like the Debye-Hückel term, depends only on ion charges and the dielectric constant of water. Remarkably good agreement was obtained for a very wide range of data for many mixed as well as pure electrolytes, extending to high molality, even to saturation in most cases, with only second and third virial coefficients. In 1980 C. E. Harvie and J. H. Weare used this model for extensive calculations of mineral solubilities in aqueous solutions at 25°C. Various geochemists and marine scientists are now using my equations in their treatments of sea water and other natural brine systems.

Although there was an extensive data base for electrolytes at or near 25°C, only a few measurements had been made at much higher temperature. In 1976 Leonard Silvester and I began a program to test the equations and extend the data base to high temperatures. For NaCl-H$_2$O there were a number of good measurements extending to 300°C. My equation fitted these data with appropriate temperature dependency of the virial coefficients. For other systems, we treated the existing data on enthalpies at 25°C, which yield the temperature derivatives of Gibbs energies. Also we built a flow calorimeter to measure heat capacities and thereby extend the temperature range of the parameters. Other laboratories are active in this area and a good data base for high temperatures now extends to most of the systems that are important geologically or industrially. Since these practical situations involve complex brines, the effectiveness of my equations for mixtures is important.

I find it interesting to work with the geochemists and the engineers in applications of our equations and data. But I have also found several theoretically interesting problems involving ionic fluids under more exotic conditions. A polar solvent continuously miscible with a fused salt is one type, for which a simple equation was found that fits the properties of many systems. For his thesis research, J. M. Simonson extended this type of equation to multicomponent systems and made measurements for the system LiNO$_3$-KNO$_3$-H$_2$O.

The critical region for an ionic fluid is also of great interest yet has been hard to study either theoretically or experimentally. It is experimentally inaccessible (above 3000K) for NaCl. Drs. de Lima, Schreiber, and I found a fused salt, polar solvent system with a critical point at 414K. In contrast to all neutral-molecule fluids, the critical exponent β has the classical value 1/2 for this system. The nonclassical behavior of neutral-molecule fluids arises from density fluctuations of range longer than that of the interparticle forces. Presumably, it is the long range of ionic forces that suppresses this nonclassical effect.

As yet no good theory has been obtained for the critical region of an ionic fluid. Monte Carlo or molecular dynamics calculations, which are satisfactory for liquid-like densities or for high temperatures, have not yet been successfully extended to critical or vapor-like densities at critical or lower temperatures. The vapor of an ionic fluid is dominated by ion pairs and other neutral clusters. Dr. Schreiber and I have extended the cluster calculations of M. J. Gillan to give good estimates of the population of charged clusters. But this method breaks down below critical density, and we could only estimate critical properties by interpolation with the Monte Carlo results for higher density. Thus there remain very interesting theoretical problems for ionic fluids. I look forward to further investigations of this and other physical chemical problems in the years ahead.

ACKNOWLEDGMENT

My research activities received generous financial support from the American Petroleum Institute in the early years, from the Robert A. Welch Foundation when I was at Rice University, and from the Atomic Energy Commission and its successor agencies, the Energy Resource and Development Agency and the Department of Energy, throughout the later years.

PHYSIOLOGY

Ann. Rev. Physiol. 1979. 41:1–24

MOSTLY MEMBRANES[1]

Kenneth S. Cole

Laboratory of Biophysics, IRP, NINCDS, National Institutes of Health,
Bethesda, Maryland 20014 and Marine Biological Laboratory, Woods Hole,
Massachusetts 02543

My first formal connection with physiology, and my first job, began in 1929 when I became Assistant Professor of Physiology at Columbia's College of Physicians and Surgeons in New York City. The association with physiology has been very happy and rewarding. Physiology and physiologists have been kind, generous, and forgiving. Now I am invited to write the Prefatory Chapter of the *Annual Review* which, in the words of the editor "is traditionally authored by a physiologist of great distinction." I am highly honored to join this group and more than a little flattered.

I switched to biology after I obtained a degree in physics in 1926; but I had been committed after the first summer at Woods Hole in 1924. I had gone to Woods Hole because I liked my first taste of biophysics at the Cleveland Clinic the summer before.

I had decided on physics research during the more than a year at the General Electric Research Laboratory, 1920–1922. There I saw electrical engineers at work and it was not for me, in spite of the fact that earlier, during a five week inspection as a merchant seaman, I had been inspired by the magnificence of the Panama Canal, and during a trip with Dad I'd felt the grandeur of the generators and the thrill of molten aluminum being cast at Niagara Falls.

As a youngster I had been lonesome but too busy to worry about it. Even then, although I spent summers as a machinist and as a deck hand on the Great Lakes, I was an electrician. I produced sparks and shocks with worn out parts from the telephone company and put together a licensed wireless station with a Ford spark coil and galena (for a detector) begged from the head of the Geology Department.

[1]The US Government has the right to retain a nonexclusive, royalty-free license in and to any copyright covering this paper.

It all goes back to my amazing parents—a Mother who, when I asked her why the big lumps came to the top when I jiggled the sugar bowl, said "Maybe they don't," and a Father who devoted his career to shaping the academic excellence of Oberlin College. As a parent and a Dean he must have been sorely tried by my pranks—especially by those he suspected but I didn't get caught at. When I asked about a couple of faculty members, he told me they could keep warm on a very cold day just by talking about him. But I don't remember ever acting against paternal advice!

COLLEGE

My scientific career really began with the year at Schenectady. S. R. Williams at Oberlin persuaded me to write for a summer job to the Bureau of Standards, which promptly turned me down, and to General Electric, where W. R. Whitney asked me to visit and promised me a job if I *had* to work there. It was a marvelous experience. I worked on my own on high frequency heating of silicon steel and evolved gas analysis in high vacuum. The techniques were all new to me. Come fall, I didn't have my first analysis.

Dad suggested I take a year off from college, and GE raised my salary so that I could quit my job as a boarding-house waiter. At the research laboratory I got to know everyone and what they were doing. I prowled the works from a battleship turbo-generator to tests of a rolling mill motor. I was properly chewed out for indiscretions and mistakes but I was really hurt when Irving Langmuir ignored me after a glance at my puzzling data. The next time he said "You *do* have a problem," and we worked on it. When I said I was interested in his octet atom, he spent the afternoon explaining and defending it, and we became friends! During this period, solo canoeing on the Mohawk out of the Edison Club and a vacation up Schoharie Creek were my principal diversions.

By far my luckiest move was to take a course at Union College on "Modern Physical Theories" by F. K. Richtmyer of Cornell. He was consultant to GE and gave several lectures on each visit. His book became the first year grad student's bible for several decades and a model for mine almost fifty years later.

My senior year was spent catching up on sophomore and junior year courses I'd missed while in the Army and on the high seas. Dad was Oberlin representative to Farrand's inauguration at Cornell. As he departed for the occasion in academic regalia and soup and fish, I told him that I didn't care about the ceremonies but that he must meet the rough-hewn Richtmyer. He was as much impressed by the man as I, and advised me to take the half-time instructorship Richtmyer had offered me. So, having considered MIT, Chicago, and Harvard, I settled on Cornell That summer my favorite girlfriend was married, and I went on the Lakes, a rough, tough high-seas bos'n—aged 22.

PHYSICS

The first semester at Cornell was so grim that I was ready to quit, but Dad told me I couldn't quit as a loser. Then a notice appeared on the bulletin board: "Wanted—Two biophysicists at the Cleveland Clinic." I knew of the Clinic and of G. W. Crile. Richtmyer said "Darned if I know what a biophysicist is, but I'll tell you something I think is biophysics." He told me of a day at Woods Hole when W. J. V. Osterhout explained the electrical conductivity of the kelp, *Laminaria.* "I think he's right," Richtmyer concluded, "and it looks like darned good fun." But Hugo Fricke wanted two physics PhDs, not a summering first year grad student. He went on vacation to Denmark with no word of what I was to do, so I had to talk myself into the job twice.

Fricke (10) measured the resistance and capacity of blood and calculated the cell membrane at nearly 1 μF cm^{-2}, to give a lipid thickness of 33 Å. He published this finding in the spring of 1923—barely ahead of the extraction and spreading experiments of Gorter & Grendel. I recalibrated his bridge and struggled futilely with his analysis, but it was a good summer. I liked Fricke and admired his history-making combination of theory and experiment. I also learned something of his principal love, the chemical effects of X rays. I knew most of the staff, including several who died in the Clinic fire a few years later. The medical atmosphere was interesting and exciting.

I spent the next summer at the Marine Biology Laboratory at Woods Hole. C. G. Rogers took me in to work on heat production of the eggs of the sea urchin, *Arbacia.* R. A. Budington told me that he was glad to see me interested in a live subject. Rogers had superb measuring equipment, but I had to design and make my first thermopile and stirring gadget while the new Lillie building was being built. The results of the *Arbacia* work were solid and spectacular, but they were first repeated and confirmed 50 years later—at my urging—by Ed Prosen of the National Bureau of Standards.

I went to most of the Physiology and Friday evening lectures as well as the MBL Club dances. I bought a decrepit sailing dory and taught myself to sail while I eased into the sailing crowd led by Ghosty Bridges, of genetics fame. I saw so many interesting and useful things that I thought I could do and had such a wonderful time that I became a Woods Hole addict and set out to try to mix biology with my physics.

Back at Cornell I turned to doing a thesis that involved chasing electrons in circles onto photographic emulsions in a brass box. Most of my fellow graduate students were in the graduate fraternity but, as I found out later, my name was always blackballed. I wasn't even curious about this; I joined a group of fellow barbarians that included a couple of my first-year students. We had some hilarious times to more than compensate for living at home during college.

In the mad scramble to finish up I applied for an NRC post-doc fellowship that would let me follow Fricke and measure the membrane capacity of sea urchin eggs. It seemed easy because they were large and beautifully spherical. But the physics division said that eggs weren't in their jurisdiction; they told me to try biology. Biology said that I didn't have the appropriate training; they told me to try physics. Richtmyer somehow persuaded the sporting biology board to play the long shot and support me at Harvard and Woods Hole under W. J. Crozier and E. L. Chaffee. When my diploma came, Dad said "It's very pretty but it doesn't mean a thing —except that now you won't have to explain why you don't have one."

BIOLOGY

So I became a biologist. Tramping the hot sidewalks of Cambridge looking for a room was discouraging until I came to the last place on my list— opposite Radcliffe—on the second day. Jack Fife clattered down, invited me in, made me a cup of tea, and sold me on the other third floor room. Then he sold his landlady on me as a roomer, and I moved in to the center of my social activities for two happy years. Lucky? Jack was an English grad student. Mrs. Williams, the landlady, had been a student of Osterhout's, was the widow of a Harvard anatomist, and was putting their two children through Radcliffe and Harvard. Her husband had done his thesis on "The Anatomy of the Common Squid, *Loligo,*" in which he discovered and described the giant nerve system. But I didn't see the monograph for ten years; I don't know why I didn't cite it in my book.

While I was trying to duplicate Fricke's bridge, my roommate at Cruft, H. B. Vincent, suggested that I use two vacuum thermocouples such as he was using for his shot-noise work. I tried the idea out on Fricke's data and it gave me his result with half of his data. So I designed and built an oscillator and an egg cell—where stirring and aerating the egg suspension were problems, as were electrodes—and got to Woods Hole before the *Arbacia.* I visited New York to meet Osterhout and Selig Hecht, who were to become lifelong friends.

The work went well with minor modifications, but the cell required the eggs from as many as five females. Preliminary calculations following M.

Philippson showed a large dependence of capacity on frequency—contrary to postulate. But I was stuck with the experiment and took as much data as possible on both unfertilized and fertilized eggs. I hoped to sort it out later. I made many friends. E. N. Harvey was a lab neighbor and listened to my woes. When Keffer Hartline complained about his experiment, my vacuum tube and circuit cured the troubles; he upbraided me some years later when his amplifier misbehaved: "It's your fault, you started me on this miserable business." I got another old sailing dory, promptly christened "Hunky"; we sailed as much and as often as possible in it and in Ghosty's beautiful Herreschoff "Virge."

The next year was mixed. In trying to make sense out of my data, I spent long periods in Widner Library. I went back into the fundamentals and tried to derive the relations that (as I didn't know) Kramer & Kronig were just then publishing. But I was very lucky. Browsing through Maxwell's text I discovered his neat derivation for the resistance of a suspension of spheres and then, a few pages earlier, his expression for a two-phase sphere. Extrapolating the outer phase to a thin capacitor and putting it into the suspension, I arrived at considerable improvements on Fricke.

From K. S. Johnson of Bell Labs, who was a visiting professor, I learned about equivalent circuits and complex plane plots. Gildemeister had plotted human skin impedance on the complex plane to give a small, short, straight line; I was able to show that for a constant phase angle impedance this became the circular arc with a depressed center that has since been widely used. I wrote up the two papers alone in Randall Cottage—where, in one of his puckish moods, G. W. Pierce designated me Director. L. R. Blinks came up from Rockefeller with a thermos full of *Laminaria* to check out their high-frequency behavior. I got some understanding of Crozier's thermal coefficients and became acquainted with Hudson Hoagland and Gregory Pincus, who later did "the pill." Crozier really went to work on my two manuscripts. He whipped them and me into shape, both grammatically and logically, for which I've been very grateful. The theory paper (2) is still useful. The experimental one only reminds me to be generous with brash youngsters.

I was intrigued by what little I knew of nerve. I got a fellowship to work with A. V. Hill, but he wouldn't have me unless I'd go back to heat. Then Osterhout called from Washington to ask if I'd like a year with Peter Debye in Leipzig. Would I ever! "Get an application in the mail today and maybe you'll get a fellowship." I missed the last mail, but a long telegram did the trick—except that Debye wouldn't have me. Then Richtmyer interceded, and I had a fabulous year. I worked over my head on Nernst-Planck theory. When I asked what one problem had to do with membranes, Debye stomped to the window muttering "You and your damned membranes."

We both laughed and that was our password. But Debye was frustrated; in a letter to Richtmyer he said, "I've enjoyed having Cole—but please don't send me any more like him." Working with Debye was a rare privilege; I was convinced that he, like Langmuir, was a different kind of mortal.

PHYSIOLOGY

Soon I was to take my first real job as a physiologist. During the winter in Leipzig I'd had a very satisfactory offer from H. B. Williams at Columbia P & S. I accepted twice—once in a letter from Athens that didn't arrive. Williams had battled for the position for a physicist and had asked Richtmyer for a recommendation once he'd gotten it. More luck! The other two PhDs on the P & S faculty, Hans Clarke and Michael Heidelberger, were chemists; I was an oddball. (But Williams himself had been a math major, had been in charge of sound ranging in World War I, and had written a couple of papers on string galvanometer theory and design.) Williams' help and understanding were nearly unbelievable. He took me to the Mens Faculty Club occasionally; usually we ate at the Attendings Dining Room of Presbyterian Hospital. My next best friend was Ross Golden of radiology. He gave me the job of calibrating his therapy machines, and soon I was consulting physicist to the hospital.

We'd had an ethylene explosion in an Operating Room soon after I arrived. Williams and I made recommendations, which fell by the wayside. A fatal cyclopropane explosion some years later was devastating. The Executive Vice President decreed there would be no more cyclo operations until I could assure him that they would be absolutely safe. When we had things under reasonable control I told him I could make the rooms safe for cyclo only by means that would make surgery impossible. I told him I would give the surgeons a probability of an explosion, which they could then consider along with all the other hazards—including those of other anesthetics. He was disgusted but convinced.

I did all sorts of odd jobs. I measured potential differences between teeth, calibrated a skin-temperature gadget, overhauled (with Williams and Hoyt) the first-year medical physiology laboratory, and gave a few of the lectures. The Department was then concentrated on circulatory physiology. B. G. King and E. Oppenheimer injected some 100 dogs with Evans Blue dye and took blood samples between a few minutes and 24 hours afterward. I found the concentrations were beautiful linear functions of the square root of time and wrote up a picture for diffusion of the serum albumin that bound the dye. But this was against the local establishment and also the journal wouldn't publish anything so absurd.

I've been very proud of my aortic aneurysm operation. After being bombarded by reports about everything that had been tried, I finally realized that a wire laid down on the wall and heated to a controlled temperature might work. We used enamelled wire, pushed the bight through a needle and heated it electrically in a bridge. After animal experiments and our first patient we had a long series of successful immobilizations. Our chief of surgery, Alan Whipple, came in once, watched the wire being inserted, looked at me and my rheostats, galvanometer, slide rule, and log-log graph paper, and, twinkling, said "I've seen strange things in operating rooms, but this is the damnedest yet." I was let down when my collaborators wrote up the operation; they gave me credit only for the circuit.

It was Ashley Weech who finally sold me to the Center staff. According to him, I could diagnose, prescribe and cure a patient by phone. He told me of his new and expensive glass electrode pH meter. I kept my reservations to myself. During the summer he reported that his meter was completely unreliable. I told him how to keep the quartz insulation dry, and that did it.

RESEARCH

My own work was slow starting. I got minimum equipment and set up a crude bridge that I used to measure everything from potato, to nerve, to a cat's diaphragm. Then I managed to approximate them on the impedance plane by depressed-center circular arcs. At Woods Hole Harvey was working on *Arbacia* eggs in the centrifuge microscope he'd built (following the design I'd sent him from Leipzig). He came up with an incredibly low value for the surface tension. This I confirmed by dropping a bit of cover slip onto eggs in a dish. I spent much of the winter building an "egg crusher" designed around specially rolled gold wire from a friend at Leeds & Northrup. The results gave me the internal pressure and elasticity I'd long wanted. Harvey hurried me into publication in the *Journal of Cellular & Comparative Physiology,* Vol. 1, No. 1, p. 1, 1932. With no little sentiment I remember that the symposium, which W. J. Adelman, R. A. Sjodin, and R. E. Taylor arranged around my 65th birthday at Woods Hole, appeared as a supplement to the last issue of the *Journal* before it changed its name. This was also about the time that M. Yoneda et al showed I'd probably been too lazy in not integrating the moment of the profile of the egg—which I had built a balance to do. They found the membrane tension independent of the area. (Recently, this question became controversial again!)

The next winter I improved the crusher. Eva Michaelis came with me to work on fertilized eggs during the summer. My brother, R. H. Cole, came to work on a fancy but stupid idea of mine to measure *Arbacia* heat. I'd shopped most of the Cape to find a knockabout I could afford; I called it "Nike." About the first thing I did was to sail it down from Monument Beach single-handed in enough of a breeze to cause Ed Norman to allow as how I must be a good sailor. He had a Herreschoff "S" and was about to rejuvenate the long-dormant Woods Hole Yacht Club. Once, when ballast shifted, my brother Bob and I were dumped in the Hole! That summer Elizabeth Roberts, a Chicago attorney whom I'd known since she was two years old, came to New York and we were married. I took her to Woods Hole. She complained that all I cared about was sailing; she got sea sick. We were to share our lives happily, through thick and thin, for a third of a century. She died with her boots on.

R. G. Harris asked me to help with a symposium he was organizing at Cold Spring Harbor in late '32. I came up with the name, Cold Spring Harbor Symposium on Quantitative Biology, and gave three papers to replace dropouts. It seems a shame to me that lately there have been so many pressures as to preclude the leisurely eight-week programs that were the rule until Reg died.

Bruce Hogg, a medical student, had volunteered to help in New York. Using a micropipette, he and C. M. Goss had explored potentials around a heart-muscle culture. The 2 μm tip killed the cells when he tried to cross a membrane, but there were one or two hot spots in some cultures where the action potentials exceeded the rest potential. He came with me to work on *Laminaria* the first Cold Spring Harbor summer in Fricke's lab. We collected the kelp by diving off Eaton's Neck at chilly dawn and repeated some of Osterhout's experiments. The kinetics were provocative but I couldn't understand them. So the graphs sat in our lab collecting Medical Center grime until we found the answers with *Nitella*. The next summer Bob Cole helped Emil Bozler and me on frog sartorius with a rather good bridge I'd put together. I calculated the resistance and capacity for parallel cylinders following Maxwell, only to find later that Rayleigh had given my resistance as his first approximation. Again we had a constant phase angle impedance, but I finally realized that this gave no information on the absolute value of the frequency-dependent capacity. Remembering my '28 trick for calculating this, I went to work before breakfast to see if the exponents checked. To my vast relief, they did. By sugar substitution, we got an early value of membrane resistance as 40 Ω cm^2.

I'd been remembering E. N. Harvey's telling me of a big white Bermuda sea urchin, then called *Hipponoë,* which he'd found with a few eggs at Christmas. When Dr. Horace Davenport told us of the delights of the Bermuda Station, Elizabeth and I decided to go on a working vacation in the fall. Will Beebe and his group were recovering from their bathysphere deep dive and included us in all their fun, from helmet diving at Almost Island to motor boating to St. Georges for dinner and moonlight dancing. Elizabeth had a rough time with her bicycle and was black and blue from running into ditches and hibiscus bushes. With the captain of the Bermuda water polo team we collected eggs from every possible source. Our last collection gave two ripe female *Hipponoë* and 100% fertilization, as Harvey had suspected. So we planned another expedition the next fall and the H. J. Curtises came along for a couple of weeks. One memory is of the four of us racing our bikes back from Swizzle Inn.

Our second experiment gave a 90° phase angle—*Hipponoë's* membrane was a perfect capacitor, altogether different from my *Arbacia*. Also the capacity increased on fertilization and decreased on swelling. (Bob Taylor recently asked about the capacity per egg—not per square cm—and it was near enough constant to suggest microscopic dimples and pimples or wrinkles and crinkles.) So we had to do *Arbacia* again. The next summer Bob and I imposed on Fricke's hospitality again; for eggs from starfish we found the same result as for *Hipponoë*. When we used urchins from Woods Hole it was the same story. Why did three echinoderm eggs have a perfect capacity while all tissues showed what I've come to call dielectric loss? And how did the capacity increase on fertilization? And what was the high-frequency dispersion at several MHz?

My ideas on a wheatstone bridge had jelled enough to justify my taking 1935 off to design and build it. Dottie Curtis and Elizabeth decided that Curtis should work with me. The Rockefeller Foundation supported the idea; he arrived at New Years, just in time to solder a few of the last joints. Joe Spencer had been dropped from Princeton, and he was volunteered to help us; we'd been invited to give a paper at the '36 CSH Symposium on nerve and muscle; Ted Jahn wanted to work on grasshopper egg cuticle during this summer: We had a feverish rush.

We knew that muscle fibers varied considerably in size and, since cell radius is a factor, whole muscles could give an apparent membrane loss. Years later, Paul Fatt showed, gently but firmly by calculation, that the contribution of the size spread was negligible. In our paper we also derived the longitudinal impedance of a single nerve fiber with narrow and also with wide electrodes. The wide-electrode result looked too difficult to be useful.

At Woods Hole, Joe Spencer did a precision job on *Arbacia* eggs. He confirmed the findings of *Hipponoë* and *Asterias* while showing that Bob and I had tried to do too much on too few eggs at Cold Spring Harbor. As Curtis worked with single eggs at Woods Hole, Rita Guttman studied frog eggs in New York. One of these was held in a cylindrical hole in the disc between the electrodes. I couldn't solve the potential theory so she got an analog solution using progressively larger glass beads as a model system. But I thought the resistance effect would be too small to measure with red cells when Coulter proposed his counter. How wrong can you be? Recently it was found that W. R. Smythe had solved the problem in an entirely different connection. Smythe was certainly pleased when I told him our data agreed with his solution. How widespread has been the use of the Coulter counter! (Then Smythe was bewildered and deeply grateful when I pointed out an old mistake in the last edition of his book!)

SQUID

The spectacular events of the summer of 1936 involved J. Z. Young and his squid giant axon. Squid were brought from the south shore of Long Island in milk cans; they arrived thoroughly inked and more dead than alive. The two axons we tried looked just like sea water but Young convinced us of their importance. "If you want to find out about nerve, you've got to work on this axon." When I asked how everyone had missed this half-mm tube as an axon, Young said he'd not done the literature until he'd mostly finished at Naples and then had found a 1912 monograph—by an American —on the giant axon system of squid. "Would that American be L. W. Williams?" I asked. I told him the story of my landlady's husband—to his utter amazement.

Elizabeth became pregnant and our lives were going to be different for a while. We went to England on an elegant small Cunarder because U.S. lines were struck, taking with us *Time* magazine with its center spread on Wallis Simpson and King Edward. This was all news to Britain and we kept current with the Paris Herald. We had a delightful lunch with J. Z. Young at Oxford, and I probably first met Alan Hodgkin in Cambridge during this visit. We took one of the first ferry trains to Paris where we visited A. M. Monnier and his wife. Elizabeth had her first, and I my second, encounter with bitter winter weather in the Atlantic in a small American freighter coming back. We were four or five days late, but home by Christmas.

The squid axon was a bigger break by far than we knew then. Our first interest was to see if 1 μF cm^{-2} and dielectric loss were characteristic of a single cell membrane. Curtis suggested that we use *Nitella* during the winter. We got the 1 μF cm^{-2} and a considerable loss—as had Blinks—but the *Nitella* had far too low a resistance. Finally when we pulled the cellulose sheath over a glass rod we found we had what came to be called an ion exchanger—the conductivity was nearly independent of that of the medium, as has been confirmed.

We arrived at Woods Hole with the squid and soon had another 1 μF cm^{-2} with considerable loss—70° to 80°. Out of plate glass Curtis ground a new cell with much larger electrodes; it made no difference to our findings. We also had a second, high-frequency dispersion, which we ignored. More disturbing was the fact that there was no change of impedance during excitation nor during deterioration until an hour or so after the axon bccamc incxcitoble (9). A constant critic kept his record clean: "I always thought you were carefully measuring something quite unimportant!"

We went back to *Nitella* in the fall of '37; using narrow electrodes we saw the transverse impedance decrease as an impulse went past. For the first time, I calculated the effect of a pure membrane resistance decrease. Our points taken from movies of the unbalance Lissajous figures during impulses could be interpreted as an average 15% decrease of membrane capacity from its resting value of 0.9 μF cm^{-2} and an average maximum resistance decrease of 500 Ωcm^2 which was independent of frequency. So the capacity change seemed negligible compared to the 200-fold decrease of membrane resistance. But the lowest resistance (500 Ωcm^2) was considerable. It was also interesting to do some analysis on the characteristics of a partial short circuit travelling with constant speed. The paper we wrote is a pride and joy to me (5). If (as we've been accused of doing) we started a new era of axonology, it was in this paper that we had to be original.

We built a new cell and a new amplifier; we modified our first commercial oscilloscope to give single, but highly nonlinear, sweeps. Then we waited, miserably, at Woods Hole for the first squid. Our first two axons showed nothing. We put in the whole nerve. With everything wide open we observed a very slight change. Curtis swore it was a decrease; I wasn't sure. But it got larger and larger as we used better and better axons, until finally we got the picture shown in Figure 1 (6). The photo was taken in the dark with my '29 Leica. I set and tripped the Lucas spring rheotome some twenty times for the impedance change, ΔZ, and half a dozen times for the action potential, V. I had a bloody thumb sometimes. Hodgkin visited when we had ΔZ on the scope. He was as excited as I've ever seen him, jumping up and down as we explained it. He also appreciated the importance of the resting membrane resistance and thought longitudinal measurements between long electrodes could give it. He assembled the equipment in New York and took the data back to Cambridge to find about 1000 Ωcm^2 (7).

On my way to the 1938 Zurich congress I stopped in London; Otto Schmitt was there; Bernard Katz had just come from Leipzig; A. V. Hill instructed me on talking to the multilingual audience I would face. Brian Matthews and I travelled together and talked sailboats most of the way. After a bad case of jitters I was truly surprised by the enthusiastic reception of my talk.

Near the end of the Congress a cable sent me to Vienna. Elizabeth's sister had married a Viennese doctor, and they wanted me to bring his mother back with me. It was very distressing that Frau Frey, her relatives, and her friends could only go to each other's homes in the evenings and talk about the Nazi restrictions and what had happened to whom. I could only get transit visas for Switzerland and France when we had to leave—two weeks before our boat sailed. The inspector on the boat train nearly had apoplexy when I showed him Frau Frey's passport. Frau Frey was the belle of the boat; she could go anywhere, do anything, and talk to everybody—which she did. Then we were met in Hoboken by the Jewish Relief, who insisted I was not an adequate escort! Frau Frey had a wonderful time in Chicago with her children, going to tea rooms and movies for the few years she had left.

Curtis had talked with Hodgkin about recording the internal potential of an axon but I wasn't enthusiastic. Why bother about an upside down action potential? We used a metal core needle. Hodgkin and Huxley got a certain overshoot, with a micropipette and reported this in a note in *Nature*. We did better the next summer (except for an overcompensation of the 100 μm glass tube).

It most certainly was a serious mistake in general for us to have directed so much attention to our single exception to the mean behavior of all our other axons. Hodgkin (11) blames it and the dextrose effect for a year's delay in proposing the "sodium hypothesis." The dextrose effect I neither understand nor remember.

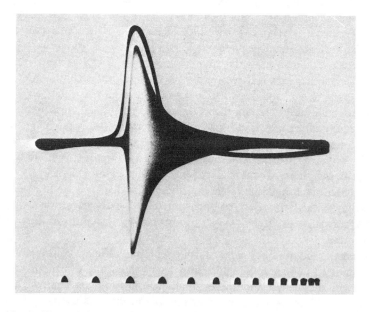

Figure 1   Oscillograms from passing squid-axon impulse. Action potential is the single line, *V*; impedance decrease, ΔZ, is the band; time mark are 1 msec apart. [After (6), 1938.]

But the Battle of Britain had begun. Curtis and I measured the potential under an external electrode and I found a neat trick to get the membrane V-I rectification curve. Hodgkin dubbed it Cole's Theorem. Years later I was disgusted to find I'd assumed linearity in one step; I'm still puzzled about why several better derivations give the same answer.

Curtis wanted to become a real physiologist and went to Johns Hopkins with Phil Bard. R. F. Baker came to us having written a mass-spectrometer thesis; we found that impedance changes by current flow duplicated those during a passing impulse at intermediate frequencies. We also measured the inductive reactance that Hodgkin and I had found in the membrane. It could be explained by the time needed to establish a nonlinear steady state and could also be a capacitive reactance.

As Baker became interested in microscopic spectroscopy and electron microscopy, George Marmont came from Cal Tech. We investigated the effects of K^+ and Ca^{2+} on longitudinal impedance. The results were striking (4) but were only published in abstract at the time. I thought the capacitive and inductive reactances might be permanent structures, seen only as allowed by shunt or series resistances, but the complexity was hideous. The time constant of nonlinearity and Mott's semiconductor theory adapted for K^+ made much more sense.

I'd studied the calculus of variations under Marston Morse at Cornell (along with I. I. Rabi) and had listened to his lectures on mechanics at Harvard. Still, I was surprised to have him suggest that I come to the Institute for Advanced Study on leave after he'd been there a while. I was tired of the medical atmosphere at the College of Physicians and Surgeons, and I got a Guggenheim for one of our very happy years. We had just moved to the Bronx, but we moved again with our very new little girl. I was set up in Fine Hall in mathematics with physics next door and the joint library upstairs. L. A. MacColl of Bell Labs had told me of V. Bush's lecture, "The Engineer Grapples with Nonlinearity." With that and Mott's semiconductor theory for starts, I chased nonlinearity and negative resistances through the mostly Japanese and Russian literature. John Tukey set up an IBM card-sort solution for a membrane, but I couldn't tell him the membrane characteristics. Riding into New York one day, I sat with Solomon Lefschetz. I told him I was planning to bring him a question.

"What is it?"

"How to tell one side of a line from the other."

"That's a good question for a topologist. Why do you want to know?"

"I want to know what makes a nerve impulse go."

"You come see me day after tomorrow and we'll talk."

Talk we did for practically a solid week. After a few days he said that because he wanted to know more, he would give a course. After the second week he had decided to translate the pertinent Russian literature. Decades later, shortly before he died, he introduced me to a friend as the man who had started him on the work he'd been doing for the past 25–30 years.

WAR

There were all kinds of visits and phone calls to enlist me in war work, but I was determined to finish the sabbatical that was starting so well. Luring me to Chicago for consultation, A. H. Compton and N. Hillberry told me the story of nuclear fission and demanded that I take charge of the biomedical problems. I knew at least how to start; and after I had persuaded them I could take no medical responsibility, it was an exciting four years. Our group grew exponentially to nearly 400 before staffing Site X (Oak Ridge). It was six months before our first laboratory was ready. It occupied the disinfected stable of an extinct ice plant on the south side of the Midway, and it was extended twice. Soon we had the first practical hunk of uranium from Spedding. We fought battles for survival with Groves; we struggled to get support from DuPont. Meanwhile the pile went critical.

Soon after R. E. Zinkle came with me, we interviewed Pat Lear for the job of animal farm manager. She told Ray she could cook. She was a tower of strength—even though she could not cook.

George Svihla and I decided to try an autoradiograph for fission product dose. An exposed guinea pig was frozen, sawed into thick sections, and reassembled with X-ray film between the sections. Svihla noticed that the machinist who had watched the procedure put on gloves to replace the band saw blade. Joe Hamilton, who supplied much of our distribution data from Berkeley, could never understand why we used thick sections.

I stole Curtis from an aviation project to head up Site X biology. In a sea of mud in the early days I had to be carried by men in hip boots. Ladd Prosser was my unhappy, able, number two at Chicago. Jo Graff was our beautiful administrator, who kept things going while her husband was interning. Only by the few times she wept on my shoulder could I guess what it cost her to be so tough.

It was all tightly programmed—except for the free 10% man days I insisted upon—but after Hiroshima and Nagasaki the place blew up. Everybody had kept his pet hates to himself until the war was won.

War is so disgusting, so futile.

PEACE

The University of Chicago set up the new Institutes. Zinkle became head of Radiobiology and Biophysics, and I resigned from Columbia to head up

Biophysics. It was a real wrench. At Columbia I'd found myself, I'd been happy. The years there were the best I've had.

George Marmont came to Chicago; we took equipment to Woods Hole for squid studies in 1946. It was not a good summer; we tried futilely to extend our prewar results on longitudinal impedance to low-frequency effects of current flow. Jimmie Savage had worked with Warren Weaver, who had thrust upon Jimmie a fellowship to work with me. Savage visited us late in the summer; after listening to our woes, he suggested putting a long current-carrying electrode inside the axon. I explained how the electrode polarization would probably defeat us; but Marmont took him seriously and soon proposed a reduced silver axial electrode, a central outside electrode with a guard at each end, and electronic control of the membrane current. I promptly added the inverse of membrane potential control.

In 1947 Carlos Chagas had invited me to Rio de Janeiro to lecture and consult at the new Instituto de Biofisica. It was a long, hard trip, but I was met by an enthusiastic delegation. During a courtesy visit to the Rector it was decided that my lectures would be published as the first of a series. There have been several reprints of the *Four Lectures on Biophysics.* I was able to help E. Leão with his spreading-depression impedance decrease; it could be interpreted as a membrane-resistance decrease. I couldn't do anything with the Instituto's favorite electric eel and I got back in time to go to Woods Hole.

After the usual start-up troubles but with Will Rall's help, we had the Mach I cell in operation in 1947. We confirmed the 1 μF cm^{-2} and 1000 Ωcm^2 findings and ran strength-duration curves with direct data. We found the initial resistance to be much too high unless there was a series resistance associated with the membrane; we had trouble with anode block of an impulse. My dream of making excitation stand still in space and time was half-fulfilled; it was not until J. W. Moore showed that excitation for the squid axon in iso-osmotic KCl and CaCl$_2$ was stable in time that I had the other half. However, Marmont was firmly opposed to my insistence on membrane potential control, and I got only a few runs—including that shown in Figure 2 (3). These I found spectacular. There was no trace of a threshold; the early inward current was a mystery, but it was a transient negative resistance that could account—at least qualitatively—for the rise and height of a spike and its propagation. The outward currents corresponded to our ideas about K$^+$.

In the fall Hodgkin told me of his Na$^+$ results with Katz; but I was more impressed by my own data, which I told him about. He wanted to visit us; still, he wasn't altogether happy to find I'd booked him for the annual biology division lecture, which he gave on the Na$^+$ work—to a full house. We went over the equipment and my results in great detail. He vigorously

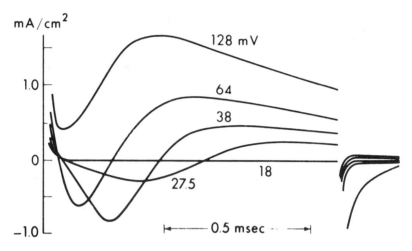

Figure 2   Squid-axon membrane currents after step depolarizations as indicated from resting potential (voltage clamps). [After (3), 1947.]

defended his and A. L. Huxley's carrier theory and blamed the slow rise of inward current on apparatus; but I believed my results, and he later confirmed them. I also pointed out troubles with electrode polarization (which they corrected) and with the membrane resistance (which they compensated for).

Through the winter, while Marmont designed and built the elegant Mach II cell, I tried all sorts of outrageous schemes to explain the potential-control data and did various Institute chores. We got a new Director, who didn't care what biophysics was, so long as it was physical chemistry. He told Savage to go back to mathematics where he belonged; Jimmie became Mr. Statistics, first at Chicago and then at Yale. The Director didn't mind having me, but he didn't want any more like me. I felt less than wanted at Chicago and our friends agreed.

DEFENSE

It was a most auspicious time for Admiral Clarence Brown to offer me the new position of Technical Director at the Naval Medical Research Institute in Bethesda and announce that he was going to stay in Chicago until I accepted. From the latter intention he was dissuaded; he sat out my indecision in Washington. Leaving the Institute was not easy—the breadth of the University was so great, and we had many good friends—but I felt I had to try the unknown again. I had not realized how discouraged Elizabeth had been in Chicago; she was delighted to be going to Washington.

I had to interrupt making friends and enemies at NMRI to go to Paris by way of London for the electrophysiology conference in honor of Louis Lapicque and his gallant war stand. I spoke for the visitors and included an April-in-Paris note. It was my first chance to present my '47 results, which I could only do briefly as an introduction, really, for Hodgkin, Huxley and Katz (14). In a little over a year they had corrected most of my difficulties, caught up and run past me. I was pleased that we had been their starting point and that they had confirmed my results, although I was not entirely happy to have the concept and the technique dubbed the "voltage clamp."

My simple plan for NMRI was to have a group of small, strong centers of research distributed as well over the medical field as possible so that any emergency would be within reach of one or perhaps two staffed working laboratories. This would avoid some of what caused our slow start up at Chicago. Manuel Morales was a fountain of good ideas (e.g. that we should invite Terrell Hill to join us). Moore, one of several students of Jesse Beams who turned to biology, came with us to work on muscle impedance; we had Dave Goldman, who started to construct an analog computer. The energetic submariner Al Behnke kept things moving as Executive Officer. Early on, Admiral Brown transferred to us his long-time right-hand man, Vic King, with instructions to get done the things we wanted done. W. E. (Bill) Kellum became CO and soon had the admiration and respect of civilians and military alike. Although he confessed it took him a year to understand what I was trying to do, he could straighten out my mistakes almost before I made them. But it was slow business, with much diversion, backtracking, and backbiting. Then we lost momentum. Ed Condon was hounded out of the Bureau of Standards; McCarthy forced out two of the most unlikely subversives of our staff—one now a full professor at Yale; A. V. Astin was only saved in the battery additive fiasco at the NBS after a slow protest from outside government and the Bureau was weakened by fragmentation; the Office of Naval Research was being forced to curtail its broad effectiveness to "man in the fleet" problems; and good rumor had it that NMRI was next. We were losing key personnel in the antiintellectual movement and couldn't get adequate replacements. I tried to get a more prestigious replacement for me but I couldn't get even medical support in my office. We seemed to be in an apparently endless decline.

Five of us flew Navy via Gander, Azores, Port Leyouty, and London to the 1951 Copenhagen physiological congress. At the Congress dinner, Lord Adrian said that if we were to rank our national preferences he was sure Denmark would be the unanimous choice for second place. One spectacular achievement reported was Ussing's electrochemical identification of an ion pump. But I was much more interested in the preliminary curves Huxley showed me of their Na^+ and K^+ conductances-vs-potential from voltage-clamp data.

The next summer (1952) there was a Cold Spring Harbor repeat run on nerve and nerve systems. Hodgkin sent me near-to-final drafts of their five *Journal of Physiology* manuscripts, and at last I realized fully what they were doing and how enormously successful they had been. Hodgkin gave the paper with his usual calm, convincing enthusiasm. After ten minutes the Chairman, Frank Schmitt, co-opted me to handle the discussion. I infuriated Hodgkin by counting the number of ad hoc analytical forms and numerical values in their equations (12). But he also said they had fully confirmed my 1947 results and had used essentially our experimental approach. Lorente de No said that it was a powerful picture and might be right but that it couldn't work for a frog axon. John Eccles dominated the rest of the all-too-short symposium and won at least grudging admiration from everyone by producing three new theories in three days.

On our way back to Woods Hole I was feeling very sorry for myself because Hodgkin and Huxley (HH) had done all that I had ever hoped to do. Consolingly, Hodgkin said they had just followed my lead. Later Huxley was to say he was only Hodgkin's student, while Eccles only claimed to apply their results. At Woods Hole, Moore and I were trying to clamp with external electrodes. Hodgkin wasn't sympathetic, but he dissected axons for us and tried to teach our other guest, I. Tasaki, to do it—even using Tasaki's favorite needles.

In 1952 the Standards Eastern Automatic Computer was in somewhat erratic operation at NBS. H. A. Antosiewicz and I explored all four quadrants of the HH equations (13) and found a saddle point that was an undoubted threshold. I passed the word to Bonhoeffer, and to his translator, Max Delbrück, that his thermochemical analogy would have to have a threshold. Huxley agreed with Richard FitzHugh's phantom saddle point, and they both thought I was wrong. Several years later I ran onto some unaccountable bumps in the curves. FitzHugh & Antosiewicz finally traced these to an absurd programming mistake that had produced the saddle point. Although important theoretically, the threshold was only 1 part in 10^8 wide and only made a percent or so change in any physiological parameter, but I had to explain and apologize as widely as possible. It was only a decade and a half later when Rita Guttman was getting much more gradual thresholds at temperatures in the 30s C, which F. Bezanilla confirmed with HH calculations.

Moore saw an advantage in using a microcapillary for the internal axon potential, and we practiced on "open chest" squid with intact circulation. Harvey had mixed feelings: "Here is a perfectly good biological experiment being done by two physicists." The undershoot recovery was delayed so much that we could expect it to disappear along with the HH leakage in an undisturbed animal. This Hodgkin and Keynes did find by boring through the mantle directly to the axon!

NATIONAL INSTITUTES OF HEALTH

Seymour Kety was intramural director for the NINDB and NIMH joint operation at National Institutes of Health and planned a biophysics laboratory. I was a very discouraged administrator after four years of war and six of defense, and the squid work I wanted to do was more than I could support at NMRI. So Moore and I moved across the street. With me I took considerable regrets that the Navy might not get what it needed and deserved for some time; I also took a pay cut!

Once again we were starting a new lab from scratch. It is not easy to sort out the apparently intertwining threads of the past twenty-three years at NIH. The most compelling strand has been the voltage clamp, with the squid axon a close second. If (as I've been accused of doing) I revolutionized electrobiology in 1947, Hodgkin and Huxley certainly took the giant step in 1952. Hodgkin (11) recounted it for the centenary of the Physiological Society in 1976. As the concept of the clamp has been accepted, many new techniques (some good, some not so good) have been developed. I guess there are well over 100 voltage clamps around the world; the published papers relying on them must be in the thousands. And now solid-state programming and data reduction are rapidly taking over manipulations far beyond my wildest dreams.

The "abominable notch" was still with us. Moore, Taylor, and I spent a couple of years on it—spurred on by Frankenhauser and Hodgkin, who ignored it, and by Tasaki and collaborators, who insisted that all good axons showed it. (Who wants to work on less than the best?) Although it almost certainly comes from poor electrodes, it is still the constant threat that I, and Hodgkin and Huxley, only narrowly avoided before 1950. Moore and I had been keeping our axons hyperpolarized between pulses to maximize the sodium currents, but all too soon the polarizing currents would begin to run away or the axial electrode would bubble. Only a few msec of prepolarization were adequate to prevent these troubles and prepotential had little effect on the initial sodium current. But after -212 mV the potassium currents were delayed by up to 0.2 msec. Didn't this prove HH were wrong? No, only limited! Our article (8) seems to have been a good start for the *Biophysical Journal,* first volume, first number, first page, 1960. The Cole-Moore delay has been confirmed for squid (Figure 3) and has been a

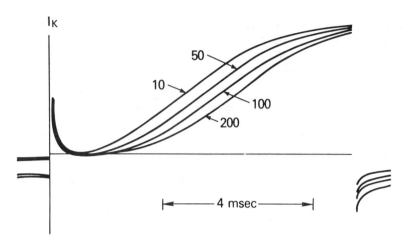

Figure 3   Depolarizations of squid axon with TTX after 3 msec hyperpolarizations to indicated potentials from holding potential. [After Keynes, Kimura & Lecar, unpublished.]

major challenge for theory. For several years there have been rumors that a frog node is different. So it may be that our simple delay is the exception rather than the rule; theorists can then let out their belts.

The work of our group on squid was hampered and finally broken up by dissension that I was unable to ease or prevent without more space. So I made everyone a section chief and went on alone.

We had only the simple analog computer—a descendant of V. Bush's mechanical differential analyzer, which I'd admired—and with it I learned the lessons I got from v. Neumann and E. Teller. The computer succeeds calculus as a way of intellectual life. Richard FitzHugh took it over—he'd always wanted one—and worked up to solutions of the HH equations before turning to keep pace with the digital developments. He worked on topology, developed the BVP (Bonhoeffer–van der Pol) analogy, and moved on to the relatively recent economics of nerve and muscle action. I've loved his teaching gadgets and his cartoons—the "Drink MyxiCola" gag, and the coy squid (which I've stolen).

We had been going along happily with *Loligo* at Woods Hole when I heard (around 1960) that the Humbolt *Dosidicus* off Chile weren't as impossible to work with as I'd been led to believe. Our first expedition didn't come off; but the second, in the fall of '63 with Dan Gilbert in charge, was a good start. Our groups had excellent seasons at Vina del Mar with E. Rojas in spite of crude collecting and laboratory facilities, fantastic logistics, and a tidal wave. But in 1971, with a capacity crowd waiting, not one single *Dosidicus* was to be found—nor have they reappeared. The unlikely troika of Keynes, Rojas, and Taylor has, however, continued to function. In Chile, efforts have been diverted to the giant barnacle muscle fiber; at Bethesda, Leonard Binstock discovered and promoted the cord of the worm, *Myxicola*, which has become a mainstay of the laboratory.

Ross Bean, whom I'd met at a bilayer meeting, came to see us at Bethesda. We persuaded him to try a very low concentration of additives; this gave impressive unit conductances. These studies, along with studies of fluctuations, have continued to expand in the lab under Gerry Ehrenstein and Harold Lecar.

For some time I'd decided that I should step aside at 65—as insurance against mistakes and to let young blood take over. I almost made it. Taylor became Acting Lab Chief in '63 and continued until '71. Adelman came back as Lab Chief and, after considerable shuffling around, moved with two sections to Woods Hole year round, leaving a section in Bethesda.

BIOPHYSICS

My first attempt at formal biophysics was a discussion group I chaired at the first summer meeting of the American Physical Society in 1927 at Cornell. During the war I tangled with Leo Szilard, who had been about to switch from physics to biology when artificial radioactivity was discovered. He dismissed my simple approach to biophysics, and it was only after many hours that I caught on: Biophysics was whatever Szilard did in biology, and that settled the argument. He started to work on phage after the war. When peace came, departments, laboratories, institutes, divisions, and branches of biophysics burst into the open. Many plans arose for national organization and publication. I managed to keep in touch with the more active people and groups, but so much time and effort were wasted in attempting definitions that I finally was able to get them ruled out of order. I was in favor of affiliating with the Institute of Physics; this notion was bitterly attacked. Alan Burton, then President of the American Physiological Society, told me what APS could offer, but I knew that a number of noisy physicists and engineers would have none of it. I had to do the best I could, even though I was a loyal member of the Society.

It was finally obvious that biophysics had to proceed without assistance or entangling alliances. An informal but considerable group at the '56 Federation meeting voted for a trial meeting to be run by the Committee of Four, E. Pollard, O. Schmitt, S. Talbot, and me. We held the first meeting at Columbus with Air Force funds from Colonel A. P. Gagge; Pollard produced the required *Proceedings*. When Hartline said that we had more people he knew and more he wanted to know than any meeting he'd been to, I was sure we had succeeded. Elizabeth and I wrote up both the

Constitution and Bylaws. I worked for their adoption, item by item. After an evening without adverse discussion at Cambridge, the Biophysical Society came into being. It has been very interesting to watch the completely independent development of a division of biological physics in the American Physical Society and the current trial of joint sponsorship of the *Biophysical Journal*. I do hope that the casual cooperation can go on.

H. B. Steinbach was commissioned to organize an ad hoc National Research Council committee to keep track of international biophysics. After considerable manipulation, the first congress was held in Stockholm. United States financing was an important factor in assuring its success, and I was relieved of one of my longest ad hoc services. The organization joined UNESCO to become the International Union of Pure and Applied Biophysics and now meets regularly.

A major effort was made in 1969 to bring MBL up to date. Steinbach and I had talked for some time about having a voltage-clamp researcher working with students at Woods Hole through the winter. Finally Adelman put together a summer program in 1969—the Excitable Membrane Training Program. He edited the book of lectures, *Biophysics and Physiology of Excitable Membranes* (1). Ernie Wright was the angel who kept it going for six years before it was ruled out. It was well worthwhile.

I was named Regents' Professor at the University of California at Berkeley for the first semester of 1963–64, and Elizabeth was determined to go with me. We had a lovely apartment near campus. It overlooked San Francisco Bay and the Golden Gate, which I'd steered a West Coast War Emergency boat in and out of long before there was a bridge. I usually got home early so we could put up our feet at cocktail time and watch the glorious sunsets. I'd decided to write up my lectures as a book, but I was not prepared for what became a five year stint.

I've been back to Berkeley almost every year for the winter quarter. I have gradually entrusted my seminar to C. A. Tobias and have added La Jolla and Galveston to my visiting list. P. E. Lilienthal of Cal Press saw me through the book *Membranes, Ions and Impulses* and its reprinting in 1972 (4). I'm almost convinced it was worth the time and effort. By the time I had recovered it was mandatory retirement time.

RETIRED AND REHIRED

My 70th birthday was ushered in with flair at the Gilberts' by the Piet Oostings' singing Happy Birthday in Dutch at midnight, and I became a rehired annuitant. As an experimentalist I've increasingly depended on Woods Hole, where for some years I've had relative peace and quiet in 150 square feet of space. I've not only gotten my hands dirty, I've cut, burned, and otherwise maltreated them on my own.

I had been somewhat querulous about various aspects of the electrical and electron-micrographic estimates of the Schwann sheath at 1.5 Ωcm^2 until I ran onto a paper Curtis and I published in '38 on the axon-impedance locus with its incomplete high-frequency tail. Could it be the Schwann sheath? Probably so; I calculated the resistance to be 1.6 Ωcm^2. Before 1970, Choh lu Li, Tony Bak, and I had run analogs to show that the low-concentration Rayleigh and Maxwell resistance equations applied to up to 100% volume concentrations of several close-packing cylinders and three dimensional forms. If cylinders made up the sheath, then the extra-cellular space constitutes about one half of one percent of cell volume. But what about the 4 MHz dispersion capacity? Might the low-frequency Rayleigh capacity equation work up to 100%? An analog said it did. Thus, if the membrane capacities are 1 μF cm^{-2}, the sheath should be six cells thick— except that the cells aren't cylinders! Membrane-covered cubes followed the Maxwell capacity up to 100%. This completes at least a sketch of Fricke's beginning in 1923. The 1935 bridge reappeared, after 18 years in hiding and minus crucial transformers, to make it possible, at least, to test guard arrangements and perhaps to find out more about dielectric loss!

IN CONCLUSION

I've had busy and exciting times with membranes since I first heard of them in 1923. In spite of my mistakes, I'm very happy to have had the good luck to participate in the development of the present widespread enthusiasm for them. I'm only slightly modest about the many and good friends who've helped so much and who share in my distinctions. These days I find it difficult to keep track of new concepts, powerful techniques, and obvious conclusions—but I keep on trying.

Literature Cited

1. Adelman, W. J. Jr., ed. 1971. *Biophysics and Physiology of Excitable Membranes.* New York: Van Nostrand Reinhold. 527 pp.
2. Cole, K. S. 1928. Electric impedance of suspensions of spheres. *J. Gen. Physiol.* 12:29–36
3. Cole, K. S. 1949. Dynamic electrical characteristics of the squid giant axon membrane. *Arch. Sci. Physiol.* 3:253–58
4. Cole, K. S. 1972. *Membranes, Ions and Impulses.* Berkeley: Univ. Calif. Press. 569 pp. 2nd printing

5. Cole, K. S., Curtis, H. J. 1938. Electric impedance of *Nitella* during activity. *J. Gen. Physiol.* 22:37–64
6. Cole, K. S., Curtis, H. J. 1939. Electric impedance of the squid giant axon during activity. *J. Gen. Physiol.* 22: 649–70
7. Cole, K. S., Hodgkin, A. L. 1939. Membrane and protoplasm resistance in the squid giant axon. *J. Gen. Physiol.* 22:671–87
8. Cole, K. S., Moore, J. W. 1960. Potassium ion current in the squid giant

axon: dynamic characteristics. *Biophys. J.* 1:1–14

9. Curtis, H. J., Cole, K. S. 1938. Transverse electric impedance of the squid giant axon. *J. Gen. Physiol.* 21:757–65

10. Fricke, H. 1923. The electric capacity of cell suspensions. *Phys. Rev.* 21:708–9

11. Hodgkin, A. L. 1976. Chance and design in electrophysiology: an informal account of certain experiments on nerve carried out between 1934 and 1952. *J. Physiol. London* 263:1–21

12. Hodgkin, A. L., Huxley, A. F. 1952. Movements of sodium and potassium ions during nervous activity. *Cold Spring Harbor Symp. Quant. Biol.* 17:43–52

13. Hodgkin, A. L., Huxley, A. F. 1952. A quantitative description of membrane current and its application to conduction and excitation in nerve. *J. Physiol. London* 117:500–44

14. Hodgkin, A. L., Huxley, A. F., Katz, B. 1949. Ionic currents underlying activity in the giant axon of the squid. *Arch. Sci. Physiol.* 3:129–50

Ann. Rev. Physiol. 1980. 42:1–16

LIFE WITH TRACERS

Hans H. Ussing

Institute of Biological Chemistry A, University of Copenhagen, Copenhagen, Denmark

OPENING

In New York in 1952 I participated in a symposium on kidney physiology sponsored by the Macy Foundation. As a part of the program we were interviewed about what had motivated us to become physiologists. The organizer of the meeting asked his first victim: "Why did you become a physiologist?" The answer came without hesitation: "I was without a job and Homer Smith offered me one." The organizer put the same question to the next participant. Answer: "I was without a job and Homer Smith offered me one." When it was my turn I answered: "I was without a job and August Krogh offered me one." I mention this incident not only to illustrate the element of sheer luck that determines one's destiny, but also to emphasize the role played by influential scientific personalities in directing the interest of young people toward particular fields. Becoming a physiologist, however, requires both motive and opportunity. I graduated from the University of Copenhagen in 1934 during the Great Depression, and opportunities for a University career were few, even when one was motivated. And motivated I was. As far back as I can remember, I had no doubt that I wanted to become a scientist. During high school my main interest oscillated between biology and chemistry. Finally biology got the upper hand.

MARINE BIOLOGIST

In 1933 I was offered the chance of participating as marine biologist and hydrographer in *The Lauge Koch's* 3 year expedition to East Greenland. My own collections from that year plus all samples of plankton collected by other zoologists of the expedition during a three year period were placed

1765

0066-4278/80/0315-0001$01.00

at my disposal and served as the basis for my dissertation on the biology of some important planktons in the fiords of East Greenland. My scientific future seemed to have been settled: I was to become a marine biologist.

But I was detoured. During my work with planktons I had been hindered by the fact that the larval stages of various species of copepods (a group of small crustaceans that, due to their number, play an enormous role in the biology of the sea) seemed virtually indistinguishable. I then got the idea that it might be possible to prepare antibodies against the various larvae and precipitate one species at a time.

ENCOUNTER WITH D₂O

Before starting on such an enterprise I discussed the idea with my professor of physiology, August Krogh. He found it attractive but advised me to contact the Serum Institute concerning the procedure. Then he switched to another subject, showing me a small glass vial that contained 1 ml of a clear fluid and asking me what I thought it was. I looked at it and shook it, judging its surface tension and viscosity. "It looks like water," I said. "It is water," he answered, "but not ordinary water. It is heavy water." Krogh anticipated that heavy water and heavy hydrogen compounds might become enormously important in the study of permeability problems, and H. C. Urey, the discoverer of deuterium, had given him this little sample. Krogh told me at some length about his plans for using heavy water in his research. Two days later he telephoned to tell me that I could start in his laboratory the following Monday to help him discover the uses of heavy water in biology. It never occurred to either of us that my answer could be anything but yes. For me, then, marine biology and the precipitation of plankton larvae were things of the past. Isotopes had entered my life to stay.

The use of isotopes in biology was not entirely new. G. Hevesy, who invented the use of isotopic tracers, had been living in Copenhagen for some time as a permanent guest in Niels Bohr's Institute for Theoretical Physics. Hevesy, a close friend of Krogh and a frequent guest at our afternoon coffee break, tried to stimulate the interest of biologists in the use of tracers. But before deuterium was discovered all available tracers had been nonbiological elements such as uranium, lead, bismuth, etc. A tracer for hydrogen was of infinitely greater biological importance.

LIFE IN THE ZOOPHYSIOLOGICAL LABORATORY

At the time I started my career as a physiologist, August Krogh's zoophysiological laboratory exhibited a whole spectrum of activities. Over the years Krogh had worked successfully on many aspects of physiology, each time attracting new students and collaborators. A sizable group worked on respiratory metabolism both in man and in marine animals. Capillary physiology was also represented. Landis had just left, but among our guests were Knisely and Beecher.

Krogh's new interest, osmotic regulation, was represented by A. Keys from the United States, R. Dean (who first proposed the term "sodium pump"), H. Koch from Louvain, and Wigglesworth from England, to mention just a few of the visitors. Among the Danish staff members I must mention Marie Krogh, August Krogh's wife; his first assistant and later successor, P. Brandt Rehberg (the inventor of the creatinine-clearance method); and the heavy labor physiologist, Hohwü Christensen.

It was a fascinating experience for me to land in that international milieu. The afternoon coffee break at 3 P.M. was a permanent institution. Invariably the gathering resulted in a lively scientific discussion in which everybody took part, though most of the time Krogh and Rehberg were the main actors. Problems of physiological interest were defined, and the possible lines of attack were mustered. In these discussions Krogh was usually the *advocatus celesti* and Rehberg the *advocatus diaboli.* Rehberg, with his enormous learnedness and keen critical intelligence, would prove that a certain experiment could not be done, Krogh would then counter by proposing an elegant approach that might make the experiment feasible after all. This benevolent intellectual fencing was an invaluable drill for us young scientists, more valuable than formal teaching. For good or bad, the atmosphere in the laboratory led us to believe that no problem was too difficult if approached with clear analysis and common sense.

The reader may wonder how the activities in the lab were financed at a time of world depression. Grants from the Carlsberg Foundation, the Rockefeller Foundation, and other granting agencies played a role, but an equally important part was played by the production and sale of equipment for use in hospitals and research institutions. Instruments like respirometers, bicycle ergometers, etc, were developed on the basis of Krogh's inventions, and the profits helped to finance the lab. This also meant that we had a first-class mechanical shop, invaluable for our research. Furthermore, inventing and producing equipment for our own research became second nature for all of us. Krogh was an expert glassblower and taught me the art.

My first task was to develop a method for estimating the toxicity of heavy water. In the meantime Hevesy, Hofer & Krogh (3) studied the permeability of frog skin to water, comparing osmotic permeability to diffusion permeability, as measured with heavy water. The data did not agree; osmotic permeability was several times larger than diffusion permeability when expressed in the same units (e.g. $cm^2 \ sec^{-1}$).

We shall return to this discrepancy later. At the time Krogh concluded that until the reason for the discrepancy was found it would not be safe to use heavy water for determining water permeabilities.

USE OF DEUTERIUM LABELLING
FOR ESTIMATING PROTEIN SYNTHESIS

My interests temporarily shifted to something entirely different: In connection with our early permeability studies on living animals we found that some deuterium disappeared from the water and had entered organic substances (mostly proteins) in the tissues. This problem fascinated me and I wanted to find out how deuterium could enter stable positions in the proteins.

Krogh felt that this study required more chemical experience than our laboratory could muster and arranged for me to do the protein work, including separation of the amino acids, at the Carlsberg Laboratory. That was a piece of good luck for me. The head of the Carlsberg Lab was S. P. L. Sørensen, a prominent protein chemist and the inventor of the term pH. He was friendly and benevolent towards me, but the problem of hydrogen-deuterium exchange did not interest him very much. However, his first assistant (and later his successor) K. Linderstrøm-Lang immediately realized the importance of using deuterium in protein chemistry and helped me in every respect to aquire the necessary knowledge of protein chemistry and other aspects of biochemistry. More than that, he became a friend to whom I could always turn when I wanted to discuss difficult physico-chemical problems.

The use of deuterium in the study of protein metabolism opened new horizons for me. In all directions lay unbroken soil. Independently of the Schoenheimer-Rittenberg group at Columbia University (and slightly earlier), I demonstrated (21) that deuterium-labelled amino acids were introduced into certain body proteins at a surprising speed so that, in mice and rats, the renewal time was only a few days. I worked out a method for quantitative determination of the renewal rate for proteins by maintaining a constant level of D_2O in the body water of the animals and following the kinetics of build-up of stably bound deuterium in various protein fractions (22). I saw clearly my future as a protein biochemist.

Then came the war, and Denmark was occupied. This hampered our research. When the Norwegian heavy water plant at Riukan was blown up by the Norwegian resistance movement our deuterium work was brought to a complete stop.

Three lucky things happened to me during the otherwise miserable years of occupation. In 1940 I married Annemarie Fuchs, in 1942 our daughter

Kirsten was born, and the same year I became lecturer in biochemistry for the science students while keeping my research position at the Zoophysiological Laboratory. The teaching job induced me to study not only biochemistry but also physical chemistry and organic chemistry. In particular, the contact with the brilliant but rather awe inspiring professor of physical chemistry, Brønsted, was valuable for my later work.

The years of occupation became increasingly grim. I lectured to my students in our home for fear that they should be rounded up by the Gestapo (some of my students were very active in the resistance movement). Krogh sought refuge in Sweden because he was known to be on the list of potential hostages. Rehberg was finally arrested by the Gestapo (and later had a very narrow escape from the burning Gestapo headquarters). Thus the laboratory limped along with a skeleton staff and no supplies.

TRACER STUDIES OF ACTIVE AND PASSIVE TRANSPORT

The war ended. Our son Niels was born the day Allied forces entered Copenhagen. I expected to continue isotope work with proteins and amino acid metabolism, but things turned out differently.

In 1945 Krogh retired as head of the Zoophysiological Laboratory and was succeeded by Poul Brandt Rehberg. Krogh moved his research to a private laboratory (in the basement of his house) where he wanted to devote his time to his new interest, the flight of insects. Before retirement he had planned to initiate and supervise a program at the Lab to study active and passive exchanges of inorganic ions through the surface of living cells, using radioactive isotopes. He developed the rationale behind the project in an important review, his Croonian Lecture of 1946 (11), and had already done an experimental paper on the giant cells of the brackish water characean, *Nitella* (5). The idea was that the constant exchange of inorganic ions between cells and their surroundings must mean an active transport of certain (or maybe all?) species in one direction and leakage in the opposite direction. If a certain ion species exhibits a sufficiently large concentration drop across the cell membrane, one must assume that the flow of ions is almost exclusively active in one direction and passive in the other. In such cases the ion exchange as measured with an isotope under steady-state conditions should simultaneously yield the rate of active transport and the leak permeability.

As we shall see in a moment this argument does not always hold, but conceptually it was a very important step forward. The work of Hevesy and other pioneers in the use of isotopes had revealed that many ionic species, thought to be nonpenetrating, exhibited a more or less lively "exchange" across cell membranes, but the mechanisms underlying the exchanges had not been rigorously analyzed. Such mechanisms were hardly discussed in the papers describing the exchanges. Now Krogh postulated that the process could be resolved into two well-defined processes: an active transport that consumed metabolic energy, and a passive leak that was a simple physical process.

Originally he planned to locate the project in the Zoophysiological Laboratory and to staff it with a few young physiologists working under his guidance. Because the start of the project was delayed until after his retirement, Krogh asked me to supervise it until it was well on its way. I had two good reasons for joining the new project. In the first place it was essential that so important a project not go astray in its initial phase. Second, as Krogh pointed out, the fact that our University had a cyclotron at the Bohr Institute and that Niels Bohr actively supported biological applications of isotopes gave us a unique chance to do pioneer work with respect to transport of inorganic ions. Carbon 14 had just become available for biochemical applications and everybody in the isotope field would soon rush to apply carbon-labelled substances in metabolic studies. In the meantime, we might have the inorganic ions for ourselves for a while. I therefore accepted the task and decided to join the project for a year or two before returning to my beloved amino acids and proteins; but I never returned. The supposedly temporary task has lasted until this day.

The team that started on the project consisted of Hilde Levi (one of Hevesy's former associates), C. Barker Jørgensen, a young physiologist who had done his thesis work at our lab, and myself. We decided to try the isotope experiments in two very different systems: live axolotls submerged in water, and isolated frog sartorius muscle.

ANTIDIURETIC HORMONE STIMULATES ACTIVE SODIUM TRANSPORT

Ten years earlier Krogh and his associates had demonstrated that frogs and many other freshwater animals take up sodium chloride from the surrounding water. This capacity is very well developed in axolotls. Krogh had assumed that the active salt uptake was regulated by a sort of mass action effect, so that low salt levels in the blood would induce an increased uptake. For the sake of argument I then proposed that the uptake might be regulated hormonally. In our storeroom I could find only one candidate (aldosterone had not yet been isolated): antidiuretic hormone. I had happily forgotten that, according to the kidney physiologists, ADH had no effect whatsoever on salt resorption. We injected ADH into the axolotls and, lo and behold, both sodium and chloride uptake were dramatically increased. This could be seen both from chemical analysis and from changes in the rate of uptake of ^{38}Cl and ^{24}Na (6). Later, as is well known, it turned out that ADH and related hormones stimulate sodium transport in many epithelia.

SODIUM EXCHANGE IN SARTORIUS MUSCLE

One thing worried me, though. Clearly sodium chloride was taken up, but the uptake was often different for the two ions. Thus they behaved like independent species. Was the uptake active for both species, or was only one transported actively, creating an electric potential difference that in turn brought about the uptake of the other? We turned to our second experimental object, muscle, where the sign and magnitude of the electric potential drop were known so that one could say with certainty which of the ions had to move against an electrochemical potential gradient.

It was well-established that muscle fibers were negative inside, that their cellular concentration of sodium was low, and that their potassium concentration was high. Now, we could easily show that ^{24}Na in the medium exchanged with all of the sodium in the muscle. By following the washout of radio-sodium from the muscle we resolved the process into two first-order processes: the wash-out of interspace sodium, and the exchange of sodium across the fiber membrane. Since sodium moved out across the fiber membrane against an electric potential gradient as well as against a concentration gradient, the process clearly involved active transport.

EXCHANGE DIFFUSION

I did a quick calculation to find out what fraction of the energy output of the muscle would be necessary to drive the "sodium pump." I found to my horror that the energy requirement surpassed the total available energy output as estimated from the oxygen consumption. This led me to ask whether any of our original assumptions could be false.

Finally it dawned on me that an exchange of an ion species across a membrane can proceed without consumption of free energy even if the ion is present at different electrochemical potentials in the two bathing solutions, but it is only possible under very particular conditions: The passage of an ion in one direction must be strictly coupled to the passage of a similar ion in the opposite direction, so that the free energy gained by the downhill transport is used for the uphill transport. On the molecular level this could be achieved if, for instance, there existed in the membrane phase a carrier molecule that could only pass from one boundary to the other in association with the ion species in question. This would inevitably lead to a one-to-one exchange of the ion, irrespective of the potential difference or concentration difference between the bathing solutions. Instead of by a diffusing carrier, the exchange might be brought about by a specific ion-binding molecule or site that could flip-flop between two positions, alternately exposing the binding site to the two bathing solutions (15, 23). Clearly, the phenomenon that I named "exchange diffusion" will also occur even for carriers that can pass the membrane in the free state if the ion in question is present in so high a concentration on both sides of the membrane that all carrier molecules are saturated at both boundaries (cf 26).

The phenomenon of exchange diffusion later turned out to be quite widespread in nature. As is well known, it is of great functional importance in physiology and biochemistry in cases where two or more molecular or ionic species share the same carrier system (counter-transport of Cl and bicarbonate in erythrocytes, antiport systems in mitochondria, etc). At the time, however, it meant that an abyss opened under the original plan of our experiments. Without additional information one obviously could not use steady-state fluxes of isotopes for measuring either active transport or leak permeability because part of the flux might be exchange diffusion. In our experiments with axolotls, active transport certainly occurred, because salt moved from a lower to a higher concentration; but unless one knew the direction and magnitude of the electric potential difference between blood and bath it was impossible to tell whether sodium, chloride, or possibly both, were actively transported. Thus we had to look for a system where all necessary parameters could be measured simultaneously.

THE FROG SKIN AS TEST OBJECT

I decided to begin with the isolated frog skin, mostly because I found it easier to skin a frog than an axolotl but also because Krogh had demonstrated that frogs can take up both sodium and chloride from very dilute solutions.

When mounted as a diaphragm separating suitable solutions (e.g. Ringer's), the isolated frog skin maintains high and rather stable potential differences, the inside bath being positive relative to that on the outside. One can then add radioisotopes of Na and Cl to the bathing solutions and measure the rates of appearance of tracers on each side. This approach soon gave a qualitatively clear answer: Sodium moved faster in the inward than in the outward direction, even when the movement took place against a concentration or a potential gradient. Hence, at least part of the sodium transport must be an active process. At least part of the chloride transfer might take place due to the electric field, but it was not possible at that time to tell which part of the chloride transport could be accounted for by "simple" electrodiffusion, i.e. the combined effect of concentration gradient and electric field.

THE FLUX RATIO EQUATION

Could an equation describing electrodiffusion of isotopic tracers through a biological structure be worked out? Having discussed this problem with Linderstrøm-Lang one Saturday afternoon, I worked hard on the problem and Monday morning called him with a solution. He, too, had worked out a solution, which was not identical to mine. In order to solve the appropriate differential equation he had used the constant concentration gradient assumption, whereas I had used the constant field assumption.

The worrisome thing was that although I now had two solutions to my problem I could trust neither. In a frog skin, or any epithelium for that matter, there must be several diffusion barriers in series: cell membranes, basement membrane, connective tissue, etc. To assume a constant field or a constant concentration gradient in such a system would be rash. However, while I was playing around with the two equations I made an odd observation: If I divided the expression for the forward flux by that for the backward flux both solutions then yielded the same equation. I suspected that the flux ratio for an ion that moves without interacting with other moving particles was a sort of state function, being independent of the properties of the barriers along its path (25).

By the time I worked out the proof for the flux ratio equation I had a temporary appointment as Rockefeller Fellow at the Donner Laboratory, University of California. It was customary for staff members at the Zoophysiological Laboratory in Copenhagen to spend one year abroad as Rockefeller Fellows after they had obtained tenure, and in 1948 it was my turn. Annemarie and I and our children Kirsten (5 year) and Niels (3 years) were installed in a one-room apartment in Oakland (later two rooms in Berkeley), and I started working in the Radiation Laboratory. Unfortunately, I had not forseen the need for a supply grant. In Copenhagen I obtained isotopes free from the Bohr Institute, but in Berkeley one had to pay. I found, however, that the clinicians at the Donner Lab used large amounts of isotopes, notably ^{24}Na and ^{131}I, for treating patients; by rinsing their empty bottles I recovered plenty of tracer material for my own work. Thus I was able to illustrate the usefulness of the flux ratio equation by showing that iodide behaved as a passive ion when passing the frog skin whereas sodium moved almost exclusively by active transport (16).

Although most of the staff at the Donner Lab used radioisotopes for one purpose or another, nobody worked on problems even remotely related to mine. Yet I was far from feeling lost. Among the senior staff Drs. Hardin Jones and Nello Pace were particularly helpful and I developed a lasting friendship with the scientists with whom I shared laboratory space. Furthermore, Berkeley was (I guess it still is?) a fascinating place. I remember that I was particularly impressed by the work of Calvin and Benzon on photosynthesis, which had just gotten underway.

THE SHORT-CIRCUITED FROG SKIN

During my stay in Berkeley I gave a seminar on sodium transport across the frog skin and one student expressed doubt that active sodium transport is the main source of the frog skin potential. He pointed out that his former teacher, Professor E. J. Lund, had shown the frog skin potential to be a redox potential. I found that assumption extremely unlikely since redox potentials cannot be picked up via a KCl bridge. Nevertheless I accepted the student's offer to lend me Lund's book, *Bioelectric Fields and Growth* (19). The book did not change my view, but I noticed something of great importance for my future work: Lund & Stapp (19) had attempted to draw electric current from frog skins via reversible lead-lead chloride electrodes. When I recalculated the currents drawn from frog skins in terms of sodium fluxes they turned out to be roughly the same order of magnitude as the isotope fluxes I had measured. Of course this similarity might have been fortuitous. Nevertheless, a plan took shape: If one could "short-circuit" the skin via suitable electrodes so that the potential drop across was reduced to zero and if the bathing solutions were identical then only actively transported ions could contribute to the current passing the skin; the flux ratio for passive ions would become one. By then, however, my stay at Berkeley was over and I had to return to Copenhagen. The idea of short-circuiting frog skins was put aside for a while.

After my return to Denmark we succeeded in showing that the fluxes of chloride through frog skin obeyed the flux ratio equation, indicating that virtually all of the chloride passed the skin passively, even when there was a net inward transport against a concentration gradient (9).

In the meantime the 18th International Physiology Congress to be held in Copenhagen in 1950 had been organized with August Krogh as President. (Krogh died before the congress began and Einar Lundsgaard had to take over the presidency; Poul Brandt Rehberg served as General Secretary.) Among the introductory talks, three were to be devoted to membrane transport problems. The speakers were to be Conway, Hodgkin, and myself. Of course I was pleased but also a bit worried. I planned to talk about the use of the flux ratio equation for distinguishing between active and passive transport, but did this justify that I, a relative newcomer in the ion transport field, was to give a main talk in front of oldtimers? I felt that I needed some striking finding that would justify my place in the program. I looked at my list of future projects and settled on the short-circuiting experiments. But time was running out. It was essential for me to find an efficient co-worker then and there. Dr. K. Zerahn, one of professor Hevesy's former associates, was working on phosphate metabolism of yeast; but when I told him about my plan and its significance, in less than an hour, he decided to join the frog skin team. He never returned to yeast.

Zerahn wired up the circuit; I did the glass blowing and, helped by our mechanic, designed the chamber. Within a week we had the first results: During short-circuiting, the entire current drawn from the skin is accounted for by the net sodium flux (influx minus outflux). We had shown that the active sodium transport alone is responsible for the electric asymmetry of the frog skin. Within two weeks we had enough data to confirm the first findings and to warrant publication of the result (31).

SOLVENT DRAG

We now had the tools for analyzing transport processes in epithelia: The flux ratio equation and the short-circuiting technique were both theoretically satisfactory as long as there was no osmotic flow through the system. I felt that as long as solutes and solvent follow separate pathways through the system the interaction could be neglected. But if there is bulk flow through narrow channels, ions and other solutes diffusing through the system will be speeded up in the direction of flow and slowed down in the opposite direction. Consequently the flux ratio, even for a passive species, would deviate from the predictions of the ideal flux ratio equation. Since such a "solvent drag" would also act on water it might explain the strange observation made by Hevesy, Hofer & Krogh (3) that the osmotic permeability of frog skin seemed larger than the diffusion permeability as measured with heavy water (see above).

A theory for the dependency of the flux ratio on solvent drag in composite systems was developed (7, 26). Predictions were verified in experiments on toad skin and frog skin: During osmotic flow, solvent drag effects do exist. These effects, however, were small compared to those bringing about active transport of sodium. Our equations for solvent drag are in complete agreement with those derived later from irreversible thermodynamics.

EXPLORING THE ROLE OF ACTIVE SODIUM TRANSPORT

In the fertile and stimulating years following the Copenhagen Physiological Congress a steady stream of visitors came to work with us. Most of the time the foreign and Danish visiting scientists outnumbered the permanent staff.

I must postpone telling the story of this period, except to mention a few developments that added to the store of concepts and tools having some general scientific applicability.

With the love for sweeping generalizations that is characteristic of mankind in general and of general physiologists in particular, I thought for a while that active sodium transport might be the charging device for all bioelectric potentials, at least in the animal kingdom. As early as 1951 we learned otherwise. Adrian Hogben (4), working in our lab, found that the potential developed across isolated toad gastric mucosa stems from active transport of chloride. In the following year we found that when stimulated by adrenaline (10) the skin glands of frog skin transported chloride in the outward direction. On the other hand we showed that active sodium transport was solely responsible for the electric asymmetry of isolated mucosa from toad large intestine and Guinea pig cecum. Of greater potential value was the isolated toad urinary bladder preparation developed by Alex Leaf (14) during his stay with us in 1954. This preparation later became the preferred test object in innumerable transport studies from many laboratories. Important developments were also triggered by Csaky & Thale's finding (1) that sodium is absolutely essential for the uptake of 3-methyl glucose through isolated toad small intestine. As is well known, the correct interpretation of this finding was later given by Crane in the form of the cotransport theory. One of my former associates, the late Peter Curran (visitor in our lab from 1959 to 1961) was to contribute greatly to our understanding of the role of cotransport for intestinal uptake of amino acids and other nutrients.

THE TWO-MEMBRANE THEORY FOR THE FROG SKIN POTENTIAL

For several years our main object was to sort out the transport processes due to active transport from those that were secondary consequences of active transport. For the exploratory stage of the studies the black-box treatment of the epithelium as a transporting entity was acceptable. After a while it became imperative to localize the mechanisms within the system. From the very beginning of my work with epithelia I had held the (rather obvious) opinion that the net transport of ions must depend on the inward and outward facing cell membranes' having different transport properties. The finding that in frog skin sodium transport was active while chloride transport was passive made it almost imperative that the sodium pump was located at the inward-facing boundary of the transporting cells (cf 24). Otherwise the cell interior would be overloaded with sodium. Another implication is that the outward-facing cell membrane must be tight to potassium. Otherwise the uptake of sodium in the cells would lead to an enormous loss of potassium.

Testing this hypothesis required a study of the ion selectivities of the outward- and inward-facing cell membranes of the epithelium. Our experimental approach was based upon the following consideration: If the skin potential results from the activity of the sodium pump, shunted by a chloride leak, it is clear that the potential measured must approach the electromotive force of the pump when the shunt conductance is reduced to zero. This situation can be approached, for instance, using the poorly permeating sulfate ion instead of chloride. Under these conditions when the compositions of the bathing solutions were varied the outward-facing side of the epithelium behaved like an almost ideal sodium electrode and the inward-facing membrane behaved like a potassium electrode.

The minimum requirements for a system that could take care of net salt transport and regulate the cell electrolyte composition at the same time would then be the following: The inward-facing membrane would have the same properties as any body cell. It would be permeable to potassium and chloride, be nearly tight to passively diffusing sodium ions, and would possess a sodium pump. We imagined that the sodium transport was performed by a sodium-potassium exchange pump (but any type of sodium pump would do). The outward-facing membrane (permeable to chloride and tight to potassium) was considered to be without active transport systems but to have a strictly selective sodium channel allowing passive sodium entry (8, 27).

This theory has had a peculiar destiny. At first it was generally accepted as a model for many epithelia. Then gradually each of its assumptions was contested. In recent years the pendulum has been swinging back: All of the original assumptions were essentially correct (28, 32). Larsen & Kristensen (12, 13) have worked out a computer program based on the assumptions of the two-membrane theory. This program predicts the steady-state electric properties of the frog skin quite satisfactorily under various conditions. Lew et al (17) have recently developed a similar program that predicts correctly even the potential and resistance transients when the system is suddenly modified. Very recently my associate Robert Nielsen has produced strong evidence (in preparation) that the ion pump is really a sodium-potassium exchange pump, with a coupling ratio of 3 Na for 2 K.

SHUNT PATHS

In our original version of the theory, based mainly on early microelectrode studies, we assumed the transporting cell layer to be the stratum germinativum in frog skin. Experiments with an improved technique (29, 30) have shown that the sodium-selective membrane must be the outward-facing surface of the outermost living cell layer, just underneath the cornified layer (which is mostly only one cell layer thick). We have modified the original model in two other respects: (*a*) Besides the transcellular chloride pathway we now assume a variable intercellular shunt pathway (via the "tight" seals) open to small ions, including sodium; (*b*) the different layers of epithelial cells are electrically coupled via a resistive cell-to-cell permeability to ions. Independently Farquhar & Palade (2) reached a similar conclusion, which also agreed with the concept of cell coupling advanced by Loewenstein (18). The idea of intercellular shunts was developed further by Windhager, Boulpaep & Giebisch (34) to describe the conductance of proximal kidney tubule. Later it was generally accepted as a property of "leaky epithelia."

COUPLING BETWEEN CELL LAYERS

The concept of coupling between different layers of epithelial cells has given rise to problems and apparent contradictions. Some studies have given clear evidence for the coupling in frog skins (20) while others have indicated very limited coupling (33). Current work by our group (R. Nielsen and H. H. Ussing, in preparation) indicates that both views may be correct, depending on the experimental circumstances. The coupling seems to vary with the degree of contact between the cells, which in turn depends on the intraepithelial pressure.

CODA

My academic position has undergone a few changes over the years. In 1951 I obtained a research professorship in zoophysiology but continued to teach biochemistry to science students. In 1960 I was asked to take over the chairmanship of the Institute of Biological Chemistry. With some reluctance I accepted the offer, because I assumed that biological transport phenomena were then ripe for study on the molecular level. As it turned out, epithelial transport studies took more time to pass from biophysics to biochemistry than anticipated. With retirement looming less than three years ahead I can no longer hope to take an active part in this development; but as long as I live I shall be an interested onlooker. In the meantime there are still plenty of tempting problems of a more biophysical nature to which I can devote my time.

Literature Cited

1. Csàky, T. Z., Thale, M. 1960. Effect of ionic environment on intestinal sugar transport. *J. Physiol. London* 151:59–65
2. Farquhar, M. G., Palade, G. E. 1964. Functional organization of amphibian skin. *Proc. Natl. Acad. Sci. USA* 51:569–77
3. Hevesy, G., Hofer, E., Krogh, A. 1935. The permeability of the skin of frogs to water as determined by D₂O and H₂O. *Skand. Arch. Physiol.* 72:199–214
4. Hogben, C. A. M. 1955. Active transport of chloride by isolated frog gastric epithelium. *Am. J. Physiol.* 180:641–49
5. Holm-Jensen, I., Krogh, A., Wartiovaara, V. 1944. Some experiments on the exchange of potassium and sodium between single cells of Characeae and the bathing fluid. *Acta Bot. Fenn.* 36:1–22
6. Jörgensen, C. B., Levi, H., Ussing, H. H. 1946. On the influence of the neurohypophyseal principles on the sodium metabolism in the axolotl (*Ambystoma mexicanum*). *Acta Physiol. Scand.* 12:350–71
7. Koefoed-Johnsen, V., Ussing, H. H. 1953. The contributions of diffusion and flow to the passage of D₂O through living membranes. *Acta Physiol. Scand.* 28:60–76
8. Koefoed-Johnsen, V., Ussing, H. H. 1958. The nature of the frog skin potential. *Acta Physiol. Scand.* 42:298–308
9. Koefoed-Johnsen, V., Levi, H., Ussing, H. H. 1952. The mode of passage of chloride ions through the isolated frog skin. *Acta Physiol. Scand.* 25:150–63
10. Koefoed-Johnsen, V., Ussing, H. H., Zerahn, K. 1952. The origin of the short-circuit current in the adrenaline stimulated frog skin. *Acta Physiol. Scand.* 27:38–48
11. Krogh, A. 1946. Croonian Lecture: The active and passive exchanges of inorganic ions through the surface of living cells and through living membranes generally. *Proc. R. Soc. Ser. B* 133:140–200
12. Larsen, E. H., Kristensen, P. 1978. Properties of a conductive cellular chloride pathway in the skin of the toad (*Bufo bufo*). *Acta Physiol. Scand.* 102:1–21
13. Larsen, E. H. 1978. Computed steady-state ion concentrations and volume of epithelial cells. Dependence on transcellular Na⁺ transport. *Alfred Benzon Symp. XI, Munksgaard*, pp. 438–56
14. Leaf, A. 1955. Ion transport by the isolated bladder of the toad. *Res. Comm.*

3ᵉ *Congr. Int. Biochim., Brussels*, p. 107.
15. Levi, H., Ussing, H. H. 1948. The exchange of sodium and chloride across the fibre membrane of the isolated frog sartorius. *Acta Physiol. Scand.* 16:232–49
16. Levi, H., Ussing, H. H. 1949. Resting potential and ion movements in the frog skin. *Nature* 164:928–30
17. Lew, V. L., Ferreira, H. G., Moura, T. 1979. The behaviour of transporting epithelial cells. I. Computer analysis of a basic model. *Proc. R. Soc. Ser. B.* In press
18. Loewenstein, W. R., Kanno, Y. 1964. Studies on an epithelial (gland) cell junction. I. Modifications of surface membrane permeability. *J. Cell Biol.* 22:565
19. Lund, E. J., Stapp, P. 1947. Biocoulometry 1. Use of iodine coulometer in the measurement of bioelectrical energy and the efficiency of the bioelectrical process. In *Bioelectric Fields and Growth*, pp. 235–80. Austin: Univ. Texas Press
20. Rick, R., Dörge, A., von Arnum, E., Thurau, K. 1978. Electron microprobe analysis of frog skin epithelium: Evidence for a syncytial sodium transport compartment. *J. Membr. Biol.* 39:313–31
21. Ussing, H. H. 1938. Use of amino acids containing deuterium to follow protein production in the organism. *Nature* 142:399
22. Ussing, H. H. 1941. The rate of protein renewal in mice and rats studied by means of heavy hydrogen. *Acta Physiol. Scand.* 2:209–21
23. Ussing, H. H. 1947. Interpretation of the exchange of radio-sodium in the isolated muscle. *Nature* 160:262
24. Ussing, H. H. 1948. The use of tracers in the study of active ion transport across animal membranes. *Cold Springs Harbor Symp. Quant. Biol.* 13:193–200
25. Ussing, H. H. 1949. The distinction by means of tracers between active transport and diffusion. *Acta Physiol. Scand.* 19:43–56
26. Ussing, H. H. 1952. Some aspects of the application of tracers in permeability studies. *Adv. Enzymol.* 13:21–65
27. Ussing, H. H., Koefoed-Johnsen, V. 1956. Nature of the frog skin potential. *Abstr. Commun. 20th Int. Physiol. Congr., Brussels.* 568 pp.
28. Ussing, H. H., Leaf, A. 1978. Transport across multimembrane systems. In

Membrane Transport in Biology, ed. G. Giebisch, D. C. Tosteson, H. H. Ussing, 3:1–26. Berlin/Heidelberg/New York: Springer. 459 pp.

29. Ussing, H. H., Windhager, E. E. 1964. Nature of shunt path and active sodium transport path through frog skin epithelium. *Acta Physiol. Scand.* 61:484–504

30. Ussing, H. H., Windhager, E. E. 1964. Active sodium transport at the cellular level. In *Water and Electrolyte Metabolism,* ed. J. de Graeff, B. Leijnse, 2:3–19. Amsterdam: Elsevier

31. Ussing, H. H., Zerahn, K. 1951. Active transport of sodium as the source of electric current in the short-circuited isolated frog skin. *Acta Physiol. Scand.* 23:110–27

32. Ussing, H. H., Erlij, D., Lassen, U. 1974. Transport pathways in biological membranes. *Ann. Rev. Physiol.* 36:17–49

33. Voûte, C. L., Ussing, H. H. 1968. Some morphological aspects of active sodium transport. The epithelium of the frog skin. *J. Cell Biol.* 36:625–38

34. Windhager, E. E., Boulpaep, E. L., Giebisch, G. 1967. Electrophysiological studies on single nephrons. *Proc. 3rd Int. Congr. Nephrol. Washington DC, 1966,* 1:35–47. Basel/New York: Karger

Alan Hodgkin

Ann. Rev. Physiol. 1983. 45:1-16

BEGINNING: SOME REMINISCENCES OF MY EARLY LIFE (1914–1947)

A. L. Hodgkin[1]

Physiological Laboratory, Cambridge, England

INTRODUCTION

Some years ago the Physiological Society published a lecture, called "Chance and Design," in which I described the experiments on nerve that my colleagues and I carried out between 1934 and 1952 (1). The present article covers roughly the same period but from a different point of view. Instead of describing experiments in detail I have tried to convey an impression of the life of a young physiologist nearly fifty years ago. I have also given a brief account of my scientific education, which in retrospect seems fairly odd. If there is a moral it is that people are more important than organizations and that universities should not streamline their administrations to the point at which it is no longer possible for the green shoots to push up through the crazy pavement.

GROWING UP 1914–1932

My father died in 1918 leaving a 26 year old widow with three boys aged 4, 2, and 0, of whom I was the eldest. This disaster might have made my mother unduly protective of her young family, but it seemed to have the opposite effect, either because she was buoyed up by some inner faith or because she made a deliberate effort to avoid mollycoddling. At all events we were allowed to wander about the pleasant country near Banbury, where I was born, or after we had moved to Oxford to spend the whole day walking, bicycling, or canoeing, often going miles from home. Later on,

[1]For a bibliography of the author's early scientific papers, see (1).

0066-4278/83/0315-0001$02.00

when we had learned to use map and compass we were even permitted to walk long distances in the snow-covered hills of the Lake District where we occasionally spent the Christmas holidays.

From an early age I was interested in natural history and was greatly encouraged by a talented but somewhat eccentric aunt with whom we used to stay in a primitive holiday cottage in Northumberland. She taught me to record my observations in a bird-diary, and although her approach to natural history was thoroughly scientific she managed to endow the subject with an exciting quality that had a special appeal for a small boy. When I was about 15 I was recruited by a professional ornithologist, Wilfred Alexander, to help him with the surveys of rookeries and heronries that he had helped to initiate. And at my second school, Gresham's, near the Norfolk coast, I overlapped with the ornithologist David Lack and spent many hours with him or another friend looking at rare birds on the salt marshes, or hunting for nightjar's nests on which Lack was making a behavioral study of some importance. All this got me interested in biology and helped to blur the distinction between learning and research.

Perhaps because there was no very suitable school at Banbury I went to a preparatory school (the Downs) near Malvern when I was nine years old. I think the parting must have been more painful for my mother than it was for me, but she consoled herself with the picture of the charming wife of the headmaster reading Treasure Island to a group of small boys in front of a glowing fire. There was nothing wrong with the picture, and bedtime reading was certainly a consolation; but small boys can behave devilishly to one another, and going away to school at nine is something I would wish to avoid if I were offered my life again. Still there was much that was enjoyable about school. We were allowed plenty of time to follow our own pursuits, which in my case involved bringing up a pet owl or hunting for birds and flowers on the beautiful hillsides that look out across the wooded Herefordshire plain to the distant Welsh Mountains. Another more sociable activity consisted in helping the headmaster build a model railway with an engine and truck large enough to carry 6 or 7 boys for a distance of several hundred yards. I do not think we can have learned a great deal because I found myself in the bottom form of my next school, Gresham's, where I went at the age of 13½.

By then I had decided that I wanted to go to Cambridge and I had to work hard in order to reach the scholarship class. There were only about two hundred boys in the senior school but it can claim some distinguished alumni, including William Rushton, Wystan Auden, and Benjamin Britten. As one might expect in so small a school, teaching was somewhat patchy. Of the subjects that later became important to me, mathematics was well taught, but I have never been able to do sums against the clock and more or less gave up the subject during the two years that I worked for a scholarship. In the lower forms physics was taught atrociously by the headmaster, with extreme rigidity, not only as to the wording but also as to the pronunciation of the laws and definitions he made us learn by heart. Thus you had to distinguish between "that" (to rhyme with hat) as an adjective and "th't" as a conjunction. I could not bear this and gave up the subject before I reached the sixth form, where I might have learned something from a different master.

For a long time I could not decide between history and biology, but in the end natural history won the day and I managed to get a scholarship at Trinity, Cambridge, in botany, zoology, and chemistry. In all these subjects I was well taught by youngish masters. We were encouraged to read widely and to work on our own, and this I think is the most important thing I learned at school. During one summer holiday I spent an enjoyable week investigating the distribution of specialized plants that grow on the sand-dunes and saltmarshes of Scolt Head Island off the Norfolk coast. This must have helped me in my scholarship examination as I was lucky enough to be set a question covering the sort of ecological work that I had done. It also may have brought home to me the powerful physiological effects of a hostile ionic environment in which only the most thoroughly adapted plants can survive.

Getting a scholarship emboldened me to visit my future Director of Studies in Trinity, Dr. Carl Pantin, an experimental zoologist of great charm and distinction. I asked him what I should do in the nine months before I came to Cambridge and he gave me some excellent advice, which I had the sense to follow. The advice was that in my last term at school I should do no more biology but should concentrate on mathematics, physics, and German. He also said "You must continue to learn mathematics," and this I have endeavored to do during the rest of my life, or at any rate until a year or two ago. For a long time my bibles were Mellor's *Higher Mathematics for Students of Chemistry and Physics* and Piaggio's *Differential Equations.* But I cannot claim to have done all the examples in Piaggio, as another admirer of that work, Freeman Dyson, (2) has evidently done.

I told Pantin that I was going to a German family in June and July but that I would like to spend May of that year at a Biological Research Station. I suggested the marine laboratory at Plymouth, where I had once been on a schoolboys' course; but Pantin thought a shy eighteen year old would be lost in a relatively large laboratory like the one at Plymouth. He suggested that I go to the Freshwater Biological Station, which had just been set up at Wray Castle on Lake Windermere and was directed by two young men, Philip Ullyott and R. S. A. Beauchamp. I jumped at the idea, not least because it would provide an opportunity (as it happened probably the only one in my life) of spending May in the Lake District.

This was my first experience of research and a fairly odd one at that. Wray Castle was a large, ivy-covered, Gothic-revival castle built in the 19th century on the western shore of Lake Windermere. It was large enough to house many scientists but was otherwise utterly unsuitable as a laboratory. There was also something strange about the work being done there. Beauchamp was carrying out fairly standard freshwater ecology, but Ullyott was trying to find out how a light-shunning planarian (*Dendrocoelum lacteum*) moved down a nondirectional light gradient (3). This sounds straightforward enough, but Ullyott had managed to build up a reputation for black magic in the quite rural community outside the castle. He worked in a cellar wearing a black cloak and mask in order to make certain that no stray light reached his apparatus, in which all rays were supposed to be normal to the plate on which the planarians crawled. At the time I admired these precautions, but I have wondered since whether they were not partly done for their dramatic effect. Later on when Ullyott moved to Cambridge after being elected to a Trinity Fellowship his colleagues in Zoology tended to laugh at his methods and had no particular respect for the problem he was trying to solve. This is sad because Ullyott's experiments were interesting and his theory of movement in a light gradient was very like that advanced some forty years later by Berg and others to explain bacterial chemotaxis (4).

Ullyott suggested that I should study the effect of temperatures on another planarian, *Polycelis nigra* and in particular should see if the animals congregated at the cold end of a temperature gradient. I spent some time building an incredible haywire apparatus but reached no definite conclusions except that the animals did congregate in the cold and that this was only partly explained by the fact that they moved faster in the warm. Six months later I tried to continue the work in the spare bathroom at home but nothing concrete came from this, apart from the disturbance to our guests.

Just before I left Windermere, Ullyott told me teasingly that I had rescued his reputation in the local community, first by playing village cricket and second by going to church. As I am almost as bad about going to church as I am at playing cricket I can only attribute these astonishing facts (which I have verified in an old letter) to the persuasive powers of the local vicar, whom I dimly remember as a most engaging old boy.

CAMBRIDGE 1932–1937

After leaving Windermere I managed to learn some German in Frankfurt but did no more science until I came to Cambridge in the autumn of 1932. I felt very young and nervous but soon started to enjoy life. I had not been seriously unhappy at school but I much preferred the holidays and it came as a pleasant surprise to find that I looked forward to the beginning of term instead of dreading it.

Carl Pantin advised me to take Physiology with Chemistry and Zoology for the first two years. He suggested that I give up Botany, which I was reluctant to do as it had been my best subject in the scholarship examination. However, after going to one or two lectures, I concluded that Pantin was right and gave up the subject quite happily. Physiology, then combined with Biochemistry, was new and exciting, particularly in the experimental classroom. The regular lectures, except for some of Barcroft's and Winton's, were not brilliant, but I remember superb lectures by Krebs on the ornithine cycle and by Adrian on referred pain. I enjoyed Zoology supervisions (tutorials) with Pantin in the evening in Trinity, and it was a great disappointment when a serious attack of tuberculosis took him away from Cambridge for nearly two years. In Physiology Roughton was less conscientious as a teacher, but if you stuck to carbonic anhydrase and hemoglobin you could learn a great deal.

The scholarship exam I had taken a year before was a pretty stiff hurdle and I soon realized that I had already done a great deal of the Part I syllabus in Zoology and Chemistry. I was too diffident of my own abilities actually to cut lectures but found that I had a reasonable amount of time for library reading, either in math and physics or in general physiology and cytology. Some of the books that influenced me particularly were those by A. V. Hill, E. D. Adrian, James Gray, and J. B. S. Haldane. I also went on a vacation course to Plymouth and learned more invertebrate zoology there than in a whole year at Cambridge. This was partly because I have always disliked seeing the delicate beauty of marine creatures transformed into the formalin-pickled relics of the museum or zoological dissecting room.

After six months at Cambridge I became a member of the Natural Science Club, a small elitist organization which was founded in 1872 and in 1981 celebrated its 2500th meeting. The active members of this society were all junior members of the university (either undergraduates or research students). The club took itself pretty seriously, having elaborate rules, minutes, and weekly meetings at which someone read a paper for an hour or so. Some of the active members in my day were Edward Bullard, John Pringle, Dick Synge, Maurice Wilkins, and Andrew Huxley. I am afraid that I have never been a particularly clubbable person but I thoroughly enjoyed this club and it did much to counter the narrowness that is necessarily associated with studying one branch of science in depth at a university. The printed record of the club's proceedings makes interesting reading. You find that the subjects chosen for student papers are often those that scientists take up many years later, probably without remembering their early interest. Here are some examples: Synge, protein structures, techniques in biochemistry, cystine, distillation; Wilkins, mirrors, watches and clocks, seeing structures; Huxley, the ear and its functions, the conduction of nervous impulses, the use of the microscope, experiments on single cells; Hodgkin, nature of cell surface, membrane theory of nervous action, the behavior of sense organs, light-reception, and nerve-muscle function.

When the Physiological Society met in Cambridge, anyone keen enough was allowed to sneak into the audience and I remember a splendid debate on humoral transmission with Henry Dale, G. L. Brown, and Feldberg on one side and Jack Eccles on the other. My scientific sympathies were wholly on the side of acetylcholine but I thought Eccles put up a good fight. In May 1934 I was lucky enough to be present on the famous occasion when Adrian and Matthews demonstrated the effect of opening the eyes or of mental arithmetic on the Berger rhythm, using Adrian as a subject. (His rhythm was unusually responsive.)

I had difficulty deciding whether to read Physiology or Zoology in Part II. I preferred the former but was advised that there was little prospect of my getting a job unless I was medically qualified. There was a good deal of force in this argument, for Cambridge, one of the few universities that accepted nonmedical physiologists, already had more than its share of these, as Rushton found when Barcroft offered him a lectureship on condition that he spend three or four years getting a medical degree. Nevertheless Roughton strongly advised me to take up physiology, and this is what I did in the end. But I am not sure that I would have had the courage to make this decision if I had not been left a legacy that brought in about £300 a year. This does not seem much now, but it was enough to live on in those days.

After finishing the Part II Physiology course in 1935, I received a research scholarship from Trinity and settled down to study the blocked impulse effect, which I had come across a year before. I have written a brief history of this work in "Chance and Design," so here I shall use the space to give a picture of life in the Physiological Laboratory as it was in the 1930s.

The building we occupied was put up in 1912 with £20,000 given by the Draper's Company. Though not a thing of beauty, the old building has proved immensely serviceable and, with additions in every direction, is still the home of Cambridge physiology. A lecture theater and a new wing were added in the middle 1930s, but I have no clear memory of timing or of the disturbance the addition must have caused.

The head of the laboratory was Joseph Barcroft, who had succeeded Langley as Professor in 1925. I learned later that he was nicknamed Soapy Joe by some and that there were those who regarded him as a tricky character. But I saw nothing of this and he treated me with great kindness and friendliness, right up to the day he died when running to catch a bus after working late in the laboratory at the age of 75. Barcroft's enthusiasm was infectious and I admired *The Respiratory Function of the Blood* (5) as well as his splendid review on "La fixité du milieu intérieur" (6) which provided me with the kind of philosophical background to physiology that I then felt I needed.

When I finished writing my first paper in 1937 I took the manuscript to Barcroft and asked if it needed his approval before I sent it to a journal. He was quite taken aback and explained first that we did not do that sort of thing in Cambridge and, second, that anything I wrote was entirely my own affair. The only time I remember his getting cross was when I rather tactlessly asked him if he wanted me to light the way to the door late one evening. He said, indignantly, that after 30 years he could find his way around the lab with his eyes shut. My concern was understandable, because the basement where we were was pitch-black and unless you were careful you were liable to fall into a large bath containing bloody saline and a dead sheep on which a caesarian operation had been performed earlier in the day. But they were Barcroft's sheep and I suppose he knew where they were. This was after the war, perhaps in 1946–47.

I do not think there were any laboratory secretaries or typists in prewar days, though I have a mental picture of the chief assistant operating with one finger on a very old-fashioned typewriter in the office next door to Barcroft's lab. His name was Secker and both his son and grandson, now head of our machineshop, have worked in the same laboratory. Old Secker was an ex-sergeant major and had a fine military moustache. He looked irascible but wasn't, except when struggling with the lab accounts. If you were sensible you did not ask him for anything at the end of the financial year when the auditors were around. ("They have me sweating," he used to say.)

There were no centralized laboratory stores and if you wanted a valve or an electrical component you went to a local wireless shop that provided components for several laboratories. You might be able to scrounge nuts and bolts or sheet metal from the workshop but it was generally better to go to one of the local ironmongers, as we call hardware stores in England. Mr. Hall, the machinist in the workshop, was often busy with class equipment; but if you could persuade him to make something for you it worked pretty well, and his ideas about design were generally better than yours. I think we were allowed to spend about £30 per annum, which was quite a lot in those days. If you wanted anything more, or needed your manuscript typed in the town, you paid for it yourself. It was some time before I realized that one might be able to get a grant from a research council or from the Royal Society.

I think that I first met Adrian in 1933 at Seatoller in the Lake District where a long-standing tradition takes a party of Trinity students and dons for a few days at the end of the summer term. I have a clear memory of him as a figure in a mackintosh cape running swiftly downhill, emerging briefly from the mist and then disappearing again as another cloud swirled up the Ennerdale valley. Three years later I had the alarming experience of being driven by him from Cambridge to the Lake District at what then seemed a terrifying speed. I am afraid that by modern standards Adrian would be considered a dangerous driver because he relied to a great extent on the quickness of his reflexes to avoid accidents. Indeed this did not always work, as I found on another occasion when he ran into a taxi in Trafalgar Square. As a young man driven by his Professor I kept my mouth shut but was surprised to find that Adrian's command of language was quite equal to that of the taxi driver.

Adrian regarded the Physiological Laboratory as a place to work and we rarely talked there for more than a few minutes. After I had given a lab seminar about my nerve-block experiments in 1936 I remember him saying "Why not work on crab nerve?" which was advice I took. On another occasion he did not actually say anything but managed to show his disapproval of the way I had soldered a joint by rushing out of the workshop with a muttered curse.

Although Adrian was not given to obvious enthusiasms you could see that he was really pleased if you told him about some technical advance such as isolating a single crab fiber or getting a microelectrode into a cell. This was very encouraging. In addition, of course, I owe Adrian a great debt for all he did to promote my scientific career. I am sure that he was largely responsible for my becoming a University Demonstrator and Teaching Fellow at Trinity in 1938, as well as for a great deal of help after the war. But it was not possible to thank him for such things and I know that if I had tried he would have choked me off at once.

As a research student I did not have a formal supervisor, but there were plenty of people one could talk to about technical matters, including Bryan Matthews and Grey Walter in Physiology and Rawdon-Smith in Psychology. On the theoretical side I learned a lot from William Rushton, with whom I did a joint piece of work in 1939. Victor Rothschild, who was then working in Zoology on fertilization, lent me apparatus from time to time and had a considerable influence on my life by softening the strongly puritanical streak I had acquired from my Quaker upbringing. The Rothschilds were infinitely hospitable and gave splendid parties, sometimes illuminated with fireworks, at their beautiful house on the Backs in Cambridge. This provided a welcome change from political discussion or gloomy contemplation of the international scene, these being two of the principal occupations of most of young Cambridge in the late 1930s.

Another person who helped me in many ways was A. V. Hill. Polly and

David, his two elder children, were close friends of mine, and I sometimes stayed with the Hills at their country cottage near Plymouth when working there in the summer. On the first of these occasions in 1935, Charles Fletcher and I camped in the garden but on the second (July 1939) I stayed in the house, as can be seen from the following extract from a letter to my mother, which I now find entertaining: "There was a rather holy atmosphere over the weekend as A. V. Hill came down bringing with him Sir William Bragg. A. V. is Secretary and Sir Wm. President of the Royal Society. They are both quite friendly and easy but all the same I felt that to have the President and Secretary of the Royal Society in one house was a bit too much of a good thing."

And while we are on the subject, perhaps I can quote from another letter from the same period in which the Royal Society gets a slightly better write-up.

Langmuir's lecture at the Royal Society was excellent. He is one of the few scientists who have been able to think out immediate practical applications of their discoveries. And his experiments are always so beautifully simple that you wonder why on earth no one ever thought of doing the same thing before. The Royal Society is always amusing. The rather pompous rooms, the royal mace and the charters going back to Charles II all give it a very dignified air. Langmuir is an American and his address was an amusing contrast to the R. S. politeness and Bragg's dignified and rather wordy presidential address. After 5 minutes of flowing sentences and sentiments about the traditions of English and American science it was nice to hear Langmuir begin "Ladies and Gentlemen, if you put a piece of camphor on water. . . ." (11 December 1938)

One of the advantages Plymouth shares with Woods Hole is that you meet people from other laboratories and disciplines. I got to know J. Z. Young at Plymouth and also Bernard Katz, with both of whom I have kept in close touch all my life. In prewar days scientific meetings meant little to me. I went to the Physiological Society from time to time but never attended a large international meeting until the Oxford congress in 1947. For this reason informal contacts such as those I made at Plymouth or later at Woods Hole were particularly important.

As Trinity College was responsible for founding the Cambridge school of physiology by bringing Michael Foster there in 1870, I may perhaps be forgiven for extending these reminiscences to include some anecdotes about the scientists who belonged there. When I came up in 1932 the Master was the legendary Sir Joseph Thomson (J. J.), who is best known for his measurements of the mass and charge of the electron. He was then aged 76 and full of life, though becoming increasingly eccentric. Mr. Prior, who until a year ago was Head Porter at Trinity, remembers that one of his first duties as a junior porter was to rescue the Master, who had been misled by the newfangled open display of goods in a well-known chainstore, from an accusation of shoplifting. The shop in question couldn't believe that anyone so shabbily dressed could be Master of Trinity, and Mr. Prior looked so young that he had difficulty convincing them of J. J.'s position. However, all ended well and the shop realized that J. J. thought 'Help yourself counter' meant what it said. Soon after I arrived at Trinity a group of new scholars were invited to dine at the Lodge. In my letter home I reported that this awe-inspiring occasion was relieved by a comic episode. After dinner the Master turned on the wireless to some "rather vulgar" dance music. Apparently he was very pleased with this and went around chuckling to himself, but Lady Thomson who was extremely conventional, was shocked. She transported as many scholars as possible to another part of the room.

The only other time that I can remember dining with J. J. was just after I had become a research fellow in 1936. One of the guests was Enoch Powell, also a research fellow but two years my senior. At that time he had no interest in politics but worked at classics with a single-minded ferocity that I have never seen equalled. One of his heroes was Bentley, the famous classical scholar and friend of Newton, who had managed to remain Master of Trinity from 1700 to 1742 in spite of the strenuous efforts of the Fellows to have him impeached and ejected. At dinner Powell insisted that Bentley was the best Master that Trinity had ever had. Lady Thomson did not like this at all, partly because she disapproved of Bentley's scandalous conduct and partly because she had no doubt that J. J. was better than any former Master. As we left she said severely to me, "Really, Mr. Powell, I do think Mr. Hodgkin's views on Bentley are too distressing for words." I was never asked to dine again, but one dinner as a research fellow was the ration so I must not complain.

I gradually got to know Trinity's great men, of whom Rutherford was the most dominating. He divided science into physics and stamp-collecting but made an honorary exception of physiology and was extremely nice to me. He had a tremendously loud voice, which could be heard all over the College dining hall, and hated anything that interfered with his weekend golf. I remember the indignation with which he spoke at a college meeting called for a Saturday afternoon to decide whether or not to give £1000 for a squash court at a women's college. "Of course we must give it," he said, "and not waste a beautiful summer afternoon debating this futile topic like a bunch of schoolgirls"—or something of the kind. This did not go down well with some of the elderly bachelor dons, but Rutherford won the day as he usually did.

When I got back to Cambridge in the autumn of 1938 after a year in America I found myself living in a beautiful set of rooms above F. W. Aston, the inventor of the mass-spectrograph and the discoverer of many isotopes. He was a complete contrast to Rutherford—a quiet bachelor with extremely regular habits but, like Rutherford, pleasant and friendly to his junior colleagues. He hated noise and, alarmed at the prospect of a young man living above him, went to the trouble of soundproofing his ceiling before my return. I thought this unnecessary but I fear he may have been justified. I remember him appearing in a dressing gown to complain gently about noise on the occasion when a friend and I entertained some of the Sadlers Wells Ballet at the end of their fortnight in Cambridge. There used to be many anecdotes about the Thomson-Rutherford-Aston trio. Someone should write them down before they are forgotten, but the *Annual Review of Physiology* is clearly not the place to do so.

NEW YORK AND WOODS HOLE 1937–1938

Fairly soon after being elected to a Trinity Research Fellowship, in the autumn of 1936, I received a letter from Herbert Gasser, then Director of the Rockefeller Institute in New York, inviting me to work in his laboratory during 1937–38. I also learned that I had been awarded a travelling fellowship by the Rockefeller Foundation. This enabled me to spend a most productive year in America, working mainly in Gasser's Laboratory, where I joined a small group consisting of Grundfest, Lorente de Nó, Toennies, and Hursch. I was able to travel and in the early summer worked with Cole and Curtis at Woods Hole, where they introduced me to the squid giant axon.

During my last few months in Cambridge I had found that it wasn't too difficult to isolate single nerve fibers from the shore crab *Carcinus maenas,* and had shown that there were transitional stages in the initiation of the nerve impulse: subthreshold responses with properties similar to those Katz and Rushton had deduced from their studies of excitation. For some time Gasser was skeptical about subthreshold activity, but he provided me with a room and equipment and didn't mind my continuing on my own. No one at the Rockefeller had heard of *Carcinus,* and I assumed the genus did not occur in America. I tried several kinds of edible crab, but in none were the fibers as robust or easy to dissect as in *Carcinus.* After some frustrating weeks I visited the Natural History Museum where I learned that *Carcinus maenas* was common on the Eastern Seaboard and that I could obtain a supply from Woods Hole. This failed in mid-winter, but by then I had arranged to have a consignment sent from Plymouth on the *Queen Mary.*

The Rockefeller Institute, which was full of distinguished people, was a somewhat formal place and I missed the free and easy life of the Cambridge laboratories. At lunchtime the great men led their teams to separate tables. You could see little processions led by Landsteiner, Carrel, Avery, or P. A. Levene. In addition to Gasser's own group the people who influenced me most were Osterhout (large plant cells), Michaelis (membranes), MacInnes and Shedlovsky (electrochemistry), as well as Peyton Rous and his family on the personal side. I met Cole and Curtis at Columbia University and they invited me to Woods Hole in the summer. In "Chance and Design" I have given an account of this visit, which had a strong influence on my scientific life. I shall not say more about it here, except to quote, again, from a letter to my mother:

> Woods Hole . . . consists of a very large laboratory and a small fishing village. It is a nice place. The village is on a little bay and you look across a smooth sea to islands, promontories and distant sand dunes. It reminds me vaguely of Blakeney or Scolt Head —mainly because of the continued screaming of terns in the harbour. I came because some scientists here (Dr. Cole and collaborators) have been getting most exciting results on the giant nerve fibres of the squid (7). As you know I spend my time working with single nerve fibres from crabs which are only 1/1000 of an inch thick. Well the squid has one fiber which is about 50 times larger than mine and Cole has been using this and getting results which made everyone else's look silly. Their results are almost too exciting because it is a little disturbing to see the answers to experiments that you have planned to do coming out so beautifully in someone else's hands. No, I dont really mind this at all [sic], what I do dislike is the fact that at present English laboratories can't catch squids so that I dont see any prospect of being able to do this myself.

In fact Young, Pumphrey & Schmitt (8) got the squid supply going at Plymouth that summer (1938), and Huxley and I worked on squid there the following year.

Someone, probably Herbert Gasser, had suggested that the Rockefeller Foundation might provide me with a grant to buy or build a modern set of electronic equipment. Toennies helped me to prepare a list of things I might need. Before I left New York I was electrified to find that I might receive an equipment grant of £300, a very large sum in those days.

CAMBRIDGE-PLYMOUTH 1938–1939

When I got back to Cambridge I joined forces with A. F. Rawdon-Smith, K. J. W. Craik, and R. S. Sturdy in Psychology. Between us we built three or four sets of equipment, some of which were still in use 25 years later. Rawdon-Smith designed the d.c. amplifier. I had help from Toennies on cathode followers, from Otto Schmitt on multivibration, and from Matthews on the camera and many other details. I also had a certain amount of teaching to do. At Trinity I gave tutorials in Physiology and had the good fortune to teach some brilliant people, including Andrew Huxley in his fourth year and Richard Keynes in his first. In the Physiological Laboratory I gave a course of lectures on the cell surface and helped with two practical classes, one on electrophysiology and one mammalian class for which I had to get up early to prepare six or seven decerebrate cats for the Undergraduates to use later. This wasn't something I enjoyed.

I got my equipment going by mid-January 1939 and started to measure the relative size of resting and action potentials using external electrodes. This led to the internal electrode experiments, carried out with Huxley at Plymouth, which showed that the action potential might exceed the resting potential by some 40 mV. There was obviously much to be done with the new technique, and it was a bitter disappointment when Hitler marched into Poland and war was declared on September 3, 1939. We left the equipment at Plymouth in the faint hope that the war might soon be over and that we could continue the experiments. In the event the war lasted six years and it was eight years before it was possible to return to Plymouth.

AVIATION MEDICINE, RADAR 1939–1945

Although I had been brought up as a Quaker, Hitler removed all my pacifist beliefs. I was anxious to do some kind of military service as soon as possible. I was pleased when Bryan Matthews offered me a temporary post at the Royal Aircraft Establishment Farnborough, where he had been working for several months on Aviation Medicine. There were two major problems to worry about. At that time aircraft were not pressurized and at high altitudes aircrew were kept going by breathing a mixture of air and oxygen. The oxygen was stored at about 100 atmospheres in small cylinders. Many cylinders had to be carried on a long flight, and their weight became a serious problem. What made matters worse was that an oxygen cylinder takes off like a rocket when hit by a bullet. After a few weeks we learned that the Germans had solved the problem by means of a very beautifully designed lung-controlled device, of which we obtained a sample early in the war. Matthews rightly felt that it would be extremely difficult to get such a device mass-produced quickly in Britain. We also thought that the RAF with its brave but individualistic traditions would not take kindly to the strict discipline required in wearing a tightly fitting mask. One of the difficulties at that time was to persuade aircrew to take oxygen at all. They felt that if mountaineers could get to 27,000 feet on Everest without oxygen, they should not have to bother with it at 20,000 feet. These considerations led us to design and build an oxygen economizer that blew oxygen into the pilot's mask when he inspired but not during the rest of the respiratory cycle. This got into service about a year later and was widely used by the RAF. A modified form was used by Hilary and Tensing in their first ascent of Everest in 1951.

The other question that concerned us was whether bubbles of nitrogen came out of the blood and produced bends at high altitudes. Matthews and I proved this the hard way by sitting in a decompression chamber evacuated to the equivalent of 40,000 feet (about one fifth of an atmosphere) breathing pure oxygen to keep us going and waiting till something happened. High altitude bends seem to come on more slowly than the decompression bends that divers used to experience, but you get them all right in the end—and very unpleasant they are, too.

I met Patrick Blackett at Farnborough. Partly through him and partly through A. V. Hill I eventually got a job with one of the teams developing centimeter radar for the RAF. This research, which occupied me for the next five years, was interesting and important; but we worked too hard, and there were too many accidents and tragedies for it to be enjoyable. The best thing that happened to me was that in 1944 I was sent for two months as a liaison officer to the MIT Radiation Laboratory. My work took me to New York, and there I married Peyton Rous's daughter Marion, whom I had first met in 1937. The authorities in Britain complained that I hadn't been sent to America to get married, but by then they'd been presented with a *fait accompli.*

GETTING BACK; CAMBRIDGE-PLYMOUTH 1945–1947

Towards the end of 1944 my work on radar slackened off and I started working on physiology again in the evenings and at weekends. I was released from military service soon after the end of the German war and my wife, small daughter, and I returned to Cambridge at the end of July 1945. Cambridge looked as beautiful as ever, but although I was mad keen to start research it was as difficult to get going in the Laboratory as it was to set up house. In six months' time universities were to be flooded with a great mass of war-surplus equipment, but to begin with there was nothing in the Laboratory and very little in the shops. I remember hunting for a piece of insulated sleeving. Eventually I found what I thought was a suitable piece in a drawer of miscellaneous electrical components. I was surprised to find that the object I held in my hand was tapered and seemed stiffer than I expected. I looked at it with a binocular microscope and was amazed at the intricate design I saw. Why on earth, I thought, should anyone go to so much trouble to make a flexible tube out of those beautifully articulated joints, and how difficult it must be to mass-produce such an object. Then I realized that what I was looking at was a lobster feeler that had fallen into the drawer and lain there since the days when I'd worked on lobsters before the war.

Some of the equipment that I had left at Plymouth was damaged in an air raid, but I managed to salvage a good deal. Fortunately I had lent the racks of electronic equipment to Rawdon-Smith and R. S. Sturdy, and they had removed them before the bad air raids began. Somehow or other I managed to collect everything and get the equipment going well enough to start experiments on *Carcinus* axons.

E. D. Adrian had obtained my release from military service on the grounds that he needed help with teaching. This was true, for the laboratory still had its full quota of medical students and many members of staff were away. One of my first jobs was to lecture on Human Physiology to student nurses. This was good practice, but not enjoyable. The nurses were in the charge of a fearsome-looking matron, and I couldn't get a flicker out of them. I felt better when Adrian, who had been giving the lectures before, told me he'd had the same experience.

Adrian let me off with a light teaching load, but I found it much harder to give tutorials than before the war. This was partly because I had forgotten a good deal and partly because I had ceased to believe in many of the principles that had once seemed to hold physiology together. The constancy of the internal environment remained as important as it had ever been, but the ways in which constancy was achieved had become much more complicated. It was also clear that much that I had read and taught before the war had been wildly oversimplified, if not downright wrong. An example is the hierarchical arrangement of respiratory centers postulated by Lumsden in the 1920s. I suppose that after five years working as a physicist I had little use for biological generalizations and always wanted to concentrate on the physicochemical approach to physiology. This doesn't go down well with most medical students.

After a rocky start my experiments on *Carcinus* began to go well. They went even better after Andrew Huxley returned to Cambridge from the Admiralty in 1945. Professor Adrian, who was head of the Department, obtained a grant of £3,000 per annum from the Rockefeller Foundation which helped to support a group working on the biophysics of nerve and muscle. The original members of the "unit" were D. K. Hill, A. F. Huxley, and myself. We were soon joined by distinguished visitors from overseas, of whom R. Stampfli and S. Weidmann were among the earliest.

I returned to Plymouth in the summer of 1947, soon after the Laboratory acquired a trawler to replace the one taken away at the beginning of the war. To begin with I was on my own as Andrew Huxley was getting married and Bernard Katz wasn't free till September. Life was not easy. Much of the Laboratory had been destroyed in the great air raids of 1941 and was being rebuilt, squid were in short supply, my hotel was squalid, and it was difficult to get enough to eat. But I was terribly pleased to be back where we had left off in 1939 and very excited when the sodium experiments started to come out so well. I don't think one ever stops trying to learn new things in science, and embarking on a new piece of work feels like starting research all over again. But in a general way Plymouth 1947 marks the end of the beginning and this was the time when I forgave fate for robbing me of what I once thought might have been the best years of my scientific life.

Literature Cited

1. Hodgkin, A. L. 1976. Chance and design in electrophysiology, an informal account of experiments carried out between 1934 and 1952. *J. Physiol. London* 263:1–21
2. Dyson, F. J. 1980. *Disturbing the Universe.* NY: Harper & Row
3. Ullyott, P. 1936. The behaviour of *Dendrocoelum lacteum.* II. Responses in non-directional gradients. *J. Exp. Biol.* 13:265–78
4. Berg, H. C. 1975. Chemotaxis in bacteria. *Ann. Rev. Biophys. Bioeng.* 4:119–36

5. Barcroft, J. 1914. *The Respiratory Function of the Blood.* London: Cambridge Univ. Press
6. Barcroft, J. 1932. 'La fixité du milieu intérieur est la condition de la vie libre' (Claude Bernard). *Biol. Rev.* 7:24–87
7. Cole, K. S., Curtis H. J. 1939. Electric impedance of the squid giant axon during activity. *J. Gen. Physiol.* 22:649–70
8. Pumphrey, R. J., Schmitt, O. H., Young, J. Z. 1940. Correlation of local excitability with local physiological response in the giant axon of the squid (*Loligo*). *J. Physiol. London* 98:47–72

Ann. Rev. Physiol. 1984. 46:1–13

A LIFE WITH
SEVERAL FACETS

Alexander von Muralt

Physiological Institute, Hallerianum, University of Bern, Switzerland

GREAT LEARNING EXPERIENCES

I had the great fortune to be in the right place, when original steps in scientific research were made, three times in my early life. My good luck enabled me to share, on a modest level, the excitement and intellectual stimulation emanating from these unforgettable experiences.

The first experience was in 1926 at the University of Zurich, where I studied physics and mathematics. Erwin Schrödinger (1877–1961), our professor of theoretical physics, gave unforgettable lectures. If the weather was cold, he delivered them in the lecture theater of the Institute of Physics; when it was warm enough, we all went to a beach on Lake Zurich. This lean man with his highly intelligent face would stand in front of us in bathing trunks, demonstrating on an improvised blackboard. With his exceptional brain he would describe, in mathematical terms, events happening at an atomic and even subatomic level, way beyond the thresholds of our sensorial perception.

Schrödinger's theoretical research closed the gap that then existed between quantum physics (introduced by Max Planck and Albert Einstein) and the classical concept of radiant energy, propagating as a continuum of electromagnetic waves (J. Cl. Maxwell). With a shy smile, he sometimes told us how his own work was progressing: a genial new theory that became known as "wave mechanics" and gained him the Nobel Prize in 1933! My second great experience came when I was fortunate enough to stay at the Harvard Medical School from 1928 to 1930. The first year I was a fellow of the Rockefeller Foundation; the second year I was advanced to a research fellow, paid by Harvard University. My luck was to find a rare combination of outstanding scientists active at the Medical School: Walter Cannon, eminent

1801

0066-4278/84/0315-0001$02.00

professor of physiology; Edwin Cohn, the great expert on proteins; Alexander Forbes and Hallowel Davis, two distinguished neurophysiologists; Lawrence Henderson, philosopher and expert in physical chemistry of the blood; and Alfred Redfield and Cecil Drinker, in comparative and applied physiology. Lawrence Henderson, whom I visited first, suggested that I start in Edwin Cohn's department, where outstanding work in the field of proteins was going on. This excellent advice was soon a major factor in my own scientific development.

Edwin Cohn asked me to work with one of his collaborators, John T. Edsall, who was busy preparing pure solutions of muscle globulin. John obtained the muscles at a Kosher slaughterhouse: because the animals had been bled at killing, their muscles were practically free of blood. This was a great help in the preparation of pure muscle globulin, because one did not have to get rid of the hemoglobin bulk in the process of muscle-globulin purification. Muscle globulin prepared by John Edsall's technique is a viscous liquid that is golden-yellow in transmitted light but bluish in scattered daylight. If a single, isolated muscle fiber is placed at a 45° angle under a polarizing microscope, the anisotropic bands appear brilliant with light and the isotropic bands appear black. The faint chance that the dissolved protein units in John Edsall's muscle-globulin solutions might have interesting optical properties encouraged us. We mounted a polarizing microscope with its optical axis in a horizontal position and fixed a vertical glass tube in the field of vision so we could let John's muscle globulin flow through it. With the solution at rest, the field of vision was pitch dark: the nicol prisms were crossed. When we let the globulin flow through the tube, brilliant light appeared—only to disappear at once when we stopped the flow!

The particles in John's solution were birefringent, an optical property only apparent when the particles were oriented by the flow in the tube, just like logs in a flowing river. In the solution in front of our noses we had the basic birefringent elements of living muscle. To summarize from our final publication in 1930: "The evidence appears to us to favor the view that the anisotropic particles are of uniform size and shape. Stuebel's work points to the existence of oriented rod-shaped particles, small compared to the wavelength of light, which are responsible for the double refraction (birefringence) in the anisotropic discs of muscle fibre. The properties of myosin solutions suggest that they may contain these rod-shaped particles." Albert Szent-Györgyi and his research group discovered in 1944 that the two principal proteins that function as the essential building stones of the muscle machinery are myosin and F-actin, which in solution aggregate easily into actomyosin. John and I in 1930 were probably dealing with actomyosin. Joseph Needham and his group repeated our experiments on birefringence of flow and showed that a small amount of ATP, added to an actomyosin solution, produced a drastic reduction of the flow-birefringence, confirming the disaggregation into myosin and actin. John Edsall and I had contributed a small building stone in the development of knowledge of the "muscle machine." I am happy to add that a lifelong friendship was the most rewarding by-product of our work together!

The research with John Edsall was only one part of my stimulating experience at the Harvard Medical School. The other benefits were the personal contacts with outstanding men, especially with Edwin Cohn. The way he directed his remarkable research group had a strong and lasting influence on me.

Towards the end of my Harvard stay, the University offered me the post of Professor of Biophysics. This generous offer forced me to decide whether or not to return to Europe to finish my studies in clinical medicine. I would have only a vague hope of obtaining an appropriate position at a Swiss university. I am now amused that I was faced with a decision between being a professor at Harvard or student in Europe. But the question, "Why did you want to finish your medical studies?" was answered by the regulations of Swiss universities. There the chair of physiology is incorporated in the faculty of medicine, and in those days the rule was adamant that a professor of physiology had to be a medical doctor! So I went back to Europe to finish the clinical part of my studies.

My third great experience brought me into an exciting phase in the development of our knowledge about the chemical background of the muscle machinery. In 1930 the research work of two great physiologists stood out in the field of muscle physiology: A. V. Hill's careful measurements of the phases of heat production; and Otto Meyerhof's studies on the chemical sources providing the necessary energy for every muscular contraction. This prompted me to write Meyerhof to ask if he would accept me as a scientific collaborator at the brand new Kaiser Wilhelm Institut in Heidelberg, with the understanding that I would like to spend some time at hospitals to finish my clinical training for an M.D. He answered positively, offering me a position at his institute and meeting my condition. Moreover, he mentioned his great interest in the work on birefringence and suggested that I continue with similar optical methods, using whole muscles.

A "Revolution in Muscle Physiology," as A. V. Hill titled a paper, happened at just about that time (early 1930). A Danish physiologist, Einar Lundsgaard, published a sensational scientific paper: "Studies on muscle contraction without formation of lactic acid." One has to realize today, what that meant at the time! The lactic acid cycle had been considered the principal energy source for all working muscles, mainly based on Meyerhof's research work. Suddenly this whole scientific edifice seemed to break down with Lundsgaard's discovery that although iodo-acetic acid (IAA) in very small amounts blocks glycolysis completely, the poisoned muscles are able to perform a considerable amount of work before ending up in an alactacid rigor.

All the previous work on the central role of glycolysis in working muscles seemed invalidated. This was the exciting scene when I entered Meyerhof's laboratory in Heidelberg in the fall of 1930. Karl Lohmann, Meyerhof's right hand man, had discovered in 1928—simultaneously with Fiske and Subarow in the United States—first the splitting of creatine-phosphate (CrP) in active muscles and one year later the presence of adenosine-triphosphate (ATP). Both compounds were considered additional sources of energy for muscular contraction, but no one had any indication of the "timetable" by which they deliver their energy. Generously, Lundsgaard came to Heidelberg to help Meyerhof disentangle the dilemma created by his discovery. In a classic example of scientific investigation, he found that the levels of CrP in poisoned, working muscles dropped exactly in proportion to the work performed, reaching zero when the poisoned muscle went into a final rigor. This showed unmistakably that CrP is one of the primary sources of energy, preceding the glycolytic sequences. But what about ATP? All attempts at that time to demonstrate a breakdown of ATP in normal or poisoned working muscles failed. Two aspects were discussed: (a) ATP is so rapidly resynthesized after breakdown that its level seems to remain constant; and (b) ATP is the primary source of energy in muscular contraction, but new methods of research must be found to prove this point.

It was again Lundsgaard who showed, somewhat later, that by very rapid cooling of IAA-poisoned muscles in liquid air one can produce a contraction, followed by a rigor without a trace of lactic acid. He also showed that the energy for this mechanical work is furnished by the splitting of ATP into ADP, which remains "frozen." All this new knowledge was very exciting.

Contracting muscles show not only the development of tension but also a transient decrease of their birefringence. I suggested to Meyerhof that I record optically this "negative wave of birefringence" in contracting muscles, an optical change that had been described by microscopists in the last century merely by visual observation. Meyerhof found the idea intriguing because he always tried to combine his chemical work with physical measurements on live muscles. I devised a spectral-optical method, independent of simultaneous changes of light scattering and possible movements of the whole muscle during isometric contractions, and yielding good optical records of the "negative wave." At the same time I saw that there were also increasing changes in light scattering, which correlated with the increase of anaerobic work performed. I measured these changes with another optical method, independent of the changes in birefringence.

A TIME OF TRANSITIONS

I stayed for five rewarding years in Meyerhof's laboratory, during which time I completed my clinical studies with the degree of M.D. One year later I became an assistant professor at the University of Heidelberg. But I remained in Meyerhof's institute, where I had the privilege of continuing my research and meeting such interesting guests and new collaborators as Hermann Blaschko, André Lwoff, George Wald, Rodolfo Margaria, and many others. But as time went by, the disintegration of the political scene dominated by the rise of Adolf Hitler, 51 years ago, became increasingly alarming. Meyerhof and his research group were left in peace until 1936, but it was sad to watch the increasing apathy and even anxiety of the academic community in Heidelberg. Fortunately, I was elected professor of physiology at the University of Bern in 1935, so I could return to my native Switzerland.

My new duties in Bern were to start in spring of the following year, so I was free to do something else for the winter and I decided to spend a few months in England to learn more about English physiology, research, and teaching. The officers of the Rockefeller Foundation heard about my plan and came to Heidelberg to ask me whether I would do a certain job for them. They wanted to know more about promising junior members in the various physiology departments in England; Would I send them a strictly confidential report? This was just the kind of task that was dear to my heart.

In London, A. V. Hill and Lovatt Evans received me most cordially. In Cambridge, Adrian, Barcroft, and Matthews spent much of their time showing me everything, including their research and teaching methods. Driving north as far as Edinburgh, in various physiology departments I was kindly received— and I learned a lot of physiology. At the end of my journey I sent the requested report to the Rockefeller Foundation: I am happy to say that four of the junior physiologists who were on my list later became winners of the Nobel Prize!

On April 1st, 1936, I started as the new professor of physiology at the University of Bern in the Hallerianum (Physiological Institute). With Walter Wilbrandt, my new assistant who became a very close friend, we reorganized the rather obsolete equipment for teaching and research. The Rockefeller Foundation came to our aid generously, enabling us to build up a modern course in physiology. At nearly every lecture we gave a demonstration. Five hours a week were reserved for practical work in human physiology, with students required to perform experiments on themselves. To aid them I wrote a book on practical physiology, which was successful in Switzerland and Germany and was translated into Spanish and Russian. Teaching physiology gave me a great deal of satisfaction. I was invited to give guest lectures in Austria, Belgium, Denmark, France, Germany, Italy, and Sweden. The outbreak of World War II put an end to this activity; I had to meet new reponsibilities.

SWITZERLAND IN WORLD WAR II

I have often been asked, in jest, "Were you an admiral in the Swiss Navy?" It shows, sadly, how little most people know about the great military efforts of my small country, especially during World War II.

At the outbreak of the war (September 1939), 400,000 Swiss soldiers were immediately mobilized. As the situation in Europe became more menacing, their number was doubled by auxiliary services—a large figure considered in relation to a total population then of 4.5 million. My personal record of military service spans five full years. I do not regret one bit of it; on the contrary, I am proud to have rendered this service to my country. I started as commander of a field artillery battery, became commander of an artillery group with the rank of major, and when the war ended I was promoted to colonel. The central figure in our army was General Henri Guisan, who was determined to defend our small country against any foreign military invasion. He was a fine leader, in whom the entire Swiss army and the majority of the civilian population had full confidence. By hard training he created a small but powerful army, ready to fight against any invader to the last cartridge in those strong defensive positions in the Swiss mountains that had been built and perfected in the course of the war years.

It is not exactly within the scope of the *Annual Review of Physiology* to cover the last war, but it might interest readers to learn how Switzerland escaped invasion, while similarly neutral Belgium, Holland, Denmark, and Norway could not. When France collapsed under the strong German attack in 1940, our small country was suddenly surrounded by the armies of the Axis powers: Austria, Germany, and Italy. They were tempted to "liquidate" and incorporate Switzerland; in order to demoralize us, maps were smuggled into our country showing Switzerland as part of the German Reich! A new and menacing situation arose when the Italian army had to retreat towards northern Italy, under the pressure of the attacking American armies advancing from the south. It was obvious that Germany had to come to the aid of the Italians. The shortest and most efficient way to bring troops and war material to northern Italy was through Switzerland, using the railroad tunnels through the Alps. But these tunnels were guarded by our army, which was prepared for and committed to destruction of the tunnels at the first sign of an invasion. The German high command knew that. They discarded their existing plans for a military invasion of Switzerland, and we remained independent and neutral.

In 1960 General Guisan died. I received the honorable order to command the military funeral procession, which went from his home on Lake Geneva all the way up through Lausanne to the cathedral. All traffic had stopped in the town. In the strange silence, one heard only the sounds of the military music and the rhythm of the marching steps of the military escort. Silent throngs of mourning people, who had come to Lausanne from all over Switzerland, lined both sides of the streets to pay their last respects to the gallant leader who had kept our country out of war by foresight and strong will.

What happened to the Physiological Institute in Bern during my long absence for military service? Fortunately, I had three able and devoted collaborators who were not obliged to serve in the army for reasons of age or nationality: Dr. I. Abelin, Dr. N. Scheinfinkel and Dr. W. Wilbrandt. They maintained admirable standards of teaching and research during my absence. The majority of male students served in the army, but during less dangerous periods they obtained sufficient leaves of absence to study and prepare for the exams. Most of them passed at the first trial! We were kept informed about important new advances in research abroad by batches of American and British scientific journals that mysteriously arrived.

During the long years of war, the European continent had suffered badly, but after the war communications were reestablished. Substantial American aid was a great help, and is not forgotten! For a program to reestablish intellectual and teaching contacts, I was asked to give physiology lectures at various universities in Holland. All the lecture students had grown up during the dark years of war; they were now eager to learn and to freely discuss the lectures. It was a great experience for me. When later I was asked to give lectures at the University of Cologne, I lectured with an improvised blackboard among the ruins under the open sky (Cologne was practically destroyed). When rains came, we all took shelter in a dugout. It was a fine occasion to talk with the young students, who were equally eager to forget the war. Since those days Holland and Germany have recovered completely; one hardly remembers the enormous civilian suffering of those years.

SWISS SUPPORT FOR SCIENCE

In Switzerland the administrative structures of education and research on the university level are entrusted to seven cantonal universities, two technical high schools financed by the federal government, and a business high school in St. Gallen. This is a great number for a small country and is a heavy load on the finances of the smaller cantons. But each school has maintained a fine level of education and at the same time kept its own, unmistakable character.

The isolation of Switzerland during World War II, coupled with the rapid development of research in the United States and Great Britain, presented serious problems for Switzerland during and after the war. It became very important for our young scientists to spend one or two years abroad, to learn about all the new advances in their specific line of research, with the hope that most of them would come back to Switzerland and find suitable positions and funds. Then they could not only continue, but also teach new research methods to the next generation. The most compelling solution to this problem was the creation of travelling fellowships for young scientists. I tried to help by two separate approaches, aided by a great number of devoted friends.

All our great industries have efficient research departments which they wish to staff with creative young scientists. The same desire exists in university departments of science and medicine because a good scientist is also, generally speaking, a good teacher and a stimulating example. To produce strong young staffs, I prepared two projects: a foundation for fellowships in biology and medicine; and, even more important, a Swiss National Foundation for the Advancement of Scientific Research. They were accepted and realized in consecutive steps.

Even in the midst of World War II, a neutral and independent foundation for fellowships in biology and medicine was created with the generous aid of leaders of our chemical and nutritional industries. This foundation, still active today, became very important when the war came to an end: our young Swiss scientists could travel abroad with fellowships for one or two years. But this raised a problem. Our universities found it difficult to meet the growing research equipment requirements of the returning young scientists.

This led to the creation of the National Fund for Scientific Research in 1952. Funded and officially recognized by the Swiss government, it is nonetheless an independent private foundation. It supports scientific research in all branches of science, and in law, literature, philosophy, and other humanities. A research council, composed of scientists from all branches of research, decides on most requests for support; very large requests are submitted to the Foundation Council. Since I helped create this fund and served as its president during the first ten years, I should not judge its success. But the financial support by our federal government has increased considerably since its origin: a reliable sign, I feel, that the National Fund not only is necessary for the advancement of scientific research in Switzerland, but also has been a success.

THE JUNGFRAUJOCH SCIENTIFIC STATION

Not far from Bern is the well-known resort town of Interlaken, situated between two lakes *(inter lacus)* in the midst of three high mountains. These beautiful snow-covered peaks are called Jungfrau, Mönch, and Eiger. Between Jungfrau and Mönch is a saddle, called Jungfraujoch, which can be reached by a cog-wheel railroad built at the beginning of this century. Rail cars take tourists up through a long tunnel, inside Eiger and Mönch, to an altitude of 3475 m (11, 389 ft). On arrival at Jungfraujoch one can enjoy, weather permitting, a magnificent alpine panorama of glaciers and mountains.

Fifty-four years ago physiologist Walter R. Hess created an international foundation to support scientific research in a research station to be built on

Jungfraujoch. He invited the leading scientific organizations in Europe to become members. Austria, Belgium, France, Germany, Great Britain, and Switzerland were represented first, followed by Holland and Italy.

The scientific station on Jungfraujoch, built in 1930, has one unique advantage for high-altitude research: the railroad. It not only delivers strong electric currents, but also can bring heavy research instruments to the scientific station. After the inauguration festivities in 1931, which I attended, Hess asked me if I would accept the directorship of the new scientific station. It was (and still is) a modern research installation with laboratories, a workshop, a library, comfortable sleeping facilities, a dining room, a kitchen and so on. This was a tempting offer, but I had to turn it down: my own research, in Meyerhof's laboratory in Heidelberg, was in an exciting phase. Five years later, when I was a professor in Bern, Hess asked me to be his successor as president of the International Foundation and at the same time direct the scientific station on Jungfraujoch. This post, second to my duties as professor, I accepted and held 37 years. Among other pleasant duties it was a welcome occasion to meet many foreign scientists, from biologists and physicists to astronomers and astrophysicists. Their principal subjects of research were cosmic radiation, UV spectroscopy of stars, the radiation of the sun, and high-altitude physiology.

High-altitude physiology is a field Joseph Barcroft, with his studies in the Andes of South America, greatly advanced. It had also its place in the Jungfraujoch station. Alfred Fleisch (Lausanne) and I developed a program of studies on hypoxia in healthy humans at Jungfraujoch, where the barometric pressure is 500 torr., two thirds of the sea-level pressure. We had enthusiastic groups of medical students, who volunteered as human "guinea pigs" during stays of 7–14 days on Jungfraujoch. They all showed upon arrival an increased ventilation rate, hypocapnia, and sometimes giddiness. All their sensorial thresholds were lowered but returned to normal by acclimatization after about five days on Jungfraujoch. In the field of cosmic radiation we had for long periods the group of P.M.S. Blackett. After the war we had a successful collaboration with a Belgian group. They installed a large sun-spectrometer, working mainly in the infrared region. Their results gave the basis for the publication of the most accurate atlas of the sun-spectrum. Among the many astonomers who came to work in the clear but very cold night air, Chalonge from the Observatoire de Paris is recognized worldwide for the precision and completeness of his work. I am happy to say that the friendships between the scientists who came to know each other on Jungfraujoch were not broken by World War II; on the contrary, many examples of mutual aid across hostile frontiers were the fruits of this international collaboration. My successor, professor Hermann Debrunner, enlarged the activities by installing two astronomical domes on Gornergrat, near Zermatt. The astronomical work and the work on cosmic radiation are now the main activities of the two Swiss high-altitude research stations.

AN EXAMPLE OF OUR RESEARCH

One research project in which we were engaged after World War II was the work with single, myelinated nerve fibers, mainly from frogs. The technique of their preparation was first introduced by G. Kato in Japan. It was considerably improved by I. Tasaki, who in 1939 developed a special "bridge-method" for recording the nodal action potentials. ("Nodal" refers to their origin in the nodes of Ranvier, which are periodical interruptions of the myelin sheath in myelinated nerve fibers.) At that time neurophysiologists were split into two groups: those who believed that the propagation of the nervous impulse in myelinated nerve fiber is "saltatory" from one node to the neighboring one; and those who refused to accept this new insight. It is surprising how many clever experiments were necessary to convince the skeptical ones.

One of the best experiments to prove saltatory conduction was by Andrew Huxley and Robert Stämpfli in Bern in 1948. Stämpfli was at that time my close collaborator, friend, and an expert in preparing long bits of single nerve fibers. The two of them developed an entirely new approach, by "threading" a single myelinated nerve fiber of the frog through a thin hole in an insulating partition. This hole was so small that the Ringer-solution on the surface of that part of the fiber that lay in the hole had a very high electrical resistance. The protruding ends of the nerve fiber were fastened onto holders on both sides of the partition in such a way that the fiber could be moved forward and backward through the hole. Each time another node had passed it, the registered action potential of the excited nerve fiber made a jump on the time-scale. This perfect and technically convincing experiment eliminated most doubts against the notion of saltatory conduction—except one. Saltatory propagation could be the result of the "isolation" of the nerve fibers, but the same fibers in situ might conduct the nervous impulses in a continuous way.

Yngve Zotterman, who came from Stockholm to work with us in Bern for a few months, planned—together with Robert Stämpfli—another fine experiment, which eliminated this objection. They used a thin, cutaneous, single nerve fiber in situ in the skin and showed with a very clever device that this single fiber conducts all nervous impulses in its natural environment with a saltatory propagation. This ended all doubts. To our great joy I. Tasaki, who had initiated electrophysiological work with single nerve fibers in Japan, came to Bern for a month to work with us. He and Robert Stämpfli improved the preparation and recording techniques of single nerve fibers to an extremely high level.

THE OPTICAL SPIKE

At 65 I retired as professor of physiology. Silvio Weidmann, who had done excellent research work in heart-action potentials, succeeded me. Suddenly I was free to escape into my laboratory to do some quiet research. I had always cherished the idea that the changes of membrane permeability for Na and K ions, which are the basis of the action potential in nerves and muscles, must be connected with a detectable optical signal or even several optical signals. One day in 1969 I saw, to my great surprise, an announcement that a scientific paper titled "Birefringence Changes During Nerve Activity" was to be presented at the imminent meeting of the Physiological Society in Hampstead. The authors were Richard Keynes and Larry Cohen. I immediately booked a flight to London, but the first announcement I heard when I reached the airport in Kloten was that the flight was delayed. Finally the plane reached London airport, but the time was short. I hired a taxi and asked the driver to take me to Hampstead as quickly as possible. Driving up Hampstead Heath, the clutch of the taxi started smoking. We both had to jump out of the car. Luckily, a kind gentleman picked me up and took me to the meeting.

When I entered the auditorium it was dark, except for the illuminated screen on which Richard Keynes was showing a slide. The transient change of birefringence in a nerve, synchronous with the action potential, was brightly revealed. I was not at all disappointed; I was fascinated that my wishful thinking had a real basis. At the end of the meeting I congratulated Richard for his great success. We agreed that I would visit Plymouth during the next "squid season" (The Atlantic squid appears in large numbers in the waters off Plymouth only from November until the new year). This was the start of a very happy six-year collaboration, during which I learned new electronic techniques and gained a close friend in Richard.

From talks I had with Herbert Gasser when I was a visiting professor at the Rockefeller Institute (University) I knew that the pike *(Esox)* has a unique olfactory nerve. It has four million unmyelinated nerve fibers on each side of the nose—more channels of communication than the entire Swiss telephone network! The isolation of this nerve is easy. Examination with a polarizing microscope shows its strong negative birefringence with regard to the fiber–axis. The average fiber diameter is two microns, with a very narrow range of diameters, so that conducted nervous impulses remain almost synchronized in all fibers over quite a distance. First experiments showed that the birefringence of the nerve changes during excitation! I built equipment that registered this optical change synchronous with the action potential at low temperatures close to 0° C. I invited Richard Keynes, Victor Howarth, and Murdoch Ritchie to come to Bern to exploit with me this new possibility to register a transient optical change, synchronous with the action potential. In the meantime Ewald Weibel, the anatomist in Bern, had made a careful analysis of the size distribution in these four million single fibers. He found that they are close to 2 μm, with only a small spread in diameter. When my friends arrived we immediately started the experiments, obtaining really fine records of the synchronized electrical and optical (birefringence) spikes at low temperature. In the Hallerianum the younger team called us "the four old foxes." Our records clearly showed that there is an appreciable, fully reversible transient change of the macromolecular structure in the excitable membrane, synchronous with the action potential. Its optical equivalent is the reversible change of birefringence, which occurs in radially oriented structural elements of the excitable membrane.

Victor Howarth had brought one of his extremely sensitive thermopiles of his own construction with him to register the heat changes, which occur synchronous with the electrical and optical spike. There is a small heat production in the rising phase of the action potential and a reabsorption of this heat in the falling phase. The optical and thermal signals can be interpreted in the following, simplified way. The excitable membrane is a highly ordered structure on the macromolecular level. In the rising phase this order increases and produces two external signals: (*a*) heat is given off to the environment; and (*b*) the birefringence increases. In the falling phase the heat is re-absorbed and the birefringence decreases. Both "signals" are extremely small: the heat changes in the order of 10^{-6} centigrades while the birefringence changes in the order of 10^{-5} μm.

We often asked ourselves how these eight million separate nerve fibers perform from the point of view of signal transmission. In any case, the pike has a remarkable sense organ. Recently I heard of an unexpected practical application of this extremely sensitive faculty of pikes to smell minute impurities in water. In Göppingen, Germany, the director of the water works keeps Nile pikes *(Gnothonemus petersi)* in an aquarium. They take turns swimming in a special pool through which small samples of the incoming drinking water for the town are deviated. The moment the slightest impurity appears, the pike "on watch" emits electrical impulses, which release an alarm system. This is a true story!

VON MURALTS PAST AND PRESENT

In 1978 Richard Keynes invited me to give a review on the transient optical changes that occur in synchrony with the action potential in excited nerves. They are changes in light scattering, in birefringence, and in artificially induced fluorescence. It was at a symposium in the lecture theater of the Royal Society in London. I gave this talk in memory of my ancestor, the well-known anatomist Johannes von Muralt, from Zürich, who in 1669 presented a paper to the Royal Society, "Concerning the Icy and Christallin Mountains of Helvetia Call'd the Gletscher." With a twinkle in the eye I would now like to remind the reader: ice is also birefringent!

The Muralt family has its origin in Italy, where they lived as the "Capitanei of Locarno" from the Twelth century until 1550. In that year they changed from Catholic to the new Protestant faith, and were forced to leave Catholic Italy.

They emigrated to the Protestant town of Zürich, Switzerland, at that time an important center for all those who adhered to the new Protestant faith. In Zürich they found a kind reception by the reformator Heinrich Bullinger. All the members of the family became citizens of Zürich, where most of my relatives live today.

My father, Doctor Ludwig von Muralt, was a well-known specialist in tuberculosis. My mother, Doctor Florence Watson, was an American, born in Philadelphia. She worked as a physician until her marriage. My American great-grandfather, Doctor John Watson, had been one of the early presidents of the New York Academy of Medicine. The ties with New York are happily renewed by the marriage of our second daughter Regula with Doctor William T. Foley of New York.

My wife Alice and I spent our first year of married life in Munich, while I studied clinical medicine. We were two years in Boston, where our eldest daughter Charlotte was born, five years in Heidelberg, where Regula was born, and we have lived in Bern, where our third daughter Elisabeth was born, since 1936. We have seven grandchildren and live now out in the country on our farm, called Arniberg.

It was an honor and a pleasure to write this prefatory chapter. I wish to thank the editors for inviting me to undertake such an interesting task.

Horace W. Davenport

Ann. Rev. Physiol. 1985. 47:1-14

THE APOLOGY OF A SECOND-CLASS MAN

Horace W. Davenport

Department of Physiology, The University of Michigan, Ann Arbor, Michigan 48109

INTRODUCTION

I was once a member of the Editorial Committee of the *Annual Review of Physiology,* so when the Black Spot arrived, I knew how I had been chosen.

SCENE: *A conference room. The editorial committee is in session.*

Chairman: . . . So it is agreed that this year we will do gating potentials in the left bundle for the heart section, and next year we will do gating potentials in the right bundle. Now, how about the prefatory chapter? You will remember that last year we wanted X, but he turned us down. We had Y and Z in reserve, and Y did a fine job. How about Z this year?
Committee Member A: Z doesn't really deserve it. We ought to get someone else. How about Davenport?
Committee Member B: Who's he?
Committee Member A: The stomach man, you know. Been at Michigan for fifty years.
Committee Member C: I thought he was dead.
Committee Member A: He was going strong writing history when I was in Ann Arbor last month. He can always be relied upon to say something amusing. We ought to get him before it's too late.
Committee Member D: I agree.
Chairman: So we'll ask Davenport. Now for the kidney section. I suggest that this year we do the first three cells of the descending loop . . .

First, I must explain my title. The word *Apology* is an allusion to Colley Cibber's *Apology for His Life.* He was a second-class dramatist and poet

1815

0066-4278/85/0315-0001$02.00

although he may have been a first-class actor, and this autobiographical essay is more like his *Apology* than like Cardinal Newman's *Apologia pro Vita sua.*

In 1937 the Examiners of Oxford University put me firmly in the Second Class in the Final Honour School of Animal Physiology. I have since lived much of my life among first-class men, and when I compare myself with Wallace Fenn or W. A. H. Rushton, for example, I know the Examiners were right. The question I shall try to answer here is, How can a second-class man accomplish anything that deserves to be commemorated in a prefatory chapter? Because there are so many more second-class men than first-class ones, perhaps the answer will encourage some of the younger ones.

After my sophomore year at the California Institute of Technology I never intended to be anything but an academic. An academic's obligations are teaching, service, and research; I shall take them up in that order.

TEACHING—AND OTHER WAYS OF LEARNING

To teach physiology, I first had to learn some. When I went to Oxford in 1935 as a Rhodes Scholar I was headed toward biochemistry—I knew no physiology. In those days Oxford had no independent School of Biochemistry, so I was required to read for the Animal Physiology School, which was three fourths physiology and one fourth biochemistry. My tutor at Balliol College parceled out the subject in one-week bits and assigned me an essay on each bit. When I had a physiological topic I read *Starling* and *Physiological Reviews,* and then a monograph if a recent one were applicable. Then I skimmed some of the primary literature. Thus I learned classical mammalian physiology with a good foundation of general physiology. It was an inefficient way to learn, and I would have done better if I had first been through the kind of survey course—a sort of physiological Chaucer to Browning—that I later taught.

The other quarter of the Animal Physiology School was devoted to biochemistry as taught at Oxford by Professor R. A. Peters. He had isolated what he then called aneurin. We now call it thiamin. He had found that aneurin is a cofactor in the decarboxylation of pyruvate, and consequently he was the first man to know the function of a vitamin. I introduced myself to him. He was kind to me and allowed me to do a little research on vitamin C in his laboratory during the summer vacation. I knew about vitamin C, for I had worked on it with Henry Borsook at Caltech, and I spent the six-week spring vacation of 1936 in a library in Munich reading all that had been published about it. One day in 1937 I got an urgent call: Would I please see Professor Peters at once? He had agreed to write an Annual Review on the water-soluble vitamins. The manuscript was already overdue, and he had just remembered that vitamin C is water-soluble. Would I help? I wrote a draft in less than two weeks, and Professor Peters revised it. Thus the 1938 volume of *Annual Review of*

Biochemistry contains an article on *The Vitamin B Complex* by R. A. Peters & J. R. O'Brien and one on *Vitamin C* by R. A. Peters & H. W. Davenport. When I last saw the professor, by then Sir Rudolph Peters and the inventor of British Anti-Lewisite, at the centenary meeting of the British Physiological Society in Cambridge in 1976, he said it had been a jolly good review. He also said that because he was 87 he would have to give up his laboratory the next year.

I was privileged to take the laboratory courses that had been developed in England beginning about 1860. When William Sharpey started to teach physiology at University College London, his only practical work was histology. In that tradition, histology was part of the physiology course at Oxford. We stained our own sections and examined muscle spindles that had been prepared by C. S. Sherrington. In 1870 Michael Foster had added work with frog heart, nerve, and muscle to the University College practical course in physiology, and John Burdon Sanderson brought Michael Foster's laboratory teaching to Oxford when he became professor of physiology there. Consequently, I spent the Saturday mornings of two terms in the "Frog Jumps" laboratory, measuring refractory periods and demonstrating decrementless conduction. Determination of the strength-duration relationship had been added in the 1920s, and I learned to use a Keith Lucas spring rheotome to obtain the data used in calculating chronaxie.

Sherrington had started a mammalian laboratory for students when he was in Liverpool, and he transferred it to Oxford in 1913. When I took that laboratory course, a young Demonstrator named Jack Eccles showed me how to make a Langendorff heart preparation by perfusing an isolated heart's coronary vessels by way of the aorta. The grand climax of the mammalian experiments was the demonstration of ipsalateral flexion and crossed extension in a decorticated cat. Carl Pfaffmann and I did that together, but when I was asked in the Practical Examination for Schools to "Demonstrate as many reflexes as possible," I was able to make that rather tricky preparation all by myself.

There was a surprising amount of human physiology. I learned to analyze my own gastric juice and to record the A, C, and V waves in the jugular vein by means of a Mackenzie polygraph. Douglas and Priestley themselves taught me how to collect alveolar air and analyze it with a Haldane apparatus. The questions I was asked on that part of the final practical examination were to determine total, free, and combined acid in a sample of gastric juice and to measure the energy cost of walking four miles an hour. For the latter I put on a Douglas bag and a nose clip and walked around the parks with a stopwatch in my hand. Then I measured and analyzed my expired air, with F. J. W. Roughton as external examiner looking over my shoulder. I had learned the equations used in the Haldane open-circuit method, and I made the appropriate calculations.

When Cuthbert Bazett gave me my first job as a physiologist in 1941, an instructorship at the University of Pennsylvania, I was of course expected to teach physiology although I had no experience as a teacher or as a physiologist. I really learned something about physiology and the technique of teaching in the student laboratory on the English model that Bazett had provided for dental and medical students.

Two years later I had the exciting experience of teaching under Eugene Landis when he revised the teaching of physiology at Harvard. He added a good bit more human physiology, so I learned about the distribution of skin temperature over the body and its changes when the body is cooled and warmed, about circulatory changes occurring when one stands on a tilt table, and about responses of blood flow through the hand to emotional stress.

In those days youngsters did what they were told. At Pennsylvania in 1941, for the first time in 40 years A. N. Richards was too busy to teach renal physiology, and I was told to do it. I started with Arthur Cushny and discovered what Richards himself and E. K. Marshall and Homer Smith had done. Those dreadful lecture rooms at Penn were entered from both the first and second floors, and the second floor door was directly opposite Vice-president Richards's office. One day while I was lecturing in my loud, self-confident voice about Homer Smith and the wonders of renal clearance I was horrified to see Richards standing in the upper door listening to me.

At Harvard I was told to teach endocrinology in both lecture and laboratory, and I had to work up all branches of that enormous subject. I remember with particular pleasure discovering the work of Fuller Albright. For the laboratory I learned to pancreatectomize a cat, to follow the estrous cycle of a rat with vaginal smears, and to measure the salt appetite of an adrenalectomized rat. When we did the Ascheim-Zondek pregnancy test, using urine I had obtained at the Women's Hospital, a couple of students, for some mysterious reason, brought samples from home.

When I went to Utah in 1945 as head of the Department of Physiology I had the responsibility of teaching the entire course, so I learned about backward and forward failure and critical closing pressure. I knew the prewar Krogh-Barcroft-Haldane respiratory physiology, but I had to learn the Fenn-Rahn-Otis-Pappenheimer kind in order to teach it. I taught acid-base physiology, as I had learned it at Harvard, to anyone who would stand still long enough. I even taught anesthesiology residents how to measure plasma carbon dioxide with a Van Slyke apparatus. Naturally, I wrote down what I taught, and the result was my book *The ABC of Acid-Base Chemistry*. An immense amount of teaching experience went into the six editions of that book. It sold over 140,000 copies in the American editions and heaven only knows how many in French, German, Spanish, Italian, Portugese, Japanese, and Korean. When I say that it has been called a "pedagogical classic" I also say that there is a point beyond which modesty becomes affection.

It has always been difficult to teach gastrointestinal physiology in the student laboratory. Frank Brooks solved the problem at Pennsylvania by assembling a kennel of chronically prepared dogs, but I never tried that. At Harvard, Eugene Landis gave me the task of devising and supervising an exercise on the radiology of the esophagus and stomach, quite appropriate for W. B. Cannon's old department. When I became associated with gastrointestinal physiologists because of my work on acid secretion, publishers nagged me to write a chapter. I always refused, reasoning that because I had no training in gastrointestinal physiology, I did not know enough.

When I later joined the Editorial Committee of the *Annual Review of Physiology*, I was asked to write about the digestive tract for the volume. I thought that as a committee member I couldn't refuse, although I learned that others didn't have the same delicate conscience. I wrote the review, and when I had finished I thought I might know enough to write an introductory text for the Year Book series that was being planned by Julius Comroe & Robert Pitts. For two years I went to the medical library every weekday evening from 7 to 9. I read and read about some topic until I knew enough to write straight off a summary of it. Then I wrote a chapter of what became my *Physiology of the Digestive Tract*. For 20 years I felt obligated to keep up with the rapidly proliferating literature on the subject in order to prepare four more editions. It was a great relief to stop!

I greatly enjoyed teaching, particularly laboratory and small group teaching, and I miss doing it. Physiology has advanced far beyond me, but the heart still beats and the blood still carries oxygen and carbon dioxide. I would like once more to be allowed to shepherd a few students through the answer to such questions as: "A man puts out four liters of urine a day. Describe in *logical order* how you would determine the cause." That question, however, reveals the nature of my teaching; it was spoon-feeding. A really first-class teacher stimulates his students to do original work and to learn on their own. Benjamin Jowett of Balliol was such a teacher; no student dared face him unless he had done the best of which he was capable. My students, on the contrary, came to me with their mouths open, ready to be fed the pap I had so artfully prepared, and I knew exactly the angle to hold the spoon. My teaching was a useful and economical method of transferring information, but it was definitely second class.

I regret what has happened to the teaching of physiology in many schools. One need only to look back to the teaching sessions of the American Physiological Society to know that in my day teachers of physiology took their job seriously. A couple of hundred persons went to the fall meetings in Madison, Wisconsin, a day early to hear Julius Comroe and his team tell the story that was eventually published as *The Lung*. Later I participated in three such sessions, one on gastroenterology, one on acid-base balance, and one on laboratory methods—for which I hauled a plethysmograph from Ann Arbor to Urbana,

Illinois. In medical schools and colleges teachers invented new ways and refined old ones, and, assisted by the Society's teaching committee, they discussed their experiences with each other. But around 1970, several things went wrong.

Students who had been through the new secondary science curriculum revolted against science as they revolted against everything else. All they wanted was a few hints on how to be humane doctors, and they refused to attend laboratory sessions. Research interests of the faculty drifted away from the whole animal and its organs to bits and pieces of cells. Lectures were given by relays of specialists who never listened to each other. "Curriculum reform" emasculated the basic science departments in medical schools and took away time for laboratory work. Teaching laboratories and small-group conferences disappeared.

The purpose of the teaching laboratory in medical schools was never to teach the scientific method. The first question asked when a laboratory exercise was planned was "Will it work?" That meant "Will it give the expected results?" The laboratory's purpose was to give the student a first-hand, quantitative knowledge of how the circulation, the respiration, and the rest work. A few years ago I taught most of a physiology course for medical students, and as I did so I thought over and over again how much the students were missing by not having laboratory work. They had never seen a triple response, felt ischemic pain, measured partial pressures in alveolar air during hyperventilation, or seen the bottom fall out of the diastolic pressure as a man is about to faint. Students are not the only losers. A young faculty member, such as I once was, is deprived of the opportunity to learn the core of physiology by teaching it. He may know a lot about down-regulation of glucose receptors, but he hasn't the faintest idea what happens to his circulation when he attempts to expire on a closed glottis while defecating.

SERVICE

I learned about the service obligations of an academic when, at the age of 32, I leaped from instructor at Harvard to professor and head of the Department of Physiology at Utah. Utah's College of Medicine had been a two-year school, and it had been told to get better or quit. It was getting better in a spectacular way, for those in authority had decided to bring in outsiders, men devoted to research, and to staff the clinical departments with full-time men. Louis Goodman, Mark Nickerson, George Sayers, Tom Dougherty, Leo Samuels, Max Wintrobe, and Hans Hecht, plus some pediatricians and obstetricians each with a Ph.D. in a basic science, were early members of the faculty. The university itself was being transformed by a physicist president who was impressed by the quality of the medical faculty and who was raising the quality of the university by persuading Mormon scientists like Henry Eyring to return to Zion.

I learned much more than I taught, and one way was by serving on many university and medical school committees. It is customary to complain of the time that committees waste, but I have found that, given a good dean or a good president, they are the best way of involving the faculty in the management of an academic institution. With good will on the part of its members, a committee can eventually reach a consensus that is often better than the original opinions brought to it, and at the end all members are committed to implementing the consensus. In committees I learned how to cooperate for the good of the institution, that you don't have to like a man to get along with him, and that the most effective form of selfishness is conspicuous unselfishness. I also learned how to deal with the Department of Buildings and Grounds and that a liberal application of soft soap helps to get the best out of the Purchasing Department. I made mistakes, but when I went to The University of Michigan as chairman eleven years later I knew exactly what I wanted to do and how to do it.

A chairman's job is to provide the atmosphere in which good men, and I use the word *men* in a neutral sense, can do good work. First, one must find the men. In recruiting I looked for *demonstrated promise*. It was a well-defined type. The man I looked for was Phi Beta Kappa from Haverford or Princeton. If he had an M.D., he was Alpha Omega Alpha from Columbia or Hopkins, but if I needed a neurophysiologist he was from the University of Washington in Seattle. He had interned at Cornell or Barnes Hospital, and he had done his national service at the N. I. H. Then followed the regulation two years in Copenhagen. If he had a Ph.D., it was from Chicago or the University of Illinois in Urbana, and he had comparable postdoctoral experience. He had published four or five papers, but he hadn't yet found himself. When he visited the department to give his seminar, I talked to him a few minutes about the job, and I told him that his full-time salary would not depend upon grant support. Then I sent him on a day-long tour of the department so that others could judge him and he could discover what they thought of the man in the front office. When he accepted the offer, I did little more than provide start-up facilities and some encouragement. After a period of quite respectable fumbling, he suddenly took off on his own with stopped flow, the role of carnitine, single-cell activity in the hippocampus, or the protective function of fever. Then the department had a man of *demonstrated performance*.

In one disasterous experience I learned it is futile to try to keep a man after he has decided to accept another job. Instead, I practiced preventive maintenance so that when one assistant professor told me he was tempted by an offer from a place with better computer facilities I was able to tell him that the week before I had proposed his promotion. He stayed, and later, by saying yes to a proposal and writing a couple of letters, I was able to advance his career materially. I did the same with others, saying yes at the right time and occasionally no, giving them teaching assignments they wanted, relieving them of teaching altogether when it seemed appropriate, proposing them for Career Development Awards, Markle Scholarships, university distinctions, and for what raises the dean allowed. The result was a stable department and a dozen or so who owe me a good bit. Whether they were happy I don't know, for I avoided personal relations. There were, of course, a few who didn't like me and who left in a cloud of acrimony. I didn't like them either, but I tried not to let that affect my behavior toward them.

Research always came first, and in that as in many other ways I tried to set an example. For years I got to work about 6:30 A.M. to work in my laboratory. If someone wanted a paper signed, he had to bring it to the room in the animal quarters where I was working with my dogs. Teaching did not suffer: it was done well at all levels. Good scientists are, in general, good teachers one way or another—science and teaching being expressions of superior qualities. Of the many ways we taught, formally and informally, I am proudest of Physiology 101–102, the course for undergraduate college students that showed them what every educated person ought to know about his own body. Through no fault of my own I had done a bad job with college teaching at Utah, and when I went to Michigan I determined to do it right. I hired a little dynamo named Virginia McMurray from Smith College whose mission in life was to spread the gospel of physiology. Within a year, working through the Extension Division, she knew more persons in Michigan than did the governor, Soapy Williams. When she left to get married, a woman from Goucher and Pennsylvania came to teach the one-term course.

In 1962–63 I was away on sabbatical leave, and the department's secretary was inexperienced. When word came from the Office of the Registrar that Physiology 101–102 was oversubscribed in advanced registration, she said: "My, my, how interesting," and threw the notice away. So when Madeline Fusco went to meet the class the first day, she found 500 students milling around in the hall. When I returned I took the problem to Vice-president Heyns, who said it was the kind of problem he liked to solve. He gave me another instructorship to be filled by a postdoctoral fellow, James Sherman, so that the course could be repeated a second term. The purpose of the course was to teach college students what they ought to know about their own bodies. A short time later Arthur Vander and Dorothy Luciano, out of missionary zeal, pitched in to help teach it. For 15 years it was second only to art history as the most popular course in the university. The Vander-Sherman-Luciano textbook, *Human Physiology,* was derived from the course, so many are familiar with the course's content.

Unlike many a first-class man who has built a monolithic department and has driven his staff to do wonders for him, I was totally incapable of dominating mine. Each member had his own independent program, and he took on one or another department job because he wanted to. A succession of men took charge of the graduate program, and all I had to do was sign the training grant application once a year. Because I was distressed by my own spoon-feeding, I was particularly proud of the rigorous course in general physiology taught by John Jacquez; when his class session was over, the blackboard was covered with integral and differential equations. Those who taught Physiology 101–102 used it to train graduate students as teachers, and they used it to recruit college students for their laboratories. Consequently, the department was always full of youngsters I didn't know. Some of them remained with us when they became medical students, so they had six or so years of research experience by the time they graduated. David Bohr scrutinized the records of students on admission to medical school, and promising ones received offers of research assistantships before their classes began. Many continued in the department for the next four years. Their presence is symbolized to me by my encounter, on the hottest Saturday afternoon one summer, with a medical student in a scrub shirt and shorts repeatedly measuring the cardiac output of a hypertensive rat by the thermal dilution method. As a result of such dedication, our department had almost a monopoly on the Borden awards for student research, and three members of one medical class are now professors of physiology at very good universities—not a random event.

RESEARCH

Is it possible to accomplish something significant in research without good luck? Certainly, I have been Fortune's favorite, although it never seemed so at the time. When my father died, my mother chose to live within what turned out to be commuting distance to Caltech. Had we lived ten miles farther away, I could not have afforded the daily journey during the Depression. Luck gave me the oportunity to serve as Henry Borsook's unpaid research assistant in the summer between my junior and senior years. I learned much more than biochemistry from Borsook: how to cook an omelet and what Cézanne had been up to, among much else.

In the spring of 1941 I was in my second postdoctoral year as a Sterling Fellow at Yale. I was desperate for a job, having haughtily turned down the University of Chicago's offer of an instructorship at $2,000 a year. At the federation meetings in the Stevens Hotel in Chicago my current chief, C. N. H. Long, went through a door just as Cuthbert Bazett of Pennsylvania was about to go through it in the other direction. Bazett said: "Hugh, do you know anyone who wants a job?" Hugh Long replied: "Yes, that tall man over there." Bazett had hired an Englishman who suddenly couldn't come to the States because of the war, so he hired me almost on the spot. Bazett was an Oxonian; I was an

Oxonian and therefore obviously qualified. So I became a physiologist. Ten seconds either way for Long or Bazett going through that door, and I would have remained a biochemist. Considering the postwar explosion in biochemistry and my inability to learn new methods, that was the luckiest thing that ever happened to me. Otherwise, I might now be an embittered associate professor in some intellectual backwater.

As I was about to return to Pasadena from Oxford in June of 1938, I discussed gastric secretion of acid with my supervisor, David Fisher. He said acid secretion might result from a sort of chloride shift, and therefore there might be carbonic anhydrase in the gastric mucosa. I said, "Pooh, pooh," for Meldrum & Roughton had failed to find carbonic anhydrase anywhere but in erythrocytes. Nevertheless, Fisher persuaded me to look, and of course I found it there. The next autumn as a graduate student at Caltech I used a version of the Linderstrøm-Lang method to determine the cellular locus of the enzyme. I punched out cylinders of gastric mucosa and sliced them on a freezing microtome. I measured the enzyme content and counted the oxyntic, or parietal, cells in alternate slices, and I found a surprisingly good correlation between enzyme content and oxyntic cell count. I held my breath for the next 20 years until others using different methods confirmed that carbonic anhydrase is indeed in the oxyntic cells. Henry Borsook, who must have been anxious to get rid of a student who was not working on his favorite problems, saw to it that I got my Ph.D. at the end of the year.

I desperately wanted carbonic anhydrase to be in the oxyntic cells and nowhere else in the mucosa. However, my data eventually convinced me that it must be in the surface epithelial cells as well. That taught me that no matter how hard I tried, I couldn't cook my data to get the results I wanted. I have often had the same experience since, and always the "wrong" result turned out to be better than the "right" one. It was fortunate I didn't leave carbonic anhydrase to be discovered in the surface cells by someone else, for bicarbonate secretion has turned out to be an important function of those cells.

That work and some papers I published in my first postdoctoral year, papers that turned out to have good data and ludicrously wrong conclusions, got me off to a flying start. Carbonic anhydrase was a fashionable subject, the prostaglandin of its day, and in addition my work was the first to provide any solid evidence, as contrasted with wild speculation, relating to the mechanism of acid secretion. As a result, I became known as a coming man. In my first paper on gastric carbonic anhydrase I had said as clearly as I could that hydration of carbon dioxide and subsequent ionization of carbonic acid could not possibly be the driving force behind the secretion of HCl at 150 mN. Others who had not paid attention erected a "Carbonic Anhydrase Theory of Acid Secretion" and fastened it on me. After an unhappy postdoctoral year at Yale—in which I failed to inhibit acid secretion with sulfanilamide, the one carbonic anhydrase inhibitor then available—I wrote a brief editorial for *Gastroenterology* called *In Memoriam: The Carbonic Anhydrase Theory of Gastric Acid Secretion.* *Gastroenterology* printed the title in suitably mournful Gothic type, and at once I earned a tremendous reputation for scientific integrity. I was amazed, for I had thought that the object of research is to learn the truth, not to provide ammunition for someone dying in the last ditch while defending an untenable theory.

There is no point in describing my next 20 years of research on the mechanism of acid secretion. I worked hard and accomplished nothing, chiefly because I didn't know enough cell biology. There wasn't much cell biology of the right kind in those days, but I didn't have the ingenuity to invent it. A few years ago I met George Sachs, the man who has come nearest to succeeding where I failed, and I am proud of the fact that I was delighted by his accomplishment rather than jealous. By 1962 I was at a dead end, and I decided to quit research altogether. The University of Michigan owed me one term's sabbatical leave. A grant from the U.S.P.H.S. paid my salary for a second term away from Ann Arbor. For a last fling at research I spent the year 1962-63 working under Charlie Code at the Mayo Clinic. I told Code to treat me like a postdoctoral fellow and to give me a research problem. He had a versatile and productive research program, and from among his many problems he asked me to explain the effect of eugenol on the gastric mucosa.

Franklin Hollander had found that eugenol, the active ingredient of clove oil and the only thing in gastroenterology that smells good, stimulates secretion of mucus by the gastric mucosa. Charlie Code, in his quest for the elusive inhibitor *gastrone* that might be a component of mucus, had bathed a dog's Heidenhain pouch with a suspension of eugenol. Acid secretion did seem to be inhibited, for immediately subsequent stimulation of the pouch with histamine caused a flow of juice that was low in acid and high in sodium. Code was too cautious to conclude that he had demonstrated an effect of gastrone, and he put the results aside until I came to the clinic. He asked me: Does eugenol affect the acid-secreting mechanism, or is there another explanation?

The pieces of the answer were all in place; all Charlie and I had to do was put them together. Years before, Code had done pioneer work on the impermeability of the gastric mucosa, and he had coined the descriptive term the gastric mucosal barrier. Juice collected from an intact Heidenhain pouch is high in acid and low in sodium, for the reason that the barrier prevents back diffusion of secreted acid into the mucosa and diffusion of sodium from mucosal interstitial fluid into the juice. Franklin Hollander had found that repeated application of eugenol to the gastric mucosa causes the mucosa to shed a "sero-sanguinous" fluid, but he hadn't quite recognized the implications of that fact. Therefore, after the barrier is broken by eugenol, acid should disappear from the juice by back diffusion, and sodium from the interstitial fluid should replace it. This could be tested by putting a buffer in the eugenol-treated pouch to trap the secreted acid before it had a chance to diffuse away.

It was easy enough to show that our idea was right. Charlie Code prepared the dogs, provided advice and technical assistance, and guided me through the splendid facilities of the clinic. Each morning I worked with one of the dogs, irrigating her pouch with test solutions for five consecutive 30-min periods, and gradually working through the elaborate protocol of controls and crucial experiments we had planned. By 5 P.M. I had finished the day's analyses and radioactive counting and could calculate the net and unidirectional fluxes. Toward the end Code had a postdoctoral fellow make the appropriate potential-difference measurements. Charlie wrote the paper and saw it through the clinic's meticulous section of publications. Mort Grossman, then the editor of *Gastroenterology*, said the paper would revolutionize the physiology of the stomach. It may not have done quite that, but, together with a little paper I published before the big one was in print, it did stir up the animals.

The work described in the little paper was done in my last few weeks in Rochester when I tackled the problem of the nature of the gastric mucosal barrier. It ought to be penetrated by un-ionized, fat-soluble, short-chain fatty acids but not by the same acids in the ionized state. I demonstrated that this is true for acetic, propionic, and butyric acids. I used acetic acid, because I knew it precipitates mucus, and even then I was worrying about the role of mucus on the surface of the stomach. I didn't use acids having longer chains, for the clinic's stock room didn't have any. Later, Code did an elegant set of experiments with longer chains, increasing their length until he ran out of water solubility.

While I was doing those experiments I read a preposterous paper by a gastroscopist describing the inflamed gastric mucosa lying beneath a fleck of aspirin on its surface. From his discussion, one would have thought that the fleck of aspirin was giving off some sort of malignant ray. I knew that Adrian Hogben and his friends at the N. I. H. had shown that aspirin, or acetylsalicylic acid, is absorbed from an acid solution in the stomach but that its ionized form, acetylsalicylate, is not absorbed from a neutral solution. I also knew that the clinical world was having one of its recurrent spasms of concern about gastric bleeding induced by aspirin. It ought to be amusing, I thought, to show that aspirin, like acetic acid, breaks the barrier. I sent to the drug store for some aspirin.

I did the crucial experiment first. I irrigated a Heidenhain pouch in a preliminary test period with 100 mN HCl to show that its barrier was intact. Very little acid disappeared from the test solution, and little sodium entered it. Then I irrigated the pouch with a solution of aspirin in 100 mN HCl. As I expected, aspirin entered the mucosa, broke the barrier, and greatly increased fluxes of H^+ and Na^+. In addition, the unexpected occurred: the pouch bled furiously and shed a large volume of fluid. Although the isotopic data showed that the permeability to Na^+ had greatly increased, my titrations showed that there was almost as much acid in the fluid within the pouch at the end of the period of incubation as there had been at the beginning. I couldn't believe that a gastric mucosa highly permeable to Na^+ was impermeable to H^+. It looked as though acid diffusing into the mucosa had stimulated the mucosa to secrete acid.

The appropriate control for that experiment was first to test the pouch with 100 mN HCl to demonstrate that the barrier was intact and then to irrigate it with the same concentration of aspirin in neutral solution. When I attempted to do that a few days later using the same dog, I found that her pouch bled during the initial period of irrigation with acid. It had not healed adequately. I debated whether to continue, and fortunately I did. When I irrigated the pouch with aspirin in neutral solution, the bleeding stopped. It took me 16 busy years to work out the implications of those two experiments.

Although I did use a flame photometer and a primitive counter, I could have done most of the work with a little glassware and an old-fashioned block comparator. I did have one enormously valuable resource: about five or six self-starting graduate students and postdoctoral fellows. Soon after I returned to Ann Arbor, Leonard Johnson—"Rusty" to everyone—came to work with me for the summer. I told him to borrow a smooth-muscle bath from David Bohr and to measure histamine. He came back a week later and told me he couldn't make the method work. I bought two airplane tickets and took Rusty to Rochester, Minnesota, where Charlie Code's assistant, Joe Kennedy, taught Rusty how to do it in one day. Thereafter Rusty took off on his own, and by the time he left me he was well on his way to becoming the best Ph.D. gastroenterologist under 40. In a similar fashion others did original work while they were nominally my students. In my more cheerful moments I think that my spoon-feeding of undergraduates is more than offset by my treatment of the others. I simply pointed them in the right direction and cheered them on.

I did my research solely to solve scientific puzzles, but because of its nature I published in clinical journals. Orthodox physiologists, like Committee Members B and C, had no idea what I was doing. That was not true of clinicians, particularly the surgeons who had to deal with gastric ulcers and with exsanguinating hemorrhages in their intensive care units. They immediately picked up the idea of the gastric mucosal barrier and the consequences of breaking it. A clinician's first paper would begin: "Davenport has shown that . . ." His second would begin: "We have shown that." His third paper would omit my name from its bibliography, a vastly amusing example of the fate of most scientists. Nevertheless, I know quite well that a substantial number of very good clinical scientists who are doing progressive work are my scientific children.

Having written this *Apology*, I must try to answer the question I posed at its beginning. If I have done nothing first class, I can say that I have been useful. My teaching, and in particular my *ABC of Acid-Base Chemistry*, has raised the general level of understanding some fraction of a millimeter. My work as a department chairman has helped the institutions and the faculties I have served. The applications of the results of my research have had some small impact on the quality of patient care. That is, I suppose, as much as a second-class man can hope for.

C. Ladd Prosser

Ann. Rev. Physiol. 1986. 48:001–06

THE MAKING OF A COMPARATIVE PHYSIOLOGIST

C. Ladd Prosser

Professor Emeritus, Department of Physiology and Biophysics, University of Illinois, 407 South Goodwin Avenue, Urbana, Illinois 61801

I spent my youth in Avon, a small town twenty miles south of Rochester, New York. There was frequent electric train service to the city, and opportunity to hike along the Genesee River and up and down the hills leading to western Finger Lakes and their outlets. My interest in nature was stimulated by weekly hikes with my father. Fossils were abundant in the slate and shale lining the many gullies. I started collecting insects with two pals with Blatchley's *Coleoptera,* and plants with Gray's *Botany* in the seventh grade. I majored in biology at the University of Rochester where a turning point in my career was a course in Physiological Psychology in which the text was Herrick's *Neurological Basis of Behavior.* During this course I decided that the neural basis of behavior could better be studied with invertebrates than with mammals. I went to The Johns Hopkins University for graduate study, and wrote a thesis on the physiology of the nervous system of earthworms. The summer after my first year there was spent as a research assistant to S. O. Mast at the Mount Desert Biological Station. My first published paper was on amoeboid movement (1). During that summer (1930) I was out at every low tide becoming acquainted with invertebrate animals.

Hopkins was a stimulating place; visitors to H. S. Jennings (protozoan genetics) and to S. O. Mast (protozoan physiology) included the geneticist T. H. Morgan from Columbia, cytologist C. E. McClung from the University of Pennsylvania, and general physiologists Newton Harvey of Princeton and L. V. Heilbrunn from the University of Pennsylvania. The summer of 1931 I spent taking the physiology course at the Marine Biological Laboratory in Woods Hole, where I learned much from the neurophysiologist Phillip Bard, respiration physiologists Lawrence Irving and A. C. Redfield, kineticist Leonor Michaelis, and membrane transport experts Merkel Jacobs and Rudolph Höber. It became evident to me that study of anatomy and ethophysiology of in-

1831

0066-4278/86/0315-0001$02.00

vertebrate nervous systems had limitations, and that electrophysiological techniques were more likely to be useful in elucidating the neural basis of behavior.

I finished my PhD in the depth of the depression; there were no National Institute of Health (NIH) fellowships and the National Science Foundation (NSF) did not exist. Phillip Bard at Harvard Medical School helped me to obtain a Harvard postdoctoral fellowship (requiring celibacy) to study in the laboratory of Hallowell Davis. These were days before Tektronix or Hewlett-Packard: a row of relay racks housed separate power supplies, vertical amplifiers, sweep circuits, and stimulators, all to serve one five-inch oscilloscope. My first recording had been at Woods Hole with a string galvanometer. From Davis I learned the basic techniques of electrical recording and some of the physiology of hearing. The Harvard Medical School Physiology Department was an intellectually stimulating place, and afternoon tea provided an opportunity to become acquainted with Professor W. B. Cannon, Arturo Rosenblueth, Phillip Bard, Magnus Gregorsen, John Edsall, and others. I participated in the physiology laboratory for medical students, and Hal Davis encouraged me to test my ideas on invertebrate nervous systems. I recorded extracellularly from crayfish central nervous systems, and everything I observed was new. I produced four papers on crayfish nervous systems during my first year. The most significant discovery was that an isolated nervous system could produce rhythmic electrical activity in single units (2). This discovery dealt a blow to the behaviorist dictum that all patterned behavior must be initiated by sensory input. Another find was a light-sensitive ganglion in the crayfish abdomen (3). I frequently visited the Harvard Biological Laboratories, and there continued my zoological interests in discussions with G. H. Parker, B. F. Skinner, W. J. Crozier, and with collaboration with John Welsh.

After a year and a half at Harvard Medical School I went to England to work with Professor E. D. Adrian, who had recently published on electrical activity in insect nervous systems. I learned not only from him but from Joseph Barcroft, Brian Matthews, William Rushton, and others. On the day of my arrival Professor Adrian took me to meet Professor Barcroft who was up to his elbows in a bathtub of warm Locke solution in which rested an anesthetized sheep from which Barcroft was removing the fetus with circulation intact. It was apparent that everyone in a British laboratory, including the Professor, enjoyed research. I attended a course of lectures on neurophysiology by Adrian, who did not seem to enjoy student contacts. His laboratory had been occupied previously by Keith Lucas, and much of Lucas' equipment—rheotomes, induction coils, a maze of overhead wires—remained. In Adrian's laboratory we used Matthews oscillographs, magnetic devices that followed fast potential changes such as nerve spikes. Shortly before I arrived in England, the German neurologist Berger had announced that rhythmic electrical waves could be recorded from outside the skull of humans. I was a subject in Adrian's confirmation of what became known as alpha waves. During my stay at Cambridge I mapped sensory fields of segmental nerves of earthworms (4).

A high spot for me as a young American physiologist in England was attendance at the monthly meetings of the Physiological Society. Dale and Feldberg had recently presented evidence for acetylcholine as a transmitter in sympathetic ganglia; Eccles maintained that transmission was electrical, and the debates were memorable. Dale's training in British debate gave him an advantage over Eccles. The best assay for acetylcholine in a perfusate was contraction of a leech muscle, and I well remember Dale's account of the contraction of this muscle to demonstrate the presence of acetylcholine in the sweat from a sock fresh off the tennis court.

I asked Eccles if I might spend a few weeks with him and in early May I rode my bicycle from Cambridge to Oxford and worked as Eccles' associate on cat sympathetic ganglia. I heard Sherrington, the most famous of Oxford physiologists, give his last lecture to the medical students, and watched him demonstrate the motor cortex of a monkey. Sherrington impressed me as a thoughtful, modest man. Eccles had many interesting stories about Oxford neurophysiologists, especially the wealthy American John Fulton. John Z. Young had recently discovered the giant fiber system in squid. I biked back to Cambridge after a useful stay at Oxford. During the spring break I travelled to the Continent, where I visited several laboratories, including that of Monnier in the Sorbonne. At that time, Lapique's idea of isochronism between nerve and muscle was under attack by William Rushton. I had long talks with the Dutch neuroanatomist Ariens-Kappers and the German comparative neurophysiologist Hans Bethe.

On returning to America in August 1934, I found that academic positions were virtually nonexistent. I was fortunate to obtain a research appointment with Hudson Hoagland at Clark University. I was then married to Hazel Blanchard who was to be my most valuable supporter and strictest critic. We had met at the MBL in the summer of 1932, and she had then spent a year studying in von Frisch's Institute of Zoology in Munich before our marriage.

At Clark I observed the political astuteness of Hudson Hoagland. The most cited of my publications was with the psychologist Walter Hunter on extinction and dishabituation in spinal rats (5). Meanwhile, I continued on crayfish nervous systems and had some teaching experience. Each summer was spent at the MBL on the staff in Physiology. I learned much from Harry Grundfest, Kenneth (Kacy) Cole, and Otto Loewi. On the recommendation of the zoologist G. H. Parker of Harvard I was offered a post at the University of Illinois, where I arrived in 1939 on the day Hitler invaded Poland. The dean of the Graduate School, R. D. Carmichael, was very supportive. My job was to give a zoology course to agriculture students, and (at my insistence) I initiated courses in comparative and cellular physiology. Then came World War II. I spent one year on a chemical warfare project developing a test with freshwater fish for war gas contamination of water supplies. In the summer of 1943, I was asked by Kacy Cole and Howard Curtis to join them at the Metallurgical Laboratory of the Manhattan Project. This was a very responsible job; I directed the internal workings of a laboratory of some 150 biologists and chemists. This was a very exciting period because all of us who knew what was going on realized that success in the war depended on achievement of the atomic bomb. Near the end of the war, after the Alamogordo test, I joined a group of scientists who signed a petition to President Truman urging that the bomb be demonstrated on an uninhabited South Pacific island rather than on a city. I learned much about radiation biology, but after the war the opportunity to return to Illinois seemed to me better than to remain with the Atomic Energy Commission (AEC). The University of Illinois was expanding in all fields.

While in Chicago I wrote a review on invertebrate nervous systems for Ralph Gerard, editor of *Physiological Reviews* (6). At about that time I concluded that the speed of animal movement is limited more by muscles than by nervous systems. On returning to the MBL in the summers of 1947–1949 I concentrated on invertebrate nonstriated muscles, and collaborated with Howard Curtis (7). During the winters I worked with mammalian smooth muscle, and have continued to do so with a current interest in rhythmicity. In 1977 I wrote a chapter for the *Handbook of Physiology* on the evolution of smooth muscle in which I indicated there was no linear evolution but great diversity. That review (8) led to several symposia and a new functional classification of muscle. Transverse alignment of thick and thin filaments has evolved many times in wide-fibered muscle, and these have T-tubules for signal coupling over the hundred micron distance between cell membrane and contractile filaments. In narrow-fibered muscles, oval, ribbon-shaped or cylindrical coupling between membrane and filaments is direct over distances of a few microns.

Meanwhile, I investigated biochemical mechanisms of acclimation of fishes to cold and warm environments. Acclimatory responses include selective protein synthesis and change in saturation of membrane phospholipids, which affect membrane fluidity. Most recently, we have found acclimatory reorganization in cultured cells much as in intact fishes. Sequential alterations in central nervous functions bring about resistance adaptations (9).

While at Clark and in my first two years at Illinois, I developed a plan for a book on comparative physiology which would emphasize evolutionary and ecological applications of physiology. The idea of such a book appealed to the W. B. Saunders editor, and the first edition of *Comparative Animal Physiology* appeared in 1951. I had several collaborators and the book set the tone for comparative physiology for many years. A second edition came out in 1961 and a third edition in 1973 (10). I enjoyed treating adaptations of animals to various environments. Comparative physiology differs from other kinds of physiology in that the comparative approach uses the kind of organism as an experimental variable, and it emphasizes the long evolutionary history leading to life in diverse environments.

Early in the fifties I served on one of the first NSF panels, which included environmental, developmental, regulatory, molecular, and systematic biologists. This was the beginning for me of two decades of service on Washington committees. These included panels at the Office of Naval Research (ONR), several at NSF, NASA, and NIH. During this period I was active in several societies: Society of General Physiology (President 1958–59), American Society of Zoologists (President 1961), A.A.A.S. and American Physiological Society (President 1969–70); and in the mid-seventies the National Academy of Sciences (Chairman, Physiology Section 1977). Experience on these committees and councils was both rewarding and frustrating. In APS the education program was implemented; my effort to sectionalize the society was voted down by the membership, but adopted ten years later.

I was the Department Chairman of Physiology and Biophysics at Illinois from 1960 to 1969. For several years as we expanded the Department toward biophysics I learned much from my colleagues, especially Bernard (Bud) Abbott (muscle) and Robert Emerson (photosynthesis). A memorable semester was when Otto Warburg spent four months in the photosynthesis laboratory. Several of us designed a degree in physico-chemical biology, but this was soon replaced by the magical word "biophysics." During several summers I participated in training programs at Woods Hole, one with Steve Kuffler on neurophysiology, and a later one with Bud Abbott on comparative physiology. During this period I jealously protected my afternoons for research, and regularly took sabbatical leaves. The most notable were in Munich and Naples, and at Monash in Australia. I have worked at some ten marine laboratories, visited others, and made three cruises on the research vessel Alpha Helix.

Official retirement in 1975 freed me for research, editing, and writing. As I contemplate my fifty years as a comparative physiologist I note that despite the Depression the thirties were golden years for training, and I was fortunate to have constructive contact with many leaders in physiology. My early discoveries with invertebrate nervous systems, later ones on diverse nonstriated muscles, and more recent findings on the biochemical and neural mechanisms of temperature adaptations of fishes have kept my interests centered in laboratory research. Writing reviews and an advanced text (three editions) provided an opportunity to contribute to the theory of comparative physiology. Frequently, physiological generalizations are discovered better in the library than in the laboratory. My research could not have been accomplished without the forty doctoral and dozen postdoctoral students whose research I have directed. Contact with colleagues and administrative duties have been rewarding, but the greatest satisfaction has come from discovering new facts and functional relationships.

Literature Cited

1. Mast, S. O., Prosser, C. L. 1932. Effect of temperature, salts, and hydrogen-ion concentration on rupture of the phasmagel sheet, rate of locomotion, and gel/sol. ratio in *Amoeba proteus*. *JCCP* 1:333–54

2. Prosser, C. L. 1934. Action potentials in the nervous system of the crayfish. I. Spontaneous impulses. *J. Cell. Comp. Physiol.* 4:185–209

3. Prosser, C. L. 1934. II. Responses to illumination of the eye and caudal ganglion. *J. Cell. Comp. Physiol.* 4:363–77

4. Prosser, C. L. 1935. Impulses in the segmental nerves of earthworm. *J. Exp. Biol.* 12:95–104

5. Prosser, C. L., Hunter, W. S. 1936. The extinction of startle responses and spinal reflexes in the white rat. *Am. J. Physiol.* 117:609–18

6. Prosser, C. L. 1946. The physiology of nervous systems of invertebrate animals. *Physiol. Rev.* 26:337–82

7. Prosser, C. L., Curtis, H. J., Travis, D. M. 1951. Action potentials from some invertebrate non-striated muscles. *J. Cell. Comp. Physiol.* 38:299–320

8. Prosser, C. L. 1980. Evolution and diversity of nonstriated muscles. *Handbook of Physiology—The Cardiovascular System*, Vol. 2, Ch. 21, pp. 635–70. Bethesda, Md: Amer. Physiol. Soc.

9. Prosser, C. L., Nelson, D. O. 1981. The role of nervous systems in temperature adaptation of poikilotherms. *Ann. Rev. Physiol.* 43:281–300.

10. Prosser, C. L., ed. 1973. *Comparative Animal Physiology*. Philadelphia: Saunders. 1011 pp. 3rd ed.

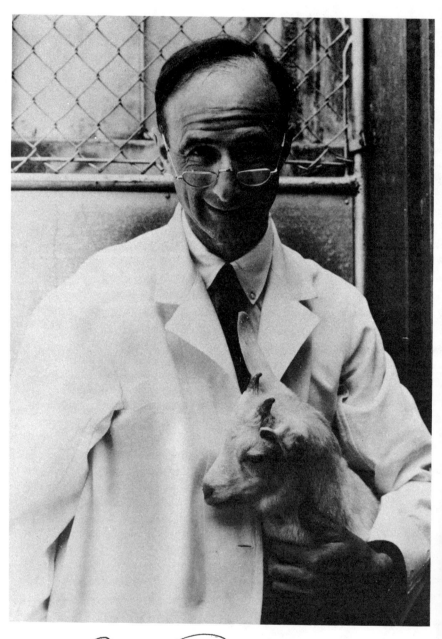

John Pappenheimer

Ann. Rev. Physiol. 1987. 49:1–15

A SILVER SPOON

J. R. Pappenheimer

Department of Physiology and Biophysics, Harvard Medical School, 25 Shattuck Street, Boston, Massachusetts 02115

An invitation to write the prefatory chapter for the *Annual Review of Physiology* provokes mixed feelings. There is no question as to the honor it confers; a glance at the list of previous invitees attests to that. On the other hand, it signals the unwanted passage of time, and it implies that the invitee, after 50 years of physiology, should have profound and useful things to say about the past and future. Alas, I must confess at the outset that the pearls of wisdom within my shell have been slow to mature, and they are still too small to spread before the sophisticated readers of this *Annual Reviews* volume. I can only use them as decorations for a personal narrative.

PRELUDE

I was born in 1915 in New York City. My father, Alwin M. Pappenheimer, Sr., was Professor of Pathology at Columbia University and one of the leading experimental pathologists of his day. He came from a well-to-do family and had every opportunity to make use of his natural talents for both science and the art of living. The summers of his youth were spent travelling in Europe, and he spoke French and German fluently. He made many drawings, etchings, and paintings, and he played the violin and viola well enough to participate in most of the classical literature of chamber music. I grew up with the sound of string quartets in the living room. My mother had a similar background, and she too was proficient in languages and music. My elder brother, who later became Professor of Biology and Master of Dunster House at Harvard, already played the violin well at the time I was started on the cello in 1922. On several occasions my scientific career has been influenced indirectly, but importantly, by music.

1839

0066-4278/87/0315-0001$02.00

My father loved his work, and the excitement of his research permeated our family life. Students and colleagues from many parts of the world came to visit, and as far back as I can remember our conversations at home included science. Sometimes I was taken to the lab to share the excitement firsthand and listen to scientific palaver over a sandwich lunch with my father's colleagues. Much of the vocabulary of biology was mine by the time I was twelve years old, and about this time, my father began taking me to meetings, including Harvey Lectures at the Academy and evening joint sessions of FASEB when they were in New York. We went to these lectures in the same spirit as we went to concerts. My father often complained ruefully that he did not know enough chemistry and physics to solve his research problems, and perhaps it was for this reason that I inclined towards physiology, which depends so heavily on the exact sciences.

On sabbatical years my father travelled extensively in Europe, Russia, and the Far East. I began my schooling at the École Alsacienne in Paris in 1921. Classes were from 8:30 to 4:30 with daily homework as well. The pace was faster than in America, and on returning home I skipped one and one-half grades at our local elementary school in Scarsdale. Later I was sent to the Lincoln School in New York City, one of the leading progressive schools of its period. I entered Harvard at 16, intellectually prepared but too young to enjoy fully the broad educational and social opportunities Harvard had to offer. My father was a devoted Harvard man, and it did not occur to any of his children to apply elsewhere. It was 1932 and we, the freshmen, were observers of the Great Depression, but few of us were a part of it. Harvard had yet to reach out for the best and brightest, and we took for granted a Certificate of Admission and full financial support from our parents. To paraphrase T. S. Eliot (3):

> We came this way, taking the route we were
>   most likely to take
> From the place we would most likely come from . . .
> Either we had no purpose or the purpose was
>   beyond the end we had figured and was
>   altered in fulfillment.

I concentrated heavily in science but did not do particularly well in formal courses and in the end failed to obtain honors. My tutor was Jeffries Wyman, Jr., who introduced me to quantitative biology in general and to the physical chemistry of proteins and amino acids in particular. In the parlance of the 1980s, Wyman would be considered a molecular biologist, and indeed, his contributions to the mechanism of the Bohr effect and the general theory of allosteric reactions of proteins are cornerstones in modern theory of protein function (27). I was in awe of his command of thermodynamics and mathematics. Our joint work on the surface tension of solutions of dipolar ions was published in the *Journal of the American Chemical Society* (23) in 1936. Jeffries (though of course I did not then address him so) wrote the first draft and gave it to me for criticism. I rewrote some of it, added a whole new section, and handed back the manuscript in fear and trepidation. He studied the changes and without hesitation or emotion said, "It is much improved, but now you must be the senior author." Not long ago I was flattered on two counts when a well-known biophysicist in England asked me whether my father had written this enduring paper.

In the summer of 1935 I took the Physiology course at Woods Hole under the direction of Laurence Irving. Our instructors included such distinguished biophysical chemists as Leonor Michaelis and Rudolf Höber; the latter taught me how to prefuse frog kidneys via the portal system and how to cannulate a frog's ureter. Later, the 1945 edition of Höber's *Physical Chemistry of Cells and Tissues* became one of the most well-worn books in my scientific library. At the end of the course, I was awarded the Collecting Net Prize ($50 was a lot in those days) to come back the next summer to work on the kinetics of CO_2 transport and carbamino formation in fish blood with J. K. W. Ferguson, who had just published important work with Roughton on carbamino hemoglobin in human blood. We did not find carbamino hemoglobin in fish bood, but we did find an anomalous distribution of HCO_3^- and Cl^- in elasmobranch red cells and duly published in the *Biological Bulletin* (5).

ROOTS IN ENGLAND

In the autumn of 1936 I sailed for England with a letter of introduction to Sir Joseph Barcroft from Jeffries Wyman. An aunt had left me a legacy of about $2000 a year, and this was more than sufficient to support a research student in pre-war England. What would I have done without it? I arrived in Cambridge not yet 21 but eager to begin work in the laboratory. I learned, however, that Sir Joseph was in the United States and would not return until October. I spent the next few weeks practicing the cello and reading old Norse poetry in the University library. When Sir Joseph returned, he was extremely kind to me, perhaps because he was fond of my brother-in-law, Will Forbes, who had worked with him ten years previously. He gave me a small research project on fetal red cells and arranged for me to take the Part II honors course in Physiology, an experience for which I am endlessly grateful. I was dreadfully self-conscious, and I admired Sir Joseph so much that I could scarcely speak in his presence. He seemed to sense this and made a point of sitting next to me at laboratory teas and trying to put me at ease by telling funny stories and anecdotes. There were eight Part II students, selected from some 300 who had taken Part I. Our instructors included E. D. Adrian, B. H. C. Matthews, F. R. Winton, E. B. Verney, W. A. H. Rushton, and F. J. W.

Roughton, and I developed life-long associations with all of them. Alan Hodgkin was a teaching assistant in the practical class; his job was to come in early in the morning to decerebrate cats for our class experiments. There were five lectures a week for three full terms, and each lecture involved analysis of significant papers in the field. We came to know the important literature in almost every aspect of physiology and in three languages. During the spring term I was introduced to F. R. Winton's experiments on isolated perfused mammalian kidneys, and this opened the door to a new world for me.

Winton was then Reader in Physiology at Cambridge and the author, with L. E. Bayliss, of a widely used elementary textbook of physiology. He had trained at University College, London, under Starling, W. M. Bayliss, Verney, Lovatt Evans, and A. V. Hill. Sir William Bayliss's "Principles of General Physiology" represented for me (and still does) the essence of all that is good in physiology, but I thought of Bayliss as someone who had lived long ago. It was amazing and thrilling for me to talk with someone who had been a close friend of both Starling and Bayliss. In Winton's hands the original Starling-Verney heart-lung preparation developed into a sophisticated technique for perfusing isolated organs from a pump-lung circulation; it was the forerunner of modern artificial heart and life-support systems. I was spellbound by the sight of an isolated dog kidney sitting on a glass plate and producing clear golden urine from thick red blood.

Winton was an expert with instrumentation, and he designed most of his own transducers, amplifiers, recording oscillographs, etc. The perfusion apparatus was itself a complex, inorganic organism—a maze of motors, plumbing, and electrical devices for measuring, recording, and controlling flows, pressures, temperatures, ion concentrations, and blood oxygen saturation. All of this appealed enormously to my scientific senses; it seemed to me to be the ideal compromise between in vitro research and the unsatisfying complexities of whole animal research. I was swayed, also, by the fact that Winton was a fine cellist and his wife, Bessie Rawlins, was one of the foremost concert violinists in England. They invited me home to play second cello in the great Schubert C major quintet (Opus 163), and this settled matters for me. I had been brought up with the Schubert but never had I played it with such accomplished musicians. Winton, together with Grace Eggleton (later to become Mrs. Leonard Bayliss), was about to start an investigation of renal oxygen consumption as a function of osmotic work of urine formation. He welcomed me as a graduate student, possibly because of my previous training in physical chemistry and blood-gas transport; or was it because of my interest in string quartets? Two foreign postdoctoral fellows were working with Winton at this time: Jim Shannon (later Director of NIH) was there to compare inulin with creatinine clearances in the isolated kidney, and Kurt Kramer from Göttingen was there to develop, with Winton and

Glenn Millikan, a spectrophotometric device for continuous recording of arteriovenous oxygen differences in flowing blood. It must be remembered that in 1936 solid-state photocells were in their infancy, and the success of this project depended on the spectral characteristics of a prototype selenium barrier cell made only in Germany. Five years later Kurt Kramer in Germany and Glenn Millikan and I in the United States used this experience to develop the ear oximeter (15). In the meantime, Winton, Eggleton, and I produced a series of papers on the mechanisms of urine formation in perfused kidneys. It was the start of a close personal relationship, which lasted until Frank Winton's death in 1985. Winton's ingenious techniques for estimating renal glomerular pressure (2, 26) provided the basis for my own first contributions to capillary physiology a decade later.

In this period, also, I came to enjoy student life in Cambridge. I played in the University Orchestra and took part in activities at Clare College. On a dark, cold winter's night we (the University Orchestra with soloists from London) performed Verdi's Requiem in King's Chapel, the most beautiful setting imaginable for this gorgeous music. William and Marjorie Rushton also played in the orchestra, and through them, I was invited to Music Camp in the Berkshire Downs. This, too, opened the way for life-long friendships. I was an avid skier and spent memorable winter holidays in the Alps and in the rugged Jotunheimen of central Norway. At Finse I raced in the downhill (Class C), coming in 89th in a field of 115 entrants. I was the only foreigner, as was duly noted by the Oslo and Bergen newspapers.

In 1938 Winton moved to the Chair of Pharmacology at University College, and he took me with him as research student on leave from Cambridge. Hermann Rein, one of the most prominent physiologists in Germany (and Kurt Kramer's chief), had just published a paper alleging that metabolism of resting muscle was regulated by the sympathetic nervous system. Winton suggested that I look into this in perfused hindlimb preparations, utilizing the methods for recording blood oxygen we had used for studies of renal metabolism. I was able to confirm Rein's surprising results, but I found a much simpler explanation for them that was based on sympathetic control of the microcirculation rather than on metabolism (17). This project gave me experience with the perfused hindlimb preparation, which I eventually used for studies of capillary permeability.

Of course we worked under the constant threat of war and with nagging inner voices telling us to stop everything worthwhile to prepare for it.

WAR YEARS

War came on September 2, 1939. I was on vacation at a music houseparty in Harvard, Massachusetts, and I intended to return to University College as

Demonstrator in Pharmacology. Passports were cancelled, along with my job. Winton became Dean of the Medical School, which was removed to Surrey before its buildings were destroyed during the Blitz of 1940. I sought help from A. N. Richards at the University of Pennsylvania. When I came for an interview, I launched into a detailed account of my work on perfused muscle, but he soon interrupted. "All I want to know," he said, "is whether you are a good experimenter." Torn between outward modesty and inner faith, I finally opted for the latter, and that was the end of the interview. He set me up with a temporary appointment in Bazett's Department of Physiology and an emergency research grant of $250, which enabled me to complete work begun in England and to send in two papers to the *Journal of Physiology*. At the same time, I joined forces with Glenn Millikan to help develop the ear oximeter. Glenn, like myself, had been cast adrift from his position in England, and he sought help from Detlev Bronk at the Johnson Foundation for Medical Physics. It seemed obvious that the US would soon be in the war, and all of us wanted to contribute what we could as scientists as soon as possible. For the next six years, Glenn and I worked closely together on oximetry, on oxygen demand valves, chemical oxygen generators, carbon monoxide poisoning in tanks and military aircraft, positive pressure breathing, and other problems in applied physiology for the military.

Let me digress for a moment to pay tribute to Glenn Millikan, who was killed in a mountain climbing accident in 1947. He was the son of Robert A. Millikan, President of the California Institute of Technology, who was famous for the oil drop experiment used to measure the charge of the electron. Glenn was known as "the little oil drop." He was probably the most well-known and charismatic young physiologist of his day in Europe and America. He exuded energy, curiosity, and *joie de vivre*. He found his way from Harvard to Cambridge, England, where he did highly original and important work on myoglobin, which culminated in a Physiological Review in 1939 (14). He was elected to a Fellowship at Trinity College, where he became a source of inspiration to students of physiology. He is still remembered with affection by colleagues in both Europe and America, and certainly his premature death deprived American physiology of one of its most promising young scientists.

In 1939, our Air Force was woefully lacking in equipment for high-altitude flying, and only a few physiologists had the knowledge and the vision to understand what was needed. Among these few were Bruce Dill, Detlev Bronk, and Walter Boothby. The Aeromedical Research Laboratory at Wright Field consisted of one altitude chamber and two or three poorly equipped laboratories staffed by a few junior medical officers and enlisted men. There was no formal provision or support for civilian research in this area, and during 1940 Glenn and I usually made the long trip out to Dayton in our own car. Operational aircraft seldom flew above 18,000 feet, and in most cases they were equipped with rudimentary oxygen equipment. Only three years later, the USAF sent more than 10,000 fighting men aloft daily in unpressurized aircraft to operate for many hours at altitudes where consciousness would fail within a few seconds without supplementary oxygen. The enterprise as a whole required vast training programs in the use of oxygen, as well as continual work to improve the design and efficiency of oxygen equipment, which competed in weight with the load of gasoline and armament.

Bronk saw the magnitude of these problems at an early stage, and he became a leader for both the military and civilian effort. Most of the time he worked directly out of the Air Surgeon's office in Washington, but he retained a small group at the Johnson Foundation to carry out applied research on oxygen equipment and on visual problems. Those of us who worked in his home laboratory as civilians had continual access to operational problems and to the flight testing facilities of the services and the aircraft industry. We spent hundreds of unhealthy hypoxic hours testing experimental equipment at simulated altitudes of 40,000 feet or more and temperatures of $-40°$ in the altitude chamber in Philadelphia.

I also participated in experimental flights to extreme altitudes in stripped down B-17s and in the first prototype B-29 heavy bomber just prior to its disastrous loss with all hands aboard in December 1942. I did a survey of carbon monoxide from gunfire in tanks and aircraft, and the Navy asked me to test a chemical oxygen generator retrieved from a Japanese Zero shot down in the Pacific. I found 0.1% CO in the oxygen. The Navy broadcast this result to the Japanese, hoping it might shake the confidence of their pilots in their equipment; at the same time, my report was classified as secret in the US, so I could no longer read it or keep it in my files.

Most of the time, at least up to 1944, we worked long hours and with a sense of urgency. Bronk was tireless; he would often arrive from Washington late in the day, talk with us until midnight and then return to Washington. I, too, led a somewhat frenetic and heady life of travel to military establishments and aircraft factories all over the country, not to mention frequent trips to Washington. Nevertheless, I remained in touch with a saner world through playing in string quartets, mostly with Catherine Drinker Bowen and with Helen Rice, who later founded the Association of Amateur Chamber Music Players, an international organization which now has more than 6000 members.

I had been brought up in an academic world that emphasized distinctions between "pure" and "applied" science. I was surprised and excited by the rapidity with which knowledge, previously considered interesting but useless, could be transformed to practical use on a large scale. Conversely, exposure to practical problems often stimulated new questions in the realm of fundamental science. Certainly, this was true of respiratory physiology, which advanced rapidly in fundamental ways under the stimulus of applied research for the military. This was a lesson for my generation of academic scientists, a lesson which in its broader context was transformed into public policy after the War.

My own experience with applied research during the War was immensely enriched by close association with Detlev Bronk and the small group of biophysicists in his entourage, including Keffer Hartline, Frank Brink, Martin Larrabee, John Hervey, Glenn Millikan, and John Lilly. They were all experts in instrumentation, especially electronics. We were a close-knit family, and one would have to be very impermeable indeed not to learn by osmosis from daily association with such alert and knowledgeable minds.

By 1945, however, the War's end was in sight and all of us were "deexhilarated" and eager to return to academic life. The project I was then working on, the storage of oxygen in the form of perchlorates, was already in pilot production in Pittsburgh, but it seemed unlikely to me that it would ever be used for its intended purpose as a source of emergency oxygen for aircraft or for portable welding apparatus. I was anxious to resume studies of the microcirculation, and after V-J Day it was natural for me to apply to E. M. Landis, an authority on capillary circulation who had just succeeded W. B. Cannon as the Higginson Professor of Physiology at Harvard. He had not yet made new appointments of faculty rank, and I had the good fortune to be selected as his right-hand man.

CAPILLARIES AND MEMBRANE PERMEABILITY (1946–1954)

I arrived in Boston in December 1945 to start a new life. There was a heavy load of teaching awaiting me. Gene Landis had organized a super course in mammalian physiology; all chapters of the fat textbooks of physiology were represented in both lecture and lab. It was a veritable *dinosaur* of a course, doomed to extinction as new branches of biology evolved to dominate the medical curriculum. For me it meant another period of intensive work and learning; I was frequently on deck at 6 AM to prepare live demonstrations or student labs prior to my 9 AM lecture.

It was not until June 1946 that I found time to resume experimental work. I had a vague plan for studying edema formation in isolated perfused muscle but without any clearly defined goal. NIH had not yet started large-scale support for university-based medical research, but even if it had, I don't think I could have written an acceptable request for a grant. Certainly I had no inkling that the project would lead rather swiftly to a logical sequence of enduring advances in the physiology of the microcirculation and the biophysics of membrane transport. After only a few experiments, I saw how to set the mean capillary pressure to known values, how to measure the effective osmotic pressure exerted by the plasma proteins across capillary walls, and how to relate these to rates of net filtration and absorption. In the following year, these preliminary experiments were verified and extended in detail with Armando Soto-Rivera; they were first presented at the 17th International Congress of Physiology held in Oxford in 1947. It was the first Congress after the War, and physiologists were happily reunited after many years of forced separation. It was a moving contrast to the grim Congress in Zurich which I had attended in 1938, just prior to the invasion of Czechoslovakia. August Krogh listened to my paper at Oxford; afterwards he came "backstage" to say nice things about it, and of course I was thrilled because his work in respiratory, comparative, and capillary physiology was (and still is) such a large part of our heritage. At the closing plenary session in the beautiful Sheldonian Theatre, Albert Szent-Gyorgi gave a speech of thanks in nine different languages.

The methods developed for measurement of capillary pressure and effective transcapillary protein osmotic pressure were soon extended to the analysis of transcapillary concentration gradients during diffusional exchange of hydrophilic solutes between blood and tissues. This, in turn, led to a general theoretical analysis of the relations between restricted diffusion, hydrodynamic flow, and molecular sieving—an analysis applicable to a wide variety of biological and artificial membranes. The theory and preliminary results were first presented (with E. M. Renkin and L. Borrero) at the 18th International Congress of Physiology in Copenhagen in 1950, under the title *Filtration and Molecular Diffusion from the Capillaries in Muscle, with Deductions Concerning the Number and Dimensions of Ultramicroscopic Openings in the Capillary Wall*. The complete paper subsequently appeared in the *American Journal of Physiology* (25). I recently wrote a history of this development and its relation to the parallel development of irreversible thermodynamics as applied to membrane transport (21). The American Physiological Society Handbook volumes on *Microcirculation* (edited by E. M. Renkin and C. C. Michel) provide full review of the many modifications and extensions of the original theory that have been made in the last 35 years.

In 1949 I married Hylie Palmer, a violinist at the Julliard School of Music and a pupil of Galamian. She has been first fiddle in my life ever since.

Alexander Forbes was still in the Department of Physiology at Harvard when I arrived in 1946, and it was a great privilege to come to know him. He was a pioneer in electrophysiology and one of its most distinguished leaders for more than 50 years (13). He was the first to use vacuum tube amplifiers in conjunction with the string galvanometer for recording nerve action potentials

(6), which set the stage for the classic work of Gasser and Erlanger on peripheral nerve and of Adrian, Bronk, Zotterman, Matthews, and others on single unit analysis in sensory systems (8). Alex was also a pioneer of oblique photogammetry from the air and an authority on off-shore navigation. In 1935 he mapped northern Laborador and Baffin Land from his own plane, thus charting the way for the northern air route to Europe during World War II (7). He was awarded the Charles Daly Medal of the Geographical Society, and his election to the National Academy of Sciences might equally well have been for his contributions to geographical science as to neurophysiology. He seemed to know every inch of the New England coast and how it had changed during the sixty years he sailed it. He was a marvelous raconteur. In the evening, after a long day's sail, he would relax over a "Chickatawbut" cocktail in the cabin of his ketch *Stormsvala* and draw on an endless repertoire of stories and anecdotes, which he always introduced by saying who had told it to him and when. There must have been dozens of young people, including my wife and me, who thought of him affectionately as Uncle Alex.

During the 1950s the American Heart Association established some fellowships designed to relieve young professors from administrative duties, and I was selected for one of these generous lifetime awards. This allowed me to continue laboratory work on a small scale in company with occasional graduate students and postdoctoral fellows. It also gave me the freedom to explore novel ideas without worrying about short-term success or failure. My first effort was indeed a failure. I had the idea that several puzzling features of the renal circulation might be explained by plasma skimming in the interlobular arteries and afferent arterioles. This idea turned out to be incorrect, at least in the renal cortex. Although my theory, which was presented as a Bowditch Lecture and elsewhere (22), failed to survive, it nevertheless stimulated lively discussion and experimentation for several years and I do not regret being wrong in this case.

GOATS

Beginning in 1957 I undertook development of techniques for perfusion of the ventricular system in unanesthetized animals. I was stimulated by the work of I. Leusen from Ghent, who had shown that perfusion of the ventricles with acid solutions caused hyperventilation in anesthetized animals even though the arterial blood became alkaline (11). Previous theories of chemical regulation of breathing were concerned solely with the composition of blood, and it seemed to me that Leusen's experiments had some exceptionally important implications for this field. No one had ever perfused the brain ventricular system in unanesthetized animals, and I had no experience with surgery for chronic experiments. My friends in the Neurosurgery Department were very skeptical about the feasibility of chronic cannulation of the cisterna magna. We found, however, that the size and shape of the skull in goats or sheep make possible the implantation of guide tubes through thick occipital bone pointing towards the atlanto-occipital membrane. The cistern could then be punctured at will through the guide tube without causing pain to the unanesthetized animals. At the same time, Feldberg, in England, described techniques for chronic implantation of cannulae in cerebral ventricles (4); so the stage was set for ventriculo-cisternal perfusions in goats. Nevertheless, it took us almost five years to perfect the surgical techniques and to learn how to carry out perfusions of the brain for long periods in healthy, confident animals. Eventually, the project was productive in three disparate fields of physiology, as follows:

1. Exchange of materials between blood and cerebrospinal fluid (CSF), including the first measurements of CSF production and absorption by clearance techniques. The results were first presented in 1961 at the International Congress of Pharmacology in Stockholm and have since been described in reviews and textbooks.
2. Central chemical control of breathing [reviewed in a Harvey Lecture, 1967 (18)].
3. Investigation of sleep-inducing factors released into CSF during sleep deprivation [reviewed for *Scientific American, 1976* (19) and in a Bayliss-Starling Lecture, 1982 (20)].

It would be difficult to initiate a project of this kind at a reputable medical school of the 1980s. My assistant, Jim Nicholl, never went to college, but he was a jack-of-all-trades who could do everything from metal spinning in the machine shop to milking goats. In fact, he became more adept than I at implanting cisternal guide tubes. Together we built a goat shed and fences amid the ventilators on the roof of the Medical School, and we did not even ask permission from Buildings and Grounds. We found farmers (within a radius of 100 miles) who would supply us with their "surplus" goats free of charge. We transported goats and hay in my station wagon. In summer the operated animals (with neckbands to protect their carotid loops and aluminum hats attached to their horns to protect ventricular guide tubes) travelled with us for vacations at our summer home in the Berkshires. Astonishment, laughter, and curiosity were plain to see in the faces of motorists who passed our station wagon with its cargo of capped goats and children. At about this time I began to serve on national committees in Washington and elsewhere, and the airplanes often passed over the Medical School, where I could look down to see my goats capering on the roof.

All of this would be illegal today, and the cost of purchase and care of more than a hundred goats in accredited facilities would be quite out of line with the potential of the project as it was viewed at the time.

SLEEP

It was my habit to spend Saturday mornings browsing amongst the journals of the week. That was how I stumbled on an article by Monnier and Hosli with the intriguing title *Humoral transmission of sleep and wakefulness; hemodialysis of a sleep-inducing humor during stimulation of the thalamic somnogenic area* (16). That article referred me to Pieron's early work on the sleep-inducing properties of CSF drawn from sleep-deprived animals. A copy of Pieron's 1913 monograph *Le problème physiologique du sommeil* was at hand in our departmental library. On several occasions I have been to the library to browse or to look up a specific reference and was led by chance to an unrelated paper that changed the course of my subsequent research. This sort of serendipity is something the computer cannot supply and I am reminded again of the quotation from T. S. Eliot's *Four Quartets* paraphrased in the prelude to this essay. In this case I realized at once that my colony of goats provided the means for collecting large quantities of CSF from sleep-deprived animals, and I resolved to "give it a fling." Jim Nicholl and I rigged up a system to keep the goats awake while Tracy Miller and Cecilie Goodrich implanted ventricular cannulae in cats to enable us to introduce CSF from the goats into the cats. To our surprise, the very first experiments corroborated the findings of Pieron: CSF from a sleep-deprived goat infused into the ventricles of a cat appeared to make the cat torporous for several hours, whereas normal fluid from the same goat had no effect. Of course we had no EEG to judge the nature of the induced sleep, but we all agreed about the behavioral effects in the three cats tested.

Our first publication in this field was in 1967 (24), and it was followed by 15 years of frustrating attempts to isolate and identify "the" sleep factor. Early in the project I was joined by Manfred Karnovsky, a distinguished biochemist who was (and still is) sympathetic to old-fashioned physiology and physiologists. Our starting material changed from CSF derived from sleep-deprived goats (six liters obtained laboriously from 25 goats over a three-year period), to the brains of 15,000 sleep-deprived rabbits, to 5000 liters of normal human urine. We had many collaborators, but much of the credit for the final isolation and identification of an active muramyl tetrapeptide from urine belongs to Dr. James Krueger (10, 12). The natural product and certain synthetic muramyl peptides have now been shown to induce high-amplitude, slow-wave sleep in rats, rabbits, cats, and squirrel monkeys. I am still not convinced, however, that the material isolated from urine is related to the sleep-inducing Factor S we originally found in the CSF of sleep-deprived animals. The latter is surely of importance to normal physiology, but there are reasons to believe that the muramyl peptides are involved in immunological reactions to bacterial infection, including fever as well as sleep. The history of this still controversial subject has been reviewed in several recent publications (1, 9, 10, 19, 20).

INTERNATIONAL

The romance of travel to international meetings has been badly tarnished by the jet. The up-and-coming young physiologist thinks nothing of waltzing off to Europe for three-day meetings several times a year, attaché case and 2 × 2 slides in hand. There is prestige value when the secretary can answer phone calls with "Dr. S. is out of the country until Thursday. Do you want him to call?" The airport at landing is the same as the one at take-off, and Hilton-Budapest is not much different from Hilton–San Diego or New Delhi.

It was not always so. Less than 40 years ago the triennial International Congresses of Physiology were about the *only* international meetings for biochemists, endocrinologists, nutritionists, and pharmacologists, as well as physiologists; and there was no such word as *neuroscience*. Biological scientists, young and old, saved their pennies for the triennial treat. It was unethical to use research funds for travel to meetings, even to local FASEB meetings. It took a week to cross the Atlantic (depending on the weather), and it did not make sense to embark on such a long trip without spending at least a month on the other side. How to compare the excitement of a midnight sailing, the sound of the bugle warning visitors off the ship, the rattle of chains as the gangplank is raised and the hawsers cast away, cutting one loose from home with no way of quick return even in emergency—how to compare this, I say, with the long line of tired, vetted passengers boarding Flight 6339, which might just as well be going to Kansas City as to Paris or Copenhagen?

The scientists I admired most, both in England and America, were internationally minded, and my generation of physiologists inherited a strong tradition of loyalty to *THE* International Congress of Physiology. I attended most of these Congresses from 1938 onwards. In 1964, when Wallace Fenn asked me to be Program Director for the 1968 Congress in Washington, D.C., I jumped at the chance. From then until 1983 I had the joy and privilege of working with Councils of the IUPS together with my counterparts from the USSR, Poland, Hungary, South America, India, Australia, and Japan, as well as those from Western Europe and Scandinavia whom I already knew well. One of my main concerns in this period was to maintain strong representation of "Neuroscience" within IUPS, in hopes that this most exciting part of physiology would not break away into a separate union. It seems to me that physiology in the United States has suffered greatly by the establishment of separate departments or divisions of neuroscience within universities, national societies, and academies. This has not occurred in other countries, where departments and societies have moved with the times to incorporate the enlarged scope of modern neurophysiology in which immunology, peptide chemistry, embryology, etc play such important roles. It may be that the size and success of developments in neuroscience within the United States made fission inevitable, but whatever the reason, it has adversely affected many departments of physiology across the land.

In addition to service with IUPS, I have had many other opportunities to participate in international meetings or projects, some in exotic places. In 1957, after a lecture tour of Swedish universities, I revisited Jotunheimen on skis to replay the adventures of 20 years previously. Björn Folkow from Göteborg accompanied me on this expedition, during which we skied to Storbreen glacier. There my old friend Per Scholander was extracting gas bubbles from the ice to estimate the composition of the atmosphere at the time the glacier ice was laid down. In 1942 Scholander and I almost lost our lives when we were caught for three days in a severe blizzard on the summit of Mt. Mansfield in sub-zero temperatures and high winds. Other memorable expeditions, all in the name of Physiology, were in Kashmir, in the foothills of the Himalayas; in the alps of Southern New Zealand; and in Valdivia, 16 hours by train south of Santiago, Chile.

On two occasions I returned to England for full years; first as Overseas Fellow of Churchill College, Cambridge, and subsequently as Eastman Professor and Supernumerary Fellow of Balliol College, Oxford. These were years for renewal of old friendships, cementing of new ones, and the sharing of both with my wife and children. Three of our children continued for five years or more in England to complete their schooling and university educations. All of them chose art, music, or literature as a life work: not an easy way to earn a living, and I admire their courage and idealism. It seems unlikely that the Arts will ever receive even a small fraction of the support given to Science in the second half of the 20th century.

ACKNOWLEDGMENTS

Parts of this essay will also appear in *Men and Ideas in Membrane Transport,* an American Physiological Society Centennial volume (21). I thank the editors for their permission to use some of the same wording in each of these two publications.

Literature Cited

1. Borbely, A., Tobler, I. 1980. The search for an endogenous 'sleep-substance'. *Trends in Pharmacological Sciences* 1:357–59
2. Eggleton, M. G., Pappenheimer, J. R., Winton, F. R. 1940. The relation between ureter, venous and arterial pressures in the isolated kidney of the dog. *J. Physiol.* 99:135–52
3. Eliot, T. S. 1943. *Four Quartets: IV. Little Gidding.* New York: Harcourt Brace
4. Feldberg, W. 1963. *A Pharmacological Approach to the Brain from its Inner and Outer Surface.* Baltimore: Williams & Wilkins
5. Ferguson, J. K. W., Horvath, S. M., Pappenheimer, J. R. 1938. The transport of carbon dioxide by erythrocytes and plasma in dogfish blood. *Biol. Bull.* 75:381–88
6. Forbes, A., Thacher, C. 1920. Amplification of action currents with the electron tube in recording with the string galvanometer. *Amer. J. Physiol.* 52:409–71
7. Forbes, A. 1953. *Quest for a Northern Air Route.* Cambridge, Mass.: Harvard Univ. Press
8. Hodgkin, A. L. 1979. *Baron Adrian of Cambridge, Biographical Memoirs of the Royal Society.* London: R. Soc.
9. Inoué, S., Borbely, A., eds. 1985. *Endogenous Sleep Substances and Sleep Regulation.* Utrecht, The Netherlands: VNU
10. Krueger, J. M., Walter, J., Levin, C. 1985. Endogenous sleep factors. In *Brain Mechanisms of Sleep,* ed. D. McGinty, pp. 253–75. New York: Raven
11. Leusen, I. R. 1954. Chemosensitivity of the respiratory center. Influence of CO_2 in the cerebral ventricles on respiration. *Amer. J. Physiol.* 176:39–44
12. Martin, S., Karnovsky, M. L., Krueger, J. M., Pappenheimer, J. R., Bieman, K. 1984. Peptidoglycans as promoters of slow wave sleep; structure of the sleep-promoting factor isolated from human urine. *J. Biol. Chem.* 259:12652–58
13. Memorial minute on the life of Alexander Forbes. *Harvard University Gazette LXI, No. 5* 1965.
14. Millikan, G. A. 1939. Muscle hemoglobin. *Physiol. Rev.* 19:503–23
15. Millikan, G. A., Pappenheimer, J. R., Rawson, A. J., Hervey, J. P. 1941. The continuous measurement of arterial saturation in man. *Amer. J. Physiol.* 133:390 (Abstr.)
16. Monnier, A., Hosli, L. 1965. Humoral transmission of sleep and wakefulness: II. Hemodialysis of a sleep-inducing humor during stimulation of the thalamic hypnogenic area. *Pflüg. Arch. Ges. Physiol.* 282:60–75
17. Pappenheimer, J. R. 1941. Vasoconstrictor nerves and oxygen consumption in the isolated perfused hindlimb muscles of the dog. *J. Physiol.* 99:182–200
18. Pappenheimer, J. R. 1967. The ionic composition of cerebral extracellular fluid and its relation to control of breathing. *Harvey Lecture Ser.* 61:71–94. New York: Academic
19. Pappenheimer, J. R. 1976. The sleep factor. *Sci. Amer.* 235:24–29
20. Pappenheimer, J. R. 1983. Induction of sleep by muramyl peptides. (Bayliss-Starling Memorial Lecture) *J. Physiol.* 336:1–11
21. Pappenheimer, J. R. Flow and diffusion through biological membranes. In *Men and Ideas in Membrane Transport,* ed. D. C. Tosteson. *Amer. Physiol. Soc.* 100: In press
22. Pappenheimer, J. R., Kinter, W. B. 1956. Hematocrit ratio of blood within the mammalian kidney and its significance for renal hemodynamics. *Amer. J. Physiol.* 185:377–90
23. Pappenheimer, J. R., Lepie, M. P., Wyman, J. Jr. 1936. The surface tension of aqueous solutions of dipolar ions. *J. Amer. Chem. Soc.* 58:1851–55
24. Pappenheimer, J. R., Miller, T. B., Goodrich, C. A. 1967. Sleep-promoting effects of cerebrospinal fluid from sleep-deprived goats. *Proc. Natl. Acad. Sci. USA* 58:513–17
25. Pappenheimer, J. R., Renkin, E. M., Borrero, L. M. 1951. Filtration, diffusion and molecular sieving through peripheral capillary membranes. *Amer. J. Physiol.* 167:13–46
26. Winton, F. R. 1931. The glomerular pressure in the isolated mammalian kidney. *J. Physiol.* 72:361–75
27. Wyman, J. Jr. 1964. Linked functions and reciprocal effects in proteins *Adv. Protein Chem.* 19:224–86

PHYTOPATHOLOGY

Philip H. Gregory

Ann. Rev. Phytopathol. 1977. 15:1–11

SPORES IN AIR

P. H. Gregory
Rothamsted Experimental Station, Harpenden, Hertfordshire, AL5 2JQ England

INTRODUCTION

My interest in aerobiology arose from discrete sources which in time converged like streams feeding a river. This attempt to identify the sources of an obsession inevitably ignores many other concurrent interests.

Interest in biology seems to me to have been my personal reaction as an asthmatic brought up in a sincere, but in my view mistaken, fundamentalist environment. But by now this is an irrelevant part of the story. Progress toward plant pathology as a specialism was slow. Improbable as it may seem, a Ministry of Agriculture advisory leaflet on potato blight was a factor suggesting a topic both interesting and useful.

At the Imperial College of Science and Technology, London, William Brown habitually set his research students two problems, one on pectic enzymes, and the other a field problem. I neglected the enzymes to concentrate on fusarium bulb rot of narcissus. George H. Pethybridge asked me whether the pathogen could spread among bulbs while out of the ground, suggesting that I should bait the fungus out of hiding by wiping cut bulb pieces over the walls of the bulb store. Not having much success, I tried to find out whether conidia of *Fusarium bulbigenum* could become airborne in the laboratory. A primitive apparatus suggested that air velocities of 4–10 mph were required to carry over the occasional *F. bulbigenum* spore from the surface of a culture, but that *Aspergillus niger* spores were removed readily at 1 mph (8). I knew no philosophy of airborne spores; neither did I know that K. M. Stepanov was doing better experiments at about the same time in Leningrad (24).

WINNIPEG

In 1931 A. H. R. Buller took me to Winnipeg to work on medical mycology with the dermatologist, Andrew M. Davidson. Winnipeg was an exciting mycological environment. Buller's work on functioning of spore liberation mechanisms, the organization of the agaric hymenium, and on hyphal fusions and cytoplasmic flow in the mycelium, dominated the University Botany Department. G. R. Bisby was studying the fungal population of Manitoba at the Agricultural College. At the

Dominion Rust Research Laboratory J. H. Craigie, W. F. Hanna, M. Newton, A. M. Brown, and others were working in parallel with E. C. Stakman and his colleagues, on genetics of fungi and on conditions for migration of *Puccinia graminis* urediniospores in winds across the 49th parallel. These were stored as impressions, but did not then form part of my philosophy. But Buller's enthusiasm did teach me, contrary to the anti-Darwinian views of the period (and long before cybernetics spread acceptable concepts of feedback), that in biology, structure and process need to be interpreted in terms of past and future events in the life cycle of an organism, as well as in terms of current happenings at the molecular level. Both the naturalist's and the biochemist's viewpoints are valid and necessary for understanding.

NARCISSUS FOLIAGE DISEASES

When I returned to England in 1935 and began work at Seale Hayne Agricultural College, Devon, the problem at once arose of spread of a pathogen from unsprayed to adjacent sprayed narcissus plots. Advising on the design of the experiment, T. N. Hoblyn of East Malling Research Station commented that analysis of data from small field plot experiments presupposes that the treatment given to one plot does not affect its neighbors—a condition substantially fulfilled in fertilizer trials but obviously not in my experiments with *Ramularia vallisumbrosae* spreading from leaf to leaf through air. Did this matter? There was no time to investigate the rules of the game. Hitler's war was threatening, and as it proved I was lucky to have three years to finish the experiment in the Isles of Scilly with the constant help of Gordon W. Gibson. But I expressed my doubts (9, 12): too diffidently as it proved (such was the prevailing enthusiasm for small plots) and made no impact until the point was taken up by J. E. Vanderplank (27), though Joyce and Roberts, working with cotton pests in the Sudan, had reached similar conclusions drawing attention to "interference" between experimental plots (23). Few workers on foliage diseases have explicitly investigated the bias introduced into their experiments by unwanted movement of inoculum or spray between small adjacent plots, until the current work by James and co-workers at Ottawa (19a) and by Bainbridge & Jenkyn (1) at Rothamsted.

POTATO VIRUS DISEASES

World War II brought the possibility that English farmers might be cut off from their traditional supply of Scotch seed potatoes which they rely on as substantially free from leaf roll and Y-virus, diseases that spread rapidly through stocks in the eastern and southern table-ware growing areas in England. We might have had to rely on locally grown seed. G. Samuel and F. C. Bawden planned research at Rothamsted Experimental Station aimed at improving the health of English-grown seed potatoes. J. P. Doncaster was seconded from the Natural History Museum at South Kensington as entomologist, and I was recruited as pathologist to study the spread of aphid-borne viruses in farmers' potato fields in eastern England. To identify promising seed stocks we studied such factors as aphid populations, early lifting, roguing out of infectors, distance of isolation from virus sources, and patterns of spread among populations of plants in the field. Geoffrey Samuel asked us to find the effective zone of influence of a single leaf roll or virus Y-infected plant. What were the rules? Walter Buddin sent us samples of seed tubers taken from sound Scotch stocks that for one reason or another had been planted alongside badly diseased local stocks. When we grew these tubers in the following year, some samples as would be expected showed decreasing incidence of infected plants with increasing distance from the source.

PLANT DISEASE GRADIENTS

In 1941, while I was working in a potato field near Holbeach Marsh, the significance of plant disease gradients struck me suddenly like a religious conversion. Again, as often happens when planting, the farmer had run out of Scotch seed and had filled in one side of the field with badly diseased seed tubers of local origin. Flying aphids had done their work and unwittingly laid out before my eyes a demonstration of the decrease of primary virus infection with increasing distance from an infective source (4).

The mechanisms underlying disease gradients in the field now appeared as a major but almost unstudied phenomenon in plant pathology, applicable not only to insect-borne viruses, but also to airborne fungal spores, and possibly also to respiratory allergens. I already knew that asthma could be associated with locality and season. Many past problems snapped into place. The subject clearly had wide theoretical implications and many practical applications. An unknown territory was crying out for exploration. This was the beginning of a concern: an unsure hunch. The difficulty was how to proceed. The problem then, as now, was how to search the literature for a nonproblem. There is still no real substitute for library browsing and talking to colleagues.

At the Mycological Institute, Kew, E. W. Mason showed me K. M. Stepanov's pioneer paper (24) on dissemination of plant diseases in air currents, evidently a major source that I needed to consult, but how to pay for a translation of 56 pages of Russian? Serving in spare time as an Air Raid Warden gave a convenient opportunity. One duty in those years was to stay awake for half of every third night, waiting to be called out on patrol when alerted for an air raid. Often this resulted in several quiet hours too good to waste. Drastic action was needed: I sold a family gold watch to raise £10 for a Russian language course on discs, and eventually translated Stepanov's paper, finding there experiments on spore liberation and patterns of spread, and much original thought. There was also the start of a bibliography.

At a British Mycological Society meeting in 1941 Sidney Dickinson provided another clue by citing the work of the Austrian meteorologist, Wilhelm Schmidt. This trail led, via D. Brunt's book on *Physical and Dynamical Meteorology* (2), to O. G. Sutton's more acceptable treatment of eddy diffusion in the atmosphere (25) with its explicit use of a diffusion constant that increased with time, unlike Schmidt's k.

Our virus studies in potato fields were producing examples of gradients of leaf roll and virus Y infections conveyed by insects. But insect-borne viruses are complicated by behavior of the vector as well as of the air. It now seemed that mechanisms underlying patterns of dispersion of airborne fungal spores might be simpler to unravel. So with Fred Bawden's encouragement I wrote a review of the literature of plant disease gradients, interlaced with interpretations (9). I believe that one has a duty to pursue one's "concern," but without allowing one's set task to suffer. I was fortunate to have a leader who encouraged this view.

In interpreting published data on airborne spores I was also fortunate in help from my wife who, as a pure mathematician, was sometimes scandalized by the assumptions necessary! The novelty here was to ask: What is the relation between the concentration of a spore cloud passing over the ground surface and the number deposited on the ground by all processes? W. Schmidt had suggested using terminal velocity of the spores to calculate distance of travel in air. At the time, we had no way of knowing which of many possible deposition factors were important. We tried to obtain an empirical answer by using data from Stepanov's experiments. Our deposition parameter, p, was designed to take account of all factors (including terminal velocity) that rob a spore cloud as it passes over the ground (except in rain). Therefore, instead of equating p with terminal velocity we estimated deposition from Stepanov's experiments with *Tilletia*. But, as pointed out later by A. C. Chamberlain, I was mistaken in supposing that p was a length, i.e. the thickness of a slice of the spore cloud (3). Chamberlain pointed out that it is a velocity which he termed v_g. It happens that our estimates were approximately equal to Chamberlain's experimental values under the wind speeds prevailing during Stepanov's experiments (approximately 1 m sec^{-1}). Our equation, as modified by Chamberlain, has been useful in estimating radioactive pollution (26). Nevertheless our treatment has obvious defects. Sutton's equations apply to stable overcast conditions, and it is uncertain how far they apply in strongly convective weather.

Some have drawn incorrect conclusions from our equations. The calculated (and observed) steepness of the gradient close to a point source has been thought incompatible with the phenomenon of wind transport of inoculum over long distances. Again, in England it has been claimed that aphid vectors coming from overwintered sugar beet crops grown for seed are of small significance in establishing sugar beet yellows infection in the newly sown crop grown for sugar. This is based on the observation that surveys show little difference between the incidence of yellows in fields less than a mile from overwintered seed crops, compared with those at 5 and 20 miles. The conclusion is fallacious, however, because the steep part of the gradient is in the first 100 yards, and crops at 1 to 20 miles distant would all be on the flat tail of the disease distribution gradients derived from many seed crops (i.e. background infection). Further, the gradient on the ground near the source gives no information whatever on the entirely different question of what proportion of the total inoculum liberated is deposited within the area where the gradient is measured. Is it 5% or 95%? We can seldom measure the "escape fraction." The sum of these escape fractions merges to form "background infection," which is of dominating importance, except near a source.

Another misconception arising from the nature of the infection gradient is to assume the presence of two different dispersal processes, "continuous" and "discontinuous," merely because near the source every plant (or leaf, etc) may appear infected, whereas at a greater distance there are gaps. There may in fact be two mechanisms, but the observation does not prove it—gappiness at greater distances is precisely what would be expected in the tail of a distribution (cf spore, sporadic).

PENICILLIN INTERLUDE

Curiosity as to the role of penicillin in fungal biology led me to a year's secondment from Rothamsted to an Imperial Chemical Industries laboratory at Trafford Park, Manchester. This curiosity was satisfied when Barbara Whinfield, one of C. W. Wardlaw's students obtained penicillin in 11–12 hr from the terminal region of spore germ tubes when only 30 μm long (28). I still wonder that a process has not been tried for removing spores from a *Penicillium* culture continuously in dry air and germinating them free from metabolites of the parent culture. As a general principle, dry airborne spores (xerospores) should be handled dry until set to germinate, if conditions are to be realistic.

As a by-product of this interlude: one fortunate evening in Manchester Central Reference Library I browsed into the now classic paper on the Cascade Impactor in which K. R. May laid the basis of modern microbial air sampling (20).

THE AIR SPORA

Back at Rothamsted after Manchester, potato virus work continued for a year or two, but fascinating as the epidemiological work on potato viruses in the field had proved, I was probably too firmly set in mycological interests to transfer permanently, and after World War II, I started an experimental program to study airborne spores, developing ideas started in the literature review of 1945. What factors controlled spore deposition? Especially how to trap airborne spores in order to find out what was in the air that bathed the aerial surfaces of crop plants, and were inhaled into the respiratory tract.

Experimental liberation and trapping of spores in open air, repeating Stepanov's tests, were done first with help from J. W. Blencowe, then with O. J. Stedman, later with J. M. Hirst while he was on vacation from Reading University.

Results from spore trapping in outdoor air with cylinders, slides, or petri dishes led to a growing dissatisfaction with existing methods. To explore these difficulties P. D. Sheppard at Imperial College suggested making a small wind tunnel. This was based on designs supplied by the Microbiological Research Establishment, Porton, and was made at Rothamsted workshops at a cost of about £175. Designed for flexibility, it is still in use after a quarter of a century. With O. J. Stedman, a long series of wind tunnel experiments measured the trapping efficiency of sticky surfaces having various shapes and orientation, using *Lycopodium, Ustilago,* and *Lycoperdon* spores at low and moderate wind speeds, both streamlined and turbulent. These showed how unsatisfactory were data obtained from the various traps in current use, including the gravity slides and sticky cylinders we were using for routine study of spore content of outdoor air. Clearly we needed volumetric sampling.

J. M. Hirst and F. T. Last came in 1950 to work in my laboratory on epidemiology of potato late blight and cereal mildews respectively. Jim Hirst's first task was to make a continuously recording sampler for airborne spores, aimed specifically at *Phytophthora infestans* sporangia, based on stage 2 of May's Cascade Impactor. His sensible and rugged instrument is worthy of his wartime service in the Royal Navy (19). My excitement at seeing for the first time a sample of the ambient air spora laid out on a slide for microscopic examination, fluctuating hour by hour, day and night, was intense. As expected, we found *Cladosporium, Alternaria,* and pollens in abundance. But there was much beside, as well as many puzzles that are still slowly being unraveled. Some, including our nicknamed "three-bar ascospores," remain unidentified still. My own part in these puzzles was helped by contacts with pure mycology through the British Mycological Society. I am convinced that one gains by exploring the margins of one's specialty. Sampling with a hand-operated portable volumetric spore trap over grass-covered downland at Juniper Top, Dorking in 1953, I was puzzled by a catch of small brown spores numbering over 1000 per cubic meter of air. After unsuccessfully consulting various experts who could not recognize them out of context, examination under oil immersion lens convinced me that they were the colored basidiospores of an agaric. We then recognized similar concentrations of basidiospores on routine Hirst trap traces, and when representative slides were sent to A. A. Pearson he was able to identify spores of some Coprini to species level. We also found vast numbers of minute hyaline spores during night hours. Remembering Buller's work on *Sporobolomyces,* I adapted the portable hand-operated sampler for drawing air direct through filter paper, and isolated the typical pink colonies of *Sporobolomyces* in abundance from night air. F. T. Last showed that these came especially from leaves of cereal crops, and he initiated the study of the phyllosphere (phylloplane).

The Hirst trap revolutionized research on many airborne plant pathogens, and on outdoor respiratory allergens as a bonus. It established for the first time the important fraction of basidiospores (including here the Sporobolomycetes) in the outdoor air spora. Yet, despite many hypotheses, the mechanism of basidiospore liberation, surely the most efficient of all fungal liberation mechanisms, still eludes us.

SOOTY BARK DISEASE OF SYCAMORE (ACER PSEUDOPLATANUS)

As a mycological plant pathologist, I have always held that one should widen knowledge of fungal biology in the field. Any fungi—pathogens, allergens, or the harmless edible ones—should be studied in their environment, amidst their competitors. What better method than to go out on fungus forays of the British Mycological Society or a local natural history society?

Out with the Essex Field Club in 1948, Sidney Waller, a somewhat sardonic retired engineer and keen amateur microscopist, introduced me to a strange fungus that he had found apparently killing *Acer pseudoplatanus* trees in Wanstead Park in east London, England. My first interest in this disease was to obtain an abundant sources of small, smooth, ovoid spores for dispersal experiments, but the fungus was fascinating, and Waller and I worked on it together until his death in 1954. F. Joan Moore suggested that the fungus was *Coniosporium corticale,* which John Dearness had discovered on maple logs in London, Ontario, in 1889. In 1950 John Dearness, then aged 99 years, wrote me an account of his original finding and sent specimens for comparison, thus establishing the identity. The fungus had been placed in *Coniosporium* by Ellis and Everhart, but was clearly an undescribed genus which we named *Cryptostroma* (18).

Earlier the fungus had been incriminated in human respiratory disease in Wisconsin and Michigan. The sooty bark outbreak was an unplanned but irresistible episode, which some years later regained interest when Emanuel, Wenzel & Lawton found that the respiratory disease in man was not typical asthma but resembled farmer's lung (5). A chance encounter in Epping Forest had led back to medical mycology.

AEROBIOLOGY

On returning to Imperial College as Professor, my first research student was T. Sreeramulu. In joint experiments with T. Longhurst we three used artificially generated spore clouds diffusing in the open air, and measured both spore concentration and deposition to ground simultaneously for the first time (17). Using specially designed miniature impactor units, 20 of which were connected to one suction pump, we obtained values for Chamberlain's v_g at distances up to 10 meters from the point source. Our data favored Sutton's eddy diffusion equations, as against Schmidt's, thus consolidating gains from the literature review of 1945.

Others in the Imperial College group studied the ambient air spora. Elizabeth Hamilton (Mrs. K. T. A. Goodhew) compared seasonal and weather changes in London (Kensington) with a rural area (Rothamsted). She later went to the Allergy Laboratory at St. Mary's Hospital, Paddington, initiating a long period of continuous sampling with the Hirst trap. T. Sreeramulu looked at mildew and smut concentrations over barley at the College Field Station, Silwood Park. Sreeramulu returned to India, eventually becoming a Professor at Andhra University, where he energetically developed an aerobiological group studying important crop pathogens. Since his early death the tradition has been continued by two of his students, A. Ramalingam, at Mysore University, and C. Subba Reddi at Waltair University.

We also started research on spore dispersal by rain splash, starting where R. F. Faulwetter (6) left off in 1917. Splash is one of the chief ways by which slime-spored fungi are launched. Gregory, Guthrie & Bunce (Mrs. J. Lacey) studied splash events in the still air of the laboratory (13). The study was later taken out into the windy field by J. M. Hirst, M. V. Carter, and O. J. Stedman at Rothamsted.

A long stay in hospital in 1956 gave an opportunity to sketch out the first few chapters of *The Microbiology of the Atmosphere* (10). E. C. Large advised on planning, and the first edition appeared in 1961 as an attempt to understand and expound microbial events in the atmosphere considered as a natural phenomenon, outlining the philosophy of airborne spores, whose lack I had noted mentally in 1931.

After returning to Rothamsted in 1958, D. R. Henden and D. H. Lapwood helped with field experiments on potato late blight gradients from an artificially inoculated line source (11). Because terminal velocity of spore fall in still air is a key factor in spore behavior, David Henden and I worked with a stirred sedimentation chamber for experimental measurements of terminal velocity using two methods: decrease of concentration, and partition between the several stages of the Cascade Impactor (14).

FARMER'S LUNG

J. T. Duncan of the London School of Hygiene and Tropical Medicine was chairman of a medical mycology committee formed by the Medical Research Council in 1943 (in 1957 this was transformed into the British Society for Mycopathology). In early meetings we discussed the baffling problem of farmer's lung, described from the English Lake District by J. M. Campbell, and eagerly championed by J. C. Fuller of Exeter. I determined to take up this problem. It appeared to be a legitimate agricultural impingement on medicine, and a problem that we had the expertise to investigate at Rothamsted. This was a major factor in my wish to return to Rothamsted in 1958 when F. C. Bawden became director. The Agricultural Research Council gave active encouragement.

The strategy adopted for the farmer's lung project, which was assisted by Maureen E. Lacey, argued that as the disease was known to result from inhalation of airborne dust from moldy hay, we would ignore the total microbial population and concentrate on dry dust blown out of batches of suspect hay in the wind tunnel. Batches of suspect hay had evidently heated spontaneously at some stage, so we concentrated on thermotolerant molds and thermophilic actinomycetes. These proved to be correct decisions: traditional examination of washings from hay would have flooded our cultures with bacteria. Cascade Impactor samples of dust from suspect hay, shaken dry in the wind tunnel, showed very few particles derived from the herbage, but immense quantities of mold and actinomycete spores. P. K. C. Austwick gave advice on culturing and suggested using the newly developed Andersen Sampler to put spores direct and dry on the surface of agar media. This provided the key to successful enumeration and isolation of the actinomycetes (15).

We sought and obtained cooperation from workers at the Cardio-Thoracic Institute, Brompton, London, to test our isolates against patients' serum using gel diffusion techniques. F. A. Skinner improved our methods of culturing the thermophilic actinomycetes on hay. It was a memorable occasion when Jack Pepys came to see me while I was again a patient for a short spell at King's College Hospital to tell me that culture extracts from one of our actinomycetes was giving strong positive reactions with sera from farmer's lung patients (22).

The study of inhaled organic dusts affecting the peripheral region of the lung has progressed far under the Brompton workers, and at Rothamsted field studies continue with John Lacey on bagassosis, suberosis, and byssinosis.

PITHOMYCES CHARTARUM

A byway led from the farmer's lung research. At the start of the project, G. C. Ainsworth sent us a reference slide of the characteristic spores of the fungus implicated in facial eczema of sheep in New Zealand (*Pithomyces chartarum*, then unrecorded in England) with the suggestion that we should look for them in hay dust. We never found them in hay. But one day Maureen E. Lacey observed a few spores in a routine Hirst trap sample from Silwood Park Field Station. Nothing very remarkable perhaps in a mere new British record: but the detective sequence, following up the concentration gradient in air, which eventually led us to find the fungus flourishing in the field, was sufficiently unusual to be worth reporting in a paper composed in a "whodunit" style (16) as more illuminating on scientific method than the thesis style which rearranges the real path of discovery to conform to a respectable standard pattern (21).

BARLEY ASTHMA

Retirement in 1967 was followed by one more incursion into the border between medical mycology and plant pathology. Two allergist friends, Kate Maunsell and R. S. Bruce Pearson, alas no longer with us, had asked me to examine the airborne spore content of the ambient environment of a group of patients whose asthma occurred in the weeks before harvest in southern England. The year 1972 proved favorable for the project. Records of symptoms were kept to correlate with continuous sampling from a Burkard model trap in the garden of a cooperative patient near Blandford, Dorset, from July to September. When I assembled records at the end of the season, without earlier comparison, it was clear that asthmatic symptoms were associated with exceptionally dense, mainly nocturnal concentrations of characteristic two-celled ascospores of up to a million per cubic metre of outdoor air. The identity of these ascospores was a puzzle for a time, but eventually at Rothamsted and elsewhere I found that they were liberated from perithecia on leaves of barley plants. After spending a couple of nights sampling air over barley with an Andersen sampler, isolates of the fungus both from barley plants and from night air were sent to the Commonwealth Mycological Institute and identified as *Didymella exitialis* by Colin Booth. Strangely, this appears to be the first record on barley in Britain of this fungus. It had been recorded only once before—on oats in Scotland. *Didymella exitialis* obviously merits further study, both as a pathogen and a potential allergen (7).

PHYTOPHTHORA DISEASE OF COCOA

In 1968, as consultant to the Cocoa, Chocolate, and Confectionery Alliance, London, my first task was to visit tropical areas where *Theobroma cacao* is cultivated, report on current research, and recommend a program for future research on cocoa black pod disease caused by *Phytophthora palmivora*. Travel to West Africa, South and Central America, and the Caribbean has brought new friends and new interests.

How else could I have visited Gambari, Ikiliwindi, Itabuna, and San Vicente de Chucurí? Eventually an epidemiological project was started with a team at the Cocoa Research Institute of Nigeria. *Phytophthora* is a large genus of crop pathogens. *Phytophthora infestans* and *P. palmivora* are two of the very few species that have successfully adapted to a career above ground. Yet any resemblance between the two turns out to be misleading. For instance, whereas the Hirst trap was successfully designed to catch sporangia of potato late fungus, despite much effort no one has yet claimed to have trapped a sporangium of the cocoa black pod fungus, which is spread largely by rain splash and insect vectors. New methods have to be devised to study and control the cocoa *Phytophthora*.

RETROSPECT

The path of aerobiological research depicted here is a fabric woven out of set tasks and duties, personal encounters, and chance problems, developed in an environment of freedom which is now disappearing. It has involved much groping for clues.

One approach to a novel problem is to let the imagination play with all of the relevant facts, always with the proviso that one or more of the supposedly well-established facts may be wrong. It is well known that an inconsistent fact may provide the clue sought after. It is equally true that an otherwise well-fitting hypothesis need not necessarily be rejected because of one glaringly incompatible fact—have another look. When a hypothesis has to be tested, as first choice take things at face value. Don't be too subtle at first, because subtlety can inhibit experimentation. One can be more subtle next time round.

Neat professional and academic demarcations, however convenient to administrators, must be brushed aside when they impede research.

Literature Cited

1. Bainbridge, A., Jenkyn, J. F. 1976. Mildew reinfection in adjacent and separated plots of sprayed barley. *Ann. Appl. Biol.* 82:477–84
2. Brunt, D. 1934. *Physical and Dynamical Meteorology.* Cambridge: Cambridge Univ. Press. 411 pp.
3. Chamberlain, A. C. 1956. Aspects of travel and deposition of aerosol and vapour clouds. *Rep. At. Energy Res. Establ. Health Phys. Rept. 1261.1953.* 35 pp.
4. Doncaster, J. P., Gregory, P. H. 1948. The spread of virus diseases in the potato crop. *Agric. Res. Counc. Rept. Ser.,* No. 7, HMSO, London. 189 pp.
5. Emanuel, D. A., Wenzel, F. J., Lawton, B. R. 1966. Pneumonitis due to *Cryptostroma corticale* (Maple-bark disease). *N. Engl. J. Med.* 274:1413–18
6. Faulwetter, R. C. 1917. Wind-blown rain, a factor in disease dissemination. *J. Agric. Res.* 10:639–48

7. Frankland, A. W., Gregory, P. H. 1973. Allergenic and agricultural implications of airborne ascospore concentrations from a fungus, *Didymella exitialis. Nature* 245:336–37
8. Gregory, P. H. 1932. The *Fusarium* bulb rot of Narcissus. *Ann. Appl. Biol.* 19:475–514
9. Gregory, P. H. 1945. The dispersion of air-borne spores. *Trans. Br. Mycol. Soc.* 28:26–72
10. Gregory, P. H. 1961. *The Microbiology of the Atmosphere.* London: Leonard Hill. New York: Interscience. 251 pp. 2nd ed., 1973. 377 pp.
11. Gregory, P. H. 1968. Interpreting plant disease gradients. *Ann. Rev. Phytopathol.* 6:189–202
12. Gregory, P. H., Gibson, G. W. 1946. The control of Narcissus leaf diseases. III. *Sclerotinia polyblastis* Greg. on *Narcissus tazetta* var. 'Soleil d'Or'. *Ann. Appl. Biol.* 33:40–45

13. Gregory, P. H., Guthrie, E. J., Bunce, M. E. 1959. Experiments on splash dispersal of fungus spores. *J. Gen. Microbiol.* 20:328–54

14. Gregory, P. H., Henden, D. R. 1976. Terminal velocity of basidiospores of the giant puffball (*Lycoperdon giganteum*). *Trans. Br. Mycol. Soc.* 67:399–407

15. Gregory, P. H., Lacey, M. E. 1963. Mycological examination of dust from mouldy hay associated with farmer's lung disease. *J. Gen. Microbiol.* 30:75–88

16. Gregory, P. H., Lacey, M. E. 1964. The discovery of *Pithomyces chartarum* in Britain. *Trans. Br. Mycol. Soc.* 47:25–30

17. Gregory, P. H., Longhurst, T. J., Sreeramulu, T. 1961. Dispersion and deposition of airborne *Lycopodium* and *Ganoderma* spores. *Ann. Appl. Biol.* 49:645–58

18. Gregory, P. H., Waller, S. 1951. *Cryptostroma corticale* and sooty bark disease of sycamore (*Acer pseudoplatanus*). *Trans. Br. Mycol. Soc.* 34:579–97

19. Hirst, J. M. 1952. An automatic volumetric spore trap. *Ann. Appl. Biol.* 39:257–65

19a. James, W. C., Shih, C. S., Hodgson, W. A., Callbeck, L. C. 1976. Representational errors due to interplot interference in field experiments with late potato blight. *Phytopathology* 66:695–700

20. May, K. R. 1945. The Cascade Impactor: an instrument for sampling coarse aerosols. *J. Sci. Instrum.* 21:187–95

21. Medawar, P. B. 1967. *The Art of the Soluble.* London: Methuen. 160 pp.

22. Pepys, J., Jenkins, P. A., Festenstein, G. N., Gregory, P. H., Lacey, M. E., Skinner, F. A. 1963. Farmer's lung: thermophilic actinomycetes as a source of "Farmer's lung hay" antigen. *Lancet* 1963 (ii):607–11

23. Roberts, P. 1960. The analysis of an experiment on interplot effects. *Inc. Stat.* 10:59–65

24. Stepanov, K. M. 1935. Dissemination of infective diseases of plants by air currents. *Bull. Plant. Prot. Leningrad, II Ser. Phytopathol.* 8:1–66 (in Russian)

25. Sutton, O. G. 1932. A theory of eddy diffusion in the atmosphere. *Proc. R. Soc. Lond. Ser. A* 135:143–65

26. US At. Energy Comm. 1968. *Meteorology and Atomic Energy 1968,* ed. D. H. Slade, TID-24190. Springfield, Va.: US Dep. Commer.

27. van der Plank, J. E. 1961. Errors due to spore dispersion in field experiments with epidemic diseases. *Rept. 6th Commonw. Mycol. Conf. London, 1960,* pp. 79–83

28. Whinfield, B. 1947. Studies in the physiology and morphology of *Penicillium notatum.* 1. Production of penicillin by germinating conidia. *Ann. Bot. NS* 11:35–39

Ann. Rev. Phytopathol. 1978. 16:1–18
Copyright © 1978 by Annual Reviews Inc. All rights reserved

PHYTOPATHOLOGY IN A DEVELOPING COUNTRY

A. A. Bitancourt

Instituto Biológico de São Paulo, Avenida Rodrigues Alves, 1252,
Caixa Postal 7119, São Paulo, Brazil

Inasmuch as the authors of the prefatory chapters of previous volumes have been from the leading nations of the world, I have thought that it might be interesting to write about plant pathology in a developing country. I am grateful to the editorial committee for this opportunity. In accord with current usage, the term *developing country* is used here instead of *under-developed country,* which has a pejorative connotation.

The northern area of Brazil is poorer and less developed than the southern portion, which presents both a better climate and greater opportunities for employment. As a result, in the past few decades there has been a considerable immigration from the north, not only of the working classes, but also of students and young professionals. The Biological Institute of São Paulo now has a number of such professionals who having come to São Paulo with a scholarship from the northern states have chosen to stay.

I was born in 1899, at Manaus, a city in the heart of the Amazon region, on the banks of the big river. When I was one month old my family moved to Paris where we stayed until 1919. All of my studies were done in Paris; I graduated as *ingénieur agronome* from the Institut National Agronomique de France. I also took two semesters each of general botany and biochemistry at the University of Paris (Sorbonne). It was only in 1947 that, on a sabbatical leave, I returned to France to complete the six semesters required for the master's degree, with two semesters of general physiology. I obtained my doctorate, also at the Sorbonne in 1954, with a thesis on auxin, the plant hormone that plays an important role in the etiology of plant cancer.

0066-4286/78/0901-0001$01.00

An *agronome* or *agronomo* is not an agronomist in the sense of soil specialist. In many Latin countries, like France for instance, the college of agriculture is not part of a university, but rather a department of the Ministry of Agriculture. Besides soil science, the curriculum covers the basic sciences applied to agriculture and animal husbandry, rural engineering, political and rural economics, and legislation. With such a broad scope there was and still is a lot of cramming during the four or five years of studies.

My training in plant pathology at the Institut Agronomique had been quite perfunctory, consisting in some twenty lectures in which little was taught about bacterial diseases and practically nothing about virus diseases which, in the minds of most pathologists, were still in a limbo. The main subject of these lectures was applied mycology, which emphasized disease-causing fungi and the symptoms they produce but which paid little attention to pathology. Laboratory exercises were reduced to drawing mounts of pathogenic fungi observed under the microscope.

In 1920, after having returned to Brazil I was appointed assistant of the Plant Pathology Section of the recently founded Biological Institute for Plant Protection of the Ministry of Agriculture. I eagerly looked forward to becoming a full-fledged plant pathologist. Unfortunately, during the first five years of my career, I was disappointed in not being able to come close to my goal. The head of the section was a mycologist who had been assistant to two former heads of the laboratory of phytopathology of the National Museum at Rio de Janeiro, Arsène Puttemans, a Belgian, and André Maublanc, a Frenchman. Both, apparently, had done mycological work only, but they had assembled a good herbarium containing specimens of many of the plant diseases known in Brazil at that time, mostly leaf spots; they had identified the fungi they found, as well as the few they received for identification from farmers, growers, and *agronomos* of the Ministry of Agriculture.

My first acquaintance with research on plant pathology came from reading articles in the *Journal of Agricultural Research*. In particular, the article by F. W. Rands on the diseases of pecan in the first volume (1913) identified the main problems in the investigation of a typical fungous disease, for instance, nursery blight caused by *Phyllosticta caryae*. The heading of each section indicated the nature of these problems: "History and Distribution; Symptoms of the Disease; Mycological and Pathological Studies: Isolation of the Fungus, Inoculations; Cultural Studies: Thermal tests, Cultural Characters; and Morphology and Taxonomy." To this I would have added "Control."

In 1925, the professor of botany, phytopathology, and agricultural microbiology at the College of Agriculture at Piracicaba, State of São Paulo, was commissioned by the Italian government for a two-year project in Libya, then an Italian colony. I was invited to take charge of the post during his absence. It so happened that phytopathology had been transferred to the third of the four years of the curriculum and I did not have to teach plant pathology at all. The college at Piracicaba was part of the Department of Agriculture of the State of São Paulo. In 1933 it was transferred to the State University, which had been recently organized by a committee appointed by the governor of the State of São Paulo and of which I was a member.

During the two years I taught botany and agricultural microbiology at Piracicaba I had no opportunity to do anything other than prepare my lectures and organize the students' laboratory which, except for a score of good microscopes and some glassware, was poorly equipped.

When the professor of the department returned from Libya I went back to Rio where I accepted the directorship of the Agrostology Experiment Station of the Ministry of Agriculture. During the three years I stayed in Rio, I had only a few occasions to become involved in research on the diseases of forage crops; most of the time I organized agronomic experiments which were performed by the two assistants (a botanist and an *agronomo*) of the station. I used practically all my spare time to improve my knowledge of genetics and statistical methods, in particular the use of the analysis of variance in experimental biology. I was to use this method in much of my work thereafter. I also did some teaching, which was more helpful to me than to my students.

However, it was not my intention to abandon my chosen career in plant pathology. An opportunity to begin work as a plant pathologist was provided when I accepted an invitation to fill the position of head of the Section of Phytopathology of the Division of Plant Biology at the Biological Institute of the State of São Paulo. The Institute was created as a consequence of the introduction to the plantations of São Paulo of the coffee berry borer (*Hypothenemus hampei*), a destructive pest of coffee in Africa. When the alarm was sounded, the government of the State of São Paulo appointed a commission of two entomologists, A. Neiva and A. Costa Lima, and an *agronomo*, A. Queiroz Telles, to recommend measures necessary to exterminate the pest. Forty years have elapsed and the pest has not yet been exterminated. For several years it caused considerable damage in the coffee plantations, but is now under satisfactory control. Two years after the commission was installed and after the publication of books, leaflets, and a colored poster that was distributed all over the state, the commission adjourned after recommending the creation of an institute for the protection of agriculture through research, extension, regulatory work, and control of diseases and pests of plants and domestic animals. The setup was not novel of course; the Rockefeller Institute at Princeton, New Jersey, had for several years associated veterinary sciences and phytopathology. The main difference, in our case, was the emphasis on control, which was not, I believe, a matter of concern at Princeton.

The Biological Institute of São Paulo was founded in 1928, with A. Neiva as director. In the beginning it had two divisions, one for animal biology and one for plant biology. The latter originally had five sections: phytopathology (later divided into general phytopathology and applied phytopathology, with seats respectively in São Paulo and at Campinas, in the Experimental Farm of the Institute), plant physiology, entomology, chemistry, and plant protection. The Section of Phytopathology of which I was then head was quartered in three small rooms of a rented building in the city of São Paulo and was staffed by R. Drummond and J. G. Carneiro, graduates of the Colleges of Agriculture of São Paulo and Recife, Pernambuco, respectively. There was also a clerk and a laboratory assistant. There were no files of any sort, the correspondence being filed in the office of the director of the Division of Plant Biology. There was, however, a remarkable, albeit small, collection of green specimens of diseased plants, preserved in cylindrical flasks by Drummond. The flasks contained a solution of sulfur dioxide in which the specimens were immersed after the chlorophyll had been fixed for several days in a dilute solution of copper sulfate. Most of these flasks are still kept in our museum and still show the fine green color they exhibited more than 40 years ago. Many more have since been added to the museum.

With a fair knowledge of mycology but only a bookish understanding of phytopathology, I did not consider myself a plant pathologist until I entered the Instituto Biologico of São Paulo. Now I had the unique responsibility of forming a research unit with the potential of becoming a school of pathology, eventually transcending the limits of the state of São Paulo and serving as a model to the rest of the country.

The problems posed by this ambitious program were many indeed. In the first place we could not remain in cramped quarters in a rented building. We needed larger rooms and a piece of land where experiments could be made with a number of plants and on which, eventually, a greenhouse could be built. All this was granted at my request and after a few weeks we moved into a pleasant building in the suburbs, some 20 miles from the city of São Paulo, and adjacent to the Forest Service in the district of Cantareira. The building had a central hall, four big rooms, and two smaller ones. An upper floor of the same size as the hall downstairs was used to initiate the museum of diseased plants. Two of the rooms were fitted with complete laboratory benches and received the furniture from our old quarters in the city. The smaller rooms served as an office and a darkroom for photography.

In the beginning, the main work of our section was to identify the numerous specimens of diseased plants sent by mail or brought personally by farmers, growers, regional *agronomos,* and quarantine inspectors of our own section of plant protection and of the Extension Department of the State Secretariat of Agriculture. We often visited the farms and experimental stations from which the specimens came and collected samples for a permanent record of our identifications in our herbarium and in the museum. In 1934, 1935, 1937, 1940, and 1941, as sole author or with the collaboration of Drummond and Carneiro, I published in the *Arquivos do Instituto Biologico* and the *FAO International Bulletin of Plant Protection* lists of plant diseases that we identified between 1931 and 1940, many of which had not been recorded before.

The herbarium has been substantially enriched by the inclusion of specimens exchanged with other institutions or purchased from individuals. A collection of 2000 specimens of European fungi was bought from the late Austrian mycologist F. Petrak. Sets of 50 specimens of plant disease fungi were exchanged with the mycological collections of the USDA and other herbariums. Ten sets of 550 specimens of the collection *Myriangiales Selecti Exsiccati* collected around the world and assembled by A. E. Jenkins and A. A. Bitancourt were sent to the foremost mycological herbariums of the world. Five fascicles of the handsome collection of Indian Fungi (*Herb. Cryptog. Ind. Orient. Exsiccati*) were received in exchange for one of those sets.

The museum now occupies a 23 by 37 foot room in the main building of the Institute in São Paulo and contains a large collection of specimens in Drummond's solution; dry specimens of roots, trunks, branches, fruits, and seeds; and 20 large 6 by 13 foot panels showing the diseases of 19 of the most important crops of the state of São Paulo in colored plates, photographs, and special 6-inch hemispherical glass containers fixed on a 8 by 12 inch glass plate. In these containers, specimens in Drummond's solution are much better displayed than they are in cylindrical flasks.

The phytopathological and mycological herbarium and the Phytopathological Museum are invaluable sources of information for the identification of casual agents and their symptoms. Whereas fungous diseases are comparatively easy to identify from herbarium specimens, virus diseases that induce typical symptoms in fresh leaves usually do not exhibit these symptoms after drying. Some time ago, however, I found that green specimens can be stored in herbaria with their distinctive patterns still evident by first keeping them in Drummond's solution and then drying them.

We have always been hindered in our work by the paucity of library facilities resulting from an insufficient annual budget, always too small as regards books and periodicals. We do subscribe to some important journals such as *Nature* and *Science,* and our files of the *Journal of Agricultural Research* and the *Review of Applied Mycology* (now *Review of Plant Pathology*) are complete. But there are a great many specialized periodicals that are lacking in our library. We are exchanging a number of journals with our publication *Arquivos do Instituto Biológico* of which the first volume was issued in 1933. Because it contains only scientific papers, it can hardly be exchanged with the numerous publications devoted to applied sciences that interest our researchers and technicians. In 1935, I proposed that a monthly magazine be published, based on our correspondence with farmers, growers, and extension agents. Our Director, H. da Rocha Lima, suggested the name *O Biológico,* an adjective used substantively which commonly designates our institute. First issued in January 1937, *O Biológico* has been published regularly for 42 years. Aimed at farmers, growers, and laymen, it has often been considered too "scientific" for that purpose. I believe, however, that our periodical is quite suitable reading for extension veterinarians and *agronomos* and also for other educated people. Through these readers are transferred to poor farmers the results of our progress in solving the problems of control of plant diseases.

To fulfill that purpose we have developed a comprehensive and accurate system of files that includes the correspondence with extension agents and farmers; the herbarium specimens on which we base our replies; photographs, photomicrographs, and drawings of those specimens; selected specimens of the museum; and reprints of articles obtained by exchange or request. The card of every item in each of these categories is numbered and filed in chronological order. Two sets of index cards, one for the hosts and one for the causal agents, have columns which list the numbers corresponding to each category.

This number system enables clerks who have no scientific training to take care of the files. Everything that has been assembled and stored in nearly the last half century—say, a leaf spot of coffee caused by *Cercospora coffeicola*—can be retrieved by its numbers in a matter of minutes. The second to the last column of each index card contains the name of the host, *Coffea arabica,* in the agent's index and that of the agent *Cercospora coffeicola* in the host index. The corresponding line of the last column names the locality from which the specimen came.

With adequate laboratory installations and a filing system already in operation, we were in a position to admit new *agronomos* to the Section of Phytopathology. These *agronomos* in due time become plant pathologists who would always be in contact with their colleagues of the extension departments; they thus serve as research pathologists and also as extension pathologists. In order to be competent and efficient, the plant pathologist must have a good knowledge of the disease he is investigating and also knowledge of the host from agricultural and economic points of view. This compels the pathologist to confine himself to one crop or at most to a group of crops ecologically and economically connected. With these requisites in mind I chose for my assistants the diseases of one crop or one group of crops. To Drummond, who had already conducted some research on grape diseases, were assigned the fruit crops of the temperate zone. Carneiro was put in charge of coffee diseases. Before he was transferred to the Forestry Department, he published, in collaboration with Dom Bento Pickel, a part-time assistant, an annotated list of the bacteria and fungi reported to cause diseases of the coffee plant.

In 1933 I was called to head the Division of Plant Biology still located in the old mansion in the city. However, my new functions did not prevent me from continuing my work as head of the Section of Phytopathology at Cantareira.

In that capacity I undertook a study of the diseases of citrus and their control, about which, in 1933, I wrote chapters for the *Manual de Citricultura,* part II. In the same book the chapters on animal pests and their control were written by two entomologists from our Institute. In 1932 oranges were beginning to be exported in increasing quantities to the European market through London and Hamburg. The first shipments were a dismal failure because of deficient packing and poor quality of the fruits. The packing problem was solved by the importation by the Ministry of Agriculture of modern equipment for two packinghouses that were installed in the two most important citrus districts at the time: Limeira, in the State of São Paulo, and Nova Iguaçu, in the State of Rio de Janeiro. The investigation of the causes of the poor quality of the fruits led to my preparation of the book *As Manchas das Laranjas (Blemishes of Oranges)* published in 1934 by our Institute as Number 53 of our miscellaneous series. The book has 135 pages, 57 black-and-white figures, and 6 beautiful color plates representing the five most important orange blemishes: leprosis, melanosis, scab, rust mite injury, and thrips injury. The sixth plate represents details of eight blemishes of lesser importance. The plates were prepared by the late J. F. Toledo, the excellent artist of our Institute, who also prepared a colored poster representing those blemished fruits and a modern 600-pound power sprayer which replaced the primitive 60-pound knapsack sprayer then in current use in orchards. A table containing instructions for the treatment of orchards against the common diseases and pests completed the poster that was hung on the walls of railroad stations, offices of extension agents, and other public buildings. After a few years Brazil became one of the chief exporters of oranges, after California and Spain, and on a par with the Union of South Africa and Israel.

In 1941, four *agronomos,* graduates of the College of Agriculture of Piracicaba, S. Arruda, Victoria Rossetti, A. Andrade, and M. Kramer were admitted as assistants after an official examination. Arruda, who was assigned the diseases of sugarcane, spent a year, in 1941–1942, at the Department of Botany of the University of Louisiana studying the diseases of that crop under C. W. Edgerton, chairman of the department. Back in Brazil he studied the mosaic disease of sugarcane; however, in 1946 he had to organize and lead the campaign to eradicate sugarcane smut (*Ustilago scitaminae*), which had been introduced into the recently developed sugarcane area of São Paulo, Brazil. The disease, which had probably spread from the sugarcane region of Tucuman, in Argentina, was destroying susceptible varieties like POJ 36 and POJ 213 and was already attacking the resistant varieties, notably Co 290 and POJ 2878. Several squads of workers in a few months uprooted and burned all plantings of susceptible varieties. Eradication of the resistant varieties had already begun when we ran short of funds provided by the Sugar and Alcohol National Institute. However, the campaign was far from a complete failure. Thanks to the eradication of the susceptible plantings, it took five years for the disease to spread to the big sugarcane districts around Piracicaba, where the susceptible varieties also had been eradicated in the meantime. Sugarcane smut is no longer a serious problem for the sugarcane plantations of São Paulo and has not yet spread to the important sugarcane districts of northern Brazil.

Victoria Rossetti became my collaborator in the work on citrus diseases. Under my direction she started her studies on phytophthora foot rot, beginning with physiological investigations of *Phytophthora* spp. This work led to the concept that growth of these fungi in solid media is determined by three factors: (*a*) factor L, which can be isolated by simple diffusion from gelidium agar gels and promotes the radial growth of the mycelium in petri dishes; (*b*) thiamin which causes the multiplication of nuclei and the production of short ramifications that elongate only in the presence of (*c*) the hypothetical growth factor R.

Inoculations with *Phytophthora* showed that growth of the fungus in the bark of citrus trees occurs only during the three or four annual flushes of growth of the host in spring and summer. Rossetti also took part in the early experiments with the tristeza disease of citrus, especially in the extensive applications of the starch test at the bud-union of susceptible stock-scion combinations. In 1952, with a fellowship from the Guggenheim Foundation she made further studies of *Phytophthora* at UC-Los Angeles, at UC-Berkeley under Machlis, and at UC-Riverside under G. A. Zentmyer. Rossetti became head of the Section of Phytopathology at the Biological Institute of São Paulo in 1954. In 1962, with a scholarship from the Rockefeller Foundation, she attended the second meeting of the International

Organization of citrus virologists (IOCV) at Gainesville, Florida. She was elected chairwoman for the period of 1962–1969 and chaired the third meeting in São Paulo in 1966.

From 1965 to 1967 Rossetti was commissioned by the FAO as assistant in its Division of Plant Protection in charge of editing the three series of its *International Bulletin of Plant Protection* in English, French, and Spanish.

Back in São Paulo, Rossetti, in 1969, became director of the Division of Phytopathology. As a result of reorganization of the Biological Institute in 1970, two new sections were added: fungicides from the Division of Plant Protection and the Section of Nematology from the Division of Plant Parasitology to make a total of nine sections devoted to plant pathology, with a staff of 45 *agronomos* and graduates of the Faculty of Philosophy of the University of Sao Paulo.

To Andrade I assigned the fungicide tests. He built a settling column after McCallan, and by controlling the elements of variation, we reduced the coefficient of variation between applications to less than 4%. In 1953 he obtained a Guggenheim fellowship, and worked at the Connecticut Experiment Station under Horsfall. Andrade is one of the plant pathologists responsible for the introduction of the dithiocarbamates for the treatment of crops against fungous diseases. Due to the difficulties of testing fungicides in Europe during World War II, the dithiocarbamates were tested first in Brazil in 1940.

Just before World War II and at the time the University of São Paulo was founded, a number of European scientists, most of whom were Jews, came to Brazil, either on their own accord or were engaged by Brazilian scientists to go on missions of the government of São Paulo on the recommendation of the founders of the University. Among them was Karl Silberschmidt, a graduate of the University of Munich who had distinguished himself as a plant physiologist but had also done some work in a few problems of plant virology as a collaborator of H. M. Quanjer, the famed Dutch virologist. Silberschmidt considered himself primarily a plant physiologist, and his appointment in 1933 at our Institute was, at his own request, that of head of a newly created Section of Physiology. One of the rooms of our pavillion at Cantareira was completely equipped according to Silberschmidt's specifications and a laboratory assistant was hired for the usual chores. Kramer, one of the four *agronomos* who passed the official examination, was appointed as his assistant, and Silberschmidt assigned him to investigate the plant growth hormone auxin, a physiological subject. In 1949 as a fellow of the Guggenheim Foundation, Kramer worked at the California Institute of Technology at Pasadena, under F. W. Went. Kramer and Went published a paper on the molecular weight of auxin.

Before the war, potato growers in São Paulo imported certified virus-free tubers from Europe. Tubers of the following first and second generations were usually sold for local consumption because they were too heavily infected with virus diseases to be used as seed. Because of the war, the annual importation of tubers for planting from Europe practically stopped, and growers had to use tubers from the southern states of Brazil, which, although less infected than São Paulo tubers, produced small, unsatisfactory crops.

With the help of a cooperative association of planters from São Paulo, Silberschmidt organized a certification program that provided a sufficient supply of potatoes for planting. In this work he was aided by his staff which had already been augmented with the admission of the late N. Nobrega, a graduate of the College of Agriculture of Piracicaba, and of unspecialized workers. In 1947, Nobrega obtained a scholarship from the British Council and stayed for a while as a trainee in Kenneth Smith's laboratory at Cambridge, England, and then at the Edinburgh Experiment Station in G. Cockerham's laboratory. In that same year I visited Nobrega and Cockerham at Edinburgh and met them again at Copenhagen where we attended the International Congress for Microbiology. After the Congress I accompanied them on an extensive tour of the certified potato fields of Sweden during which Nobrega and I learned a great deal about the routine of certification used at the experiment stations we visited.

In the meantime Silberschmidt was working with *agronomo* A. Orlando who had joined the staff of the Section of Plant Physiology in 1946. With the collaboration of Silberschmidt, Orlando demonstrated that the hitherto unknown insect vector of infectious chlorosis of Malvaceae is the whitefly *Bemisia tabaci.*

Silberschmidt was the initiator of research on plant viruses in Latin America. When he died in 1973, he was known the world over and held in high esteem by his colleagues. He published comparatively few papers as sole author; his numerous collaborators were mostly members of his staff but also colleagues from other institutions in Brazil and abroad.

In 1949 I was invited to fill the position of Director General of the Institute. A few weeks later I published in our periodical *O Biológico* an article entitled "Research, Extension and Teaching" in which I emphasized that as the Institute was a department of the Secretariat of Agriculture of the State of São Paulo, our chief duty was to help farmers and breeders to control the pests and diseases of their crops and cattle. But for us to be able to give valuable advice on those matters, the diseases and pests required research much more advanced than what was taught in our colleges of agriculture and faculties of veterinary sciences. Considering that postgraduate courses were not taught in those institutions, we had to provide them in our Institute itself. As Director General of the Institute I had now a chance to execute a plan that I had cherished for several years. Avoiding the meddling of bureaucracy by making sure not to use the slightest fraction of our yearly budget, I established three yearly courses of phytopathology, economic entomology, and animal pathology. The teachers were selected from the staff of different sections of the Institute who gave lectures on the subject of their choice, and usually pertaining to their own research. During the three and a half years of my tenure as the director of the Institute, the courses were maintained for a period of 11 months each year. A total of 37 students were admitted, 13 of them in the phytopathology series. A monthly stipend was paid during 12 months to the students of the first two years, the funds coming from donations of farmers and breeders and from commercial and industrial concerns dealing with pesticides and veterinary products. Bureaucrats found this irregular and during the third year a special sum was included in the budget of our Institute to pay the stipend of the students. So much red tape was involved, however, that no payment could be made.

In 1952 we invited Professor J. C. Walker to give a series of lectures on general phytopathology to our students. Attendance was free and besides the plant pathologists of our staff those of other institutions were welcome. Some came daily from the Instituto Agronômico at Campinas, 60 miles away.

At the conclusion of the course, the students were given a short examination, and most of them were later employed by our Institute. The courses were not continued after 1953, when I moved from the position of Director General back to that of Director of the Division of Plant Biology. Victoria Rossetti returned to the post of head of the Section of Phytopathology and twice a year she gave a one-month course during the holidays of the state college of agriculture, with the purpose of testing possible candidates for an appointment to her section.

When the University of São Paulo was founded in 1933, the College of Agriculture, formerly a division of the State Department of Agriculture, was incorporated into the University. A graduate course in phytopathology was established in 1964 by Professor F. Galli, head of the Department of Plant Pathology. Since the institution of that course, nine of our *agronomos* have obtained the title of Doctor of Agronomy under the direction of Professor Galli and his staff. Two biologists, graduates of the Faculty of Philosophy, Walkyria Moraes and Marly Oliveira, are PhDs. In collaboration with some members of his staff, Galli wrote a comprehensive treatise on phytopathology, which was published in 1968.

Our Section of Phytopathological Mycology is now headed by *agronoma* Regina Amaral. Since her admission in 1953 to the former section of

Phytopathology of which she became head in 1970, she has worked on rice diseases, especially rice blast caused by *Piricularia oryzae,* of which she has distinguished several races that occur in Brazil. In 1966 she received a scholarship from the government of Japan and spent five months in that country visiting experiment stations. In 1971 she presented a paper on diseases of rice in Brazil at the meeting of the International Commission on Rice Diseases at Pelotas, RGS, Brazil.

There are two phytopathological societies in Brazil: The *Sociedade Brasileira de Fitopatologia* founded in 1966, with its office in Brasilia DF., has published since 1967 the periodical *Fito-patologia Brasileira* and its supplement *Notícias Fitopatológicas.* It had its tenth congress in 1977, in the city of Recife, PE. The *Grupo Paulista de Fitopatologia,* which has its office in São Paulo, SP and publishes a quarterly *Summa Phytopathologica,* met at Botucatu, SP, in January this year. In January 1977, the Brazilian Society of Nematology had its third meeting in Mossoro, RGN.

Brazilian phytopathology owes a great debt to many foreign scientists and institutions. Outstanding in this respect are numerous American colleagues, philanthropic institutions, and organizations dealing with scientific agriculture.

My first collaboration with an American plant pathologist was in 1934 with Anna E. Jenkins, mycologist of the Mycology and Disease Survey of the Bureau of Plant Industry of the US Department of Agriculture in Washington DC. When I started working on citrus diseases, I prepared a mimeographed write-up of my observations and research on the rots and blemishes that affected the oranges that Brazil was beginning to export. I sent a copy of my report to P. H. Rolfs, former plant pathologist and director of the Citrus Experiment Station of Florida, and also founder and director of the College of Agriculture of the State of Minas Gerais, at Viçosa. Rolfs sent me a letter with valuable commentaries and criticisms and the suggestion that I write to Jenkins about it. Jenkins had published several papers on sour orange scab, a severe disease in the nurseries of Florida and other parts of the world, including Brazil. She called its causal agent *Sphaceloma fawcetti,* a fungus belonging in the Myriangiales and to which the sweet orange is immune. Jenkins had just published a note on sweet orange fruit scab which occurs only in South America and described its agent under the name *Sphaceloma fawcetti* var. *viscosa.* Years later we found the sexual stages of both fungi, which we named *Elsinoë fawcetti* and *E. australis* respectively. The latter does not occur in the United States. Thus we started our collaboration which was to last 38 years, until her death in 1972. In the course of all those years we exchanged well over one thousand letters. But more important than our correspondence were the numerous occasions in which we worked together either in São Paulo or in

Washington DC. During my visits to the States, I never failed to visit her in the South Building of the Department of Agriculture in Washington where I worked one full month in 1939 and in Beltsville, Maryland, when her office moved to the new quarters of the Bureau of Plant Industry.

Jenkins and I assembled the best specimens of Myriangiales in our herbaria into a collection that we called *Myriangiales Selecti Exsiccati.* It was issued between 1940 and 1951, by the Bureau of Plant Industry in 10 sets of 11 fascicles of 50 specimens each. The sets were given to the many important mycological herbaria in the world. In exchange for one of the sets, we received from the Indian Agricultural Research Institute five handsome fascicles of Indian fungi, three of Uredinales, and one each of Ustilaginales and Cercosporae.

At our request to the USDA, Jenkins was commissioned to continue her work in my laboratory during nine months in the period between 1935 and 1936. After she retired she was again invited by the Government of the State of São Paulo and commissioned for a period of four years in 1952–1956 by the Office of Foreign Agricultural Relations of the Department of State. During our collaboration we published numerous papers on the Myriangiales, especially on the genus *Elsinoë* and its imperfect stage *Sphaceloma.* Many of the infected hosts had been collected during my travels in South and North America, and trips to Europe, Africa, and Asia; these hosts showed two types of symptoms, scab and an anthracnose of a distinct type that gave to the group the general name of spotted anthracnoses. Although these papers were mostly descriptive and taxonomic, some, like those on economic plants such as citrus and the rubber plant *Hevea brasiliensis,* involved more detailed investigations: culture work, temperature relations, somatic variation (saltations), and inoculations.

In 1937, H. S. Fawcett, head of the Department of Plant Pathology of the Citrus Experiment Station of the University of California at Riverside spent his sabbatical leave working in our Institute and traveling in my company, through the citrus districts of Brazil, Argentina, Paraguay, and Uruguay. I met him in Rio de Janeiro and took him directly to the states of Pernambuco and Bahia in northern Brazil where we visited experiment stations and orchards. Later we published an account of our observations, and the specimens we collected were filed at the *Instituto Biológico.* In São Paulo where he spent three weeks, Fawcett worked with me on the footrot fungi of citrus caused by *Phytophthora* spp. We made many inoculations with the efficient technique he had developed in Florida and we published a joint paper on the results of this work. At the end of his stay in São Paulo, Fawcett, always in my company, made an extensive trip to the citrus districts of southern South America. Starting in the State of Rio Grande do Sul, in Brazil, we traveled to Argentina where we were met by H. Speroni, an *agronomo* of the Argentinian Ministry of Agriculture. Speroni had been sent from Buenos Aires by plant pathologist Marchionatto, director of the Plant Protection Service of Argentina, to accompany us in the Argentinian territory. It was in the province of Corrientes in the citrus district of Bella Vista that we saw for the first time the unforgettable scene of big orchards of sweet orange completely destroyed by tristeza. Two years later, commissioned by the government of the province of Corrientes to advise growers on how to control tristeza, I found that it had spread all through Bella Vista, practically covering the whole province and already was beginning to invade the neighboring territory of Misiones. This was, in my opinion, sufficient proof that tristeza is an infectious disease.

On the night boat on the Paraná River from Posadas in the territory of Misiones back to Corrientes, Fawcett, Bitancourt, and Speroni, as representatives of three countries, collaborated in the preparation of a report on tristeza to the Argentinian authorities. Of the six hypotheses we presented on the cause of the disease, the last one was that it might be due to a virus infection.

Fawcett returned to the States from Tucuman, by way of the west coast of South America, where tristeza appeared only several years later, and I returned to São Paulo. Around 1938 tristeza appeared in the orchards of São Paulo and I started experiments to determine whether it was caused by a virus. In 1944, swabbing the bark of diseased trees with an iodine solution, I showed that translocation of carbohydrates had been blocked at the bud union so that starch accumulated in the sweet orange above and did not show at all in the sour orange stock below. In 1945, Fawcett, in California, applied the same test to the roots of sour orange in quick-decline affected trees of sweet orange on sour stock, and found that the young roots were deprived of starch, like the sour orange bark of trees affected with tristeza. I considered that both results were acceptable evidence that tristeza and quick-decline were the same disease and probably caused by a virus. It remained to M. Meneghini in our Institute to prove in 1947 that tristeza is transmissible by grafting and by aphids previously fed on diseased trees.

The US authorities were alarmed by the news of this destructive disease that had already killed millions of trees in Argentina and that was soon to spread to a still larger number of trees in Brazil. In 1945 the US Department of Agriculture sent the entomologist A. C. Baker, head of the USDA fruitfly laboratory at Mexico City, to Brazil, to observe tristeza and to suggest measures to prevent its introduction into the United States. I suppose Baker was convinced by my evidence that tristeza is in fact a virus disease; therefore, C. W. Bennett, a virologist-pathologist of the beet diseases laboratory of the USDA at Riverside, California, was sent to São Paulo. He established his quarters at the Instituto Agronômico at Campinas in 1946. Bennett and

A. S. Costa, head of the Section of Virology at the Institute, conducted a cooperative study of tristeza chiefly in regard to its transmission by grafts and by insect vectors and to the susceptibility of various scion-stock combinations. After Bennett returned to the United States in 1948, the work on tristeza was continued cooperatively by T. J. Grant, pathologist of the USDA, and Costa at Instituto Agronômico. Their work laid special emphasis on susceptibility of different stock-scion combinations to identify suitable stocks for sweet orange and grapefruit, to determine the reaction of scions of a number of citrus varieties on sour orange rootstock.

In 1941, I was granted a Guggenheim fellowship, which allowed me to spend a seven-month leave in the Department of Plant Pathology of the Citrus Experiment Station at Riverside, where Fawcett had demonstrated that psorosis of citrus is transmissible by grafts. I met him and A. E. Jenkins at the annual meeting of the AAAS at Dallas, Texas. From there, Jenkins and I were taken through Texas by Fawcett. After Jenkins left us at El Paso and returned to Washington DC, we continued our trip to Riverside by way of New Mexico and Arizona, visiting Fawcett's colleagues at universities and experiment stations along the way. Arriving at Riverside in January 1942, I met J. M. Wallace, a virologist of the beet diseases laboratory who had recently transferred to the Citrus Experiment Station. Until July I collaborated with Fawcett and sometimes also with Wallace, in the work on psorosis. From this work three papers were published in *Phytopathology* in 1943 and 1944. In July, I returned to Brazil from Riverside by way of Mexico where I attended a Pan American Congress of Agriculture.

The Rockefeller Foundation, which had established agricultural experiment stations in Mexico, Colombia, and Chile that were staffed with North American scientists of the highest caliber, promoted meetings of plant pathologists, usually with the cooperation of local authorities, in Mexico City, São Paulo, Bogota, Santiago, and other cities. In 1963, on the occasion of the celebrations of its fiftieth anniversary in New York, the Foundation sponsored three lectures that I gave, in the Department of Plant Pathology of the University of Wisconsin at Madison. Other plant pathologists from Brazil received scholarships to complete their training in postgraduate courses in the USA and other countries.

The Guggenheim Foundation gave fellowships to Kramer, Bitancourt, Victoria Rossetti, and Andrade of our Institute and to Costa of Instituto Agronômico to work in American Institutions.

The French Government, through its Ministère des Affaires Etrangères, gave a grant to Victoria Rossetti to work on plant tissue culture as substrate for the cultivation of obligate parasites such as rusts in the Uredinales in G. Morel's Laboratory at the Centre de la Recherche Agronomique at Versailles.

Among the foreign pathologists that have contributed to the development of phytopathology in Brazil I must cite A. S. Muller, the first professor of phytopathology of the College of Agriculture of Viçosa, MG, when P. H. Rolfs, former director of the Florida Experiment Station, was its director. E. E. Honey, a graduate of Cornell University, became professor of phytopathology of the College of Agriculture of the State of São Paulo in Piracicaba. Three of his students were later plant pathologists at the Instituto Agronômico in Campinas, SP: A. S. Costa, who collaborated with C. W. Bennett and T. J. Grant in their research on tristeza; A. P. Viegas who obtained a PhD after four years of study under H. H. Whetzel, chairman of the Department of Phytopathology, Cornell University; and H. P. Krug.

The simultaneous and independent development of methods for the culture in vitro of plant tissues by R. Gautheret in France and P. White in the United States roused the imagination of plant pathologists everywhere. Some saw in Gautheret's method the possibility of using the callus tissue of healthy plants as substrate for the isolation, in aseptic conditions, of obligate plant parasites such as the rust fungi and the powdery mildews. Others perceived the advantages of studying in vitro the tissue of crown gall. The bacteriologist E. F. Smith had already described the similarity between crown gall and animal cancer, and White's cultures offered a convenient technique for extending the investigations into plant cancer to nutritional and biochemical studies.

Both methods, which used essentially the same technique, appealed to me as a plant pathologist, but that of Gautheret, who started his cultures from healthy plants, seemed to be more interesting in view of the economic importance of obligate parasitic fungi compared to that of crown gall. I therefore decided to go to Paris to learn the technique of Gautheret. In Paris an unexpected event made me change my attitude toward crown gall, or rather plant cancer. I found that Gautheret, who had just returned from the United States where he had met White and seen his setup at Princeton, was already working on plant cancer in collaboration with one of Gautheret's assistants, a young Dutch student named Zoyza Kulescha.

Kulescha was making a comparative study of the content of auxin, i.e. indoleacetic acid (IAA) in callus tissue and in tumor (cancer) tissue. Ten times as much auxin were found in tumor tissue than in callus tissue. These results had already been found or confirmed in other laboratories, and they were generally interpreted as resulting from an enhanced power of biosynthesis of IAA in tumor tissue. I chose the opposite interpretation according to which the increase of auxin is due to a block in the biosynthetic pathway of other metabolites that compete with IAA for the pool of common precursors, i.e. amino acids and dicarboxylic acids. This places the cause of cancer in the category of Garrod's errors of metabolism, merely a problem of genetics. I went one step further when I found a similarity in the structural formula of IAA and that of the pyrrole unit of protoporphyrin, the precursor of the active cofactors of the biological porphyrins: chlorophyll, the cytochromes, peroxidases, and catalase. A block in the synthetic pathway of protoporphyrin would inhibit the synthesis of porphyrins.

At the same time, the pool of precursors of the porphyrins in its entirety would be available to IAA while the porphyrins and their multiple effects on the metabolism would be suppressed.

Known facts about tumor tissue support that interpretation of carcinogenesis. For instance, the galls produced by the inoculation of *Agrobacterium tumefaciens* in a tomato plant are pure white in contrast with the unaffected green chlorophyll color of the surrounding tissues. According to the late biochemist Otto Warburg, animal cancer is the consequence of the lack of what he calls grana, obviously meaning mitochondria, the seat of activity of most of the porphyrins. His respiratory experiments showed that the lack of grana was the cause of the shift of the respiratory metabolism of normal cells to the fermentative metabolism of cancer cells, a consequence of the suppression of terminal oxidations by the cytochromes in the mitochondria. He also stated that the lack of catalase in the cancer cells results in the lethal accumulation of peroxides whereas normal cells are spared because of the presence of catalase that destroys the peroxides. This, according to Warburg, is the reason for the beneficial effect of the irradiation of tumors in the treatment of cancer.

All these cogitations point to auxin as the pivotal factor in plant carcinogenesis. We postponed the use of tissue culture in the study of plant cancer, and since 1954 our research have been almost entirely devoted to indoleacetic acid and its metabolism.

The first result that we obtained was the finding that IAA is not directly active in the coleoptile straight growth test. To become active, it must be oxidized and converted to a dimer, which I have named deuterauxin. Autoxidation of solutions of IAA diluted to physiological concentrations is very rapid, and it is doubtful that it participates directly in experiments in which these solutions are used. Solutions of deuterauxin polymerize and aggregate as we have demonstrated by chromatography and are thus inactivated, but when they are boiled for a few minutes, activity is restored. Oxidation and tautomerism, which have been demonstrated by several authors, explain the sometimes erratic behavior of what was supposed to be IAA.

What has not yet been explained is why a gas like ethylene can have the same physiological properties as auxin. We have sent samples of deuterauxin and IAA to Varian to be run in their mass spectrograph. The results were surprising because oxidation had completely changed their patterns of ionic fragmentation exhibited in the spectrograms. The pattern of IAA is easily explained by decarboxylation resulting in the production of a skatole ion; the pattern of deuterauxin shows that the greatest relative abundance is from clusters of alkanes and alkenes whose fragmentation finally yields ethylene. Our attempts at proving that this also occurs in plant tissue have so far been unsuccessful. If it is eventually proved, we will finally have a unified theory of auxin activity.

There is no doubt that there has been a great development of phytopathology during the past half century, parallel to the growth in the whole country in Brazil. As part of the curriculum of the colleges of agriculture, phytopathology has increased with the number of colleges and universities. The number of universities with departments of phytopathology has increased so that there is at least one department in nearly every unit of the Brazilian Federation. The leading southern states have more than one college, and in São Paulo there are new federal, state, and private colleges of agriculture. There is also a department of agriculture, usually with a section of plant protection, in the administration of some of Brazil's largest cities. The number of plant pathologists has followed the same trend. The Biological Institute of the State of São Paulo, which in 1920 had one head and two assistants in a single section of phytopathology, now has nine specialized sections with 45 chief scientists and their assistants. There are two professional societies (*agronomos*) in São Paulo and one society of phytopathology in São Paulo and one in Brasilia.

The number of teachers, students, and workers in any institution, however, is not an indication of the quantity and quality of their achievements. The late plant pathologist Gerold Stahel, director of the agricultural experiment station of Surinam, after visiting our pavillion in the woods, 20 miles away from São Paulo, and the new institute under construction in the city of São Paulo, told me, rather prophetically: "Remember that the quality of the production of a scientific institution is inversely proportional to its size."

Ann. Rev. Phytopathol. 1980. 18:1–9

RECOLLECTIONS OF A GENETICAL PLANT PATHOLOGIST

Adrian Frank Posnette

East Malling Research Station, Maidstone, Kent, England ME19 6BJ

An invitation to write a prefatory chapter with an autobiographical theme is especially difficult to decline when one is on the verge of retirement and contemplating a period of retrospection. Having accepted the honor, one then faces unexpected problems of synthesis not previously encountered in writing research accounts for scientific journals. Where to begin, what seemingly trivial matters might or might not interest the readers, what sort of person, if any, will read it—these are among the questions that encourage procrastination.

My education did not fit me in the conventional way for becoming a plant pathologist. My chief interests as a boy were more zoological than botanical, and I went to Christ's College, Cambridge University in 1932 with the intention of learning how to earn my living as some kind of a zoologist. The Tripos system demanded that three subjects be studied in Part I; I read zoology, botany, and animal (mainly human) physiology and spent much time playing college football and flying with the University Air Squadron before concentrating on zoology for Part II. Finding genetics particularly interesting I hoped to be employed as an animal breeder.

The Colonial Agricultural Service attracted me for several reasons: It provided unrivaled opportunities for travel, it paid for two years' postgraduate study, and the salary was considerably higher than could be obtained in research in Britain. Consequently entry was highly competitive and I was genuinely surprised to be offered one of the two research studentships available in 1935; one was in soil science and the other, mine, was in *plant* genetics. My college tutor, Dr. C. P. Snow (later better known as a novelist than as a scientist) allayed my doubts about my mathematical ability by assuring me that I would always find statisticians willing to analyze my

1889

figures. There followed a year at the Cambridge University School of Agriculture attending lectures on the theory and practice of most aspects of crop and animal husbandry, and my first exercise in plant genetics (of barley) under the guidance of Professor F. L. Engledow and Dr. G. D. H. Bell, who gave me an enduring appreciation of the potential as well as the limitations of crop improvement by breeding.

The second year (1936–1937) was at what was then the Imperial College of Tropical Agriculture in Trinidad, now part of the University of the West Indies. Besides learning some of the differences between temperate and tropical agriculture I was allocated two research problems: One was a genetic analysis of many varieties of cowpea (*Vigna* spp.) that had been collected from most countries where the crop was grown. As a lesson in the genetic diversity to be found in a crop the exercise was illuminating but I know of no other way in which it was useful. The second research project, which was to investigate the natural pollinating agents of the cocoa tree (*Theobroma cacao*), I was able to continue for many years because my next assignment was to the Gold Coast, now Ghana, where cocoa was and still is the main crop and the most valuable of its exports.

It is a common fallacy that British colonial administrations concentrated on the development of export crops, such as cocoa in the Gold Coast and oil palms in Nigeria, and neglected food crops. While this may have been the tendency after the Second World War, when the economic stability of the soon to be independent colonies became increasingly important, the reverse was true in West Africa during the interwar years. When I arrived in the Gold Coast in 1937, very little research effort had been devoted to cocoa there and the first experiment station for it was only then about to be started. The country's plant breeding programs were confined to food crops. I was to be the first plant breeder for cocoa, and was instructed to breed for resistance to whatever was killing a great many of the trees.

Cocoa Research in West Africa

The extent to which the cocoa trees were being lost only became apparent when several agricultural officers were deployed in planting shade trees, mainly *Saman samanea,* among the cocoa in an attempt to ameliorate the environment. The cocoa was believed to be suffering primarily from physiological die-back due to lowered humidity following the degrading of much of the rain forest. Stem swellings had been reported as a condition new in cocoa but it was regarded as a minor problem of scientific interest, irrelevant to the major one of "drought die-back."

Another serious problem was the damage to cocoa trees caused by two species of mirid bugs (then called capsids). Severe attacks on mature trees frequently followed injury by fallen forest trees that had provided shade for the cocoa; but more general, even widespread damage occurred periodically on trees of all ages. Many young ones were killed; hence the planting of new cocoa farms was often unsuccessful. As S. H. Crowdy proved later, fungal infection followed the capsid damage, weakened the trees, and caused a chronic die-back condition.

After spending most of my first year in cocoa farms on preliminary selection work in many parts of the country, I concluded that capsid damage and "drought die-back" were easily distinguished, and that the latter condition was almost invariably accompanied by leaf symptoms in the form of chlorotic mosaic or vein banding, and usually by stem swellings, however inconspicuous. The density of shade trees seemed irrelevant; much of the affected cocoa was unshaded, but even where shade trees were dense the cocoa beneath them was also severely affected or dead.

Experiments to test for transmission by grafting established that the pathogen was probably a virus and, more significantly at the time, that it was capable of killing trees after inducing a sequence of leaf patterns and stem swellings. Later, the variety of syndromes encountered in different regions was shown to be due to variation in pathogenicity among a multiplicity of virus strains. An intensive search for resistant trees and propagating the survivors where all but a few had died revealed that some were infected with avirulent virus which could confer resistance when transferred to clones known to be sensitive. Protective inoculation seemed a possible method of control, and I shall return to this aspect later.

Attempts to eradicate the disease from individual outbreaks of different sizes showed that small ones consisting of fewer than about 50 affected trees could be controlled by cutting them down to ground level together with a ring of neighboring symptomless trees. This treatment usually had to be repeated two or three times at about six-month intervals as trees with latent infection developed symptoms. The larger the outbreak the more protracted the operation needed to be; nevertheless the treatment was sufficiently effective, when done by observant and conscientious men, to be adopted nationally, compensation being paid to the farmers in various ways by successive governments.

The benefit of those control measures, involving the destruction of many millions of trees, is difficult to assess. In the eastern region of Ghana where swollen shoot disease was most prevalent, it still remains epidemic, but cutting out accelerated replanting and some production of cocoa has been maintained, albeit at a low level. In other areas, and especially in Ashanti where the disease had not become widespread, outbreaks were effectively controlled and undoubtedly most of the cocoa farms there were saved from the devastation that would presumably have occurred otherwise.

To revert to the research theme—having established that swollen shoot disease was a primary cause of the cocoa dying, my next objective was to find the vector. As acting entomologist I was responsible for the first tests with likely insects: Thrips, aphids, and leafhoppers gave no transmission but psyllids did, once and not repeatable! The first clue came from gauze-caged trees that had been used for pollination studies and afterwards inoculated with swollen shoot to see whether they would die from the virus infection if shaded and protected from capsids; they did. Some cocoa seedlings became infected while growing in a cage with an infected tree, and on the seedlings were mealy bugs identified as *Pseudococcus njalensis.*

Several species of *Pseudococcus* and related genera proved to be vectors, transmitting the virus in a nonpersistent manner but requiring feeding periods of hours to become infective. The study of virus-vector relationships (and other aspects of cocoa virus research) was greatly facilitated by the discovery that mealy bugs will feed on the cocoa bean cotyledons from which the testa has been removed. A block watchglass with a glass cover serves as a cage for each bean so that large numbers of test plants can be accommodated on a laboratory or glasshouse bench. When the required feeding time has elapsed, the insects can be removed or killed by an insecticide and the beans planted. If infection has occurred symptoms develop in the first or second pair of leaves of sensitive varieties, or in later leaves of tolerant ones or with avirulent virus strains. The ideas behind this method were thought out while contemplating the problem during my third voyage to the Gold Coast—an advantage of a sea passage over the thought-obliterating discomfort of traveling by air! As most of the vector species are polyphagus, the possible alternative hosts of the virus were legion. The first to be identified, *Cola chlamydantha,* was found because my collaborator, J. M. Todd, saw vein banding on the leaves of a regenerating stump near infected cocoa trees in the extreme western part of Ghana, the only area of the country where *C. chlamydantha* occurs. Mealy bugs fed on the cola leaves transmitted swollen shoot virus when transferred to cocoa beans.

Although *C. chlamydantha* trees appear to be important reservoirs of swollen shoot virus where this species is indigenous, it does not occur where cocoa is intensively grown in the eastern region and Ashanti. The other species of forest trees shown to be susceptible to the virus are generally tolerant; with the exception of the baobab, *Adansonia digitata,* they show transient symptoms or none, and support only a low level of virus available to vectors. Spread of virus from some of these hosts to cocoa can be demonstrated only rarely, although the frequency of isolated outbreaks in cocoa, even that planted in forest reserves, the multiplicity of virus variants, and the extreme tolerance of the indigenous hosts to virus infection suggest that they have an important role in the epidemiology of swollen shoot and have probably been infected for a much longer time than the cocoa which was introduced to Ghana little more than a century ago.

The occurrence of avirulent strains of the virus in cocoa trees that had survived an epidemic and the evident ability of these survivors to resist the effects of the most virulent strain when inoculated clearly indicated an alternative control measure. Cross-protection tests to study relationships between virus isolates of different pathogenicity had already shown that those from outbreaks in the same locality were often antagonistic whereas those from different regions usually showed no interference.

Besides many small-scale tests, a field trial using about 1500 mature cocoa trees demonstrated that mild strain protection would greatly delay the rapid deterioration caused by the prevalent virulent strain. The mild strain used to "protect" half the trees was taken from a "resistant" tree; it had no apparent effect on yield over a period of 5 years, by which time the virulent strain had reduced the yield of trees inoculated with it by 90% and killed 77% of them. Only 14% of the protected trees had developed symptoms typical of the virulent strain after 5 years compared with 76% of the unprotected trees that became infected by natural spread.

Despite these promising indications, protective inoculation has not been used as a control measure against swollen shoot disease. There are obvious practical difficulties in implementation and considerable preliminary work would be needed to select the mildest strain that was both stable and adequately protective. Also the massive campaign of cutting out diseased trees would have to cease in any area where protective inoculation was being practised. I have no doubt, however, that if swollen shoot disease had been allowed to spread unchecked and replanting had continued without the removal of surviving trees infected with avirulent virus, in the course of time (perhaps a century) most of the cocoa trees in Ghana would be infected with mild strains because virulent strains should be self-eliminating in cocoa, while trees infected with avirulent strains would be protected from the effects of virus occasionally transferred from the wild host plants.

Selection for virus resistance or tolerance among the cocoa types available in West Africa continued while this work on insect transmission and alternative host plants was in progress. Genetic diversity was very limited because most cocoa trees were derived from limited introductions from Brazil toward the end of the nineteenth century. The prevalent Amelonado type is self-fertile and considerably inbred; the Trinitario types introduced in 1905–1912 period, mostly from Trinidad, had left comparatively few descendants, some of which were more tolerant of virus and capsid damage than the Amelonado. The need for a wider genetic base was evident if resistance breeding was to be successful.

F. J. Pound's material collected in 1938 from the Upper Amazon region of Peru was established in Trinidad and was already known to contain genotypes resistant to witches'-broom disease, *Crinipellis* (*Marasmius*) *per-*

niciosa. By a series of fortunate circumstances I was able to return to Trinidad in 1943. In the course of hand-pollination to prepare seed of the Upper Amazon types in Trinidad for dispatch to quarantine in Accra, I found that all were self-incompatible but with a high degree of cross-compatibility controlled by a series of sterility genes different from the single gene system in previously studied cultivars (Criollo and Trinitario).

Among the progenies sent as seed from Trinidad were crosses between representatives of different Upper Amazon populations and between them and Trinitario clones. Some of these seedlings showed hybrid vigor, growing much faster than Amelonado and Trinitario cocoa, starting to crop in the second or third year after planting instead of in the fourth or fifth, and soon producing yields much above those expected from other types of cocoa. Because of ease of establishment on degraded soils and their precocity, such Upper Amazon hybrids now form the basis for most cocoa improvement programs worldwide.

In reaction to swollen shoot virus some of the Upper Amazon introductions proved generally to be more tolerant than any previously tested cocoa, and resistance to infection via the vector was found in accessions from near Iquitos in Peru. This resistance has been the subject of detailed investigations over the past 10 years by a U. K. technical aid team with the objective of producing planting material combining virus resistance with early cropping and high yield. Because immunity to infection has not been found in cocoa, the effectiveness of resistance must be judged by its influence on rate of spread—a long-term assessment.

Virus Diseases of Temperate Fruit Plants

In 1949 I moved to East Malling Research Station in Kent, England, primarily to work on virus diseases of strawberry and other berry fruits. Ian Prentice had already separated the three components of yellow edge and crinkle diseases of strawberry on the basis of their persistence in the aphid vector. My efforts were first directed to obtaining virus-free plants by thermotherapy and then to the etiology of two new diseases, green petal and mosaic. In collaboration with N. W. Frazier on sabattical leave from Berkeley, the first of these was shown to be transmitted by leafhoppers from clover in which the "virus" caused flower phyllody previously regarded as a physiological condition. We were concerned with establishing the relationship with aster yellows which was not known in Britain; the "yellows-type" viruses being established as not sap transmissible, we concentrated on vectors and host range. We failed to question the virus nature of the pathogen and so missed the opportunity to find mycoplasma in green petal plants and spiroplasma associated with our clover witches'-broom.

Strawberry mosaic (or yellow crinkle) also presented a vector problem, which was resolved in collaboration with my graduate student Asharfi Jha and the nematologist R. S. Pitcher. The pathogen, identified as arabis mosaic virus, was shown to be infective in soil collected from a strawberry field at my home where the disease was endemic; the vector proved to be *Xiphinema diversicaudatum,* the second nematode species proved to transmit a virus after Hewitt and Raski had pioneered with *X. index* and grapevine fan leaf.

Most of my research at East Malling was on virus infection of stone and pome fruit trees and on thermotherapy for producing virus-free clones of established varieties, in collaboration with Roy Cropley. Knowledge of what viruses might be latent in each species and of efficient methods of testing plants for infection are prerequisites for establishing that all viruses are *probably* absent from the particular plant under test. Infection is not difficult to prove, but the absence of infection cannot be defined as more than a probability, with confidence growing with increasing knowledge of how to detect more and more viruses and strains of differing pathogenicity.

By comparison with virus diseases of stone fruits, which had been investigated in the USA and Canada for many years, those of apples and pears had not received much attention before 1950. Stony pit of pear and mosaic, rubbery wood, and chat fruit of apple were known to be graft-transmissible; these three apple diseases had become prominent in England because a new variety that was sensitive (Lord Lambourne) had been grafted onto trees of other varieties in which infection was latent. It soon became evident that commercial varieties of pome fruits and their clonal rootstocks were infected, without showing symptoms, by several other viruses that had severe effects on either related species or noncommercial clones of the same species. English plum and cherry varieties also proved to have some previously unrecognized diseases that were transmissible by grafting.

Identification of the pathogens was virtually impossible without transfer to herbaceous hosts, partial purification, serological tests, and tests of Koch's postulates. This has so far been done with lamentably few of the graft-transmissible diseases of fruit trees; nevertheless the number of possibly distinct viruses has been diminished by establishing that the same virus causes several diseases, e.g. plum line pattern by a strain of apple mosaic virus, ring russet on apple and pear fruits and pseudo-pox blemishes on plums by apple chlorotic leaf virus, cherry ring mottle by prune dwarf virus, and a myrobalan/plum graft incompatibility by necrotic ring spots virus.

Not the least rewarding aspect of fruit tree virus research in the 1950s and 1960s was the international collaboration that developed from the founding of an international committee to organize frequent symposia in

European countries. The eleventh of these meetings was held in 1979 in Budapest, previous ones having been in most European countries that grow deciduous fruit crops commercially. (The group is now under the aegis of the International Society for Horticultural Science, and the next meeting will be held in North America.) The rapid application of research to practice, especially the provision of virus-free clones, owes much to the free exchange of information and virus-indicators. In most countries virus-free fruit trees have replaced virus-infected ones in many orchards planted in recent years, following the example of the EMLA scheme organized by East Malling and Long Ashton Research Stations in England.

Research Administration

While my responsibilities for research programs other than my own widened at East Malling Research Station, where I was Head of the Pathology Department for 15 years and then Deputy Director and Director for 10 years, I was frequently reminded of the advantages I gained from my experience of being largely self-sufficient during my first eight years in West Africa and after that of developing collaborative research with the graduates who came in 1945/1946 on their first posts to the newly established West African Cocoa Research Institute (which had developed from the Cocoa Experiment Station that I had helped to start in the Gold Coast). These young men did not stay at the institute many years, but their enthusiasm, ability, and dedication enabled them to make valuable contributions to their individual fields of research. Not surprisingly their talents and success in their cocoa research afterwards led them to further achievements so that their names are now well known internationally—H. Owen, the mycologist; N. F. Robertson and J. M. Todd, the virologists; and A. H. Strickland, the entomologist. From them, other colleagues, and experience I have learned many things, including the following:

1. The synergistic effect of collaboration that makes teamwork so much more effective than go-it-alone efforts. A corollary is the value of travel to see one's own research problems in perspective of those of others. Cocoa research sent me to Brazil; the Ivory Coast; Nigeria; Papua, New Guinea; Tanganyika; Venezuela; West Indies; and Zanzibar. Research into fruit viruses took me to most European countries, Australia, Canada, India, Iran, South Africa and the United States.

2. The fundamental importance of techniques. The most valuable advances often depend on developing a new method for investigating the subject.

3. The unimportance of sophisticated equipment for solving some problems in plant pathology, how much can be done with the proverbial string and sealing wax. My first official letter was requesting very modest equip-

ment but, I pleaded, it was an "urgent necessity" because the cocoa flowering season would soon end. The letter in reply from the Director of Agriculture in Accra, stating that "in this country nothing is urgent and very little is necessary," was indeed half true. If one is studying a plant virus disease with the first objective of minimizing its damage to the crop, then the identity and ecology of its vectors, alternative host plants, and sources of resistance are more relevant than the morphology of the virus particles —valuable though that information can be in another context.

3. Ensuring, as best one can, that others are well equipped with facilities and a congenial environment appropriate for their research can be as rewarding as doing experiments oneself.

4. The importance of having a wife who is both tolerant and critical. Mine has not only tolerated uncomfortable and unhealthy climates, the torture of sand flies at their worst, periodical family separations, and the primitive conditions of a medieval house, but also has been ready to give a constructively critical appraisal of my theories, deductions, and research papers. She has been the most cogent of my collaborators.

Ann. Rev. Phytopathol. 1981. 19:1–19

RECOLLECTIONS AND REFLECTIONS

L. M. Black

Department of Genetics and Development and Department of Plant Pathology, University of Illinois, Urbana, Illinois 61801

INTRODUCTION

When I was invited to write the prefatory chapter for this volume I much appreciated the honor and I must confess that I never considered the possibility that I might not have anything worthwhile to say. I accepted at once and worried about the reader later. During my career, I have been fortunate enough to work virtually full time on problems of my own choosing without regard to practical applications. My bent was to do "fundamental research" as I conceived it. Most plant pathologists, of necessity, must be concerned with the practical value of their work in controlling plant diseases. At times I have had the impression that some plant pathologists feel that this is the only justifiable motivation for their research. Moreover, many fundamental discoveries have been derived from applied research. And vice versa. But it was because of plant pathology's major emphasis on the practical applications that I was quite surprised and much pleased by the editorial committee's invitation to one in the profession who has contributed so little directly to control.

It is extremely improbable that this chapter will contain any statement of a universal principle that has not been enunciated before; on the contrary, it is quite likely that it will contain more than one generalization, without attribution, that has been stated previously and more clearly. Nevertheless, it seems possible to make a worthwhile contribution in describing some of the unique experiences that any person has during a lifetime and how things were viewed before and after important discoveries. Those who arrive on the scene later and do not have first-hand experience of an important change

1899

0066-4286/81/0901-0001$01.00

may have real difficulty in appreciating what the prior outlook was. This approach also appeals to me because I have found that learning is much more likely to be derived from history than from prophecy; I also fear that there is some evidence for the latter's being a gerontological disease.

I have never regretted my choice of plant pathology for a career in science. My initial area of study was agronomy but after I had worked one summer as an inspector of certified seed potato fields, I was converted to plant pathology. I still remember a field trip in which a number of potato inspectors were accompanied by the Dominion Botanist. He was pointing out leaf spots on various plants other than potatoes and calling them diseases—I thought rather indiscriminately. Of course, he was right and the gradual realization of my abyssmal ignorance about plant diseases at that time has kept me from later being contemptuous of others in a similar state. In this connection one subsequent incident made a particularly strong impression on me: I was fortunate enough to be a member of L. O. Kunkel's group from 1936 to 1946. The Board of Directors of the Rockefeller Institute for Medical Research was visiting Kunkel's laboratory in that Institute. One MD member of the Board—he must have been new—asked Kunkel if it was true that plants have disease. An admirable man, not afraid to learn by a question that might expose his ignorance! Most of us scientists are laymen outside our area of expertise.

Actually plant pathologists have certain advantages over students of human diseases just because of the nature of the material we work with. For example, one cannot be a student of plant pathology without being impressed with the importance of heredity in determining susceptibility and resistance to disease. The basic observation that different diseases afflict different species of plants is everywhere apparent and itself bespeaks the importance of heredity. But in plant pathology, the widespread use of clones for many agricultural crops such as potatoes and apples (a practice long antedating modern agriculture) makes it obvious that many differences in disease are hereditary, even small differences in incidence, symptoms, timing, etc. It is not uncommon to see in the field thousands of plants of one clone alongside thousands of plants of another clone of the same species. Less commonly, clones may be interspersed in a field. Thus under identical conditions, one may, for example, determine with certainty that the incidence of a rapidly lethal disease on a genome of one heredity may be only as small as say, 5% less, but definitely 5% less, than that for another genome with a slightly different heredity. Students of man are obliged to work long and hard to acquire information on a statistically significant minimal number of identical twins to be able to exploit clonal material relevant to a problem.

On the other hand students of animal disease have the great advantage that particular functions are often performed by highly specialized animal organs large enough to readily provide material for separate physiological or biochemical study. Although organs are not lacking in plants they are not so generally available and specialized function is commonly performed in an organelle of a cell with more general capabilities.

The large number of plant diseases and the many different kinds of plant disease provide abundant and challenging opportunities for research. In

"Plant Pests of Importance to North American Agriculture," published in 1960, about 1250 genera of plants are listed. On some species only one disease has been recorded but on others, there are many—for example, the potato, *Solanum tuberosum,* has more than 150 diseases, depending on how one counts them. During the six years I worked as an inspector of certified seed potato fields in British Columbia and the state of New York, the disease picture of this crop frequently changed in interesting ways. Was it any wonder that at almost every annual A.P.S. meeting I attended, some exciting new development was described!

I well remember one bright young botanist remarking in a departmental staff meeting that plant pathology was "only applied mycology anyway." It is difficult to know where one should begin to challenge such an overwhelmingly pejorative declaration. Of course, in those days, and for many years later, even the most complete abstract journal covering plant pathology was still called *The Review of Applied Mycology.* And even today, probably most PhD degrees in biology, including some in botany, are awarded without the candidate even attaining the level of understanding implicit in the description "applied mycology." Obviously, it is impossible to study in formal courses everything relevant to one's scientific discipline and perhaps certain biologists can make better use of their time than in attaining some basic understanding of plant pathology. My own repeated suggestions that plant pathology be included as one of the essential core courses for botany rarely elicited any response whatever and never had any effect that I can remember. In fact, plant pathology not only involves other core biological areas, such as taxonomy, anatomy, ecology, genetics, and physiology, as they affect the diseased plant host, but also similarly as they affect the pathogen, be it fungus, parasitic flowering plant, bacterium, mycoplasma, spiroplasma, nematode, virus, viroid, plasmid, insect, protozoan, or physiological agent or condition. And the interactions between pathogen and host form in many ways what is like a third separate biological entity. Additionally, plant pathology is broadened because it is a science that cannot afford to ignore knowledge from sciences other than biology proper.

But to end the digressions; my main purpose is to relate my personal impressions of four events that radically altered conceptions in plant virology. These four are the isolation of tobacco mosaic virus, the discovery of multiparticulate viruses, the discovery of the viroids, and the discovery of plant mycoplasma-like organisms. I shall give minimal citations to specific papers, with apologies for the inevitable inequities that result, but shall end the chapter listing a few of the most comprehensive and well-indexed works that provide entry to the literature.

ISOLATION OF TOBACCO MOSAIC VIRUS

In 1928 at the University of British Columbia, my teacher, F. Dickson,[1] suggested that I write a term paper on tobacco mosaic. By the time I submitted it I was near exhaustion and, even so, was more aware of the inadequacies of my reading on the subject than of the questionable substantiality of the paper. The mystery about the nature of viruses was what attracted me to their study. In 1930, when I went to Cornell as a graduate student teaching assistant to H. H. Whetzel,[2] there was no formal instruction in plant virology as such. The properties of viruses were unknown except for their ability to cause disease, to reproduce themselves in the host, and to pass through filters that would hold back bacteria. For a long time the two words "filterable virus" were used almost as one.

Before I obtained my PhD from Cornell in 1936 one of the most important papers clarifying the virus mystery was published. L. O. Kunkel had included a chemist W. M. Stanley among the scientists he chose to do research on the very stable, very infectious tobacco mosaic virus at the branch of the Rockefeller near Princeton, New Jersey. Some years earlier F. O. Holmes, in Kunkel's group at the Boyce Thompson Institute, had introduced primary local lesion assays for plant viruses. Such assays greatly increased the accuracy of bioassay while at the same time greatly reducing the time, labor, and greenhouse space required. Previously each of a number of dilutions of an extract containing the virus was tested in 10 or more tobacco plants by scratching inoculum into leaves or pricking it into stems. Several days later all the plants inoculated with the more concentrated solutions would develop systemic symptoms at about the same time. Of course solutions could be made dilute enough to give no infections but there would be few dilutions with intermediate numbers of infections. Stanley adopted Holmes' assay from the outset and it was a great asset to his work.

About a year before Stanley's isolation of tobacco mosaic virus, W. J. Robbins had speculated, on the basis of a number of assumptions, that it might take somewhere between 10 and 10^8 liters of sap from diseased tobacco leaves to yield 0.1 g of virus. Thinking on the subject was then

[1]Other Dickson students who subsequently became plant virologists include J. B. Bancroft, W. B. Raymer, R. Stace-Smith, N. S. Wright, and C. E. Yarwood.

[2]Whetzel, the founder of the department at Cornell, was an inspiring teacher who was continually and freely dispensing ideas. One of his most important was the conviction, acquired from a lifetime of observation, that the supposedly functionless microconidia of the sclerotinias were really male gametes. His persistent persuasion of my fellow graduate student, F. L. Drayton, to try to demonstrate this, led despite many obstacles to the identification of functional sex organs in these fungi.

dominated by ideas that the virus was exceedingly small, that the virus was present in very low concentration in diseased plants, and that very little virus was required to produce an infection. Some earlier claims that tobacco mosaic virus had been isolated in crystalline form were shown a little later to be based instead on crystals of inorganic substances contaminated with small amounts of the active agent. These experiences did nothing to ameliorate such extreme expectations and must have tended to induce caution in others.

In this context Stanley's papers of 1935 and 1936, which boldly asserted that he had isolated a crystalline protein possessing the properties of tobacco mosaic virus, were a crucial contribution and had a tremendous impact. His claim was based on several critically important demonstrations. His isolated product was more infectious than the juice from diseased plants and formed needle-like protein crystals which were infectious at a dilution of 10^{-9}. Unlike the products of earlier workers his preparation suffered no reduction in activity through 10 successive crystallizations. The dissolved protein was filterable through the bacteria-proof, porous stone Berkefeld W filters of the period, but would not pass through collodion membranes. He estimated it to have a molecular weight of a few million daltons. Its further characterization as a specific precipitin antigen different from any in healthy plants, its formation of opalescent solutions at certain concentrations, and certain other properties made the product readily identifiable by subsequent workers despite Stanley's failure to find the phosphorus and carbohydrate fractions which Bawden et al (1936) Bawden & Pirie (1937) showed to be present in the nucleoprotein of tobacco mosaic virus.

Stanley's paper resulted in a period of intense experimental activity, worldwide, to determine whether his isolate was really the causal agent of tobacco mosaic. It was easy to show that plants with tobacco mosaic always yielded this product whereas healthy plants did not. But the critical procedures devised by Koch for isolating and growing a pathogenic organism in a medium, using the growth to inoculate the host to produce the specific disease, and reisolating the same organism could not be used to prove pathogenicity of the virus.

In spite of a qualifying reservation in Stanley's paper itself, namely that it is "difficult or impossible to obtain conclusive proof of the purity of a protein," the fact that his isolate would form crystals caused me, and probably others, to think of it as pure. And if a pure protein, and of such infectivity, it must of necessity be the virus.

Very soon another scientist (K. S. Chester) in Kunkel's group showed by a very sensitive serological technique that isolates prepared by Stanley's method contained some host protein.

Later Stanley and his colleagues subjected his product to many different tests that might separate the infectivity from the nucleoprotein. Infectivity could be destroyed in certain tests without destroying the antigenicity of the nucleoprotein. With formaldehyde treatments the nucleoprotein could even be inactivated and subsequently appreciably reactivated by dialysis at an appropriate pH. However, none of the results actually disproved the pathogenicity of his isolate and the weight of much evidence from other tests convinced most people that Stanley had isolated the first virus. In sum, although Stanley's original description of his isolate was incomplete and contained some errors, his isolate was essentially the virus. Deficiencies were almost inevitable in the description of a hitherto unknown kind of entity. The papers of other investigators (and Stanley's associates), completing the description of the virus and correcting some inaccuracies, mostly traced back to Stanley's paper.

It should be remembered that during this period researchers were working "in the dark" because it was not until 1939 that single virus rod-like particles were visualized by electron microscopy of purified tobacco mosaic virus. Later the electron microscope was to reveal other plant viruses with their great variety of shapes and sizes, many of them nearly spherical.

The scientists involved during this period were not entirely unemotional or completely objective observers and participants. My consistent impression was that the biologists hoped that the tobacco mosaic pathogen would prove to be an organism. No bacterial viruses infect plants but I remember the excitement over the first pictures showing tails on bacteriophages. This fitted very well with the concept of an organism and the more the microscopy improved the better it fitted. Although plant viruses do not have such obvious organ-like structures we know that their surfaces must have a great variety of highly specialized properties that have evolved to enable some viruses to fuse with the membranes of the cells of their insect vectors, others to lodge in favorable spots of the buccal cavities of round worms that transmit them, others to make them very stable in extracellular environments and so on. The chemists on the other hand seemed to hope that the causal agent of tobacco mosaic would prove to be a substance. Stanley (1936) discussed the possibility that the increase of virus in plants is like the conversion of the inactive protein trypsinogen to the active enzyme trypsin simply by the addition of the latter to the former; he considered that the evidence indicated that the virus was "not alive." Later, Stanley (1947) wrote about the virus being an entity between the world of the living and the world of the nonliving. At the time, I thought such writings were somewhat sensational and I doubt that I was alone in this. But as it happens the virus is actually something between what we had known previously as living and chemicals we had regarded as nonliving. With my biological bias, the virus still seems more like an organism than a chemical. But more of that later.

Electron micrographs of purified preparations of tobacco mosaic virus showed rigid rod-shaped particles about 18 nm in diameter. Many of the largest proportion of particles were about 300 nm long and these were thought to be infective. However, from earliest days to the present, shorter particles of the same diameter but of various lengths, sometimes in large numbers, have always occurred in purified preparations. A question soon arose over the possibility that some of the shorter particles were infective. Extensive studies of several kinds convinced investigators that infectivity resided exclusively in the particles about 300 nm long, a conviction that still holds. Subsequently the concept of a virus species consisting of one kind of particle was dominant for many years. People tended to think of a virus as a macromolecular particle which was capable of replicating itself and producing disease in a host. This replication was conceived as producing large populations in which all the infectious particles were identical—like molecules of a substance. When bits and pieces of these particles were found they tended to be regarded as breakdown products formed in purifying preparations of virus or as residual surplus components required for the production of virus in the host.

The discovery by Markham and Smith of two kinds of particles in turnips with yellow mosaic, one with nucleic acid which was infective and another without nucleic acid which was not, also fitted the idea that the latter was an incompleted particle. These interpretations still stand. The occurrence of such empty or nucleic acid free particles as top zones in density gradient centrifugation tubes turned out to be rather common. After density gradient centrifugation of extracts from plants or insect vectors infected with wound tumor virus a zone of soluble antigen much smaller than the virus but serologically related to it was found at the top of the tube whereas a separate virus zone was found below. This observation could also be explained on the basis of breakdown products from virus or residual surplus components required for forming virus.

MULTIPARTICULATE VIRUSES

New questions were raised when several different plant virus infections each yielded two kinds of particle with nucleic acid, but with obviously different sedimentation rates. In general, the fastest sedimenting zone of such particles was the most infectious. After centrifugation in density gradient tubes, the particles in these zones were commonly called middle and bottom components because usually there was a top component particle without nucleic acid. Sometimes in the electron microscope all three particles were similar in size and near spherical shape although staining often was different. Was the middle component a partially completed virion or a breakdown product?

Around 1960, these phenomena were studied in several laboratories, one of the most active being that of J. B. Bancroft. The zones of middle and bottom components often appeared to be clearly separated when examined by eye. However, optical density measurements throughout the density gradient column in a centrifugation tube, and other evidence, indicated that the separation of the zones was not complete and it was possible that each contained a mixture of particles. Both middle and bottom zones were infective but addition of the middle zone to the bottom zone produced a mixture which was more infective than either alone. Why? Did the middle component enhance the infectivity of the bottom component in some physiological way? It was well known that various physiological conditions affected the infectivity of virus preparations. Or did the middle component particles restore the infectivity of some defective bottom component particles that had lost an essential part? Interparticle reactivation of bacterial viruses inactivated by ultraviolet light had been demonstrated earlier.

Wood & Bancroft (1965) used the near spherical comoviruses extensively in their studies of this problem and related the infectivity of various fractions to the optical density at a wavelength used for measuring the concentration of nucleic acid. On the basis of the same nucleic acid content the bottom zone was still the most infective of the zones but they clearly demonstrated that the infectivity of a mixture of middle and bottom components, was greater than that of either of the separate zones from which it was prepared, when comparisons were made at the same nucleic acid content. They found that the purer the zones the greater the enhancement of infectivity in mixtures. Furthermore, in some cases the enhancement was restricted to mixtures of fractions from the same virus species even when a mixture of middle and bottom components came from different species of the same group. They favored the hypothesis that the bottom fraction contained fully infectious virus particles and also others which were defective but which could be rendered infective in some way by the middle component particles.

Although further efforts at purification with this system and other similar ones decreased the infectivity of the separate zones and increased the enhancement effect, no physical-chemical technique was available that could provide convincing proof that a zone containing one kind of particle did not contain any particles of another kind. In any case at this time, physical-chemical efforts at purification did not result in completely noninfective bottom component preparations that could be rendered infective by the addition of noninfective middle component. Moreover, such an experiment would not have provided a conclusive answer because it is easy to imagine purified preparations of middle and bottom components which, after each had been diluted just beyond the end-point for infectivity, would produce an infective mixture when recombined. One should have in mind the fact that the most sensitive infectivity assays of plant viruses require a concentration of at least a million particles per milliliter at end-point dilution and that solutions diluted beyond the point where infectivity could be detected might contain many infective particles detectable by reconcentration.

Light was to be thrown on this problem by research, carried out during the same period in several laboratories, on the very different tobacco rattle virus (TRV). This virus produces rod-like particles of two very different lengths, but with the same diameter. When preparations of tobacco rattle virus near limiting dilution were used to inoculate plants, or when single primary lesions were used as inoculum, some of the infected plants exhibited a syndrome strikingly different from that of the usual infection. Such plants contain no detectable virus particles but they do contain free virus ribonucleic acid (RNA) which can be used to inoculate healthy plants. The RNA is unstable and for this reason the infection has been called unstable, although the unstable RNA inoculum from such sources always results in the same syndrome with new virus RNA but no virus particles. There is never reversion to the original unmodified syndrome or stable inoculum during an indefinite number of passages in series. However, purified preparations of short particles at dilutions that will not produce any infections in plants will, in combination with RNA from the unstable infection, produce the original unmodified disease and virus particles of both lengths. Therefore, one can be sure of the complete absence of any short rods in the right biological source of inoculum, that is, the unstable infection. The occurrence of any short rod in such an infection would soon result in the infection reverting to the stable form with the normal syndrome and particles of both lengths. This biological source of one infective component, absolutely free of any short particles, or their RNA, was something unattainable by physical-chemical purification techniques of the time.

To explain these results, Lister (1966) hypothesized, "Present theory suggests that the simplest explanation of this system would be that the RNA of 'long' particles of the TRV type is deficient in the information required for some stage of the process leading to the enrobement of viral RNA with virus protein; possibly the coding for the virus protein itself. Hence, inocula containing only long particles give rise to infections only of the unstable type. The RNA of 'short' particles, on the other hand, though containing at least, or perhaps only, the information lacking in that of the 'long' particles, is inadequate to mediate other stages in the infection cycle." This constituted an explicit statement of a viral genome normally divided and having different parts encapsidated separately in different particles. Implicit in his statement is the idea that the two different particles are not defective. This hypothesis was a crucial challenge to the dominant concept that the infectious particles of a virus species were all the same. The correctness of the hypothesis was supported by additional experiments on tobacco rattle virus by Lister and by others.

Later work with near spherical viruses, and others, and more and more highly purified preparations of top, middle, and bottom virus particles, some of them noninfective alone, has shown that the phenomenon of such multiparticulate virus species possessing two or even more normal and essential virus particles is a widespread phenomenon among plant viruses. This concept had many important consequences such as correct determinations of genome molecular weights and convincing hybridization experiments with plant viruses. In fact, the latter hybridizations, with marker genes for symptoms or other traits, often provided crucial evidence for other viruses having genomes divided between two or more particles.

Actually the phenomenon has proved to be so widespread among plant viruses that we cannot now be sure, even when all the virus particles of a species appear the same in the electron microscope or in density gradient tubes, that they are in fact genomically the same. The outlook of plant virologists underwent a revolutionary change. Sometimes the electron microscope reveals very slight differences between two kinds of particles in a preparation, or a preparation with apparently one kind of virus particle reveals several nucleic acid species by gel electrophoresis. However, conceivably, one might have a virus with a genome divided between two particles containing different essential genes but which nevertheless might appear identical under the electron microscope and not have detectably different nucleic acid base ratios or electrophoretic migration rates. Whether such particles have different genes which are together essential for infection may be tested by the curve relating virus concentration to infectivity. Sometimes in the current period of high technology such a simple, unsophisticated test as the use of the "dilution curve" seems not to be sufficiently regarded to be utilized even when appropriate. Wound tumor virus has 12 separate double-stranded ribonucleic acid genome components but it gives a convincing single hit curve relating virus to infectivity. This, more than the approximate equimolarity of the 12 genome segments in virus preparations, provides convincing evidence that it is a monoparticulate virus with the same multipartite genome in each particle.

There were some premonitory indications of this complex situation—a phenomenon not so unusual in science. As a result of studying certain plant viruses with steep dilution curves, Fulton (1962, p. 484) had recalled the reactivation of phage inactivated by ultraviolet light, and had suggested that two or more virus particles "at one site might provide a complete complement of genetic units, one or more of which were lacking in single particles." He also had suggested that (p. 477) a "dose requirement may be satisfied partly by virus that is unable to infect." Bawden (1964, p. 371) had speculated that the long particle of tobacco rattle virus had the abilities to infect and to synthesize coat protein but that the short particle lacked the ability to infect. Apparently he regarded the long particle as a complete virus particle. Looking back, it is clear that the density gradient centrifugation technique played an important part in making it apparent that extracts from diseased plants contained virus-related components—soluble antigen and more than one kind of virus particle—in many plant virus infections.

THE VIROID

Viroids turned out to be a mystery within a mystery. Of those diseases that later proved to be viroid, the first to be described was potato spindle tuber. It was identified as a distinct transmissible disease of *Solanum tuberosum* in the early 1920s and was considered to be caused by a virus. One should recall that the physical nature of viruses was unknown until about 15 years later. Since 1935 more than 200 distinct plant virus species in at least 23 genus-like groups have been well described and the mystery of their physical nature resolved. However, it was not until the early 1970s that this could be claimed for any viroid. Then, because they were basically so different from viruses they were given the new name *viroid.* PSTV once meant potato spindle tuber virus; now it means potato spindle tuber viroid.

In 1952 Brierley found that chrysanthemum stunt virus (now viroid) survived heating in boiling water for 10 min. When he asked me what I thought of this, I had to confess that I did not know what to think; tobacco mosaic virus, which was regarded as the acme of stability, is inactivated by heating at about 90°C for 10 min. The unusual properties possessed by the "viruses" causing chrysanthemum stunt and potato spindle tuber and citrus exocortis diseases remained anomalies over a period of decades during which more and more plant viruses proved to be nucleoproteins of specific size and shape that could be isolated and visualized by means of electron microscopy.

Little progress was made in revealing how PSTV differed fundamentally from other plant viruses until Diener and his colleagues began to study it extensively. Economically important on *Solanum tuberosum,* spindle tuber was difficult to work with. As a student inspecting certified seed potato fields I had more trouble in recognizing and identifying symptoms of this infection in the tops of plants than those of any other. They form a gradient of imperceptibly gradual and often vague changes from the healthy condition. K. H. Fernow, under whom I worked as a potato inspector and who supervised my thesis research on potato yellow dwarf, was much better than I at detecting plants with spindle tuber. Of course, Fernow had been among the few plant pathologists who first described and transmitted the disease. I was often unsure of my diagnosis unless I had the help of tuber unit plantings (groups of four consecutive plants in a row, all derived from pieces of the same tuber). Such a planting made it easier to detect a difference between the overall appearance of spindle tuber plants and adjacent healthy ones. However, even Fernow had trouble diagnosing the mildest strains of spindle tuber and he devised a cross protection test in tomato plants. Preinfections by suspected mild strains in this plant were revealed by the interference they offered to subsequent challenge inoculation with a readily recognized strain.

During much of the period of active research on the nature of PSTV no local lesion host was known. Raymer & O'Brien had been the first to infect tomato plants and in these systemic symptoms appeared in much less time than in potato plants. This led to a unique bioassay by Raymer & Diener (1969) in which the proportion of plants infected and the lengths of incubation periods at each of a number of inoculum dilutions were all incorporated into a single numerical infectivity index. This bioassay permitted relative measures of the concentration of PSTV in extracts. Subsequently, in the long period of prerequisite purification work of Diener and his collaborators this infectivity assay was the only means for identifying PSTV in various fractions and eventually in zones in gels in which preparations from diseased plants had undergone electrophoresis. After many years of work Diener (1972, 1979) obtained an infective zone in a gel electrophoresis column that he could identify by untraviolet light absorption and clearly distinguish from other absorption zones caused by small normal host RNA components. This identification was crucial to subsequent studies. In this successful work Diener obtained only 80 μg of potato spindle tuber viroid RNA from 5 kg of infected tomato leaves, or about 1 part of PSTV in 10^8 parts of infected host tissue. Even with inevitable losses during purification this result provided a tangible basis for understanding earlier failures to detect ultraviolet absorption even when infectivity was readily demonstrable.

Semancik & Weathers (1972) had been studying the pathogen (CEV) causing citrus exocortis disease for a number of years and obtained results indicating that the cause was a similar viroid. Subsequently, Diener and his collaborators, Semancik with his, and a number of other workers have added much to our understanding of viroids.

In prior studies there had been many surprising results in addition to the unusual heat stability. The behavior of viroids in experiments with ribonuclease, with gel electrophoresis, and with other treatments was often puzzling or anomalous and of uncertain interpretation. There was wide disagreement in determinations of the approximate molecular weight from gel electrophoresis studies. In addition to the early failures to detect any ultraviolet light absorption that could be identified with infective preparations that were susceptible to inactivation by ribonuclease, there were other negative results: No virus particles could be detected in infected tissue and no new proteins or antigens could be demonstrated in infected plants. My impression of reactions to reports of such findings ranged from quizzical to skeptical—reactions that were not lessened by the tentative evidence that viroids were much smaller than any pathogen previously known. How could a naked viroid RNA, less than one third the size of the genome of the smallest known virus, replicate without parasitizing an infection by a helper virus? The aforementioned smallest known virus requires such a host virus infection. How could a single viroid particle infect and replicate? PSTV is not multiparticulate. It has a single hit curve relating concentration and infectivity and this was demonstrated clearly when a local lesion host became available. How could different strains of something so small maintain separate identities unaffected by the host in which they replicated? Amazingly enough, viroids do possess all these implied properties.

The progress that has been made in characterizing viroids is scarcely less amazing. Viroids have been visualized by electron microscopy and it is clear that a particle of a single viroid species may take the form of a single-stranded, circular, covalently closed RNA or a single-stranded linear RNA. Because of the high internal nucleotide base complementarity, the circle may be zipped together in such a way and amount that it appears as a short double-stranded RNA rod. The double-stranded RNA segments of the rod are assumed to be interrupted by small single-stranded RNA loops. The linear and circular forms assumed when the rod is denatured can be separated by gel electrophoresis; all forms are infective. The entire sequence of the 359 nucleotides in PSTV has been determined and from this a precise molecular weight has been calculated. Digestion of a viroid with ribonucleases and chromatography of the nucleotide fragments yields a specific fingerprint. The fingerprints of PSTV and CEV are different and although both viroids produce spindle tuber symptoms in potatoes and the same striking disease in a laboratory host plant called *Gynura aurantiaca,* the specific fingerprint of each is not changed by the experience.

Investigations of the attributes of viroids are still proceeding at an astonishing rate. Already the identification of viroid infections in individual plants of foundation potato seed stocks by gel electrophoresis is more reliable than examination for symptoms. Cadang-cadang, a devastating disease of coconut palm trees, which had long been fruitlessly investigated as a possible virus disease, seems to have yielded positive results when examined for a viroid etiology. It seems quite likely that some long-standing mysterious diseases of animals and man will prove to be caused by viroids.

The discovery of viroids is undoubtedly a contribution from plant virology which has importance for other scientific disciplines. It is too early, of course, to know if it ranks in importance with certain other contributions from plant virology such as the discovery of viruses or the discovery of multiparticulate viruses.

MYCOPLASMA-LIKE ORGANISMS

The resolution of the virus and viroid problem and, to a lesser extent, the resolution of the puzzle of multiparticulate viruses were long drawn out affairs. In contrast, the considerable group of diseases like aster yellows, for many years considered to be virus diseases although virus had not been demonstrated, were suddenly shown to be associated with mycoplasma-like organisms (MLOs). Although Koch's "rules of proof" have even yet not been satisfied for plant MLOs, Doi et al (1967) and Ishiie et al (1967) provided convincing evidence that they cause diseases like aster yellows. This came both as a shock and a tremendous stimulus to research. It was not long before about 50 plant diseases were removed from the category of viral diseases to a new group caused by MLOs or other cellular parasites.

Kunkel's 1926 paper was the definitive paper on the insect transmission of plant diseases of the aster yellows type. People seem not to realize the implications of the interval of almost 10 years between his publication and that of Stanley on tobacco mosaic virus. It is interesting to reexamine Kunkel's paper to see how he wrote about the pathogen causing aster yellows. He thought that the incubation period of at least 10 days in the insect vector was probably "due to a development and multiplication of the *causative agent* in some tissue of the leafhopper," page 701 (my italics). Elsewhere he wrote that aster yellows "belongs to that group of obscure plant maladies known as virus diseases," page 646, but "it has not been shown to be due to a filterable virus," page 699. In the same paper, on page 700, he writes about "the condition of the virus" in the insect vector.

Between 1926 and 1935 the nature of viruses was itself obscure and during this time it became conventional practice to write about pathogens of the aster yellows type as viruses. The clearing of the leaf veins in early stages of the diseases they cause is almost universally the first distinctive systemic symptom of known plant virus infections. Transmission only after a week or more of incubation following acquisition by leafhoppers also occurs in some plant virus diseases.

Between about 1940 and 1970, F. C. Bawden was, in my opinion, the most lucid and rigorously critical writer on plant virus diseases. A fine example of such writing is the following excerpt from page 13 of his 1950 textbook:

> ... in spite of the wide differences between the properties of the individual viruses so far purified, it would be rash to assume that they form a random and representative sample of the whole group. Because these are chemically similar, it can almost be taken for granted that some other viruses are also nucleoproteins, but to assume that all are would be decidedly premature. The methods of isolation so far used may be acting as a selective agency, succeeding only with those that are nucleoproteins. If this is so, and that it may be is perhaps suggested by the failures of attempts to isolate some viruses, the apparent chemical uniformity of plant viruses may be a temporary phase, reflecting merely the limited range of procedures yet applied to their study, and new methods may identify some that are chemically different. This needs stressing, but the identification of a plant virus as anything other than a nucleoprotein will be a major event, calling for much supporting evidence, whereas a claim to have isolated a new one as a nucleoprotein will be accepted as merely adding another to the growing list.

It is true that this cautionary note does not explicitly mention the idea that some of the "viruses" might not even be viruses but it could have been logically extended to do so. However, on page 104 of the same book, Bawden, the persistent critic of evidence for multiplication of plant virus in insect vectors, accepted my evidence that the clover club leaf virus multiplied in leafhoppers transmitting it. He did not question the idea that the pathogen was a virus any more than I did. Nevertheless we now know that it is not a virus but a fastidious phloem-limited bacterium. Actually, after critical evidence was available on what was virus and what was not, the first plant pathogen proved to be a virus and to multiply in the vector turned out to be wound tumor virus.

Obviously Bawden was not alone in such matters. In 1943 I reported that my filtration and ultracentrifugation studies on aster yellows suggested that activity was "associated with a particle that is of large size relative to plant viruses heretofore studied." I made no suggestion that it might be other than a virus.

These facts provide a good measure of the universality of the usage of *virus* for various pathogens at that time. It had become conventional to write of these pathogens as viruses and no doubt to think of them as such —at least most of the time—but obviously not all of the time. It should not be forgotten that during this same period many more viruses than mycoplasma-like organisms were termed *viruses* before they had been proved to be such.

The identification of mycoplasmas as agents of disease in animals had resulted in knowledge of the distinctive features of fine structure in these simplest of cellular pathogens as revealed by electron microscopy. The recognition of similar organisms in the phloem of plants with diseases of the aster yellows type probably could not have occurred until after the techniques of embedding, ultrathin sectioning, staining, and electron microscopic resolution had attained a very high level. I feel certain, for example, that at least as late as 1950 the techniques would not have been adequate to the task. The Japanese discoveries of the plant mycoplasma-like organisms led directly to the discovery of the phloem-inhabiting spiroplasmas, the phloem-limited bacteria, and the xylem-limited bacteria as the pathogens of other diseases transmitted by leafhoppers. Because clover club leaf, caused by a phloem-limited bacterium, exhibits a virescence of the petals in *Vinca,* long considered a very distinctive symptom of the "yellows" diseases, the discovery of its pathogen in the phloem in a search for mycoplasmas was inevitable.

AFTERTHOUGHTS

Many scientific "firsts" can be attributed to the great expansion of research on tobacco mosaic virus—one of them being the clear evidence that RNA as well as DNA can provide the basic hereditary substance. It seems obvious that if there had been a bandwagon development on tobacco mosaic virus to the extent that basic research became its exclusive province many important discoveries in virology would have been delayed. This appears to have been avoided by the economic necessity of investigating many other plant virus diseases. On the other hand, no economic motive initiated the research on Brakke's[3] invention of density gradient centrifugation. It resulted from an effort to purify wound tumor virus, which had no known economic importance, and the technique was worked out with a second virus, potato yellow dwarf, of minimal economic importance at the time, because it was thought to be better than wound tumor virus for the purpose. The importance of density gradient centrifugation has spread far beyond plant virology.

Some of the discoveries on viruses and viroids were among those that at least blurred, if they did not obliterate, any nice distinction between the living and nonliving. They also shook some of our neat conceptions about major biological categories. Originally I had learned to think of organisms as being comprised of one or more cells. Now, some of the more compli-

---

[3]The invention of rate zonal density gradient centrifugation is correctly attributed to M. K. Brakke (1951) *J. Am. Chem. Soc.* 73:1847. K. R. Porter's article in *Science* 186:517 (1974) on the Nobel laureates A. Claude, C. de Duve, and G. E. Palade may have inadvertently created a different but erroneous impression. This footnote is based on my correspondence with all five men.

cated viruses, such as the phage with its tail, I tend to think of as an organism below the cellular level, with specialized noncellular organs for performing definite functions. Even when the structures of a virus cannot be visualized, the apparently simple surfaces may exhibit very specialized properties as with the very select transmission niches of the apparently simple plant viruses. Viruses may show no anatomical features obviously connected with transmission but consider their capabilities for physiological effects on the host, such as the aggressive biting behavior of rabid animals. Can one doubt that these capabilities are also subject to change based on mutation and evolutionary selection? Characters directly involved in transmission would seem to be especially sensitive to selection because they are so crucial to perpetuation of the species.

The surprising properties of the viroid seem unlikely to have a simple explanation. A viroid rod or ring looks very simple in the electron microscope but if the resolution were such that we could see the position of each atom and understand its part in viroid endowments a long evolutionary selection might well appear to be a necessity rather than a surprise.

Like classical cellular organisms, all of these entities have a double-stranded nucleic acid phase for ensuring their hereditary continuity, and four nucleotides essential to the double helix are susceptible to the same mutations and evolutionary selection. When the organism concept was developing, would we have separated viruses and viroids off as nonorganisms if we had known what we know today?

ACKNOWLEDGMENT

This is the first time that my wife, Helen Wilhelm Black, has accepted any acknowledgment in my publications. She has been my most honest critic and I welcome this opportunity to express my appreciation for her long and generous assistance.

SELECTED REFERENCES

Research Papers

Bawden, F. C., Pirie, N. W. 1937. The isolation and some properties of liquid crystalline substances from solanaceous plants infected with three strains of tobacco mosaic virus. *Proc. R. Soc. London B* 123:274–320

Bawden, F. C., Pirie, N. W., Bernal, J. D., Fankuchen, I. 1936. Liquid crystalline substances from virus-infected plants. *Nature* 138:1051–52

Diener, T. O. 1972. Potato spindle tuber viroid. *Virology* 50:606–9

Diener, T. O. 1979. Viroid discovery. *Science* 206:886

Doi, Y., Teranaka, M., Yora, K., Asuyama, H. 1967. Mycoplasma- or PLT group-like microorganisms found in the phloem elements of plants infected with mulberry dwarf, potato witches' broom, aster yellows, or Paulownia witches' broom. *Ann. Phytopathol. Soc. Jpn.* 33:259–66

Fulton, R. W. 1962. The effect of dilution on necrotic ring spot virus infectivity and the enhancement of infectivity by non-infective virus. *Virology* 18:477–85

Ishiie, T., Doi, Y., Yora, K., Asuyama, H. 1967. Suppressive effects of antibiotics of tetracycline group on symptom development of mulberry dwarf disease. *Ann. Phytopathol. Soc. Jpn.* 33:267–75

Kunkel, L. O. 1926. Studies on aster yellows. *Am. J. Bot.* 13:646–705

Lister, R. M. 1966. Possible relationships of virus-specific products of tobacco rattle virus infections. *Virology* 28:350–53

Raymer, W. B., Diener, T. O. 1969. Potato spindle tuber virus: A plant virus with properties of a free nucleic acid. 1. Assay, extraction, and concentration. *Virology* 37:343–50

Semancik, J. S., Weathers, L. G. 1972. Exocortis disease: Evidence for a new species of "infectious" low molecular weight RNA in plants. *Nature New Biol.* 237:242–44

Stanley, W. M. 1935. Isolation of a crystalline protein possessing the properties of tobacco-mosaic virus. *Science NS* 81:644–45

Stanley, W. M. 1936. Chemical studies on the virus of tobacco mosaic. VI. The isolation from diseased Turkish tobacco plants of a crystalline protein possessing the properties of tobacco-mosaic virus. *Phytopathology* 26:305–20

Wood, H. A., Bancroft, J. B. 1965. Activation of a plant virus by related incomplete nucleoprotein particles. *Virology* 27:94–102

Books and Reviews

Bawden, F. C. 1950. *Plant Viruses and Virus Diseases.* Waltham, Mass: Chronica Botanica. 335 pp.

Bawden, F. C. 1964. Speculations on the origins and nature of viruses. In *Plant Virology*, ed. M. K. Corbett, H. D. Sisler pp. 365–85. Gainesville: Univ. Florida Press. 527 pp.

Diener, T. O. 1979. *Viroids and Viroid Diseases.* New York: Wiley. 252 pp.

Diener, T. O. 1981. Viroids. *Sci. Am.* 244:66–73

Gibbs, A., Harrison, B. 1976. *Plant Virology, The Principles.* New York: Wiley. 292 pp.

Harrison, B. D., Murant, A. F., eds. 1980. *Descriptions of Plant Viruses.* Farnham Royal, England: Commonw. Agric. Bur. 14 sets.

Jaspars, E. M. J. 1974. Plant viruses with a multipartite genome. *Adv. Virus Res.* 19:37–149

Semancik, J. S. 1979. Small pathogenic RNA in plants—the viroids. *Ann. Rev. Phytopathol.* 17:461–84

Stanley, W. M. 1947. At the twilight zone of life. In *The Scientists Speak*, ed. W. Weaver pp. 164–68. New York: Boni & Gaer. 369 pp. 3rd ed.

Whitcomb, R. F., Black, L. M. 1981. Plant and arthropod mycoplasmas: A historical perspective. In *Plant and Insect Mycoplasma Techniques*, ed. M. J. Daniels, P. G. Markham. London. Croom Helm. In press

Kenneth E. Baker

Ann. Rev. Phytopathol. 1982. 20:1–25

MEDITATIONS ON FIFTY YEARS AS AN APOLITICAL PLANT PATHOLOGIST

Kenneth F. Baker

Professor of Plant Pathology Emeritus, University of California, Berkeley, and Collaborator, Ornamental Plants Research Laboratory, USDA-ARS, Oregon State University, Corvallis, Oregon 97333

After due reflection it has seemed best to use the pleasant but awesome privilege of preparing this Prefatory Chapter to discuss some general developments that have long been on my mind—a memorandum on observations and views of plant pathology in the last half century.

PLANT PATHOLOGY IN FLUX

Plant pathologists in the last two decades have indulged in a surprising amount of self-criticism (e.g. 9, 10, 24, 25, 34, 35, 36, 41), indicating a confused uncertainty about our objectives, and concern about our public image. Horsfall (24) felt in 1959 that the science of our discipline was underdeveloped, and Stevens (41) thought in 1961 that the art was thriving but that the science was essentially nonexistent. Pound (35) in 1965 thought that we must use the technology of molecular biology or "settle into a lower category of eminence." Three years later, from the vantage point as Dean of Agricultural and Life Sciences, University of Wisconsin, he (36) was concerned that "We are experiencing a steady retreat from the applied research field. . . . we dare not forsake production research. We must do both concurrently, and in balance! Plant pathology must maintain an agricultural orientation." Bateman (9) in 1980 also felt that the "empirical approach to problem solving . . . has tended to decrease . . . development of the theory and science of pathology." All of these writers excessively emphasized a spurious distinction between basic and applied research, and

1919

0066-4286/82/0901-0001$02.00

did not mention philosophy, perhaps reflecting the deficiency of synthesis of our knowledge into principles and generalizations.

The Effects of Grantsmanship

Plant pathology had its inception in man's need to protect his crops against disease, and until at least 1950 this was regarded as the raison d'être for our profession. However, there was a strong trend away from applied or field research following the end of World War II in September, 1945 (25). Because it was felt that the United States was deficient in basic research, the generous war-time funding of university and industrial laboratories was continued. The National Science Foundation was established in 1950 to advance fundamental scientific research,and began giving grants to individual scientists for basic studies. The National Institutes of Health, research arm of the long-established Public Health Service, also began giving research grants about 1950 for basic studies in plant pathology, often only marginally related to medical research. The Atomic Energy Commission was established in 1946, and made grants for studies relating to peaceful uses of atomic energy.

The research funds available to plant pathologists from these and other federal, philanthropic, and industrial sources were greatly augmented after 1957. On October 4 of that year the USSR put Sputnik 1 into orbit as the first artificial satellite, followed by five more in the next year. Although the US launched Explorer 1 in January, 1958, followed by nine others within two years, and landed men on the moon in July, 1969, there was a widespread feeling that the US was falling behind in basic research and in training of investigators. It is interesting to ponder why the investigations of the National Aeronautics and Space Administration were considered to be basic, since they certainly were directed toward a practical goal.

This feeling of intellectual unease was translated into large sums of money pumped into the universities and centers for fundamental research untainted by utility. This brought about major changes in American universities, and began erosion of the idea that the prime goal of plant pathology is to be of service to agriculture.

EFFECT ON THE INDIVIDUAL   The historically accepted concept that the individual should select his research relative to the needs of agriculture, within the limits of his ability, interests, and insights, and that he would be judged by his success, changed radically within a few years. Research projects of plant pathologists in many universities came to be selected on the basis of potential financial support, that was determined in the last analysis by an anonymous panel of basic scientists. This led to grantsmanship, the submission of "safe" projects likely to appeal to panels who wished to invest the funds in fundamental studies that they thought had the best chance of success, a decision that the National Academy of Sciences recently found, at least for the National Science Foundation, perhaps depended as much on the luck of the reviewer draw as on merit (33a). The subject needed a tinge of newness, but should not be so close to the frontiers of knowledge as to appear to be a risky investment. While the project had to appear to the local Experiment Station Director to fit into reasonable station activity and presumably to have potential practical value, it had to appear to the panel to be entirely basic. This naturally led to distortions, pressures, and "fashionable" research.

Investigators were frankly told by university administrators to seek outside funds, since state funds were drying up. When this was done, these funds often were given top priority by the investigator. There was a good deal of lip service to agriculture. Complaints were heard from university administrators that problems of their state's agriculture were not receiving attention, from growers that station staff were too busy on their priority grant projects to give them help and that they were expected to contribute money to get work done on their problems, and from Agricultural Extension workers that they were losing their sources of practical information and were becoming second-class citizens. Investigators explained that they had been told to seek funds, and having successfully done so, the grant project must receive top priority to insure continued support. Financial and priority control had effectively shifted from station administrators to anonymous panels in Washington.

These pressures often produced insidious strains on ethical standards in the profession, ranging from prematurely releasing inadequately checked data, to reporting the same data to different granting agencies, to presenting a colleague's data as one's own, and even using administrative blackmail to get one's name on a paper or project. As publication pressures on individuals increased, there was greater conscious or unconscious borrowing of ideas from the oral communication network.

Devious maneuvering often was required to gain approval of legitimate projects funded by granting agencies. For example, the National Research Council Committee on Biological Control of Soilborne Plant Pathogens wished to organize an international symposium to stimulate work in that field. Funds were sought from the National Science Foundation, who insisted that the subject be less applied—control was a utilitarian term. The title was changed to Factors Determining the Behavior of Plant Pathogens in Soil, and the topic Crop Rotation became Absence of the Host, Tillage Practices became Altered Physical Condition of Soil, and Cover Crops became Plant Sequence. More than two years of negotiation were required to sanitize our agricultural language before the project was approved.

Graduate training was also affected by the grant system, since most American and foreign students in plant pathology in this country were supported by such funds. The thesis subject was likely to be selected, therefore, to comply with the criteria of the anonymous granting panel in Washington. Several generations of students were thus directed away from the art of plant pathology for more than 20 years. They were confronted with the well-equipped laboratories and numerous assistants of the efficient grant-promoter, and perhaps led to confuse these symbols with true scientific merit. It was particularly unfortunate that students from developing countries were led to devote their time to using expensive specialized equipment rather than learning the art of field pathology, so needed when they returned home.

EFFECT ON THE PROFESSION   The very essence of plant pathology has gradually been altered by these various stresses, so that the field has listed heavily toward abnormal physiology. Thus, McNew (33) emphasized the biochemical approach as the hope of the future: "The primary fact in all plant pathology is the one of nutrition or nourishment of the pathogen." This ignores the important nonpathogenic diseases and microorganism interrelationships, inter alia. Bateman (9) took a somewhat broader view, but still emphasized "impaired physiology." Molecular biology and molecular genetics have now become a new focus of interest. Biology exhibits waves of bandwagon enthusiasm or fads, useful or harmful (39), the same as other human activities. Are we on the crest of a biochemical and molecular wave? There is a very real danger that aspects of plant pathology currently unfashionable (e.g. diagnostics, biology of the pathogen, ecology of plant disease, pathological anatomy, biological control) will be neglected. Perhaps the real essence of our field, the areas that are peculiar to plant pathology and that tend to be ignored by biochemical and molecular advocates, are diagnostics (20) and the ecology of plant disease. With these factors included, the concept of plant pathology as "defender of plants" will continue; with pathogen nutrition or molecular studies as the essence we may become a satellite of plant physiology and molecular biology. Biochemical and molecular investigations must be continued, but not at the expense of applied studies.

"Theoreticians, with mathematical ability that most of us can only envy" have attempted to reduce some aspects of our field to mathematical models. Hirst (23) continued, "we must strive to ensure that, in the course of their intellectually satisfying work, they are not tempted to diverge entirely from the ways in which the organisms really operate. Inevitably models portray averages; equally inevitably every occasion . . . is peculiar. . . . In aerobiology . . . dispersal models have not provided more valuable information than observation." Models have contributed little beyond what is implicit in the actual data on which they are based, and it is frequently said that one of their main values is in indicating significant missing data (e.g. 9, 30). Kranz (31) commented that "a model is . . . an abstraction of the real world . . . but by no means the reality itself or its replica. This implies that a model is rarely complete, final and an objective in itself." Perhaps the major hazard from model building is in "mistaking a blueprint for a going concern" (8). As Grogan et al (21) stated, "biological interpretations should be based on biological observations and not on tests for linearity of transformed data."

Vocal advocates of the "new pathology," coupled with the pressures of grantsmanship, have made many older pathologists feel the creeping malaise of obsolescence, and efforts at rehabilitation are fairly common. There are negative aspects in such a trend. People relatively untrained in biochemistry may dress up a paper by grafting on little-understood or poorly designed experiments, even as their elders blossomed out with proba- ble error data in their tables when statistics came into its own. Others, in attempting to master the new pathology, may use time that might be more profitably spent in working within the limits of their training, since there is much that still needs to be done on the biology of plant disease. This process of obsolescence promises to accelerate, with each generation facing the problem at a younger age.

Forty years ago students took several courses on diseases of specific crop plants. These are now very rare, and therefore preparation of syllabi or collections of information that lead to reference books on the subjects has almost ceased. Furthermore, since these courses are no longer given, and plant pathologists have a reputation of not being book buyers, publishers are reluctant to produce books on crop pathology. Unless other sources can be developed, one can foresee a time when there will be no modern detailed reference books in these areas comparable to those available to the medical profession. Semipopular publications like the Compendia published by the American Phytopathological Society hardly fill this need. Will we wait for some publisher to discover this area of need or will we supply the leadership rightly to be expected from our profession?

Extension Plant Pathology

The Agricultural Extension Service is undergoing a rapid change of orienta- tion. The percentage of the population in agriculture has declined in the last 80 years from 60 to less than 5, while the number of both plant pathologists and extension workers has greatly increased. The growers today are far better educated and informed than ever before, and the extension worker's best source of tested information often is the most successful grower in his area.

The service concept has sometimes come to mean doing something for the grower that he would better do for himself. Experience in Australia convinced me that our detailed extension assistance may actually be undermining the ingenuity and initiative of our growers. In the very early stages of development of aerated steam treatment of nursery soils, for example, I told a group of Australian nurserymen how we hoped it might commercially be accomplished; a few weeks later several of them independently had successfully worked out methods and had them in commercial operation. A similar group of California growers were later given much more specific details, but waited for detailed blueprints before attempting it.

The new graduate in agriculture is hardly trained to advise an educated and experienced grower on production problems. Recognition of this has led to upgrading the level of training of extension workers. There are more extension pathologists with an M.S. or Ph.D. each year, and sabbatical leaves often are available to update the training of county agents and farm advisers. With better training and less service work, these men are increasingly participating in research, particularly in developing fungicidal and culture control measures, in obtaining pathogen-free plant propagules, and in field experiments. With the expected future increase in applied research, the extension plant pathologists will have a more important role in investigations, and extension specialists will participate even more in applied research. Unfortunately, farm advisers sometimes come to regard their county as a private fiefdom, and attempt to exclude information not in accord with their personal beliefs.

There are several types of growers that may be grouped as to their reactions to the occurrence of plant disease in their crops. They are arranged here in ascending order of their merit as cooperators in extension research. (a) The Pollyanna Type isn't aware of or denies the existence of disease in his crop, through indifference or ignorance. (b) The Walter Mitty Type thinks that nothing can be done about disease, blaming it on the weather, and citing as an alibi other growers that are even worse off. (c) The Mañana Type appreciates help and intends to do something about it, but just doesn't get around to it. (d) The Las Vegas Type operates on an expected percentage-loss basis, and sets prices accordingly; he changes crops frequently to reduce loss. (e) The Conscientious Type recognizes disease, seeks help, and carefully follows instructions if he has the necessary capital. (f) The Investigator Type contributes to and pioneers new ideas. (g) The Misplaced Researcher Type may get so involved in developing new ideas that he doesn't get around to the exacting business of growing plants well. Types e and f (and possibly g) are the best cooperators; they usually are better educated and financed, and are followed by other growers. As Dimock (18) pointed out in 1950, if one works with the better cooperators there is no need to carry the word to every grower, nor to keep hounding a convert to prevent his backsliding. In my experience, cooperative tests with growers of types *a-d* are apt to be more harmful than helpful, as the plots are apt not to be properly cared for and the failure of the test is likely to be widely publicized by the grower.

Make no mistake about it, even though some plant pathologists may think that they have outgrown the earthy aspects of our profession, this is not in accord with the needs of a hungry world. Demand generates supply, and if we become too elitist to heed these demands of practical need, then someone else will do so. It is at our peril that we abdicate our birthright. There is at least as great a world need today for answers to plant-disease problems as there was in the days when plant pathology began. If this need is translated into the lush green of available funds for the necessary applied work, will there be a rapid reversion to our original purpose? I am concerned that the accumulated indifference to and prejudice against applied work will not easily be forgotten, even though history bears witness that money fosters respectability in science as elsewhere. The animosity between basic researchers and extension workers in the last 15 years shows that much healing still is needed. A heartening trend toward balance is shown by new students coming into the field. Just as Sputnik 1 fueled the trend toward basic research, so the empty Surplus Food Storage bins may reverse it. The future of plant pathology is bright if we maintain balance and harmony in our work.

ON GRADUATE TRAINING

Graduate study should be one of life's most pleasant and rewarding experiences. At no other time will the student have so many people ready, willing, and eager to advise, help, instruct, and even assist physically, financially, or psychologically. The world *is* the student's oyster, but there is the usual price tag: "From one to whom much is given, much is expected." One must make full use of this period. The experience should resemble a walk down a long hall of a well-stocked museum. Each of the many rooms that open on either side contains a different world of fascinating things. One cannot explore each and still reach the exit at the far end before closing time, but neither should one remain in a single room. This is an unequaled time to peer into other cubicles, stretch one's mind, and learn something of other disciplines and cultural subjects. One should get sufficient familiarity with a wide range of subjects to know how to go about getting information on each if future need arises. However, it is not necessary to take a formal course in each subject to gain such a brushing acquaintance with it.

Study for the qualifying examination for the doctorate is an opportunity to synthesize accumulated knowledge, rather than simply to memorize a mass of facts. One should seize this opportunity to construct a mental framework of the generalizations and principles of biology as well as plant pathology, on which to hang new facts encountered in the years ahead.

Four qualities necessary for success in science were outlined by Hull (28) in decreasing order of importance and increasing order of alterability by training: character, aptitude, attitude, and knowledge. These qualities also determine the suitability of a student for graduate study.

Character, as expressed in scientific honesty, integrity, ethics, patience, and persistence is essential and largely nontrainable. Sensitivity and perception are related to ability. The more sensitive and able a student, the more care and judgment must be exercised in bringing him on, as with a champion racehorse or a fine wine. N. E. Stevens, our best phytopathological essayist, pointed out (40), "for a life of real research, character, that is the ability to stand up under disappointment and frustration, is at least as important as intellectual ability."

Aptitude or ability, as shown by imagination, analytical powers, curiosity, and judgment, is crucial. Aptitude can be evaluated, but should not be confused with quickness of mind or brilliance. Since biological materials and processes develop slowly, there is ample opportunity for reflection and thought. The brilliant mind may become bored and discouraged, and turn to abstract theories poorly supported by facts. Imagination, coupled with persistence and patience, is invaluable; it can be encouraged but not learned. Judgment rests on thoughtful observation and analysis sharpened by experience, the balance wheel of imagination.

Attitude toward one's work, and one's work habits are important. Interest in one's field can be stimulated, and is a catalyst to success. There should be developed what Barzun (8) called "the sense of good workmanship, of preference for quality and truth, which is the chief mark of the genuinely educated man." Work habits established during graduate study are likely to persist. Older pathologists will have seen several *undesirable characteristics* of graduate students, and followed the impact of them on their careers: The Eight-to-Five Type is self defeating; the student may build up an interest in a problem by 5 o'clock, then puts it out of his mind until morning. Graduate students who avoid evening work in laboratories miss the pleasant and effective education from exchanging information, concepts, and philosophies with their peers—an excellent means of mutually acquiring the knowledge and experience in research to which each is exposed. The Mercenary Type views his profession merely as a job, and will do little more than is required; the wonder is that some of this type are able to continue this attitude until retirement, but they have missed the fun in the meantime. The Personality Type learns to trade on his pleasant manner to get ahead and to avoid work; it may take him quite far up the status ladder, with little benefit to his profession. The Brilliant Escapee uses his facile mind to outwit others and avoid work. These types tend to miss the real excitement and fascination of research.

Knowledge of the fundamental principles of the field, and expert skill in a subject, are highly communicable. Plant pathology is a derivative subject, bordering on crop fields (e.g. agronomy, horticulture, forestry) on the one hand, and source fields (e.g. botany, soils, plant physiology, biochemistry, genetics, meteorology, mathematics) on the other; it must be firmly rooted in the source fields, but must solve problems in agriculture. One must accumulate in an accessible form a large body of knowledge. This is the fertile soil into which clues and insights fall, growing into fruitful investigations or even a new discipline. As Chandler (14) commented, "Too much . . . is said about teaching students to think; for thinking about living things is dangerous unless that thinking is based on a detailed system of coordinated facts." The problem is to keep alive the powerful stimulant of individual thought while acquiring a background of facts. The larger the overall viewpoint and background of the individual, the more nearly correct his interpretations from research will be, because of a lessened chance of overlooking some fact or factor.

It is fortunate that the mass of knowledge does not increase proportionally with time. Studies often displace previous facts and relegate them to the historic scrap pile. New facts clarify comprehension and decrease necessary explanations and qualifications. Many isolated facts of 25 years ago have vanished into concepts and generalizations. Thus, the early theories on the nature of viruses and the function of pycniospores of rusts have been reduced to historical general knowledge. Once a principle or concept has been stated, it often becomes self-evident, and may therefore be difficult to publish (34). As Barzun (8) said in his inimitable *Teacher in America,* "The only thing worth teaching anybody is a principle. . . . just as a complex athletic feat is made possible by rapid and accurate coordination, so all valuable learning hangs together and works by associations that make sense."

The faculty establishes the reputation and atmosphere of a department, and this is reinforced by the students who sustain a sense of standards in a graduate program by competition and comparison, rather than discipline or force.

It is unfortunate that the beginning course in plant pathology is regarded in some departments as an unpleasant or unimportant duty, often fobbed off on a young or inexperienced person. Actually it may well be the most important course offered, as it is the only contact most students have with our discipline. If it is poorly taught, the subject will be held in low esteem by the students, one of whom may become a Dean, Director, University President, Legislator, or Congressman, and this low opinion then may affect the profession as a whole. A Dean under whom I once worked had such a view of plant pathology, and this was reflected in department support. The reason for this became clearer when he gave me his notebook prepared for the only course he had had in plant pathology; it was a course in mycology, emphasizing drawings of fungus structures, with nothing on epidemiology and almost nothing on control. The beginning course should be taught by the best qualified and most inspiring individual in the department, and this should be a respected position.

A graduate student's major professor has a very important role, being a sort of father- or mother-figure to inspire, counsel, or even to reprimand if necessary. The thesis problem selected should have been explored sufficiently to demonstrate that its completion is possible with the available time and facilities, and it must be of stimulating interest to the student. The thesis should be part of a continuing program, but should not be used to provide cheap labor on the professor's project. The student should be encouraged to develop a broad training and outlook. These things require that the professor be readily accessible to the student; thus, C. E. Allen at the University of Wisconsin used to make the rounds each morning to give every student opportunity to discuss or demonstrate his research progress.

ON RESEARCH

Every investigation begins with an *observation* of a phenomenon and some question concerning it; in plant pathology this usually is a field problem. It is then taken to the laboratory for *analysis* and investigation to determine the factors involved, and their interactions. The facts discovered are returned to the field for *verification,* and this usually will reveal a means of control or diminution of the disease. By reflection and *synthesis* the facts are then fitted into the body of knowledge, as exemplified in the publications of J. E. Vanderplank. It is perhaps as pointless to argue which of these stages is the most important as to debate which is the most important leg of a chair. Where in this sequence is the line separating basic and applied? The boundary between them clearly is more illusory than real, each grading into the other. That work which today is regarded as analytical and basic tends to become tomorrow's matter of casual observation. The first observation and analysis is the difficult one. It has correctly been said that there is no distinction between basic and applied research, only between trivial and significant or between good and bad research.

Research workers who make the greatest contributions to agriculture pursue a coordinated program for extended periods; "grasshopper research" may produce publications, but few accepted commercial practices. The most successful biological controls of plant pathogens certainly have resulted from such persistent studies. The unity of a program continued over many years may not be obvious. Thus, my studies over 40 years on *Phytophthora cinnamomi* on pineapple and avocado, on seed pathology, thermotherapy of plant propagules, steam treatment of soil, *Rhizoctonia solani,* and on biological control were parts of an overall program to obtain pathogen-free propagative materials and soil, and to support the program with sanitation. This was based on an early insight that the ultimate sources of plant pathogens are living plants and the soil, including its organic matter and water.

Alert intelligent growers play an important role in observations and in developing the working hypotheses with which all research begins. Because of their intimate daily contact with the crop and intense personal interest in it, growers have often made early perceptive observations and some have conducted far-sighted experiments that defined fruitful areas for investigation. Experienced research workers pay close attention to grower observations and ideas, and encourage free expression of them.

Philosophical vs Pragmatic Research

The need for distinction between fundamental (basic, prospecting, pioneering) and applied (mission-oriented, production) research is primarily an administrative convenience for allocation of funds. Too frequently, however, biologists have come to equate fundamental with difficult and applied with easy, although actually the reverse often is true. One wonders how many of the vocal advocates of fundamental studies have ever devised an effective control of a plant disease, or could do so. One does not hear medical research aimed at control of human disease, or engineering research on the relative strength of various metals, deprecated as applied.

Most of the easy answers in plant pathology have been given. Many of the remaining problems involve integration of many practices (e.g. manipulation of culture techniques, environmental modification, host resistance, biological control, chemicals). The complexities of biological control of plant pathogens are as great as any facing man today, even though the goal is practical.

Perhaps a better and less pejorative terminology can be devised: *philosophical studies* (pertaining to principles underlying a branch of learning) and *pragmatic studies* (pertaining to application rather than principles). The former tends to be analytical or synthetical, comprehensive, and long-term, the latter to be descriptive, specific, and short-term. Durant (19) defined science as analytical description, and philosophy as synthetic interpretation. Art is defined by Webster's Third International Dictionary as the systematic application of knowledge or skill in effecting a desired result. Plant pathology, as indeed all science, is moving toward the philosophical type of research, underemphasizing application.

Principles of Plant Pathology

Plant pathologists generally are uneasy and distrustful when one speaks of principles, although courses are given and books written that carry this label, usually erroneously. A principle is a fundamental proposition or generalization that aids in the interpretation of related facts, and has prediction value. It is a concept rather than a law (a statement of a relation of phenomena that, as far as is known, is invariable under given conditions), but must be stated definitely and concisely, rather than left to reader inference. Statement of true principles fortunately is becoming more common (e.g. 3, 7, 37, 43). Some socalled principles that have been published are really indices (27), or mere statements of fact (44) or sometimes even of prejudice.

Leach (32) stated the results of his damping-off studies, "In all combinations of host and pathogen preemergence infection was most severe at temperatures that were relatively less favorable to the host than to the pathogen as measured by the ratio of their growth rates." This statement of fact has restricted utility, but combining this work with other studies (e.g. 11, 12, 17) permits statement of a more general principle: Environmental conditions relatively more favorable to facultative types of pathogens than to the host are conducive to pathogenesis.

Charles Singer, the English historian of science, commented (38), "Amidst the multiplicity of phenomena, order must be sought if knowledge is not to lose itself in detail. . . . The progress of science is to be measured by the success men have in applying these general formulae to the knowledge they have collected. . . . But it is remarkable how far a science may advance before coherent expression is given to its principles. . . . many . . . scientific principles . . . [are] applied long before . . . [they are] adequately formulated." While books and courses in plant pathology frequently announce that they are concerned with principles (and usually are not), chemistry and physics books and courses seldom make a point of this, even though they are organized around laws. Subjects thus tidily arranged appeal to analytical minds and attract highly qualified students. Although a good deal of revision of even scientific laws has proved necessary, such statements provide a useful framework on which to hang ideas, extrapolations, and hypotheses, as well as serving as challenging targets for those who try to find exceptions to them.

Prestige

Among the rewards for research, one of the most invidious is prestige. Appropriately, the word is from the latin *praestigium*, meaning conjurer's tricks. When an honor or award unexpectedly comes to one for work well done, it generally is regarded as good. Even then, however, it is questionable whether it may not, in the total picture, do more harm than good. If it is truly deserved, the merits of the man already are well known to him and to others in his field. He probably has passed the time when it would encourage or aid him, and he often is indifferent to it. However, to other equally deserving workers it may be discouraging and lead to diminished effort. The distinction of the award is also lessened by the knowledge that, in a subject studied by many investigators, if a discovery is not made by one scientist it soon will be by another. Since the award is determined by a committee who often must choose between people in different and unfamiliar fields, there is always the chance that the award is not deserved. If it is not, it is even more discouraging to those who are more deserving but passed over, and it elevates the winner beyond his merit. Awards can become prime objectives, and intense political pressure often is exerted on the selection committee by a candidate's friends and colleagues. This of course debases the award. The publicized story of the race to resolve the double-helix structure of DNA in order to win a Nobel Prize is an example of the undesirable effects of the award system.

In this era of grantsmanship the obtaining of grants has become a mark of superiority, and is often advanced as justification for academic promotion, even when there is scant evidence that it has led to productive research.

As pointed out by Bawden (10), expensive laboratory equipment has become a status symbol, and younger workers come to feel that they cannot do worthwhile work without it. The effect of this atmosphere on graduate students is unfortunate, for they are then often ill-trained in applied plant pathology. There is a great deal of important work yet to be done on soilborne plant pathogens that does not require elaborate equipment, and Bawden pointed out that the same is true for plant virus diseases.

With grantsmanship and promotion systems has also come a tendency to measure research results by the number of papers produced, rather than by the benefit to agriculture or society. This is particularly unfortunate in plant pathology, whose objective should be the control of plant disease. To slightly paraphrase John Carew (13), "when the technique becomes more important than the plant; when a scientific paper is a greater reward than the answer to a problem or the gratitude of mankind; and when we are content to have our knowledge gather dust in a library and are no longer interested in communicating with those who produce or use our plants, . . .

[plant pathology] will be dead." It is sad to see a plant pathologist reach retirement, knowing that his work has not resulted in or improved the control of a single plant disease. As S. D. Garrett cogently commented, "The best monument to a plant pathologist is a forgotten plant disease." Carefully to study a plant-disease problem, which usually involves several disciplines, and see the results translated into improved quantity or quality of food for the world's populations is to find purpose in research. The proximity of the final judges, the intelligent grower and the completely objective plant, encourages cautious statement. There are, therefore, few journalistic prophets, ivory towers, and brass bands in phytopathology.

Enthusiasm and Faith

Ralph Waldo Emerson's dictum, "Nothing great was ever achieved without enthusiasm," certainly applies to research. If one is not enthusiastic about his studies, does not find them exciting fun, then they will become heavy going, and unlikely to be very productive. Perhaps 90% of research is routine drudgery; the other 10% is the stimulus that keeps one fired up. Research progress is apt to come in spurts, and to take advantage of these periods one must devote long hours of mental concentration. To quote Barzun (8) "The momentary glimpse that shows a relation, a truth, or a method of proof does not come at will. It is watched for like big game."

Research, like writing, is an act of faith. One must believe in what he is doing, and that he can produce an answer to the question that has been asked. Science is conservative, and new ideas or concepts spawn opposition that must be overcome. H. P. Agee [in (15)] outlined the stages in the adoption of new concepts: "The first stage . . . says the idea is crazy, it will never work. . . . The next stage . . . [is] admitting that there might be something to it but that it will take so long to put it into effect that the industry by that time will have folded! The third stage says it probably is good, but it is so academic and impractical that it would cost huge sums of money and labor to operate. . . . [In] the fourth stage . . . the critics begin saying 'Oh sure, it will work! There's nothing new in it. I used it 50 years ago! I know all about it!' " I have gone through these stages for the U.C. system for producing healthy container-grown plants (1), for soil and seed treatment with aerated steam (3), and for biological control of plant pathogens (7), and know the discouragement of initial nonacceptance.

On the other hand, I wish to acknowledge the encouragement received from Australian plant pathologists concerning biological control of plant pathogens. In the 1960s and early 1970s most pathologists in the US were very skeptical that biological control would ever be more than a novelty studied by petri-dish biologists. However, the attitude of Australian plant pathologists when I was there on leave in 1961–2, 1969, 1972, and 1976 was,

"Sounds beaut; let's have a go at it." Had it not been for this open-minded encouragement, which led to studies on the successful biological control of *Phytophthora cinnamomi* on avocado there, it is doubtful that the book *Biological Control of Plant Pathogens* (7) would have been undertaken. It has correctly been said that minds, like parachutes, only function when they are open.

There is frustration in that most experiments are failures; these can, however, supply valuable clues if the reason for failure is studied and resolved. An early experiment with seed bacterization of carrots in Australia gave a 48% increase in marketable roots one year, but no increase the following year. An administrator stopped the work because of the variability, rather than encouraging careful study to find the reason for the inconsistency. Pathologists (29) at the University of California, Berkeley, have, through careful studies since devised methods to minimize such variability. Seed bacterization appears to be primarily a form of biological control; the inoculated bacteria diminish the injurious activity of noninfecting pathogens or exopathogens on plant roots. In addition to being a very promising means for increasing crop yields, this technique expands the scope of plant pathology beyond its traditional concern with parasites that penetrate the host to produce disease. The discipline must now also include microorganisms in the rhizosphere that decrease plant growth but rarely or never penetrate the root (6).

In our early work on steam treatment of soil we observed that a small flow of steam injected into soil produced a temperature gradient ranging from 100°C at the point of injection, to perhaps 60°C 2 inches away, and ambient temperature at 3 inches. Since steam moves as an advancing front through soil, condensing on particles until they reach 100°, then flowing to the next cool particles, this broad 3-inch band was puzzling. We discovered that air was displaced from the soil pores as the temperature rose, gradually increasing the content of steam as the air was displaced. It then occurred to me that perhaps injecting air into steam might lower its temperature. An engineer specializing in thermodynamics was consulted; his calculations showed that the scheme would not work. Although discouraged, we nevertheless combined a steam line and a line of compressed air into a ⟩── tube. Varying the air flow readily and precisely controlled the temperature of the steam-air mixture. When this information was reported to the engineer, he found that a factor had been omitted in his equation.

It had been recognized for many years that soil is damaged by extended heating to 100°C, and that moist heat at 60°C will kill plant pathogens, but there was no commercial means of achieving that temperature with steam. Thus, studies on aerated-steam started with a philosophical study of steam movement through soil pores, and led to a new method of soil treatment. It was then learned that engineers in England and Norway were investigating the use of aerated steam to save fuel in soil steaming; the saving was not considered by them to be economic. Our interest in the method was as a means of selective thermal manipulation of the soil microbiota for improved disease control, and this approach has led to its commercial use. This is an example of the benefit from faith and persistence in one's work, of the unwisdom of accepting authority rather than experimental data, and of the evaluation of a new technique solely from one viewpoint. In the words of Charles Fillmore, "Only those ideas in which we have faith ever reach complete expression." One must have confidence, faith, and enthusiasm for what he is doing, and this is most easily maintained if the research is fun.

People gifted in analytical synthesis, and with disciplined imagination and judgment, become increasingly essential as our discipline increases in complexity. Unfortunately these are premium characteristics, and their total supply does not increase with need, training, or available funds. We must try to attract such gifted students to our profession, but there is little evidence that their relative numbers can be greatly increased. We therefore must encourage and fully support the gifted minds that we are able to attract. One of the major problems of our times is to find enough superior minds to keep the wheels of our technological civilization turning, and yet have enough able people to conduct business as usual, let alone to find enough of the best minds to advance science.

Employment of pathologists who are able to "grow" with a developing, broad, integrated, long-term project is necessary in these competitive times, as opposed to hiring persons for some restricted practical objective, leaving them at loose ends when that problem is solved. Men are stifled quite as much in being strictured by the boundaries of their research area as by being placed in a position that is much too big for them.

ON PUBLICATIONS

Graduate students in plant pathology in an earlier day were encouraged to find and use publications of their subject in the original. In the 1930s they were expected to keep abreast of the world literature of their field, and were able to do so without undue effort by reading abstracts in the *Review of Applied Mycology,* and papers in a relatively few journals. The occasional review or summary paper in the *Quarterly Review of Biology* and *Botanical Review* was welcomed. Because of the steadily increasing number of published papers in plant pathology (2, 4), investigators soon found it impossible to read all material published in their specialty, let alone that in other aspects of plant pathology. For a time the abstracting journals enabled workers to keep in touch with the world literature, but the increasing flood of publications forced such a reduction in abstract size, and in the percent-age of papers covered, that they gradually became little more than truncated annotated bibliographies. The multilingual *Bibliographie der Pflanzen-schutz-Literatur* has provided extensive annotated bibliographies for 1914 to 1954, in 1958, and from 1968 to date. This excellent service is too little known and used, possibly because its multilingual nature is not recognized.

As review journals increased in number and coverage after 1950 investi-gators came to depend on critical authoritative reviews to keep abreast of subjects outside their specialty. There was a remarkable increase in number of reviews in the next 15 years, but we still are, to a large extent, "unprinci-pled" collectors of facts. Some abstracting journals, such as the *Review of Plant Pathology,* have begun including reviews. Probably the most effective of our review journals is the *Annual Review of Phytopathology,* started in 1963.

More recently, information retrieval services have been started. They produce the equivalent of an annotated bibliography, requiring one to read the paper for details. With the widespread retreat from the graduate re-quirement of a reading knowledge of foreign languages, one wonders how the pathologists of tomorrow will be able to effectively use this service.

The great increase in international exchange and travel by plant patholo-gists since the 1950s (36), made possible by more plentiful travel funds and by increased speed of travel, has affected publication practice. The number of, and attendance at, international congresses has increased, and symposia have multiplied, leading to many published syntheses and reviews. Labora-tory visits and visiting lectureships have become common, and the tele-phone is increasingly used. Oral communication of research results has become so rapid that a new development usually is known around the world by at least a few investigators well before publication of the work. The published paper has now become the printed record of information known months before; its appearance often receives only a brief scan to see if the story still is the same as reported earlier. Even the purported value of seeing the supporting data may be diminished through independent verification by visitors. Thus, Ralph Cooper (16) was sent from the Rothamsted Experi-mental Station, England, to the USSR in 1958 to appraise studies conducted there on seed bacterization ("bacterial fertilizers"), and Alexander Holla-ender was sent by the National Research Council, USA., to the USSR in 1933 to evaluate claims for mitogenetic radiation by A. Gurwitsch (22). The published records in these cases were insufficient for adequate appraisal. Papers are no longer the built-in device "to keep scientists honest" that they once were.

Paradoxically, the significance of full-length scientific papers has thus diminished as their number has increased. It has been said that man is slowly burying himself in paper, and plant pathologists are doing their best to follow this trend. The time is near when something must be done to lessen the volume of detailed data published. Publication of extensive abstracts of not more than one or two pages probably would suffice for the casual reader, and the specifically interested reader might obtain a copy of the complete edited typescript or microfiche from the National Auxiliary Publications Service, Washington, DC. Each abstract would carry a NAPS Document Number, and would indicate the cost of the complete paper. Annual Reviews Inc. has for several years used this service for extended bibliographies and supplemental information. Arrangements might be made by which each subscriber of a journal would be given tickets for several such complete papers to be supplied at no additional cost each year. Such an arrangement is now used for communications in *Behavioral Biology*, for example.

It has become apparent to me that there is nearly as much useful phytopathological information filed away in offices as is ever published. Numerous mature scientists [e.g. reference (30)] have agreed that this is so. The number of abstracts from meetings that are not followed by complete papers also supports this conclusion. The increasing oral communications between workers in a given subject probably makes this loss to the field less apparent than it formerly was. Some of this information is salvaged in abbreviated form in review papers, books, symposia, and in teaching, but there still is a significant loss of information.

ON EDITING AND REVIEWING

My editorial background for these comments is 16 years with *Annual Review of Phytopathology*, 3 years with *Phytopathology*, 5 years with *Plant Disease Reporter*, 8 years with *Phytopathological Monographs*, and as editor of three books. On the other side of the fence, I have published with more than 70 different journals and publishers.

In this time of "publish or perish," with its attendant flood of papers and increasing emphasis on peer reviews, pathologists frequently become involved in the mechanics of publication. Many are thus expected to judge a paper that may be on a subject outside their area of competence, one that differs from their experience, work, or opinions, that is boring and of no interest to them, or that is written by someone they dislike, distrust, or regard as a competitor. It is common, therefore, for one reviewer to reject a paper, another to praise it, and a third to decide its fate. The more significant and original the subject, the greater is the likelihood that the third reviewer will be crucial. Our experience with the publication of The U.C. System manual for nurserymen (1) suggests that, as the number of reviewers increase, their comments tend to cancel each other. Because of professional infighting and the breadth of the subject matter involved in that manuscript, it was reviewed by 32 scientists over a 2-year period. As a result, every criticism or suggestion made was cancelled by someone else's review! One reviewer stated that it was "too complicated; no grower will be able to understand it." However, some 50,000 copies were sold to growers before it went out of print, and it has since been twice reprinted by a nursery association. The peer review system has been critically examined by others (e.g. 33a–34a). As Theodore Roosevelt is said to have observed, scientists have a quarrelsome interest in the work of their colleagues.

Experience demonstrates that the review system does winnow out papers that should not be published. All too frequently, however, these rejected papers are submitted to and published by another journal. Certainly the peer review system has not been very effective in restricting the quantity of publications, and it may even be questioned that it has improved the quality. Regarding Nelson's suggestion (34) that names of reviewers be published with the paper, it should be noted that *Current Anthropology* as early as 1964 carried this suggestion even farther. A controversial paper was sent to as many as 26 reviewers, who supplied written comments, to which the authors replied, and all of this material was published together. Thus, a true cross-section view by recognized authorities was presented with the concept in a stimulating forum. Only the occasional paper would stimulate such extended comment, but it could well be worth the journal space when used.

Editorial and Publisher Faults

Principal causes of difficulties that arise in reviewing or editing range from assertive inexperience, presumptive intolerance, excessive rigidity, and unfortunate tactlessness, to overbearing arrogance. These may lead to a flat statement that a point is not true, rather than requesting supporting data or phrasing a question concerning its validity. Surprisingly, if a reference is given for a statement, even if there is no supporting data in the reference, the statement will go unchallenged by most reviewers.

Perhaps the most common irritation is the presumption by editors or reviewers that their style of writing is superior to that of the author. Unquestionably, there are many different ways something can be said, and some are clearer than others. However, the paper is the author's, and his reputation, not the editor's, is at stake. His writing ability is one of the things on which he will judged, and his thoughts in someone else's words distort the record. An editor should modify a sentence only to prevent its being misinterpreted, not to make it sound better to him.

When the lengthy editorial procedure for The U.C. System manual (1) was completed, our excellent editor commented that her writing style then resembled ours. That should be the aim of an editor of scientific matter—to adapt to the author's style.

There is a school of thought that holds that short sentences make for easy reading (42), and some editors follow this dictum. In the Prefatory Chapter by F. C. Bawden (10), the average sentence length was about 38 words. There were numerous sentences of 60–85 words that were completely lucid and that retained one's interest; very few sentences had less than 16 words. On the other hand, the Prefatory Chapter by J. G. Horsfall (26) had an average sentence length of about 14 words, with many sentences of four to six words, and very few of 30–50 words. Clearly, nothing as mechanical as sentence length determines clarity and interest. As E. B. White (42) has said, "writing is an act of faith, not a trick of grammar."

There are more irrational and capricious editorial policies concerning references than almost anywhere else in publication. Author alphabetical sequence there, when it is followed, is highly variable and whimsical. One would expect that so esteemed and well-known a name as Anton de Bary would be used consistently. Although he always is spoken of as de Bary, never as Bary, frequently he is listed under B as Bary, A. de, or de Bary, A. Why not simply follow the accepted oral usage familiar to the reader, and list him under D as de Bary, A.? Similarly, an alphabetical arrangement following the sequence of letters in names, instead of various artificial devices (e.g. grouping MacDonald and McDonald together) would simplify matters for everyone. Listing references in the order in which they are cited in the paper, rather than alphabetically, also makes it difficult to find a given paper, and fortunately is not common.

Title of the paper is deleted in many journals, fortunately few in plant pathology. When the policy to be followed on this by *Annual Review of Phytopathology* was being discussed, Glenn Pound expressed the view of most plant pathologists when he pointed out that a reference without a title was of little use because one would have to examine the paper to see whether it was one he had already read.

Indication of the number of pages in an article gives a valuable clue to the completeness of the work. It is inexcusable to list only the first page of a paper; a book thus becomes a single page. Another irritation is to restrict book citation to author, date, and title; try to get an interlibrary loan with that information! The inconsistency of abbreviation of journal names is legendary and confusing. The recent practice of the *Annals of Applied Biology* in giving the unabbreviated journal name certainly is the best solution of this problem.

Most of these journal or editorial policies aim at saving space, but objective examination suggests that the economy often is minimal. For example, why truncate and abbreviate references, only to leave a half page empty at the end of the paper? If the journal name is given in full, it often simply fills the space left blank by the abbreviations.

One who has done literature searching appreciates the usefulness of indexes. Without them some facts can only be found with great difficulty, even when the specific volume is known. *Phytopathology* has annual indexes, and published 10- or 30-year indexes covering 1911 to 1950. The decision of the Publications Committee to then abandon publication of the Ten-year Index because it "was not useful" is more a reflection on the Committee than a statement of fact. As Mrs. A. R. Jaffa, former Agricultural Librarian of the University of California, commented, "To get specific information from an unorganized collection of books is like trying to get a drink from a revolving lawn sprinkler." One of the values of *Annual Review of Phytopathology* is the excellent indexes provided for each volume. A detailed subject index is provided by a plant pathologist, and an author index and a cumulative chapter index for 10 volumes are provided by Annual Reviews Inc.

Another source of confusion is the capricious changing of content of a journal. For example, publication of abstracts of papers presented at meetings has long been a matter of controversy in the American Phytopathological Society, and this is reflected in placing them at various times in *Phytopathology, Phytopathology News,* and *Annual Proceedings.* Similarly, *Phytopathology News* has been published as a separate journal, then in *Plant Disease,* and now again separately. Even though flexibility is desirable in changing times, these rapid fluctuations suggest poor judgment and insufficient thoughtful planning, and certainly do not endear us to librarians.

Author Faults

A pathologist once told me that he never bothered with references until the paper was written, and then added a few "to dress up the paper for the editor." Some authors do not cite or mention earlier workers on a subject, perhaps hoping that the reader will think that the paper is essentially new work. All work builds on the studies of others, and failure to acknowledge this is a measure of the stature and ethics of the individual. Giving credit to others doesn't cost anything; not to do so, may!

My editorial experience has indicated that references are held in low esteem by many authors. Many citations are incorrect, often with the wrong volume or pages, sometimes in the wrong journal, or by the wrong author. An author's own publication has even been cited in the wrong journal. The only reason for citing references is to enable the reader to find the paper, and if the information is incorrect it isn't worth printing. This is particularly true in a review journal because of its frequent use as a bibliography. The *Annual Review of Phytopathology* has for this reason been particularly careful in checking for accuracy of citation.

ON HISTORY

Every scientist is an instrument of history; everything he has learned and everything he attempts to do are products of history, and everything he does or publishes becomes history. In many fields of science, especially the older and more mature ones, history of the subject holds a respected position. This is hardly the case with phytopathology in this country, even though some of our most honored leaders (e.g. L. R. Jones, G. W. Keitt, J. G. Horsfall, E. C. Stakman, J. C. Walker, H. H. Whetzel) have emphasized the subject in their teaching and writing. As a pathologist phrased it recently, "the interjection of a broad historical perspective and a humanistic philosophy . . . [is] an essential leavening . . . in the Doctor of Philosophy degree . . . [it] makes a profession out of what is too much a technological trade." My interest in the history of plant pathology began while a graduate student, and has continued over the years (5). A course on the History and Literature of Plant Pathology was taught at the University of California, Berkeley, from 1963 until my retirement in 1975.

Benefits From Knowledge of History

Understanding the background of one's discipline should impart a sense of belonging to a long-flowing intellectual current. The relative size of this stream is indicative of the number of tributaries and sources, and it eventually flows into the common pool of knowledge. Some of the tributaries are small, noisy, and riffled, contributing little to the main current; others flow steadily and deeply, and contribute much. The stream may abruptly change its course as obstructions are encountered or new channels develop. There may be quiet intervals of little forward progress, followed by excitingly rapid flow and turbulent cataracts, or even the occasional breakthrough of waterfalls.

It helps, in the loneliness of intellectual effort, to be aware of the majesty and beauty of the cumulative efforts of one's predecessors and peers, and this awareness may supply a focal point for the sense of dedication necessary for outstanding work in any time and place. Perhaps the failure in our time to stimulate man's enormous capacity for dedicated service through a sense of belonging to some large and significant movement is related to lack of interest and appreciation of history. A sense of belonging, an esprit de corps, is surely needed to bolster the scientist today in his isolated specialized compartment. Charles Singer (38) gloomily commented on this isolation of the individual, "As knowledge advances it must become more divided. . . . it will be appreciated by fewer and fewer. . . . the body of science will be so fragmented that each investigator will be intelligible only to himself. He can thus leave no scientific heir."

Concern frequently has been expressed about the fractionation of biological science into ever-smaller units. Numerous attempts to reassemble these scattered units on the basis of some common denominator, such as evolution, molecular biology, RNA, or ecology, have not been very successful. The sequence and manner in which the concepts of biology have developed (i.e. history) could provide the necessary cement to bind the diverse subject matter of biology together, and to maintain lines of communication between disciplines.

Knowledge of some of the landmarks in our field, from the standpoint of the background of the times and how the results were obtained, can only be helpful. Examination of some of the "shipwrecks" in our field, in an effort to understand where the investigations went wrong, should give experience in recognizing shoddy work, and stimulate an appreciation of how vulnerable we all are, encouraging a tolerant view of all investigations. It is important to a plant pathologist to know about the origin and development of agriculture and the principal cultivated crops. Frequently knowledge about the native home of a plant, and its ecological requirements there, will supply valuable clues for avoiding or reducing stress diseases of the crop, abiotic or biotic.

History provides a sound basis for evaluation of one's worth, one's work and decisions, one's department and its work, compared with others, past and present. Experience (history) is the stuff of which judgment is made.

Research is an organized way of discovering some practical course of action that probably would be found anyway by trial and error, although more slowly. It is unfortunate that man, who so hastens discovery by the experimental method, and quickens evolution by selective breeding and genetic engineering, should be so tardy in aiding, through the study of history, development of the necessary judgment to cope with the accelerated pace of science and technology.

Knowledge of the history of one's subject is an effective antidote to arrogance, since it clearly shows that one's contributions are small in the total picture, and that their importance bears no relationship to the notice that they may attract at the time. As in photography, where one enlarges the field of view by moving further away, so knowledge of history increases the breadth of one's scientific outlook. The less specialized and the broader a scientist's viewpoint, the greater his interest in the history of his field is apt to be.

Decision making can be improved by historical background, replacing emotions and loyalties that stand in one's light. As W. H. Chandler said, "Do what is best for your field, and it usually will prove to be best for the Department, the University, and for you." We should seek a profession-oriented rather than an egocentric or institution-oriented approach.

MEMORIALS

The four people who exerted the greatest influence on my professional career may be chronologically listed. They were all singularly apolitical, highly ethical, totally lacking in sham, hard-working, and completely dedicated to their subjects.

F. D. Heald, my graduate professor, stimulated industrious work habits by his example and by his statement that his worthwhile contributions had been done on overtime. His encyclopedic knowledge of plant science instilled the desire to be well-informed in a wide subject area. H. F. Clements, plant physiologist and anatomist, and perhaps the most successful practitioner of crop logging and crop control [on sugarcane in Hawaii (15)], instilled a lasting conviction that the goal should be synthesis of laboratory and field work for the good of society. M. B. Linford, scholarly plant pathologist and meticulous technician, was a superlative colleague and supervisor in my first professional position. W. H. Chandler was one of the grand old men of plant science and a strong friend of plant pathology; his rugged honesty, humility, courage, philosophy, and enduring faith in the essential goodness of man have profoundly affected my view of the world.

In a less personal way, two administrators have influenced my career. C. B. Hutchison, principal architect of the California Agricultural Experiment Station, was the ablest and most humane administrator I have known. The concern of Robert Gordon Sproul, President of the University of California, for even the lowest staff member, was legendary. His personal encouraging note, written at home at Christmas, when I was in the hospital with a serious eye problem, made a lasting impression of his humanity. To have served under both of these men was a privilege for which I am grateful.

Literature Cited

1. Baker, K. F., ed. 1957. The U.C. system for producing healthy container-grown plants. *Calif. Agric. Exp. Stn. Manual 23.* 332 pp.
2. Baker, K. F. 1958. The development of floricultural pathology in North America. *Plant Dis. Reptr.* 42:997–1010
3. Baker, K. F. 1962. Principles of heat treatment of soil and planting material. *J. Aust. Inst. Agric. Sci.* 28:118–26
4. Baker, K. F. 1970. The changing pattern of APS and Phytopathology. *Phytopathol. News* 4(6):2–3
5. Baker, K. F. 1980. Developments in plant pathology and mycology, 1930–1980. In *Perspective in World Agriculture,* pp. 207–36. Slough, England: Commonw. Agric. Bur. 532 pp.
6. Baker, K. F. 1980. Microbial antagonism—the potential for biological control. In *Contemporary Microbial Ecology,* ed. D. C. Ellwood, J. N. Hedger, M. J. Latham, J. M. Lynch, J. H. Slater, pp. 327–47. London: Academic. 437 pp.
7. Baker, K. F., Cook, R. J. 1974. *Biological Control of Plant Pathogens.* San Francisco: Freeman. 433 pp.
8. Barzun, J. 1951. *Teacher in America.* Boston: Little, Brown, 321 pp.
9. Bateman, D. F. 1978. The dynamic nature of disease. In *Plant Disease—An Advanced Treatise,* ed. J. G. Horsfall, E. B. Cowling, 3:53–83. New York: Academic. 487 pp.
10. Bawden, F. C. 1970. Musings of an erstwhile plant pathologist. *Ann. Rev. Phytopathol.* 8:1–12
11. Beach, W. S. 1949. The effects of excess solutes, temperature and moisture upon damping-off. *Penn. Agric. Exp. Stn. Bull. 509.* 29 pp.
12. Bliss, D. E. 1946. The relation of soil temperature to the development of Armillaria root rot. *Phytopathology* 36:302–18
13. Carew, J. 1966. The composition of horticulturists. *Proc. Am. Soc. Hortic. Sci.* 89:766–72

14. Chandler, W. H. 1942. *Deciduous Orchards*. Philadelphia: Lea & Febiger, 438 pp.
15. Clements, H. F. 1980. *Sugarcane Crop Logging and Crop Control: Principles and Practices*. Honolulu: Univ. Hawaii Press. 520 pp.
16. Cooper, R. 1959. Bacterial fertilizers in the Soviet Union. *Soils Fert.* 22:327–33
17. Dickson, J. G. 1923. Influence of soil temperature and moisture on the development of the seedling-blight of wheat and corn caused by *Gibberella saubinetti. J. Agric. Res.* 23:837–70
18. Dimock, A. W. 1950. Extension work with commerical florists—some points of policy. *US Dep. Agric., Ext. Serv. Mimeogr. Publ.* 3 pp.
19. Durant, W. 1926. *The Story of Philosophy*. Garden City, NY: Garden City Publ. Co. 592 pp.
20. Grogan, R. G. 1981. The science and art of plant-disease diagnosis. *Ann. Rev. Phytopathol.* 19:333–51
21. Grogan, R. G., Sall, A., Punja, Z. K. 1980. Concepts for modeling root infection by soilborne fungi. *Phytopathology* 70:361–63
22. Gurwitsch, A. 1932. *Die mitogenetische Strahlung*. Berlin: Springer. 384 pp.
23. Hirst, J. M. 1965. Dispersal of microorganisms. In *Ecology of Soil-borne Plant Pathogens*, ed. K. F. Baker, W. C. Snyder, pp. 69–81. Berkeley: Univ. Calif. Press. 571 pp.
24. Horsfall, J. G. 1959. A look to the future—the status of plant pathology in botany and agriculture. In *Plant Pathology Problems and Progress, 1908–1958*, ed. C. S. Holton, G. W. Fischer, R. W. Fulton, H. Hart, S. E. A. McCallan, pp. 63–70. Madison: Univ. Wis. Press. 588 pp.
25. Horsfall, J. G. 1969. Relevance: are we smart outside? *Phytopathol. News* 3(12):5–9
26. Horsfall, J. G. 1975. Fungi and fungicides. The story of a noncomformist. *Ann. Rev. Phytopathol.* 13:1–13
27. Horsfall, J. G., Cowling, E. B. 1980. Cumulative index of major principles, volumes I–V. See Ref. 9, 5:519–34
28. Hull, A. W. 1945. Selecting and training of students for industrial research. *Science* 101:157–60
29. Kloepper, J. W., Schroth, M. N., Miller, T. D. 1980. Effects of rhizosphere colonization by plant growth-promoting rhizobacteria on potato plant development and yield. *Phytopathology* 70:1078–82
30. Kramer, P. J. 1973. Some reflections after 40 years in plant physiology. *Ann. Rev. Plant Physiol.* 24:1–24
31. Kranz, J. 1974. *Epidemics of Plant Diseases. Mathematical Analysis and Modeling*. New York: Springer-Verlag. 170 pp.
32. Leach, L. D. 1947. Growth rates of host and pathogen as factors determining the severity of preemergence damping-off. *J. Agric. Res.* 75:161–79
33. McNew, G. L. 1960. The training of students for research in plant pathology. *Phytopathology* 50:511–16
33a. National Academy of Sciences. 1981. *Peer Review in the National Science Foundation. Phase Two of a Study*. Washington: National Academy Press. 106 pp.
34. Nelson, R. R. 1980. Some thoughts on the current and future health of Phytopathology and the American Phytopathological Society. *Phytopathology* 70:364–65
34a. Öpik, E. J. 1977. About dogma in science, and other recollections of an astronomer. *Ann. Rev. Astron. Astrophys.* 15:1–17
35. Pound, G. S. 1965. The Position of Plant Pathology in Current Administrative Changes in Biology and Agriculture. *Presented at Ann. Meet. Am. Phytopathol. Soc., 57th, Miami.* 7 pp.
36. Pound, G. S. 1968. A Midstream View of Plant Pathology. *Presented at Ann. Meet. Am. Phytopathol. Soc., 60th, Columbus.* 21 pp.
37. Savile, D. B. O. 1955. A phylogeny of the Basidiomycetes. *Can. J. Bot.* 33:60–104
38. Singer, C. 1959. *A History of Biology to About the Year 1900*. New York: Abelard-Schuman. 580 pp. 3rd ed.
39. Stevens, N. E. 1932. The fad as a factor in botanical publication. *Science* 75:499–504
40. Stevens, N. E. 1949. Fun in research. *Am. Sci.* 37:119–22
41. Stevens, R. B. 1961. Is plant pathology a fake? *J. Wash. Acad. Sci.* 51:129–31
42. White, E. B. 1954. Calculating machine. In *The Second Tree From the Corner*, ed. E. B. White, pp. 165–67. New York: Harper. 253 pp.
43. Wilhelm, S. 1959. Parasitism and pathogenesis of root-disease fungi. See Ref. 24, pp. 356–66
44. Yarwood, C. E. 1962, 1973. Some principles of plant pathology I, II. *Phytopathology* 52:166–67; 63:1324–25.

Kohei Tomiyama

Ann. Rev. Phytopathol. 1983. 21:1–12

RESEARCH ON THE HYPERSENSITIVE RESPONSE

Kohei Tomiyama

Plant Pathology Laboratory, Faculty of Agriculture, Nagoya University, Nagoya, 464, Japan

INTRODUCTION

When I was invited to contribute this prefatory chapter, I deemed it a great honour, but doubted whether I could write a chapter worth reading. Nevertheless, I thought that it might be worthwhile to write about how a phytopathologist of Japan entered our profession in his youth and how he advanced his life work.

When I started my studies in phytopathology at the Hokkaido University in Sapporo, the professors of phytopathology were S. Ito, Y. Tochinai, and T. Fukushi. These professors were students of professor emeritus K. Miyabe, who was about 90 years old and still worked on his research in the herbarium every day. He initiated one of the first courses in phytopathology in Japan in 1889. (Kotaro Shirai started a course at Tokyo University in 1886.) Thus, I belong to the third generation of phytopathologists in Japan.

In the feudal age of Japan, peace was maintained for about 250 years from 1615 to 1867 by the adroit policy of Tokugawa Shogunate. However, dissatisfaction with government policy grew during the long term of peace. The visit to Japan of the American squadron commanded by Admiral M. C. Perry in 1853 triggered the Meiji Revolution in Japan. The visit of the Americans frightened the Japanese people, who had never seen a steam ship. The Tokugawa Shogunate could not cope with the beginning of the new age (7).

The young leaders of the new government made it their policy to modernize Japan by introducing western culture. They sent many capable young students to Europe and the USA to learn western culture and science. Kingo Miyabe was one of those students, and he studied botany under W. G. Farlow at Harvard University from 1886 to 1889. During the long, peaceful reign of the Tokugawa Shogunate, while Japan was closed to western influences, Japanese phi-

1945

0066–4286/83/0901–0001$02.00

losophy, knowledge, technique and arts became extremely refined. Thus the young students who went abroad had a strong desire to rapidly learn western science and to modernize Japan. However, the success of the revolution also produced a group of people with an ultranationalistic spirit. The colonialism of the advanced western nations that prevailed in Asia at that time may have increased this ultranationalism, which grew stronger with time. Finally, although a majority of Japanese people did not want it, Japan became more militant. In 1936, while I was still a University student, a coup d'etat by a small group of fanatical soldiers took place in Tokyo. The next morning, students met to express their anger against the oppression of the military, but the protest was not put into action at our university. This coup d'etat was not successful, but eventually Japan fell under the control of militarism, and the disasters of World War II became inevitable.

I was a student at Hokkaido University just before World War II. In Japan at that time most major plant pathogens besides the viruses had already been identified and classified, although many scientific problems remained to be solved. It was a turning point for plant pathology in Japan. In Hokkaido University some physiological and ecological research was being carried out, but most of the educational program was concerned with the identification and classification of pathogens. I graduated from the University in 1940.

SNOW BLIGHT DISEASE

Shortly before the start of World War II I obtained a position in the Hokkaido Agricultural Experiment Station in 1941. The Station concentrated its effort on the promotion of food production on the farms of Hokkaido Island. My main duty was to identify the pathogens on diseased samples sent by farmers and to advise them of methods of protection from disease. The identification was, in most cases, laborious and difficult, and advice to the farmers was usually simple and repetitious: "Burn the diseased plants," "bury in the soil," "spray with bordeaux mixture," "disinfect the seeds," and so on. As a result, I was greatly discouraged and disappointed with my profession. Before long, I had to perform my military duty. My first work after the war was research on snow blight disease of winter cereals. Both during and after the war, winter cereals in northern Japan suffered serious damage from snow blight disease. Later, the results of our research disclosed that this damage might be caused in part by a shortage of phophorous fertilizers during the war. I attacked this problem in the hope that I could alleviate the doubt I felt during the war about the value of my profession. Thus, I worked very hard.

The research was extensive, including classification, distribution, and ecology of the pathogens; physiology of wheat under the snow; resistance of wheat to pathogens; protection by cultivation methods; protection by spraying chemicals. Improvement of cultivation methods and spraying of organic mercury compounds, which had been used as seed disinfectants, effectly protected cereals against snow blight. After about five years, I terminated my research on the snow blight disease. Later, however, the use of mercury compounds as spray chemicals was banned because of the possibility that mercury may accumulate in humans and animals. My study on the snow blight disease was finished, but I was not satisfied with the results. Improvement of fungicides appeared to me to be most desirable, but I was not a chemist. Therefore I directed my research toward the physiological mechanism of disease resistance in the hope of applying the information to plant protection. My scientific interest was most important in this decision.

EFFECTS OF PLANT JUICE ON PARASITES

The research on the snow blight diseases required all of my research time during autumn, winter, and spring. In the summer season, however, I could find time to carry out experiments on a research theme that I myself thought important. Although the period of time available was short, I carried out experiments to test the effects of juice obtained from various species of *Graminae* on germination of spores of different fungal species parasitic on each of the tested cereals. These experiments were my first trials in a series of studies on disease resistance physiology. Many researchers seemed to consider that preformed toxic substances might have nothing to do with disease resistance. I thought that I should determine whether the preformed toxic substances play a role in disease resistance before starting the research on the defense response.

We examined the effects of nontreated juice obtained by pressing leaves of oat, wheat, barley, corn, rice, and millet on germination of spores of *Ustilago avenae, U. tritici, U. nuda, U. zeae, Helminthosporium avenae, H. sativum, H. turcicum, H. oryzae, Piriculatia oryzae,* and *Sorosporium manchuricum.* The results showed that, when the juice obtained from cereal species had fungitoxicity, all of the fungal species parasitic to that species were tolerant to the toxic substance (14). In contrast, the fungal species nonparasitic to that host were sensitive to that toxic substance. These results indicated that preformed toxic substances play some role in the resistance of nonhost plants to pathogenic fungi, possibly in cooperation with defense responses. The tolerance of pathogenic fungi to the phytoalexins of host plants has been considered to be the most convincing evidence for the role of phytoalexins in disease resistance.

The author presumes that preformed substances mainly play a role in the resistance of nonhost plants, but may not in host plants. On the other hand phytoalexins are believed to constitute a factor in the resistance of plants to pathogens that can invade into plant tissue to some extent. In both cases, cooperation of a series of responses of plant tissue including papilla formation, oxidation of cell contents, suberization, lignification, and so on may be indispensable.

DISEASE RESISTANCE: A BASIC PRINCIPLE OF RESEARCH PLAN

In 1953, The Ministry of Agriculture and Forestry set up a research group to study resistance of potato late blight at the Hokkaido Agricultural Experiment Station. I was installed as chief of this group. N. Takase, R. Sakai. M. Takakuwa, T. Takemori, and K. Kitazawa took partial charge of potato late blight problems, but I investigated mainly the physiological mechanism of disease resistance. Many researchers, especially K. O. Müller, had already done extensive and brilliant work on resistance to potato late blight. Thus we had a good starting point.

During my student tenure, M. Sakamoto told me almost every day that research on disease resistance performed without microscopic observation is often meaningless. My experience after graduation convinced me that this is true. Dr. Sakamoto also greatly influenced N. Suzuki and Y. Takahashi in their excellent work on violet root rot of sweet potato and infection process of rice by *Piricularia oryzae,* respectively, as well as other researchers.

We made it a principle to observe quantitatively the process of infection and response of host to infection using living tissue in all physiological experiments. A combination of morphological and physiological studies became a characteristic feature of research throughout my life.

In the case of intact surfaces of leaf and stem tissues, the difficulty of penetration and the degree of host-cell response in resistant and susceptible types usually differs widely, even in the same tissue. Often only a few of many incipient infections produce typical symptoms. Accordingly, quantitative and statistical measurements of the behavior of infecting hyphae and responses of the infected cells in an intact tissue surface give only the average of various types of cell response. To avoid misleading results, the experimental model system, therefore, must possess the following characteristics:

1. Insofar as possible, all of the cells in the tissue should be uniform with regard to their response to the infection.
2. Almost 100% of the inoculated spores should pentrate synchronously and at almost the same rate.
3. Existence of races of pathogens having similar characteristics except for pathogenicity to the host concerned is necessary.

Model systems meeting these requirements were cut surface cells of the cortex of potato-leaf petioles and disks of potato tuber. The intact surface of midribs of young unfolding leaflets was also suitable for observation of the infection process of intact cells. These systems, of course, do not completely satisfy the requirements but statistical treatment of the observations on many cells make up for their incompleteness.

To determine chemical changes in infected cells, in adjacent cells, and in cells at a definite distance from infected cells, tissue blocks cut from tubers of potato cultivar Rishiri are suitable. The cells of this tissue are relatively uniform in their hypersensitive response and can be cut into thin slices with a microtome for chemical analysis. In this system, growth of an incompatible race of *Phytophthora infestans* is stopped after infection of 1, 2, or 3 cells. When the spores of the incompatible race are inoculated heavily onto the surface, a brown tissue zone appears in a flat plane parallel with the cut surface, which allows the separation of the healthy cells from infected ones with a microtome.

The use of cut surfaces of the tissue block was often criticized because of possible inteference of wound-healing responses with the infection responses. Of course, this criticism is reasonable, if the aim of the experiment is to determine how a pathogen invades intact tissue. However, when the aim is to determine the basic principle of the host-parasite interaction, the cut tissue block is not only very useful but also sometimes indispensable if necessary attention is paid to the interference from the wounding response. Because the cut surface of tissue block is highly resistant to incompatible races and sufficiently susceptible to compatible ones, the block is a good model system for investigating host-parasite interactions. Progress in physiological research could not be achieved without using such a simple model system.

DISEASE RESISTANCE: CHEMICAL DEFENSE

The initial and most impressive result from the quantitative observation of the infection process was the fact that the earlier the infected cells died, the more strongly the infecting hyphae were restricted, regardless of the presence or absence of resistance genes. This observation strongly suggested that hypersensitive cell death is an important phenomenon in disease resistance. Accordingly, almost all of the experiments involving chemical analysis were related to the occurrence of early hypersensitive cell death.

The first research on biochemical changes in the infected tissue was done on respiration and polyphenol metabolism. These were prevalent themes for research in those days. In Japan, the research groups of I. Uritani and N. Suzuki had been working actively in this field. The objective of our experiments was to characterize the metabolic changes in the infected cells, and also changes in nearby healthy cells at various distances from the infected ones, in relation to the infection process observed under a microscope (11, 16).

One of our findings was that the acceleration of metabolic activity, including respiration and polyphenol metabolism, was greatest in the adjacent cells and decreased exponentially in proportion to the distance from the infected cells.

This and other results suggested that the accelerated metabolism in the neighboring healthy cells may supply substances for repair of the infection lesion. Results from a series of experiments suggested that polyphenol compounds synthesized in the tissue near the infection lesion may be transported to the lesion. Subsequently the compounds become oxidized and are deposited on the cell walls of both host and parasite and also on the cell contents. Before phytoalexins accumulate to some degree, branching of the hyphae in the hypersensitive dead cells and penetration into adjacent cells is inhibited, but growth rate in the total length of the intracellular hyphae is not inhibited. This finding strongly suggested that oxidation and deposition of polyphenol compounds is a highly possible mechanism for inhibition of the branching and penetration of infection hyphae. It is noteworthy that, in this stage of browning of the dead cells, total hyphal growth rate is not inhibited, so that the main axis of hyphae in dead cells grows faster than that in living cells. This implied that the oxidation and deposition of polyphenols might not affect the metabolism of intracellular hyphae, but plays an important role of confining the parasite within the lesion. Thus polyphenols may cooperate with phytoalexins in inhibition of disease development. Recently, inhibition of hyphal development in the tissue has often been attributed exclusively to phytoalexins, but it is obvious that polyphenols also play a role.

Four years after we started our research on potato late blight, K. O. Müller visited us in Sapporo. He gave a lecture on the phytoalexin theory. We were very much impressed by both his achievement and his personality. Meanwhile, on a fellowship from the Rockefeller Foundation, I went to University of Wisconsin to study physiological plant pathology under the guidance of J. C. Walker. He introduced me to M. A. Stahmann because of my desire to learn biochemistry. I carried out experiments in Stahmann's laboratory, where R. Heitefuss, I. Uritani, and C. Hiruki were also visiting at that time. My stay in Madison with these scholars was exciting and I learned very much. While I was in Madison, I was invited by R. P. Scheffer to give a seminar at Michigan State University. It was an impressive experience for me. One week after I left the University, Dr. Scheffer went to the Rockfeller Institute on sabbatical where he advanced his brilliant research on host-specific toxins.

In 1963, I. A. M. Cruickshank visited us in Sapporo. He delivered a lecture on his discovery of pisatin, the phytoalexin of *Pisum sativum* (2). Talking with him was very stimulating. In Japan, ipomeamaron was isolated by M. Hiura in 1940 (3) from sweet potato roots infected by *Cereatostomela fimbriata* as a toxic substance that was newly produced after infection. Its significance in disease resistance was studied by I. Uritani and his group (17) and by N. Suzuki (10) before pisatin was isolated. However, the experimental results were not elucidated as a general concept in disease resistance in the plant kingdom.

In 1940, K. O. Müller and H. Börger (6) presented the phytoalexin hypothesis as a general concept based on their observation of resistance of potato to *P. infestans*. But it was not then known whether phytoalexins were really produced in potato tissue. I thought it might be impossible to elucidate the inhibitory mechanism of hyphal development in infected potato tissue without phytoalexins. Thus, after Dr. Cruickshank left Sapporo, we began research to isolate phytoalexins from infected potato. T. Sakuma, N. Ishizaka, and N. Sato cooperated with me in this research. After a few years, we could isolate a toxic substance as a single spot on thin layer chromatograms. We named this compound rishitin, because we isolated it from the cultivar Rishiri. We then cooperated with organic chemists at Hokkaido University, T. Masamune, K. Katsui, and later N. Murai, in an attempt to determine the chemical structure of rishitin. They eventually elucidated the chemical structure of rishitin and related compounds. Thereafter, many related terpenoid phytoalexins were isolated from solanaceous plants by Metlitskii's (5), Kuc's (18), Stossel's (9), and our groups. Gradually, a general outline of the biosynthetic pathways of solanaceous phytoalexins and their pathological and physiological roles were revealed.

In 1966, the first United States-Japan Seminar on plant-parasite interaction was held by intergovernmental agreement at Gamagori, Japan. Subsequently, we had four more US-Japan Seminars prior to the most recent one, which was held in 1981 at Brainerd, Minnesota, and chaired by W. R. Bushnell and Y. Asada. In 1965, the first Summer School for physiological plant pathology was held in Sapporo. Since then the Summer School has been held every year under the leadership of T. Hirai and N. Suzuki. Exchanges of ideas and discussions with distinguished phytopathologists in the US-Japan Seminars and Japanese Summer Schools were invaluable in promoting our field of science in Japan.

In 1971, I moved to Nagoya University as professor of Phytopathology, where I continued research on disease resistance in potato in cooperation with N. Doke. A succession of graduate students including T. Nakajima, N. Nishimura, T. Horikawa, M. Nozue, N. Matsumoto, N. Furuichi, Y. Ishiguri, K. Sato, S. Sakai and others worked with us in these studies. At Nagoya, we continued to study the biosynthetic pathways of rishitin in cooperation with the organic chemists T. Masamune, N. Katsui, and A. Murai. In another series of experiments we studied the pathological role of phytoalexins. From first to last, these studies were aimed at answering the question of "when and where" rishitin is produced. This simple question led us to unexpected results. Contrary to our expectation, the rishitin synthetic pathway was found to operate in metabolically active young intact tissue at such a low level it was detectable only with radioisotopes.

Our most recent results from the experiments on rishitin and related compounds demonstrated the following ideas (13):

1. Rishitin and related compounds play an important role in hyphal growth inhibition.
2. The synthetic pathway of rishitin operates in metabolically active healthy tissue at a level detectable with radioisotopes.
3. Rishitin is synthesized via farnesol → solavetivone →→ lubimine → oxylubimin → rishitin. Then rishitin is metabolized to rishitin-M-1 and rishitin-M-2 and other more hydrophilic compounds.
4. When local necrosis occurs in the infected tissue, or the infection site becomes highly permeable to rishitin and its related compound, the compounds are exuded from the living cells, excluded from further metabolism, and accumulate in the dead host tissue, in intracellular spaces, or on the surface around the infected cells.
5. Possibly by exclusion of product from the normal metabolic pathway on the one hand and metabolic acceleration as wound response on the other hand, rishitin synthesis may be accelerated in the neighboring healthy tissue.
6. The synthetic pathway of rishitin may not be an abnormal metabolic pathway induced by stress.

To resolve the ambiguous points of phytoalexin production, it seems necessary to know the site in a cell where phytoalexin is synthesized and stored and also to know the path for transfer and the details of its synthesis.

DISEASE RESISTANCE: CELL DEATH

From the beginning of our research on potato late blight, we continued experiments on the infection processes and host responses using the microscope. Our observations strongly suggested that hypersensitive cell death is a key phenomenon in the hypersensitive type of disease resistance (11). We presumed that hypersensitive cell death may constitute a general and basic phenomenon in the response of a living organism to an incompatible microorganism.

Thus, research on hypersensitive cell death became the central theme of our group. Recent results are summarized as follows:

1. The host cell membrane adheres to the hyphal surface in a very early period of infection, possibly almost immediately after penetration. The adhesion seems to be mediated by a host lectin. Adhesions occur in both incompatible and compatible combinations. The adhesion appears to be indispensable for the occurrence of rapid hypersensitive cell death, but does not decide specificity.
2. Cells in intact tissue seem to have little or very low potential to react hypersensitively to infection by an incompatible race. Wounding or infection per se increases or induces the potential. This process requires protein synthesis.

3. ATP is involved in rapid hypersensitive cell death in tissue with fully developed hypersensitivity potential. Protein synthesis is not involved in this rapid cell death process.
4. Presumably hypersensitive cell death can be regarded as a kind of "suicide response" of the infected cell to save the rest of the tissue from disease. As a result, the infected cells as a whole are isolated from the normal healthy tissue by formation of periderm or browned cells.
5. Phytoalexin cannot be the cause of hypersensitive cell death. Accordingly, there is no evidence as yet indicating that infection converts the normal metabolism to an abnormal one leading to death of host cells.
6. There is some evidence that hyphal wall components may be the cause of the hypersensitive response, but there also is evidence that the effects on the host cell of hyphal wall components are different from those of infection. Further investigation is necessary to understand this matter more fully.
7. Infection by the compatible race strongly reduces the hypersensitive reactivity of host cells to the incompatible race.
8. Soluble glucan isolated from a compatible race appears to inhibit the hypersensitive response more strongly than that from incompatible races. Further examination is necessary to determine whether the soluble glucan is really the cause of host-parasite specificity.

Hypersensitive cell death is followed by periderm formation involving renewed meristematic activity. It is known in some cases that cell death constitutes an important part of morphogenesis (1). These facts tempt us to postulate that the hypersensitive response in which cell death plays a key role is essentially a kind of morphogenesis.

DISEASE RESISTANCE: ELECTROPHYSIOLOGY

In 1980, I retired as professor of phytopathology from Nagoya University. Two years before retirement, I started to conduct electrophysiological experiments on the infection process of potato by *Phytophthora infestans*. I intended to investigate the first electrophysiological sign in the infection process. However, the experiments were not easy. We had to develop a new device and new techniques. It was laborious work. After retirement I have continued this research in cooperation with H. Okamoto and K. Katou (4, 15), electrophysiologists in the faculty of science at Nagoya University. Electrophysiological measurement is thought to be an irreplaceable research tool for some kinds of cell response, since the proposal of the chemiosmotic theory of energy coupling and the discovery of the cotransport system linked to the activity of the electrogenic ion pump via a H^+ electrochemical potential gradient (8, 19).

By setting up equipment to measure the membrane potential of an infected cell under a microscope, we found that there was no signal such as action potential at the time of penetration by either the incompatible or the compatible races. Infection by an incompatible race appeared to produce the first detrimental effect on the mechanism responsible for the maintenance of the passive component of the membrane potential, but the electrogenic component was increased and seemed to compensate for the deficit of the passive component in an early period of infection. Later, however, the electrogenic component also decreased.

PERSPECTIVES

My views on disease resistance in plants have been summarized in a book written in the Japanese language (12):

> During the process of evolution, pathogens acquired the ability to break through innumerable obstacles in order to invade into and live upon a host plant. For instance, the host cell wall is a sufficient barrier against many microorganisms which do not have the ability to break through the wall. Host plants susceptible to pathogens with efficient armaments have been selectively eliminated. Thus, both the plants and parasites now surviving have developed various means for attack or defense against each other; sometimes one or the other wins or is defeated in the desperate fight. Pathogens continue to devise new armaments by mutation and selection. Besides these armaments, all of the morphological and physiological factors of both hosts and pathogens may affect disease development in one way or another.

Thus, disease resistance seems to be disordered and haphazard. However, disease resistance is not an incoherent pathological phenomenon, but a defense response to invasion by a parasite that occurs step by step in perfect order according to a definite rule.

The pathogens, at first, have to break through the preinfectional chemical and physical barricades including the dynamic processes of host response which accompany this invasion. Even if the initial invasion is successful, various kinds of defense responses, such as papilla formation, hypersensitive cell death, production of toxic aglycones by hydrolysis of glucosides, metabolism and oxidation of polyphenols, accumulation of phytoalexins, and so on, must be overcome. These responses to invasion by the pathogen take place in succession according to a definite program. Some pathogens may be inhibited by defense factors at an early step and others may be trapped by other factors at a later stage, depending upon the strategy of the host plants. Thus, each pathogen causes specific symptoms which are the aftermath of an old battle. It is presumed that the basis of these defense responses may be the recognition and rejection by plants of foreign organisms, which is required if they are to maintain their inherent characteristics.

Results from research in the field of host-parasite interaction provide much useful information for agricultural technology. Further advances in biotechnology require more precise knowledge of the host-parasite relations, which will involve analysis of the mechanism of disease resistance at the molecular level. This field of science may also be indispensable to understanding the general principle of immunity of living organisms and its relation to the process of evolution.

EPILOGUE

As described above, during the World War II, in my youth, I was disappointed with the profession I had chosen. After the war, I worked hard on various research projects to resolve my doubt concerning my profession. Now, after a long research life, I have no doubt of the importance of my profession.

I should like to express my sincere thanks to the colleagues with whom I have worked. I also express my appreciation to my wife, Toshiko, for her generous assistance for nearly 40 years since our marriage.

Literature Cited

1. Bowen, I. D., Lockshin, R. A., eds. 1981. *Cell Death in Biology and Pathology*. London: Chapman and Hall. 493 pp.
2. Cruicksank, I. A. M. 1963. Phytoalexins *Ann. Rev. Phytopathol.* 1:351–74
3. Hiura, M. 1943. Studies in storage and rot of sweet potato. *Gifu Norin Semmon Gakko Gakujutsu Hokoku* 50:1–5
4. Katou, K., Tomiyama, K., Okamoto, H. 1982. Effect of hyphal wall component of *Phytophthora infestans* on membrane potential of potato cell. *Physiol. Plant Pathol.* 21:311–17
5. Metlitskii, L. Ozeretskovskaya, O., Vulfson, N., Calhova, L. 1971. The role of lubimin in potato resistance to *Phytophthora infestans* and its chemical identification. *Mikologia i Fitopathol.* 5:439–43
6. Müller, K., Börger, H. 1940. Experimentelle Untersuchungen uber die *Phytophthora*-Resistenz der Kartoffel. Berlin: Arb. *Biol. Reichsanstalt. Land-u Forstwirtsch.* 23:189–231
7. Norman, E. H. 1940 *Japan's emergence as a modern state—political and economic problems of the Meiji period.* New York: Institute of Pacific Relations. Translated into Japanese by Soji Okubo. Tokyo: Jijitsushin. 318 pp.
8. Novacky, A., Karr, A. L., Van Sanbeek, W. V. 1976. Using electrophysiology to study plant disease development. *BioScience* 26:500–4
9. Stossel, A. 1980. Phytoalexins—a biogenetic perspective *Phytopathol. Z.* 99:21–72

10. Suzuki, N. 1957. Studies on the violet root rot of sweet potato caused by *Helicobasidium mompa* TANAKA. *Bull. Natl. Inst. Agric. Sci. Japan.* Ser. C. 8:69–123
11. Tomiyama, K. 1963. Physiology and biochemistry of disease resistance of plants. *Ann. Rev. Phytopathol.* 1:295–324
12. Tomiyama, K. 1979. *Physiology of Plant Infection.* UP-Biology Series No. 35. Tokyo: Tokyo Univ. 149 pp. (In Japanese)
13. Tomiyama, K. 1982. Hypersensitive cell death: its significance and physiology. In *Plant Infection,* ed. Y. Asada, W. R. Bushnell, pp. 329–42. Tokyo: Tokyo Univ. 467 pp.
14. Tomiyama, K., Akai, J., Washio, T. 1952. Studies on the specific affinity between the host plant juice and parasitic fungus. I. Effect of host plant juice on germination of the spores of parasitic fungus. *Res. Bull. Hokkaido Natl. Agric. Exp. Stn.* 63:1–12
15. Tomiyama, K., Okamoto, H., Katou, K. 1983. Effect of infection by *Phytophthora infestans* on membrane potential of potato cell. *Physiol. Plant Pathol.* In press
16. Tomiyama, K., Sakai, R., Sakuma, T., Ishizaka, N. 1967. The role of polyphenols in the defense reaction in plants induced by infection. In *The Dynamic Role of Molecular Constituents in Plant-Parasite Interaction,* ed. C. J. Mirocha, I. Uritani, pp. 165–82. St. Paul, Minn.: Am. Phytopathol. Soc. 372 pp.

17. Uritani, I., Akazawa, T. 1955. Antibiotic effect on *Ceratostomella fimbriata* of ipomeamarone, an abnormal metabolite in black rot of sweet potato. *Science* 121:216–17

18. Varns, J. L., Kuc, J., Wiliams, E. B. 1971. Terpenoid accumulation as a biochemical response of potato tuber to *Phytophthora infestans*. *Phytopathology* 61:174–77

19. Wheeler, H. 1978. Disease alteration in permeability and membranes. In *Plant Disease* III, ed. J. G. Horsfall, E. B. Cowling, pp. 317–47. New York: Academic. 488 pp.

Ann. Rev. Phytopathol. 1984. 22:1–10

EXPLORING TROPICAL RICE DISEASES: A REMINISCENCE

S. H. Ou

The International Rice Research Institute, Los Baños, The Philippines, and National Science Council, Taipei, Taiwan, Republic of China

INTRODUCTION

I never expected to have the privilege and honor of writing a prefatory chapter for the *Annual Review of Phytopathology*. I appreciate the editorial committee's broad perspective, which extends over the people and problems of plant pathology in nations all around the world, and I treasure this opportunity.

Much has been written by eminent scientists on the philosophy, concepts, principles, and directions of plant pathology, as well as on the attitudes of plant pathologists; I have little to add. It is easier to say something of one's past. More than half of my 50 years as a plant pathologist was the turbulent period of civil wars and World War II. During the early years research programs were disrupted and often changed. As a result, the most peaceful, interesting, and rewarding time in my career has been my years of association with the International Rice Research Institute (IRRI) in the Philippines. I want to recount some of the events that I believe are significant in the development of rice pathology in the tropics, the details of which are largely unknown to most people. Hindsight is often helpful; L. M. Black said: "Learning is much more likely to be derived from history than from prophecy" (1).

The recent history of rice pathology in the tropics reflects the general state of plant pathology in many tropical nations. The number of plant pathologists has been insufficient to cope with important disease problems. Mistakes have been made through prejudice, misconception, and blind faith in old ideas and untested new ones. Rice is an important crop to the hungry people of the world, yet it was long neglected by the world scientific community.

Rice is a remarkable plant. It has great antiquity and a very broad and diverse genetic base in its many thousands of cultivars and landraces. The genetic

1959

0066-4286/84/0901-0001$02.00

diversity of rice provides numerous genes for resisting many pathogens. We are convinced that most rice disease problems in the tropics can be resolved with genetic resistance. In its long history of cultivation in broad regions, rice has been associated with all types of pathogens. Many deter rice production and until recently were very little known. Here are the results of our explorations into some of these problems.

MENTEK AND PENYAKIT MERAH

A rice malady called mentek was known in Indonesia as early as 1859, and severe losses were reported between the 1930s and 1950s. For many years Dutch scientists studied the disease intensively from the soil chemistry and nutritional point of view, working on the hypothesis that it was a physiological disorder. However, to their credit, in spite of detailed studies they never concluded that it is physiological. Later they studied nematodes as a possible cause of the disease.

Penyakit merah is another disease known in Malaysia since 1938. Several foreign experts were invited to study it for several years. Perhaps influenced by the study in Indonesia, it was also considered physiological and was reported to be due to a metabolic deficiency of nitrogen. Similar but less known is so-called suffocation disease, seen in Taiwan in the early 1960s, said to be due to a lack of oxygen in the soil. All of these diseases turned out to be due to virus infections. Mentek and penyakit merah were later known as tungro disease; suffocation disease was called transitory yellowing.

Many plant pathologists know about cadang-cadang disease of coconut in the Philippines, but very few know there is also rice cadang-cadang disease. Rice cadang-cadang disease was first reported in 1939 and caused some 30% loss of rice nation-wide in the 1940s. Researchers suspected that it had a viral nature and thought it similar to Japanese dwarf disease. It was later shown to be due to the tungro virus.

Even though virus diseases of rice had occurred in the tropics for a long time in different places, almost no reliable information on them was available when the IRRI began to operate. When the first large batch of several thousand rice germ plasm collections was planted at an IRRI farm for seed multiplication in 1962, there was a great display of reddish, orange, yellow, and dark green leaf discoloration. Many of the entries were stunted. Because the physiological disease hypothesis was still strong, some thought that these were physiological disorders. However, extensive field observations showed that the discoloration occurred in upland rice as well as in the lowlands. Sometimes among seedlings in the same hills some were healthy while others showed symptoms. Thus, the disorder was not likely to be physiological. We therefore turned our attention to its possible viral nature and during the next two years identified three new virus diseases, which we called orange leaf, tungro, and grassy stunt (4). Even after positive identification of the viruses, some continued to think of the diseases as physiological. One day the leader of a rice team from an international organization in Manila invited us to see an experiment developed by their plant physiologist, who claimed that it proved that tungro in his potted plants was due to a nutritional disorder. This was an interesting challenge to our report. Several plant physiologists and pathologists, as well as reporters and photographers, gathered in the developer's laboratory. What we found in his pots, however, were seedlings showing simple nitrogen deficiency symptoms, not the tungro disease of the field. After adding a little nitrogen fertilizer, the plants returned to normal. That plant physiologist left Manila soon afterward.

In 1964 I visited Malaysia and had an opportunity to see penyakit merah in the field firsthand. It struck me as being very much like tungro disease. We dispatched one of our staff to live in a small agricultural experiment station, taking with him a few small insect cages and equipment for transmission studies. In less than two months, he obtained proof that penyakit merah was indeed the same as tungro. The symptoms, the vector, and the manner of transmission were the same. The experiments did not convince everyone, however, and for many years some pursued evidence that penyakit merah is a physiological disorder.

In 1965, we set up a simple field test that served as a demonstration plot as well. In a diseased field found early in the season, a dozen microplots of 2 × 2m were set up side by side. The diseased young plants were removed and replaced with new seedlings, some of which were healthy (raised in a greenhouse) and some of which were artificially inoculated with tungro virus. Some of the plots were covered with insect cages and some were exposed. At the end of the season, the results were dramatic. Plots with healthy seedlings covered by the insect cages grew normally and gave normal yields. Those inoculated or those in the exposed plots showed typical symptoms of penyakit merah (tungro) and yielded only a few hundred kilograms per hectare. This changed the attitudes of many people both in and outside Malaysia.

The identification of penyakit merah of Malaysia as a viral disorder led me to suspect that mentek of Indonesia was also a virus disease similar to tungro. This proved to be true in 1967, when we were able to travel to Indonesia. Transmission experiments were again conducted and positive evidence was obtained.

In 1965, one of our students found tungro when he returned to Thailand, and in 1966–1967 it affected one-third of the rice in that country. India began to survey for virus diseases in 1967 and eventually found them to be widespread in many states. It became apparent then that virus diseases were major pathological problems of rice in the tropics.

Many other virus and virus-like (mycoplasma) diseases of rice have been reported since 1970; e.g. necrosis mosaic (Japan), yellow mottle (Kenya), giallume (Italy), waika (Japan), wilted stunt (Taiwan), ragged stunt (Indonesia, Philippines, Thailand, India), bunchy stunt (Mainland China), gall dwarf (Thailand), chlorotic streak (India), and crinkle (West Africa). More will probably be found in the future. The discovery of these new virus diseases and the fact that they appeared in different locations led me to think that the viruses might come from wild graminaceous hosts.

In the early 1950s, Dutch rice breeder J. G. J. van der Meulen developed several cultivars (e.g. Tjeremas, Bengawan, Peta, Intan, Sigadis) from a well-known cross between Latisail and Tjina in Indonesia. These cultivars were resistant to mentek, even though the nature of mentek was not known at the time.

In rearing the green leafhopper for tungro transmission studies at IRRI, K. C. Ling hypothesized that the cultivar Peta would be a better host for feeding and ovipositioning the insects because its leafsheath is longer and larger than that of the conventionally used Taichung Native 1 (TN1). However, the results were the contrary. Insect populations on Peta plants were much smaller than those on TN1. This was the first recognition of leafhopper resistance in rice, now an accepted fact of great importance in rice insect control. Many other discoveries were incidental. The diverse genes with which rice resists both viral pathogens and vector insects should enable us to protect new cultivars from heavy damage.

BACTERIAL LEAF STRIPE AND KRESEK

In the early days bacterial diseases of rice in the tropics were considered unimportant and reports on them were very scarce. O. A. Reiking in 1918 reported a bacterial leaf stripe disease from the Philippines that produced narrow stripes with a watery, dark green, translucent appearance. He did not identify the causal organism. Many authors have erroneously referred to the disease Reiking described as bacterial leaf blight. It was not until 1957 that this bacterial leaf streak disease, as it is now called, was differentiated from bacterial blight during a study in Mainland China.

In 1950 J. Reitsma & P. S. H. Schure described a bacterial disease of rice in Indonesia called kresek, and in 1953 Schure identified the organism that causes it, *Xanthomonas kresek*. *Kresek* is a local word, we were told, that describes the dead rice plant with its leaves floating on the water in the field. The symptoms were very different from those of bacterial leaf blight known in Japan.

In 1964, Masio Goto came to the IRRI as a visiting scientist. In this capacity, he visited many countries in Southeast Asia, including Indonesia, where he had an opportunity to closely scrutinize kresek disease. He noticed typical bacterial leaf blight symptoms mingled with kresek symptoms and suspected that the two might be the same disease. He was excited by this observation and the evening he returned from Indonesia said to me that we might find kresek right in the IRRI plots. We found it the next morning. The symptoms that all along we thought were due to injuries from stem borer, so-called dead heart, were actually symptoms of kresek. Later he proved experimentally that leaf blight bacterium caused the kresek symptoms. Since his report, kresek has been found in many countries, not only in the tropics but also in Korea and Japan. Presumptions, prejudices, and belief in the pronouncements of others are so often impediments to finding the truth!

In addition to kresek, Goto described another distinct symptom of bacterial blight in the tropics, the pale yellow syndrome. One or two young expanded leaves of normal size on some of our field-grown plants had partly or entirely turned a pale yellow color while the other leaves stayed green. A few of the lowest leaves often died. We had had the pale yellow plants brought to our laboratory before Goto described the symptoms. There was one post-doctoral visiting scholar who wanted to see if the disease was viral in nature. Since he had experience with tobacco mosaic, he comminuted the leaves and rubbed the juice on leaves of healthy seedlings, just as he did when inoculating tobacco with tobacco mosaic virus. A few weeks later he showed me the inoculated plants, saying he thought his inoculation experiment had succeeded in transmitting a virus. I was not convinced. After three repeated trials, he got the idea that I was preventing him from publishing a new discovery. Soon afterward, Goto produced results with bacterial inoculation. Goto's reports opened up new horizons in tropical bacterial disease research. Cultivar IR8 was becoming popular in the late 1960s, and its susceptibility to bacterial blight aroused further interest in the disease.

For a long time bacterial blight was known as an Asian disease. Only recently has it been reported in Africa, Latin America, and, a little earlier, from Australia. The origin of the disease in areas outside Asia is of interest. Seeds have been reported as the method of transmission in temperate regions, but we failed to find seed transmission in the hot and humid tropics. In the tropics the bacterium dies quickly, and even in an active leaf lesion living bacterial cells seem to exist only in the peripheral areas. Isolations cannot be made from the center or older portions of the lesion. Seeds of rice cultivar IR8 and others have been sent to Africa and Latin America for several years, sometimes in large quantities, but the leaf blight disease has never been reported. In 1976 I visited Central America and found a few rice plants showing symptoms of bacterial blight in an isolated, newly developed rice area. I also observed many typical bacterial leaf blight-like symptoms on at least four species of wild graminae in large areas nearby. I speculated that the bacterium on rice may come from the wild graminae. The presence of the disease on rice was confirmed later by Lazano in Latin America (4), but he believed that the disease was probably introduced through imported seeds. Yet relatively few rice seeds are imported into Australia and the strictest quarantine probably prohibits introduction of the disease. For several years I. W. Buddenhagen did not observe the disease in Africa but reported its presence recently. He also suspects that it is of local origin.

PHYSIOLOGIC RACES OF *PYRICULARIA ORYZAE*

The presence of pathogenic races of rice blast fungus, *Pyricularia oryzae,* was noticed as early as 1922 in Japan; the reactions of cultivars to the fungus differed on the plain and in the hilly areas. The phenomenon was demonstrated easily by inoculation experiments. During the 1950s and 1960s, Japanese and US workers studied physiologic races intensively, and many other countries began to identify their races in the 1960s. Imitating the procedures that began with cereal rusts, a set of 10–12 differential cultivars were inoculated with pure culture isolates of the fungus, producing a table of resistance and susceptibility grouped into races. Several investigators have spent 15 or more years on the subject. A great deal of effort has been expended trying to find out what races exist in this or that country.

To establish a rice cultivar as resistant (R) or susceptible (S) to an isolate of the blast fungus is not always easy. During earlier studies, some investigators complained that certain differential cultivars were not good because they produced different types of lesions, from resistant to susceptible on a single leaf. Others found some fungus isolates inconsistent, producing one pattern of reaction (race) on test cultivars in the first inoculation but another pattern in the next inoculation.

We started the race study as everyone else did. We also found the different lesion types on the same leaves and very different lesion numbers among the cultivars. In our inoculation experiments, among 20 seedlings of each differential cultivar there may have been numerous lesions on each seedling of some cultivars, while other cultivars showed only a few lesions on a few of the seedlings. In these cases, the designation of cultivar as resistant or susceptible depended on the discretion of the investigator. In many instances, a leaf may have had few large S-type lesions but many more R-type lesions. If typical S-type lesions were present the cultivar would be conventionally considered susceptible, but this did not seem to be logical. The expression of reaction was often quantitative. Simple R and S signs to indicate the reactions between the differential cultivars and the isolates seemed grossly incomplete and arbitrary.

We began our race study by employing the Japanese and the US as well as the Taiwanese differential sets. After a hundred inoculations using the US set, we identified some 20 races in the conventional manner. Based on the US set we found a race, say race X, consisting of 17 isolates. To six of them a local popular cultivar, Peta, was susceptible and to the other 11 isolates Peta was resistant. If someone asked whether Peta was susceptible or resistant to that race X, what would be our answer? The existing sets of differentials were not useful in the Philippines.

To search for a new set of differential cultivars, we started with a broad base, selecting 110 rice cultivars, including the Japanese, Taiwan, and US differentials as preliminary candidates. After inoculating these 110 cultivars with 50 isolates, two things among others were apparent: (*a*) some cultivars were susceptible to none or only a few of the isolates, while others were susceptible to as many as 90% or more of the isolates, and all gradations from zero to 100% were found, (*b*) no two isolates had the same reactions on all the cultivars; they were all different.

It became clear that conventional race grouping is an oversimplification that had led to a common misconception among rice workers that the isolates belonging to a race are identical. Often investigators have selected one of many isolates as representative of the race for testing the resistance of new cultivars. In fact, it may not be representative, as shown in the case of the isolates of race X and Peta above. The results also showed that the number of races identified depends not only on the number of differential cultivars, but also depends on what cultivars are used, as we all know. If one selects the cultivars with the highest resistance or susceptibility reactions, there will be few races. If one selects those with both resistance and susceptibility in approximately equal numbers, there will be more races. Not only are the isolates of the same race not identical, each isolate within itself may show variance.

P. oryzae co-evolved with rice for a very long time. It must be extremely diverse to adapt to the wide genetic base present in rice. The results of the inoculation experiments mentioned above suggest the complex nature of the pathogenicity of the fungus. Studying single-spore progenies from a common origin would be an easy and efficient way to determine diversity. When we inoculated about 50 single-spore cultures, each from two lesions, onto a set of 12 Philippine differentials, we found 14 different reaction patterns (races) from one lesion and eight races from the second lesion. In a further step, we inoculated cultures of 25 daughter conidia from each of two monoconidial cultures. Again each family showed nine and ten different pathogenic patterns. Even conidia produced from hyphal-tip cultures of single cells of one conidium differed in pathogenic patterns. This new dimension of variability was a great surprise to us and more so to other people. Some did not believe or did not wish to believe it was possible, since it rendered all previous race studies almost meaningless. A year later an independent study confirmed this extreme variability. In fact, such variability can be found in several earlier reports, but its significance was not recognized or emphasized. It seems that the physiologic races of *P. oryzae* are temporary or transitory phenomena. We appreciate the words of W. B. Hewitt: "If we merely follow the beaten path or just sit in our canoe and float with the stream relying on the imagination of others . . . and/or doing more of what has been done, then life is but an empty dream" (3).

THE HORIZONTAL RESISTANCE BANDWAGON

Plant pathology, like other professions, has its changing fashions. Horizontal resistance (HR) has been popular in the past 20 years. Many rice pathologists have been trying to find horizontal resistance to rice blast. Others have offered hypotheses on how it should be done. Most of their theories sound good on paper but are difficult to achieve in practice. Some plant pathologists have had strong opinions on studying horizontal resistance that discredited conventional resistance study. Unfortunately, despite these efforts, we have found no practical way to measure horizontal resistance to rice blast. Some believed that measuring the rate of disease development was a sure method. By measuring the rate of development, several rice cultivars were considered to have good horizontal resistance in West Africa. At least two of them, Blue Bonnet and Tainan 8, were well-known susceptible cultivars in their respective homelands. The races of *P. oryzae* have patterns of geographical distribution, and in the absence of major pathogenic races the susceptibility of the cultivars could not be shown.

Yorinori & Thurston attempted to find out if certain characteristics that had been shown to be associated with horizontal resistance to late blight in potatoes existed in rice blast (7). They reported that one of the most difficult problems encountered in studying rice blast was the diversity of lesion types observed on the varieties and selections used. Even using a single isolate, wide variation in lesion types was observed on the same variety in both detached leaf and greenhouse inoculations. Reactions varied from highly resistant to highly susceptible. The authors came to no conclusions about what might represent horizontal resistance to *P. oryzae* in rice. Their study confirmed the extreme variability of the organism and illustrated the difference between theory and practice. Finally, many of the so-called field resistant cultivars selected in Japan were very susceptible in the Philippines (6). As yet the answer to the question of horizontal resistance remains illusive.

In our studies on resistance to rice blast, we found something like vertical and horizontal resistance but we described them as qualitative and quantitative. Qualitative indicates the reaction of each rice cultivar to fungus isolates, resistant or susceptible. Quantitiative is the total amount of disease, i.e. the number of lesions produced. Qualitative resembles vertical, and quantitative horizontal, reactions.

In one of our studies, isolates were inoculated on the cultivars from which they were isolated. The results showed that on the cultivars that had a broad spectrum of resistance, i.e. resistant to most isolates, small numbers of lesions developed. On the other hand, cultivars with narrow spectrums of resistance developed large numbers of lesions. For instance, when 37 isolates from Tetep, a cultivar of broad-spectrum resistance, were inoculated back onto Tetep, five produced no lesions, 17 produced less than one lesion (average) per plant, others produced a few lesions, and two produced 14 and 16 lesions. The average was two lesions per plant. In the same experiment, a susceptible cultivar, Khao-teh-haeng 17, used as a control, produced 10–67 lesions on each seedling, with an average of 33. Similar studies with other rice cultivars disclosed that the resistance spectrum and the number of lesions were always negatively correlated.

We further found that cultivars with broad spectrums of resistance as identified in the international blast nurseries (i.e. resistant in most test locations) had fewer lesions, while those with narrow spectrums of resistance had the reverse. Regardless of whether the isolates were from the field or from single-spore cultures or whether the amount of disease was measured in the field or in artificial inoculations, there was always a close negative correlation between the spectrum of resistance and the number of lesions.

Some 240 pathogenic races have been identified in the Philippines, and we know exactly the number of races to which each differential cultivar is resistant. The spectrum of resistance of the differential cultivar varies from 10–90%. Whether they were inoculated artificially or naturally infected, the amount of disease was again negatively correlated with the spectrum of resistance.

From these many results, we assumed that cultivars resistant to most isolates (races) have more genes for resistance, and they appear to have horizontal resistance. This seems to fit very well with R. R. Nelson's hypothesis that the accumulation of vertical resistance may confer horizontal resistance. The reactions basically are vertical, but the end results appear to be horizontal. In other words, if you see only the end results, resistance is horizontal (amount of disease), but when the resistance is analyzed isolate by isolate it is vertical or racial. We prefer to call this stable resistance rather than horizontal resistance. Horizontal resistance in this case seems to be another artifact (2).

THE INTERNATIONAL APPROACH

P. oryzae is capable of producing many pathogenic races. However, the survival and prevalence of a race depends on the availability of a host cultivar. After their long association, *japonica* rice cultivars are readily infected by the prevailing races in Japan, while many *indica* cultivars are resistant. The reverse is true in the tropics. Pathogenic races differ from locality to locality and also from season to season. One of our studies showed that the composition and frequency of races in the IRRI blast nursery differed from month to month, although there were one or two prevailing most of the time. Tests conducted in one country or only a few times cannot assure us of the resistance levels of any cultivar. Obviously, the best test of resistance is to expose rice cultivars to all existing races at different regions over a period of years through international cooperatives tests.

After a symposium on rice blast disease in 1963, the IRRI assumed the responsibility of coordinating the international blast nurseries (IBN) program, from multiplying the seeds to compilation of the data. Some 30 countries and some 50 or 60 testing stations from all the rice-growing regions of the world have participated. Each year several hundred cultivars are tested. These experiments are the most extensive and efficient way available to test blast resistance, and they are now in their twentieth year. One of the most significant results is the identification of rice cultivars with broad-spectrum resistance. The test cultivars have shown a wide range of resistant reactions, from 20–98% of the tests. Some rice cultivars, such as Tetep and Carreon, have been resistant in 95% or more of the tests and have been consistent year after year.

Many of the cultivars with broad spectrums of resistance were crossed singly or in combination with cultivar IR8, which has a desirable plant type. In some cases nine or ten back crosses with IR8, as well as the progenies in different generations, were screened continuously for many years at the IRRI blast nurseries. The best 50 lines were put into the IBN test in the late 1970s. In the test results over the last five or six years, many lines showed a level of resistance as high as the parents, such as Tetep, Carreon, and *Oryza nivara*. These include lines of IR1416, IR1905, IR3259, IR4227, IR4547, IR5533, IR13429, and others. Some, such as IR4547, are also resistant to other diseases and to some insects. Now we have lines with both high levels of resistance and good plant type. These should be better donors for breeding blast resistance. This exercise also suggests that rice cultivars with a high level of stable resistance can be developed. At present, this can be done most effectively by exposing the selections of one country to the *P. oryzae* strain of a number of cooperating countries over a period of years, as is being done in the IBN.

Literature Cited

1. Black, L. M. 1981. Recollections and reflections. *Ann. Rev. Phytopathol.* 19:1–19

2. Ellingboe, A. H. 1975. Horizontal resistance: An artifact of experimental procedures? *Aust. Plant Pathol. Soc. Newsl.* 4:44–46

3. Hewitt, W. B. 1979. Conceptualizing in plant pathology. *Ann. Rev. Phytopathol.* 17:1–21

4. International Rice Research Institute. 1969. *The virus diseases of rice plants.* Proc. Symp. Intl. Rice Res. Inst. Baltimore: Johns Hopkins Univ. Press. 354 pp.

5. Lazano, J. C. 1977. Identification of bacterial leaf blight in rice caused by *Xanthomonas oryzae* in America. *Plant Dis. Rep.* 61:644–48

6. Ou, S. H. 1979. Breeding rice for resistance to blast: A critical review. In *Proc. Rice Blast Wrkshp*, pp. 81–137. Los Baños, Laguna, Philippines: Intl. Rice Res. Inst.

7. Yorinori, J. T., Thurston, H. D. 1975. Factors which may express general resistance in rice to *Pyricularia oryzae* Cau. In *Horizontal Resistance to the Blast Disease of Rice*, pp. 117–35. Cali, Columbia: CIAT Ser. CE-9

PLANT PHYSIOLOGY

Photograph by Debra Biale

Jacob Biale

Ann. Rev. Plant Physiol. 1978. 29:1–23
Copyright © 1978 by Annual Reviews Inc. All rights reserved

ON THE INTERFACE
OF HORTICULTURE
AND PLANT PHYSIOLOGY[1]

Jacob B. Biale

Biology Department, University of California, Los Angeles, California 90024

CONTENTS

[1]Abbreviations used in this chapter: ABA, abscisic acid; IAA, indoleacetic acid; NAA, naphthalene acetic acid; 2,4-D, 2,4-dichlorophenoxyacetic acid; GA, gibberellin; GA_3, gibberellic acid; Ethrel, 2-chloroethanephosphonic acid; Amo-1618, 4-hydroxyl-5-isopropyl-2-methylphenyl trimethyl ammonium chloride 1-piperidine carboxylate; B-9 or Alar, 1,1-dimethylamino succinamic acid; CCC, 2-chloroethyl trimethyl ammonium chloride; Phosphon, 2,4-dichlorobenzyl tributyl phosphonium chloride.

1971

0066-4294/78/0601-0001$01.00

To My Family

PREFACE

The extension of the invitation by the editorial committee of the *Annual Review of Plant Physiology* to prepare this introductory chapter is a distinct honor; to live up to the challenge is no easy task. In contemplating a subject to be covered, I searched for a topic that has not been in my mind for some time. I felt that I could discuss the relationship between plant physiology and horticulture with some degree of detachment as it is now 15 years since my transfer from the College of Agriculture to the College of Letters and Science. The advantages of taking a look as an outsider may be counteracted by insufficient insight on my part into problems encountered recently by investigators in horticulture.

The choice of the term "Interface" was intended to narrow down the discussion to fields in which there is no sharp demarcation line between horticulture and plant physiology. I shall use the term "horticulture" to designate that body of knowledge concerned with understanding the behavior and culture predominantly of trees bearing consumable fruit. Studies with fruits borne by annual plants, vines, or shrubs have also contributed considerably to certain "Interface" topics and they, too, will enter into the total picture. In some quarters the term horticulture also includes trees, shrubs, and flowering plants used by society for ornamental purposes.

The realization that basic knowledge is necessary to explain and advance the operations of gardening goes back to Hale's *Vegetable Staticks*. In a treatise on "The Theory and Practice of Horticulture" (London, 1840), John Lindley stressed the need of physiological principles on which to base horticultural practices. It was almost a century before sufficient knowledge in plant physiology and in other disciplines, such as soil science, entomology, plant pathology, and genetics, was accumulated to furnish a basis for better understanding and for improvement of practices in fruit culture.

The dependence of the advancements in horticulture upon plant physiology and the interactions between these two fields can be clearly seen in the pages of the *Annual Review of Plant Physiology* since its inception in 1950. The earlier volumes placed emphasis on mineral nutrition, growth substances, carbon assimilation, respiration, and water relations. Within the framework of these fields a number of reviewers from agricultural institutions contributed topics dealing with nutrient deficiencies, salt uptake, soil moisture, foliar applications of fertilizer, dormancy in trees, growth substances in horticulture, and postharvest physiology of fruits and vegetables. With greater stress on biochemical aspects of plant physiology, the reviews delved deeper into mechanisms of action, formation of biologically important substances, the nature of the major plant processes, and metabolic pathways.

Special sections in the series devoted to development, environmental physiology, and special topics continue to cover subjects directly applicable to research in agriculture including horticulture.

In the first part of this discussion I wish to cite two important fields of physiological horticulture, indicating underlying ideas for experimental approach. As part of one of the quoted studies, a tribute is included to two late teachers and colleagues of mine who personified an effective approach to the integration between basic and applied science. In the second part of this presentation, the more personal section, emphasis is placed on the training desirable for work on "Interface" problems. In addition, I hope that drawing some thoughts from my own field of research points to the attractiveness of being engaged in a field in which there is no distinct demarcation line between the basic and the "bread-and-butter" problems.

FACING UP TO COMPLEXITIES IN HORTICULTURAL RESEARCH

During the end of the third decade and the beginning of the fourth, while I was enrolled as an undergraduate student in the College of Agriculture of the University of California, certain studies with fruit trees, which were in progress at that time, exerted considerable influence on the approach to research in physiological horticulture.

The Problem

The fruit industries of certain districts in California, as well as in other states and countries, were faced with physiological disorders that caused drastic cuts into yields of both deciduous and evergreen orchards. These disorders were known as "little leaf" on peaches and other stone fruits, as "rosette" on apples and pecans, and as "mottle leaf" on citrus. In citrus the mottle leaf syndrome is characterized by yellow streaks between the veins, while the veins themselves retain chlorophyll. In the other species the symptom manifests itself in the spring by the opening of tufts or rosettes of leaves that are very small, frequently no more than 5% of the normal leaf area. When it was recognized that the causal agent was neither fungal nor bacterial, the attention was directed to a possible nutritional deficiency.

Interdisciplinary Approach

The problem was undertaken by W. H. Chandler, professor of pomology, D. R. Hoagland, professor of plant nutrition, and P. L. Hibbard, analytical chemist. They decided that the effort should be spread out through many localities rather than concentrating on one spot, thus assuring a better chance of success on some of the many plots under treatment. Massive applications of fertilizer were ineffective except for one site which had a sandy soil with low fixing power for applied inorganic chemicals. In that plot, recovery of peach trees from little leaf was observed when large amounts of commercial grade ferrous sulfate were applied, but not when the chemically pure compound was used. Analysis showed considerable quantities of zinc in the commercial $FeSO_4$. Heavy applications of zinc sulfate worked only in orchards of one district with a special light soil. After a series of experiments in more than 50 sites throughout the state of California the conclusion was reached that the most effective and universal cure could be obtained from foliage sprays and in some cases from zinc nails driven into the wood or from zinc powder plugged into tree trunks and allowed to be carried to the foliage by the transpiration stream. It turned out that zinc deficiency was the cause for the several disorders occurring in a number of horticultural species listed above. In greenhouse experiments with sand culture it was possible to reproduce the disorder and to cure it by supplying low levels of zinc. Subsequent biochemical studies led to a tentative conclusion that the role of Zn is that of a cofactor in the synthesis of the amino acid tryptophane, the precursor of auxin. I do not know whether these biochemical studies were substantiated, whether auxin application was tried to correct the disorder, or whether the levels of the auxin were found to be low in the foliage of little leaf or mottle leaf. Answers to these questions are less revelant here than the implications of the work under consideration.

Take-Home Lessons

The studies with little leaf and mottle leaf made a very important contribution in solving a pressing problem, and at the same time they are indicative of the benefits derived from interaction between different disciplines.

The concurrent pursuit of nutritional requirements by the Division of Plant Nutrition was helpful in focusing attention on possible need for micronutrients by the tree. The laboratory effort with rapidly growing plants to show that certain elements are required in minute quantities demanded a high degree of purification of chemicals and the use of special laboratory apparatus free of any contamination. Realizing these demands, the laboratory investigator might have concluded that there must be in the soil sufficient impurities of such elements as B, Mn, Cu, Mo, and Zn to make it highly dubious to be able to demonstrate that these elements were deficient in trees. Some investigators with interest in pure science might have argued also that the understanding of basic processes of plant responses is essential prior to any attack on problems of an applied nature. The cooperators on this project did not adopt this philosophy, realizing full well that the trees would die before the mechanisms of plant processes were unraveled. They felt that complete understanding of the responses is not necessary for carrying out field trials and that answers to perplexing problems in a relatively short time are feasible even with long-lived fruit trees.

How to conduct the tests under orchard conditions was another important lesson derived from the zinc studies. Serious doubts were cast on the orchard fertilizer experiments in which large tracts of land were divided into plots and the effects of additions of each element or combination of elements were tested. With trees the experimental error is large because few can be planted per acre and uniform acreage is not too abundant. A greater degree of uniformity can be expected in deep rather than shallow soils because of better development of the root system in the soil with greater depth. Trees with an extensive root system on one plot may absorb more nutrient even if it is present in lower concentration than do trees on another plot with restricted roots. Under such conditions soil analysis would not tell the story, while deficiency symptoms are more reliable in indicating the need of supplying the unavailable nutrient. A grower may not wish to delay in applying fertilizer until the deficiency is manifested because yield and quality could be affected prior to the appearance of deficiency symptoms. In many cases the need for a particular nutrient can be demonstrated by leaf analysis for the element in question. Foliar analysis is certainly useful where a nutrient is in excess or where it is definitely deficient. As more knowledge is gained on physiological responses the horticulturist will be able to rely on the leaf to tell him about the needs of the tree.

Great and Humble Men of Science

I have chosen to discuss the zinc deficiency work not only because of the merit of this study for the subject under discussion but also because of my reverence for the two cooperators of this project: William H. Chandler (1878–1970) and Dennis R. Hoagland (1884–1949). I valued them as inspiring teachers and I continued to maintain close contact with them long after my student days in Berkeley. The impact of these men on a generation of students and on scholarship in their respective fields, coupled with their qualities as human beings, deserves special consideration.

Chandler was unique as a scientist, administrator, man of ideas, and man of character. There is no question in my mind that he earned to the fullest the title of horticultural scientist. He had the ability, the insight, and the experience of integrating the anatomical features, the functions, and the responses to the environment of fruit trees. In his books on both deciduous and evergreen orchards he succeeded in systematizing horticultural knowledge, but at the same time he questioned horticulture as an independent discipline as he questioned the existence of an independent science of agriculture. A research man, in his opinion, to be effective in horticulture must have thorough training in botany, chemistry, and plant physiology. When a student was preparing to do horticultural research, Chandler would recommend that he take a minimum number of courses in horticulture, since the practical aspects can be picked up later on the job. At the same time he felt that work with rapidly growing grasses cannot take the place of tests and observations of trees in the orchard. Neither can laboratory results be transposed into recommendations for practice without field trials. Frequently the problems of the farm do not seem to be closely related to available scientific principles; neither can one wait for a system of knowledge to evolve complete enough to supply answers to all farm problems. In such situations the farm adviser has fulfilled a crucial role. He brings field problems to the attention of the experiment station man, helps him in setting up trials, evaluating results, studying limitations, and drawing up conditions for restudy. In his personal discussions and in his speeches, Chandler advocated extra caution in interpreting data from field tests because of the inherent complexities involved in working with trees. To him the outstanding virtues of a good experimenter are curiosity, diligence, patience, tentativeness, and integrity in reporting his observations.

Chandler himself was a patient man and a man of sturdy character; his sturdiness seemed to come from the trees he worked with. He was an humble man, but not a timid one. He tended to see the good in people, but he did not mince any words when he noticed abnormal selfishness and capriciousness. Egocentrism, excessive ambition, self interest, and lack of sharing of credit among academic people irritated him particularly. He felt that for everyone in the institutions of higher learning who considers himself important and indispensable there are hundreds on the outside who could do as well had they been given the opportunity for education. He referred to these people as the dormant buds and he stated that the "emblem of his faith was the tree and its system of dormant buds." He hoped that man would cultivate his desire for prestige somewhat less and instead develop his group instincts and acquire a greater appreciation of his heritage and cultural values of life.

It is no surprise that the cooperation between Chandler and Hoagland was so successful. They both shared the philosophy of harmonious interaction between the applied and basic in agricultural research. They also had similar broad outlooks on university and world affairs. They were both men of unusual personal qualities which inspired their associates and students. While they were both aware of the complexities of plant-soil interrelationships, their research trends and research programs differed considerably except for the cooperative zinc problem. Chandler had chosen to stand by his trees despite the long years required for results and the cumbersome nature of his experimental material. Hoagland went in the direction of adapting more simplified systems in order to gain insight into mechanisms of ion accumulation by plant cells. The selective action in absorbing certain ions such as potassium against the concentration gradient fascinated him. He came to the conclusion that problems of this kind must be studied in a less complex medium than the soil, and therefore designed a suitable culture medium known to this day in plant physiological laboratories everywhere as the "Hoagland solution." He also realized that in order to obtain reproducible results it is essential to use relatively simple systems such as *Nitella* or isolated roots and constant environmental conditions. Together with A. R. Davis, he designed two plant chambers for the control of temperature, humidity, and light intensity. This equipment was probably one of the first phytotrons introduced to work with plants. Hoagland was very much admired by members of his department and by the faculty of the university as well as by plant physiologists everywhere. Despite his preoccupation with many departmental and campus duties, he found time for students and staff. Whenever I went to Berkeley I knew that I would be welcome in his office. Conversations were always lengthy and on many subjects; he always asked about the wellbeing of his friend Chandler. Hoagland, who was ailing in the last years of his life, would remark repeatedly that Chandler was wise in selecting good genes. If Hoagland was lacking in genes for physical health, he made up for this lack with a brilliant mind and a generous heart.

CHEMICAL REGULATION IN PHYSIOLOGY AND HORTICULTURE

The curiosity of plant physiologists about unilaterally illuminated coleoptiles turning toward the light and about horizontally placed roots bending downward has resulted in discoveries with a far-reaching impact on horticulture as well as on plant physiology. I wonder whether the horticulturist of 50 years ago realized the full significance of Went's successful isolation in agar blocks of minute amounts of a substance which moved from the tip downward in the oat coleoptile and caused cell stretching. It took less than a decade from the time of that discovery for investigators in horticulture to embark on research and use of IAA and synthetic auxins on a variety of responses in fruit trees. With the introduction of a number of other naturally occurring as well as synthetic growth-regulating substances, new opportunities have opened up to look for agents with specific functions and for actions of combinations of regulators. The field of chemical regulation of flower and fruit development selected here for demonstrating the interrelation of horticulture and plant physiology is of particular relevance to investigators whose aim is improvement in yields and quality of fruit crops.

Flower Bud Differentiation

The chemical regulation of floral initiation has been studied directly by the application of naturally occurring or synthetic substances and indirectly by the use of growth retardants which effect the levels of the endogenous regulators. The results of these studies are variable and the conclusions are complicated by interactions and changes in the nutritional status of the tree.

A spectacular result with flowering of pineapple and of other bromeliads was achieved with external application of NAA, 2,4-D, ethylene, and Ethrel. Ethylene is the product of Ethrel decomposition and is probably produced also as a result of auxin application. Flowering success with ethylene was reported also for apples and mangos, but not for other species. In some plants cytokinins were observed to promote flowering and in others a combination of benzyladenine and one of the gibberellins was more effective than either one alone. The promotion of floral initiation by GA_3 is well established for rosette plants with a long-day photoperiod, but gibberellins act as inhibitors of flowering in most woody dicotyledons.

The role of gibberellins in the control of vegetative and reproductive development has been the subject of several studies by Anton Lang and associates. In brief, gibberellins appear to have different effects at different concentrations. By the use of morphactins such as CCC, B-9, and Amo-1618 it has been possible to lower the content of endogenous GA_3 to a level suitable for flower initiation. The idea was advanced that flower formation is suppressed above a certain maximal level of GA. As to the endogenous controlling mechanism, it is not known what takes the place of the morphactins. Conceivably, abscisin occurring universally might turn the switch on GA synthesis. Another possibility is a dynamic equilibrium between the various gibberellins, each with a different specific function. Whichever the case may be, interactions play an important role whether between different forms of one group or between different groups of regulating substances. Some of the workers in this field ascribe an ultimate flowering role to florigen, postulated from photoperiodic studies as a substance produced in induced leaves and capable of moving through a graft union to a noninduced branch. The attempts to isolate and identify florigen met with little success.

In light of these complexities the work of the horticulturist has to be largely empirical as he explores the various treatments suggested by the laboratory people.

Normally the fruit investigator is not faced with problems of floral initiation. Failure to produce sufficient flowers for a good yield is rarely a problem in the highly selective clones of orchard species. The control of time of flowering to avoid spring frosts or other hazards is probably of greater importance, especially in deciduous orchards. Under these conditions flower buds are formed in mid- to late summer following the flush of growth and remain dormant until the next spring. The concern in this case, especially in warmer climates, is that there should be enough days at temperatures below 6–8°C to break the rest and allow for abundant and uniform bud opening. In evergreen trees, differentiation of flowers takes place from apical meristems terminally or at axils of leaves of current season's shoots or, as in deciduous trees, from shoots formed in the previous year. In both cases day length does not appear to play a role in the reproductive process. Flower formation in fruit clones can be influenced indirectly by nutrition. Insufficient flowers may be produced when the nitrogen deficiency is so severe that small leaves with low chlorophyll content are formed. Another cause, though less likely, for reduced flowering is when a great excess of nitrogen results in succulent growth. Micronutrient deficiencies could affect flowering, but only when the deficiency is uncommonly severe. The horticultural literature tends to stress the relationship between nitrogen level and carbohydrate accumulation as a factor in flowering. On this point there may be a meeting of minds with plant physiologists who postulate that the role of the endogenous growth regulators is to mobilize the photosynthates from the leaves.

The Control of Fruit Set and Growth

To insure a proper fruit set in relation to foliage surface and to make certain that the fruits develop to maturity is of special concern to the horticulturist. Plant physiologists within agricultural organizations have contributed significantly to the clarification of the role of growth substances produced during pollination, fertilization, and seed development. Some of the most notable contributions were made by J. P. Nitsch, L. C. Luckwil, and J. C. Crane. A brief summary of their findings and of others will serve as background for drawing a few general conclusions and implications.

Fruit set commences with pollination, thus supplying a stimulus to the ovarian wall in the form of auxin or auxin-like substances and possibly also in the form of GA or GA-like substances. The stimulus which continues with syngamy, the fusion of male nucleus with the egg cell, is ascribed to these substances as well. The action of pollination and fertilization in bringing about cell expansion in the tissue of the ovary wall can be replaced in some but not all species by external applications of growth-regulating substances. Parthenocarpic fruits were set by auxin in tomatoes and in strawberries, while in stone and pome fruits, in blueberries and in cranberries, only gibberellins were effective. In cherry a combination of the two substances was required. Either cytokinin or GA_3 was responsible for setting seedless grapes. In Calimyrna figs the role of pollen was replaced by an auxin-like chemical, 4-amino-3,5,6-trichloropicolinic acid, as well as GA and cytokinin. Both growth substances and growth retardants were found to increase yield in several species. Apricots responded to auxins but not to GA, the level of which is high in this fruit. GA_3 was effective in pears and in cranberries. Alar, CCC, and phosphon, as well as ABA applications, resulted in increased crops of grapes and tomatoes. Apparently the level of cytokinins was raised by the use of growth retardants. A spectacular increase in the rate of growth of figs and consequent acceleration of maturity was attained by use of ethylene, Ethrel, and by high concentrations of auxin, resulting in ethylene production by the tissue.

In order to translate the effects of external applications into mechanism of action, analyses were made of the various substances at several stages of fruit development. In apple seeds and pericarp tissue, known and unknown auxins as well as GA_4 and GA_7 were identified. In the banana, cytokinin-like and gibberellin-like substances in addition to IAA were discovered. A special non-indolic auxin was reported in citrus fruits. The content of growth substances is always much higher in the seed than in the mesocarp, but when changes in concentration were followed in relation to growth of the entire fruit no correlations were observed. In some cases it was shown that peaks in growth substance content coincided with peaks of nucellus or endosperm or embryo development. In general, the fluctuations in auxins or gibberellins do not correspond to trends in the rate of growth of the mesocarp.

As to the role of endogenous growth substances, the most widely held theory is that they fulfill a mobilization function. By maintaining relatively high levels of the regulators in the seed a sink for products of photosynthesis is established between the reproductive organs and the vegetative tissue. This theory does not explain the need for a variety of substances and the reason for widely differing chemical structures. Neither does it explain the differences among different fruits in response to applications.

Hormonal Effects in Mature Fruit

The complex problems arising from interactions and antagonisms between different growth-regulating substances can be clearly demonstrated in the studies dealing with fruit maturation. Horticulturists have used auxins, growth retardants and other substances to control abscission and to prevent premature fruit drop. In some instances, advanced maturity and ripening were ascribed to treatment, while, in fact, delay in harvesting could have made the difference. There was also the possibility that sufficient substance penetrated the abscising zone to prevent fruit drop, while permeation into the bulky fruit tissue was too slow to have an effect. By infiltrating thick banana slices (McGlasson and associates) with IAA or 2,4-D it was shown that ripening was delayed, while ethylene production and respiration were stimulated by treatment. Similarly, C. Frenkel and associates, working with intact pears, found that auxin can restrain the ethylene action. Infiltrating the tissue with the antiauxin p-chlorophenoxyisobutyric acid promoted ripening and reduced ethylene production, though it is not clear whether the level of ethylene was below the threshold concentration shown to be required for inducing the processes leading to ripening. An attempt to eliminate the ethylene action was made by placing antiauxin-infiltrated fruit under low pressure of 0.1 atmosphere; ripening was not retarded under this condition. In other studies it was found that auxin pretreatment delayed both normal and ethylene-induced ripening. As ripening commences there appear isozymes of peroxidase and IAA oxidase, suggesting that conceivably degradation products of auxin play a role in the process. The overall picture that emerges is that the auxin level in the fruit must be reduced and IAA oxidation products formed before the action of ethylene can be manifested. However, before this controlling role can be ascribed to auxin, it is necessary to obtain changes in the endogenous levels of the phytohormone.

Reports were made that GA also counteracts ethylene effects. This is based mostly on the retention of chlorophyll by treatment with GA as is the case in citrus peel. Some significant delays in ripening by GA were observed in stone fruits, especially in sweet cherries. In the case of apricots, cytokinins such as benzyladenine as well as GA treatment delayed color changes. The delay in chlorophyll destruction in oranges by cytokinin was ascribed either to the maintenance of GA levels or the retardation of the build-up in abscisic acid or ABA-like substances.

ABA levels have been investigated in apples and pears with a well-defined pattern of respiration related to ripening, the climacteric, as well as in nonclimacteric fruits such as citrus, grapes, and strawberries. In both cases endogenous levels of ABA increased preceding and during ripening. External applications of ABA accelerated the onset of ripening, and treatments which hastened senescence increased the level of ABA in the tissue. Some fruits exhibit a shift of sensitivity to ethylene during an advanced developmental stage when an increase in ABA concentration takes place. ABA was found to stimulate ethylene production in tissue slices. As a result of these studies the question arises of the nature of the ABA-ethylene relationship with respect to fruit ripening. Is it possible that ABA alone can be the triggering agent or is there a requirement for both substances coupled with disappearance of auxin and appearance of IAA oxidation products? The control of fruit ripening appears to involve complex hormonal interactions.

The State of the Art

The studies in my laboratory were not in the field of growth substances except the work on ethylene before it was recognized as a regulator. In order to systematize in my own mind the state of knowledge in this field, I made an attempt to select some of the major horticultural applications and relate them to one or more physiological roles. The tabulation presented here is a rough approximation. I am confident that there must be somewhere in the literature a more complete and more precise presentation of this kind. The inclusion of the effects of interactions would be of added importance. From the standpoint of the horticulturist, the classification of the numerous synthetic substances could be highly useful. This task would be quite difficult because of the great number of papers on growth regulators that have appeared in the horticultural literature and that have dominated the meetings of horticultural societies. The tendency to test each substance for a variety of activities as well as for yield and quality leaves one with the impression that the applications are more of an art than a science. This state of affairs is not surprising inasmuch as the plant physiologists and biochemists have not as yet unraveled the mechanism of action of the regulating substances at the cellular and at the tissue level.

Table 1   Some physiological roles and horticultural applications of growth regulators

Regulator	Physiological role	Horticultural application
Auxin	Cell elongation Tropisms Apical dominance	Rooting of cuttings Fruit set Tissue culture Abscission control

Gibberellin	Stem growth
	Flower formation in long-day
	  plants
	Enzyme induction
	Enhancement and retardation
	  of senescence
		Breaking dormancy
		Fruit growth
		Ripening retardation

Cytokinin	Cell division Inhibition of elongation Bud differentiation Regulation of foliar senescence	Delay of senescence Vegetative propagation Callus culture

Ethylene	Inhibition of cell elongation
	Diageotropism
	Suppression of bud growth
	Stimulation of formation of
	  abscission layer
		Fruit ripening
		Flower induction
		Leaf, flower, and fruit
		  shedding

Abscisic acid	Dormancy factor Inhibitor of seed germination Role in fruit abscission	Prolongation of dormancy in woody species

ABA, as the last of the growth substances discovered, has been credited with the regulatory role in several phenomena such as dormancy, abscission, ripening, and senescence. The control of these processes was ascribed previously to ethylene. The appearance of new substances results in increasing search for interactions and for mode of action. Kende and Gardner reviewed recently the literature on binding sites and came to the conclusion that "no single receptor protein for any of the plant hormones has yet been isolated." The search for receptors has been motivated by the situation in the animal field. It is possible that the plant regulators do not have the high degree of specificity of animal hormones. Conceivably, the mechanism of action involves a dynamic interrelationship between two or more of the regulators in which the balance between endogenous levels plays a role.

EMERGING RESEARCH FRONTIERS

Control of Basic Processes

The worldwide crisis in energy and in water supply is demanding drastic changes in orchard practices as well as in growing field crops. Research in all branches of agriculture will be increasingly directed toward utilization of atmospheric nitrogen as the fertilizers manufactured by energy from petroleum become too costly. In some vegetables and in ornamental horticulture notable success has been achieved in establishing symbiotic relationships between roots and nitrogen fixing microorganisms. We are far removed from accomplishing this with fruit trees. The development of new cultivars with this capability may hinge on achievements in modern methods of hybridization at the cellular level. There will be increasing demand for varieties requiring minimal quantities of water per unit dry matter produced. Transpirational losses can be reduced by antitranspirants and by regulation of stomatal opening, but care has to be exercised not to lower the photosynthetic efficiency. In horticulture the utilization of special substances for controlling transpiration, photosynthesis, and photorespiration will have as its goal not only economic use of water but also regularity of bearing behavior, increased yields, and improved fruit quality. For rational experimentation along this line more basic knowledge is needed on the mode of action of activators and inhibitors in plant processes.

Crop Regeneration and Improvement

The advances in plant cell and tissue culture coupled with the great strides in molecular genetics promise a bright outlook for the development of new varieties and new hybrids with an array of highly desirable agricultural traits. The dream may materialize when cultivars will be able to fix atmospheric nitrogen, have a high degree of resistance to drought, salinity, and disease, and produce a crop of high quality and high yield under a wide range of environmental conditions. There are many requirements for this achievement, but notable success has already been attained in the field of vegetative propagation. The principles of development of clones are well established and are universally applicable to all plants. Commercial production by rapid clonal propagation is now widely used in foliage plants, woody ornamentals, flowering plants, etc. The tissue culture method has been adapted also for the production of pathogen-free clones.

For crop improvement by cell culture the following stepwise procedure will be required: 1. growth of somatic and haploid cells, 2. isolation and culture of protoplasts, 3. intra- and intergeneric hybridization by protoplast fusion, 4. selection of desirable variants, and 5. vegetative regeneration of plants. For this technique to be successful in crop improvement certain conditions must be met. A well-developed tissue culture method is essential, and the modified trait has to find expression in the culture. Once an effective system is selected for the identification of the variant, it must be established that it is capable of regenerating and transmitting the desirable trait. Thus far, the first critical limitation for crop plants is the development of the tissue culture techniques. The analysis of the agronomic or horticultural trait under consideration at the cellular level presents another serious problem. Yield and quality are the usual goals of the horticulturist, but these phenotypes are highly complex end products of a series of biological processes and reactions. It would be helpful if use could be made of assays involving intermediary steps. For fruit trees the time saving factor is of special significance. A great deal of biochemical work involving metabolic pathways will have to be done to relate key reactions to horticultural goals.

Old Problem—New Approach

Some of the perplexing horticultural problems may be solved through studies at the cellular and subcellular level. I doubt whether the workers with apple scald or other chilling injury disorders back in the third or fourth decade of this century dreamed that the structure of cellular membranes might hold the key to this phenomenology. Chilling injuries are nonpathogenic disorders affecting horticultural produce at low temperatures, but not so low as to cause freezing of the tissue. The critical temperatures below which symptoms of physiological disorder appear may be as high as 12°C for the banana, 10°C for the lemon, 5–6°C for avocado, and 4°C for certain varieties of apple. The symptoms show up as spotting, discoloration, necrosis, and pitting on the skin of fruits. Surface cells collapse and the tissue becomes susceptible to invasion by decay organisms. Abnormal ripening follows chilling injury.

A great deal of physiological and chemical work has been done on this problem, but only recently a promising new approach was undertaken. Comparisons were made on the oxidation rates at different temperatures of mitochondria prepared from healthy tissue of both chilling sensitive and chilling resistant materials (Lyons and Raison). The results were presented as changes of Arrhenius activation energies plotted against temperature. In mitochondrial oxidations from chilling-resistant tissue the activation energy was constant over the entire range of temperature. On the other hand, in chilling sensitive material there was a break in the Arrhenius value at the temperature at which chilling occurs. The oxidative pattern was correlated with a physical phase transition in membranes from liquid-crystalline to solid-gel structure. This transition is considered to be responsible for conformational changes in the membrane-bound oxidative enzymes. An analysis of lipid composition revealed that the phospholipids from chilling-sensitive species had a tendency toward a higher ratio of saturated to unsaturated fatty acids. If this proves to be the case, the question arises whether it is possible to modify the composition and thus alter the response to temperature.

TRAINING FOR INTERFACE TEACHING AND RESEARCH

An Immigrant in California

My arrival in the USA from Poland was during the year prior to the stock market crash, but it took some time before I felt the impact of that event and never to the extent that the millions of unemployed did. To me America meant freedom and opportunity. When I mastered English sufficiently well to read novels, I enjoyed Louis Adamic's "Laughing in the Jungle" and "A Country Full of Nice People," which gave a fascinating insight into immigrant life in those years. But before I was privileged with this type of enjoyment I had to face up to the problem of language.

Since the fall semester had already started, I was taken directly from the transcontinental train to the campus in Berekeley. I was immediately confronted by a special examination in English for foreigners. The examining professor disclosed to my aunt what was totally obvious to me: I did not know any English. He said he would pass me because otherwise I could not register. My next problem was to decide which courses to take without the knowledge of the language of instruction.

Upon the advice of some fellow students, I enrolled in analytic geometry, scientific German, general botany, and Subject A (English). Mathematics had been a favorite subject with me in high school; therefore I had no difficulty following the symbols and solving the problems. Neither was there a serious problem with reading scientific passages in German, since I had studied the language for several years. The dilemma was to convince others that I understood what I was reading. The rest of the class experienced no difficulty in expressing themselves in English once they could understand the German. A sympathetic teacher excused me from class participation and trusted my competency. The story with botany was more complex. I had to take the introductory course as it was required for other courses. The evidence of the struggle is on all the pages of my copy of the textbook by Holman and Robbins. Nearly every sentence was translated into Polish, but that was of little help in following Professor Holman's lectures. The consequence was grim one day when I came to class and sensed an atmosphere different than usual. Everybody grabbed a paper and started writing without exhibiting the slightest surprise that an examination was taking place. I did not even make any attempt, whereupon I was called in by Holman who tried to talk to me in German. He excused me from examinations and instead suggested that I write two term papers. With the European notion of what a scholarly paper should be, I spent long hours in the small library of the botany department, removed the dust from the books by Sachs, Pfeffer, etc, and produced two papers entitled "Photosynthesis" and "Sexual reproduction of flowering plants." These papers more than fulfilled the requirement.

By the end of the fall semester of 1928, my English was sufficiently good to leave me free to contemplate the purpose and course of my training in agriculture. My goal in coming to California was to become a farmer, and later to join a cooperative settlement (kibutz) in Palestine and help establish a national home for the Jewish people. The studies in the agricultural curriculum in Berkeley did not hold out much promise to transform a city boy into a man of the soil. David Appleman, whom I met shortly after my arrival and who has been one of my closest friends to this day, tried to convince me that mathematics, chemistry, and physics are essential for training in agriculture. This made me dubious about continuing in Berkeley. About this time I discovered that Davis offered the kind of course that appeared to suit my purpose.

The Nondegree Program

The two-year course at Davis was designed to teach a wide variety of farm practices. You could learn how to plant, prune, raise chickens, use the tractor, and, if you could muster up the courage, arrange a date between a bull and a cow. There was little reference to principles, and use of books was kept to a minimum. Entrance requirements were very lax so as to make it possible for the sons (and rarely daughters) of farmers to enjoy college life. Since nearly all of the students came from farms, they did not need to learn the art of farming. I believe that later in life they would have valued their college years more had they learned to appreciate a book, analyze a play, or gain insight into the history of their country and of other countries. For anyone preparing to run the family farm or to manage a commercial orchard, courses in economics, chemistry, and biology might have been more beneficial for facing up to problems that could not be predicted a priori than learning the art of farming.

As for myself, I certainly did not belong there, though in later years administrative officers of the nondegree course cited my academic career as evidence of success for the program. Soon I came to the realization that I was learning very little about the principles of agriculture and the acquisition of farming practices was negligible. This state of affairs was brought home to me during the first winter recess when, following a full semester's course on pruning, I fancied myself an expert in this art and looked for a job. When I declared proudly to the man in the employment office in Sacramento that I came with experience from Davis, he promptly rejected me with: "We don't want college boys." I did succeed in landing a job pruning peaches that paid 50¢ per tree rather than by the day. My expertise in pruning resulted in pruning one tree the first day, when I should have done eight in order to make a living wage. Despite my realization of the inadequacies of the nondegree program, I did obtain a certificate of completion, utilizing at the same time the opportunities at Davis for taking as many degree courses as I could possibly schedule. One of these was in pomology in which Chandler's book on *Fruit Growing* was used. I sensed that this man had a great deal to offer and realized that this was so when I returned to Berkeley.

Prior to my resumption of studies toward a BS degree in horticulture at Berkeley, I took an eight week summer session course in citriculture which was given at Riverside by R. W. Hodgson and staff from the Division of Subtropical Horticulture in Berkeley. The course attracted students from nearly all citrus producing regions of the world and was famous in particular for the weekly all-day field trips. Moving in a long caravan of cars and stopping in groves, packinghouses, and storage plants, the student learned about various problems not so much from the staff as from the farm advisers, extension specialists, growers, and managers. Hodgson's lectures presented a masterpiece of organization, easy for students to follow and reproduce, but with little attention to the unknown and the problematic aspects of fruit growing. This type of presentation was in striking contrast to that of Chandler, who stimulated the student to question and analyze.

Horticulture and Plant Physiology at Berkeley

The plant science curriculum of the statewide College of Agriculture specified requirements in English, mathematics, physical sciences, biological sciences, with minimum exposure to the major. Thus, a student wishing to obtain a BS degree in horticulture was required to enroll in a few courses in his specialty covering about 10% of total semester units (12 out of 120). He could take more if he wished to since provision was made for electives. The curriculum had considerable flexibility and was subject to administrative interpretation. The assistant dean, who served as student adviser, took the view that rules can be enacted by a secretary and it was his function to make allowances for individual cases. This attitude coupled with credits for my European high school enabled me to graduate in one year after the Davis experience.

The point that I wish to stress is that the plant science and other four-year curricula in the College of Agriculture, in contrast to the nondegree program, prepared the student for various agricultural activities such as teaching, advising, appraising. As farm advisers in the numerous counties of the state and as extension specialists, the graduates of the college not only serve the growers, but they serve as a bridge between the investigator and the orchardist; often they assist in setting up field experiments to test out new findings. With the decrease in rural population, few of the graduates returned to the farm, but those who did with the BS degree were prepared to follow new developments. Years later when some of these people returned as alumni, they complained that the faculty should have insisted on their taking more science and humanities; they were never concerned about insufficient courses in agriculture. Some, if they had to do it again, said they would have majored either in social sciences or in humanities.

In the postgraduate plant program leading to the PhD degree in the College of Agriculture, no specialization in any of the crops was available. The degree had to be in plant physiology, administered by an interdivisional committee on an individual basis even though certain requirements pertained to everybody, as, for example, reading knowledge of two foreign languages, calculus, physical chemistry, biochemistry, and advanced plant physiology. There was a great deal of freedom in the choice of a topic for research, but one could expect interest and helpful suggestions from any member of the teaching and research staff. The doors of some of the very busy professors were open to students not only for professional inquiries but also for a chat on politics, economics, literature, etc.

The atmosphere in Berkeley was conducive to graduate studies, for free exchange of ideas, and for a great deal of latitude in the choice of the dissertation topic. The doctoral candidate was not required to tailor his thesis research to the activities of a particular laboratory as long as significant physiological questions were raised and a reasonable experimental approach selected. Students associated with crop departments could easily formulate a problem that fitted well in the "interface" of agriculture and plant physiology. My own doctoral research was of this kind. The selection of a study of water relations in citrus was motivated partially by my affiliation with the Division of Subtropical Horticulture in which I held the appointment of research assistant. My work was sponsored by this division as well as the Division of Plant Nutrition. The control chambers of Davis and Hoagland were essential for an investigation of the saturation vapor deficit as a factor in the rate of transpiration of rooted citruc cuttings. Fick's law of evaporation was found to be applicable within a narrow range of temperatures, while outside that range significant changes took place as a consequence of physiological effects. Subsequently, when I transferred to UCLA, I continued these studies for a few years, stressing in particular the relationship between the absorbing system and the transpiring surface. An interesting and unexpected by-product of this work was the observation of periodicity in the rate of transpiration of rooted lemon cuttings under constant environmental conditions. The water loss from plants grown on 4-hour light periods alternating with 4 hours of darkness depended on prior light history. The light period which coincided with previous day or artificial light caused a greater rate of transpiration than an equal light period which corresponded with previous night or darkness. In other words, the plants remembered for several days when there was supposed to be day and night. This problem was not pursued further because of my transfer to a new field of research dealing with fruit physiology.

POSTHARVEST PHYSIOLOGY—MODEL FOR RESEARCH AT INTERFACE

The transfer from studies of transpiration to respiration and fruit physiology took place some three to four years after I came to UCLA and was working on a temporary basis during the height of the depression of the 1930s. The opportunity for a regular academic position opened up when I was offered and agreed to take over a project initiated by the California Fruit Growers Association (Sunkist) through a $2000 donation. This was probably one of the first, if not the first, extramural grant received by UCLA. The citrus growers were interested in certain physiological disorders which caused serious losses in oranges and in lemons subjected to low temperatures. The regents accepted the fund without strings attached, but it was understood that the research would concentrate on postharvest problems.

A Fortunate Break

Lacking experience and having only scanty knowledge of the literature, I followed a lead from work with apples and bananas which showed that ripe fruit produced a gas that accelerated the ripening of unripe fruit. Accordingly, a simple experiment was designed. Air was passed from jar A with yellow lemons to jar B with green lemons. In the control, air was passed from jar C (green) to jar D (green). Each of the jars contained 50 lemons. Measurements of CO_2 evolution were made on the comparable jars B and D. For weeks no significant differences were recorded and the experiment was about to be terminated. Then an unexpected result was observed as the respiration rate rose in jar D, but not in the expected jar B. The possibility that mold on fruit had developed in jar D, thus contributing to the increased rate of respiration, proved to be negative. Instead we found one lemon partially covered with mold in jar C only. Since no respiration determinations were made on jar C, the suspicion arose that the fungus was producing a potent emanation which affected the rate of CO_2 evolution of the sound fruit. After extensive studies with pure cultures, it was established that the common green mold *Penicillium digitatum* and not some of the other storage fungi, not even the closely related blue mold *Penicillium italicum,* produced an emanation which caused a rise in CO_2 production, destruction of chlorophyll, and abscission of stem tissue. The potency of the fungal gaseous product was demonstrated by passing air from a single infected fruit over 500 green lemons and observing a respiratory rise in the entire sample.

The actual identification of the gaseous emanation as ethylene took several years. This was possible only after studies of the nutritional requirements of the fungus and after the development of a manometric method for the determination of low concentrations of ethylene. It was more than two decades afterwards when S. F. Yang traced ethylene biosynthesis by *P. digitatum* to glutamic acid involving the citric acid cycle. The mechanism of ethylene formation in the fungus is distinctly different than in fruits in which methionine was shown to be the ethylene precursor.

Going Out on a Limb with Ethylene

As the manometric method became operational, a survey was undertaken of 14 species of fruit of different climatic zones with respect to the relationship between ethylene evolution and CO_2 production. As a result of this study "the hypothesis was advanced that native ethylene is a product of the ripening process rather than a causal agent" (from Biale, Young & Olmstead, *Plant Physiology*, 1954). We decided to make this statement even though we realized our methodological limitations, which did not allow for collection of information on the relationship between external ethylene evolution and the internal concentrations in fruit tissue. I am pleased that this daring statement was made because it served as a stimulus for a great deal of work during the ensuing decade. Stanley Burg at first, and others later, challenged our hypothesis on the basis of results obtained by the newly introduced gas chromatographic procedure which was much more sensitive than the manometric method. Ethylene was proposed as the ripening hormone because concentrations higher than threshold were detected prior to the climacteric rise in respiration. The fact that physiologically active amounts were present in some fruits without inducing ripening was countered by the contention that the effectiveness of ethylene depends on the sentitivity of the tissue to it. In present-day thinking the resistance of the tissue might be ascribed to auxin, suggesting the need to follow changes in auxin along with ethylene prior to the respiratory rise. Recent precise studies by Roy Young and associates on avocado and cherimoya indicate that the rise in O_2 uptake either precedes or follows closely the increase in ethylene. Reservations about ethylene as the sole triggering agent emanated also from the work with grapes by Coombe and associates. Burg supplied evidence for the role of ethylene as ripening hormone by suppressing banana ripening under hypobaric conditions; contradictory results were reported on tomato by Bruinsma and associates. Obviously, the field is still in a state of flux though the onus of proof is on those who question the role of ethylene. The complications resulting from interaction with ABA and other growth-regulating substances were discussed in a previous section of this paper.

Unfinished Research

Much more can be written about the problems not completed and about studies that went out on a tangent than on the published reports of work done. I shall limit myself here to a few examples to indicate the spectrum of interests of those who have been associated with the laboratory and possibly to stimulate interest in pursuing some of the unresolved questions.

By working simultaneously with avocados and citrus fruits we were exposed continuously to the differences between the species. To obtain an orange fit for consumption one waits until it ripens on the tree, while the avocado, at least those grown in California, does not ripen on the tree. If a Fuerte avocado is picked in December and kept at 20°C, it will be in proper edible condition in 8–15 days. The same fruit may be left on the tree until April or May in unripe condition. The number of days for ripening of the late-harvested fruit are reduced, but the pattern of ripening is the same as that of the early pick. Thus far, we have no inkling of what this inhibitor of ripening present in attached avocados is, despite many attempts to pinpoint its nature.

The differences between avocados and citrus in postharvest respiratory behavior prompted us to differentiate between climacteric and nonclimacteric classes of fruits. Can the difference be explained simply by differential rates of ethylene production, or should we look for metabolic patterns that might be different? In the avocado, apple, and banana a great deal has been learned about protein and nucleic acid metabolism and about energetics, while there is only scanty work along these lines with fruits classified as nonclimacteric.

When fruits of either class are harvested and placed under controlled conditions, one may predict certain responses to temperature and to the gaseous composition of the atmosphere which do not differ too significantly between species. An important exception was discovered when citrus fruit were subjected to different levels of CO_2. The respiration rate in terms of oxygen uptake was stimulated rather than suppressed as the CO_2 level was increased, and more so in air than at lower oxygen levels. In following up this problem with labeled CO_2, it was established that radioactivity was incorporated rapidly into malate, citrate aspartate, suggesting oxaloacetate as common precursor. We left the problem at this point and did not establish whether increased CO_2 tension raises the rate of respiration by increasing the content of Krebs cycle acids.

An attractive feature of postharvest physiology is the opportunity and desirability of working at various levels of complexity. In attempting to understand the events in the detached fruit, we looked into reactions taking place in tissue slices and in subcellular fractions with concomitant studies of responses of the whole fruit. Investigating metabolism at several levels of organization also raised a number of perplexing questions up to now unresolved. I want to list one which has to do with the response to cyanide. We observed years ago that the oxidation rate of avocado tissue slices is not only insensitive to concentrations of cyanide inhibitory to the cytochrome oxidase, but is, in fact, highly stimulated by the inhibitor. On the other hand, mitochondrial oxidations are highly sensitive. HCN applied to the intact fruit was shown by T. Solomos to bring about a climacteric rise in respiration and to stimulate ethylene production. These discrepancies are amenable to experimentation in terms of studying the relative roles of the cytochrome system versus the alternate pathway of oxidation. It is gratifying to me to see that George Laties and his associates are following up these problems.

FRUIT RIPENING AND PLANT SENESCENCE

As more knowledge is acquired on the metabolism of the detached fruit, the question arises whether ripening is typically a senescence phenomenon or a prelude to it. Plant senescence has been viewed as a phase in the development of an organism or an organ in which irreversible events lead to cellular deterioration culminating in death. The essence of the stated question is to weigh the various processes taking place during ripening in terms of catabolic or anabolic reactions. A comparison of ripening with foliar senescence might be enlightening inasmuch as extensive post-maturation studies were conducted on leaves and on fruits.

In comparing these two organs, certain striking differences and certain similarities are evident. In the leaf both total protein content as well as incorporation of labeled amino acids decline with yellowing, while in fruits there is enhanced protein synthesis at the preclimacteric minimum as well as during the respiratory rise. In the avocado, incorporation of ^{14}C valine and leucine proceeded actively until close to the peak of the climacteric. Puromycin inhibited protein synthesis but not the respiratory rise, thus suggesting an independence of the two processes. A conclusion can also be reached that no ADP generated by protein synthesis is required to initiate the rise. As far as specific proteins are concerned, the indications are that hydrolytic enzymes are formed in leaves during yellowing and in fruits during ripening. There is perhaps a more striking rate of formation of cell wall degradative enzymes such as cellulase, pectinmethylesterase and polygalacturonase in fruits than in leaves. We know more about synthesis of proteolytic enzymes in the latter than in the former. The observations on increased activity of certain oxidative and peroxidative enzymes with the onset of the climacteric are indicative of the operation of the machinery for transcription and translation during senescence.

The differences between fruit ripening and leaf senescence extend also to changes in nucleic acids. In most cases reported for fruits there was little upward or downward trend in the content of total RNA or of the various components of RNA. In foliar senescence, on the other hand, total RNA declined and nucleic acid metabolism was suppressed compared with activities in mature green leaves. During fruit ripening incorporation of radioactive inorganic phosphate into nucleic acids was progressing at an active rate. In the avocado, ^{32}P was incorporated actively into the regular and into the heavy ribosomal fractions during the preclimacteric stage as well as during the period which corresponded to the first half or two-thirds of the rise. The rate of incorporation fell off drastically when the respiration of the fruit was close to the peak. On the other hand, labeling in the soluble RNA fractions progressed more actively in the fruit close to the fully ripe stage than in the unripe. It is not known whether the soluble RNA fraction represents transfer RNA or RNA degraded from ribosomes. In some fruits it was found that uridine was incorporated into nucleic acids and into ribosomal RNA in particular. Of special interest are the findings that ethylene enhanced the uridine incorporation and that inhibitors of RNA synthesis inhibited ripening. It appears, therefore, that from the standpoint of nucleic acid metabolism there is a wide divergence between leaf senescence and fruit ripening.

The differences recorded thus far might be correlated with differences in function and structure of the two organs. In leaves the photosynthetic activity declines with age along with yellowing and loss of insoluble nitrogen. The respiration proceeds at a constant rate until the stage of advanced senescence when, in a few species, a slight rise in oxygen uptake was recorded. At the ultrastructural level it can be observed during leaf senescence that endoplasmic reticulum and ribosomes deteriorate first, followed by disorganization of chloroplast structure. It appears that chloroplasts are more vulnerable than mitochondria. In green fruits prior to yellowing, or in those fruits which retain chlorophyll, no significant function can be ascribed to the chloroplasts. In avocados no Hill reaction could be demonstrated with isolated chloroplasts. On the other hand, mitochondria retain intact structure at least until the climacteric peak. Their oxidation activities are more pronounced during the rise and at the peak than during the preclimacteric stage. The oxidation rates of succinate may not be significantly higher in mitochondria from avocados in the process of ripening than unripe avocados, but they are certainly higher in the case of malate, α-ketoglutarate and pyruvate. Another important difference between mitochondria from preclimacteric and climacteric fruit is the dependence of the oxidation of the former and not the latter on the addition of thiaminepyrophosphate to the reaction medium. The hypothesis was advanced that α-keto acids limit the respiration rate of preclimacteric fruit. Support for this idea may be secured from studies on apples in which malic enzyme activity rises with ripening.

In drawing comparisons between fruit ripening and leaf senescence it was taken for granted that in fruits exhibiting the climacteric pattern the beginning of the ripening process coincides with the onset of the respiratory rise. This is assumed to be the case by fruit physiologists generally. If so, one must come to the conclusion that the events occurring during ripening prior to the climacteric peak are in many respects different from those in foliar senescence. The temptation is therefore to suggest that the major part of the rise reflects a signal to senescence rather than senescence itself. Ripening is not a form of senescence, but a prelude to it. According to this view, the logical conclusion to reach is that in ripening we find an interplay of synthetic and degradative processes.

FROM SUBTROPICAL HORTICULTURE TO BIOLOGY

Research at the "interface" of agriculture and plant physiology demands constant interest in additional disciplines such as genetics, biochemistry, cell physiology, and cell ultrastructure. Consequently, one is prepared for teaching in a variety of subjects. After the early years of teaching fruit physiology and respiration I had the opportunity of participating in cooperative courses in plant physiology. The program was initiated by Sam Wildman when he came to UCLA, and was joined later by Anton Lang, George Laties, and Roy Sachs. It included an advanced undergraduate course in which the students had the opportunity of working in the laboratories of the staff members. Of special interest to graduate students, postdoctoral fellows, and plant physiologists in several departments was a three year seminar-lecture course that covered the advances in the major fields of plant physiology. This course served also as a forum for discussions by visiting scholars. A large percentage of students participating in these courses came to UCLA from distant countries because of interest in citriculture and other subtropical fruits. The program toward the PhD degree broadened their training and interests. A good number of the doctoral recipients went into fields other than horticulture and made notable contributions in plant physiology and biochemistry.

As the College of Agriculture started to be phased out from the UCLA campus because of the low undergraduate enrollment, the number of graduate students from abroad also decreased. The orientation in the graduate program shifted from inter- to intralaboratory. At that stage I redirected my teaching interest toward an intro- ductory course in cell and molecular biology. For the past decade I have been enjoying the contacts with younger and less sophisticated students. I feel strongly that the more senior members of departments should teach the elementary courses. I have been trying to convince my colleagues, without too much success, that it is highly rewarding to direct and stimulate students early in their career.

ACADEMIC FREEDOM

An item in a recent issue of *Newsweek* brought back to me the problem of academic freedom which has come up from time to time. The report was entitled "To Shrink a Scientist" and dealt with the dismissal of a biophysicist in the USSR. This man was not a dissident, but neither was he a comformist. He preferred not to attend prescribed political meetings and did not choose to include the name of the chief as co-author on a paper. Our academic system certainly has more safeguards against arbitrary decisions by administrators. I recall that at least on two occasions I did not hesitate to oppose deans on matters of considerable importance. I knew that the tenure system gave me a great measure of protection. We have had our share of bitter struggles for academic freedom. We know, of course, of sufficient instances of suppression of unpopular ideas and we realize the need to be on guard. We remember only too well the stormy period of 1949–50 when the regents demanded of the faculty a special loyalty oath. A number of professors, mostly from the Berkeley campus, resigned rather than take the oath. Eventually the oath was rescinded and the nonsigners were reinstated.

The hysteria of those days touched the lives of many people in Academia. I, too, had my own encounter with the FBI. In 1958 I was invited by the government of Israel to assist in the establishment of a laboratory for fruit physiology at the Volcani Agricultural Research Institute in Rehovot. Since the assignment was sponsored by an agency of the United Nations (FAO), FBI clearance was required before appoint- ment could be made. After a long period of uncertainty I was presented with a bill of particulars. The FBI had a full list of my activities no matter how trivial. A serious charge was that on a certain date in 1947 my name appeared among others on telegrams to President Truman and to the Speaker of the House urging the abolition of the House Committee on Un-American Activities. They wanted to know the circumstances that induced me to take this step. They also inquired about the nature and extent of my association with members of the faculty who lectured before the People's Education Center.

I promptly replied that I trusted their information about the date and content of the messages, though I myself didn't keep a record of such protest telegrams. The reason that I gave my signature was that I felt the activities of that House Committee were undemocratic and un-American. I told the FBI that having immigrated from a country in which there was no freedom I was particularly sensitive to the witch hunt of the Congressional Committee. As to my friends on the faculty, I replied that I saw these individuals almost daily and I was proud of my association with them.

I considered them fine members of the faculty and good citizens. The clearance promptly arrived. Mine was an insignificant case but it is indicative of the times. It reminds me that other members of the academic community had more serious encounters with the authorities without any justification and that we must be constantly on guard.

NO BETTER LIFE

The occasional inconveniences caused by extra- and intramural politics do not subtract from the great attractions of the academic way of living; any other life style can hardly compete with it. While serving on various committees of the academic senate I heard a great deal of complaining about the low salaries compared to the incomes of other professionals. Only too often do we forget that we recieve a good living wage to do what we enjoy doing. Not too many people can say that. Here you have the privilege of determining your own research program and of teaching in your own style. You have the opportunity for professional associations with colleagues throughout the world and of traveling to meetings held in countries far and near. In some places, such as California, you have other advantages which cannot be measured in money. The lofty mountains are close enough for back-packing trips in summertime and for skiing expeditions in winter. When you finally retire after a lifetime of service you are supported by a good pension, you get an office or you share an office, and, if you can justify it, you frequently can count also on laboratory space. You can continue doing what you did before or you can take advantage of the numerous offerings by the university and start as a student all over again. For the privileges extended to me I am deeply grateful to the University of California and to the people of California.

ACKNOWLEDGMENTS

My appreciation to graduate students, postdoctoral fellows, and associates who contributed materially toward this chapter. In chronological order of their association with the laboratory—PhD recipients: H. K. Pratt, H. W. Siegelman, J. M. Tager, M. Avron, C. T. Chow, R. J. Romani, S. Ben-Yehoshua, S. H. Lips, C. Carmeli, D. L. Ketring, and H. A. Schwertner. Postdoctoral fellows: M. Lieberman, S. Uota, E. K. Akamine, W. O. Griesel, J. P. Nauriyal, R. Kaur, J. T. Wiskich, P. K. Macnicol, G. G. Dull, H. A. Kordan, M. Gibson, G. E. Hobson, C. Lance, A. E. Richmond, J. E. Baker, and A. D. Deshmukh. My thanks to G. G. Laties and D. Appleman for their ever-ready assistance to the students. A very special role in our program was fulfilled by R. E. Young, whose contributions were indispensable.

I wish to express gratitude for arrangements to do research or to lecture during sabbaticals to: H. A. Krebs, A. A. Benson, D. Koller, and B. Jacoby. My appreciation also to F. Lattar and M. Schiffmann-Nadel and their associates for the opportunity of working with them at the Volcani Research Institute.

Special thanks to Marjorie Macdonald for superb work on the manuscript.

C. S. French

Ann. Rev. Plant Physiol. 1979. 30:1-26

FIFTY YEARS OF PHOTOSYNTHESIS

C. Stacy French

Department of Plant Biology, Carnegie Institution of Washington, Stanford, California 94305

CONTENTS

INTRODUCTION

I first went to a scientific meeting as an undergraduate about 1928. The nature of the scientific community as well as the subject matter of plant physiology has changed so much in the past 50 years that it may be of some interest to make comparisons of then and now. I want to speak particularly about what might be called the sociology of science—the personal relations of scientists to each other that greatly affect their scientific output. The change as I see it has been essentially the transformation from a few small groups of near-amateurs with enthusiasm and interest in a wide range of

1995

0066-4294/79/0601-0001$01.00

disciplines to a very large number of groups all intent on a small segment of science and seemingly with less interest in followers of other subjects than typical scientists used to have.

Possibly the easiest way to become a successful scientist is to cultivate the right people and to follow their way of life. By this means almost anyone, even of limited ability, may be channeled into more or less productive enterprises. Everyone's career, in science or any other field, depends largely on one's interactions with other people. Very few of us can develop a significant intellectual life from within ourselves. Therefore I will mention with appreciation some of the many people who have served as models, have been excellent guides to life in general, or who have been helpful with the details of particular scientific investigations. I hope to be forgiven for using the autobiographical format because it is by far the easiest way to string together a series of reminiscences and comments.

I was fortunate in getting a good start and continued guidance to age 25 from suitable parents. My father, Charles Ephraim French, born in Berkely, Massachusetts, in 1864 on a one-acre strawberry farm, put himself through the University of Maryland Medical School by carpentering and painting; then after European hospital experience, had a good practice of ear, eye, nose, and throat work in Lowell, Massachusetts. His comment on scientific research was "Well, every once in a while they give us something we can use." I was born in Lowell December 13, 1907. My mother, Helena Stacy, born in Colebrook, New Hampshire, in 1867, grew up in a small business and lumbering family and later lived in Bathurst, New Brunswick. She had a year in Europe, then two years at Radcliffe, followed by kindergarten teaching in Massachusetts.

My early education in Lowell public schools was not much good. Fortunately, respiratory infections made me lose so much time that a private tutor became necessary. This well-educated spinster, Flora Ewing, lightened the day's dose of Latin, English, and algebra with kitchen-scale chemical experiments. Before going away to school, I learned from my father how to do simple carpentry, wood turning, house painting, and general repair work that formed a basis for later laboratory life. I entered Loomis in Windsor, Connecticut, in 1921, failed the first year, repeated it the next and later, when the time came, was refused permission to take the Harvard entrance exam so as not to spoil the school's record. Fortunately, my mother put me in Lowell High School for a year. From there, I got into Harvard via a summer school course in botany given by Carroll W. Dodge. During the first year in college I started to prepare for engineering with math, physics, engineering drawing, and languages. In those days engineers could both draw and write. In the second year of college I discovered W. J. Crozier's course in general physiology and L. J. Henderson's introductory biochemistry, including some lectures by John Edsall, now a valued family friend 50 years later. The laboratory work in general physiology first brought contact with real science done with good apparatus and of sufficient significance to have an unpredictable outcome. From then on engineering was forgotten.

In early 1928 a few lectures on photosynthesis by Robert Emerson, who had recently returned with a PhD from Otto Warburg's laboratory in Berlin, got me interested enough to take Emerson's course on photosynthesis the following year, and I have stayed with the subject ever since.

When Harvard was somewhat smaller and under the direct personal care of President A. Lawrence Lowell, the faculty was composed largely of Boston gentlemen. Now that Harvard has become more like the large public universities, some of the local flavor that made it such a pleasant environment for quiet scholarship has been displaced by academic competitiveness. I have long been interested in observing whether academic excellence can be promoted without running into the destructive aspects of intense competition.

As an undergraduate I was lucky to have Professor George D. Birkhoff as an adviser. A prominent mathematician, he well understood the peculiarities of a creative academic career. He and his family guided me into the good life of Cambridge and Boston.

An undergraduate thesis on the temperature coefficient of catalase action under the guidance of W. J. Crozier and A. E. Navez very nearly kept me from graduating with my class in 1930 through lack of attention to the math and organic chemistry courses. My first paid scientific job was to help move the laboratory of general physiology from under the glass flowers in the Agassiz Museum to the new biology building guarded by the rhinoceroses. All went well in the new building after a pair of workmen's shoes were removed from the distilled water system.

GENERAL PHYSIOLOGY AT HARVARD

The years of graduate work and teaching assistantships in general physiology were excellent, not only because of the faculty but equally for the association with outstanding fellow graduate students and postdocs. I watched the slightly older men like Edward Castle experimenting with *Phycomyces'* response to light, Ted Stier measuring yeast respiration, Gregory Pincus being concerned with the sex life of rats, Fred Skinner studying snail behavior on sloping panels, Bob Emerson shaking manometers full of *Chlorella,* Doug Whittaker doing the same with *Fucus* eggs, and George Clarke getting equipped to measure light penetration in the ocean. Morgan Upton and Hudson Hoagland were somehow concerned about psychology —all this in one department.

My contemporaries felt the department seminars were too dominated by the professor, so after a year or two we organized the "Chlorella Club" meeting weekly in secret session to educate each other. Bill Arnold had just come to Harvard from Caltech, where he and Bob Emerson had measured the size of the chlorophyll unit. Bill and Caryl Haskins, later President of the Carnegie Institution, led most of the discussion, while Henry Kohn maintained a skeptical attitude. The Chlorella Club gave us more education than any of the biology courses. Continuing interaction with Bill Arnold and Caryl Haskins has been a lifelong pleasure. The only formal courses I remember much about during the graduate years were George Shannon Forbes' "Photochemistry" and James B. Conant's "Natural Products" and "Physical Organic." My major conclusion from all this is that exposure to great men is more valuable than the subject matter acquired from them.

Two summers at Woods Hole in the early 1930s, one to take the invertebrate course, the other to study the effect of light on respiration of a red sponge, led to a broader acquaintance with American biologists. In graduate school I eventually did some experiments on the rate of respiration of *Chlorella* at various temperatures. Having accumulated too much data, some of doubtful significance, it became nearly impossible to write a doctoral thesis. My life was saved by Pei-Sung Tang, a postdoc from China, who spent much time helping me organize the data. We also collaborated on measurements of the rate of *Chlorella* respiration as a function of oxygen pressure, which resulted in my first scientific paper appearing in the *Chinese Journal of Physiology.* It was a great pleasure to be in correspondence with Pei-Sung Tang again about 1974 after his long leadership of both academic and practical plant physiology in China. In 1934 the good times came to an end with the completion of the degree requirements. Due to the depression, few academic positions were available. My applications to teach at Williams and at Howard were turned down and there was no way to stay on at Harvard. Bob Emerson however, allowed me to go to his lab at Caltech to work on photosynthesis in purple bacteria.

WITH VAN NIEL AND EMERSON IN CALIFORNIA

On the way to Caltech I spent the summer taking Kees van Niel's famous course on microbiology at the Hopkins Marine Station at Pacific Grove, California. No one else I ever knew could lecture for four to six hours without losing the student's attention. That summer I visited the Carnegie Institution's laboratory at Stanford and met Herman A. Spoehr, James H. C. Smith, and Harold Strain. Spoehr was a most helpful and kind adviser whose guidance I relied on until his death in 1954. I became his successor in 1947.

As an undergraduate, under the stimulus of A. E. Navez, I had developed an interest in purple photosynthetic bacteria. Through van Niel's course and his personal interest, I learned to grow and work with bacteria more efficiently. At Caltech I tried to measure their photosynthetic efficiency, but the excellent skiing in the mountains near Pasadena left little time for science, so that year was not productive. Bob Emerson was justifiably disgusted with my performance and we were barely on speaking terms for the academic year. Some years later we became friends again. However, in spite of my poor performance, he arranged for me to spend the next year in Berlin with Otto Warburg, which was what saved my scientific career.

At Caltech I consulted the optics professor, Ira Bowen, later a C.I.W. colleague, about how to measure the absorption spectra of purple bacteria in the near infrared. He referred me to Theodore Dunham, who said it could be done with the astronomical spectrograph on Mount Wilson. We made a date for a particular night on the telescope with the understanding that if it was clear he would work on a star, but if it was cloudy we would photograph spectra of bacteria. The weather favored the bacteria. That was probably the only time astronomical equipment has been used for the study of microscopic objects. On another occasion I wanted to calibrate a thermopile and was told that a young assistant professor, Arnold O. Beckman, had a setup for that purpose in the chemistry department. He generously helped me out. There is an uncorroborated story that he had a market survey made when preparing to manufacture the Beckman D.U. spectrophotometer about 1940. The survey report is reputed to have been that 100 instruments would saturate the market. About 1976 production was discontinued after something like 30,000 of these spectrophotometers and a very large number of competing instruments had been sold. These figures must bear some relation to the explosion of scientific research in general as well as of biochemical plant physiology.

WITH WARBURG IN BERLIN—DAHLEM

The Cold Spring Harbor Symposium on Photosynthesis in June 1935 was my introduction to the world of professional photosynthesis investigators. The Physiological Congress in Leningrad and Moscow that summer followed by the Botanical Congress in Amsterdam provided some acquaintance with European science.

While in Leningrad I had the pleasure of a visit to Professor N. N. Lubimenko, a corresponding member of the American Society of Plant Physiologists. He was the first as far as I know to believe in the existence of different forms of chlorophyll a although the evidence then was not clear. His laboratory in the Botanical Garden of Leningrad was very simple—all the equipment visible was a microspectroscope and his only collaborator seemed to be one old lady. He was extremely kind and helped me revise and clarify the speech I was to give at the Physiological Congress. This was before biochemistry split off from physiology to have its own international congress, so papers on photosynthesis were a reasonable part of a physiology program.

In Stockholm I visited the biochemist, Professor H. von Euler, who treated me very well initially. After some pleasant and informative conversation, he asked about my plans. Two minutes after telling him I was on the way to work with his enemy, Professor Warburg, I was out in the street. That kind of intense personal competitiveness seems to be less common now than it was in those days—perhaps because of the greater number of people now involved in every field. By contrast with the visit to von Euler, Professor John Runnström, the zoologist, was most kind and gave me a fine country weekend complete with lessons on the details of expected behavior at Swedish dinner parties, which I still find formidable.

The year with Professor Otto Warburg at the Kaiser Wilhelm Institut, (now the Max Planck Institut) was one of intense concentration on the efficiency and action spectrum of photosynthesis in *Rhodospirillum rubrum*. Living in the laboratory building and eating at Harnack House just around the corner in Dahlem was most convenient. The laboratory was excellently supplied with optical equipment left over from Warburg's determination of the action spectrum for dissociation of the CO complex with the respiratory enzyme, which had brought him the Nobel Prize. I was treated very well by the professor and all his staff. The rigid discipline and long hours without outside distraction were just what was needed to convert an easygoing freewheeling academic type into a professional scientist.

Training in constructing apparatus specifically for particular research purposes was of great value. At Caltech I had experimented with a Christiansen filter to produce monochromatic light. The principle is that powdered optical glass suspended in an organic liquid will scatter all wavelengths of light unless the refractive index is identical for the glass and the liquid so the light can therefore pass straight through. Since the refractive index for the liquid and the glass vary differently with temperature, the wavelength transmitted can be selected by temperature control. In Dahlem I constructed several of these devices but only later found out that the optical glass powder has to be annealed to remove strains produced by the pulverizing.

The unusual intellectual climate of Dahlem under the Weimar Republic several years before my time there has been well described by Nachmanson (3). I saw little of Dahlem science outside the laboratory except for a few seminars on photosynthesis at Max Delbrück's house with Hans Gaffron and Eugene Rabinowitch. During the 1935–36 year that I was in Berlin, the good days of the Weimar Republic had largely faded away. The Hitler government was busily working up war spirit through well-planned propaganda. It was disillusioning a few years later to see the same propaganda tricks being played at home by our own government. That the well-established procedures for inciting any population to enter a war are not immediately recognized by the people being manipulated is most unfortunate.

At that time quantities of horse blood were being processed in the basement of Warburg's Institute to isolate and identify a redox substance then called coenzyme II, later TPN, and now NADP. Daily association with the small group of the professor and the assistants—Negelein, Kubowitz, Haas, Lütgens, and Gerischer—made it possible to learn many biochemical procedures by watching the experts. As far as I remember, these people, with a janitor, a secretary, two mechanics, and a courier, were all the inhabitants of the most famous biochemical laboratory of the time. (Apologies to Cambridge biochemistry where a different scale of values may prevail.) Fortunately this was before all German scientists spoke English, and I am grateful to Warburg for refusing to speak English though he could do it better than I. There one saw the advantages of a small laboratory free from the distractions of students, seminars, and committee meetings. In later years, however, that sort of isolation showed its negative effect when Warburg, no longer at the top of the field, refused to consider the advances in photosynthetic research made in other laboratories.

WITH HASTINGS AT THE HARVARD MEDICAL SCHOOL

Having lived so far on a modest inheritance, I returned to the United States in the late summer of 1936 to look for academic employment. I was rescued by Baird Hastings, who gave me an Austin Teaching Fellowship ($1000 per year) in biochemistry at Harvard Medical School for two years. This allowed half time for research, so it was possible to continue with purple bacteria photosynthesis.

One lesson I learned at the Medical School was to believe and to publish experimental results. I found that light completely stopped O_2 uptake by purple bacteria, but never having heard of the effect, I thought something must have been wrong with the experiment, so I dropped that work. Many years later the effect was discovered and taken seriously by others.

Another valuable experience was to see the difference between "big science," or at least what was big for those days, and the small scale one-man research then more typical of academic life than it is today. My objective at the time was to measure the absorption spectra of various strains of purple bacteria and to compare them with water extracts of the organisms. Attending a meeting of the Optical Society at MIT, I heard one of the older members, Frederic Ives, or one of his contemporaries make the following remark: "The time has about gone by when one can get a PhD in physics for publishing the absorption spectrum of a single substance." The only commercial spectrophotometers available then were the Koenig-Martens visual-balance monstrosities that worked only for clear solutions and in the visible spectrum. However, these bacteria have their most interesting bands in the near infrared. I wanted to build a near infrared photoelectric spectrophotometer suitable for use with scattering suspensions. The facilities at Harvard Medical School were not adequate for such a deviation from standard biochemistry, so I went over to MIT for a talk with Professor George Harrison. He had a fine laboratory for all sorts of spectroscopy with many collaborators. This was "big science" for that time. His program was to remeasure the emission lines of all the elements. He agreed that my plan was feasible, but insisted that one of his graduate students be paid $200 to follow me around to be sure I didn't spoil any of his equipment. Perhaps he had had unfortunate experiences with someone else having medical connections. However, neither I nor anyone else in my situation had $200 for research expenses. That was my first and only contact with the "big science" of that day. After a visit to Professor Theodore Lyman at the Harvard Physics Department, I was given all the equipment and help necessary. Several kind people, knowledgeable in electronics and in optics, helped me put together a workable instrument, and the measurements were completed without difficulty in a very pleasant atmosphere.

Breaking the bacteria, however, was a problem. I tried grinding them with abrasive powders between rotating glass plates and also by a cheap and simple procedure with a hypodermic syringe rotating in a lathe. The outlet of the syringe was plugged and the stationary plunger was forced in slowly. This squeezes the bacteria out between the ground glass surfaces. Also I had heard that Stuart Mudd and Alfred Loomis had broken bacteria with supersonic vibration. Luckily a magnetostrictive supersonic generator was in the physics building, the creation of Professor G. W. Pierce, who kindly let me use it. As far as I know, this was the first use of supersonic vibration to break photosynthetic organisms.

While at the Harvard Medical School, I had the good luck to watch a demonstration by O. A. Bessey of his method for titrating ascorbic acid with dichlorophenol-indophenol. Ascorbic acid had recently been isolated from cabbage by Szent-Georgy. Several years later the memory of this demonstration led me to try indophenol dyes as Hill reagents for detection of chloroplast activity. Among the people in the Hastings laboratory then were John Taylor, a former student of W. Mansfield Clark, Kenneth Fisher, and

Oliver Lowry, later to become the most often cited man in science because of his protein determination procedure. It was a stimulating group. These two years were great training in biochemistry, but for a longer period it would not have been appropriate to try to make a career of photosynthesis research in a Medical School. I am most grateful to Baird Hastings for his support while I was looking for a suitable position.

ASSISTANT TO FRANCK AT CHICAGO

In the fall of 1938, the day before the great hurricane, I left for Chicago with an appointment as instructor (research) in chemistry ($2400 per year). This was to help James Franck set up a photosynthesis laboratory with support from the Fels Foundation. The unspoken thought that seemed to be in the air was that a Nobel Prize winner in physics would easily be able to take care of the photosynthesis problem in a few years with some simple critical experiments. My first assignment was to study the time course of chlorophyll fluorescence from green leaves while the other assistant, Foster Rieke, a physicist, was to measure quantum yields of photosynthesis. In December I returned to Cambridge briefly to marry Margaret Wendell Coolidge, daughter of the mathematician, Professor Julian L. Coolidge, first Master of Lowell House at Harvard. We have had a very happy home life ever since. In addition to her work as a counselor, she has taken complete responsibility for household affairs and family life, thus leaving me free to concentrate on science. Through Margaret I again became on good terms with Bob Emerson since their fathers had been college friends.

It soon became evident that Franck and I had very different ideas about the conduct of research. Given the phenomenon of complex time courses of chlorophyll fluorescence, I wanted to explore systematically the influence of temperature, CO_2 and O_2 concentration, previous light regimes, etc on the effect. His plan was to think about the situation, develop a theory, and plan a critical experiment that would prove or disprove the theory. According to that view, my job was to do the critical experiment. I survived this unhappy situation for three years before a better position became available. After the first year, Hans Gaffron, a former Warburg collaborator, joined the group. He was better able to deal with the conditions and managed to work more independently. For many decades Gaffron remained in the small group of a dozen or so leading investigators of photosynthesis. One of his major contributions was the idea of "photosynthetic units." I learned some practical physics from Rieke to combine with the biochemistry picked up during the past three years.

At Chicago the studies of kinetics of fluorescence intensity changes with time of illumination brought the effects discovered by Kautsky in Marburg to the attention of photosynthesis workers in this country. A graduate student, Ted Puck, joined in this project. He later became famous by culturing human tissue cells. A freshman student was assigned to help in the work of the Fels group through some sort of student aid program about 1939. His first assignment was to prepare a set of neutral absorbing filters for attenuation of light beams. This set of filters is still in daily use at the Carnegie Institution and has played a significant part in experiments leading to a large number of publications by many investigators at Chicago, Minnesota, and the Carnegie Institution. This was Roderick Clayton, now a Cornell professor and member of the National Academy, well known for his studies of the photosynthetic reaction centers and his excellent books on photosynthesis. About 1940 Robert Livingston, already a distinguished photochemist, joined the group. His later work as a professor of chemistry at the University of Minnesota on spectroscopy of chlorophyll in solution free from water vapor led to some of the present ideas about chlorophyll aggregation. After I escaped from Chicago, Warren Butler did his graduate work with Franck and continues as a leader in the biophysical side of photosynthetic research at the University of California, San Diego.

My life changed greatly for the better one day when Mortimer L. Anson, of the Rockefeller Institute in Princeton, on his way home from an Arizona vacation dropped in at the Chicago chemistry department to tell James Franck about Robin Hill's discovery of oxygen evolution by isolated chloroplasts. He stayed about a month while we repeated Hill's experiments with some variations. In spite of Franck's opinion that all this had nothing to do with photosynthesis, he did allow me to continue to play with isolated chloroplasts. Tim Anson and I worked on chloroplast isolation, stabilization, and O_2 evolution kinetics. We eventually prepared a manuscript for the 1941 American Society of Plant Physiologists meeting in Texas. Neither of us could attend, so Jack Myers (2) read our paper at the meeting. Anson was trained in the customs of organic biochemistry, so he proposed to call the effect the "Hill Reaction." Following Jack Myer's presentation in Texas, the chloroplast reaction has been so known, except by Robin. Our first objective was to try all the common biochemical tricks of homogenization, freeze drying, fractionation, etc on chloroplasts.

TEACHING PLANT PHYSIOLOGY AT MINNESOTA

I continued these chloroplast projects after being brought to the University of Minnesota by George Burr to be assistant professor of botany in 1941. There the memory of Bessey's ascorbic acid titration by dichlorophenolindophenol led to use of that blue dye as a Hill reagent that could be followed photoelectrically. Fortunately, I then had at Minnesota the one and only doctoral student of my whole career, Stanley Holt. His vigor and stamina developed as a long distance runner was applied to studies of chloroplast activity and later, on his own, to chlorophyll chemistry. Together we investigated the stoichiometry of the reduction of various dyes and inorganic salts by chloroplasts. We also found by use of $^{18}O_2$ that the oxygen evolved by isolated chloroplasts comes from water, as was already known for whole cell photosynthesis.

At the Minnesota botany department I inherited a spectroscopy laboratory set up by Elmer Miller as a university-wide service department. This was intended to take care of both absorption and analytical emission spectroscopy for all comers. Fortunately only a few agronomists wanted chlorophyll determinations and some analyses of plant material for inorganic nutrient deficiencies. The service aspects were soon forgotten, so I had access to much useful equipment.

In the days before large-surface photomultiplier tubes or before Shibata's opal glass techniques for measuring the absorption of scattering material like leaves or cell suspensions, the Ulbright sphere seemed more essential than it does now. Glenn Rabideau, Stanley Holt, and I put together a large white sphere with a homemade monochromator using a large replica grating. With this we measured the absorption spectra of various leaves and algae.

For chloroplast disruption I used a piezoelectric quartz supersonic oscillator at the University of Minnesota chemistry department. This machine, imported from Germany by Freundlich, the colloid chemist, was the typical Hollywood idea of scientific equipment. It comprised a large table covered with electronics of the 1920–30 era. The two-foot high vacuum tubes were spectacular when in operation. The apparatus was made available and its use explained by a former collaborator of Freundlich. Thus chloroplasts were first intentionally broken into smaller pieces in the hope of isolating the active particles.

At Minnesota I was exposed to real botanists like Ernst Abbe, Lawrence Moyer, Orville Dahl, William S. Cooper, and Don Lawrence, the plant physiologists George Burr, Allan Brown, and Albert Frenkel, and such biochemists as Harland Wood and R. A. Gortner. One unexpected pleasure was to meet Leroy S. Palmer, whose book on carotenoids I had studied long before.

Before the second World War, I had two worries about research in photosynthesis. One was that far too little biochemical work was being done in relation to the physical thinking of the time. The other worry was the apparent shortage of young scientists interested in the subject. It is now obvious that these two concerns need not have been disturbing in view of the predominantly biochemical nature of the present work on photosynthesis. The number of people now engaged in photosynthetic research is overwhelming.

During the war years I was registered as a conscientious objector, so I had to find some occupations more acceptable to the draft board than merely teaching plant physiology. When will freedom from conscription take its rightful place in the list of basic human rights? Draft dodging led me at various times into teaching elementary physics, researching chlorophyll-containing paint for camouflage purposes, and a long project on mold selection for penicillin production in E. C. Stakman's Department of Plant Pathology. There I enjoyed an association with Clyde Christensen.

THE CARNEGIE INSTITUTION

In the summer of 1946, Herman A. Spoehr invited me to spend a month visiting the Carnegie Institution of Washington, Department of Plant Biology, at Stanford with the thought of joining the group. During the following winter he arranged a meeting for me in Chicago with Vannevar Bush, President of CIW, and with Alfred Loomis, a physicist and trustee of CIW. As a result I became Director of the Department of Plant Biology to succeed Spoehr on July 1, 1947. Spoehr remained on as a staff member while we worked closely and happily together for the rest of his life. There were two groups in the department: Biochemical Investigations (essentially limited to photosynthesis), and Experimental Taxonomy, consisting of Jens Clausen, Bill Hiesey, David Keck, and Malcolm Nobs. The latter group worked largely together on transplant studies to separate and identify the relative contributions of heredity and environment in plant development, originally under Harvey Monroe Hall, a professor at Berkeley and a research associate of the Carnegie Institution.

With the hope of producing an improved strain of range grass, the experimental taxonomy group was making crosses of different grasses that had probably not previously met in nature. This was done on a very large scale with help in testing the progeny from many experiment stations in various parts of the world. The history of that enterprise shows some of the advantages of research support by a private organization with sufficient patience. In the first place, cooperation with universities and experiment stations in various countries was easily arranged by Jens Clausen, the group leader. Secondly, the work was continued for several decades without pressure to publish prematurely. Annual reports and occasional papers on certain aspects of the results were published, but only now after 35 years is the final survey of the whole range grass project nearing completion by Hiesey and Nobs.

The photosynthesis group, mainly working individually, was made up of Spoehr, James H. C. Smith, Harold H. Strain, and Harold Milner, all basically chemists. Fergus D. H. Macdowall and Violet Koski came with me from Minnesota as graduate students. Spoehr and Milner were working on large scale algal culture. *Chlorella* culture continued with many collaborators for six more years and eventually resulted in the Institution's all time best seller, *Algal Culture from Laboratory to Pilot Plant.*

Once I asked James Smith if he was going to the next American Society of Plant Physiologists meeting. He replied, "When I was a boy I lived on the bank of the Ohio River. A small steamboat worked its way up against the current but when blowing its whistle lost steam and drifted down again. No meetings for me this year."

Until about the middle of this century, the justification for spending time and money on research in photosynthesis was that knowledge of its basic mechanism might lead to increased food production. Of course, that thought is still valid, but now the expectations are less specific and photosynthesis research itself is seen more as an essential part of the larger scientific enterprise from which practical human benefits may arise in many unpredictable ways. The *Chlorella* culture work was an attempt to force some practical value out of existing information about photosynthesis. The actual results, however, were of more value in educating scientists into the basic facts of economics than in feeding hungry people. At least we tried.

We had a small shop for wood and metal working where I usually spent more time than at the desk or in the laboratory. The shop was open to all scientists and we have had excellent mechanics, first George Schuster, who left after a few years to head the shop of the chemistry department at UCLA, then Louis Kruger, great-nephew of Oom Paul Kruger of South Africa, and now Richard Hart, who with Frank Nicholson has set high standards for construction and care of laboratory equipment. All the accounting and secretarial work of the laboratory was done when I first came by Wilbur A. Pestel, who had worked previously in the CIW administration building in Washington, at the Desert Laboratory in Tucson, Arizona, and in the department's laboratory in Carmel.

In 1902 the Carnegie Institution founded the Desert Laboratory in Tucson, Arizona, and in 1905 Dr. Daniel T. MacDougal became its director. Spoehr joined the group in 1910 after research experience in Berlin and at the University of Chicago with Ulrich Nef. Spoehr, Otto Warburg, and H. O. L. Fischer, Emil Fischer's son, all had been postdocs together in Emil Fischer's laboratory. A summer laboratory at Carmel was used by the department until early in the 1940s. The present laboratory at Stanford was constructed two years after Spoehr became chairman in 1927. Robert Emerson and Charlton Lewis had worked here for three years from 1938 and while measuring action spectra had discovered the "red drop." The later explanation of that effect and of Lawrence Blinks's chromatic transient experiments revolutionized the ideas about photosynthesis by making obvious the existence of two separate photochemical systems.

Until he retired in 1956, Dr. Vannevar Bush, President of CIW, visited the laboratory for several days once or twice each year. These visits were extraordinarily stimulating because of his intense interest in the details of our activities. His great personal warmth and enthusiasm always left us invigorated for long periods. This tradition was continued by Dr. Caryl Haskins, who succeeded Dr. Bush in 1957 as president of the Institution. It was also a pleasure to serve under Caryl, who was an old friend from graduate school times at Harvard. In addition to these two good presidents, the routine affairs of the Institution in Washington were always easily and pleasantly handled by the most tolerant and understanding executive officers, Paul A. Scherrer and later Edward A. Ackerman. Because of the small size of our department and the kindly attitude of the CIW officers in Washington, the administrative work involved in being director was simple and left most of the time for research.

RESEARCH FELLOWS AND VISITING INVESTIGATORS AT CIW

About the time I came to the Institution, Dr. Bush was starting a fellowship program largely for postdoctoral training. This program has the long-recognized values of the apprenticeship system. It also brings to each laboratory knowledge and techniques developed in other places. By careful selection of stimulating fellows it is possible to keep a small permanent staff near the advancing edge of scientific progress. To select the most promising young people for fellowships, James H. C. Smith and I traveled extensively and tried to keep in close contact with the leading laboratories in photosynthesis and related subjects. We were helped a great deal by Dr. Spoehr, who had previously spent a year as Director of Natural Sciences for the Rockefeller Foundation and knew all the photosynthesis people. In 1950 and again in 1955 Smith and I visited 50 European laboratories. After that we decided to go oftener and to cut the number of laboratories visited to a much smaller list of more specialized places. Much of our hunting later was done at meetings even though we continued to believe that laboratory visits are far more definitive in getting a picture of a man's ideas and working habits. We were continually appreciative of the wisdom of the old adage: "Never ask a professor what he is doing because he will probably tell you." This frequently takes a long time. Our systematic visiting turned up many extremely competent investigators and in looking back on the list of fellows, my major regrets concern only those we could not take for one reason or another. Of the Institution's particularly successful "graduates," I expect most of them would have done equally well regardless of their time here and that their contribution to our life exceeded whatever they may have acquired by coming. In addition to the young fellows, we also had visits of various lengths from well-established investigators on sabbatical or shorter periods of leave. The intangible residue left at a laboratory through which a number of great men have passed is significant in itself. Lists of the Institution Fellows and other associated scientists up to 1962 are given in Year Book 61, "Report of the President," p. *76* and p. *102* and were brought up to 1973 in Year Book 72, pp. 321–27. The latter article also gives an abbreviated survey of the department's history. Details of each year's work are reported in the Annual Report of the Director of the Department of Plant Biology in the CIW Year Book.

The Biophysical Research Group at Utrecht, founded by A. J. Kluyver and L. S. Ornstein with a Rockefeller grant, operated for many years under the direction of E. C. Wassink and later of J. B. Thomas. The work of that group on photosynthesis was of great interest to all researchers on the subject. James Smith and I were frequent visitors, and a succession of scientists from that group worked at the Carnegie Institution. These are Bessel Kok, L. N. M. Duysens, Joop Goedheer, C. J. P. Spruit, Cornelis Brill, and currently G. van Ginkel. Related to this group is also Jan Amesz, a "second generation Utrecht scientist" trained by Lou Duysens, now Professor of Biophysics at Leiden.

From Japan we had very productive visits from Hiroshi Tamiya, Atusi Takamiya, and Kazuo Shibata of the Tokugawa Institute, Norio Murata of Tokyo University, and Tetsuo Hiyama. All these Japanese people became close friends of our group.

From Scandinavia came various other delightful friends, now distinguished leaders in the field of plant physiology or related subjects. These were: Hemming Virgin, Göteborg; Per Halldal, Oslo; Lars Olof Björn, Lund; Axel Madsen, Diter von Wettstein, and Erik Jorgensen, Copenhagen; Hedda Nordenshiöld, Uppsala; Axel Nygren, Uppsala. Continuing friendship with Hemming and other Scandinavian colleagues lead to a very pleasant lecture tour and a Göteborg degree in 1974. Among our German Fellows or Visiting Investigators were Günter Jacobi, Alexander Müller, and Wolfgang Wiessner from Göttingen; August Ried and Eckhard W. Gauhl, Frankfurt; Friedrich Ehrendorfer, Vienna; Wilhelm Menke, Cologne; Eckhard Loos, Regensburg; Ulrich Heber, Düsseldorf; Helga Ninnemann, Tübingen; Ulrich Schreiber, Aachen via Vancouver; Carl Soeder,

Dortmund; and Wolfgang Urbach, Würzburg. Contact with German photosynthesis workers such as Karl Egle, André Pirson, and others resulted in an invitation in 1964 to speak at a meeting in Halle and become a member of Leopoldina. Kenneth Thimann was also in the group and on this occasion we celebrated our third penetration of the Iron Curtain together. With regret I omit a complete listing of our Fellows and Guest Investigators from other countries and the USA. It was a great pleasure to have Gordon Gould, a fellow biochemical sciences college classmate appear later on the Stanford faculty so that our friendship since freshman days could be resumed.

The laboratory was so small, usually about 15–20 people all together, that all visiting scientists became close personal friends of the whole group. We rarely had more than 3–4 visiting scientific workers at a time and usually all the staff followed each other's activities and those of the visitors with considerable interest. It is sometimes said that the department's Annual Reports are not very good publications because of the lack of outside reviewers. This comment is mildly amusing to those of us who lived through more than one of the Annual Report seasons when each member of a group, including visitors, independently edited each other's work in far more critical detail than would have been expended in reviewing any journal article referred by an editor.

SOME OF THE EXPERIMENTAL WORK AT CIW

Chlorophyll is clearly the most obvious organic material in the world. Nevertheless, more is known about the spectroscopy of the rare earth elements than about chlorophyll in its natural state as it occurs in leaves. This situation was mentioned in my first annual report in 1948 and much time since then has been spent on spectroscopy of in vivo chlorophyll. The revival of Lubimenko's ideas and their great expansion by Krasnovsky about 1945–1955 stimulated me to concentrate on the various forms of chlorophyll *a*. It is now clear that there are four major forms of chlorophyll *a* with absorption peaks at 662, 670, 677, and 684 nm in all green plants. Much of James Smith's work when I first came to the Institution was on chlorophyll formation and the transformation of the protochlorophyll by light. Smith, Koski, and I collaborated on the action spectrum for this photochemical effect in dark-grown seedlings. During this time Harold Strain was perfecting his chromatographic methods of pigment separation, an art only slowly being rediscovered since the pioneer work of Tswett long before. Unfortunately for plant physiology, Strain was called away to the Argonne laboratory to apply chromatography to a wider range of chemicals.

Continuing with Harold Milner and others the chloroplast fractionation experiments started in Minnesota, I was always looking for better ways to disrupt chloroplasts. My attempts to do so with detergents were never successful for lack of work with finely graded series of different concentrations. My crude experiments either did nothing with low concentrations or dissolved and ruined the whole mixture with high concentrations. Dr. Bush discussed our individual problems in considerable technical detail. He told me about some company that could make very small holes in sapphire and suggested that extruding a chloroplast suspension through such a small hole might do a good job of disruption. This sounded very promising, but I worried about getting the hole plugged up by fibers and bits of dirt. To have a small hole that could be opened up when plugged and then constricted again led me to the idea of a needle valve instead of a fixed hole. I threaded an ammonia needle valve into a steel cylinder with a hole in the center to contain the chloroplast suspension. The suspension was forced through the valve by a steel piston with a seal. This extrusion principle worked well and has since found much use for disruption of bacteria and other cells as well as for chloroplasts. When I told Dr. Bush that the device was being made commercially, he suggested that I ask the manufacturer to attach my name to the device and to give a free one to the laboratory. The first request was acceded to at once, and about 20 years later the American Instrument Company gave the department a "French Press." We still drive it with an old hydraulic jack made for automobiles.

Combining an interest in chlorophyll fluorescence from the time at Chicago with a taste for building spectroscopic equipment, I put together two homemade grating monochromators. One of these was used to irradiate the sample such as a leaf or an alga with blue light while the other monochromator had its wavelength setting swept by a synchronous motor to record the emitted fluorescence as a function of wavelength. A synchronized drum carrying a hand-drawn curve corrected the photomultiplier output to give the true emission spectrum of the sample. This was the first automatic recording fluorescence spectrophotometer ever built. For its construction and for many similar purposes I found the part-time help of students at Stanford's electrical engineering department to be indispensable.

In the summer of 1949 we enjoyed a visit from Richard H. Goodwin of Connecticut College, who used fluorescence spectroscopy to identify traces of insoluble uroporphyrin particles in the cells surrounding the guard cells of *Vicia*. These particles have a brilliant orange-red fluorescence which Goodwin duly recorded by color photomicrography and sent in for development. A package of tourist photos of the Matterhorn came back.

Shortly after I came to the laboratory, Shao-lin Chen, a research fellow, measured the action spectrum for di-chlorophenol indophenol reduction by spinach chloroplasts, confirming for this reaction the "red drop" of Emerson and Lewis. The carotenoids present did not appear to be active.

In 1953 we had a visit of a month more or less from Robin Hill. With James Smith we started to measure the absorption and fluorescence spectra of protochlorophyll. The stimulation from these experiments and the subsequent discussions were out of proportion to the results published from this brief collaboration, but they led to further related studies and to life-long friendships.

After Violet Koski and I had measured some fluorescence spectra of red algae, we found it necessary to resolve these spectra into their overlapping components. To do this we wanted to add together, in adjustable proportions, curves of the individual substances present in the mixture. With help from an engineering student, George Towner, we experimented with vertically movable tables carrying heavily inked curves. The curves were followed by a light beam and photocell assembly attached to a potentiometer. This arrangement produced a voltage proportional to the Y-axis of the curve as the X-axis was driven along. For four years I worked on this device with various engineering students. Eventually we had a so-called "Curve Analyzer and General Purpose Graphical Computer." This consisted of five curve follower tables, a recorder, two integrators, numerous adding amplifiers and servo drives. In final form it was much more versatile than the original plan. This device served us well until Stanford's IBM computer and associated plotting system displaced it about 15 years later. One of the early experimental curve followers later was converted to a curve digitizer. It is now used to tabulate numbers from plotted curves and enter them into the digital computer.

In 1953 I started to build a first derivative spectrophotometer to detect small changes in the slopes of absorption spectra. This was based on a slit vibrating over a small wavelength interval in the spectrum. The construction and subsequent modifications took several years. With this, as well as with the curve analyzer, the continued technical interest and support by Dr. Bush was a great help. At first the machine plotted the derivative of transmission against wavelength but later it was altered to plot the derivative of absorbance. We surveyed the absorption spectra of various algae. When we got to *Euglena* the strange results made us suspect trouble in the apparatus. However, the irregularity turned out to be an unexpected form of chlorophyll a with an absorption band at 695 nm. With Dr. Jeanette S. Brown and others, work on the different forms of in vivo chlorophyll was continued for many years.

A simple but useful technique for removal and readdition of lipid components of chloroplasts was developed in 1957 with Victoria Lynch. Extraction of dry chloroplasts with petroleum ether greatly reduced their Hill reaction capacity, but readdition of the extract or of carotene solutions in hexane to the dry material followed by evaporation of the solvent caused reactivation. The remarkable effect was that the readded substance went back to the right place. Subsequent work by others has implicated plastoqui-nones as well as β-carotene in the reactivation.

In that same year we enjoyed Per Halldal's first visit. He arranged a spectral projection apparatus with a vertical intensity gradient on one side of a glass vessel. On the other side uniform light from the opposite direction fell on the same vessel. Within the vessel some motile algae in suspension plotted their own action spectrum for phototaxis by swimming to one side of the vessel or to the other side.

To study the growth of algae at various temperatures and light intensities, Halldal and I fixed up a thick aluminum plate with a thin layer of inoculated agar on its surface. One side of the plate was cooled by circulating cold water; the other side was heated to produce a temperature gradient across the plate. At a right angle to this temperature gradient a light intensity gradient was established. Thus combinations of temperature and intensity were provided at different places on the plate and the resulting growth pattern gave a visual picture of the intensity and temperature effects on growth rates. Later Ruth Elliott and I used the same apparatus to study lettuce seed germination as influenced by light at various temperatures.

A new technique was brought to the group by Francis Haxo, who worked at the laboratory for several months about 1957. He showed us how to measure oxygen exchange easily. The platinum electrode arrangement originally developed by Blinks and Skow had been used by Haxo and Blinks for measuring photosynthetic action spectra of algae. About this time Jack Myers came for a sabbatical year and with this platinum electrode measured action spectra for the Emerson enhancement effect and also for Blinks' chromatic transients. Furthermore he found that the two wavelength en-hancement effects persisted even though the two beams were given consecu-tively instead of concurrently. This established the nature of the interaction as being due to accumulation of a chemical substance produced by one light reaction and used up by another.

The rate-measuring O_2 electrode was also used in an apparatus for auto-matic recording of action spectra as the wavelength was continuously var-ied. With this device the rate of O_2 evolution as measured by the electrode output was used to control the light intensity so that the rate of O_2 evolution was kept constant as the actinic wavelength was swept through the spec-trum. The reciprocal of the light intensity, so adjusted to give a constant rate of O_2 evolution, was then plotted automatically against wavelength. Later this system was improved by Per Halldal during a return visit in 1969. A variety of Blinks O_2 electrodes were used by many different investigators over a long period of years. Among this group were also: Guy McCloud, David Fork, Govindjee, Martin Gibbs, William Vidaver, Carl Soeder, Yaroslav de Kouchkovsky, August Ried, James Pickett, and Eckhard Loos. After David Fork returned from Witts laboratory in Germany and set up equipment for measuring absorbance changes, the O_2 electrode was less in demand. This was because the absorption changes can be used to tell which particular substances in the cells or chloroplasts are actually changed by a light exposure rather than merely measuring overall metabolism.

This absorption change technique was first used for photosynthesis investigations by L. N. M. Duysens at Utrecht. He came to the laboratory in 1952 and extended his previous work with absorption changes in purple bacteria to *Chlorella* and other photosynthetic species. He thus discovered the "515 change" later found to be caused by shifts in the carotenoids.

Before 1947 Harold Strain had been interested in the possibility that changes in the absorption spectrum of leaves might be caused by light. He looked at leaves with a visual spectroscope, left them in light for minutes to hours, then took them downstairs to the spectroscope to look for changes. None were observed. About that time I took a flask of *Chlorella* culture to Britton Chance's laboratory in Philadelphia where he had excellent optical equipment and techniques. We pumped the *Chlorella* suspension slowly through his apparatus. The suspension first went through one measuring chamber, then through a chamber where it was exposed to light, then through a second measuring chamber. The transmission difference for various wavelengths between the two measuring chambers was plotted automatically. No difference! One reason for Duysen's success was that in addition to the improved apparatus sensitivity, he illuminated the sample during the measurement.

In 1960, while our new house was being built in the country, I renewed a previous, very primitive interest in land surveying. Experience with simple optics applied to the illumination of plant preparations and the work on the curve analyzer led me to think about combining a range finder with a mechanical computer and alidade to partially automate plane table plotting. With help from Dick Hart, a functional but cumbersome model was built. Later Charlton Lewis, Bob Emerson's former collaborator, wrote a fine application for a patent, which was eventually granted. Part of our thinking, largely inherited from Dr. Bush, was that an occasional by-product of scientific work might be of enough use to produce some income to further facilitate research. That this idea can be productive had been very successfully demonstrated by Dr. Cottrell, a Palo Alto resident in his later years, who frequently visited the laboratory and whose dust precipitator patents had been the foundation of the Research Corporation. The patent for the automatic plotter was placed in the hands of the Research Corporation with the intent of splitting profits equally between CIW and RC. However, there weren't any. About that time laser range finding was being developed which made optical range finders obsolete. The patent never became of interest to an instrument manufacturer in spite of considerable effort. When the development was well along, I found out that a somewhat similar device had been built before and was located in the geography building at Harvard next door to the biology building where I had worked 25–30 years before. While I have no conscious knowledge of having seen that device, I often wonder if some previous but forgotten exposure to the basic idea had been behind my "invention." The main lesson learned from this adventure was to appreciate the cost in time and money of getting too interested in a sideline not directly concerned with the normal affairs of the laboratory.

About 1967 we tired of the complexities of first derivative spectroscopy and converted that instrument into a more conventional absorption spectrophotometer. The device was, however, made for convenient use at liquid N_2 temperature. We then were able to see irregularities in the absorption spectra of chloroplast preparations due to the presence of different forms of chlorophyll a. Through a discussion with Glenn Bailey I heard of the computer program developed at the Shell Development Laboratory by Dr. Don D. Tunnicliff, who kindly gave me a card file of the program. With the help of the Stanford computer group and of Mark Lawrence, that program was adapted to the Stanford IBM computer. Numerous modifications were made and the program has been widely distributed with permission of the original author. With this computerized curve analysis, spectra can be resolved into Gaussian or Lorentzian components. Glenn Ford has adapted a simplified version of the program to the department's Hewlett-Packard computer. Jeanette Brown and I used it to characterize various forms of chlorophyll a. Other specific components were similarly identified with Hemming Virgin in natural protochlorophyll spectra and with Atusi Takamiya and the Muratas in water-soluble chlorophyll protein complexes.

COMMENTS ON VISITS TO OTHER LABORATORIES AND ON LABORATORY ADMINISTRATION

Until after the 1950s, the average scientist seemed to take more interest in fellow scientists in other fields and in their work than is now common. As a biologist looking for new techniques and different approaches to photosynthesis research in the 1930–1970 period, I frequently visited and was treated hospitably by specialists in many different fields. Many of these kind people, though they may have thought of biology as butterfly collecting and naming the wild flowers, nevertheless would willingly give their time to an interested visitor. I have always valued contacts with some of the great men in such diverse and to me largely unknown fields as spectroscopy, physical chemistry, photochemistry, bacteriology, algology, mammalian physiology, photocell technology, lamp manufacture, biochemistry, servosystem theory, optics, electrical engineering and, of course, various phases of plant physiology. For arranging many of these visits, I am particularly grateful first to my major professor, W. J. Crozier, and later to Dr. Vannevar Bush, both of whom realized the potential value of cross-fertilization of fields through scientific visits without an immediate specific purpose. I hope the increase in numbers of scientists and attendant specialization will not make such contacts between men in different fields less common in the future. I fear that the increase in size of the National Academy may, or already has, led to a decrease in its potential for developing friendships or at least speaking acquaintances between practitioners of different subjects.

A particularly memorable visit was to Robert W. Wood, Professor of Physics at Hopkins, author of *Physical Optics* and of *How to Tell the Birds from the Flowers.* I showed him a curve that I had measured for the transmission of a replica grating for various wavelengths. He said he would give me a grating in exchange for the curve. This grating is still in use for fluorescence spectroscopy of photosynthetic material by Bill Hagar. I also remember having lunch at the Rockefeller Institute with Leonore Michaelis, who pointed out that mathematical analysis of kinetic experiments is of great value in very simple systems but may become nearly meaningless if the system is too complicated. Several visits to James B. Sumner at Cornell were particularly helpful when I was first trying to separate out chloroplast components by biochemical means. To see a man with only one hand pour liquid from a flask into an unsupported test tube shows the value of experience and familiarity with the tools of the trade. The take-home lesson from Sumner was to keep the protein concentration high and the liquid volume low when trying to crystallize a protein. Unfortunately my fractionation experiments never got to the point where crystallization of anything but ammonium sulfate seemed likely. George Shannon Forbes once pointed out that many famous chemists have rediscovered ammonium chloride.

Sometime during the late 1930s or early 1940s Jack Myers and I were talking about the future of photosynthesis research. The upper limit of our thinking was to establish a laboratory with a million dollar endowment. That, we thought, would enable all the interested people to get together and solve all the problems we could think of at that time. Since then many people have each run through much more than that amount of money and the same unsolved problems along with some new ones are still around. This is progress in research. Myers (2) has recently written a remarkable review of the development of the basic concepts of the process of photosynthesis.

It has been a particular pleasure to have had the stimulation of knowing some of the photosynthesis investigators of the previous generation. In addition to the long and close association with Spoehr, I appreciated several visits with Professor Harder in Göttingen and enjoyed his visit to CIW accompanied by his students, von Denfer and von Witsch. At a Physiological Congress in Switzerland Emerson introduced me to Professor Arthur Stoll, who had worked with Willstätter on photosynthesis and on chlorophyll before and during World War I. After I gave a short paper at the Congress, Stoll kindly stood up and pointed out the significance of the work which I had not made clear. He visited the CIW laboratory some years later. James Smith and I had a memorable visit with Stoll at Basel when he was President of Sandoz. Other contacts with photosynthesis pioneers were several visits with René Wurmser in Paris, with G. E. Briggs at Cambridge, W. O. James at Oxford, H. Kautsky in Marburg, J. Buder, early investigator of photosynthetic bacteria, in Halle, A. Seybold in Heidelberg, William Duggar and Farrington Daniels at Wisconsin, W. J. V. Osterhout at the Rockefeller Institute, M. G. Stolfelt at Stockholm, and various others of later dates. Having known many of the active people in the first three quarters of this century gives a sense of continuity in the field of photosynthesis research.

Throughout my time as a scientist I have been amazed at the progress that can be made by combining the techniques of different branches of science. Furthermore, a facility with the simple elements of assorted trades such as carpentry, metal machining, plumbing, painting, electrical work, and all that sort of thing can give an investigator great freedom to improvise unusual equipment. The very small amount of knowledge from other fields that is needed to significantly promote one's main line of investigation seems almost unbelievable. Applications of simple mathematics, elementary optics, primitive electronics, and basic metal working was about all that I found necessary to build various pieces of apparatus that were new to plant physiology. Using the principle of combining assorted techniques is great fun and with very little effort can easily break new ground when the problem at hand is clearly visualized.

Most everyone who is responsible for more than one or two assistants has to face the question of laboratory organization and how to develop the customs and the atmosphere that best facilitate research. It seems that the answer to this question depends primarily on whether the laboratory is basically a one-man affair with everything devoted to furthering his program or whether it is a place to develop independent scientists. Whatever attributes of a good laboratory are listed, it is easy to find a clearly successful place that drastically violates anyone's preconceived ideas of good organization. This makes the subject difficult to discuss. However, by keeping in mind two contrasting types of laboratories, some useful attributes of each type may reasonably be described.

For the one-man show, organized very likely to capture a Nobel Prize for its head, it is essential to have all the staff personally loyal to and preferably afraid of the boss. This means that independent thinkers with unconventional approaches must be suppressed. Travel of the staff to meetings and other laboratories is very dangerous and must be strictly limited. Visitors must be urged to talk about their own work and their questions evaded as much as possible. Of course, it is forbidden to talk about the work of the group to outsiders, even nonscientists. Working hours and vacation times are strictly kept. Many great discoveries have come from such groups but one hears very little from most of the workers after they leave that sort of an organization. Surprisingly enough, that kind of structure occasionally flourishes within universities as well as in more business-like research organizations. Some places of this type may be patterned on older European models or on customs in the business world. Such places are not necessarily all bad because that otherwise abominable system may provide the optimum environment to take advantage of the particular abilities of certain rare geniuses.

The contrasting kind of a laboratory would abhor all the prescriptions for success of a director-led group as just described. Perhaps its main characteristic would be the encouragement of "enthusiasm without stress" Tiselius (4). There the individual scientist usually thinks and acts independently but with frequent friendly discussions with colleagues both in his own and in outside groups. In that environment cooperation predominates over competitiveness and the workers are happier. Perhaps only in this way can true academic excellence be encouraged without danger from the almost inevitable degradation that follows from the spirit of too much competition. This good laboratory environment fortunately seems to be nearly universal in plant physiology. Strangely enough, I have learned as much or more from visiting bad laboratories as from studying good ones. It is a great help to see clear illustrations of things to avoid.

The question of the optimum size of a laboratory is entirely different from that of its form of government, i.e. dictatorship, democracy, oligarchy, or anarchy. In theory the optimum size depends on the nature of the people there, on the nature of the problems being investigated, and on the teaching responsibilities, if any, of the group. In actual fact, the size is usually determined by the available budget or space on the principle that bigger is better. The critical size below which the effectiveness of a group evaporates may be very different for trying out unusual ideas with little probability of success or for pursuing a well-defined objective. The optimum must be somewhere above the critical size but below the point where in the words of Paul Kramer (1): "Increasing numbers cover both talent and mediocrity."

THE PLEASURES OF RETIREMENT

After 26 years at CIW I retired in 1973 with the great good luck of being succeeded by Winslow Briggs. He has kindly provided me with a comfortable office and a laboratory. This is an excellent opportunity to try to refine methods of measuring action spectra of the separate steps in photosynthesis. The hope is to compare action and absorption spectra of photosynthetic material with enough precision to see if action spectra for the two photosynthetic systems can add together to match the absorption spectra accurately. It may be possible to establish definitely the presence or absence of other reactions than the well-known two photosystems.

A photostationary state, steady-deflection method for measuring DCIP oxidation has been found to give more precise measurements than direct rate measurements. The principle is that with an excess of reductant present the dye oxidation rate by light is just balanced by the chemical reduction rate. The resulting steady-state dye concentration can be measured easily because the noise level can be averaged for a long time period since nothing changes after equilibrium is established.

Strangely enough, this dye oxidation reaction seems to be driven by all the forms of chlorophyll in the preparation (*Nostoc* particles kindly provided by Arnon and Hiyama) while the dye reduction is driven, at least preferentially, by the usual system II components. I look forward to a happy old age devoted to clarifying these questions. Working alone has for me always been a pleasure and is particularly desirable as the usable length of a working day becomes shorter.

The slow decline in competence with old age is fortunately compensated by an increasing delight in the completion of simple jobs. It is a great satisfaction to be able to continue with scientific work in a renovated laboratory and with many highly skilled and modern investigators available for discussion and often needed enlightenment.

THOUGHTS ON THE FUTURE OF RESEARCH IN PHOTOSYNTHESIS

"Solving" the problems of photosynthesis really means describing the process in terms of currently understood concepts of molecular interaction. Many aspects of photosynthesis have been well described already. Probably the most comprehensive interpretation of one part of the photosynthetic problem is the path of carbon through the various intermediates from CO_2 to carbohydrate as described by Calvin and his associates in terms of organic chemistry. Also, the moderately clear pictures of the process of light absorption followed by energy transfer through various pigments to a reaction center have been widely accepted. Furthermore, the electron transport processes from the reaction center to reduced NADP and the accompanying formation of ATP are believed to be reasonably well understood and the chemical nature of the transporting substances are mostly known.

Where then are the remaining problems? In the first place many details remain to be filled in to complete and verify the current concepts of electron transport, and no doubt some major revision of the present concepts of the two-reaction scheme and associated carriers may be expected.

If past experience is any guide to the future course of development of the understanding of photosynthesis, we may confidently expect some discoveries that would make radical changes in the present concepts. It may be that some new enlightenment comparable in significance to the discovery of the two separate light reactions and to the involvement of photophosphorylation in photosynthesis may again appear.

In any case, as the relevant chemistry and physics develop further, the language in which presently known effects are described will certainly be very different in the future.

We may hope that future research on photosynthesis and related processes will show ways to increase agricultural production, possibly through influencing the path of carbon fixation.

Literature Cited

1. Kramer, P. J. 1973. Some reflections after 40 years in plant physiology. *Ann. Rev. Plant Physiol.* 24:1–24
2. Myers, J. 1974. Conceptual developments in photosynthesis, 1924–1974. *Plant Physiol.* 54:420–26
3. Nachmanson, D. 1972. Biochemistry as part of my life. *Ann. Rev. Biochem.* 41:1–28
4. Tiselius, A. 1968. Reflections from both sides of the counter. *Ann. Rev. Biochem.* 37:1–24

Ann. Rev. Plant Physiol. 31:1–28

SOME RECOLLECTIONS AND REFLECTIONS

Anton Lang

MSU-DOE Plant Research Laboratory, Michigan State University, East Lansing, Michigan 48824

CONTENTS

DIPLOMATIC CORPS, OPERA, OR BOTANY?

I have never regretted becoming a botanist, specializing in plant physiology and not evolving into a plant biologist, developmental biologist, or new biologist, but the decision to become one was not entirely easy. It is true that curiosity in and outright fondness for plants is the first persistent interest I can remember having. It began when I was at the age of 10 or 11 and living in Labes, a small city in Eastern Pomerania, then part of Germany, now Lobez in Poland, attending its Private Boys' High School.

I am a native of Russia, although of German ancestry on my father's side, but at age 4 I left Russia, in early summer of 1917, to go with my mother (my parents had separated and Father remained in Russia) from our place of residence in Petersburg (now Leningrad) to the family *dacha* in Sestroryetsk, a resort place near the Finnish border. The trip was a yearly event, so our departure from Russia was in no way as adventuresome as that of many later emigrés. At that time the Czar had abdicated and Russia had its brief period of a government that endeavored to follow genuinely demo-

2023

0066-4294/80/0601-0001$01.00

cratic principles. However, the situation in the country was tense and uncertain, and the family council decided we should move to Finland, which at that time was about to acquire independence, and wait until things in Russia had settled down, with the Czar back on his throne and the good old times returned. After waiting four years, Mother accepted an invitation from her sister and brother-in-law and we moved to Labes. Otherwise, I might still be in Finland, waiting for those good old times.

In Labes, we lived outside the city in a house that had been the residence of the owner of a combined flour mill and noodle factory and therefore was locally known by the unusual name of *Villa Nudelmühle* (Villa Noodle Mill). School was about three-quarters of an hour away. In winter one had to fight one's way through snowdrifts, but in spring, summer, and fall there were ample opportunities for nature watching and "botanizing," and these activities were gently encouraged and guided by our natural history teacher.

Those who planned to attend a university in Germany had to complete four years of grade school and nine of high school. The high school of Labes was quite modest. It occupied one half of a building (the other half housed the parallel establishment for girls, and the breaks between classes were carefully scheduled so that boys and girls would never meet), consisted of the first four high-school grades and two teachers. The natural history teacher was not a member of this regular staff, but was borrowed from another school for two periods a week. Although at first impression a rather dry man, he knew how to challenge his pupils' curiosity and make them look closely at nature. Each summer he had each student choose some creature, plant or animal, study it on his own, and write a report. For my first assignment, I picked the pea and the cucumber, discovered that the flowers of the former had stamens and pistil while those of the latter had only one or the other, and at the age of 11 or 12 wrote a monograph on this subject. It was only much later that I learned, with some chagrin, that the phenomenon had been quite well known to botanists for some time.

In 1926 we left Labes and moved to Berlin. There was much less opportunity for getting into nature, and the teacher at my new school preferred to do his instructing with books, drawings, and charts, yet my interest in plants persisted. However, as time drew nearer for making a choice of subject for further studies, two challenges arose, each one championed by an uncle. One, the squire of Villa Noodle Mill, a former German officer, had for some reason the ambition to make me into a diplomat. For this career it was very useful to have a degree in law, and he offered to pay the tuition. This was an offer quite hard to refuse. We were far from wealthy, nor was he (our money as well as his had been made worthless by inflation in Germany), so the offer meant a great sacrifice on his and my aunt's part. I went as far as attending some law classes in the university during my last years in high school, but decided that this was a form of human endeavor I could neither master nor enjoy. The challenge to botany, championed by the other uncle, was more serious. This uncle (Mother's brother) was quite an unusual person. By profession he was a banker; by avocation an opera enthusiast, and I suspect he was a far better opera enthusiast than banker. In fact, he had helped to found and manage a small but progressive opera company called the Musical Drama Theater in Petersburg. Now, I was (and am) quite an opera fan myself. During my last years in high school and my first ones in the university, I was an important member of the Berlin State Opera, namely, an extra, and was sometimes performing three times a week, undoubtedly to the detriment of my studies and with no material reward either, since extras were not paid. However, it was a delightful experience. Performances were allotted by seniority; junior extras were assigned the tougher ones while the most senior extras could choose almost any performance they liked, as long as extras were needed. I enrolled with *Tannhäuser* and graduated with *Aida*. *Tannhäuser* was not popular with extras because it kept them very busy. Of course, we were not allowed to take part in the orgies of the Venusberg, but in Scene 2 of Act I we were reinforcing either the ranks of the pilgrims about to depart for Rome or the Landgrave's huntsmen and other followers. My first chore, with another unfortunate extra, was to lug a wild pig which was part of the huntsmen's bounty. It was made of wood and incredibly heavy, and to add insult to injury, earned me the name of *Wildsau* (wild sow) even though there was no proof whatever that it was a sow and not a boar. In Act II, we were part of the Landgrave's court, and in Act III again pilgrims, now returning from Rome. With changing costumes and makeup, there was hardly time for standing in the wings, as we were allowed to do when not on stage, and enjoying the performance. *Aida* was quite a different matter. The only scene where extras were needed was the triumphal scene, where we were either victorious Egyptian soldiers or Ethiopian prisoners. Being one of the latter was a particularly coveted part since they came on stage rather late and left early. The only problem was getting washed and changed in time to be ready in the wings for the Nile scene. Changing was a minor matter, for the costume designer's idea of Ethiopian warriors was to have them wear nothing but what seemed to be very old, very worn, and very dirty underpants. The washing, however, was something different for we were painted from face to toe with a highly repulsive brown paint which was slopped on us with a huge brush, and the dressing area for extras was quite cramped, with few showers and sinks.

In 1933, after Hitler had come to power, the Berlin State Opera, like all Prussian state theaters, was assigned to Hermann Göring, who had become

Prime Minister of Prussia. In many respects, this was a blessing. All other German theaters, and cultural institutions in general, became the charge of Joseph Goebbels, and of these two top Nazi leaders Göring was less rigid, at least in areas such as the arts which were relatively marginal to ideology and politics. The Germany of the Weimar Republic, while wracked by endless internal political strife and handicapped by severe external strictures—the Versailles peace treaty with its unrealistic and humiliating provisions—had an incredibly active and diverse cultural life, and Berlin was its undisputed center. There were nearly 40 theaters and three or four opera houses; musicians, writers, poets, painters, and sculptors seemed to be working as if paid overtime (some may have been, too), their work ranging from the conservative to extreme experimental, their *Weltanschauungen* from the right to the far left.

Thanks to the efforts of Doctor Goebbels, all of this excitement vanished overnight. It is true, one could still see excellent theater performances and hear outstanding concerts and recitals, but the scope had become quite narrow and parochial. Opera, and the Berlin State Opera in particular, however, was somewhat less affected, undoubtedly partly because of Göring's attitude. The Berlin Opera lost some excellent Jewish singers, like Alexander Kipnis and Richard Tauber, and several fine conductors, like Otto Klemperer and George Szell, but the standards remained very high. Göring supported "his" theaters generously, and the extras now were paid *Reichsmark* 2.50 per performance.

Soon money became the main purpose of being an extra, the old enthusiam disappeared, and after a year or so, although I could use the money, I preferred paying my own RM 2.50 for the cheapest seat (or RM 1.00 for standing room) rather than earning Göring's RM 2.50. However, when my uncle began to build up an opera career before my mind's eye, I was still very much involved in my duties as an extra and quite susceptible to his blandishments. He decided, for reasons I no longer recall, that I had the makings of a great singer. He was a very persuasive person, and when he started describing the future, I soon could see myself acknowledging thunderous applause before the curtain, with him, my coach and manager, beaming from a box near the stage. I had no voice training whatever, but when he decided to test my voice (bass naturally) I sang with considerable abandon (and probably many a sour note). After I had finished, he sat for awhile saying nothing, his hands still on the piano and a quizzical look on his face, then finally pronounced, "You know, Anton, your voice is very small—but extremely disagreeable."

So in 1931 I enrolled in Berlin University, with botany as my major subject, and graduated with the equivalent of a PhD in spring 1939.

BERLIN UNIVERSITY, 1932–1939

I have often been asked how high school, university studies, and an academic career in Germany compared with the United States. It almost seems as if the two systems were designed to represent opposite approaches. In German high school, although there was quite a variety of types— *Gymnasium, Reformgymnasium, Realgymnasium, Reformrealgymnasium, Realschule, Oberrealschule,* and perhaps more—which differed mainly in the degrees of emphasis on classical languages and lack of emphasis on sciences, the schedule was firmly established for five or six hours daily, six days a week, every week of the school year for 9 years, with no electives, no study periods, no driver's education or other "nonacademic" subjects. However, once one entered a university and did not study toward a particular profession, such as high-school teacher, physician, or pharmacist, for which definite curricula were prescribed, but rather went for an academic degree, one was entirely on one's own. I could take or leave any course I liked. Moreover, there were no midterm or final examinations, no tests, no quizzes, no term papers, and no grades. For lecture classes, all that was needed was the instructor's signature at the beginning of the term; for laboratory classes, another signature at the end; but whether one actually attended classes or studied at home, in the library, or elsewhere was one's own decision. (In science, this was obviously less so than in other disciplines, but I knew some fellow students who had not attended a single lecture; they even got that necessary signature by asking another student who did attend classes.) The only mandatory formal examination I took was the final PhD examination, which consisted of the thesis defense and three hours each of general and taxonomic botany, plant genetics, plant physiology, and chemistry.

As far as thesis research was concerned, there was usually little leeway. The custom was to find a professor working in one's area of interest and have him accept you as his student. He would give you the *Thema,* i.e. the problem you were expected to solve, but from then on you were again very much on your own. Studying toward the PhD was a matter of swimming or sinking.

In retrospect, I would have preferred having had more opportunities for checking my choice of classes and my research progress (just as I would have liked to have some options which my children had in American high school). There is no doubt that quite a number of German students who had fancied they were vigorously swimming toward the doctor's degree discovered much too late that they had sunk. However, those who did reach the goal had acquired a degree of independence and self-reliance which most of their American counterparts acquire only during graduate or postdoctoral work, if at all. I also cannot help being concerned that this trend in American universities seems to be growing even stronger, with increasing efforts at removing any problem or obstacle from the student's course. I have far too often seen that people realize their full potential only when faced with problems and challenges which they had to master largely on their own to believe that the smoothest, fastest superhighway is the best route for reaching one's goal—if that goal is one, like a career in science, which requires imagination and initiative.

But back to the young scientist who has obtained his PhD and has perhaps acquired some years of additional postdoctoral experience. In America he is now ready to find a faculty position and is expected—apart from teaching, some committee work, and perhaps some public service obligations—to develop a research program of his own. Not so his German equivalent, whose options, at least at my time, were quite restricted. He had to find a position as *wissenschaftlicher Assistent,* the first rung of the German academic stepladder, and as the structure of German academic institutions was very hierarchical, his opportunities depended very much on the ideas and intents of the *"Chef,"* the director and sole full professor of the institute (department). The assistant usually had to run laboratory courses, often had to prepare exhibits and demonstrations which were part of the professor's lectures, and while he was expected to carry out research, this could mean participation in the director's research program or could at least be greatly influenced by the latter's ideas of science. I have known several German scientists who, at the start of their "independent" careers, were told by their directors in which problems areas they could or couldn't work. In one case, one of the forbidden areas was plant hormone research since the *Chef* did not believe that plants had hormones. The fact that the *Chef* usually was in control of all funds in his unit made arguments quite difficult, too.

These remarks should not be taken to mean that all German institution heads were petty tyrants. The great majority were fine, often outstanding scientists who had reached their rank through a rigorous selection process. Most were utterly devoted to their discipline and very willing to raise competent students to continue their work. But once at the top they were usually exposed to little if any professional criticism and developed very firm views about what should be done in their science, and how it should be done, sincerely believing that progress was not possible any other way. I have heard it said that in the American academic system one has to be different in order to be accepted; in the German system, one has to conform. Now that is an overstatement, to be sure, but one of the greatest and most attractive differences between the two "systems" which I discovered after

coming to the "New World" was how much easier communication is between students and faculty and between junior and senior scientists in
American and Canadian universities. Things seem to have changed in this
regard in Germany also, but I am not quite certain to what extent. Students
do not address professors as "Herr Professor" any more, but simply as Herr
So-and-So—which would have been lèse-majesté in my time. However,
when on a recent visit there I commented on this fact to a young German
colleague, his response was, "Yes, but otherwise not much has changed."

The two instructors who had the greatest impact on me in my university
years were the plant physiologist K. Noack and the geneticist Elisabeth
Schiemann. Noack had not done very much research of his own and was
a poorly organized lecturer, but he was very alert to any new developments
in experimental biology. It was in his plant physiology course that I learned
for the first time of such classical work as Warburg's research on cell
respiration and photosynthesis, or of Krebs' early experiments which ultimately led to the establishment of the TCA cycle. Noack's was very probably the only botany class at that time in which these topics were treated.
Dr. Schiemann was my thesis adviser. Her principal research interest was
the history of cultivated plants, and she approached the problem not only
genetically but became also quite an expert in archeological and paleonto-
logical techniques.

My own thesis "theme," however, consisted of genetical and cytological
studies in native plants of the genus Stachys (Labiatae), with the objective
of elucidating the relationships between species, tribes, etc. It is a large and
rather poorly defined heterogeneous taxon, and I doubt that I made any
startling contributions to the theory of evolution, although I was able to
identify two groups where further work might have been interesting, mainly
on the role of species hybridization in evolution.

However, I learned from Dr. Schiemann the person at least as much as
from Dr. Schiemann the scientist. She was one of the first women to make
her way into the German academic establishment, and like most of them
did not have much of a career. When I was working as her student, she had
a small office-laboratory in one of the botanical institutions of Berlin University, the Botanical Museum, and was a Privatdozent, i.e. was giving
courses in her specialty which were not required of any students and thus
taken only by those interested in exactly the same specialty. A basic academic rule in German universities was Tres faciunt collegium, but it was
not clear whether the instructor counted as one of the three, and some of
Dr. Schiemann's classes were indeed attended by herself, me as her oldest
graduate student, also doubling as projectionist, and one or two more
students. She was not the easiest person to work with, somewhat rigid in
her outlooks, and with little sense of humor, I fear, but her dedication to

science was absolute and incorruptible. It was, of course, a period when all Germans, unless they were able and willing to emigrate, had in some manner to adjust to the Hitler regime. Many had no particular problems in this regard and became, whether by conviction or design, quite faithful followers if not active supporters. Not so Elisabeth Schiemann. I remember a discussion on the genetic bases of culture, as late as 1937 or 1938, in which she got up and stated that we should not forget the contributions of the Jews to German culture. This could not go on for very long. A year or two later, soon after I had finished my degree, her *venia legendi*—the authorization to teach—was withdrawn and she was left without a position. It was thanks to the efforts of F. von Wettstein, the leading plant geneticist in Germany, that she was awarded a stipend by the *Deutsche Forschungsgemeinschaft*, the approximate equivalent of the much more recent National Science Foundation of the USA, and was capable of continuing her research. Only after the end of the "Third Reich," and near the end of her own active career, was she given a professorship and relatively better support.

Another question I have frequently been asked is: What was it like to be a student in Hitler Germany? As far as I was concerned, the answer was very much "business as usual." Germany, or I should say those parts of Germany where I happened to be, or those groups of German society with which I was mainly associated, were very nationalistic throughout the entire period between World Wars I and II. Pomerania was among the first regions of Germany to give the Nazis large majorities in elections. Berlin as a whole was far more liberal, in fact the most liberal place in all Germany. However, all of my high school teachers, the great majority of my fellow students in high school (most but by no means all Jewish ones excepted), and especially those in Berlin University, and at least the politically more vocal ones of my professors were on the right of the political spectrum.

I realize quite well that nationalism exists in other countries, too, the USA being no exception, nor do I feel that attachment to and pride in one's country is something unnatural or reprehensible. On the contrary, I am nowadays getting very impatient with Americans—college teachers, students, writers, film makers, newsmen, etc—who seem to see only the negative features of American society and some of whom even claim that fascism and racism in America are not different from what existed in Hitler Germany. Having seen the latter ones in action, I can assure anyone who makes such claims that if these were indeed true, he would not be able to make them. American nationalism is permissive, the German was not.

Over the years I had—and still have—many very good individual German friends, but among my fellow students in Labes and Berlin, as a group, I remained an *Ausländer,* and never felt at home in Germany nor ever had the desire to become a German citizen, while I never, from the very start, had similar feelings in the United States. The narrow nationalistic attitude prevailed in virtually all German educational institutions, from grade school through university. There is no doubt that the entire German educational "community"—the educators and those who were being educated—bear a very heavy share of responsibility for the advent of National Socialism in Germany. Yet I have sometimes wondered what the world might be like if education in the schools of all nations, instead of starting with one's own nation and its accomplishments, would start with those biological, sociological, and cultural features that are common to all people, and only then would deal with the different ways by which different people and nations developed this common property.

In any case, Hitler's advent to power was neither a particular surprise for me, nor did it have major effects on my studies. If anything, I was in a better position to pursue these than my German fellow students who had to join the Party's student organization, attend "schooling courses" and political rallies, and in an increasing measure undergo paramilitary training. It is true that there were depressing and sickening developments: the appearance of yellow benches in the parks, designed for Jews; the "crystal night" in November 1938, in which all Jewish stores were smashed and all synagogues burned, "spontaneously" all at the same hour of the same night throughout the entire "Reich"; the disappearance of people one had known for years, with inquiries countered with a stern "None of your business; you are advised not to pursue the matter."

As to direct encounters with the Nazi authorities, however, I had only two, and while they caused much personal anxiety, they were nothing compared with the experiences of others, nor did they result in permanent problems. The first happened when I had completed my thesis and registered for final examination. My thesis and the registration fee were returned —the latter minus the postage—with a letter stating that I had not been admitted to the examination. No reasons were given but it was clear that they were my lack of citizenship and of involvement in political activities expected from students. This set me back for a year, and it was a difficult time, for financial if no other reasons. However, a number of friends rallied to the rescue, in particular a physician for whom my mother had worked on one occasion, and who was an old Party member. He wrote a glowing letter claiming that Mother had raised me as an ardent National Socialist. Mother almost died of shock, but the letter had its effect: I was admitted and obtained the degree. However, after the end of the Third Reich, when everybody in Germany was screened for Nazi affiliations, I was somewhat worried that that letter might surface and I would be called upon to prove that I hadn't been a Nazi after all.

Authorities can be hard to please. Many years later, when I was applying for permanent residency in the United States and had completed the necessary questionnaire, answering, truthfully and with some satisfaction, with

"No" all the questions about membership in the NSDAP, SA, SS, etc, back came a stern letter from the immigration service stating that all men of my age in Germany had had to belong to some Nazi Party organization. The fact that I claimed that I had not was highly suspect, and would I please explain. Again, some friends, this time American ones, came to the rescue, in particular the late Ralph E. Cleland and the late Harry A. Borthwick. Harry Borthwick in particular proved a tremendous asset for he had two veritable US Senators among his relatives, one a Democrat and the other a Republican. Thus, whatever party was in power in Washington, Harry couldn't go wrong. In this case, the offices of both Senators wrote to the immigration authorities, and my application was approved. Still, for awhile I felt a little like a character in a play by Kafka: guilty if one had belonged to The Organization, and guilty of one had not.

The second encounter with the Hitler regime—although it really never developed into anything serious—happened in 1943 or 1944, when Germany's situation had become very critical and the authorities, desperate for manpower, made aliens of German extraction subject to the draft. My then administrative superiors promptly requested exemption, but in case this should not be granted, I made preparation to go underground. This sounds far more heroic than it was at this time, particularly in a large city like Berlin. The administrative structure of the country had so far deteriorated that it was quite possible to "disappear" and with some luck and circumspection survive and escape capture by the authorities. I had several friends who for one reason or another were living underground and who were ready to help if necessary. But the deterioration of the administrative structure I just mentioned had evidently progressed so far that I received neither deferment nor the summons for registration and physical examination which preceded the actual draft.

One may ask why I did not make efforts to leave Germany. I admit that at the beginning of the Hitler period I saw no urgent reason to do so. My intent was to finish my education and then seek work in another country. And later it became increasingly difficult to emigrate, at least unless one had relatives or friends abroad, which I did not happen to have. When I had been barred from completing the doctorate, Dr. Schiemann wrote urgent letters to colleagues and friends in Sweden, The Netherlands, and Switzerland asking them to admit me to their schools and enable me to finish my studies. The responses were sympathetic but all had to state that the authorities of their countries or schools had passed laws or regulations preventing the admission of more students or scholars from Germany. The Swiss authorities—who even now have not made public the archives on the country's refugee policies during the Nazi period—not only forced many German Jews who had fled to Switzerland to return to Germany and a very uncertain fate, but even requested the German authorities in 1937 or 1938 to place a special mark in the passports of German Jewish citizens so that they could be identified and turned back right at the border (a valid passport was all that was needed to travel from Germany to Switzerland; no visa was required). The Germans complied, but in this case with some reluctance because they were not yet ready for quite so blatant a demonstration of their antisemitic attitudes.

I must also admit that although those manifestations of the Hitler regime were depressing and their number and variety increasing, they were not my main daily concern because all the time I had left after classes or thesis research was needed to earn some money. Not having German citizenship, I was not eligible for a scholarship or similar support (which was quite scarce anyway) nor for any regular job because for this a work permit was required. Thus, I worked in various odd jobs, and after entering the university started writing abstracts of papers for various journals. Abstract writing at that time was paid but the pay was low, so it had to be compensated by sheer mass, and in all humility, for some years I must have been one of the world's busiest abstract writers. I abstracted papers not only in botany and plant genetics, but also in zoology, animal genetics, breeding and husbandry, and in tropical agriculture—the latter for the *Tropenpflanzer,* a small journal trying to maintain some of the German colonial tradition. I remember abstracting a lengthy article on jelly-making from tropical fruits, which seem generally unsuitable for this purpose, but at that time I could have instructed all housewives in the tropics how to cope with this domestic problem.

Pay, apart from being low, was irregular; it usually came when a volume had been completed, and any delay was likely to cause a financial crisis. However, when pay did come, I splurged by having a full-course meal in one of the little restaurants in suburban Berlin which served good food at modest prices. In summer, my favorite one had some tables on a terrace, and this was separated from the sidewalk by a nicely trimmed privet hedge. Once, when I was consuming a pork chop, a dachshund came trotting along that sidewalk. The odor of my chop must have hit the beast's nostrils, for he stopped in his tracks, sat down on his haunches, and looked at me as only dachshunds can look. His expression was one of utter *Weltschmerz,* and it was clear that nothing but a share of that pork chop could restore his faith in life or mankind. So, having finished, I tossed the bone to him; he departed with it happily while I ate my dessert with the feeling of having done my good deed for the day. When I came to the restaurant next time, I saw the young waiter, who had been serving me on the previous occasion, nudging his elder colleague and saying, in what he thought was a whisper but which I could hear quite clearly, "Look, look, Robert, that's the fellow who ate the chop with the bone." To add to the humiliation, the German language has two words for "eat": people *essen* but animals *fressen,* and I heard that fellow distinctly say *"aufgefressen."*

Abstract writing, apart from helping to keep the wolf from the door, had other advantages. It undoubtedly helped in learning how to focus onto the "message" of a paper, and some of it, although perhaps not the jelly-making paper, helped in keeping up with literature which was useful for my own work. But the biggest bonus by far was that abstract writing turned out to be instrumental in helping me get a job after graduating.

At this time (1935–1936) work by several investigators, including G. Melchers, had led to the concept of flower hormones. I was fascinated by this discovery, and abstracted a number of the pertinent early publications, among these one by Melchers. Melchers, who was working at the Kaiser Wilhelm Institute for Biology in Berlin, read the abstract and must have felt that it did reasonable justice to his work, for some time later he invited me to participate in this work as his scientific assistant, and early in 1939 I joined his laboratory, thus starting on one of the most exciting and cherished periods of my career.

KWI/MPI FOR BIOLOGY, BERLIN/TÜBINGEN, 1939–1949

Melchers was—and still is, of course—a scientist with great and bold ideas, and always ready to fight for them fiercely. But at the same time he was open to ideas of others and had none of the idiosyncrasies which constrained relations with many German scientists, as I have already mentioned. We had plenty of arguments, but he never "pulled" his seniority, and in the end we usually worked out a strategy that satisfied us both. Unfortunately, we had only a short period of unhampered work, for the political situation in Europe was rapidly worsening, and in July 1939 Hitler attacked Poland, starting World War II.

One other question I have often been asked since coming to the US is, "What exactly did you do in Germany during the war?" In turn, it was one of my great surprises after coming to this country to discover to what an extent many American scientists had been involved in the war effort. In World War II Germany, for most of us who were not drafted and except for those who were working in problem areas of immediate military relevance, like solid state physics, it was once again largely "business as usual." We tried to carry on our research, and those in universities to continue teaching. It is true that a large and ever-increasing number were drafted, many never to return (only a few in key positions, like Melchers, were exempted), while supporting personnel, both men and women, were put into factories, and the "business" we were trying to carry on became increasingly unreal. In 1943 I happened to pass through Heidelberg, and having some time between trains decided to visit the Botanical Institute of the University, which was then located quite near the railroad station. The door was opened by an elderly gentleman with a broom and dustpan. When I asked whether the director, Professor Seybold, was in he replied, "Yes, I am Professor Seybold." He was practically the only member of the entire staff that was left, and having no *Reinemachefrauen* (cleaning women—normally a fixture in German institutions of that time) nor other help, had decided to do some housecleaning of his own. Still, the answer to above question is that throughout the war I was working with Melchers on the physiology of flower formation (and occasionally in Melchers' other area of interest at that time, plant virology).

Our warhorses were short-day and day-neutral varieties of tobacco and the long-day plants *Nicotiana silvestris* and *Hyoscyamus niger*. The latter —Shakespeare's cursed hebona by means of which Hamlet's uncle had murdered Hamlet's father—was Melchers' pet material, adopted from Correns, who had used it to demonstrate the inheritance of a physiological character, cold requirement in flower formation, the first case ever of this kind (5). By grafts between these plants we showed that the hypothetical flower hormone was neither species- nor type-specific, a strong indication that it was identical in long-day, short-day, and day-neutral plants (14). We also undertook an extensive kinetic analysis of the photoperiodic response of *Hyoscyamus,* showing that this response was primarily determined by the length of the dark period, and that darkness had an inhibitory action in flower formation in long-day plants (10). Finally, we made a valiant effort at isolating that flower hormone and were among the first in a long series of distinguished scholars who failed in this endeavor; only very recently some results that look promising have been obtained (3). Not that we conceded defeat meekly. Having soon discovered that plants do not like to be treated with their own extracts, responding by dying and rotting—and rotting *Hyoscyamus,* to continue with Shakespeare, is among the vilest smells that ever offended nostril—we tried various ways of obtaining less noxious preparations. We tried to drain leaves and shoots by placing them in containers with water, but found that plants do not like to be drained either; only many years later were King & Zeevaart (7) able to overcome the problem, at least partially, by treating the cut ends with chelating agents. We even sucked water through whole plants under vacuum, and that did yield some relatively clear liquid with, however, no flower-inducing capacity whatever.

In the short time I had there before the war and even during the first years of the war, the "KWI" for Biology was one of the most stimulating places to do biological research, worldwide. The KWIs as a whole represent a unique attempt at organizing and administering research. The institutes date back to 1910. At that time German industry, trade, and commerce were flourishing, and this success was to a very large extent based on advances and discoveries made by German science. In a remarkable "turn-about," one that has happened neither before nor since, the German industrial and commercial enterprises were persuaded to contribute large amounts of money not only to build a number of research institutions but also to underwrite their continued operation. The emperor, Wilhelm II, lent his prestige to this undertaking and became the patron of the institutes. Nor were institutes only created for disciplines such as chemistry, physics, metallurgy, or plant breeding research whose relevance for industry or agriculture was obvious (the breeding institute was founded later, in the 1920s); there were institutes for sciences where such a relevance was less apparent—the Biology Institute being one—and even for law.

The institutes were highly elitist and no apologies were made about this fact. The directors of the institutes and of the major subunits (divisions) were selected for eminence in their fields and given free hand in their research. If one of them died, retired, or moved to another place, his division would often be phased out and he would be replaced by someone of equal distinction but working in quite another problem area of the same broad discipline. The institutes were engaged only in research, but many of the senior members had adjunct appointments in universities and technical colleges, and many of the junior ones later transferred to university positions so that science in the entire country benefited from the KWIs. After German industry and commerce had been severely set back by World War I, inflation, and depression, support of the institutes passed largely to the German federal government, but their setup and *modus operandi* did not change. And when Hitler assumed power the KWIs lost some of the most distinguished senior members—the physical chemist F. Haber and the biochemist O. Meyerhof, both Nobel Prize winners, to name but two—as well as many junior ones, but they still remained formally independent, being administered by the Kaiser Wilhelm Society for the Advancement of Sciences and one important step removed from direct government control. It was only because of this unusual setup that I could be employed in one of the institutes and work without becoming a German citizen or being involved in Party activities which were mandatory for all younger staff members in universities and other institutions directly dependent on the government.

When I entered the KWI for Biology it had three divisions, headed by M. Hartmann, A. Kühn, and F. von Wettstein. These names mean probably little to American biologists of our day, but they were outstanding scientists who made profound impacts in (to use current terminology) cellular, developmental, and genetical biology. They all also had younger, brilliant associates who became excellent scientists in their own right: H. Bauer, J. Hämmerling, K. Pätau (who, like myself, later settled in the USA, in Madison, Wisconsin), J. Straub, H. Stubbe, and of course Melchers. In addition, there were many visitors who either came to present one of the regular Dahlem Biology Colloquia—so named for the Berlin suburb where the KWI for Biology and several other KWIs were located—or to spend periods of work in the institute. Even after the war had started some of this tradition continued for some time. I can recall seminars by T. Caspersson from Sweden, R. Collander and H. Federley from Finland, and A. Buzzati-Traverso from Italy, while B. Györffy from Hungary and F. Resende from Portugal came to work in the institute, the latter even after life in Berlin had become not very comfortable and not entirely safe. On another level, the KWI for Biology and some of the other KWIs in Berlin—there were at least half a dozen in sciences alone—were able to help some young scientists from the occupied countries (France, The Netherlands, even Russia), who were either POWs or had been drafted as forced labor, by getting them assigned to the institutes and enabling them to work in their profession rather than in munitions factories.

As the war went on, however, and especially after Hitler started his Russian campaign, these entire activities gradually withered, and our research also became increasingly inefficient and unproductive. Equipment deteriorated and could be neither replaced nor repaired for lack of parts, and even the most common supplies became hard or impossible to obtain. At the same time, the air raids on Berlin intensified, work became increasingly difficult physically, and one never was sure whether one could continue at all next day. Thus, sometime in 1943, the decision was made to transfer the KWI of Biology, and several of the other KWIs in Berlin, to other locations. I have no conclusive information but assume the entire move was part of the effort of Albert Speer, Hitler's Munitions Minister, to salvage as much as possible of the German scientific establishment and to prevent it from falling into Soviet hands. The Biology Institute was relocated to the southern part of the State of Württemberg. Melchers' unit was assigned quarters in the botany department of the University of Tübingen, while other parts of the Institute were housed in cities and villages not far away in factories that had been closed because of lack of raw materials and of labor. The excellent library found a home in the old city hall of Hechingen. It was in these places that we experienced the end of the war and, after Germany's surrender in May 1945, found ourselves in the French occupation zone.

From the viewpoint of sheer physical survival, this was not the happiest location since the food situation was distincly worse here than in the American and the British occupation zones, but from the professional viewpoint it was an advantage. The French occupational authorities placed great priority on cultural affairs and kept scientific institutions operating or tried to restore them to operation without delay. The various factions of the Biology Institute, wherever they were located, remained intact and their work, modest as it may have become, was not disrupted as was that of some institutions in the other occupation zones. The University of Tübingen also reopened quite soon after the end of the war. And we were visited by French scientists. They came under the auspices of the French Military Government, and some acted somewhat in accord with this status, issuing rather peremptory summons to their German colleagues. One big difference was Professor P. Chouard, who behaved as a fellow scientist, giving our morale a substantial boost. He even gave me a ride home in his official car. I have had a soft spot for him ever since, and we became good friends even though I never learned to follow his rapid-fire French.

Other contacts with the former enemy countries also began to emerge; some books and journals began to arrive; many colleagues were very generous in sending reprints of their work for the whole war period, and quite a number also sent CARE food parcels, which certainly helped us survive the first lean post-war years. Altogether, we felt as if we had been released, if not from prison at least from a strict house arrest during which we knew very little of what was happening even in the house next door. At first, the traffic was somewhat one way as we were not yet allowed to travel freely, but in summer 1948 Melchers and I were given permission to attend the International Genetics Congress in Stockholm—the barriers were indeed disappearing. Making use of the renewed access to literature, Melchers and I decided to write a review of the physiology of flowering (15), partly to catch up with developments in this field and partly to present our own ideas.

We also were able, both before and after the end of the war, to continue some research. One project involved kinetic studies of vernalization in biennial *Hyoscyamus;* they showed that the process proceeded in two stages, the first reversible and the second not (11). And having been defeated in our earlier efforts at extracting florigen, we decided on another approach, one that would not only coax this elusive material out of the plant, at least for a fleeting moment, but also reveal something about its nature by giving us an estimate of the molecular size. We made grafts between flowering donor and nonflowering receptor plants but inserted membrane filters— filters with different pore sizes manufactured by a German company— between the graft partners. The idea was much too good to work. We did get flowering in some of the nonflowering receptor plants, but with every single filter, regardless of pore size; and a closer inspection showed that the tissue had disrupted the filter and a nice vascular connection had been established. The score was still 1:0 for florigen (16).

Otherwise, the years at Tübingen were a holding and later a planning period. During the first years after the end of the war the German states (*Länder*) which had been established by the Allies functioned very much as separate units; any central power rested with the Allied Military Government, and the Western Allies on the one hand, the Soviet Union on the other, were already pursuing quite different policies in their respective occupation zones. Thus, scientific institutions were dependent on the *Land* in which they happened to be. The KWI of Biology and some of the other KWIs that had been relocated from Berlin were now wards of the *Land* of Süd-Württemberg, neither one of the biggest nor the wealthiest of the *Länder*. Nevertheless its government decided to offer these institutes permanent homes, and gradually plans were made to find or construct the necessary facilities.

Incidentally, the KWIs were no longer KWIs but MPIs, the Allied authorities had objected to the Kaiser's name as too suggestive of Germany's military past, and the institutes had been renamed in honor of Max Planck. Planck was a great scientist and a great man; however, the KWIs were perhaps the single greatest peaceful accomplishment of the last Kaiser, and his name deserved to remain with the institutes—from some of which one now can see the castle of the Hohenzollerns in the distance.

I participated with Melchers in planning the building for his Division, but did not see the results until half a dozen years later, on my first visit back to Germany. For soon after contacts with abroad had become possible, I started making inquiries about possibilities for continuing work elsewhere. In retrospect, I find it not altogether easy to explain the reasons. I may have underrated the drive and energy of the German people: in 1947–1948 it was quite difficult to foresee that conditions in the country, including opportunities for science, would become normal and in fact prosperous as soon as they actually did, at least in West Germany. The Cold War had already become quite hot, and there was concern that all of Central Europe would come under communist domination. However, the main reason was an admittedly vague desire to be in a part of the world where there were not the many things which reminded one of German nationalism and National Socialism.

Contacts with the United States did not result in anything definite, but K. C. Hamner, one of the American scientists working on flowering, with whom I had been able to establish contacts, drew my attention to the Lady Davis Foundation of Canada. I applied and in 1948 was fortunate enough to obtain a fellowship to McGill University. There followed a period of red tape. The Allied authorities tried to screen carefully anyone wishing to leave Germany, to prevent the escape of war criminals and others on their "wanted" lists. But with the German adminstrative structure still thoroughly disrupted and many records destroyed or not yet accessible, reliable information was very difficult to obtain. Sometime after applying for a travel document that would permit me and my wife Lydia to move to Canada I was visited at our home by a police official of the county of Tübingen—we would say, from the county sheriff's office—and a French officer of the local military government, and subjected to searching questions about name, parents, other relations, education, jobs, any political affiliations and activities. Some more time later I was summoned to the Central French Military Government in Baden-Baden and there interrogated by a higher officer. From his questions it was quite apparent that all the information he had on me came from my own answers to those two local officials. In February 1949, a few weeks before I would have been with Dr. Melchers for 10 years, came a summons to report at the UNRRA office in Rastatt, and after some days in a *Sammellager* (collecting center) we traveled by train and boat to London and then by air to Montreal.

CANADIAN INTERLUDE

My career in the "New World" can be recounted briefly since it was similar to that of a native of the United States or Canada except that it started 10 to 12 years later than usual. The stations were: about a year at the genetics department of McGill; eight months at Texas A&M College (now Texas A&M University); two years (1950–1952) at the California Institute of Technology as research fellow; seven years on the faculty of the botany department, University of California at Los Angeles; 1959–1965 back in Caltech as professor. of biology and head of the Earhart Plant Research Laboratory as successor to F. W. Went; since then at the Plant Research Laboratory at Michigan State University.

That my first professional home in the New World was a genetics department did not mean a return to my graduate student days as a geneticist. No opportunities in my area of research happened to be available in Canada at that time, and in turn there were no geneticists among the crop of Lady Davis Fellows. My assignment was a compromise, but one I did not have to regret at all since, as things turned out, it permitted a gradual transition to life and work in the New World. One of the most important and difficult experiences during this transition was learning to understand the English as it is spoken in the New World but not taught in German high school. It was not difficult to make oneself understood, but it did take some time to realize that when somebody told one of having been in Yerrp and having had a marvelous time nittly he meant that he had been in Europe and had had a marvelous time in Italy. And when I first heard of what sounded like "cicology" or maybe "sicology" I seriously thought this was altogether new science, invented while we had been cut off from most of world science during the war, before realizing that it was good old psychology which I of course had pronounced *P'sükhologhee,* the p definitely not being silent. However, as the Lady Davis Fellowship year was drawing to its close, there were still no definite prospects for a more permanent job in Canada. Canadian universities were at the time quite strongly teaching-oriented while, after those experiences with Yerrp and sicology, I did not feel quite ready to teach and was also anxious to continue, or rather resume, some research. Luckily, there arose at this time not one but two opportunities for periods of work and study in the USA. Texas A&M had received a grant enabling it to invite three visiting scholars to participate in its cotton research program. At the Genetics Congress in Stockholm I had met Meta S. Brown, who was heading the cytogenetics part of this program, and it was thanks to her recommendation that I was offered one of these positions, for research on regulation of flowering in cotton and a special topics course in the physiology of flowering and related problems. At the same time, James Bonner offered me a research fellowship in his group in Caltech, funded by a fellowship from the Lalor Foundation. I accepted both offers with alacrity, and traveled by Greyhound bus at the end of 1949 from Montreal to College Station and in the fall of 1950 on to Pasadena. As things turned out, I did not return to Canada except as a visitor, but I remember my time at McGill with great fondness, especially the chairman of the genetics department, J. W. (Wally) Boyes, and his wife Bea, who were marvelous in helping Lydia and me adjust to the "New World."

TEXAS A&M, CALTECH I, UCLA, CALTECH II

Beginning my career in the USA on a southern campus was a fascinating experience. Meta Brown and many others were a great help in furthering the adjustment process, and my morale was undoubtedly enormously boosted by the title I had—the most impressive one of my life: Distinguished Visiting Professor! Some of you, dear readers, may be visiting profs, some may even hold a distinguished professorship, but how many can claim being or having been a dis-vis-prof? This exalted status couldn't last; when moving to Caltech I dropped precipitously to a humble postdoctoral research fellowship, and it took 7 years to climb back to a professorship, and that one was neither visiting nor distinguished.

All the institutions in which I have worked since coming to the US have added to my continuing education, but the one that added most was Caltech. While different in many other ways, intellectually it compared with the Kaiser Wilhelm Institutes at their best: a superb and at the same time remarkably open-minded faculty, with numerous outstanding postdoctoral and graduate students, many of whom have developed into leaders in their respective disciplines. No one who has "passed" through Caltech has left it quite the same person, and probably retains a trace of regret at having left. The sole drawback from my personal viewpoint was that, particularly during my second period at Caltech, research with plants, once one of the strongest programs in the biology division, had been greatly reduced because the faculty members who had worked with plants had either left or retired or had turned to other problems and other materials. I was the only faculty member still interested in plants as distinct organisms. I had generous support and could get advice on many technical problems, but had nobody outside my own group with whom to discuss specifically plant-related problems in depth. Now plant research has completely disappeared at Caltech—in my opinion a very regrettable development in an institution that has otherwise been distinguished by a remarkable insight, even foresight, into the needs and promises of all areas of experimental biology.

The lack of fellow plant scientists was one of the reasons that caused me to leave Caltech—even though with an aching heart—and move to Michigan State. Another reason was a growing uncertainty in my mind whether the phytotron was a genuine asset to my research. When Went built this facility—I had read about it while still in Tübingen—it had seemed a unique and very useful installation, but as I worked in it myself, various limitations became increasingly apparent. The principal one was that the facility—in essence, a large, highly sophisticated item of equipment, fundamentally not different from an ultracentrifuge or a computer—tended to be the prime factor determining the type of work to be carried out. In addition, being designed with the idea of providing a large variety of environmental conditions for a large number of investigators, the facility proved insufficiently flexible; on quite a number of occasions, when special needs arose in my own work or that of an associate, it became necessary to improvise or go outside the phytotron. I do feel that even by the strictest criteria the Caltech phytotron has served a useful purpose (the facility was closed down for research a few years after my departure and has meanwhile been torn down; before that it served for awhile as a studio for budding artists, with nude models who, hopefully, enjoyed the conditioned climate—an unusual transition from science to art). However, if more phytotrons should be built, they should contain ample "uncommitted" space to provide for new needs as these may arise. At the Plant Research Laboratory at MSU neither I nor, I believe, the other members of the research staff have seriously missed the availability of a full-fledged phytotron.

To return to my own research, the time at McGill and Texas, the first period at Caltech, and the first years at UCLA were a phase of search and readjustment. My prime interest remained in the physiology of flowering, particularly the hormonal regulation of flower formation and its dependence on specific environmental factors. Some work along the "old" lines, i.e. those pursued with Melchers, continued: at McGill and Caltech studies on the kinetics of vernalization; at Texas A&M, in line with my obligations, work on the photoperiodic responses in cultivated and wild cotton which, I'm afraid, I never managed to write up for publication but which showed the validity of the florigen concept in another genus and family. However, it became clear that these types of experimenation were increasingly exhibiting the operation of the law of diminishing returns and that new, perhaps less direct approaches were needed. Together with J. L. Liverman, we tried to use auxin as a tool in studying flower induction. We were able to obtain some flowering response by auxin treatment in some long-day plants (12), but somewhat later I found, with considerable dismay, that substances like some chlorinated phenoxyisobutyric acids which in other assays behaved as auxin antagonists elicited the same response—a result I did not publish as I found, and still find, it difficult to rationalize. However, the effect was in either case quite small and apparent only under marginal photoperiodic conditions; it provided no indication that auxin was a major factor in flower regulation.

At about this time the gibberellins had been recognized as another, powerful type of plant growth regulators, and the first cytokinin, kinetin, had made its appearance. But while the effect of gibberellin on shoot growth had been investigated in numerous plants, they had all been caulescent, i.e. possessed an elongated (although in some cases dwarfed) stem. No one seemed to have had the simple-minded idea of testing for this effect in "stem-less," rosette-type plants. I did and found, this time with boundless delight, that treatment with gibberellin elicited first stem elongation, by greatly increasing cell-division activitiy in the subapical region of the terminal shoot meristem, and subsequently flower formation in rosette plants maintained under strictly vegetative conditions: nonvernalized biennial (cold-requiring) *Hyoscyamus* and carrot, and *Samolus parviflorus* and other long-day plants kept on short days (8, 18). As to kinetin, some information in literature and preliminary experiments in a plant physiology class supplied indications that this regulator had properties of Chibnall's "root factor" which prevented the aging of leaves, and in follow-up experiments conducted with A. E. Richmond (17), we showed that kinetin treatment prevented loss of protein and chlorophyll in detached *Xanthium* leaves. These various results provided so many leads for further work that it became necessary to decide which ones to follow, and as the gibberellin effects were closer to the problems I had studied before, I concentrated on these.

Results similar to those with biennial *Hyoscyamus,* carrot, *Samolus,* etc were soon reported by other investigators in numerous other plants in which flower formation is dependent on a cold treatment or on long days, making the gibberellins the first chemical compounds known to induce flower formation in two large physiological groups of plants. This finding immediately led to two questions: 1. Were the gibberellins a factor in the endogenous, physiological regulation of flower formation? 2. What was their relationship to the other flower-inducing factor demonstrated earlier by means of grafting experiments, i.e. to florigen?

The effect of gibberellins on flower formation was demonstrated by applying the compound to the plants, i.e. by pharmacological experiments. These were soon supplemented by studies indicating that the content of gibberellin-like substances in a plant may increase when the plant proceeds to flower initiation or at least stem elongation, i.e. by correlational evidence. A very large part of our knowledge about the function of plant hormones rests on pharmacological and correlational data. However, such data, while they may be suggestive, are not conclusive evidence that the material in question—in our case, gibberellin—plays a physiological role in the process in question—in our case, flower formation. Pharmacological data may show us the scope of the potential functions of a hormone, and because of the great multitude of effects of the major plant hormones they have been very important in work with these hormones, but because of this very multitude of effects, i.e. the lack of a marked functional specificity, the data must be interpreted with great caution. And correlational data are notoriously difficult to interpret in terms of cause and effect unless one has an independent, experimental tool permitting one to interfere with the one component of the correlation, here the hormone, and register the result of this interference on the other, the response.

More generally, we should whenever possible adopt the principles which Robert Koch proposed 100 years ago for the identification of the etiological agent of a disease: the syndrome should disappear when the agent is removed; it should reappear when the agent is reintroduced; and the second-named effect should be specific, i.e. causable only by this one agent. Auxin and cytokinin were discovered, and much of the early work with these hormones, particularly auxin, was conducted on the basis of these principles, even if this may not always have been clearly recognized. With auxin it was possible to remove the source of the hormone in one of the major assay materials, the grass coleoptile, by simple surgery. With cytokinins, hormone-free tissue—the tobacco stem pith—was obtained by cutting it free from all surrounding tissue which might produce or transport the hormone. But when the effects of applied gibberellin on flower formation had been discovered, no analogous procedures were available for this class of hormones (later it was shown that the decapitation technique can also be used in research with gibberellins). As to the relationship of gibberellin and florigen, the latter seems to be the same in long- and short-day plants, but while gibberellin treatment induces flowering in many long-day plants maintained on noninductive photoperiods, it does not, except in a few instances, have the analogous effect in short-day plants. Thus, gibberellin and florigen cannot be simply equated.

The question of whether the gibberellins were a physiological element of flower regulation could be answered by making use of another discovery in plant growth regulation made at this time, namely, the discovery of the first growth retardants. As is now well known, these are compounds that affect plant growth in a manner seemingly opposite to that of the gibberellins, and as Lockhart (13) showed by kinetic studies, they are indeed specific antagonists of gibberellins. The cellular basis of their action—inhibition of gibberellin biosynthesis, enhancement of gibberellin degradation, competition at the site of action in the cell—remained unknown. In joint work with several associates and colleagues (e.g. 2, 6), we could show that several of the retardants inhibited gibberellin synthesis both in *Gibberella fujikuroi* and in higher plants. These retardants can thus serve as "chemical knives" for obtaining gibberellin-deficient plants, and have been useful in determining whether gibberellins participate in the endogenous regulation of growth or development in various plant organs and tissues. With respect to flower formation, experiments conducted with B. Baldev and J. A. D. Zeevaart showed that treatment with the retardant 2-chloroethyltrimethyl ammonium chloride (CCC, Cycocel) inhibited flower formation in *Samolus* and *Bryophyllum daigremontianum* without affecting growth, and that this inhibition was completely overcome by application of gibberellic acid, thus satisfying the first two principles of Koch (1, 20); later, unpublished work proved that this effect of gibberellin could not be duplicated with auxins, a cytokinin, vitamins, sugars, or organic acids, i.e. was evidently specific for gibberellins.

Insights into the relationship of gibberellin and florigen were obtained in other experiments with *Bryophyllum,* also conducted jointly with Zeevaart (19). *Bryophyllum* is a long-short-day plant, and the long-day part of its induction can be substituted by gibberellin application but the short-day part cannot. However, when a *Bryophyllum* plant was treated with gibberellin under short-day conditions, it became capable of inducing flower formation in a grafting partner that had been and continued to be maintained in long days throughout. Thus, the gibberellin treatment had caused, in the former plant, the appearance of a graft-transmissible material which was capable of inducing flower formation in *Bryophyllum* under a photoperiodic regime under which direct gibberellin application did not cause this response. We assume that this second hormone-like material is florigen, and that gibberellin thus functions in some manner as a physiological precursor of florigen, at least in *Bryophyllum*. The chemical relationship between the two materials obviously remains unknown as long as the chemical nature of florigen is not known.

THE AEC→ERDA→DOE/MSU PRL

In 1965 I moved once more, to the newly founded Plant Research Laboratory at Michigan State University, first as director, then since 1978, having reached 65, the retirement age for administrators at MSU, as professor.

The PRL was built and started to operate under a contract between MSU and the US Atomic Energy Commission. A prime mover in its establishment was Jim Liverman, with whom I had worked when he was a graduate student at Caltech. He had later joined the AEC and now became my boss —somewhat remote in Washington, D.C. but sympathetic and helpful in various "crises." AEC was after some years absorbed by the US Energy Research and Development Administration, and the latter in turn by the new US Department of Energy. Each move entailed a change in the laboratory's initials. We may well hold the world's record for a scientific institution in this regard, but its purpose has not changed; the PRL is functioning as a center for broadly based, experimental research and related training in plant science. Increasing emphasis has been placed on studying the role of plants in the overall energy budget of the world.

To develop a research unit, not too large but not too small, in one's own area of interest and "from scratch" was a challenge which would have been very difficult to resist, and I am not sorry for having accepted it. But I am also not sorry for having relinquished the directorship, the reason being that throughout my tenure as director the "red tape" increased steadily and became stifling. This was the result of a continuous stream of new rules and directives, and as the PRL has to abide both by university and federal regulations, our share has been larger than the average. Most of these regulations were intended to protect rights of certain groups: the faculty, the students, monthly and hourly employees, ethnic and other minorities. Each of them meant completing more forms, keeping more records, and writing more reports, leaving less and less time for productive work, and since the unit administrator is responsible for compliance with the regulations, he cannot delegate these chores entirely to an administrative assistant.

No less and in the longer range, even more disturbing are two other consequences of this increasing protectionism. First, it encourages an increasing, unhealthy polarization in academic institutions. Second, it runs counter to other rights of a different and more general order: the right of students to obtain the best possible education; the right of society—which directly or indirectly is paying for all of this—to obtain the best possible research. Education and science, in the widest sense, contain an important elitist element. For progress it is not enough to provide "average" education for all or to train "average" scientists. The student who can absorb it deserves the *best* education since only this will enable him to realize his full potential; the scientist who is capable of keeping ahead of his ever-changing discipline will be discouraged by an atmosphere of egalitarianism.

Another reason which reduced my personal research activities in the PRL, and in fact caused their complete suspension for a period of time, was the study of the effects of the use of herbicides in the Vietnam war, a study which I headed, as chairman of a committee of the National Academy of Sciences, from 1971 to 1974. To do justice to this undertaking would, however, require another prefatory chapter, and I must reluctantly pass over it here.

I was able to return to "the bench," or more accurately to the greenhouse, after the Vietnam study had been completed and during a terminal sabbatical leave which I spent under the scientific exchange program of the National and the Soviet Academies of Sciences in Moscow, August 1975 to February 1976, at the K. A. Timiryazev Institute of Plant Physiology, working mainly in the Laboratory (Division) of Growth and Development headed by M. Kh. Chailakhyan. Chailakhyan is the "father" of the florigen concept. But since florigen had proven so recalcitrant to isolation, we decided to make a turn of 180 degrees and to search for flower inhibitors. The literature on the physiology of flowering contains a number of reports on inhibitory phenomena in flower formation, including some that have been interpreted as evidence for transmissible, hormone-like materials. However, the number of this kind of report is small compared to those on which the florigen concept is based; the effects, while suggestive, are still open to other interpretations and are often quantitative (delay rather than suppression of flowering). We decided to study the problem with the technique which has proved so useful for establishing the existence of flower-promoting materials, i.e. by grafting, and in plants where there was already good evidence for the presence of florigen so that the relationship of promoters and inhibitors of flowering could be assessed in the same plants. By grafting the long-day plants *Nicotiana silvestris* and *Hyoscyamus niger* on a day-neutral tobacco or on the short-day tobacco Maryland Mammoth and keeping the grafts on short days, we were able to suppress flower formation in the latter plants, and found in addition that their growth habit was modified in the direction of rosette growth (shortened internodes, thickened axis) which is quite characteristic of long-day plants maintained on noninductive photoperiods (9; and unpublished data). The entire syndrome was found to persist for periods of at least one year. Thus, the inability at least of some long-day plants to flower under short days seem to be not only the result of the absence of flower promoting material(s)—florigen—but on the presence of potent flower-inhibiting and growth-regulating material(s)— "antiflorigen(s)." Like florigen, antiflorigen seems to be neither species- nor type-specific since antiflorigen from two long-day plants belonging to differ- ent genera was effective in a day-neutral and a short-day plant belonging to another species. Whether short-day plants, in particular Maryland Mam- moth tobacco, also produce antiflorigen requires further study. In our experiments with day-neutral tobacco (9), we did not find much evidence that this was the case, but in grafts between Maryland Mammoth and *Nicotiana silvestris,* Chailakhyan and coworkers (4) observed marked inhi- bition of flower formation in the long-day partner when the grafts were kept on long days.

A FEW THOUGHTS IN CONCLUSION

I would like to close these rather unsystematic recollections with some thoughts on the factors which had the greatest influence on my research career. I shall not carry this very far for the temptation of rationalization and hindsight is severe. Still, as I see it, some factors can be singled out.

The first and foremost were fellow scientists. I have mentioned Noack and Elisabeth Schiemann among my university teachers. At the KWI/MPI for Biology, there were Melchers and the three Division heads, Hartmann, Kühn and von Wettstein—very different personalities and for this reason alone, fascinating to watch. At UCLA there were K. C. Hamner and S. G. Wildman; at Caltech, Bonner, Went, G. W. Beadle, M. Delbrück, and many others; and still others whom I met during visits at their institutions or at conferences and symposia. Not that I talked "shop" with all of them. As a matter of fact, Melchers was the only one with whom I spent a lengthy period of time in genuine team work, and to get into too close an association with all of them could even have been hazardous. Most were very strong personalities, with very definite ideas about research, while I have always felt that one learns far more from one's own mistakes than by proving that somebody else's ideas are right. It was mainly by watching these various people "in action," in formal and informal meetings, discussions, etc that I had some of the most valuable insight into science and research. And all of those people were very kind and helpful when help of some sort was needed.

Most of my research has been done with simple, well-established tech- niques—varying photoperiod and temperature, grafting, and the like. Somebody, who shall be nameless here, has claimed that if one uses a simple apparatus to answer a question, the answer will be a complicated one, and vice versa. I think nothing can be further from the truth. A technique is useful not because it is simple or complicated, nor because it is old or new, but because it is applicable to the given problem—when the phenomenology of the latter has been worked out to such a degree that the technique can be exploited to yield answers and not merely data. As already mentioned, I felt increasingly uncomfortable with the phytotron at Caltech because, while being an instrument, it tended to determine the problems to be studied. There's also another very important aspect to this story: a technique is useful only *as long as* it is applicable to the given problem. This has been probably the most difficult lesson to learn, but I hope I have learned some of it: when to stop working on a problem because the available opportunities and approaches are beginning to be exhausted or, as I have said earlier, exhibit the operation of the law of diminishing returns.

And this is where one more factor comes in which I would like to mention and which has helped me a great deal through my research career—serendipity. Serendipity is usually regarded as the propensity of having nice things happen to oneself, i.e. essentially as good luck. Good luck is a very fine thing, and I have had my share of that too. But serendipity is somewhat more; it is the ability to see properties of things that have not been seen by others. The "thing" itself can be something old or new, and can be close to one's own experience or quite remote from it; it is important to recognize some feature that has a definite bearing on the problem or problems one is studying. On many occasions throughout my research work, when that law of diminishing returns was clearly operating, it was this kind of insight, of recognizing the significance of perhaps a very minor observation, that permitted resumption of efficient and satisfying work, although perhaps in a somewhat different direction than originally intended. I wish I could say in conclusion that these insights always came promptly, in sudden, blinding flashes, but that would be stretching truth quite far. In hindsight—if I may be permitted one piece of hindsight—Chailakhyan, Melchers, and I should have done the antiflorigen grafts at the same time as we did the "proflorigen" grafts. Looking back at those ancient experiments and examining some of the old photographs, I see that those inhibitory effects were staring right into our faces from some of our very own controls. However, being fascinated by having obtained flowering in nonflowering plants, we paid no attention to the obvious possibility that nonflowering may be the consequence of an inhibition. The "flash" did finally come, but with a lag of 40 years, and this may be as good an insight as any with which to conclude these reminiscences.

Literature Cited

1. Baldev, B., Lang, A. 1965. Control of flower formation by growth retardants and gibberellin in *Samolus parviflorus,* a long-day plant. *Am. J. Bot.* 52:408-17
2. Baldev, B., Lang, A., Agatep, A. O. 1965. Gibberellin production in pea seeds developing in excised pods—effect of the growth retardant AMO-1618. *Science* 147:155-57
3. Chailakhyan, M. Kh., Grigoryeva, N. Ya., Lozhnikova, V. N. 1977. Effect of extracts from leaves of flowering tobacco plants on flowering in seedlings and plantlets of *Chenopodium rubrum. Dokl. Akad. Nauk SSSR* 236:773-76 (In Russian)
4. Chailakhyan, M. Kh., Yanina, L. I., Lotova, G. N. 1979. Inhibition of flowering in plants of the long-day type by grafting them on plants of the short-day type under conditions of long day. *Dokl. Akad. Nauk SSSR.* In press. (In Russian)
5. Correns, C. E. 1904. Ein typisch spaltender Bastard zwischen einer einjährigen und einer zweijährigen Sippe von *Hyoscyamus niger. Ber. Dtsch. Bot. Ges.* 22:517-24
6. Kende, H., Ninnemann, H., Lang, A. 1963. Inhibition of gibberellic acid biosynthesis in *Fusarium moniliforme* by AMO-1618 and CCC. *Naturwissenschaften* 50:599-600
7. King, R. W., Zeevart, J. A. D. 1974. Enhancement of phloem exudation from cut petioles by chelating agents. *Plant Physiol.* 53:96-103
8. Lang, A. 1957. The effect of gibberellin upon flower formation. *Proc. Natl. Acad. Sci. USA* 43:709-17
9. Lang, A., Chailakhyan, M. Kh., Frolova, I. A. 1977. Promotion and inhibition of flower formation in a dayneutral plant in grafts with a short-

day and a long-day plant. *Proc. Natl. Acad. Sci. USA* 74:2412-16
10. Lang, A., Melchers, G. 1943. Die photoperiodische Reaktion von *Hyoscyamus niger. Planta* 33:653-702
11. Lang, A., Melchers, G. 1947. Vernalisation und Devernalisation bei einer zweijährigen Pflanze. *Z. Naturforsch. Teil B* 2:444-49
12. Liverman, J. L., Lang, A. 1956. Induction of flowering in long-day plants by applied indoleacetic acid. *Plant Physiol.* 31:147-50
13. Lockhart, J. A. 1962. Kinetic studies of certain antigibberellins. *Plant Physiol.* 37:759-64
14. Melchers, G., Lang, A. 1941. Weitere Untersuchungen zur Frage der Blühhormone. *Biol. Zentralbl.* 61:16-39
15. Melchers, G., Lang, A. 1948. Die Physiologie der Blütenbildung. *Biol. Zentralbl.* 67:105-74
16. Melchers, G., Lang, A. 1948. Versuche zur Auslösung der Blütenbildung an zweijährigen *Hyoscyamus-niger*-Pflanzen durch Verbindung mit einjährigen ohne Geweberverwachsung. *Z. Naturforsch. Teil B* 3:105-7
17. Richmond, A. E., Lang, A. 1957. Effect of kinetin on protein content and survival of detached *Xanthium* leaves. *Science* 125:650-51
18. Sachs, R. M., Lang, A. 1957. Effect of gibberellin on cell division in *Hyoscyamus. Science* 125:1144-45
19. Zeevart, J. A. D., Lang, A. 1962. The relationship between gibberellin and floral stimulus in *Bryophyllum daigremontianum. Planta* 58:531-42
20. Zeevart, J. A. D., Lang, A. 1963. Suppression of flower induction in *Bryophyllum daigremontianum* by a growth retardant. *Planta* 59:509-17

Birgit Vennesland

Ann. Rev. Plant Physiol. 1981. 32:1-20
Copyright © 1981 by Annual Reviews Inc. All rights reserved

RECOLLECTIONS AND SMALL CONFESSIONS

Birgit Vennesland

Forschungsstelle Vennesland der Max-Planck-Gesellschaft,
1 Berlin 33, Germany

CONTENTS

The Preacher sought to find out acceptable words - even words of truth

Ecclesiastes 12, 10

On seeking inspiration from the prefatory chapters of my distinguished predecessors, I began to wonder what I had done to deserve inclusion in this series, beyond surviving until retirement. I represent a scientific generation that lived through a developmental explosion when scientific research came of age. Maybe I am typical of those biochemists with plant physiological interests who trained in the 1930s, when refrigerated centrifuges and spectrophotometers and isotopic tracers were only available in homemade versions, fractions were hand-collected, and jobs were rare. My generation has seen greater scientific change in our lifetime than any generation before us. All I can do is describe what it was like.

2053

0066-4294/81/0601-0001$01.00

LIFE HISTORY

I was born together with my sister Kirsten on a stormy November day in 1913 in Kristiansand, a small town on the southern coast of Norway. Ever since, when I return to Norway and look at it, my nose tickles a warning of impending tears. I had forgotten how beautiful it was. This was one of the reasons that I moved, in 1968, from Chicago to Berlin, but that is a later part of the story.

My ancestors, so far as I know, were all Norwegian farmers. My father grew up a country boy, but my maternal grandparents settled in Kristiansand, a town of about 20,000 population. Both my parents attended the normal school in Kristiansand with the object of becoming school teachers, but my father made frequent trips to the USA and Canada, where he functioned variously as a school teacher (in a small Norwegian settlement in Iowa), a bookkeeper, a lumber dealer, a real-estate agent, and finally a DDS, graduate of the Chicago School of Dental Surgery.

In May 1917 my parents got tired of waiting for the end of World War I, and my mother bundled my twin sister and me onto a ship to rejoin my father, who had been caught by the war on the other side of the Atlantic. I grew up on the north side of Chicago and attended Waters Grammar School and Roosevelt High School, before entering the University of Chicago in the fall of 1930. Except for two exciting years in the Department of Biochemistry at the Harvard Medical School (in 1939–41), I lived continuously in Chicago until I returned to Europe in 1968.

My parents were liberal-minded middle class citizens, glad to be in the United States—Europe was a rather unhappy part of the world in the 1920s, but life in the USA was comfortable during this decade. We grew up bilingual, speaking Norwegian at home, and our house was always overflowing with books, both English and Norwegian. My mother had an intense interest in women's rights, without any urge to participate in politics directly. She maintained an emotional attachment to Norway which was apolitical, but ensured that my sister and I returned to Norway for three extensive visits, in 1921-22, in 1930, and in 1935, so we would not "forget."

I am a product of the public schools of Chicago. In my recollection, they were not so bad. Standards may not have been very high, but we were encouraged to read extensively and to try our hand at writing or debating or acting or singing, or whatever other hobby might interest us. For a period, I was an ardent girl scout, and spent quite happy summers outside the city earning badges for flower, bird, and star finding. The science involved was nonexistent. One acquired handbooks that enabled one to attach a name to what one could see: that was all. But I enjoyed these activities very much and began to think about being some kind of "naturalist." My picture of a naturalist was of a person who spent most of his time collecting things that grew or flew and learning all about them. The museum of natural history (then the Field Museum) strengthened this interest. For years I spent most of my Saturdays haunting the Field Museum, the Art Institute, and the Public Library in downtown Chicago. I read hungrily and quite indiscriminately. My first favorite fictional characters were probably "Tarzan of the Apes" followed by Jim of "Treasure Island." Robert Louis Stevenson was a little too realistic. I found it cruel of him to give a sore throat to the hero of "Kidnapped."

My father had been prevented by strong pecuniary pressure from entering the occupation he thought he would have liked best, which was civil engineering: bridge building to be exact. In the eyes of the small boy in a wild and mountainous countryside, the man who designed and built bridges was the great hero-figure. My mother had become a school teacher by passionate choice and looked for the same inclinations in her offspring. The result was that when I was asked what I wanted to be, at an age when most boys opted for fireman or locomotive driver and most girls expressed a preference for nursing or acting, I lisped that I wanted to be a teacher. This conviction never left me, but it is hard to say how much was instilled from outside. My mother was a rather compelling educator of the very young. If I search my heart in this matter, I would say that it was the subject matter to be taught that interested me. But what subject matter? Here choice was a little difficult because I was interested in almost everything.

I suppose that under other circumstances I could have ended up in any of a large variety of occupations. But fate took a hand. The University of Chicago offered tuition scholarships to graduating high school students on the basis of competitive examinations. In those years the examination was given in a considerable variety of standard high school subjects, but the examinee picked only one. When invited to take the exam in physics, I was only too glad to say yes. I placed first, with the result that I entered the University of Chicago a little before my seventeenth birthday in the fall of 1930.

From this course of events you might infer that I had a special interest in physics, but that wasn't true. I preferred to take an examination in physics because our high school physics course was taught with a little rigor and was therefore considered "hard." Even in grammar school I had learned that it was much easier to "shine" in a "hard" subject. Later I learned that it had been my great luck that our high school physics course had been a very simplified version of the first year physics sequence taught at the University of Chicago. Had I taken the examination in chemistry, I would have laid an egg. Our high school chemistry course was not taught as a science but as an odd mixture of facts about chemical technology.

In high school I had taken what was then known as a general language major, which was considered, probably rightly, the best college preparatory course. I entered the University of Chicago in the very early days of the reign of R. M. Hutchins, President, then Chancellor, of the University. Hutchins was planning a big reorganization of college education. The class in which I entered was the last class subject to the much-maligned "old" plan. Our successors flourished under the "new" plan, which underwent so many subsequent revisions that one quite lost track of it. But I don't want to get involved in the subject of specialist (old) vs general (new) education. For myself the main point was that I had, so to speak, the benefits of the best of both systems. I was enrolled for the first two quarters in a general science course entitled "The Nature of the World and of Man." This course was staffed by top science professors, each of whom gave a few lectures in his specialty. Most of them were good lecturers. I was spellbound. Since the subject organization was partly alphabetical, we began, I believe, with astronomy and ended with zoology, and I fell in love with each subject in turn, which meant that I began my college career with a decision to become an astronomer and shifted to a committment to zoology, having changed my mind repeatedly over the course of six months. The result of this series of changes was that I chose to follow a premedical program, since it provided for a maximal mix of both physical and biological sciences, at least at the beginning.

I had no real commitment to medicine. Though I dreamed a little of medical research, it was in terms of *Arrowsmith* and *Microbe Hunters* (my major as a premed was bacteriology). I took the first course in biochemistry in the beginning of my junior year, and found myself in a dilemma. From now on it would all be biology. I was finished with the physical sciences, if I continued with my original plan. Though committed to biology, I felt that I needed much more chemistry to think successfully about biological problems. The obvious compromise was biochemistry. There were few BS degrees given at that time with a major in biochemistry, mainly because all of the necessary prerequisites plus a year of biochemistry could not easily be crowded into a four-year program.

The credits on my college transcript showed an appalling concentration on science. As a horrifying example of the uneducated specialist, I was exhibit A. Presumably, I had never read a book except for science texts. This presumption was not true. I had actually done all the assigned reading for the "survey" courses in the humanities and social sciences plus most of the assignments that Hutchins and Adler gave in their "Great Books" course. It was my exceeding good fortune that I went through college with my physical wants provided for in the University dormitories and no need to earn my way by taking part-time jobs. Thus, I had plenty of free time to indulge in hobbies, and I have always been a voracious reader with rather catholic tastes. It only enhanced my pleasure that I could read St. Augustine and Dostoevsky and Democritus and Hume for fun, without subsequently being subjected to the indignity of a multiple choice examination. The use of such examinations as a measure of the well-rounded human being is in my opinion a barbarism. The important thing in the humanities is the exposure to the works of great thinkers and artists. For this the University of Chicago in the early thirties was an excellent place. One could attend or cut any lectures one pleased, and the libraries were open til ten in the evening.

Becoming a Biochemist

I took my BS in biochemistry in the spring of 1934 and spent five subsequent months working as a chemical technician in the department of medicine of the University of Illinois Medical School on Chicago's West Side. That was my first taste of earning my own living. The work was not uninteresting, because I was attached to a research project, but I realized that I had still only scratched the surface of what I needed and wanted to know if I was going to do research in biology. In the fall I returned to the University graduate school in the same department of biochemistry.

The decade of the thirties was not a happy time in the graduate schools. It was the decade of the great depression. Most students had to live on a shoestring. The department of biochemistry happened to be better off than most. It provided its graduate students with teaching assistantships: $ 200 for each 12-week quarter = $ 800 per year, from which a minimum of 4 X $ 33 = 132 was repaid as tuition. A single person could almost live on the remainder, though not with any frills. We took outside jobs wherever we found them. I remember doing pH determinations for the meat packers, with a homemade glass electrode. The necessary special glass had just become available commercially.

In *The Adventures of Augie March*, Saul Bellow has described the milieu around the Midway in the thirties. His time there coincided apparently in part with mine and our paths probably crossed. Though I have no clear recollection of having met him, I have suspicions. But the picturesque squalor described by Bellow wasn't so picturesque (or squalid either) for the graduate science students. We spent most of our time, morning, noon, and night, in the laboratory. Meals were cooked there, and romance sometimes flourished there.

Throughout my years as a student of biochemistry at the University of Chicago, Fred C. Koch was Departmental Chairman. Koch's research interests were in steroids, both in the form of vitamins and of sex hormones —a very ambitious program. In addition, he was interested in the development of analytical methods suitable for use in the clinical laboratories attached to the University hospital. In the student training there was a heavy emphasis on blood and urine analysis. This had good practical reasons. Hospital laboratories provided jobs for biochemists, and jobs were a very precious commodity.

The biochemistry of the thirties was mainly concerned with the chemistry of small molecules. Vitamins, hormones, and growth factors were being identified. From the chemistry of carbohydrates, fats, and amino acids, we made a conceptual leap to the physiological aspects of diabetis, ketogenesis, and glyconeogenesis. In between there was the very mysterious process of metabolism which took place in something equally mysterious called protoplasm and was catalyzed by agents called enzymes.

Enzymes, we were taught, were of unknown chemical nature. This may seem surprising, since Sumner had long since identified urease as a crystalline protein, and Northrop and Kunitz were crystallizing the proteolytic enzymes of the digestive tract. But Fred Koch was a great admirer of Willstätter, and adhered to Willstätter's view that peroxidase was not a protein. It has been pointed out that Willstätter, who was a first-rate chemist, had the sheer bad luck of selecting for purification an enzyme with a very high turnover number. He purified it to a point where the available protein tests were negative and the material left no ash, though it still gave a good enzyme test. (This business about the ash was the origin of a certain friction between Willstätter and Otto Warburg, as described in a later section.)

The opposition to the conclusion that enzymes were proteins was not irrational. The prevailing view was that "proteins are big polypeptides. Such molecules do not have catalytic properties. They can absorb other substances, however. The isolation of a few crystalline proteins with enzymatic properties is the result of the absorption of the enzyme on the protein." What was actually demanded was the isolation of the active center of the enzyme. Later the identification of special prosthetic groups for various reaction types made the notion that enzymes are proteins easier to accept. But what really clinched the point was not one single discovery but a massive accumulation of findings with many different enzyme reactions, showing that no enzymatic activity could be separated from protein.

For my PhD I picked my own problem with Martin Hanke's sponsorship and help. By 1938 I had completed my thesis on the *Oxidation-reduction requirements of an obligate anaerobe.* This thesis didn't, in my later opinion, amount to much. I was trying to approach the problem of what it was that distinguished obligate anaerobes from other bacteria, and I didn't select a very good approach. But the work taught me two things: It taught me that bacteria require CO_2 for growth, and it taught me the importance of collaborators. If you have to do absolutely everything all by yourself, progress can be very slow, especially if you have no money.

Boston and Radioactive Carbon

After working for a year as research assistant for E. A. Evans, Jr., I had the colossal good luck in 1939 to get a fellowship from the International Federation of University Women. This fellowship carried the stipulation that it had to be used in a foreign country. Arrangements were made for me to join Meyerhof, who was at that time in Paris. A summary and evaluation of Meyerhof's scientific achievements can be found in Nachmansohn's autobiographical chapter (5), in which the latter also outlines the problems faced by Meyerhof in 1939.

I was to leave for Paris in the middle of September. On the outbreak of World War II, all passports of American citizens were cancelled. You had to reapply for a new one in order to go to Europe, and you had better have a good reason. This was at first a bitter blow. In retrospect, however, it is clear that the fates had been kind, because I managed to find sanctuary in Baird Hastings's department of biochemistry at the Harvard Medical School. When I arrived, he assigned me to his ^{11}C-glycogen project, which was just getting organized. As a departmental chairman and group leader of the ^{11}C project, Baird Hastings was one of the kindest and most considerate human beings that I have ever encountered. He ran a happy department in an unhappy time. The following two years were for me depressing in terms of the continual escalations of the war; and at the same time they were intensely stimulating and exciting as far as scientific work was concerned.

The utility of deuterium for studying metabolic processes or paths had recently been demonstrated at Columbia by Schoenheimer and his coworkers, but the usefulness of D was limited. ^{11}C was the first radioactive isotope of carbon made available to biochemists. It had a half-life of about 20 minutes—a severe limitation which required speed in all operations. There was a substantial group involved. To achieve the necessary velocity, four or five people actively participated in each labeling experiment. We studied the synthesis of glycogen by rat liver in vivo, with the object of learning something about how glycogen was made from precursors of low molecular weight. The most interesting discovery was that the carbon of CO_2 got into the carbon of the glycogen. There were several laboratories working on related reactions at this time, more or less unaware of each other. The full story has been told elsewhere (2) and calls for no recapitulation. Earl Evans, who had been working with Hans Krebs in Sheffield, came back to the University of Chicago in 1940, got hold of some ^{11}C from Louis Slotin, and showed that the α-ketoglutarate, made during the Krebs cycle by pigeon liver brei, was labeled by ^{11}CO$_2$. In the fall of 1941, I returned to the University of Chicago as an Instructor of Biochemistry with the intention of continuing ^{11}C experiments with Evans and Slotin, but we didn't get much accomplished before the United States was also involved in the war. During the following year or so, the chemists who could make and handle isotopes were swallowed by war projects, and research unrelated to the war effort ceased almost completely. I was attached to a Malaria project, but my teaching assignments were so heavy that there was time for little else.

In retrospect, when I revive my recollections of my two years at Harvard, it seems to me that there was an electrical intellectual excitement in the Boston air. This was partly due to ^{11}C, but there was also something else. It was the time of the great migration of scientists from Europe to America. They came first to the eastern seaboard and often gathered in Woods Hole, which was within easy weekend reach of Boston. Through Gertrude Perlmann, I had an opportunity to read Fritz Lipmann's paper, "Metabolic Generation and Utilization of Phosphate Bond Energy," when it was still in press in Volume 1 of *Advances in Enzymology*. This manuscript seemed to have the power of a revelation. Now I could understand why Meyerhof had suggested that I work on the "oxidizing enzyme" of glycolysis. Now I felt that I understood glycolysis for the first time. Of course it wasn't until the postwar years that the biochemical seeds sown by the immigrants grew and bore fruit a hundredfold.

β-Carboxylations and Stereospecificity

When things began returning to normal after the war was over, I turned my attention to plants, with the object of looking for dark fixation of CO$_2$, Krebs cycle, and other enzyme reactions. The known facts about the metabolism of *Crassulacean* plants suggested strongly that a set of reactions similar to those of the Krebs cycle plus the dark carboxylations known as Wood-Werkman reactions, would probably be found in plants also. I believe that what attracted me to higher plants as experimental material was the paucity of information available about their metabolism, plus the fact that they were the seat of one of the most fascinating processes in nature: photosynthesis.

James Franck and Hans Gaffron were located in a neighboring building, and I often went over to listen to their seminars and discussion groups, in this way learning also a good deal about the history of the photosynthesis problem and the development of current experimentation. Since different research groups at the same institution were not supposed to compete, it was agreed that I would limit myself to dark reactions, unless it was possible to find an area for collaboration. My group had invested much time and effort in the preparation of the pyridine nucleotides, and Eric Conn gave some of our TPN (NADP) to one of Franck's students (Tolmach), who was overjoyed when it worked as a reagent for eliciting O_2 evolution. But no real collaboration developed here, mainly because James Franck didn't like enzymes. I suspect that the root of this antipathy was similar to the root of Otto Warburg's unreasonable antipathy for radioactive isotopes. After a certain age, it becomes difficult if not impossible to master a new technique. It happens to all of us.

It was about 1949–50, as I recall, that Frank Westheimer of the Chemistry Department and I discovered that we were both interested in β-carboxylations, except that he, of course, was interested in reaction mechanism, whereas I was looking for enzymes that catalyzed such reactions. When Harvey Fisher asked if he could do a joint PhD with Westheimer and me, I rather expected that what would emerge would be a carboxylation project. But no, what Westheimer wanted to work with was pyridine nucleotides. He had been using deuterium to study the mechanism of oxidation of alcohols by chromates. (That problem, Harvey informed me, was known as "how to make better cleaning solution.") Now Westheimer wanted to know how diphosphopyridine nucleotide (NAD) and alcohol dehydrogenase oxidized ethanol. Hydrogen must either be transferred directly, or the extra hydrogen in the reduced pyridine nucleotide must have its origin in water. The problem was beautifully formulated from the very beginning. By great good luck, Pabst put diphosphopyridine nucleotide on the market at just the right time, so that the preparative work required was simplified. The first serious experiment was successful. We soon knew that there was direct transfer of one hydrogen atom. Since Ogston's celebrated paper (1) had appeared, and we had sweated out the significance of Ogston's paper in connection with citrate synthesis, as told elsewhere (7), the possibility of using deuterium to do stereospecificity studies was apparent. All in all, this was a happy and very successful collaboration, perhaps because Westheimer and I brought different past experience and skills to a problem of common interest. What little I could contribute to mechanism studies rests on the tutoring I received from Westheimer.

A summary of known information about the stereospecificity of hydrogen transfer in pyridine nucleotide dehydrogenase reactions has appeared (8), and there the reviewers give me a little more credit than is my due for getting this work started. That isn't quite fair to Westheimer, though I feel a flush of gratitude to the reviewers. But when I think of the long line of students and postdocs who worked on this project and contributed all the hard labor, I feel very humble indeed. I do deserve credit for durability. Perhaps I stuck with it longest.

Photosynthesis and Otto Warburg

After Gaffron left Chicago, I began to work seriously with chloroplast preparations, trying to learn something about photophosphorylation and the Hill reaction. A fact that particularly caught my fancy was the catalytic effect of CO_2 on the latter. Warburg was putting a strong emphasis on this point because it fit his theory of photosynthesis. Gaffron said Warburg's experiments were irreproducible. I decided to try it myself, and found that one could get quite good stimulation of Hill reactions with CO_2, provided one picked the right conditions. This was the background for my initial visit to Warburg's laboratory in 1961.

A succession of visits ensued and culminated in my accepting a position as a director at Warburg's institute in West Berlin in 1968. There I began to work on nitrate reduction by *Chlorella.* There were complex reasons for the selection of this problem. One was that Warburg regarded nitrate as the "natural" Hill reagent. Later I gradually developed a strong suspicion that the reason Warburg got such fantastically low values for the overall quantum requirement of photosynthesis was mainly that he had nitrate in the medium and excess carbohydrate in the cells. Better methods have long since superseded those used by Warburg, and the problem of the quantum requirement is no longer cogent.

The editors made it clear that I was not to write a scientific review, so I will exercise heroic self-restraint and refrain from saying much more about nitrate reduction.

Warburg's theory of photosynthesis was sensible and internally self-consistent, though it paid no attention to any experiments that he hadn't done himself. There were some assumptions in his theory and he was quite aware of them. I asked him once whether he didn't think there might be more than one kind of light reaction. "Of course, there might be. But you should keep everything as simple as you possibly can. Never introduce any more detail into a theory than you must." Me: "But, Herr Professor, I think there is evidence that there are two different kinds of light reactions." Warburg: "What is your evidence? Have you seen it?" Me: "Well, not exactly—uh. There is lots of evidence from different labs." Warburg: "Don't tell me about anybody else's experiments. I want to see yours. Show me. That's the way one should argue. I'll show you my experiments. Then you show me your experiments." Me: "All right."

When I last looked at Warburg's desk, it looked as though he was doing energy calculations for an equation which I had suggested to him:

$$H_2 + HNO_2 \xrightarrow{h\nu} NH_3 + O_2 \hspace{4cm} 1.$$

I had thought of this formulation when I was reading his earlier work on nitrate reduction. The advantage of Equation 1 over Warburg's earlier formulation was that the new equation is really a one-quantum reaction, almost. Now one didn't need a back reaction of O_2 to make up the energy deficit, and CO_2 was pushed over to the Calvin cycle where it belonged. Had it been possible to convince him that there was a second light reaction that caused dismutation but led to no O_2 evolution, he might have added:

$$NH_3 + HOH \xrightarrow{h\nu} NH_2OH + H_2 \hspace{3.5cm} 2.$$

and

$$NH_2OH + HOH \xrightarrow{h\nu} HNO_2 + 2H_2 \hspace{3.5cm} 3.$$

All three equations add up to look like water-splitting, but in principle they represent an attempt to divide the overall process into three chemical steps, each of which might be energized by one light quantum.

In my opinion, the apparent naivete in Warburg's theories was studied and intentional. The rules seemed to be: keep maximal simplicity and stick to minimal numbers. Make changes only when you must. The advantage of writing balanced equations is mainly that it is an indispensable approach to correct energy calculations. The chemical identity of the components need not be taken literally. One knows that the entire process is far more complicated than the symbols suggest. The very starkness of the symbols protects one from the easy semantic error of confusing the picture with the phenomena that the picture represents. To me, this approach to theory building seems preferable to the currently accepted practice of beginning with a diagram of an hypothetical electron-transport chain. This latter term, invented to represent a sequence of reactions in time, has been converted in the minds of many to a real chain along which electrons flow. Arguing about the properties of such chains is a little like counting angels on the head of a pin, though it seems to pay.

Lest the previous paragraphs give the impression that my relations with Warburg were totally harmonious, I hasten to add that the harmony was mainly confined to scientific questions at a fairly elementary level. Warburg was a man of violent emotions. Anger was predominant, and he wasn't very rational when his emotions took control. We all have a rational self and an irrational, emotional self. Mostly, the rational self is in control. Warburg's emotional self was unusually powerful. Science was his recipe for staying rational.

A mutual acquaintance—not a scientist—was reported to be so disappointed in an emotional relationship that he got sick. Warburg's advice was, "Tell him not to think about anything but science—think about absolutely nothing else—only science." Since the sick man was not a scientist, this prescription probably didn't help him much. But I regard it as a clue to the reason that science, for Warburg, was such an obsession.

On a visit to Warburg's laboratory, I had asked him once for his opinion about the claim in an earlier biographical sketch that Warburg had never made any mistakes. He pondered awhile and said, "Of course, I have made mistakes—many of them. The only way to avoid making any mistakes is never to do anything at all. My biggest mistake—", here he paused a long time before continuing, "my biggest mistake was to get much too much involved in controversy. Never get involved in controversy. It's a waste of time. It isn't that controversy in itself is wrong. No, it can even be stimulating. But controversy takes too much time and energy. That's what's wrong about it. I have wasted my time and my energy in controversy, when I should have been going on doing new experiments, and now—". In the course of the conversation, Warburg also tried to explain to me why he got so mad at James Franck. Warburg: "He said—he said—I couldn't measure light. He was a theoretician. By himself he couldn't measure anything, and he said I—I couldn't—measure—." Something curious was happening. Warburg was getting a little incoherent. In the course of telling me about why he got angry, Otto Warburg got angry all over again. He got as angry as he would have been if James Franck had been sitting there in the same room, now, telling him how sorry he (Franck) was, that his (Warburg's) measurements couldn't possibly be right. At that time, however, I hadn't learned to recognize Warburg's anger. He didn't usually signal it the way most of us do, by raising the pitch and/or the volume of his voice. He had perfect control over his behavior.

Warburg's views on photosynthesis could be considered to be a further development of Willstätter's views. The well-selected library of the Institute for Cell Physiology included Willstätter's works on chlorophyll and photosynthesis. But there was an antagonism to Willstätter on Warburg's part. What was its origin? Warburg told me about this as follows:

> Willstätter gave a lecture about peroxidase and the audience was very big. When he had finished and it was time for questions, I asked him if there was any ash. That wasn't such a stupid question for a young man to ask, was it? Willstätter said, "No, Herr Warburg, there isn't any ash." It wasn't what he said, but the way he said it. He was a good speaker. Because of the way he said it, five hundred people laughed.

In his fine biography of Warburg (3), Hans Krebs cites two sentences concerning the 46-year war over bioenergetics as an example of Warburg's often very aggressive polemical style. I translated the original German version of that manuscript into English. Warburg asked me to do this for him. Now I know perfectly well that the translator isn't supposed to try to improve on the original. He is supposed to be accurate. I did the translation as conscientiously as though I were taking an examination for a translator's certificate, except at one place. After pondering carefully, I translated the "war" into "argument," knowing that this would flunk me. Except that I thought the examiner wouldn't notice. The examiner, I assumed, would be Warburg himself. At first, this guess seemed right. Warburg concentrated on the science and didn't notice the omitted war. He was pleased and complimented me on how well I understood his experiments and theories. Since he was never prodigal with compliments, I felt pretty good. But there was a second examiner, a Warburg No. 2, who acted as a censor. I hadn't known about him. The censor didn't know any science, but he could read English. So the day after getting the nice A, I was told my grade had been changed to an F. "I won't use any of it," said Warburg, in anger. The baleful glare of his blue-gray eyes could be uncannily frightening. Of course it didn't help to try to point out that I was only giving his own advice back to him. Advice is a commodity that can readily be transmitted from senior to junior, but seldom in the other direction.

Warburg was a stubborn man, and apparently something of a rebel in his youth. In German universities there is a procedure called habilitation, which qualifies a person to hold a professorship. This procedure is a kind of final examination, in which the candidate demonstrates his teaching competence as well as his mastery of a research area. Warburg was proud of the fact that he had never submitted to the habilitation process.

Krebs has told how Warburg became a professor by sending lamb chops to Fisher (3). Warburg told me once how he first managed not to become a professor. Warburg: "I have never given a lecture to students. Never. Not one. When I was young, my Professor said, 'The time has come. You have to give some lectures.'
I said, 'I won't.'
He said, 'You must.'
I said, 'I won't.'
He said, 'Oh come, just a few, not many, you can easily do it.'
I said, 'I won't.'
So they held a meeting and decided it must be schizoid-paranoia." Warburg smiled when he got to the last sentence.

In a book review, I once wrote a kind of sketch of Warburg (6). His comment, on reading this review was, "How nice, how very nice. But I am not like that. Really not. You are mistaken." Indeed I was mistaken, as I later discovered. Up to that time, I had seen only the sunny side of a very complex character. The brightest sun casts the darkest shadow.

I hope Warburg approves of the present manuscript, as he once did of another one. This latter paper described some work that had been done in his laboratory, and was dutifully mailed to him for his approval. That manuscript crossed the Atlantic six times. After two revisions I finally got the hoped-for letter: Dear Dr. Vennesland: Imprimatur! Warburg.

SOME PREJUDICES AND OPINIONS

The editors have given me laissez-faire. I can express my prejudices and opinions without editorial censorship. After a lifetime subjected to the discipline of science editors, I find this new freedom heady stuff. Here we go! Let me apologize for the next section by explaining that it is not completely factual; but it contains some grains of truth, nevertheless.

The Problems of Scientific Publishing

These are so great that I am going to have to switch to parables. The journals, some of them, are making brevity such a virtue that clarity has become a sin. It isn't that I have anything against editors. Editors work hard for little or nothing but the glory of having their names appear in public; and the job is usually sheer drudgery. Editors deserve our gratitude for the time and effort they invest in preserving high standards. OK. But there's more to the job of an editor than that. I demand that the editor use his head before I'm going to give him any gratitude.

Here is parable 1. I once got a rejection slip from a well-known journal saying that there might have been some interest in this material if the work had been done with a purified enzyme, but since all of the work had been done with nothing but crude extracts, it was totally unsuitable - oh - you know the usual lines. They were accompanied by a more soothing suggestion that I try another journal. Now, Ladies and Gentlemen, imagine my feelings! I know you can. Half of that paper dealt with experiments with purified enzyme! Imagine my feelings - pardon - my becoming a little incoherent. That bastard hasn't even taken the trouble to read my paper. Imagine? *My* paper! That's what my feelings were. Now it turned out all to the good, because the paper probably *was* more suitable for the second journal, and it was also revised for the second journal and became a better paper. This time we put the last half in front and the front half in the back. I couldn't resist that final slap at the anonymous referee who wasn't interested in my paper.

My second parable is inspired by a horrifying suggestion that I recently read somewhere—perhaps in TIBS—that referees should be named and authors should be anonymous. When I act as referee, I try to make my hands soft as silk. I am not an intemperate critic when I am writing anonymously to a defenseless author. I know who he is, but he doesn't know me. He can't hit me back because he doesn't know whom to hit; and I have never enjoyed slapping defenseless children. But an editor once had the nerve to write to me, to ask me for my opinion of a paper by an unidentified author. This isn't the normal kind of request made to a referee. The referee is usually told who the author is. Now, I'm not to be told who the author is, but of course the editor knows who he is, and the editor wants my honest opinion, so the wise editor can protect the author from injustice, can't he? So what did I do? I blasted that author—oops that was a Freudian slip— I mean editor—no—I mean author—who do I mean anyway?—with the roar of all the cannon I have in my artillery! "I don't know who the author is, but in my opinion the paper shows that he doesn't know what he is talking about. The following two examples should suffice, etc. I hope that these are written in a form suitable for transmission to the author." I never received the reply of the outraged author. It was probably unprintable.

Women in the Universities

I have already mentioned the debt I owe to the International Federation of University Women. It seems obligatory to comment on the women's rights movement with its attendant publicity, which has been louder in the USA perhaps than in Europe. First of all, I feel a genuine sisterly sympathy for any young woman who is determined to make her way in a research career, and I am glad that it has been recognized in principle, at least, that women should be considered for assistant professorships and promotion on the academic ladder on the same basis as men. Reverse discrimination, however, bothers me very much indeed. I do not approve of it in any form. In the long run, such a practice will guarantee a lower average quality in academia. In other words, I don't think that one should try to establish a particular sex ratio for academic appointments in too short a time. And I don't even think we should try to calculate what an appropriate "fair" ratio of sexes should be. Maybe the men really do have brains better suited to perform best in hardware subjects like physics. Maybe the girls have brains better suited to perform best in biology? Let's let the optimal mix demonstrate itself.

The entry of women into public life in more than occasional numbers is a fairly recent phenomenon in human history. The cause undoubtedly lies in underlying economic changes, the same ones that have caused a large population shift from country to city. If we go back several hundred years to colonial America or to Europe in the eighteenth century, most of our female ancestors were living on farms and engaged in a form of housekeeping that was a full-time job. The raw materials often came straight from the field. It was the man's job to provide them and the woman's job to convert them into daily meals and clothing. The upper classes had servants, true, but these rulers were a tiny minority. The great bulk of the population lived like Adam and Eve, delving and spinning, and no one questioned the importance of the woman's job for the family prosperity. What has housekeeping become now? Every single task has been lightened with labor-saving devices to such a point that fewer and fewer able-bodied women can take pride in their housekeeping performance as a difficult job well done. With the exception of the care of small children, which I grant is an all-consuming task, women are left with no satisfying occupation except for the period of child-rearing. The consequence is that women move into the market place in increasing numbers, demanding a share of the more interesting positions and society is adapting to this change. My impression is that most men view this movement with sympathy and understanding. Their wives and daughters are involved. And take note, you blacks and women, who demand instant action. Take note: If you get all the reverse discrimination you want, this will provide what some will regard as proof that you really are "inferior." And you ought to be able to figure out what that proof will be. Most of the intellectual lightweights on the staff *will be* blacks and women.

Research Financing

The support provided for the individual scientist by the Max-Planck-Society is ideal. I have nothing but gratitude for this support and for the manner in which it is administered. Since my move to Germany preceded the crunch in research funding in the USA, my experience of research poverty is limited to the thirties, which is a long way back. Nevertheless, one's earlier recollections are often the most vivid. My generation could have accomplished much more if we hadn't been forced to spend so much of our time on do-it-yourself activities. Though this was not totally bad training, it seems reasonably clear that one individual cannot aspire to becoming an expert glass blower and instrument maker, bench chemist and lecturer, as well as an authority on the theoretical aspects of his subject. Even if the talents for these occupations are all there, human life is too short to acquire the necessary expertise in all crafts and disciplines. The provision of positions with other positions attached to make a unit recognizes this need for cooperative productivity. The problem of hierarchy (who is the boss?) remains, and is solved in the Max-Planck-Society by providing real tenure only for the top positions. Isn't there room here for an adjustment?

In some areas, seven independent groups of three or four scientists may be more productive than one group of 25. This is a plea for more intermediary tenure positions at an independent level within the Max-Planck-Society. I grant that there are problems which can only be handled by very large units, but only a fraction of biological research is of such a nature.

Science: A Personal View

I have never before attempted to articulate my view of the natural sciences. When I was still quite young, formal philosophy appealed to me, but its attractions gradually vanished as I immersed myself in the study of living things. The contents of this section reflect this transition, and are personal in the sense that they express tastes or preferences which are bound to vary with the individual. But I have attempted also to express as clearly as possible my own rational view of the relationship of a self-conscious human being to the universe of changing things.

Scientific thinking is thought about the real world. A mind trained (or untrained) in classical logic has a tendency to confuse scientific thinking with the thinking of classical logic because the logic is the same. But scientific thinking does not permit the use of symbols, concepts—call them what you will—unless they stand for things that are real. Therefore, it is very important to be especially careful to examine the real meaning of the words we are using.

Logic is an internally consistent system of thinking. Classical logic is expressed in terms of symbols and speech. These entities need have no correspondence to the things of reality. Classical logic is attractive to the young mind which knows speech but doesn't know much more. The young mind likes to dance. Logic is the dance of the mind. Classical logic is the art of reasoning about nothing. As symbols are not real things, they are no-thing. There is no progress in logic. It is the process the mind uses; therefore it need not be taught. It is there in the mind as the process the mind uses to think.

I prefer to work with plants and to think about humans, which seems foolish, but that's the way it is. For example, one thing I have often wondered about is why most (not all) of the smells emanating from the plant world are experienced as a pleasant feeling, whereas the opposite is true for the smells emanating from the animal world. But "why" is not the right word with which to question nature. The small child often uses "why" in the sense of "tell me more." The adult uses "why" in the sense of "tell me your motives." "Why" is a word that applies only when two brains are communicating with the language of human speech. "Why" is a nonsense syllable in the system man uses to communicate with nature. This system is the scientific system. Don't ask why? Ask what it is? And think of it as things like chemical compounds or as a process involving molecules.

The word revolution is being applied right and left to the development of modern biological sciences—without justification in my opinion. Kuhn apparently started this custom with *The Nature of Scientific Revolutions,* which I once read with real pleasure. Kuhn has applied himself to the history of science by applying the scientific method of observation to the process whereby scientists arrive at concensus (4). His method is a welcome change from that of the philosopher-historians. But I have been surprised by the fact that many of my fellow scientists didn't seem to read Kuhn the way I did. Some of them wanted to claim kinship with Galileo every time they made a new discovery.

New discoveries are very nice things, everybody is happy, progress is great, business is good, we are going to heaven fast. That's right. It is fun to make discoveries. But that is not the nature of revolutions. Revolutions hurt, because something we believed in has to be rejected. That's how you can tell the difference between a discovery and a revolution. If the molecular biologists want a big word to describe a big thing—the surge of new information in the field of genetics and cell differentiation—let them use the word explosion.

The basic premise of biology—that it can be understood in the terms of the physical sciences—has in no way been altered by the molecular biologists. Nevertheless, modern biology has given us a deep sense of kinship with all living things. Molecular biology is the Saint Francis of the sciences. I used to think that speculation about the course of evolution was pointless because it was impossible to duplicate the process. But if the nucleic acids carry within themselves the records of their origins, as Eigen suggests, then we may in fact be able to deduce the chemical course of our origins from dust. The prospect is wonderfully exciting. I regret being born too soon.

But this basic assumption of the biological sciences that all biological phenomena can be explained in terms of the laws of physics and chemistry doesn't always seem to apply. The mind-body problem defies such an explanation. Our awareness, our consciousness, our capacity for thought reflect complex chemical processes occurring in our brains. Free will implies a capacity to control these chemical events—albeit to a limited extent. We cannot deduce the experience of consciousness from physical and chemical laws. At this level, biology endows chemistry and physics with a new set of properties.

Mind is not a real thing, but a different entity, associated in a curious relationship with processes in which only things are involved. The natural scientist can examine the things involved and note their correlation with subjective phenomena.

The scientific method cannot answer all our questions. We insist on raising senseless questions. We want purpose and design in the universe. Our notions of purpose and design are based only on our own built-in natural drives.

THE SERIOUS AND THE NOT-SO-SERIOUS

I appreciate the logic of the theologians who said, "God's name is secret." But they weren't theologians as defenders of dogma. They were searchers for truth. I cannot subscribe to any dogma. I consider myself religious, but many religious people wouldn't. Man is an animal, endowed with a brain that tells him he is going to end in Nirvana, and the same brain gives man a set of drives that makes him fight his own extinction with all his might. This is the human dilemma. The great religions provide comfort by either telling us there isn't such a place as Nirvana, there is a Heaven and a Hell, or by telling us that Nirvana is really very pleasant, which comes out in a curious sense to the same thing. The "Hell" of the Christian is the "Rebirth" of Buddhism.

I believe that the best thing a human being can do is to try to increase understanding of the entire reality—the all—around us. This can only be done by the techniques of the natural sciences. One can't figure it out by meditation. One must observe nature. In this sense, the natural scientist isn't creative at all. Perhaps this word should be applied only to the artist. Still, there is an aspect of the work of the scientist that is very similar to the work of the artist. The scientist puts questions to nature and nature answers. No speck of imagination here. The art comes in the process of figuring out how to put the questions, and the thrill comes when you get the answer—the thrill is the *signal* that tells you you have the right answer.

I enjoy great literature. There are passages in Shakespeare and the Bible that have moved me in a way impossible to describe. One has to be in the mood for that kind of reading, and one's taste changes with circumstances, so I can't tell what my favorite passages are. It might be this one this week and next month it will be another one. I like fiction that can be read with several layers of meaning, but I have never been able to wade through Joyce's monumental later work. I don't like nonsense. I cannot understand that anyone can be interested in the meaning of symbols that represent nonsense.

Ariel's song is a sample of the kind of poetry that gives me the same kind of thrill as the thrill of scientific discovery.

Full fathom five thy father lies;
Of his bones are coral made;
Those are pearls that were his eyes;
Nothing of him that doth fade
But does suffer a sea-change
Into something rich and strange.
Sea-nymphs hourly ring his knell:
Burden. Ding-dong.
Hark! now I hear them—Ding-dong
bell.

Ariel's song—The Tempest (Shakespeare)

The above lines may have been inspired by the exhilarating excitement of the naval explorations of Elizabethan England, but they can be applied equally to describe the excitement of scientific discovery. The scientists aren't taking the mystery out of life. Even if they wanted to, they couldn't do it.

I have borrowed the title of this article from a book by Barbra Ring, a Norwegian novelist and newspaper woman who was born in my home town, Kristiansand, a good many years before I was. After getting a divorce from an insanely jealous husband, she made her living by writing. She provided the original version of my favorite Kjutta joke. Kjutta, I must explain, was once a real character who lived in Kristiansand, but he has become a legend there in the form of a hundred varieties of Kjutta stories.

Kjutta was in a fight and was beaten up. He is describing the fight: "First, I hauled off and slammed him on the nose and didn't hit him, so I slammed him again the same way. Then he began to run, but I got ahead of him."

Literature Cited

1. Bentley, R. 1978. Ogston and the development of prochirality theory. *Nature* 276:673–76
2. Krebs, H. A. 1974. The discovery of carbon dioxide fixation in mammalian tissues. *Mol. Cell. Biochem.* 5:79–94
3. Krebs, H., Mitarbeit von Schmid, R. 1979. *Otto Warburg: Zellphysiologe—Biochemiker—Mediziner,* 1883–1970. Stuttgart: Wissenschaftliche Verlagsgesellschaft. 168 pp.
4. Kuhn, T. S. 1977. *The Essential Tension.* Chicago/London: Univ. Chicago Press. 366 pp.
5. Nachmansohn, D. 1972. Biochemistry as part of my life. *Ann. Rev. Biochem.* 41:1–28
6. Vennesland, B. 1963. Review of *New Methods of Cell Physiology. Perspect. Biol. Med.* 6:385–88
7. Vennesland, B. 1974. Stereospecificity in biology. *Top. Curr. Chem.* 48:39–65
8. You, K. S., Arnold, L. J. Jr., Allison, W. S., Kaplan, N. O. 1978. Enzyme stereospecificities for nicotinamide nucleotides. *Trends Biochem. Sci.* 3:265–68

Ann. Rev. Plant Physiol. 1982. 33:1–26

A PLANT PHYSIOLOGICAL ODYSSEY

Philip F. Wareing

Department of Botany and Microbiology, University College of Wales, Aberystwyth, Dyfed, United Kingdom

CONTENTS

To be invited to contribute to the *Annual Review of Plant Physiology* is always a compliment, but to be given a *carte blanche* to write as one chooses is indeed a privilege. At the age of retirement, the urge to reminisce and pontificate is strong and normally to be resisted, but to have a positive invitation to indulge in these activities is too great a temptation to refuse. I am not sure what benefit readers can be expected to derive from this exercise, unless it be as a warning to the young so that they may avoid falling into the same errors, but this seems highly unlikely. Probably the editors

2075

0066-4294/82/0601-0001$02.00

intend it to be a form of light relief from the more serious and weighty subjects treated in the remainder of the volume. If it fulfills this function I will be content.

BIOGRAPHICAL

Roots and Early Years

My paternal grandparents came from the Preston area in Lancashire, in northwest England, where the economy was based on the cotton industry. Indeed, an essential step in the Industrial Revolution regarding production in mechanized factories started in Lancashire in the late eighteenth century, following the invention there of the spinning jenny and the spinning mule and their exploitation by Arkwright, who first developed the factory system. The industry was dependent upon the supply of American cotton, the production of which increased greatly at the same time, following the invention of the saw-gin, to separate cotton seed from the fibers, by Eli Whitney in the United States. My grandfather emigrated to Boston, Massachusetts about 1884, shortly after my father's birth with the intention that his wife and child should follow him there as soon as he had settled in. But in a comparatively short time news came that he had been killed in circumstances which were never made clear.

My father started work as a "half-timer" in a cotton mill at the age of 11, working in the mornings and attending school in the afternoons. Later he started working full time but continued to attend evening classes, and when he was 15 or 16 he passed a competitive examination to enter the Civil Service in London, where he spent the rest of his working life. Although he was largely self-educated, he was quite widely read in literature, history, and especially in the complex theological works of Swedenborg, so that he was a natural scholar and in many ways a more educated person than many who have received an expensive formal education. Working men with intellectual interests were not uncommon in Britain, and in Wales it is still not considered unusual for quarrymen and miners to be poets and scholars. There was an unsatisfied hunger for education among working people which to some extent was met by evening classes provided by various types of institutes and by organizations such as the Workers' Educational Association, but the greatly improved opportunities for higher education which have become available since World War II have gone far to satisfy the need, although there is still a great demand for part-time degree courses offered by the Open University.

My mother was the daughter of a Methodist minister and had received a boarding-school education, no doubt in an atmosphere of strict middle-class respectability characteristic of late Victorian times. Her life was more than fully occupied in bringing up a family of six, of whom I was the eldest. My father served overseas in the British Army for over four years during World War I. I was born in Essex, near London, shortly before the outbreak of war in 1914, and I lived with my mother and aunts at a seaside town on the Thames estuary, where some of the first air raids on civilian targets were carried out by German zeppelins (airships), and I still have a few hazy recollections of standing with the family under the stairs, singing hymns, during an air raid!

After my father returned from war service, we lived in the country for two or three years, and for me it was a delight to roam the countryside "footloose and fancy-free"; it was then I developed an abiding interest in the wild plants and animals of the countryside. This idyllic interlude was brought to an end when my parents decided that they ought to move to some place where a better education could be provided for me and the other members of the growing family than could be had in the small village school. They both attached the greatest importance to education and were anxious that their children should have the best they could afford; ultimately, four of us ended up in the teaching profession.

In 1923 we moved to Watford, north of London, which was chosen by my parents because of the high reputation of the Boys' Grammar School there, which I started attending. Although at that time it was not recognized as a "public school" (which in Britain means private!), it adopted the traditional public school "ethos." Discipline was fairly strict, with caning by the Headmaster as the ultimate deterrent. The teachers remained somewhat aloof, and American school students and even contemporary British ones would have found the deference and respect we paid to the teachers very strange. At school, "games" (rugby, cricket, cross-country running) were compulsory and were regarded as an essential part of the "character building" process ("Waterloo was won on the playing fields of Eton"). There were separate grammar schools for boys and girls, situated at least a mile apart, and any interest shown in the opposite sex was severely dealt with—it was an offense to be seen in the company of a girl! However, times have changed and I understand that the school is now mixed.

One of the characteristics of the British educational system was and still is the very early age at which pupils (students) start to specialize. Although at that time all pupils had to study English, mathematics, and a modern language (normally French), at the age of about 13 they had to choose whether they would enter the Classics, Modern (Arts), or Science streams. If they chose Science, they would take additional mathematics, physics, and chemistry, but biology was not taught at any level, presumably because it was not regarded as academically respectable. However, Latin was studied from the ages of 10 to 16 even by science students, since it was required for entry to Oxford and Cambridge. After the age of 16 the number of subjects was reduced to three or four—mathematics, physics, and chemistry for the scientists. Nowadays there is less specialization up to 16, but the same specialization prevails after that age. As a result of this, students who continue at school until 18 can reach a high standard of knowledge in three subjects, which makes it possible for them to be exempted from introductory first-year courses at a university and hence to graduate in three years.

At the time I chose to go into science I had no great enthusiasm for any of my school subjects, but later I became increasingly absorbed with physics and chemistry. Apart from the formal school lessons, I started reading a considerable number of semipopular expositions of nuclear physics. In the 1930s there were a number of semipopular accounts of the remarkable developments in physics during the first three decades of the present century. As a result, it was possible for the informed layman to get a grasp of the concepts and ideas which were exciting physicists at the time, including the Niels Bohr model of the atom, the Heisenberg Principle of Uncertainty, and the wave properties of electrons. I was excited by this ferment of ideas and became convinced that the study of these problems was the most fascinating and worthwhile pursuit that anyone could follow. I realized that to achieve my ambitions I would need to gain a university scholarship, and the Headmaster and teachers encouraged this idea. Great prestige was attached at school to winning competitive scholarships to Oxford and Cambridge, but other universities were considered beyond the pale and hardly mentioned.

My father was sympathetic to my aspirations but decided that even if I could get a university scholarship, the additional financial burden this would entail was beyond his means. There were no grants from public funds to cover living expenses, as there are now, and he felt, quite rightly, that with a large, young family he would not be justified in spending an undue proportion of his limited resources on my further education. Consequently, it was decided that I would have to leave school at 17 and start earning my living. Accordingly, I took and passed a competitive examination for entry into the Civil Service and found myself in a section of the Inland Revenue (Income Tax) in London. I had no interest in any type of office work, but in the difficult times of the 1930s, when jobs were hard to come by, I was prepared to accept it as a means of earning a living. However, my ambitions for scientific research had been aroused and I had no intention of staying in the Civil Service all my life. But what could be done about it?

Apart from my primary interest in physical science, my hobby for some years had been gardening. At home we had quite a large garden, and I developed a keen interest and satisfaction in growing plants. Since I couldn't pursue my first love of physics, I decided to explore the possibility of becoming qualified for a research post in horticulture.

Birkbeck College

Although it was, and still is, much less common for a student to "work his passage" through college or university in Britain than in America, nevertheless there were opportunities to obtain a degree through part-time study at London University. For many years London University has offered "external" degrees to students who have attended part- time or full-time courses at colleges and polytechnics outside the university itself. Moreover, London University has a unique institution, Birkbeck College, which offers evening courses leading to "internal" degrees of the university. Birkbeck College has had many distinguished scientists and scholars on its staff, and during my time there P. M. S. Blackett, discoverer of the positron and Nobel prize winner, was appointed Professor of Physics, and later J. D. Bernal was Professor of Biophysics there. The courses offered at Birkbeck compared favorably with the full-time courses at the other colleges, and the same academic standards were applied to both part-time and full-time students, who took the same university examinations.

Although Birkbeck College did not offer courses in applied science such as horticulture, it offered an honors degree course in botany and I decided to enroll for this. During the first year I attended courses in botany and zoology, as well as physics and chemistry. Not having studied biology at school, this was my first acquaintance with the biological subjects, and at first I found the purely descriptive accounts of structure and life histories very strange after the strictly quantitative and analytical approaches of physics and chemistry. I tried to understand biological phenomena in terms of physicochemical principles and found it frustrating not to be able to do so. The practical experiments in first-year plant physiology were interesting but seemed crude in the extreme. However, by the end of the first year I had made the necessary intellectual readjustments and had learned to accept biological phenomena for what they were, even though it was not yet possible to explain them in physicochemical terms.

The Professor of Botany was Dame Helen Gwynne-Vaughan, a well-known cytologist and mycologist, who had been Commander of the Women's Royal Auxiliary Corps during both World War I and again at the beginning of World War II. She was a charming but somewhat formidable person and a good teacher. Under her influence I developed a special interest in cytology and genetics. London University has, for many years, had a system of "Intercollegiate" lectures which are attended by students from all of the 6 or 7 Colleges of the university, and I was especially thrilled to able to attend extensive courses of lectures in cytology and genetics from C. D. Darlington and J. B. S. Haldane. I also had the opportunity in 1934 of attending a summer school on cytology and genetics at the famous John Innes Institution, which was at its heyday at that time, and again had lectures from Darlington, Haldane, Mather, and several other recognized authorities in their fields. Darlington had revolutionized thinking in cytology, which previously had been mainly a descriptive subject without unifying hypotheses. Haldane was a rather poor lecturer, but what he had to say was always interesting and stimulating, and he had a great influence on thinking in genetics, evolution, and even enzymology. One of the ideas which I picked up from him was that genes have specific times of action during the development of the organism, a concept which antedated by many years the current dogma that development is a matter of selective gene expression, an idea that seems self-evident now, but 50 years ago it was new and striking. Haldane also saw the need for biochemical approaches to genetics and was responsible for initiating research on combined biochemical and genetical studies on flower pigments.

By the end of my second year as a student I was so attracted by genetics that I had decided that this was the field in which I wished to work later. However, at this time (1934) Dr. F. C. Steward joined the staff of the Botany Department at Birkbeck after spending several years in the United States, during which he had established a good reputation for his work on salt uptake. I had previously been mildly interested in plant physiology but the teaching in this subject had been dull and pedestrian. Dr. Steward's rigorous approach appealed to me, so that physiology was added to cytology and genetics as my main areas of interest in botany, but I was also attracted by parts of zoology, which I took as an accessory subject, and I was particularly interested in animal embryology and development. By contrast, the botany course included almost nothing on plant development, and I saw here a fascinating and challenging potential field of research.

I graduated four years after starting my biology course, the normal time for a student who had not studied biology at school. It had been fairly tough and had meant attending lectures and classes five evenings a week and also studying a good deal in the evenings during the vacations, leaving little time for other activities. In common with other young people of the 1930s, I was deeply disturbed by the advance of Fascism, but I had no time for political activities, or even for any social life. However, I was by no means unique in obtaining a degree by part-time study, and indeed three of my friends did so, one graduating in engineering, another in law, while a third obtained not only a first degree in classics but also a PhD in philosophy by part-time work and later became a university professor in the subject. The total number of students graduating each year by part-time study must have been quite large.

Soyabeans and Other Matters

By the time I had graduated I was impatient to get started on research. Dr. Steward was prepared to take me on as a graduate student if I could get a studentship, which were very hard to obtain. But this would have meant giving up my Civil Service job, which seemed a risky step to take since even people with PhDs were out of work at that time. In the end I decided to play safe for a time and do research on a part-time basis. Being imbued with youthful idealism, I wished to undertake research which would be of immediate benefit to mankind, although problems of development still attracted me strongly. It so happened that at the time there was considerable interest, both in Britain and in Nazi Germany, in the possibility of growing soyabeans under European conditions. There was talk, even at that time, about the potential of soyabeans for helping to ease the shortage of proteins in what we should now call the Third World.

I discovered that a retired plant breeder, Professor W. Southworth, who had previously worked on soyabeans at the University of Manitoba, was actually working on this problem at Rothamsted Experimental Station at Harpenden, near London, and I was given permission to work with him on a part-time basis. Progress was slow, although I tried to speed up the breeding work by having my F_1 hybrids grown in South Africa during the English winter. The work came to an abrupt end with the outbreak of the World War II in September 1939, and I remember harvesting my last experimental plants on the morning that war with Germany was declared. We had achieved almost nothing, although I had gained three years' research experience, and while living at Harpenden I met my future wife, Helen, so that my stay there brought other rewards, if not in research.

Helen's grandfather Charles Clark, was an English metallurgist from Middlesborough who had gone to work in Russia in the latter part of the nineteenth century. In 1869 a Welsh iron master and entrepreneur, John Hughes, had been commissioned by the Russian government to develop an iron and steel industry in the Ukraine as one of the first steps in the industrial development of Russia, and Charles Clark later joined the enterprise. The industry developed rapidly and led to the growth of a new town, formerly known as Yuzovka ('Hughes-ovka'—now Donetsk), named after John Hughes. Both Helen and her father were born and brought up near Yuzovka but had British citizenship. She returned to Britain with her parents in 1933, after Stalin had staged a show trial of British engineers. In his memoirs, Kruschev, who lived in Yuzovka as a young man, states that after the Revolution he was Commissar for the village in which Helen then lived with her mother and maternal grandparents and he may, therefore, have been responsible for taking over their house as a Workers Club in 1922. A vivid description of Yuzovka is given by the well-known Russian writer, Konstantin Paustovsky, in his autobiography.

During the war I served in the Army as a Telecommunications Officer, with responsibilities mainly for radar equipment used by coast artillery. A number of young biologists who later became well-known were engaged in this type of work in the Services, including Professor John Heslop-Harrison. I finished my Army service in the West Indies, where I made contact with the Imperial College of Tropical Agriculture in Trinidad and learned something about tropical plants in my spare time.

At the time of my release from the Army, my wife was living in Manchester, and one of the botany professors there was C. W. Wardlaw, who was carrying out his exciting work on morphogenesis in ferns and studying exactly the sort of problems of development which strongly interested me. I spent my demobilization leave in his laboratory, but during this period I was successful in obtaining a junior post at Bedford College, London University, which was at that time a women's college, being among the first colleges in the world to provide a university education for women, beginning about 1850. Although I could have gone back to my old Civil Service job, I decided to burn my boats and change to the academic life.

THE ACADEMIC LIFE

Bedford College, London

The Botany Department at Bedford College had been destroyed during the bombing of London, and the Departments of Botany and Zoology were housed in the former mansion of a member of the gentry on the edge of Regents Park. My office had been "Lady Anne's bathroom," with the bath still there! The department had had to start again from scratch and there was very little equipment beyond the minimum required to run classes. There were only four academic members of the department to teach all phases of botany, and during my three years there I was responsible for teaching cytology and genetics, mycology, ecology and the gymnosperms, as well as developmental physiology. However, there were advantages to working in a small department; there were fewer distractions and very little administration, so despite the heavy teaching load, I found I had more time for research than I have ever had since.

Head of the department was Professor Nielson Jones, who had been interested in a number of developmental problems, including chimeras. His wife, Dr. M. C. Rayner, was a well-known mycologist with a special interest in mycorrhizas—she had been stressing the importance of mycorrhizas, both ectotropic and endotropic, in the nutrition of a wide range of plants some 50 years before the recent resurgence of interest on the subject. Her research at that time was concerned with mycorrhizas associated with coniferous forest trees and was supported by the British Forestry Commission. Professor Nielson Jones suggested that I should work on a tree problem for a PhD and drew my attention to an observation that they had made, namely that the earlier in the season that Scots pine seeds were sown the better was the growth of the seedlings in the *second* year. This was the problem they suggested I should investigate.

It occurred to me that the effect of sowing date on Scots pine seedlings might be related to seasonal changes in photoperiod. Accordingly, I set up experiments to investigate the effects of a range of different photoperiods on the growth of seedlings of Scots pine in the first and second years. Since I depended on manual covering and uncovering of the seedlings, these experiments were set up at our home and my wife had the job of attending to the covers a various times during the day when I was away at work, the lengths of the photoperiods being chosen to fit in with the baby's feeding times!

The original problem proved to have a relatively simple solution, since after the first year, the number of needles and internodes of the annual shoot is predetermined by the number of primordia laid down in the resting bud formed in the previous year; consequently, if first year seedlings are sown late in the season, the number of needles and internodes laid down in the resting buds is fewer than for earlier sown seedlings and this is reflected in the reduced growth of the seedlings in the second year.

Although the results of these early experiments on Scots pine seedlings were not exactly epoch-making, they drew my attention to the large number of interesting problems associated with tree growth waiting to be studied, and which occupied my attention for a number of years. My experiments with Scots pine also included observations on cambial activity and aroused my interest in the general question of the control of cambial activity in trees. During my second year at Bedford College, Nielson-Jones retired and Professor Leslie Audus was appointed in his place, and he started up his well-known researches on the hormonal control of root growth, geotropism, and other topics.

In 1950 I was approached by Professor Eric Ashby (now Lord Ashby) as to whether I would be interested in moving to Manchester University; he himself was about to leave Manchester to become Vice-Chancellor of Queen's University, Belfast, and this created an opening to teach the courses in plant physiology for which he had been responsible. I decided fairly quickly to accept, since the Botany Department at Manchester was much larger than at Bedford, and although it was far from well equipped, it appeared to offer new opportunities, and I was not disappointed.

Manchester

The new Professor of Botany at Manchester was S. C. Harland, a distinguished geneticist and cotton breeder, while C. W. Wardlaw continued as Professor of Cryptogamic Botany. The intellectual atmosphere (as opposed to the heavily polluted physical atmosphere) was buoyant and exciting, especially for those interested in development. One of my closest colleagues was Dr. Herbert Street, with whom I was responsible for the teaching of plant physiology. Street rapidly built up a very active research group in root and tissue culture, and his cheerful enthusiasm infected the whole department. His dynamism and energy continued to play a major role in the development of tissue culture until his untimely death in 1979, when he was Professor at Leicester University. Another of my close friends was Dennis Carr, with whom I shared a common interest in photoperiodism. We had differences of opinion regarding the validity of Bünning's hypothesis about the role of endogenous rhythms, about which I was very skeptical and carried out a few experiments to test the hypothesis, but later I had to concede that endogenous rhythms in photoperiodic sensitivity do exist, and probably are involved in the time measuring process in photoperiodism.

There was a lively group of graduate students working under Wardlaw on various aspects of morphogenesis, several of whom later became leading workers in the field themselves, including Taylor Steeves, Ian Sussex, and Elizabeth Cutter. There was also an active group under Dr. Joyce Bentley working on auxins, in association with Professor E. R. H. Jones of the Chemistry Department, who was attempting to re-isolate auxins "a and b", but who finally isolated indole-acetonitrile from cabbage. I was able to draw on their expertise in hormone work and to start our own work on endogenous hormones in relation to bud dormancy in woody plants. The main lines of our future research were established in Manchester and continued in Aberystwyth, and I will describe these under separate headings below.

Aberystwyth

In 1958 I was asked whether I would be interested in the Chair of Botany at Aberystwyth. In Britain the Professor in a small department is normally also the Head and carries the bulk of the administrative work. Having given up an office job in the Civil Service, I had no wish to become desk-bound again, and for this reason I had always said that I did not wish to become a Professor. However, on visiting Aberystwyth I was impressed with the facilities it could offer, since a new biology building was being constructed and there were good greenhouse and garden facilities for growing plants. Moreover, the sea, and the unrivaled countryside around the small town were an attraction I could not resist. My predecessor was Professor Lily Newton, a distinguished algologist.

The 1960s and early 1970s were years of unprecedented university expansion in Britain and elsewhere, and it was relatively easy to get faculty appointments, studentships, and research grants. Whereas previously our work had been handicapped by inadequacies in modern equipment, suddenly it became possible to obtain expensive items of equipment which only a short time earlier seemed well beyond our means. In this climate it was possible to build up sizeable research groups and to maintain the impetus of research projects. The department expanded greatly and we were able to establish a Chair of Microbiology, occupied by my good colleague, Professor Gareth Morris. However, this Golden Age lasted only about 10 years, and a slightly chill wind began to blow with the first energy crisis in 1973, and we have now moved back into the Ice Age. My scientific career was delayed by 10 years because of the economic depression of the 1930s and the War, but I was fortunate to be able to build up an active department and research group at a time when all things seemed possible. As always, any success we achieved depended upon the efforts of the whole department, including a succession of very able graduate students, many of whom have established themselves as leading researchers throughout the world.

MEANDERINGS IN DEVELOPMENTAL PHYSIOLOGY

Inhibitors, ABA, and Dormancy

I was interested in the fact that it had been known since the observations of Klebs over 60 years ago that dormant buds of beech (*Fagus sylvatica*) are apparently sensitive to photoperiod even in the leafless condition, and I was able to confirm this for birch (*Betula*) and larch (*Larix*). This observation seemed incompatible with the well-known fact that in the photoperiodic control of flowering the shoot apex is relatively insensitive and that it is the daylength conditions to which the mature leaves are exposed which determines the response. (It was primarily this observation which led Chailakhyan to postulate the existence of florigen in 1936, of course.) Consequently, I decided to investigate the locus of photoperiodic perception in sycamore (*Acer pseudoplatanus*), the buds of which do not respond to photoperiod in the leafless condition, and in birch, in which they do. It turned out that the situation in sycamore is exactly analogous with that in herbaceous species, in that the response is determined only by the photoperiodic conditions under which the mature leaves are maintained. In birch, however, both the leaves and the buds respond. This finding led to the demonstration that if birch seedlings are first rendered dormant by exposure to short days, the buds can be induced to expand and resume growth if either (*a*) both buds and leaves are exposed to long days, or (*b*) if the buds only of defoliated seedlings are maintained under long days; but if the buds are exposed to long days and the leaves to short days, then the buds remained dormant, indicating the transmission of some inhibitory influence from the leaves under short days, whereas "long day" leaves had no such inhibitory effect.

These observations suggested that an inhibitor of bud growth may be produced in the leaves of woody plants in higher amounts under short day than under long day. This idea was fully in accord with Hemberg's hypothesis that bud dormancy is regulated by endogenous inhibitors, and Bennet-Clark and Kefford had recently demonstrated the occurrence of a β-inhibitor fraction in many plant extracts. We therefore decided to initiate a program of work to investigate the role of growth inhibitors in bud dormancy and to determine the effects of photoperiod on inhibitor levels in seedlings of woody plants.

This problem was tackled by David Phillips and myself in sycamore. We used simple techniques of purification including paper chromatography, and inhibitory activity was determined by bioassay, using the *Avena* straight growth test. We found the expected β-inhibitor fraction in extracts of buds and leaves of all three species, and we showed that there is a decline in the activity of the β inhibitor levels in the buds during the winter when they are exposed to natural chilling. We also found significantly higher levels of β-inhibitor in the leaves and apical buds of sycamore seedlings under short day than under long day.

In 1957 I attended a meeting on photoperiodism at Gatlinburg, Tennessee, at which I met Dr. Jean Nitsch and found that he had been investigating the effects of photoperiod on inhibitor levels in dogwood and sumac, and that he had obtained similar results to those we found in sycamore. Moreover, similar differences were obtained later by several other graduate students in our group in sycamore, birch, and blackcurrant. This was the point which our work on inhibitors had reached at the time I left Manchester in 1958.

On moving to Aberystwyth we decided to make a direct attempt to identify the component of the β-inhibitor in sycamore that appeared to be varying with photoperiod. Two graduate students, Peter Robinson and Peter Jones, were involved in the early phases. Varga had demonstrated that a number of phenolic substances, including salicylic, ferulic, cinnamic, and coumaric acids, are present in the β-inhibitor fraction in many types of plant tissue, and since these substances are known to inhibit growth, it was concluded that the activity of the β-inhibitor fraction was due to the cumulative effects of these weakly inhibitory phenolics. However, we decided to investigate whether it was possible to separate the inhibitory activity of the β-inhibitor zone from the phenolics, and we were very quickly able to do this using a chromatographic solvent based on n-butanol. We therefore proceeded to concentrate on the nonphenolic inhibitory fraction, and we were able to purify it sufficiently to show that it was significantly more active than coumarin, the most active natural inhibitory substance then known. We felt we needed the assistance of an organic chemist for further progress that would lead, we hoped, to final identification.

We approached Dr. J. W. Cornforth, who was then at the Medical Research Institute in London, who agreed to collaborate. On moving from Mill Hill to the Shell Agricultural Research Centre at Sittingbourne, near London, Dr. Cornforth and his associates, Dr. B. Milborrow and Dr. G. Ryback, began further purification of the inhibitory fraction from sycamore leaves. In 1963 I reported progress on work at the International Plant Growth Substances Conference organized by Dr. Jean Nitsch at Gif-sur-Yvette, Paris, and it became apparent that the properties of our material corresponded closely with those of abscission-accelerating fraction (abscisin) isolated from cotton fruits by Addicott's group and from yellow lupins by Rothwell and Wain, and which also showed high growth inhibitory activity. It seemed very possible that the active substance was identical in all three fractions. In 1965 Addicott's group published the structure of Abscisin II, and Dr. Cornforth was able immediately to confirm this structure by synthesis and to show that the active fraction from sycamore leaves was identical with Abscisin II.

While work on the isolation of the inhibitor from sycamore leaves was going on, we also investigated the effect of a crude preparation of β-inhibitor on the growth of birch seedlings. The inhibitory fraction was applied to the leaves of birch seedlings growing under marginal day lengths, and we were successful in causing the cessation of growth and the formation of normal-looking terminal resting buds. Thus, we seemed to have evidence that (a) there were higher levels of inhibitor in the leaves and buds (and also the phloem sap) of woody seedlings growing under short days than under long days; (b) that crude preparations of the inhibitor would induce resting bud formation in seedlings under long days; (c) that the major part of the activity in the β-inhibitor appeared to be caused by the presence of ABA; and (d) that there is a mutually antagonistic interaction between the effects of gibberellic acid and ABA on bud growth. Although not conclusive, the evidence seemed to suggest quite strongly that the short-day induction of bud formation in woody plants is mediated via photoperiodic regulation of ABA levels, although the possibility that gibberellin levels are also modulated was always in our minds.

However, subsequent work has shown that this was an oversimplified view, and the role of ABA in bud dormancy is still not clear. First, after the development of GLC techniques for direct determination of ABA levels, Lenton and Saunders at Aberystwyth showed that the levels of ABA in sycamore leaves and apices were no higher under short days than under long days, but indeed were sometimes *lower* under short days. This result was totally unexpected in view of our earlier observations on β-inhibitor levels determined by bioassay. To add to our consternation, it was reported by Hocking and Hillman that application of R,S-ABA to seedlings of birch and alder growing under long days does not induce bud formation. Thus, two of the main props of the hypothesis that ABA is involved in the photoperiodic induction of bud dormancy were demolished, leaving uncertainty about the role of ABA. This problem has been reviewed by several authors (6, 7) and I have myself discussed the problem recently in relation to both bud and seed dormancy (10, 12).

Hormones and Seed Dormancy

Although my interest in dormancy originally developed from studies on the effects of photoperiod on bud formation in woody plants, it became clear to me that seed dormancy presented a number of fascinating problems and that certain aspects of dormancy could be studied more conveniently in seeds than in young trees.

We investigated three types of seed showing different forms of dormancy: birch (light requiring), cocklebur (the classical seed for oxygen effects) and ash (chilling requirement). One of our main interests was the possible role of endogenous inhibitors in seed dormancy, to complement the work on bud dormancy. The role of inhibitors in preventing premature germination of the seeds in succulent fruits such as tomato had been recognized since the early work of Molisch and his associates (8), so the concept of growth inhibitors antedates the discovery of auxin. However, the idea that inhibitors may be important in preventing germination after shedding of the seed (i.e. in dormancy) had hardly been considered, although it was discussed by Barton in 1945 and discounted by her, but following Hemberg's suggestion that inhibitors play an essential role in bud dormancy, it was logical to extend the concept to seeds.

Needless to say, our experimental results were inconclusive as to whether inhibitors are involved in the dormancy of the seeds we investigated, but with each type of seed, leaching reduced the dormancy and was shown to result in lower inhibitor levels in the embryos or seed coats, while application of the inhibitor to leached seeds reduced germination, thus providing some circumstantial evidence for the involvement of inhibitors.

In the studies on ash (*Fraxinus excelsior*) carried out by Trevor Villiers, it was found that chilling resulted in an increase in the levels of a germination promoter in the embryos, but whether or not this promoter showed gibberellin activity was not determined. However, these studies on ash produced some of the earliest evidence in support of the hypothesis that seed dormancy and germination may be regulated by an interaction between promoters and inhibitors.

These latter studies on ash were paralleled by work on the seed of two other species showing a chilling requirement, via hazel (*Corylus avellana*) and beech (*Fagus sylvatica*), in which the effects of chilling on endogenous gibberellin levels was studied by Barry Frankland. Only relatively slight increases in gibberellin levels during the actual chilling period were found, but this work has been developed extensively by our former colleague, Dr. J. W. Bradbeer, and his former graduate students, N. Pinfield, B. Jarvis, and J. Ross, and has resulted in substantial advances in our understanding of the effects of chilling on gibberellin metabolism in hazel seeds.

Since exogenous cytokinins are effective in releasing dormancy in some seeds, such as sugar maple, which do not respond to gibberellins, P. Webb and J. van Staden investigated the effects of chilling on endogenous cytokinin and gibberellin levels in sugar maple (*Acer saccharum*) seeds and showed that there are successive peaks in the levels of both types of hormone. Similar results were obtained in sugar pine (*Pinus lambertiana*).

There have been several reports that exposure to red light leads to rapid increases in endogenous gibberellin levels in both etiolated tissues and seeds. We were able to show that similar peaks in both gibberellins and cytokinins occur in seeds, but whether these effects are important in the light responses of the seeds remains uncertain. Indeed the whole question of the role of hormones in the regulation of the various forms of seed dormancy remains obscure (1), but it would be surprising if they turned out to be of no significance, especially in seeds with responses to chilling (10).

Hormones and Cambial Activity

My initial interest in the seasonal pattern of growth and cambial activity in Scots pine seedlings brought to my attention several other intriguing problems of growth and flowering in trees which seemed to have received little attention, especially during the war years, and which I was tempted to explore. The first of these topics was the control of cambial activity in trees.

Priestley and others had shown earlier that there is a close correlation between cambial activity and the activity of the shoot apices, and the presence of swelling buds had been demonstrated to be essential for renewal of cambial activity in several species. Moreover, the "cambial stimulus" assumed to arise in the shoot apices was apparently transported in a strictly basipetal direction. Following the prewar work of Snow and others in which it was shown that application of auxin to pea seedlings and shoots of woody plants stimulates cambial division, it seemed very probable that the major component of cambial stimulus was auxin. When supplies of gibberellic acid became available I was able to show that it acted synergistically with IAA in stimulating cambial division, but gibberellic acid alone allowed phloem formation but not xylem differentation. These observations were followed by studies on endogenous auxins and gibberellins in seedlings of several broad-leaved woody species, which gave support to the hypothesis that their photoperiodic responses were mediated via variations in endogenous hormone levels. However, these latter studies involved relatively simple purification techniques in association with bioassays, with all the limitations to which these techniques are subject, and our interest in the subject has recently been reawakened. A reexamination of the role of endogenous hormones in the regulation of cambial activity in conifers was started by Dr. C. H. A. Little of the Canadian Forestry Service, Fredericton, Canada, while on sabbatical leave in our laboratory and is being continued by Rodney Savidge, a Canadian graduate student, using the full range of modern techniques for hormone purification and estimation; his results are leading to a drastic reappraisal of current thinking on hormonal control of cambial activity and xylem differentiation.

Growth and Flowering in Trees

While at Manchester, I decided to investigate the statements in the horticultural literature that training or pulling down the branches of fruit trees into a horizontal position checks vegetative growth and promotes flowering. Through the work of T. Nasr and K. A. Longman, we were very quickly able to confirm these reports, not only for fruit trees, including apples, plums and cherries, but also for forest trees, including birch and larch. All the species we studied showed reduced extension growth when the shoots were trained horizontally, and they were accompanied by a loss of apical dominance, with lateral buds on the upper side of the main stem, which normally were partially or completely inhibited, growing out strongly. However, the effects on flower initiation were not found in some species, such as birch and beech, whereas the effect on "flower" initiation in larch was dramatic. Since these were not covered by any existing term, we called the phenomenon "gravimorphism." From the results of subsequent studies by Harry Smith and other graduate students, it seems that the effects of shoot orientation on apical dominance probably result from a redistribution of auxin within horizontal or inclined shoots, so that the buds on the upper side are released. However, studies on the effects of shoot orientation on cambial activity and xylem differentiation suggest that redistribution of other factors, possibly cytokinins and/or inhibitors, may also be involved.

My studies on growth problems in forest trees brought me into contact with John Matthews (now Professor of Forestry at Aberdeen), who drew my attention to the phenomenon of juvenility in forest trees. The breeding of fruit and forest trees is greatly hampered by the inability of seedling trees to flower below a certain size, and considerable practical benefit would accrue from methods for inducing early flowering. The phenomenon also presents problems of considerable general biological interest, since once the adult, flowering condition is attained it is a remarkably stable state and can be transmitted by vegetative propagation by cuttings or grafting. Thus, the phenotype of woody plants can exist in two distinct, stable phases, but the transition from the juvenile to the adult state clearly cannot involve an irreversible change in the genome since it is the adult phase which gives rise to the juvenile phase of the next generation. The phenomonon has been referred to as "phase-change."

We addressed ourselves to the problem of what determines the stage at which the juvenile-adult transition occurs—is it chronological age, the number of annual cycles of active growth and dormancy, or size? Direct experiments to test these alternatives gave unequivocal results in favor of the last, since birch seedlings grown continuously under long days started flowering about 13 months after sowing, whereas under normal conditions they would not flower until the age of 3–5 years. Similar results were obtained with larch. The formula for rapid attainment of flowering in seedling trees therefore seems to be to grow them as rapidly as possible to a certain size to attain the adult condition and then to apply specific treatments, such as long days, short days, or horizontal training, which are conducive to flower initiation in the species in question.

Apart from its practical importance, phase change is of considerable theoretical significance. I have argued that phase-change is an example of "determination" in shoot apices (9), and that other phenomena such as vernalization and the sporophyte-gametophyte alternation in lower plants are analogous to phase change since they also involve the capacity of shoot apices to exist in alternative states. The phenomenon of habituation in tissue cultures appears to provide a further example of the capacity of meristematic cells to exist in alternative stable states, as Meins & Binns (5) have pointed out. I believe that the phenomenon of cell determination is more widespread than has hitherto been recognized and that it deserves much more attention (11).

Apical Dominance and Stolon Development

Another line of our work began at Manchester in the 1950s and which still continues concerns apical dominance, and especially in relation to stolon development in the potato plant. We have chosen to work with clones of *Solanum andigena,* which show marked photoperiodic responses with respect to stolon development and tuberization, rather than cultivated varieties of *S. tuberosum,* which do not have an obligate requirement for short days for these processes. The work was started by Alan Booth, as a graduate student, who made the interesting discovery that whereas application of IAA or GA alone to decapitated potato shoots either inhibited or stimulated the growth of the uppermost lateral bud, respectively, application of a combination of IAA and GA caused it to develop into a stolon-like aerial shoot. This finding suggests that natural stolon development depends upon high endogenous gibberellin levels, a problem which we are still pursuing. However, to develop as a natural stolon at the base of the plant a bud needs also to be subjected to apical dominance, because if the aerial shoot is decapitated and all lateral buds removed from it, the subterranean stolons turn up and become leafy shoots. This conversion of stolons into orthotropic leafy shoots is promoted by the presence of roots (although we now know that it will also occur in absence of roots if high levels of inorganic nutrients are supplied). A stolon can be converted into a leafy shoot while still subject to apical dominance if kinetin or benzyladenine is applied directly to the stolon tip. Thus, the conversion of a stolon into a leafy shoot is promoted both by roots and by exogenous cytokinin. Since roots are known to be a source of endogenous cytokinins in the plant, these observations led to the hypothesis that the role of IAA in the development of a lateral bud as a stolon is in some way to prevent the access of root-produced cytokinins to the lateral bud, either by diversion to the main shoot apex or by blocking the movement of the cytokinin into the bud itself. The work of Thimann's group had, of course, also led to the concept that apical dominance is controlled by an interaction between IAA and endogenous cytokinins, but they had suggested that IAA acts by inhibiting cytokinin biosynthesis in the bud itself.

It was clear that the solution of the problem would require more precise information on (*a*) the identification of endogenous cytokinins, (*b*) their sites of production within the plant, (*c*) the nature of the transported cytokinins, (*d*) their mode of transport and distribution within a plant, and (*e*) the metabolism of cytokinins in different tissues, especially in relation to the regulation of their levels. Although this is a formidable program, we decided over 10 years ago to initiate a long-term project that is still continuing. The first task was to develop the necessary techniques for the purification, identification, and quantification of endogenous cytokinins, using HPLC, GCMS, and the various other opportunities that modern physical methods offer. This has been achieved very successfully by my colleague, Dr. Roger Horgan, so that we have been able to identify a new natural cytokinin and several new cytokinin metabolites, thus complementing the work of Letham and his associates.

This is not the place to describe in detail the results of our recent biochemical and physiological studies, but although the pathway of cytokinin biosynthesis in higher plants is proving singularly difficult to elucidate, more progress is being made on the fate (through catabolism and conjugation) of cytokinins in different tissues of the dwarf bean (*Phaseolus vulgaris*), which will ultimately help to build up a model of cytokinin relations in the whole plant. It is clear, however, that the cytokinin relations within the whole plant are far more complex than might have been expected. In fact, in plant hormone research it is usually safe to assume that things will turn out to be more complicated than you originally expected.

Hormone-directed Transport

The "Diversion Theory" of apical dominance, suggested by Went, postulated that nutrients are diverted from lateral buds and shoots toward the dominant apex by the high auxin levels it contains; that is to say, it was held that tissues containing high auxin levels tend to "attract" nutrients and assimilates. There was indeed some early evidence that assimilates and proteins tend to accumulate in localized regions of stem tissues to which exogenous auxins had been applied, but these were long-term effects during which growth and "sink activity" could also have been stimulated. With the availability of radioactive sugars and nutrients such as ^{32}P phosphate, it became possible to carry out much shorter term experiments during which growth effects would be minimized. Over the past 20 years, we have carried out a series of investigations on "hormone-directed transport." We have worked mainly with young dwarf bean seedlings decapitated at the top of the second internode after it has ceased extension. We have deliberately chosen nongrowing tissue in order to avoid confusion between any direct effects of hormones on the mobilizing ability of a tissue and indirect effects resulting from the stimulation of growth and hence of demand for assimilates. Mothes and his co-workers had, of course, shown that applied cytokinins increase the mobilizing ability of mature tobacco leaf tissue, but the distances involved were relatively short and apparently did not involve long distance transport. However, we have shown that application of auxins to decapitated, mature bean internodes results in increased accumulation of ^{14}C sucrose from a point of application several centimeters distant. The effects can be detected within a few hours of hormone application and do not appear to involve stimulation of growth or metabolic demand. These effects can be demonstrated for auxins, cytokinins, and even gibberellins, although these various types of hormone appear to differ in their mode of action. The responses to auxin appear to be mediated through effects upon the process of unloading from the phloem, although this is held by some workers to be a passive process determined by diffusion gradients and therefore unlikely to be affected by hormones.

The subject remains controversial, but I am convinced that applied auxins and other hormones can stimulate the long-distance mobilization of assimilates by tissues independently of any effects on growth or metabolic demand, and the only doubt I have relates to the magnitude of the effects, since the nongrowing tissues with which we work have very low "sink capacity" and the effects involve only minute amounts of ^{14}C-sucrose or other assimilates. However, these hormone effects may be important in the intact plants in conferring increased competitive ability for assimilates in growing tissues, such as young fruits and shoot apices, which contain high hormone levels. This work is being continued by Dr. John Patrick in Australia, and his elegant and critical experimentation has already greatly contributed to the development of the subject.

HINDSIGHT AND FORESIGHT

The Problem of Hormonal Control

Most of the problems on which I have worked have been concerned with hormonal control of some aspect of development—in seed and bud dormancy, cambial activity, gravimorphism, apical dominance, hormone-directed transport—and it will be only too apparent that we seem to be no nearer to an understanding of the processes than we were 30 years ago. This is true for almost every physiological process in which we believe hormonal factors are important. Indeed, in some areas the situation is almost more confused than it appeared to be 40 or 50 years ago—in phototropism and geotropism, for example. What are the reasons for our slow rate of progress and what should we do about it?

Until the recent remarkable improvements in techniques for purification and estimation of plant growth substances, including especially HPLC and GCMS, the impurity of our extracts, with no estimation of recovery, and the variability inherent in bioassay techniques were major stumbling-blocks. We can now achieve a standard of precision which was unthinkable 10–15 years ago, and although the cost of the hardware required is likely to reduce the number of laboratories able to work in the hormone field, we now have very adequate techniques for the determination of endogenous levels of all classes of hormones, although the gibberellins continue to present difficulties because of their large numbers.

However, why have problems of hormonal control not immediately become greatly clarified with the availability of the new techniques? Probably we have to allow more time for them to be applied, but in a number of cases where hormone levels have been determined with high precision, our understanding of the physiology of the system studied has not advanced significantly. Why should this be? Generally, samples taken for hormonal analysis, such as stems, leaves, and roots, include a variety of tissues, whereas it is likely that the hormone levels vary considerably between different tissues, e.g. between cambium and cortex, and determinations of average hormone levels for various tissues may not be significant. However, it is also likely that the hormone concentration will vary within the different compartments of the individual cell, e.g. in cytosol, vacuole, and cell wall. Fractionation of cell components prior to hormone analysis has not yet been widely used, but probably it will be necessary to do this before we can obtain fully meaningful results from hormone analysis. It is generally assumed that hormones bind with receptor molecules, and where this is shown to be so, we shall ultimately need to know the partition of "free" and "bound" hormone.

It has been suggested that the mode of action of some hormones, such as gibberellins, involves their inactivation and that the significant parameter is not the steady state concentration but the rate of turnover. If this is so, then attempts to correlate steady state levels with developmental processes are likely to be fruitless. In any case, we need to know more about the biosynthesis and regulation of hormone levels in the plant, and the present activity in this area is timely and appropriate. So also are the increasing efforts to isolate and identify receptors for each class of hormone. The information which will come from these studies will provide a much sounder basis for understanding the functional control of growth and development than we have had hitherto.

Yet another large lacuna in our knowledge relates to the modes of action of plant hormones. After some 40 years of study we may be gaining some insight into the mode of action of auxin, but for the other classes of hormone our knowledge is virtually nil. However, this lack of knowledge may reflect the more general lack of understanding of how cell division, cell enlargement, and differentiation are controlled. Our ignorance of these subjects is not confined to plants but extends also to microorganisms and animals cells. Possibly we shall not be able to determine the mode of action of plant hormones until we know more about the control of gene expression in eukaryotes.

At another level of organization there is a dearth of information on hormone relations within the plant as a whole. As we realized when we came to consider the possible role of endogenous cytokinins in apical dominance, we need to know much more about the sites of biosynthesis within the plant for each class of hormone and about their transport, distribution, and regulation within the various parts of the plant.

As though all these difficulties and limitations were not enough, there are also inherent difficulties in determining whether a given process is under hormonal control, so that even where a correlation can be established between variations in hormone levels and developmental or metabolic events, it is usually difficult to establish the causal relations between the two sets of observations. Jacobs (3) set out a list of criteria (the "PESIGS" rules) which need to be met to establish that a process is under hormonal control, and Dennis (2) has developed this approach further, specifically in relation to regulation by growth inhibitors. There is clearly no simple answer to this problem, and final solutions are unlikely to be forthcoming until we know precisely the mode of action of a hormone in a given process.

Phenomenological and Molecular Approaches

Our main lines of work were established during a relatively short period at Manchester, and these we have continued to pursue for 25 years with a few minor diversions. In the early phases of the work we were mainly engaged in investigating the phenomena, using relatively simple experimental approaches, and in this phase it was fairly easy to keep a number of separate lines going and still to make reasonable progress. Subsequently we have tried to investigate the mechanisms involved, and this is a much more difficult task, particularly where it has involved studies on endogenous hormones, for the reasons already discussed.

Progress in plant physiology as a whole over the past 20–30 years has, of course, been less rapid and less spectacular than in molecular biology and microbiology. This is no doubt partly explained by the fact that there are far fewer researchers working on plant problems, including plant biochemistry, and it also reflects greater difficulties in working with plant tissues, arising from the presence of cell walls and vacuoles and the difficulty of obtaining large quantities of homogeneous plant material. However, progress in physiology, both plant and animal, is inevitably going to be slower than in molecular biology or microbiology because of the complexity of the systems under study.

Obviously, there are basic differences in the nature of problems within plant physiology itself. In areas such as water relations or electrolyte uptake, one is dealing with problems to which basic principles of physical chemistry and thermodynamics can be applied more or less directly, but in developmental problems, such as flowering or apical dominance, we are dealing with systems showing a complexity which is greater by at least an order of magnitude and it is impossible at present to interpret the phenomena directly in physicochemical terms, so that initially they have to be studied and described at the phenomenological level. The differences in approach, which might for convenience be referred to as the "molecular" and "phenomenological," respectively, are highlighted in studies on hormonal control of growth and differentiation. On the one hand, where we have a specific hormone of known chemical structure, it is possible to study the metabolism of the hormone and ultimately to determine its mode of action at the molecular level. On the other hand, a phenomenon such as apical dominance exists only at the whole plant level of organization, and then the complexity inherent in the system is superimposed upon the events occurring at the molecular level. It is perhaps not surprising, therefore, that nearly 50 years since Skoog and Thimann first demonstrated that auxin probably plays a primary role in correlative inhibition of buds, we still do not fully understand the mechanism of this phenomenon.

However, apart from the general problem of complexity of whole plant systems, developmental physiology involves additional complexities arising from our lack of understanding of development at the molecular level. The dogma that differentiation involves selective gene expression seems to me to be valid, but since we are only now beginning to get an understanding of the structure of the eukaryote genome and still have little or no understanding of the mechanism of selective gene expression in eukaryotes, there is clearly a greater lacuna in our knowledge at the molecular level than is the case in other areas of biology. This lack of knowledge of the molecular biology of development is probably a major obstacle to progress in developmental physiology, both in the hormone field and in other areas, such as the mode of action of phytochrome. However, the very rapid progress in molecular biology of eukaryotes now made possible by the techniques of "cloning" and sequencing will probably result in a transformation of the situation within a few years, and this will undoubtedly lead to progress in our understanding of plant development.

Until it is possible to investigate more effectively the mechanisms of developmental processes, in many cases it will remain possible only to observe phenomena and to attempt to put forward speculative interpretations of the observations. To those who have been schooled in the physical sciences this type of approach is alien and futile, and they emphasize that the only worthwhile approach is a rigorous one at the molecular level. While I applaud all attempts to gain information at the molecular level, I consider that both types of approach are valid and complementary. F. G. Gregory used to emphasize the importance of fully investigating the developmental phenomena before plunging into detailed biochemical work, and his classical studies with Dr. O. N. Purvis on vernalization illustrate his point admirably, but it must be admitted that we are still almost totally ignorant of the molecular basis of vernalization. On the other hand, there comes a time in studies on hormonal control, for example, when further progress from the 'phenomenological' approach becomes impeded by lack of detailed biochemical information on hormone metabolism. Therefore, both types of approach are valid and must be kept in step as far as possible. Indeed, all molecular studies must ultimately be related to the functional situation in the intact plant, otherwise they are likely to acquire a momentum of their own and to proliferate without direction or justification.

The complexity of developmental processes makes it peculiarly difficult to formulate unifying hypotheses that can be put to crucial experimental tests. Though others may be more successful in this respect, I usually find when studying developmental problems that my original hypothesis turns out to be invalid and has to be abandoned or modified. However, in the course of the investigation one is somewhat unlucky if at least one new and interesting observation does not turn up, which is worth pursuing for its own sake. In other words, serendipity plays an important part in research in developmental physiology, as Anton Lang has pointed out (4). On this basis, perhaps a good maxim for research in developmental plant physiology might be "Don't just sit there, do something!"

Some Afterthoughts

The criticism could probably be made that I have tried to keep too many different lines of work going and that it would have been more profitable to concentrate on fewer. However, experiments with trees tend to be long-term, sometimes lasting several years, so that it is easier in this area to keep several lines of work going at the same time. Moreover, although it is probably better, in the short run at least, to concentrate one's efforts over a fairly narrow front, there are also advantages in having several fields of interest, although not necessarily in pursuing them all at the same time. There is a danger, if one is too specialized, in getting bored with a single problem over 10 to 20 years, whereas with several areas of interest this is less likely to happen, and there is a better chance of maintaining one's enthusiasm and hence research momentum. One of the advantages of academic life is that it is possible to switch to a new problem if progress on a research program is slowing down and momentum is being lost, and in my opinion it should be made easier for workers in research institutes to do the same, since success in research is obviously entirely dependent on the researcher's excitement and enthusiasm for the problem. In other words, if the fun goes out of research, one might as well give up. However, sometimes it may be sensible to give up a particular line of work until better techniques are available or newer approaches become possible—this has certainly been the case with work on endogenous hormones.

During the course of a lifetime of science one is obviously continuously influenced by a host of people, young and old, and whereas the scientific giants of the past were frequently able to advance almost single-handed, today the existence of large numbers of scientific workers and the rapidity with which experimental results are disseminated through the world mean that advances in discovery and understanding are the result of the interplay of many ideas and of interactions between many individuals. However, I would not agree with the assertions of some sociologists that scientific knowledge is not objective but is a product of the political and economic climate in which it is pursued. On the contrary, I would stress the international nature of science, which cuts across natural and political boundaries, and which constitutes one of the rewards of the scientific life. It is true that the idea that scientists are engaged in a disinterested pursuit of the truth is an overidealistic fiction, since what most scientists are seeking is recognition and this is a powerful factor in providing their motivation—"fame is the spur" for most of us. However, to gain this recognition we have to conform to the rules of the game and to subject ourselves to the criticism of our peers, so I maintain that scientific knowledge is to a high degree objective, although the individual scientist is far from disinterested.

Whereas it is frequently said that the ability to pursue original research in mathematics or physics declines rapidly with age, I take comfort in the fact that age seems less of a handicap for biologists. I am inspired and encouraged by the examples set by many of my senior colleagues, including Dr. Kenneth Thimann and Dr. Sterling Hendricks, who have continued to be excited by their research and to make valuable contributions, although it is a great loss that Dr. Hendricks is not longer with us. I look forward to the opportunity that retirement offers to tie up some of the loose ends left in our own research. However, the task is no doubt endless, since each advance opens up new problems.

The prospect for the future in developmental biology is exciting, and the next 10 or 20 years will undoubtedly see major advances in our understanding. One looks forward to participating in a minor way to this new phase in the unraveling of the old mystery of development.

The scientific life is indeed an Odyssey, a voyage of discovery, in which the Golden Fleece is a better understanding of the natural world, and some of the thoughts that go through one's mind seem admirably embodied in the words of Tennyson's Ulysses:

I am a part of all that I have met;
Yet all experience is an arch wherethrough
Gleams that untravelled world, whose margin fades
For ever and for ever when I move.
How dull it is to pause, to make an end,
To rust unburnished, not to shine in use,
As though to breathe were life.

ACKNOWLEDGMENT

I am indebted to my colleague and successor, Dr. Michael Hall, for helpful comments on the manuscript of this chapter, and I wish him bon voyage on his own Odyssey.

Literature Cited

1. Black, M. 1980/81. The role of endogenous hormones in germination and dormancy. *Isr. J. Bot.* 29:181–92
2. Dennis, F. G. 1974. Growth inhibitors: correlations vs causes. *Hort. Sci.* 9:180–83
3. Jacobs, W. P. 1959. What substance normally controls a given biological process? Formulation of some rules. *Dev. Biol.* 1:527–33
4. Lang, A. 1980. Some recollections and reflections. *Ann. Rev. Plant Physiol.* 31:1–28
5. Meins, F., Binns, A. N. 1979. Cell determination in plant development. *Bioscience* 29:221–25
6. Noodén, L. D., Weber, J. A. 1978. Environmental and hormonal control of dormancy in terminal buds of plants. In *Dormancy and Developmental Arrest,* ed. M. E. Clutter, pp. 221–68. New York: Academic
7. Saunders, P. F. 1978. Phytohormones and bud dormancy. In *Phytohormones and Related Compounds: A Comprehensive Treatise,* ed. D. S. Letham, P. B. Goodwin, T. J. V. Higgins, pp. 423–45.
   Amsterdam: Elsevier/North Holland Biomed. Press
8. Wareing, P. F. 1965. Endogenous inhibitors in seed germination and dormancy. *Encycl. Plant Physiol.* 15:909–24
9. Wareing, P. F. 1978. Determination in plant development. *Bot. Mag. Spec. Issue* 1:3–18
10. Wareing, P. F. 1982. Hormonal regulation of seed dormancy—Past, present and future. In *The Physiology and Biochemistry of Seed Development, Dormancy and Germination,* ed. A. A. Khan. Amsterdam: Elsevier/North Holland. In press
11. Wareing, P. F. 1982. Determination and related aspects of plant development. In *Molecular Biology of Plant Development,* ed. D. Grierson, H. Smith. Oxford: Blackwell's. In press
12. Wareing, P. F., Phillips, I. D. J. 1982. ABA in bud dormancy and apical dominance. In *Abscisic Acid,* ed. F. T. Addicott. New York: Holt, Rinehart & Winston. In press

Ann. Rev. Plant Physiol. 1983. 34:1–19

ASPIRATIONS, REALITY, AND CIRCUMSTANCES: The Devious Trail of a Roaming Plant Physiologist

Pei-sung Tang

Institute of Botany, Academia Sinica, Peking, China

CONTENTS

OF EMPERORS, REBELLIONS, INVASIONS, AND INDEMNITIES (Personal and historical background)

I was born in the county of Kishui (now Xishui) in Hupeh province, China, on November 12, 1903. Times were hard for the Chinese people in that period under the decaying rule of the Manchu (Ching) dynasty. During the few decades prior to the turn of the present century, the doors of the then self-imposed "Central Kingdom" had repeatedly been pried open by Western invaders, and the nation under Manchurian rule was in danger of, if not actually being, dismembered by the preying powers. Faced with the constant threat of invasion from without and the rising wrath and discontent of the impoverished populace from within, the tottering Manchurian rulers promised political reform by taking steps toward constitutional monarchy.

2103

0066-4294/83/0601-0001$02.00

Socially they were forced to introduce modern (European) education by sending students abroad for new knowledge. In short, I was born at a time when the old (imperial rule) was giving way to the new, and when East and West clashed and also mingled, in a land of turmoil and contradiction.

I came from a family of classical Chinese scholars. My father, Tang Hua-lung, received most of his training in Chinese classics from private tutors. Upon passing the imperial (civil) examination, he was appointed a junior clerk in the Board of Punishments (equivalent to the ministry of justice). Because he believed that the acquisition of new (Western) knowledge was essential in effecting political reform of Manchurian rule, he asked to be sent to Japan for further studies in law and political science (3).

When the Manchurian rulers, in an attempt to quell popular discontent generated by the Boxer Uprising, took their first steps toward constitutional reform in 1907, they established the system of government by election. Soon after my father's return from Japan he was elected to the newly formed Hupeh provincial assembly and became its speaker in 1910.

On October 10, 1911, revolution against the Ching (Manchurian) dynasty broke out. My father supported the revolution and functioned as the civil administrator in the provisional military government of the revolutionaries. I was then 7 years old and can still recall vividly the noise of fighting all through the night of that day which quickly led to the end of centuries of feudal rule and the beginning of the modern era in China. As an irony of fate, the last child-emperor, the overthrow of whose dynasty my father and I witnessed in 1911, worked for a short period of time in the Botanical Garden of the Botany Institute of Academia Sinica in the 1960s as a caretaker of the gardens. I was then a director of the institute and hence his boss! Acually, of course, he was "employed" there to provide him with a quiet place to write his memoirs.

Because of my father's connections, I was sent to Tokyo for grammar school education. Had it not been for World War I, and also my mother's death in 1915, I would have remained there and been educated as a law and political science student. I returned to Peking before the war's end and took the entrance examination to enter the then much coveted Tsing Hua College, actually a preparatory school for teenagers who would be sent to study in the United States after eight years of training. They were then sent to enter American universities as juniors, all on government scholarships, the so-called "Boxer Indemnity Scholarships." This is a part of the "Indemnity Fund" extorted by the eight invading powers in 1900 to "punish" the Manchurian government on the pretext of "property damage" to their nationals on Chinese soil due to the Boxer Uprising. The indemnity fund came from heavy taxation on the then 450 million Chinese people, some 90 percent of whom were peasants. The few candidates (ca 50 per year) were selected from hundreds of applicants on open examination, and the competition was keen. I was among the five students from Hupeh province who qualified for the 1917 examination. The remainder came from quotas assigned to each province proportional to the amount of tax levied on them. The great majority of these students had very good school records and excellent grades. I had a good father. He was by then Minister of Education in the Central Government. My father was assassinated by his political adversaries in 1918, a year after I entered Tsing Hua College, and I was entirely supported by that scholarship which came through taxation. I remained at the college until graduation in 1925, when I left for America. To this day I have a strong feeling of deep indebtedness to my country and my people who nurtured and nourished me.

IN QUEST OF KNOWLEDGE

I went to the University of Minnesota because it had one of the best agricultural colleges known to us at that time. But I soon transferred to the School of Science, Letters and Arts because I discovered what I really needed was a solid general foundation in science and liberal arts before specialization. I filled my credit hours with a full load of the general and intermediate courses in physics, chemistry, biochemistry, and botany, graduating with an AB degree magna cum laude, majoring in botany and minoring in chemistry and physics. My first plant physiology instructor was Rodney Harvey, who translated Maximov's textbook into English; the second instructor, also at Minnesota, was George O. Burr, who later almost discovered the C_4 (or Hatch and Slack) pathway of carbon metabolism in photosynthesis. But the most unforgettable classes I attended then were the lucid and inspiring lectures on physical chemistry by Professor Frank McDougall. His lectures sparked my interest not only in thermodynamics, but through it, the energetics of living matter in general.

Johns Hopkins Days

On recommendation from my major adviser, and because of the reputation of Johns Hopkins, I went to study with Burton E. Livingston in the fall of 1928. His doctoral thesis on osmotic pressure and his treatment of the "three-salt mineral nutrient solution" in terms of physical chemistry interested me. But by then Livingston's interests had shifted to the more ecological aspects of plant physiology which did not appeal to me. Against his wish to work on physiological ecology problems, I started to perform some exploratory experiments on photosynthesis and seed respiration. Our relationship became somewhat strained. Maybe because I was the first and perhaps the only Chinese student he ever had, and perhaps also because I was sent to him by his good friend, William S. Cooper, he did not quite have the heart to dismiss me as he did some of my contemporaries. Behind his back we students called him "Uncle Burt," and on occasions, "the old crank." Finally we compromised on doing a PhD thesis on "the influence of temperature and aeration on the germination of wheat seeds under water." This research topic itself is meaningless if not absurd. Who would germinate wheat seeds under water, even for "pure" research? And what is the meaning of the redundant effort of bubbling air through the water to overcome anaerobiosis? Livingston's aim was to obtain some experimental data to refute Blackman's "theory of limiting factors." My aim was to finish getting my degree. By 1930 I had the experiments finished and written up. Neither of us paid much attention to the physiological meaning of this research, but Livingston was very impressed with the graphical analysis of the data, a method which I learned in the physical chemistry course at Minnesota. He was pleased and elaborated my thesis into a long paper which appeared in 1931 (9). I was also very pleased, not so much because he had conferred on me the PhD degree, but also because I learned much from his writing abilities. I admit that whatever scientific writing ability I have acquired is due in large part to Livingston's legacy. More importantly, the idea of multifunctional relations in physiological processes which was employed in the analysis of the data was the germinal beginning of my later concept of multiple pathways in respiratory metabolism and their relations to other physiological functions.

It was also at Johns Hopkins that I had the good fortune of befriending Robert Marshall, who shared the laboratory with me. I do not quite remember whether he finally got his degree after I left Hopkins, but I do well remember that it was he who introduced me to the writings and views of H. L. Mencken, Norman Thomas, and Oliver Wendell Holmes. It was he, too, who introduced me to the plays of Eugene O'Neil and the superb performances of Lynn Fontanne and Alfred Lunt, among many others. The biology seminars organized by H. S. Jennings and his student Tracey Sonneborn opened my eyes to the then rising science of genetics. It was Jennings who greeted and congratulated me upon hearing that I was awarded the same fellowship he was awarded some years earlier when he was at Harvard. These friends and many others of that period taught me aspects of American life other than that found on the campus. In my days at Johns Hopkins the laboratories were modest and austere, but exchange of ideas was rich and fruitful.

I left Johns Hopkins in the summer of 1930. Little did I know that not so long afterwards, when Livingston retired, that same small and austere laboratory would become the very famous McCullum-Pratt Institute of Biological Science. Tony San Pietro, when I met him at the Columbus meeting in 1979, told me that no less than eight brilliant stars in biochemistry were crowded in that very building where I once worked. Much excellent research on enzymology flowed out from that small red brick building, including the now classical volumes of *Methods of Enzymology* edited by Colowick and Kaplan.

When I returned to visit the United States in 1979, I revisited Johns Hopkins—after a lapse of exactly half a century. That small laboratory was still there, intact, but with its greenhouse abandoned. When I palmed and turned the ever-so-familiar brass doorknob and shouldered open the laboratory door, nobody greeted or stopped me. Rows of animal cages took the place of laboratory benches, and the bleak walls had been enameled white. The once austere laboratory where we sweated in the summer has been turned into a deluxe air-conditioned animal house! And the basement of Gilman Hall, where we had seminars, is annexed to a much bigger biological laboratory where molecular biology thrives. Such is the march of time for Biology!

Woods Hole Summers and the MBL

My early interest in the physical chemical aspects of living phenomena naturally led me to seek for an apprenticeship with W. J. V. Osterhout. His earlier work on the use of relatively simple plant parts or cells for respiratory and photosynthetic studies impressed me. I wrote to inquire if I might come to work with him at Harvard. He wrote back that he had already left for the Rockefeller Institute, but I might contact his successor, W. J. Crozier. I wrote to Crozier and informed him of my acquaintance with the Arrhenius equation and chemical kinetics in general. I got an unexpectedly quick response and was offered the Parker Fellowship from Harvard University to work in his recently established Laboratory of General Physiology.

In the meantime, the praise from my colleagues at Johns Hopkins for the Marine Biological Laboratory at Woods Hole intrigued me. I applied for and was granted a space there for the course in general physiology for the summer of 1930, before I was to report for work at Harvard.

Those three months at Woods Hole, followed by a second visit in the summer of 1931, were memorable ones—both because of the course which I took the first summer and the numerous great biologists whom I had the good fortune to meet or to study with. The two summers I spent with the biologists there, especially with cellular and general physiologists, set the theme for my life work. Among the many friends who had a direct influence on me are: Ralph S. Lillie and Ralph Gerald of Chicago, Leanor Michaelis and W. J. V. Osterhout of the Rockefeller Institute. I learned in the general physiology course the then "modern" method of Warburg microrespirometry, the Van Slyke and Haldane gas analysis machines, and oxidation-reduction titrometry, among other things. Very important were the lectures given with the course and the evening lectures and discussion afterwards. Among the instructors of the general physiology course, Ralph Gerald influenced me the most. His brilliant lectures and witty remarks impressed me. He told us about A. V. Hill's research on heat production in muscle, electric action potentials in nerves, and especially the work of Otto Warburg on the rapid rise in respiration of sea urchin eggs upon fertilization, as well as his use of the light-reversible property of CO inhibition on cell respiration to obtain the action spectrum of the *Atmungsferment*. These and Otto Meyerhof's attempt to measure "energy of organization" enthralled me. I decided then and there to take up energetics of cellular and plant respiration and of photosynthesis as my life work. Fortified with such "modern" techniques, and more importantly with such lines of thought, I was well prepared to do some serious work as a "post doc," as it is termed now, at Harvard.

Besides the many friends I made at Woods Hole and the knowledge and skill in general and cell physiology which I learned from them, I also made some contribution of my own to them and to all my colleagues at the old MBL. I entertained them with almost daily after-lunch tennis exhibitions between myself and Hallowell Davis of Harvard on the court next to the mess hall. These matches were remembered by my fellow students of those days, such as Emil L. Smith and David Green, when I met the former in China in 1976 and the latter during my visit to the United States in 1979. My other contribution was the publication of three little papers: The rate of oxygen-consumption by *Asterias* eggs before and after fertilization (11); the oxygen-consumption curve of unfertilized sea urchin eggs (10); and the oxygen-consumption curve of fertilized *Arbacia* eggs (12). These three papers have been generously quoted by Joseph Needham in his book on *Chemical Embryology* (5), and by Rashevsky in his book on *Mathematical Biophysics* (8).

Harvard Days

I went straight to Harvard from Woods Hole in the fall of 1930 and reported for work at the Laboratory of General Physiology, perhaps the first institution in the country to bear that modern name. I must confess my spirit was a little dampened when I was ushered to the basement of the age-worn Peabody Museum, world famous for the glass flowers it housed, but incompatible to the name of a modern laboratory which it sheltered. I went up the squeaky stairs to a large room on the second floor where Crozier shared his office with his secretary, separated only with a placard on a wooden stand which read: "Please do not disturb" between their desks. After a brief introduction, I was returned to his second in command, Albert Navez, in the basement. While going down the stairs, I ran into Bob Emerson and had a brief exchange of greetings. Bob had recently returned from Warburg's laboratory in Germany, and was in a rush to gather his things for departure to Pasadena to join T. H. Morgan at Caltech.

Navez, who first came to Harvard as a fellow on the Belgium Relief Program, arranged for me to share his dark and crowded basement laboratory, together with two or three others, including Trevor Robinson, who was then Navez's graduate student. Navez took me on a tour of the other basement laboratories and introduced me to all the junior members and graduate students. In touring the laboratory rooms, one outstanding feature came to my attention. That was the presence of at least one or two thermostat tanks of all sorts and sizes in each and every room: the hallmark of Crozier's trade. After a few days of brief visits with the colleagues at the laboratory to acquaint myself with them and their work, I quickly settled down to work.

Since I knew exactly what I wanted to do and had my project planned ahead, I had no difficulty getting my work started. I took over the then newly designed Warburg respirometers that Bob Emerson left behind, grabbed the then perhaps most versatile refrigerated thermostat tank to go with the set, and started experiments on the temperature characteristics of seed respiration during germination. I worked very hard and very efficiently. Not infrequently I had to make continuous observations at 15 minute intervals for 24 or even 48 hours at a stretch. On those occasions I took a camp cot to the laboratory for momentary naps between temperature adaptations. I made plans for the next series of experiments before the first ones ended. In this way my research program went off with the efficiency of a one-man production line at an average output of one paper published every three months. This pace of research activities was kept up essentially throughout my period of studies in Crozier's laboratory with a slight letup after the first year due to my marriage and having a family, but the general tempo was essentially the same. Both Crozier and I were pleased during the first two years, but for different reasons. He was pleased by the additional evidence for the universality of his temperature characteristic idea, while I was mainly concerned with repetitive rappings at the door of Science's opportunity.

This happy coexistence did not last long, however. As with many of my more independent colleagues in the laboratory at that time, I began to deviate from the boss's assembly line and set up a little side business of my own. The goods from this unauthorized sideline are: the publication in 1932 of the now well-documented report on the discovery of cytochrome oxidase in plants (13), and a thorough survey of the then existing literature on the relation of oxygen tension and rate of oxygen consumption in cells and tissues (14). I summarized the findings with an equation describing the hyperbolic relation between the two. That equation had been used and mentioned in some biochemical and physiological textbooks at that time, e.g. in C. Ladd Prossors's *Comparative Animal Physiology* (7). Similarity between this and the equation of Michaelis was pointed out in that paper, but the true meaning of the constant was not understood until years later. The constant of that equation has since been used to express the affinity between oxygen and cytochrome oxidase in the cell. In other words, the constant is actually the Michaelis constant for oxygen affinity of the substrate cytochrome oxidase in vivo.

With these two side products from the benches of the temperature characteristic shop, together with some half a dozen of the "standard ware" type, I edged into the threshold of science, at the expense of my relations with Crozier. So, in the summer of 1933, with the shocks of the stock market crash still reverberating, and when my first wife lost her eyesight, I accepted an invitation from Wuhan University to return to China.

On hearing of the difficult position I was in, and my desire to leave the United States for home, my very close friend of Johns Hopkins days, Bob Marshall, made a special trip from New York to Boston to persuade me to remain in the States and continue with my scientific research. He promised to provide a private endowment at a New York university where I would be accommodated with full research facilities. He came from a very wealthy family of lawyers in New York City. Knowing his intellectual inclinations in Hopkins days, I knew his offer was sincere, and not out of philanthropy. But with my personal misfortune and my self-imposed commitment to my country, I regretfully declined.

In spite of all that I have said about my "grind mill" life in Cambridge, it was certainly not "all work and no play." The social life among many of the friends at Harvard, the numerous spontaneously organized and unscheduled discussions and seminars, were not only stimulating, they somehow banded us "dissenters" into close unity. One of these is the "*Chlorella* club" that Stacy French mentioned in his account of Harvard days in Volume 30 of the *Annual Review of Plant Physiology.* (4). I can never forget the friends I made in Cambridge (Mass); the vacation at the summer house of Walter B. Cannon and his family in the New Hampshire hills; Bill Arnold of the Emerson-Arnold photosynthetic unit fame, who constantly harassed me with such "nonsensical" (then, but not now) questions such as whether carotenoids play a part in photosynthesis; B. Fred Skinner, who introduced me to the philosophies of Ralph Waldo Emerson (who by the way was an ancestor of Bob Emerson whose respirometers I took over at Harvard), and especially of his hero, Henry David Thoreau; and finally my very close and lifelong friend and colleague, Stacy French, of derivative spectroscopy and of the French press fame, a sequoia among the few old giants of American photosynthesis.

Unlike at Johns Hopkins, I have been in touch with friends at Harvard off and on through the decades, first through Stacy French and B. F. Skinner, and more recently through Lawrence Bogorad. I came to know Lawrence during his visit to China in 1978 when he led a group of ten botanists sent over by the Botanical Society of America, sponsored by the head of its China Relations Committee, Peter Raven. Lawrence learned of my desire to revisit my old laboratory, room 208 of the Biological Laboratories on Divinity Avenue. So when we paid a return visit to the USA, he very thoughtfully arranged a grand "homecoming" for me. After a tour of the various departments at the Biological Laboratories, I was first taken to the section to which the glass flowers were moved and housed in modern display cases. To further link me with the olden days, I was taken to visit the octogenarian Ralph Wetmore. He was professor of botany when I was busily grinding out my temperature characteristic papers in his neighboring General Physiology wing. Then Lawrence took me to his own office and I immediately recognized it to be the expanded and refurnished office that Crozier once occupied. The climax came when he took me to the laboratory next to his office: room 208 where I worked half a century ago! The door opened. Standing before me was its present occupant, Daniel Branton, a young scientist in his 40s. My mind immediately shifted back 40 years. I was of that age, too, 40 years ago, and working in the same room! But the similarity ends there. The change has been enormous. Whereas it was a sparsely equipped large room with a barren bench on which stood only a set of Warburg respirometers, it has now been completely renovated and sectioned, filled to every corner with modern sophisticated instruments for molecular biology and biochemistry. Dan politely gave an account of the brilliant work he was doing, but most of his account escaped my attention. I caught little of what he told me, not only because of my inadequate knowledge of modern biology, but more because my mind was occupied by something much deeper than the subject matter itself. What I was thinking was between Dan and me and how much water had passed under the biological bridge! And more importantly, how wide is the gap between this science in my country and that of the scientifically more advanced countries!

To top the "homecoming," Lawrence kindly arranged for a reunion between me and many of my old friends at the Harvard Club. Among them were Stacy French, who came specially for the occasion from his trip to the Caribbean, B. F. Skinner, John and Wilma Fairbank, Ralph Wetmore, and Martin Gibbs. And to wind up the occasion, they arranged for a dinner at the Pier Four Restaurant. They told me they chose this place because of its food and its name. But I doubly appreciated this location for another reason. Nearly half a century before it was from these wharfs that I embarked and left America for home.

CHINA RETURNED: THE CALM BEFORE THE STORM

I accepted the invitation of and chose to remain at the National Wuhan University because they were good enough to offer me time and funds for research, a privilege rare in Chinese universities at that time. They allotted $2000 (US) to set up a small laboratory for cell and plant physiology. I spent the first two years designing and building the laboratory and purchasing the equipment and chemicals needed for studies in cellular respiration and photosynthesis. The first of a series of seven papers on the kinetics of cell respiration appeared in 1936 (15), intermingled with another series of short papers on practical aspects of biochemistry—a survey of the iodine contents of seaweed off the South China Coast in collaboration with the phycologist C. K. Tseng.

The happy and productive years at Wuhan University did not last long. At the time when the experiments for the seventh paper of the respiration series were in progress, the Japanese army invaded North China—soon the entire country—the beginning of eight years of holocaust and devastation, the effects of which were felt years afterwards. My feelings at that time were expressed in the paragraph which ended the last paper (16) published together with my close friend and colleague Dr. Mao-i Woo, thus:

> It was the original plan of this series of studies to investigate the respiration of unicellular organisms in simple physical chemical systems such as those used in the present account so that we might obtain some knowledge of the physical chemistry of cell respiration like that of the kinetics of reaction in simple chemical and in enzyme systems. Circumstances, however, do not permit the completion of this plan. It is our earnest hope that investigations along this line of thought may be continued by more fortunate workers who are at the present enjoying the blessings of tranquility.

Nevertheless, I did not give up. I turned my laboratory into manufacturing active carbon for gas masks. When Nanking fell and Wuhan was in immediate danger, I dismantled my laboratory as well as the active carbon furnace in the factory and took up a new task: that of establishing a medical school in the more interior city of Kweiyang.

THE WAR YEARS; THE ROAMING BEGAN

At the time when Wuhan (Wuchang then) city was crowded with refugees and soldiers retreating from all the Japanese occupied areas, I accepted an emergency appointment on a five-men committee to establish a medical school in the interior city of Kweiyang for training physicians to supplement the needs of the army. The first batch of students were those who fled to Wuhan from the medical schools of the occupied areas. Since I was the only nonmedical member of the staff, I was entrusted with the admission and transportation of the first batch of some 40 students to the remote city of Kweiyang. We spent 16 days on a steamer overcrowded with refugees of all sorts and went up through the treacherous rapids of the Yangtze River to Chungking. From there we had a further two days' journey by bus over the mountainous staircase highway to Kweiyang, passing many overturned vehicles down the steep cliffs below the narrow paths.

I remained with the Kweiyang Medical School for six months. After fulfilling my assignment of organizing the preclinical teacher staff and after designing and building the teaching laboratories, I moved farther to the interior stronghold of Kunming, Yunnan province, in answer to the call of my alma mater, which had long since attained the full status of a renowned university.

While I was a founding member of the medical school in Kweiyang, I was concurrently a staff member of the Chinese Red Cross (war area) Medical Relief Corps, headquartered in the same city. I was called back to that organization on numerous missions during the years I was in Kunming as an adviser on nutrition problems of the soldiers at the front. In fact, a part of the research of my plant physiology laboratory at Tsing Hua University in Kunming was on the Red Cross Medical Relief Corps program, the soybean milk nutrition project.

Kunming Days: Tapuchi, a Sanctuary for Wartime Chinese Plant Physiology

I spent practically the entire period of the war (1938–1946) in Kunming. Those were eventful and of course difficult years. During the eight years in Kunming, my laboratory was bombed out three times, moved to four different locations, and rebuilt as many times. The last location, where we worked the longest, was in the small village of Tapuchi, about 10 kilometers in the city's northern suburb, where we set up adobe laboratories. This village of not more than a score of dwellings was made widely known during and after the war by the Needhams' account of its scientific "immigrants" in their book *Science Outpost* (6).

During the years at Tapuchi, my laboratory served as a sanctuary as well as an assembly post for youthful and dedicated physiologists of all kinds: microbial, cellular, plant, and animal. That was the major goal which I set myself to attain when I accepted the call to establish a laboratory of plant physiology (actually general physiology) at the Institute of Agricultural Research of Tsing Hua University soon after the war broke out. On the aims of my laboratory, Joseph Needham in his introduction to my book of essays

(17) had this to say: "Yet at Tapuchi. .... Tang Pei-sung created laboratories of general physiology, not ill equipped, though constructed of mud bricks and wood work; and above all knew how to surround himself with eager young scientific workers in the true atmosphere."

No less than 40 scientists passed through the doors of that humble laboratory during the eight difficult years in Kunming (Tapuchi). Almost all of them were later sent to work or study in America or in Europe. Many of them attained success in their own respective fields but never returned to China. But many of them did return and are the backbone of Chinese plant physiology today. Some of these are: H. C. Yin, director of the Institute of Plant Physiology, Academia Sinica; C. H. Lou, Professor and vice president, Peking Agricultural University; S. W. Loo, also of the Institute of Plant Physiology; T. S. Tsao, Professor at the Peking University; Y. L. Hsieh, Professor at Futan University; B. L. Cheng and T. Y. Hsieh, Professors at the Shantung College of Oceanology.

The research projects of my laboratory had been diverse, but the main theme was very clear: to do what we could to serve the nation at war, to utilize what knowledge we had to improve agricultural production, to exploit the abundant and varied plant resources in the rich subtropical regions of southwest China, and finally of course to train a young crop of scientists.

In war work, besides what we did for the Chinese Red Cross Medical Relief Corps, I served for a short period as a technical adviser with the rank of colonel to the then only armed division of the Chinese army which thwarted a Japanese offensive to the southwest from occupied Indo-China. I was trying to persuade them to use castor bean oil as a substitute for hard-to-get mineral oil lubricant.

In the exploitation of natural resources, we made a general survey of the vegetable oil resource in the southwest provinces and experimented on castor oil, sumac wax, and laurel oil for civil as well as for military use.

In the field of agriculture, we worked extensively both on the application of plant hormones for plant growth and the use of colchicine to induce polyploidy. We even worked on the physiology of the silkworm.

Looking back, those were difficult, eventful, but rewarding years; rewarding because of the gratifying feeling that I have done something, however insignificant, for my country in time of her great misfortune, and also because to some extent it was adventuresome. In performing the many varied missions during these years, I traveled all over the southwestern part of the country by bus, pushing the bus sometimes. On occasions when it was dark and no lights were available, I held two torches in my hands and ran before the bus so we could find our way to the city for the night. The athletic prowess of my college days won its belated applause.

As the war dragged on, I began to write a series of essays on my scientific activities through the 15 years since I first started my work at Johns Hopkins in the 1930s. These essays were meant to be both an account of what I had done during those years and a collecting of my scientific line of thought, a philosophical look at living phenomena in general. When Joseph and Dorothy Needham came to visit me again in Tapuchi at war's end in 1945, I showed them these essays. At their suggestion and persuasion, I agreed to have the essays published in London. Joseph wrote a generous introduction to the book and christened it *Green Thraldom* (17) from the title of the first essay dealing with my views on the conversion of solar energy through photosynthesis. This book marks the maturation of my views on the bioenergetic and metabolic concept of life phenomena, a concept which I had been closely attached to all through my later research. The theme sounded in the opening paragraphs of the book:

> Life connotes activity, and activity implies the expenditure of energy. In man, the energy for his activities is derived from the food and fuel he consumes, and these are manufactured for him, directly or indirectly, by the green plant. Primitive man, living in the forest, feeding on its fruits and deriving warmth from burning its timber, was a parasite. When gradually he learned the usefulness of certain kinds of plants and animals for food and fuel, and singled them out for cultivation, propagation and protection, he initiated the art of agriculture and thereby passed from a state of parasitism into a state of symbiosis. . . .
>
> Civilized man, long having abandoned his worship of fire and sun, is scarcely further advanced than his ancestors, since the fundamental relation between the sun, the green plant and himself is still shrouded in mystery. Not until the time comes when he can reproduce the apparently simple process of photosynthesis in his laboratory, independent of the green plant, can man claim to be free from the vestiges of his ancestral worship of sun and fire. When that day comes the energy awaiting his disposal is enormous.

This theme is continued throughout the book and the crescendo is reached in a passage in its final chapter where my metabolic views on living organisms are emphasized thus:

> A living organism is a state of aggregation of matter which in a limited range of physical environment can utilize relatively unorganized matter: a part of the material is used for the maintenance of itself and for the reproduction of its kind, while the other part is converted into energy necessary for the performance of these processes.

This is my metabolic concept of living matter, or my concept of bioenergetics. It has been the main theme on which my later work was based. We shall come back to this later. As an aside, I am grateful that the above passages were quoted and used as epigrams to the appropriate chapters in two of Bladergroen's books (1, 2).

INTERMEZZO

With the completion of these essays, my scientific activities of the prewar and the war years came to an end. At the close of the eight-year Sino-Japanese war in 1945, I, like all the scientists and intellectuals of my country, had to start from scratch again, richer with experience gained through the years, but impoverished by the soaring inflation. I left my mud-brick laboratory at Tapuchi in the summer of 1946 and returned with the university to its devastated home in Peking to build laboratories again.

We expanded the Institute of Agricultural Research into the College of Agriculture in the university. It was my goal to make the college both a research institution and a teaching school to train students of such quality that they could do advanced teaching and research in other colleges and in the experiment stations, as well as doing biological research. This was the influence of the Johns Hopkins system. We had only a dozen or so senior members on the faculty in the beginning, but these were selected from institutions all over the land on merit of their research and teaching achievements.

After a year's preparation, we began to admit students in 1947, and again in 1948. The dream and aspiration of an advanced school of agriculture, somewhat similar to the schools of biological sciences in some Western countries of recent years, did not last long. Just after the first batch of students entered the respective departments of their choice after finishing two years of training in the college of arts and science, Peking was under siege in the fall of 1948 by the People's Liberation Army, and activities of Tsing Hua as well as other universities were suspended. Very soon Peking was liberated, followed by the rapid collapse of the entire old regime, and New China was founded by the Chinese Communist Party. This change was much more profound than the 1911 revolution which overthrew the Manchurian dynasty. It was the dawning of a new era in an ancient country comparable only to the "October Revolution" of the USSR which Sidney and Beatrice Webb described as "A New Civilization" (21).

NEW CHINA

Plant physiology, like all other branches of science, grew rapidly in New China, but not without its ups and downs. Since I have given an account of the general state of this science in China at the 1979 Columbus meeting of the ASPP (18) and more recently at the Thirteenth International Botanical Congress (19), I shall only give a very sketchy account of some of my personal experiences during this period.

Soon after the founding of New China, my college of agriculture was merged with two others to form the Peking Agricultural University. I was one of three deputies on the governing board. The chief of the board was also the chief proponent for "New Biology" of the Lysenko version. He turned the university into a stronghold for propaganda of Lysenkoism. I was open minded and willing to learn anything new, but when scientific argumentation degenerated into idealogical polemics, I supported the then underdogs who adhered to facts and theories of modern genetics, however "decadent." I left the Agricultural University for the Institute of Plant Physiology, Academia Sinica, in the fall of 1952, and I have been with the Academy ever since, with a concurrent post as professor of plant physiology at Peking University from then to the present. Although my position was changed to director of the Institute of Botany in 1977, I have continued with teaching and research in plant physiology throughout the 32 years since the founding of New China in 1949. In spite of frequent disturbances and interruptions mentioned in my earlier account (18), I have managed to continue with my research in both bioenergetics and basic problems in relation to practical needs of the country, e.g. on biological nitrogen fixation and on bioenergetics of solar energy conversion.

Soon after the founding of the People's Republic of China, fundamental changes in the social and political structure of the nation took place—a complete overhauling, so to speak. It was not merely a change in the form of government, but in ideology and in all systems, including the educational system. For these changes we naturally sought the assistance of the Soviet Union. During the early years of the newly established republic, literally thousands of advisers were sent to assist us in every phase of our national reconstruction, including, of course, those who came to renovate the educational system down to the actual texts to be used for teaching the courses.

The old system of university teaching was abandoned, and new ones were introduced. Plant physiology, for example, was being taught in agricultural "universities" which had been detached from the professional colleges of the original universities. The original universities still maintained their biology departments in which there was a special faculty for plant physiology! This duplication was caused by the segregation of the professional colleges from their main universities. The regular universities retained their departments of biology with faculties of plant physiology, animal physiology, biochemistry, zoology, botany and some others. In this way, when a student graduated, he was trained as a "specialist" in his own narrow field, say of plant physiology. Even in this field he could be further "specialized" in such topics as growth, mineral nutrition, or photosynthesis. This system of teaching plant physiology was then new to us. In addition, there was a lack of teachers for the "advanced" courses. There was also the very difficult problem of obtaining the "progressive" texts even for the course of introductory plant physiology.

Faced with these difficulties, the teachers of plant physiology as well as those for the other fields were rather at a loss of what to do. Fortunately, the government foresaw this need and invited "specialists" in the various fields as advisers to teach us how to handle the situation. I was at the time fortunate enough to be a professor of plant physiology at the Peking Agricultural University which was specially favored because it was entrusted with the task of indoctrinating the "Progressive Russian New Biology" to the country. Because of this we were privileged in two aspects: first, we could receive outlines and syllabi for the biology courses from Russia; and secondly, we had a strong staff of translators to render the materials into Chinese. In spite of these advantages, it took a long time to have all the materials translated, checked, and tried out before they could be distributed to the plant physiologists at large who were anxiously awaiting intellectual food to feed the hungry students. Unable to resist their cry for help, I, with the collaboration of my junior colleagues of the plant physiology faculty, mimeographed hand-printed copies of parts of the translated outlines and syllabi in the form of "Plant Physiology Newsletters" (from 1951 on) as quickly and as frequently as they were available. These newsletters were so well received that after the first four mimeographed issues (1951–1952) and beginning from the fifth issue in 1953 they had to be changed to regular printed pamphlets. Finally, in 1955, beginning with the twenty-first issue, they were published as a regular journal which now bears the respectable name *Plant Physiology Communications,* with a wide circulation within the country.

The availability of the translated outlines of Russian plant physiology relieved the anxiety of the country's plant physiologists only for awhile, for we still needed the all-important specific materials to fill in the outlines. There was further confusion because of the then denunciation of the backward, "decadent" old biology and the indoctrination of the "Progressive New Biology" of Lysenko vintage. To help us solve this very important problem, our Russian friends came to our rescue again. They sent in the nick of time a couple of very nice young damsels to teach us how to teach the "progressive" plant physiology as it was done in Russia. They were very dedicated and pleasant young teachers who had perhaps just received their "Kandidat" from some teacher's college back home. We discussed in detail the syllabus and outlines of the introductory course. At first they were a little apprehensive about whether we were ready to accept the "new" plant physiology in place of the old. They were soon relieved and perhaps even pleased to find that I, among a few other old souls, not only could accept the material and the contents, but embraced them wholeheartedly! In fact,

I could even supply the necessary experimental evidence or factual and theoretical documentation at the right places. And, in a few instances where the translation appeared to be inexact, I supplied my alternative interpretations. The training class went on very smoothly and came to a successful conclusion, with the teachers marveling at my quick grasp of the "new progressive" plant physiology. Little did they know that out of scientific courtesy and of bourgeois etiquette, I refrained from telling my teachers that the syllabus was that of Maximov's *Text Book of Plant Physiology* with which I had been teaching my students for at least a decade. Only it was the English rendition by Rodney Harvey that I used, instead of the Chinese translation which my Russian teachers brought to us. Science, like the sun, begets its light and warmth to all people on Earth, irrespective of country, race, or belief.

To this day I still have fond memories of that training class and of the many accomplished Russian plant physiologists whom they sent us later, among them B. A. Rubin of Moscow University and P. A. Henkel of the Timiriazev Agricultural College. And it was my pleasure indeed to have the chance to serve on the same committee with such an eminent Russian scientist as M. Kh. Chailakhyan at the recent Thirteenth International Botanical Congress in Sydney.

With the initial worries over, plant physiology, together with other similar courses, began to be taught throughout the country. But there soon arose another difficulty: the unification of the proper interpretation of many of the points raised when the syllabus was taught in the various colleges. In addition, there were the contents of the laboratory courses, especially of the General Laboratory (the "Gross Prakticum" in the old European countries). Then there was the problem for the training of teachers for the advanced courses. I was then with Peking University as the head of its plant physiology faculty. Under pressure from colleagues of other colleges and universities, I sponsored and organized the first training course for plant physiology teachers of the entire country in the summer of 1956. With the collaboration of C. H. Lou of the Peking Agricultural University and C. Tsui of Nankai University and our younger colleagues at these and some other universities, we gave a comprehensive program in plant physiology in which laboratory experiments were offered, together with lectures and seminars on advanced topics. Discussions on points needing clarification or unification were conducted. A total of 130 plant physiologists from every institution in the country attended, with about a dozen or so of the senior members acting as lecturers. Fortified with what they learned from this course, all, especially the younger teachers, went back to their teaching posts with confidence. From this group that received their training from us emerged the new crop of younger generations of plant physiologists in New China.

This training course, the initiation of the journal *Plant Physiology Communications,* and the sanctuary for Chinese plant physiologists at Tapuchi which I established during the war years are perhaps the only three pebbles which I gathered on my wandering trail as a roaming plant physiologist.

During the some 30 years in New China, I had on the whole enjoyed stability in my academic life as compared with the turbulent days in old China. I have essentially been left to do research on basic problems. But with a nation of nearly one billion people, over 80 percent of whom are agrarian, aspiring to lift itself from austerity, one cannot shut one's self in an ivory tower. All through the past three decades there had been many occasions when we had to pack our equipment and move to the communes or to the factories to work with the peasants and the factory hands. I call these the "Open-door Laboratory" days. Many of the things we did may appear unorthodox, but they had their educational value to me and to my younger colleagues. They enriched me both professionally and morally through contact with life's realities. In these practical activities I have profited as much as the children in some industrially more advanced countries who had to be taken to the zoological garden or the botanical garden to learn that chickens, pigs, monkeys, or cabbages really existed as living things, and are much more sophisticated and different from what they appear on the dinner table or in the picture books.

Aside from such interruptions, I was left fairly alone to do my own research. Since a general account of what my colleagues and I did were given in my recent talk (18), and since a summary of my research on plant respiratory metabolism has just appeared (20), I shall not dwell on specific items. But I would like to recapitulate the main theme of thought underlying these research efforts to close my account on the eve of my exit from the stage of plant physiology. I shall quote from selected passages from that article (20):

> When I resumed my research on plant respiration in the early fifties. . . . I initiated the series on the theme of treating it as a physiological function which furnishes energy and material for the performance of the plant as a living organism. In other words, the theme deals with the functional and regulatory aspects of respiratory metabolism in higher plants: integrating metabolic changes (material and energetic) with the other physiological functions in the living plant.
>
> This concept, formulated in the early fifties, was presented in 1965. . . . After an interruption of another decade, . . . it can now be summarized as follows:
>
> "Respiratory metabolism is the process whereby a part of the material stored in the plant (organism) is converted into biological work (function) for maintaining its state of being alive, while another part of the same material is converted into substances of higher degrees of orderliness (negative entropy) in the form of structure and organization. Within limits imposed by the genetic potential, these processes are regulated by internal and external factors."

This last paragraph summarizes the entire line of thought which has been the central theme of my research on bioenergetic aspects of living organisms since the early 1930s, gathering form in my book of essays in 1949 (17) and leading to the present mature form. And in concluding that article (20), which summarized my 50 years of research in this field, I made a final assessment of my accomplishments:

... the decrease in entropy is the unique function of metabolism and of life. This is perhaps the inner sanctum of biology and which remained stubbornly invulnerable to assaults up to the present ... This was the aim which I set myself to attain when I performed my first experiments on oxygen-consumption by starfish eggs before and after fertilization as a young and innocent novice, inspired by the elegant works of Otto Meyerhof and of Otto Warburg. After almost half a century, I have not even approached the fringe of the fortress, but neither has any one else, I think.

These, in summary, are my aspirations and the realities. I published much, but contributed little; I labored hard, but I accomplished even less. Looking back over the years, I have left many things undone which I should have done, and I have done many things which I should have refrained from doing. Such are circumstances and reality!

Literature Cited

1. Bladergroen, W. 1960. *Problems in Photosynthesis,* p. vii. Springfield, Ill: Thomas
2. Bladergroen, W. 1955. *Einführung in die Energetik und Biologischer Vorlängen,* p. 330. Basel: Wepf
3. Boorman, H. L., Edward, R. C., eds. 1967–71. *Bibliographical Dictionary of Republic of China.* 4 vols. New York: Columbia Univ. Press
4. French, C. S. 1979. Fifty years of photosynthesis. *Ann. Rev. Plant Physiol.* 30:1–26
5. Needham, J. 1931. *Chemical Embryology.* Cambridge: Cambridge Univ. Press
6. Needham, J., Needham, D. M. 1948. *Science Outpost.* London: Pilot Press
7. Prosser, C. L., Brown, F. A. 1961. *Comparative Animal Physiology,* p. 167. London: Saunders. 2nd ed.
8. Raschevsky, N. 1938. *Mathematical Biophysics.* Chicago: Univ. Chicago Press. 1st ed.
9. Tang, P-S. 1931. An experimental study of germination of wheat seeds under water. *Plant Physiol.* 6:203–48
10. Tang, P-S. 1931. The O_2-consumption curve of unfertilized *Arbacia* eggs. *Biol. Bull.* 60:242–44
11. Tang, P-S. 1931. The rate of O_2-consumption by *Asterias* eggs before and after fertilization. *Biol. Bull.* 61:468–71
12. Tang, P-S. 1932. The O_2 tension-O_2 consumption curve of fertilized *Arbacia* eggs. *J. Cell. Comp. Physiol.* 1:503–13
13. Tang, P-S. 1932. The effects of CO and light on the oxygen consumption and on CO_2 production by germinating seeds of *Lupinus albus. J. Gen. Physiol.* 16:65–73
14. Tang, P-S. 1933. On the rate of oxygen consumption by tissues and lower organisms as a function of oxygen tension. *Q. Rev. Biol.* 8:260–74
15. Tang, P-S. 1936. Studies on the kinetics of cell respiration I. The rate of oxygen consumption by *Saccharomyces* wanching as a function of pH. *J. Cell. Comp. Physiol.* 8:109–15
16. Tang, P-S., Wu, M. 1938. Studies on the kinetics of cell respiration VII. Respiration of *Saccharomyces* wanching in acetate, lactate and pyruvate buffer solutions. *J. Cell. Comp. Physiol.* 11:495–502
17. Tang, P-S. 1949. *Green Thraldom.* London: Allen & Unwin
18. Tang, P-S. 1980. Fifty years of plant physiology in China: A prelude to the new long march. *Bioscience* 30:524–28
19. Tang, P-S. 1981. Aspects of botany in China. *Search* 10:344–49
20. Tang, P-S. 1981. Regulation and control of multiple pathways of respiratory metabolism in relation to other physiological functions in higher plants: Recollections and reflections on fifty years of research in plant respiration. *Am. J. Bot.* 68:443–48
21. Webb, S., Webb, B. 1945. *Soviet Communism: A New Civilization?* London: Longman's. 3rd ed.

C. Ralph Stocking

Ann. Rev. Plant Physiol. 1984. 35:1–14

REMINISCENCES AND REFLECTIONS

C. Ralph Stocking

Department of Botany, University of California, Davis, California 95616

CONTENTS

Sitting in a mountain cabin in the Cascades in front of an open fire, I look out across the meadow, snow covered still even though it is now nearly mid-May. Although it is an honor to be asked to write a prefatory chapter for the *Annual Review of Plant Physiology,* it is difficult to decide what to include in such a chapter. Perhaps at various periods in life there are certain times that are particularly appropriate to contemplate where one has been and what direction one should take in the future. A particularly fitting time for reflections, though certainly less important for planning ahead than earlier evaluations would be, follows retirement, when the pressures of work are things of the past and life is less hectic and more relaxed.

Over half a century has passed since I had my first course in plant physiology at Riverside Junior College in California. I reminisce about my involvement in plant physiology as a student, teacher, and scientist. I go over in my mind some of the changes that have taken place in our understanding of the chloroplast and its cytoplasmic environment, my major area of research interest for many

2123

0066-4294/84/0601-0001$02.00

years. It is tempting to try to identify some contributions that I may have made to the ever increasing body of knowledge regarding chloroplast structure and function. I think of the joys of research, of the privilege of teaching and of knowing students, of the challenges of departmental and college obligations, of committee meetings, of stimulating friendships with other scientists both here and abroad, of retirement, and of the future.

The beauty of the crystal clear air, the snow-covered mountains with their evergreen trees, the blue sky dotted with cotton-white clouds, and the peace and solitude are conducive to reflection, but the realization is forced on me that such islands of beauty and peace are rare indeed. The air is scarcely clean, the sky hardly visible, and strife and contention prevail in many parts of the world. These things, too, are part of my reflections.

Young plant physiologists today, in academic positions at least, appear to spend much more energy worrying about attaining promotion and/or tenure than we did 40 years ago. This worry is understandable since competition is now much keener and permanent positions more difficult to obtain. But we too had our concerns and our difficulties; the path was not always smooth. It is more to these younger plant physiologists than to my contemporaries that I direct most of these reflections in the hope that they may find a few thoughts of encouragement to pursue their chosen areas of study without fruitless, wasted energy. As is frequently the case with such thoughts, my reminiscences and reflections do not constitute one cohesive theme, but are rather centered on several more or less related ideas.

IN SEARCH OF BEAUTY

"Beauty is the splendor of truth" wrote the English economist Schumacher. All of us are searching for truth, consciously or unconsciously, in our lives, in our associates, and in the world around us. What is plant physiological research but a search for truth in the living plant? The poet and artist are free to express their quest for beauty through a variety of media. Should not the research scientist also feel the joy and splendor of truth in a biochemical reaction, in a physiological process, or in the intricate design of a membrane that can be unraveled, comprehended, or probed? Surely one of the most satisfying personal rewards that I have experienced as a research scientist has been to see the relationship between the complex architecture of the plant cell and its organelles with the precise regulation of metabolic processes as they relate to this ultrastructure. Marvelous as is the extreme miniaturization of our most advanced computers, it hardly compares with the intricate miniaturization of the living, reproducing cell with its information centers, energy trapping and releasing sites, and its precise machinery for synthesizing the numerous compounds that it uses, transports, or stores. To identify a physiological problem, to formulate a specific set of questions to be answered, to anticipate the possible physiological reactions involved, to design a series of experiments to answer the questions posed, and to see the data fall into place as the experiments are carried out, this, to me, is the "splendor of truth—beauty."

CHLOROPLAST-CYTOPLASM INTERACTIONS: FIFTY YEARS OF CHANGING CONCEPTS

After a brief early immersion into the physiology of plant water relations, my major research efforts have dealt with the relationship between chloroplasts and their cytoplasmic environment. Because these interactions are so intimately related to the structural as well as the biochemical nature of the interface between chloroplasts and cytoplasm, it was my good fortune to work closely with a skilled cytologist, T. E. Weier. It is interesting to follow the highlights in the advance of our knowledge in this area over the past years. As in many areas of science, early advances were slow and frequently limited by inadequate instrumentation and methods. Only with the development of precision instruments and improved methods did our knowledge in this area move into the exponential phase of growth.

As early as 1676–1678, Anton von Leuwenhoek, in letters to the Royal Society, reported observing green globules in leaves. However, because of limitations of the microscopes of that day, it was almost two and a half centuries before a hazy image of even the gross structure of the chloroplast began to emerge. In 1883, Meyer coined the term "grana" in his detailed description of the granular nature of chloroplasts. Little additional advance in the understanding of chloroplast structure was made during the next 50 years. Thus, by 1933, when I took my first course in plant physiology, the English translation of N. A. Maximov's *Textbook of Plant Physiology* which was used in that class had only about one page devoted to a description of chloroplast structure. Since the work of Engelmann, von Mohl, Sachs, and others, the chloroplast was known to be the site of photosynthesis and starch synthesis; however, arguments still occurred as to whether chloroplasts were formless masses of green cytoplasm, vesicles, optically homogeneous bodies, or fibrillar or granular structures. The stroma was described in Maximov's book as a hyaline plasmatic base in which the green pigment chlorophyll was found. The primary importance of the stroma was considered as "being the material basis through which the particles of chlorophyll are distributed and which, in a way not yet understood, makes their functioning possible." No reference was made to possible interactions between chloroplasts and cytoplasm.

In 1938, E. C. Miller's monumental effort to bring all aspects of plant physiology into one comprehensive volume resulted in the publication of over 1000 pages and several thousand references. This book became the major reference for many young plant physiologists. In it Miller devoted about two pages to a consideration of chloroplast structure. Much of this discussion centered on the idea that "the ground substance or stroma of the chloroplast is in the form of a hollow, flattened prolate spheroid surrounding a large central 'vacuole.' The stroma has a granular appearance that is due to numerous pores which connect the central vacuole with the cytoplasm surrounding the chloroplast. The pigments of the chloroplasts are intimately mixed and evenly distributed throughout the protein ground substance." In the same year, T. E. Weier, in a review of chloroplasts, stated that there was "probably some sort of membrane surrounding the chloroplast." With the paucity of accurate information concerning chloroplast structure at that time, it is not surprising that essentially no effective research was being conducted on the movement of solutes into and out of chloroplasts.

Development of the electron microscope and then its subsequent application to the problem of chloroplast structure in 1947 by Granick and Porter stimulated a new and rapidly expanding interest in chloroplast structure. This also was a period of intensive investigation of the physiology of carbon metabolism and energy trapping. Consequently, little attention was paid to the intracellular movement of solutes. Such studies awaited the perfection, in the late 1950s to mid-1960s, of techniques for the immobilization of solutes prior to chloroplast isolation, the rapid isolation of intact chloroplasts, and the successful isolation and chemical characterization of the chloroplast envelope. To have been able to share in the development of our knowledge of the interactions between chloroplasts and their cytoplasmic environment; to have seen our understanding of this important area of the physiology of plants grow from a state of complete ignorance, through a rapidly expanding state of knowledge, to our present understanding where even today so many unanswered questions still exist; to have had the opportunity to know most of the scientists who have made major contributions to our knowledge in this area; to have followed their work and to see how it contributed to our increasing knowledge—these are some of the highlights that come to mind as I reminisce about the past 50 years of searching for beauty in the chloroplast.

THE LAW OF INDEPENDENT DUPLICATION

One point of frequent worry to young scientists especially is the possibility that their research will be duplicated and the results published before their own work is completed. Early in my research career, I learned about the unwritten law of plant physiological research which states that if your research is worth doing, there is a better than fifty percent chance that someone else already has done it, is doing it, or plans to do it. Perhaps a few geniuses are exempt from this law, but for the rest of us, it is in full force. I first became aware of this phenomenon after I had completed my dissertation on water relations under Alden Crafts. On browsing through an old copy of FLORA, I chanced to see an article that described experiments that were quite similar to a section of my own thesis. The approach was different, but the basic idea was the same. In fact, I recall at least ten instances in which I independently either duplicated or completed research prior to the publication of a similar study done by others. In the last instance, a graduate student in my laboratory investigated the biosynthesis of ribulose bisphosphate carboxylase in barley leaves during chloroplast development. Her research was carried out quite independently and prior to the publication of a similar study done in Rachael Leech's laboratory in England. Although the techniques differed, some of the results and the conclusions were similar. In a way, this can be quite satisfying, because it shows you have been on the right track.

Perhaps one of the most amusing instances whereby this law almost spelled disaster for me occurred when I was preparing an outline for a proposed project to be submitted as part of a fellowship application in anticipation of a sabbatical to be spent at King's College in London. (A previous sabbatical had been spent at Imperial College in London.) Just before mailing the fellowship application, I was visited for a day by M. D. Hatch and C. R. Slack of C_4 fame. In the course of our discussion, I mentioned that I proposed to determine the intracellular distribution of the carbon cycle enzymes in the C_4 species, maize, using the nonaqueous method of isolating chloroplasts that I had developed previously. One of my visitors smiled and said, "It just so happens that we have done that, and a paper is in press." He reached into his briefcase and produced page proof describing the exact experiments that I had intended to include in my fellowship application. Slightly taken aback, I gulped and said, "Well, the second phase of what I propose to do on sabbatical is to conduct a kinetic study of the intracellular distribution of a series of critical metabolic products in the C_4 plant." Imagine my dismay when, after a moment of silence, one of my visitors said, "I'm sorry, but here is page proof of our manuscript covering our study of these products." He showed me another article, to appear shortly, that was essentially identical to what I had proposed to do. There went the second and last section of my proposed research!

After my visitors had left me to recover from the shock, several things became very apparent to me. My proposed research would, of course, have to be changed entirely. Time for a fellowship application was also running out. Had I been secretive about what I proposed to do on my sabbatical, I would have missed learning, until too late, that the research already had been done. After a concentrated effort to get a new research project outlined, I was fortunate enough to receive a Senior Postdoctoral NSF Fellowship and spent a very profitable year at King's College working with Bill Bradbeer in the Botany Department, chaired at that time by F. R. Whatley. A sequel to this episode was that about a year later, as I was reporting some of the results of my sabbatical research concerning the effects of iron deficiency on the structure and function of maize chloroplasts, a person in the audience commented that they had tried an approach similar to the one I had used.

Unfortunately, the possibility of having one's research duplicated and published first constitutes, especially for some young plant physiologists, a constant threat. It is a sad commentary that some scientists stoop to parasitize others' ideas, but it seems to me that one should not be unduly alarmed by the prospect of being "scooped" by independent workers. Obviously, no one field of science "belongs to" or is the "exclusive territory of" any one scientist just because he or she happened to be one of the first in the field. Young scientists should be undaunted in their search for beauty and truth in what appeals to them. Friendly competition among scientists can have a very positive, stimulating effect. It is understandable that young scientists who are trying hard to make a name for themselves may dread the possibility that a paper will appear describing research that is similar to that in which they are involved. Although it is disappointing when this happens, in general it is not a tragedy. After all, scientists should be searching for truth rather than competing for personal recognition. Granted, the criteria for academic advancement, with the emphasis upon publication, often stimulates the "competitive race" rather than encouraging creativity per se. If one's research is truly independent, then it is highly likely that the approach used will be somewhat different than that first published by others, and the results may be more comprehensive or differ in other ways. Often the second research publication confirming and possibly extending the conclusions in the first is fully as significant and sometimes more important. One may not be aware of the influence of a particular publication until years later.

For example, after publishing a paper describing a nonaqueous method for isolating chloroplasts that I developed while in Bob Burris's laboratory in Wisconsin, I received a reprint from Ulrich Heber in Germany in which he had described a similar method. Although both of our methods were based on earlier work by Berhen, in which he isolated nuclei from animal cells, Heber and I used the method for different studies. I had looked at the intracellular distribution of enzymes, and Heber had studied the intracellular location of some sugars. Had I known of his simultaneous effort, I might not have continued research along these lines. Quite to the contrary, Heber told me some years later that, had he not seen my paper dealing with enzymes, he probably would not have continued his studies using this method. However, as a result of his continued investigation of the intracellular distribution of metabolites, our knowledge of the exchange of materials across the chloroplast envelope was greatly advanced. At that time, little information was available concerning this very important field, and reliable aqueous methods of isolating intact chloroplasts had not yet been developed.

ON THE VALUE OF EXPOSURE TO "WHOLE" PLANT PHYSIOLOGY

It is well known, at least among scientists, but I'm afraid not many politicians, that major advances, not only in our understanding of the world around us but also in our ability to solve difficult practical problems, come from fundamental and theoretical research. However, today's young plant physiologists, in the unavoidable pressure to become experts in a relatively narrow area of plant physiology, too often develop the idea that research involving the whole plant as a unit is less sophisticated or less important than the investigation of a single set of plant functions or metabolic reactions. Actually, research that attempts to integrate the many facets of plant activity into a comprehensive unified picture is often more difficult and certainly no less important than research on a more narrow aspect of plant function. It is necessary to take a system apart in order to study its individual components, but we also need those whose skills can help us to reassemble the parts and show us how they work as a unit.

Some physiology students and professors lack the comprehensive training that would give them an understanding of the living plant as a unified, functioning organism. They can speak with authority on biofeedback, enzyme modulation, DNA or RNA, gene splicing, photorespiration, or other complex topics, but may be less knowledgeable about the interrelationships that keep the plant functioning as a unit. Interactions such as those between nitrogen absorption by roots and the photosynthetic rate in leaves, effects of leaves on the rooting of cuttings, or the delicate balance in a plant affected by biosynthesis, water availability, mineral supply, shifts in temperature and in light quality or duration may be only vaguely appreciated by the more narrowly trained plant physiologists. I believe that a period of involvement with whole plant physiology would benefit even the specialist so that a comprehensive appreciation of the plant as a functioning unit would be acquired.

Some involvement in applying physiological principles to food or fiber production also can be very instructive. Perhaps this belief results from my own experiences with commercial food processing and in a Forest and Range Experiment Station. Prior to returning to college to study for my PhD degree, I worked for 2 years at the Great Basin Forest and Range Experiment Station in Utah. My major activity, in addition to dishwashing, involved the collection and analysis of photosynthetic and storage organs of range grasses and herbs, taking growth measurements, and attempting to determine the effects of varying time and intensity of clipping (simulated grazing) on food reserves, growth, and survival of these plants. This experience not only stimulated my interest in photosynthesis but gave me insight into the complex physiology of the entire plant throughout its life cycle.

After receiving my PhD, I spent about a year and a half as a chemist in a food processing plant that dehydrated carrots, potatoes, cabbage, and onions. One episode that emphasizes the fact that what works in the laboratory may not always be applicable to the production line, stands out in my mind. Cabbage that we were processing tended to stick to the wooden trays used in the dehydration process. This could be overcome by lightly coating the surface of the trays with vegetable oil. However, this process had to be repeated each time a tray was used. Since one aspect of my graduate thesis dealt with the removal of air from xylem elements of cut stems by a vacuum, I thought that I would apply this in a practical way. Consequently, I suggested that the trays might be vacuum infiltrated with the oil. This should make the process more efficient.

A small slat of one of the dehydration trays when dipped in vegetable oil and submitted to a vacuum in a laboratory test absorbed the oil very satisfactorily when the vacuum was released. Consequently, a large bath was constructed in which the 6 by 3 foot wooden trays could be dipped in oil. A stack of trays on a trolley was then run into a large vacuum chamber. Upon removal from the vacuum chamber after being submitted to a vacuum, the trays looked very good and the oil had penetrated into the wood. I thought that I should get a salary increase. The trays were put into the production line and after passing through the dehydration tunnel, they were shaken against a rack so that the dehydrated cabbage would fall onto a continuous inspection belt. All went according to plan, except that in addition to the cabbage falling off of the tray onto the belt, the trays fell apart! The oil had lubricated the nails holding the trays together, allowing the slats to pull free. I was afraid that I might need to look for another position.

My position also involved testing the food during processing to see whether it had been adequately steam-blanched prior to dehydration. This testing necessitated testing for the presence of catalase, peroxidase, and phenol oxidase. I had worked with all of these enzymes in a plant physiology laboratory course, but there were many new and challenging problems involved in setting up adequate sampling and test procedures in a production line. This experience also made me appreciate some of the differences between laboratory work and applying the theory to an actual production situation.

Now, with increased emphasis on bioengineering, it is more than ever important that physiologists with a comprehensive knowledge of the functioning of the entire plant work closely with biochemists and geneticists involved in these studies. The gene manipulator should know or have advice on the physiology of the plant on which he or she eventually may be working, whether it be maize (a C_4 plant), sugarbeet (a sucrose-storing plant), wheat (a C_3 starch storer), soybean (a C_3 plant and nitrogen fixer with some chloroplasts that appear to store starch more than others), alfalfa (a nitrogen fixer), onions (where starch is not stored), or any of the many other plants that may be under study. To attempt to increase productivity without a knowledge of the physiology of the plant studied would be illogical.

ON MANUSCRIPTS, REVIEWS, AND PUBLICATION

After long hours of intensive research, careful data analysis, and painstaking writing, it is generally satisfying and also somewhat of a relief to complete a paper. However, it is frequently with a feeling of concern that the manuscript is submitted to an editor for review. What kind of review will it receive? Will the reviewer by unnecessarily critical? Will the paper be accepted for publication? If the research has been carefully conducted, the data critically evaluated, the conclusions fully substantiated by the results, and the paper carefully written, then I would urge young plant physiologists not to be discouraged if the manuscript comes back with critical reviews. Although reviewers are generally experts in at least some aspect of the field covered by the manuscript, often they may not be experts in all parts. Nevertheless, each comment should be carefully evaluated. Reviewers are human too. Unfortunately, a reviewer, particularly if he or she has evaluated very few manuscripts, may feel that the review will not be considered adequate unless it contains major critical comments. The reviewer may be embarrassed to admit an inadequate knowledge of a segment of the work. This weakness may result in an attempt to make critical comments on that portion of the manuscript, when instead the manuscript should simply have been returned to the editor with the comment that part of the manuscript was not evaluated. It might be of some encouragement to realize that you are probably more knowledgeable than a reviewer about research that you have been involved in for weeks or months. Be sure of your position, and when you know that you are right, don't hesitate to defend it with facts.

How soon should research results be written up for publication? This is a problem that confronts all research scientists, but I suspect that more err by attempting to publish too soon before a comprehensive piece of research has been completed. On the other hand, research data that fills notebooks and lies on office shelves or in file drawers does little good and represents essentially a waste of time and resources. To try to solve all of the questions that inevitably arise during the course of a piece of research before publishing anything that might have been discovered generally will be impossible. Excellent researchers do not always wait for completion of the definitive experiment before publishing research progress.

Much of the research on chloroplasts from the early 1930s to the mid 1950s centered on the elucidation of the carbon-reduction pathway of photosynthesis and the quantum conversion process and photosynthetic electron transport. By the mid to late 1950s, the essential features of the C_3 carbon assimilation pathway had been worked out. As a young plant physiologist, I followed each new scheme for the movement of carbon from CO_2 through various proposed pathways. Many of these were published from M. Calvin's laboratory in Berkeley and appeared in the *Journal of the American Chemical Society*. It seemed that almost monthly there would be a modification of a proposed scheme as new data became available. Finally in 1952, after some 15 papers and numerous schemes and modifications had been proposed, the essential features of the currrently accepted pathway for carbon flow in photosynthesis in C_3 plants was worked out.

A further point that perhaps should be kept in mind with reference to publication is that if one's research results lead to conclusions that are different from the usual current dogma, if one is confident that they are correct, then even repeated negative criticism should not be overwhelming. I recall the protracted running controversies that took place in the mid-1950s at a Gordon Conference, in scientific publications, and in laboratories concerning the concept of photophosphorylation. In these controversies, D. I. Arnon successfully defended his position against severe criticism.

THE DEVELOPMENT OF A BOTANY DEPARTMENT

Academic departments frequently have growing pains, so perhaps some of my reflections on the factors influencing the growth of the Department of Botany at the University of California at Davis will be of interest. At the time I came to the department as a graduate student in 1934, the faculty consisted at Kathrine Esau (plant anatomist), W. W. Robbins (plant physiologist and chairman), Alden Crafts (plant physiologist), and Elliot Weier (cytologist). It has been a privilege to have been associated with these outstanding scientists and teachers and to have been a member and to have been of help in the development of this department as it grew from a very small service unit in the College of Agriculture to its present stature as a major department of the University of California. Of the many factors that influenced the development of this department, including the excellence of the original core of faculty and the sympathetic support of college deans, two unique factors affected the course of departmental growth. First, the department has had a continuous close affiliation with the College of Agriculture and its plant science departments; and secondly, the emphasis in the recruitment policies on the cooperative ability of new faculty members may have been somewhat unique.

Since the faculty of the department was at first too small to administer graduate programs effectively, graduate groups were established in botany and plant physiology. The faculties of these groups were responsible for the establishment of academic requirements, the administration of examinations, and the granting of graduate degrees. They were made up not only of the appropriate members of the faculties of the botany and plant science departments of the College of Agriculture at Davis but also of similar departments at Berkeley. Thus, students studying for advanced degrees had the opportunity to choose a major professor from a large and diverse faculty, and the establishment of guidelines for what constituted a desirable training for prospective botanists or plant physiologists was the result of ideas contributed by a diversified but effective faculty. In addition, the Department of Botany at Davis, although small, functioned as an integral part of this comprehensive program of graduate instruction. When the department had grown to sufficient size, the Berkeley members of the graduate groups withdrew. However, the graduate groups have been retained at Davis, and the many outstanding botanists, plant physiologists, biochemists, and geneticists who are members of other departments in the College of Agriculture have been a very important factor in the development of graduate instruction in plant physiology at Davis. There was a very rapid growth of the Department of Botany between the mid 1940s to the mid 1950s. At this time, as many as 12 plant physiologists and physiological ecologists became members of the department. Consequently, they played a major role in the graduate instruction in plant physiology on the campus.

A further factor that had a significant effect on the direction of development of the Department of Botany at Davis was the department's rather unique position. First it was a small department in the College of Agriculture, and later, with the establishment of a liberal arts college, it became two combined departments, the Department of Agricultural Botany in the College of Agriculture and the Department of Botany in the College of Letters and Science. The department became responsible to two deans in two different colleges, had two budgets, and had faculty members, some of whom had appointments in one college, some in the other college, and some split between both colleges. This was a situation that would seem especially designed to foster the maximum internal disruptive friction among faculty members. However, over the years the department has operated very smoothly so that it has been a pleasure to be a part of it. Instead of wasting energy with internal friction, faculty members have directed their energy along productive lines.

This relatively smooth interaction speaks to the character of the individual faculty members, but I believe that it also reflects a conscious effort in the selection of new faculty. Such points as excellence and originality in teaching and research obviously were of prime consideration. However, an additional factor on which more stress was placed than may be customary was that each prospective candidate for a given position was evaluated with respect to his or her ability to work cooperatively with the rest of the faculty. Even when outstanding established scientists were being considered, if it appeared that they would have a tendency to demand space and support to build their own little empire at the expense of other faculty members, their applications were considered less favorably. Undoubtedly this philosophy resulted in the department losing some potentially very productive scientists. However, I believe that the overall result, especially for this double department, was very positive in terms of the smooth functioning of the entire departmental unit. It has resulted in a balanced department that is recognized for excellence in the broad field of botany rather than in a few narrow areas of specialization.

SECOND THOUGHTS AND UNFINISHED BUSINESS

It is stimulating to let one's mind wander over the past, to contemplate the present, and then to project one's imagination into the future. However, we have sufficient futurologists and fantastic model builders extolling the roles of computers, bioengineering, space stations and colonies, etc. I shall not attempt to speculate on the role of the plant physiologist in helping to bring about this incredible future. Instead, having survived many years of academic life, contact with hundreds of students in lectures and laboratories, committee meetings, grant proposals, and research problems, and now having finally reached the status of Emeritus Professor, I can look back over the past and try to imagine what I might have done differently. I don't mean all of the false starts, blunders, and downright goofs, etc. I have had my full share of these! Would I have chosen a different profession, or a different research program, or are there certain research problems that I would have liked to tackle but did not?

Certainly to have been involved for almost 50 years in the general area of plant physiology, to have been associated with and been stimulated by students at all levels from freshman to graduate and postdoctoral, and to have known and worked closely with so many outstanding scientists and admirable people from many different countries, and to have been free to explore the structural and functional beauty of plants, I have been richly rewarded.

I owe much to the many students who were a constant stimulus with their helpful criticisms both of my teaching and research. Their questions and suggestions often led to significant experiments. For example, I remember a lab meeting held in my home one evening in 1969 and the discussion concerning light-stimulated cellular activities that are linked to photosynthetically trapped energy but not to CO_2 reduction. We tossed ideas back and forth, dragged out a blackboard, and outlined some of the possible shuttle systems that might account for these phenomena. As a result of this discussion, specific experiments were initiated in the laboratory. These experiments showed the importance of a triosephosphate-glyceraldehyde shuttle system that could effectively transfer photosynthetically generated reducing power across the chloroplast envelope. I'm sure that discussions like this have been constant sources of stimulating and innovative ideas in many laboratories where plant physiologists openly exchange thoughts on research problems.

Why did I become a plant physiologist? During my high school days, I thought that I would be a chemist. I would skip lunch to save money to buy chemicals to stock the lab that I had fashioned in the basement at home. I suspect that we would need a most sophisticated and detailed analysis of innumerable small, seemingly insignificant occurrences to gain any under-standing of the factors that led to my decision to become a plant physiologist. Some factors, however, are more apparent. Among these are a friend's gift, on my twelfth birthday, of Paul de Kruif's book, "Microbe Hunters"; an opportunity to pay my way through junior college by working in a cooperative program in which I alternately spent six weeks in E. C. McCarty's (a plant physiologist) laboratory and then attended class for six weeks; and a teaching assistantship under A. R. Davis at the University of California at Berkeley. Although my choice of being a plant physiologist was influenced by many small factors that occurred over a number of years, if I had it to do over now, I would not change my choice. Undoubtedly this reflects my years of effort in the area, but it also emphasizes the satisfaction that this work has had for me.

On reflecting on the direction of my past research, I believe that if I had it to do over, I would include some research of a more immediate applied nature but still retain my major effort in fundamental or basic research. Recently, I heard the statement made that all of the research on photosynthesis over the past 50 years has contributed very little to an actual increase in food production. Most increases have resulted from modified cultural practices and breeding work. Of course, this may change entirely in the next 50 years. Perhaps by then what I half jokingly would tell my students will come true. When asked by them why I did research on chloroplasts, I replied that some day when we knew enough about all of the intricate and structural and biochemical characteristics of the plastids, we could devise a system utilizing these principles to trap sunlight, pass water and carbon dioxide across an artificially pigmented membrane, and have sugar coming out of the spigots at the other end. Will such a system be found in space stations within the next 50 years?

If I were to continue doing research on interactions between chloroplasts and cytoplasm, what projects would I like to do? Our knowledge of the chloroplast envelope, particularly of the intramembrane location and orientation of the membrane components, is still very incomplete. What are the intracellular environmental factors that might cause a reorientation of the phospholipid and intrinsic protein constituents of the envelope membranes? Are there fundamental shifts in membrane composition or orientation of membrane components during plastid development that might have an influence on the permeability properties of these envelopes? Since the bundle sheath and mesophyll plastids of C_4 plants are enzymatically different, are there fundamental differences in the molecular architecture of the membranes of the envelopes? I would like to test the feasibility of isolating these two kinds of chloroplasts and studying the effects of various enzymes and membrane probes on their properties, including their ultrastructure and their electrophoretic mobilities.

Although we now know a great deal about the transport of metabolites across the chloroplast envelope, especially the export of metabolites from the chloroplast into its cytoplasmic environment, with the possible exception of one or two enzyme components, particularly the small subunit of ribulose bisphosphate carboxylase, we know relatively little about the transport of large molecules into plastids. This important aspect of cellular physiology still holds many intriguing questions.

ESCAPE FROM REALITY?

One of the temptations of being a scientist is to use scientific endeavors as an excuse to escape from the unpleasant and sometimes frightening realities of life. Scientific research can be an island of tranquility, but it is too easy to rationalize by thinking that the research that one is doing is so important that no time is left even to write a grant application or a research paper, much less to become involved in any way with local, national, or international problems.

When the shadows of nuclear catastrophy, environmental degradation, and widespread starvation are now so startlingly clear, one should not wait for retirement before reflecting on what are the most important problems to solve today. I have been accused of being an optimist. I do believe that in spite of all of the signs of insanity that seem to point toward world calamity, the human spirit, will, and intelligence can overcome these difficulties, which, after all, are also products of human thought, emotions, and actions. Yes, I am optimistic. The optimist has an advantage in that he or she experiences the joy of life day to day, while the pessimist may be continuously fearful of the future. Nevertheless, the optimist should be fully aware of the realities of the world around and strive in whatever way possible to justify this optimism.

In spite of their contributions to science, some of the best-known scientists have taken very active roles in expressing their concerns for the future of the human race. Eminent Nobel Laureates such as Einstein and Pauling have been outstanding in their efforts to promote world understanding and peace. It is no accident that 48 of the 95 sponsors of the Federation of American Scientists (an organization of 5000 natural and social scientists and engineers who are concerned with problems of science and society) are Nobel Laureates. It seems axiomatic that we, as trained scientists, have an obligation not to let our scientific endeavors mask our responsibilities as citizens. We may not be experts in foreign affairs or nuclear disarmament, in controlling toxic substances, in the resolution of nutritional needs of Third World citizens, in the resolution of social inequities, in slowing population growth, or in reducing the depletion of global resources, but at least we can become reasonably informed and concerned. This awareness and knowledge on our part should, in itself, inevitably lead to some action, however minor. My plea is not for plant physiologists to become activists, but for all scientists not to spend their entire intellectual effort in that island of scientific tranquility.

Michael Evenari

Ann. Rev. Plant Physiol. 1985. 36:1–25

A CAT HAS NINE LIVES

Michael Evenari

Department of Botany, Hebrew University, Institute of Life Sciences, Givat Ram, Jerusalem 91904, Israel

CONTENTS

YOUTH

When I was 13 years old I knew that I was going to be a botanist. I lived at that time in Berlin with my sister and brother-in-law, the poet Gerson Stern. One day he presented me with a book by R. H. Francé, *Die Welt der Pflanze: Eine populare Botanik* (8). The book hit me like a ton of bricks. I had studied in a humanistic high school, "Humanistisches Gymnasium," a school in which at that time we were taught mainly Latin, Greek, and with the exception of German, no modern language and no biology. The book, therefore, opened up a new world before me.

Up to that time, a plant was for me simply a "thing," and now it was revealed to be a complex entity provided with a great number of astonishing structures and mechanisms which were all *"zweckmaessig,"* i.e. adaptations enabling the plant to exist in very different environments. Two of the book's chapters aroused my particular curiosity. One was called "Invisible technicians," dealing with what we know since Haberlandt's classical book (8a) as physiological

2139

plant anatomy. Another was on seeds, their dispersal and their germination. Francé called the chapter "The plant as mother." In Metz, where I was born as Walter Schwarz and where I lived until my thirteenth year, we had a garden that was cared for by our factotum Johann, who every year had sown some flower seeds. I had never paid any attention to this. The idea that a plant could be a mother with children, as described so romantically by Francé, therefore seemed to me not fully believable. I started to germinate some bean seeds in pots and was deeply awed when the hypocotyl of the seedlings broke out of the soil. I confess that even today germinating seeds still affect me the same way. At that time I could not analyze my feelings, but I now know what struck me then. My awe apparently emerged from my collective subconscious (J. C. Jung). It was the same spiritual experience which made germination for many ancient civilizations and religions the symbol of death and resurrection and of the yearly recurring rebirth of nature in spring after its "death" in winter.

Shortly after I had received Francé's book, my brother-in-law presented me with a children's microscope. It looked like one of the first microscopes invented. It had a substage mirror but no condenser, the three objectives had to be screwed on since it had no revolving nosepiece, and the focusing had to be done by elevating or lowering the tube by hand. I immediately collected water from any puddle or pond I could find. Observing my first *Euglena* flagellating through the microscope's field of vision and my first *Volvox* majestically rotating was a revelation. My reaction was twofold. My curiosity to see (and know) more was awakened, and at the same time I was overwhelmed by the elegance and pure aesthetic beauty of these organisms.

In reviewing my scientific life now at the age of 80, I see that this primeval reaction to nature at age 13 has remained the mainspring of my scientific work—continuously stimulated by curiosity and, perhaps even more important, wonder, admiration, and love for the beauty of nature and its creatures. From then on animals and plants were for me far more than objects of study. In trying to find out what made these creatures tick I was bound to them by the common bond of life, a feeling which elevated them from study objects to brother living beings. I am convinced that most good biologists, even if they claim objective detachment, react, mostly unconsciously, in a similar way to their objects of research.

While I was still in my thirteenth year, my brother-in-law gave me another booster shot when he presented me with *Die Lebensgeheimmiss der Pflanze* (*The Secrets of Plant Life*) by Adolph Wagner (26). The title page of the book depicted two growing bean plants, one climbing up a pole, and the other, having climbed a broken pole, now reaching from its pole to that of the other bean plant. When writing this paper, I thought of this book, which I had not seen for more than 50 years. I had forgotten its title and only remembered the author's name Wagner, the fact that the book had been published before World

War I, and most of all, the title page picture. I asked friends in Germany to find the book, and they located it on the basis of my description of the climbing bean plants. I think the picture imprinted itself so deeply on my mind because, for the first time, I became aware that a plant has sensors enabling it to orient itself in space according to its needs. When I first received the book, I read it in a few days, and grew a bean plant in a pot to see if Wagner told the truth. This was my introduction to plant physiology.

Having devoured France's and Wagner's books and looked down the microscope, my fate to become a botanist was sealed. From then on I called myself (naturally only secretly in my mind) in good Latin style, "discipulus scientiae amabilis."

As stated earlier, I was born in Alsace-Lorraine, the two border provinces between France and Germany. After the Prussian-French war of 1870–71, the Germans had annexed both. In 1918, the provinces became French again, and in 1940 returned to the Germans, only to become French yet again in 1945. In contrast to Alsace where the people spoke a German dialect similar to the "Schwytzer Deutsch" of the Swiss, the language of the indigenous population of Lorraine was French. Since in Metz after the occupation of 1871 German was the official but French the unofficial language, I was bilingual from early youth. Everybody spoke at least some German, often intermixed with French. You could, for instance, hear a sentence like this one: "Maman, komm à la fenetre. Jean ne croit pas dass Du schielst" ("Mom come to the window. Jean does not believe that you are crosseyed").

My bilinguality as well as my knowledge of Latin and Greek helped me later in life to acquire other languages such as Hebrew, Spanish, Italian, English, and Russian with comparative ease.

It is most regrettable that today in science we do not have one common language, in the way that Latin was the language of scholars and scientists up to the eighteenth century. Since a return to Latin is most unlikely, a modern scientist should be able to at least *read* English, French, and Russian. If, as in most cases, he cannot, the results are sometimes bizarre.

For instance, a modern textbook of ecology cites 1695 papers written in English, 15 written in German, 7 written in Russian, 3 written in French, and one paper written in Dutch. Students will certainly get a lopsided view of ecology because they will believe that nearly all the important papers in ecology were written in English. Lack of knowledge of literature written in foreign languages also leads to objective mistakes. The same textbook can serve as an example. In it the term "allelopathy" is ascribed to Muller (15), whereas the phenomenon was first described and the term first coined by Molisch in 1937 (14).

The Germans introduced into Alsace-Lorraine a school system modeled along that which was prevalent in Prussia. The school had 12 grades numbered in Latin, as befitting a humanistic gymnasium that required seven years of Latin. One started at age five or six in the "nona" (ninth grade), went through Octova, Septima, etc, and finished in upper Prima with the "Abitur" which entitled one to enter a university.

Up to the beginning of World War I, school discipline was very strict. The teachers were tyrants, and we were often punished for small misdeeds with hard strokes of a cane on the inner hand surface. This changed completely after the beginning of the war. The young teachers joined the army and were replaced by old retired ones who could not master the unruly crowd. We learned very little, especially since we spent much time in the large underground air shelters of our school, a former French monastery.

Metz at the time of World War I was a strange place. The city was surrounded by a double ring of fortifications kept by the German army, and the front line was only about 20 km away at Pont à Mousson. From my window I could see the cannon fire there and hear the continuous rumbling of the guns at Verdun, about 50 km west of Metz. In the first days of the war I had my first taste of aerial warfare. French planes attacked Metz and used for the first (and I think the last) time, *Fliegerpfeile* ("aviator's arrows"), aerodynamically constructed metal arrows which the pilots released in bundles. One pierced the visor of the peaked cap of the gatekeeper of my parent's department store. We children fought for the possession of such a rare souvenir. Later French planes often bombarded the city, once hitting an ammunition train. The explosion kept us for hours in our cellar, where we sat huddled together, frightened to death.

The combination of inefficient teachers and the atmosphere of war not far off had a strong effect on me. I stopped doing my homework and spent most of my time playing trench warfare with a band of other rowdies in unused trenches. The result was that in one year I went down the ladder from the first to the last place in my class. The patience of my parents broke, and they sent me off to my sister and brother-in-law in Berlin.

In Metz I had already had my first experience with anti-Semitism and with Zionism. When I was seven, coming home from school, some children called me "*sale juif.*" I reacted promptly by jumping on them and a fight ensued.

Another anti-Semitic event made an even more profound impression upon me. My elder sister and my brother, who fell as a German soldier in World War I, belonged to the "Wandervogel." This German youth movement decided in 1916 to expel its Jewish members. This brought home to us that the social environment to which we thought we belonged apparently didn't want us. At about the same time, Jewish soldiers serving in the nearby German frontline visited us during the Jewish holidays, speaking to us about Herzl (the founder of modern Zionism) and the Zionist aim to build a Jewish homeland in Palestine. After my experience with anti-Semitism, the idea appealed to me, and I became, and still am, a Zionist.

When my parents moved in 1920 from Metz to Frankfurt/Main, I joined them. During the two and a half years of high school in Frankfurt I dedicated most of my time to the Zionist youth movement which was the Jewish-Zionist counterpart of the Wandervogel.

STUDENT DAYS

I entered Frankfurt's Johann Wolfgang Goethe University in 1923. I naturally took botany as my main subject but told my parents that I was studying chemistry (one of my secondary subjects) because at that time it was unthinkable that a good Jewish boy would take up such an outlandish profession as botany. I continued this pretense until the day I received my PhD.

Three years after I had received my PhD, my parents were still so worried about my professional future that my mother once visited Moebius, my professor of botany, and asked him: "Herr Geheimrat, has my son as a Jew really a chance as a botanist at a German university?" Moebius, knowing that at the time I was working on the physiological anatomy of the fruit stalks of heavy fruits, gave her a metaphoric answer citing a German proverb, "Es sind die schlechtesten Fruechte nicht, an denen die Wespen nagen," meaning, I was a good "fruit" and that she should not worry about my future.

My professor of zoology was Otto zur Strassen. He and Moebius were both heads of their respective departments and, as usual at that time in Germany, absolute bosses. Both carried the splendid title of "Geheimrat" (Privy Councillor) and had to be addressed as "Herr Geheimrat" even after years of professional contact. To call the professor by his personal name as is usual in American universities was unthinkable.

Moebius was a gentle man who treated his students like a benevolent father and took an interest in the personal life of each of them. He had an unusually broad and humanistic cultural background. During the Nazi period he also displayed unusual personal courage. In 1935 he dared to write in a letter to the mayor of Frankfurt: ". . . Presumably you have seen . . . the giant antisemetic placards on the fence of the Gontard house. We Frankfurters should be ashamed . . . that here Jew baiting is done in such a hateful and disgusting way while Frankfurt especially should be grateful to its Jewish citizens for a multitude of endowments . . . I remind you that Frankfurt would never have been a university town if Jews in particular had not donated the necessary means . . ." Somebody leaked the letter to the "Sturmer," the notorious anti-Semetic hate journal edited by Julius Streicher, which published it. As a consequence, Moebius's successor as head of the Botany department banned him from ever entering again the department which he had founded.

The botanical fields in which Moebius was most active were taxonomy, developmental and physiological anatomy, nature and origin of plant colors

(12), and the history of botany (13). I am proud to say that in his book (13) he twice mentions my own work. His interests stimulated my own. The elements of taxonomy that I learned from him came in very handy later in Palestine, when I worked in Aaron Aeronsohn's herbarium. The effect of environmental factors on anatomical structures had interested me throughout my botanical career. I was also much taken by Moebius's occupation with the history of botany. In his lectures he always tried to trace the origin of concepts or terms back to the men who had first conceived them. In my own lectures, I have always tried to follow Moebius in this regard because I feel that educationally it is important that students know that, with few exceptions, we in science are "standing on the shoulders of giants."

Moebius was not a good lecturer. We used to say that listening to him was like taking a sleeping pill we called "Moebiol." In the lab he was an excellent teacher. He had prepared a hand-written manual in which each consecutively numbered paragraph described what we had to do. We called it the "Fahrplan" (railway time table). We proceeded from number to number in our own time. I spent every free minute in the laboratory which was always open to us. I often worked there until the early morning hours. Moebius visited us every day and discussed our work with us, sometimes for hours. If he found that we were much interested in a certain matter, he encouraged us to enlarge upon it and to do some additional work. This very personal way of teaching was possible because only about eight students participated in the lab. It was ideal because one was completely free from any time limit and constraint. When he noticed that my special interest was in physiological anatomy, he permitted me to deviate from the "time table" and to do some special work.

In the department's botanical garden, I had observed that in the mature petioles of *Heracleum pubescens,* the central vascular bundles were arranged in a wave-like pattern whereas they were straight in the young petioles. When I showed this to Moebius he said, "Why don't you investigate the reason?" And so it came about that my first paper was published in 1926 (18).

From Moebius I received my basic training in general botany and from Fritz Overbeck, then the one and only assistant in the department, the basis of plant physiology. In his course I did my first scientific experiments on germination, a theme which later in my scientific career became very important to me. Under his influence I also carried out my first physiological investigations, one on the so-called "mitrogenetic rays" of Gurwitsch, which proved to be nonexistent (20); the other one on the etiology of variegation in *Coleus* (21). The phenomenon of variegation occupied my interest for many years.

Geheimrat zur Strassen, the zoologist, was a type very different from Moebius, an extrovert, aristocrat, and a brilliant lecturer. Not only students attended his lectures but half the intellectual elite of Frankfurt, especially women attracted by this impeccably dressed, charming man. I, as did all the others, listened spellbound to his lectures on evolution of the animals, where, according to him, every step was well known and documented. Only after the lectures, I asked myself if everything was really so easily explainable. I still have some doubts today about the *mechanism* of evolution, as proposed by Neo-Darwinism.

In zur Strassen's seminars we had to lecture on a zoological paper he had supplied. He then not only criticized what we said but also *how* we said it. We should never wander around the podium, but stand still facing the audience. We should never read from a prepared text, nor talk when writing on the blackboard with our backs turned to the audience. And above all, we should prepare our lectures carefully and, before giving them, practice them and time them. When listening today to some of my colleagues, I pity them that they did not have such teachers!

After four years at the University, I graduated with a PhD (19) at the age of 22, having passed an oral examination in botany, zoology, physics, and chemistry. In contrast to today's praxis, this was the only examination I had to go through during the four years of my university studies. I am sorry for the students of today who are plagued by so many consecutive examinations which, I suspect, obstruct the learning process more than they promote it.

My PhD examination was more a friendly conversation than an examination. Moebius and zur Strassen asked what I knew about the evolutionary process of animals and plants. When they saw that I had my own slightly unconventional opinions about the mechanism of the evolutionary process, they discussed these with me for over an hour.

ASSISTANT IN EUROPE

A few days after receiving my PhD, I married and worked for two terms as an assistant to Moebius. Then Ernst Pringsheim offered me a similar position in the department of plant physiology of the German University in Prague. I accepted and stayed there from 1927–1930. Prague at that time had two universities, one Czech and one German, and therefore also two departments of botany. There was practically no contact between them, an evil omen for things to come some ten years later after the takeover of Czechoslovakia by Hitler.

In Prague I worked on two problems, variegation and the physiological anatomy of the fruit stalks of heavy fruits. I found that mechanically the fruit stalks were much overdimensioned, and that with much less mechanical tissue there still would be no danger of the fruits falling off the tree. My investigation taught me to be very careful in relating structures to functions. Often certain structures seem to fulfill a certain function, yet in reality are only a necessary functionally neutral by-product of physiological events that are not directly related to the observed structure.

In my work on variegation I was lucky to find in the botanical garden a form of *Selaginella* in which the variegation was temperature dependent. In the white parts of the variegated leaves the originally normal plastids degenerate until they disappear completely. I found stomata in which one guard cell contained a normal green plastid whereas the plastid in the second guard cell degenerated. Both plastids derived from the apparently normal plastid of the stoma-mother cell. Here I was confronted with a basic problem of all developmental processes: physiologically unequal cell division.

My stay in Prague widened my botanical horizons considerably. I attended many of the lectures of the four professors and their assistants. In Prague I also became interested in floristics and plant sociology.

The three years I spent in Prague belong to my most pleasant memories. Prague was at that time, when Thomas Masaryk was the president, the cultural center of central Europe, abounding with writers like Franz Kafka, Karel Čapek, Jaroslav Haček (the good soldier Schweik), Franz Werfel, Stephan Zweig, Ernst Brod, and many famous musicians. I enjoyed the best of two cultures, the German and the Czech slavic. I decided that an educated person should know at least one slavic language and started to learn Russian. This opened for me a new cultural horizon.

While I was in Prague, an event occurred that was of the greatest importance for my future life. In 1930 Heinz Oppenheimer came to Prague in order to work in Pringsheim's department. After finishing his PhD under Molisch, he had immigrated to Palestine where he worked in the herbarium of Aaron Aaronsohn, the man who in 1906 had found one of the wild ancestors of cultivated wheat *(Triticum dicoccoides)* in Galilee, and over the years had collected a large herbarium of plants from Palestine, Syria, and Jordan, containing a number of new species. After his untimely death the family asked Oppenheimer to classify and publish the new species and a list of all the plants collected, as well as Aaronsohn's very interesting travel journals. At that time Oppenheimer had already published the *Florula Transjordanica* (16). In Prague we lived in the same house and became friends. After some time he asked me if I would be willing to come to Palestine and work with him on editing the *Florula Cisjordanica*. I promised him that I would seriously consider his offer. In the meantime, I was offered an assistantship in the botany department of the Technical University in Darmstadt. I accepted because the job offered the possibility of becoming a lecturer.

My first boss in Darmstadt was Friedrich Oehlkers. He was highly intelligent and had a special interest in modern philosophy. He was also a difficult personality and most excitable. Once when I disturbed him, he threw a chair at me, which for a professor was quite an extraordinary thing to do.

In Darmstadt I continued my work on the physiology of variegation (22). In the youngest cells of apical shoot mainstems I observed in vivo mitochondria and proplastids and found that mitochondria and plastids are physiologically and developmentally completely independent. This was an important new statement, because some authors then doubted the existence of mitochondria and others believed that plastids developed from mitochondria.

At the end of 1931, Oehlkers left Darmstadt and Bruno Huber became head of the department. The cool aloofness of Oehlkers toward his subordinates was replaced by the warm, affectionate, and friendly attitude of Huber. He encouraged me to write the thesis needed to become a lecturer, and in 1933 I gave my probation lecture before the faculty. He did all this, knowing that I was a Jew and knowing that the new Nazi regime was fanatically anti-Semitic. He later had considerable trouble with the regime because of what he had done for me.

In the department I was responsible for the laboratory exercises and the floristic excursions of the students of pharmacology and of the future biology teachers. When I met them for the first time I noticed that all of them carried the badge of the national socialist party. I immediately made my position clear, telling them that I was a Jew and a Zionist. From this point on an interesting relationship developed between us. The students knew that as a Zionist I wanted to go to Palestine. This made sense to them. I, on my part, was curious to know what attracted them so forcefully to Nazism. It came out that they strongly believed in the *socialist* part of the national-socialist movement. Hitler at that time had a strong competitor for leadership in the party, Gregor Strasser, who proclaimed socialism as one of the main aims of his party. He based his "socialism" on the ideas of Gottfried Feder, who in 1919 had published a booklet entitled, "The breaking of the bonds of interest slavery" (Die Brechung der Zinsknechtschaft). In this booklet he proposed a noncapitalistic social system in which everybody could borrow money without paying interest. This impressed my students very much. I had read Feder's pamphlet and tried to explain to them that the whole idea was an impossible bluff. I still hear myself telling them that Hitler used their naive idealism for his very different political purposes and that they were dupes. How right I was they must have seen later when, after he came to power, Hitler immediately got rid of Strasser, and Feder and his whole "socialist" program just vanished. Looking back after so many years I am still astonished that we could discuss those explosive matters freely and still remain on good personal terms.

Hitler's book, the discussions with my students, and the opinions expressed by the man in the street convinced me that Hitler would soon come to power, sealing the fate of the Jews. I had kept up my contact with Oppenheimer, and in October 1932, I signed a contract with the Aaronsohn family, which stipulated that I should come to Palestine in October 1933.

My nearly two years with Huber were, professionally speaking, quite productive. I had observed earlier that the leaves of a variegated form of *Coleus* rooted easily without forming buds. Their petioles thicken considerably and the white parts of the variegated leaves become green. I found that leaves of many other species can be induced to root in the same way as *Coleus*. The changes that take place when the petioles form a cambium and anatomically turn into stems, and their possible physiological reasons, became the theme of my *Habilitationsschrift* (23).

Under the influence of Huber I began to be interested in ecophysiology. Huber was the first to use a fast-weighing torsion balance for measuring transpiration in the field, and he also constructed a cumbersome instrument for the field measurement of photosynthesis. He turned my attention to Maximov's book on water relations (11), subtitled "A Study of the Physiological Basis of Drought Resistance," a topic which keeps me busy to this day. When in 1932 I told Huber of my intention to leave for Palestine, he encouraged me to take with me a torsion balance and his photosynthetic apparatus and to start ecophysiological work in the desert there. He also gave me the books by Volkens (25) and Stocker (24) on the physiological anatomy and water balance of desert plants. To this day, both are for me a kind of desert bible.

Otto Stocker, who much later became my personal friend, was the first ecophysiologist to work in the desert of the old world. I was thrilled by his book and decided to walk in his footsteps.

On April 1, 1933, the day Hitler declared a boycott against the Jews, the university sacked me. In the morning, the rector of the university summoned me. When I entered his office he said, after some mumbling and stuttering, "Herr Doktor, you were denounced to me as a confirmed Jew *(bewusster Jude)* and I should dismiss you on the spot. But I am personally willing to give you four weeks to leave the university." I had expected something of that kind and answered rather brashly, "Your magnificence (this was the official title of a rector), in times to come you may remember this day as the beginning of the downfall of Germany. As to the four week's grace that you want to give me—you can keep it. I am leaving this afternoon." With these words I stood up, turned my back on him, and walked out banging the door. I immediately sent a cable to the Aaronsohns telling them to expect me in Palestine in April instead of October. They agreed.

In the afternoon of that memorable day I packed my things in the department. Interestingly enough, my Nazi students told me how sorry they were at my leaving and then helped me to pack. The father of one, the owner of a transport company, moved all of my meager goods to Frankfurt and refused to be paid for his service. We left all of our furniture in our apartment, inviting everybody to take what they wanted. Three weeks later we arrived in Haifa. Thus began a new life.

PALESTINE

Oppenheimer and the Aaronsohns received us with open arms. They had prepared a flat in Jerusalem for us, and for the next four years I went every month for a few days to Zikhron Yaakov to work in the herbarium, returning with a bundle of herbarium species which I could not properly identify in Zikhron. In Jerusalem I could compare them with specimens in the herbarium of the department of the Hebrew University. I always tried to find the species that I had identified alive in nature and collected my own herbarium. By the time this taxonomic work ended and Oppenheimer and I had written the manuscript of our book, *The Florula Cisjordanica* (17), I had acquired a good working knowledge of the flora of Palestine, Syria, and Lebanon. This helped me greatly in my ecophysiological field work.

I think it would be good for science if all physiologists would have some knowledge of the flora surrounding them. This is very often not the case, and it shocked me when I heard a famous plant physiologist say that he was proud of not knowing the difference between a rose and a carnation.

A few days after our arrival in Jerusalem, Heinz Oppenheimer, who was the first head of the plant physiology and anatomy section in the botany department of the Hebrew University, brought me in contact with Alexander Eig, the head of the department. In 1933 the Hebrew University was still very young. It was officially opened in 1925 as a research institution and was only opened to students in 1928. Otto Warburg was its first professor of botany, followed by Eig, a taxonomist, phytosociologist, and phytogeographer. I became Heinz's successor when he left for the Agricultural Research Station in Rehovot, and gave my first lectures in plant physiology and anatomy in 1934.

In the meantime, much had happened to me. A few days after our arrival in Jerusalem I met, through Oppenheimer, Richard Richter. He was (and still is, at 90) a very colorful character. He is half Jewish and served during World War I in the German army as a fighter pilot. I asked him to cooperate with me in my desert research, and a few days later we went to the Judean desert to find an appropriate spot for our work. I had brought with me the torsion balance for measuring transpiration and Huber's instrument constructed for measuring photosynthesis.

On our first trip to the desert, on a day when the dry hot desert wind (Khamsin) was blowing and the temperature rose to 42°C, it soon became evident that under these conditions the photosynthesis apparatus refused to work. It was also so heavy that Richter, carrying it on his back in that heat, collapsed with a slight heart attack and we had some trouble returning to Jerusalem. We had no money for a car and had to drag ourselves and our equipment to the faraway bus station. From then on we had to restrict our work to the measuring of transpiration. Thus began my first ecophysiological work in the desert, the results of which were published in 1937 and 1938 (4–6).

The moment I had my first glimpse of the Judean wilderness, I fell in love forever with the desert. I was spellbound by its somber and sublime beauty. It moved me emotionally, spiritually, and intellectually: emotionally and spiritually because in it man in all his tragic loneliness is confronted with nature in the raw; only here could God have spoken to Moses, Jesus, and Mohammed; and intellectually, because its faunistic and floristic structure is, in comparison with a jungle or even a prairie, comparatively simple, comprehensible, and researchable.

In 1933 I also became acquainted with the Jordanian, Syrian, and Iraqian desert. The department of botany of the Hebrew University was invited by the Iraqian government to make a survey of the forests of Kurdestan. Eig asked me to participate in the expedition, together with my instruments, in order to get an idea of the water balance of the Kurdestanian forests. We traveled by car from Jerusalem to Amman and via Kasr el Asraq to Bagdad, then via Kirkuk to Suleimaniyeh, and from there on horseback up into the mountains of Kurdestan. I had never mounted a horse and felt very romantic riding one, an emotion enhanced by an episode which earned me the respect of our military escort. Before riding out from Suleimaniyeh, we had to choose our horses. The Iraqi soldiers wanted to give us the tamest ones because of our inexperience. Out of silly pride I asked for a "normal" horse. As we started out, my horse, an old cavalry steed, immediately began to run and I was soon at the head of our long column. The soldiers apparently got the impression that I wanted to challenge them and raced after me. My horse must have been accustomed to races and galloped faster and faster, jumping over ditches and trenches. I held my arms around its neck, holding on for dear life. I was deadly afraid and tried to slow it down but did not find the right brakes. Miraculously, I was not thrown off and was the first to reach the forest. From then on I was the racing champion of our party.

For the next two months we traveled through the forests along the Iraqian-Iranian Turkish border in an arc of scenic mountains from Suleimaniyeh, Rawanduz, Amadiyeh, Zakho, to Dihok. Wherever we stayed for a few days, the Kurds built an airy hut from tree branches in which I measured evaporation, transpiration, and stomatal opening of the main forest tree, *Quercus brantii*. From this tree the Kurds collected the galls for tanning and ink production. They also gathered its leaves, which are covered by a layer of sugar, and put them into large vessels where hot water dissolves the sugar that later is used for making Turkish Delight (Lokoum). The sugar is produced by aphids which tap the phloem, use the protein, and excrete the surplus of sugar.

During our journey we also detected in the mountains far off the beaten track two heretofore unknown villages inhabited by Jewish peasants speaking Aramaic, i.e. the language spoken in Palestine at the time of Jesus. They claimed to be in Kurdestan from the time of the Babylonian exile (7th century B.C.). However this may be, it is certain that these Jewish peasants were there for at least more than 1500 years.

When we returned to Jerusalem, Eig offered me a part-time position as "external teacher" in the department which I accepted.

My narrative so far may have given the impression that my transition from Germany to Palestine was smooth and painless. This certainly was not the case.

Palestine in 1933 was very different from the Israel of today. Large parts of what is today good agricultural land was unproductive swamp, steppe, and desert. Living conditions were harsh. We often had no water in Jerusalem; we cooked on small petrol stoves (Primus); there was no central heating in winter and Jerusalem can be very cold; electricity often failed. Daily life was further complicated by the Arab-Jewish conflict. Since the British police were unable to protect us, we had to defend ourselves. In 1933, at the time of one of the recurrent outbreaks of violence of Arabs against Jews, I joined the Jewish self-defense organization, the Haganah. This meant that I had to do guard duty two or three nights a week in one of the Jewish suburbs of Jerusalem. In the beginning this seemed to be quite romantic, but when it went on year after year it became quite a physical and emotional burden. It was not easy to lecture and give labs after a night of guard duty without sleep. But we felt it our duty to carry on with our research and teaching as if times were normal.

These physical discomforts were only minor nuisances in comparison with the language problem. The transition from Germany to Palestine had robbed me of my mother tongue. From now on I could neither lecture nor publish anymore in German. I had to do both now in English and Hebrew. I felt like a man deprived of air. In school I did not learn English and my Hebrew language was most scanty. In some way I acquired the rudiments of both languages. I had to because in 1934, less than one year after leaving Germany, I already had to lecture in Hebrew on plant physiology and anatomy, for which there were no Hebrew textbooks. When I knew Hebrew better I remedied the lack of text-books when I published, together with my assistant Konis, two Hebrew books on plant physiology and one on general botany and translated, together with Konis and Michael Zohary, the famous popular book by Timiriasew on "the life of plants" from Russian into Hebrew.

The fact that Nazism had driven me out of Germany and deprived me of German as a means of scientific communication persuaded me to get rid of my two German names. When in 1935 I took out Palestinean citizenship, Walter Schwarz officially became Michael Evenari. I choose Michael because that is the name of one of the guarding angels, and Evenari is the Hebrew for Loewenstein, the maiden name of my mother. Schwarz (black) in Hebrew did not sound nice to me. To change names was not an easy decision. The botanist

Walter Schwarz, who had already published 17 scientific papers, had buried himself, and the new Michael Evenari had to start ostensibly as a newcomer to science.

Besides my own work on the water balance of desert plants, I also induced some of my students to do ecological and ecophysiological research. These included Alexendra Poljakoff, E. Shmueli, and Ephraim Konis. This was the work I had already planned to carry out in Palestine when I was still in Germany, but the special conditions of my new homeland turned me toward new lines of research.

When Jews started farming in Palestine they did so without any agricultural tradition behind them. This had its drawbacks and advantages; drawbacks because they lacked experience; advantages because they were not bound by old agricultural practices. Since the new Jewish farmers were not peasants and were often university graduates, they tried to farm scientifically. Thus one day David Zirkin, a member of Kibbutz Ain Kharod, came to me and asked if there was a scientific method to stimulate the rooting of cuttings. This was the beginning of a cooperation between Zirkin, my assistant Konis, and myself, which lasted for many years.

We were the first in the Middle East to use plant hormones for root formation on cuttings of grapevines, figs, etc for stimulating the union of stock and scion in grafts. We found new ways to force early flowering of *Iris* bulbs and to break the rest periods of *Gladiolus* corms. But most important for my future research was the fact that Zirkin turned my attention to the difficulties he was having in germinating apple and plum seeds. Thus started our research on germination inhibitors and germination physiology which occupied me and the plant physiology section of the Hebrew University's botany department for the next 25 years. The first paper was published by my student Gershon Mosheov in the first volume of the *Palestine* (now Israel) *Journal of Botany,* which was founded and paid for by the staff of the department. Mosheov found that a water extract of wheat grains first inhibits and then stimulates the germination of the grains. His two papers on germination inhibition were his only publications since this most able and promising student was killed in 1936 while on guard duty in a kibbutz which was attacked by Arabs.

I enjoyed the cooperation between science and agriculture. It stimulated my scientific curiosity and led to a number of unexpected results, and it made me feel that I contributed my share to the development of the country.

At this point I must say something which many of our colleagues may not accept. The intellectual and emotional satisfaction we get from our profession in the automated world of today is a great privilege for which we owe society a debt. One of our duties should be to apply our knowledge to the solution of practical problems. At the same time we should force society to give us a decisive role regarding the way our knowledge is going to be applied. In specific cases, we should have the moral courage to refuse to divulge our knowledge if we feel that its application would or could be catastrophic.

One of the many departmental excursions was of special importance for my future work, though at the time I did not realize it. In 1936, on the way from Amman to Aqaba, we visited Petra, the capital of the Nabateans, one of the seven wonders of the world. Besides the breathtaking beauty of the many tombs hewn into the red-rose Nubian sandstone, the many waterworks, channels, and cisterns aroused my curiosity. What was the water source of this desert city? Who were these Nabateans? I started to read out of curiosity about the culture and history of this forgotten nation, but the problem of their water source must have sunk into my subconscious and remained there until it popped up much later.

WORLD WAR II

In 1940 I volunteered for the British army. The unit I joined was called the "Palestine Light Anti-Aircraft Battery" and was composed only of Palestinian Jews, mostly members of kibbutzim, all members of the Haganah. Apart from wishing to show Hitler that Jews were not just victims of his persecution but could fight back, Rommel was nearing Egypt, and were he to occupy Palestine its Jewish population would be in danger. We were ready to turn our unit with all its weapons, together with the Haganah, into a guerilla fighting unit in order not to be slaughtered without resistance. Later we were also to fulfill another function: to seek out in Europe Jews who had survived the concentration camps and to smuggle them into Palestine. I thought that under the circumstances all this was much more important than to continue scientific work as if everything were normal.

This is not the place to tell of my experience during 5½ years in the British army. I only want to mention that even as a soldier I found time to botanize. During a period when I was stationed in Cyprus I systematically explored the flora of the island, collecting plant specimens everywhere. Floristically, Cyprus is very interesting because it contains a great number of endemic species. These I hunted specifically: on the igneous alpine top of the Troodos mountains above the beautiful forest of the Cyprus cedar (*Cedrus libani* var. *brevifolia*) and in the limestone mountain range stretching from the magnificent ruins of the Bellapais monastary to the eastern tip of the island at Rhizokarpas. I collected a whole herbarium which I sent to my colleagues at the Hebrew University.

Of my other experiences in Cyprus I mention only that there my knowledge of ancient Greek paid dividends in the material sense of the term. I had only to enter one of the many Greek-owned taverns and to cite some verses of the

Odyssey, the Iliad, or from a drama of Euripides or Sophocles to be treated to a free glass of cognac or wine.

I also botanized in Italy when the Jewish brigade, into which our unit was incorporated as a regiment of artillary, was sent from the Eastern desert to Italy.

One interesting event of our campaign in Italy occurred while I was in a military hospital in Rome. When I was able to leave the hospital for some hours I visited the Vatican, by chance on a day when Pope Pius XII gave an audience to allied soldiers. After a short speech he went round the first row of the audience where I, heavily bandaged, had been given a place. Everybody whom he passed knelt down and kissed the fisherman's ring. Being non-Catholic, the English officer next to me and I remained standing, bowing slightly. The Pope stopped and asked me in English, "From where are you my son?" On the spur of the moment—I think I just wanted to show off—I answered him back in Latin: "terrae sanctae civis sum Judaeus. Tibi gratias ago nomine populi Judaei quia salvabas vitam Judaeorum tam multorum" (I am a Jew from the holy land. I thank you in the name of the Jewish people for saving the lives of so many Jews). The pope looked at me, slightly taken aback and asked me in *Hebrew:* "Then you speak Hebrew. Let me bless you." He then extended his hands over my head and with spread fingers like the ancient priests of the Jerusalem Temple, gave me the priestly blessing in Hebrew, "The Lord bless thee and keep thee. The Lord make his face shine upon thee and be graceous unto thee. The Lord lift up his countenance upon thee and give you peace" (Numbers 6:24–26).

I thanked the pope for saving Jews because during our campaign in Italy we met with many Jewish families who were saved from the Germans, finding, by the pope's order, shelter in monasteries.

AFTERMATH OF WAR

When the armistice was signed in May of 1945, we found ourselves in Palmanova, south of Udine in Northern Italy. We asked our commander to permit us to prepare for our return to civilian life. Since the majority of the soldiers in our unit were farmers or agriculture students, we wanted to do agricultural work. I was asked to organize this venture, and we established ourselves in Fagagna, a small, romantically situated village in the Italian alps not far from the Austrian border. We lived with the peasants in their homes. The "students" worked in the morning in the fields, and in the afternoons I returned happily to my real profession, teaching them botany as the basis for agriculture. In my free time, I botanized in the mountains, collecting another herbarium. My most interesting trip was to the Triglav on the Italian-Austrian-Jugoslav border, where at a height of about 2800 meters, I had the incredible experience of seeing a whole field of Edelweiss, ordinarily a very rare alpine flower. Apparently because of the war, nobody ascended the mountain and the Edelweiss was able to expand unhampered by man.

I returned to Palestine after my demobilization in August 1945. I continued with my physiological-agricultural work but spent most of my time in the service of the Haganah because the country was in turmoil. The British did not permit the entry of the refugees from the concentration camps, so we had to bring them in secretly. In 1947/48, open war broke out when the state of Israel was established and the armies of six Arab states attacked us. We lost the access road to the Hebrew University on Mount Scopus. In spite of the fighting and the loss of our equipment and our books, we continued to teach in private houses, but our research came to an end. Since the finances of the Hebrew University were in a catastrophic state, I was sent to the United States and South America to collect money from the Friends of the Hebrew University for the university's upkeep. This work kept me and my wife Lieselotte, whom I had meanwhile married in New York, busy for the next ten months. I was then granted a year's leave of absence and spent this sabbatical at Caltech, which offered me a visiting professorship. The time I spent there was the happiest and most productive period in my whole scientific career. I can only repeat what Anton Lang wrote: "No one who has passed through Caltech has left quite the same person, and probably retains a trace of regret of having left" (10). At that time Beadle, Frits Went, and James Bonner were the leading biologists, together with Arthur Galston and George Laties. I soon became friends with Frits Went, but the man who had the greatest impact on my work was James Bonner. He advised me to investigate the factors affecting the germination of lettuce seeds, paying special attention to the effect of light and to germination inhibitors. I followed his advice and for the next 20 years germination of lettuce became the main theme of my work and that of the Hebrew University's physiology department.

I had now turned full circle back to my juvenile infatuation with the phenomena of germination. When late in the 1960s somebody at an international meeting talked about the "Jerusalem School of germination research" I was filled with secret pride.

I may be permitted here to intersperse a general remark. At the time I am speaking of, all the members of my department concentrated all their efforts on the elucidation of one problem, seed physiology. Today, this situation has changed completely. There is no unifying topic anymore. While each member may be quite excellent in his field, I doubt that this is really the most productive way for a university department to proceed in research. There is less mutual intellectual cross-fertilization, and there is also a material disadvantage—more and more expensive equipment is needed which cannot be bought since money for research is so scarce. This difficulty would be minimized if the department had a common topic of research.

When I started to write this story I did not realize that I let myself in for a soul-searching process. Therefore I have to confess here that I myself am at least partly to blame for what happened to my department. Years ago I had propagated the idea of having an institute of life sciences at the Hebrew University, incorporating botany, zoology, and biochemistry. The idea was good, but the way it was executed was disastrous for biology since it led to the deterioration and internal dissolution of zoology and botany. We no longer have one integrated research program, no longer one team working together, which still seems to me the best way to achieve optimal results.

When I left Caltech and returned to Jerusalem I found that the various university departments were housed in 55 different buildings all over town because we had lost access to our Mt. Scopus campus. Botanical taxonomy was located in a hut about 3 km from the physiology section, housed in a building of the former British police. Our students were continuously on the run from lecture to lecture. Our equipment was very poor. To break a Petri dish was a disaster because it was so difficult to replace.

ISRAEL

I became vice president of the Hebrew University in 1953, and my archaeological colleague and friend, Benjamin Mazar, became rector and president. Our first decision was to build a new campus in town. I served as vice president from 1953 to 1959. During this time Mazar and I, aided by an able administrative staff, were able to build from scratch a new campus in town. When I felt in 1959 that the main work was done and that a longer stay in that office would cut me off for good from scientific work, I resigned.

During my vice presidency, an event occurred which guided my work into a completely different direction. My then PhD student, Dov Koller, invited me on a trip to the Negev Highlands, to show me the impressive remains of ancient desert agriculture in a wadi (dry river bed) near the ruins of the ancient Nabatean city of Avdat. Looking at the ancient terrace walls, water conduits, channels, and fields, I remembered Petra and what I had seen there. All of a sudden my interest as to how this could have functioned was rearoused and I decided on the spot to try and find out. Somebody told me that I should meet one of our students, Naftali Tadmor, who, as he said, was himself bewitched by the Negev. We met and were immediately drawn to each other as if by some magnetic force. He introduced me to the hydrological engineer Leslie Shanan, and the three of us decided that we would dedicate ourselves to the investigation of the ancient agriculture of the Negev Highlands. For the next 20 years until the untimely death of Tadmor, we worked together as a team.

The late archaeologist, Johanan Aharoni, helped us in the historical-archaeological part of our work. From 1954 to 1958 we spent at least one week each month in the Negev surveying and excavating ancient farms and villages. In 1958 we published the first of a series of papers on this topic, and in 1971 we summarized our experiences in a book (7).

But we were not satisfied with theorizing about how the ancient farms functioned. The instigation to do more than theorize came from my wife Liesel. In August 1956 we were eating our lunches near the ancient Nabatean town of Shivta, and hotly exchanging our opinions about the working of the ancient farms, when my wife said, "Why don't you test your theories and reconstruct an ancient farm?" And so we did. We first rebuilt a farm near Shivta, then one near Avdat, and later a much larger one in Wadi Mashash.

Our chief motivation was to find out if the water source of the ancient farms was really the runoff the ancient farmers collected from a catchment area and led through channels into the terraces of their farms. When we saw that this worked in our reconstructed farms, we went one step farther and wanted to know if we could grow agricultural crops in the farms and if this could be important for turning part of the desert into agriculture land. As a result we now have three flourishing runoff farms in the Negev and various runoff projects in a number of developing countries [see the second edition of our book (7)].

The Avdat farm, which started as an agricultural project, also became a research station for the ecophysiology of desert plants and cultivated plants growing under desert conditions, and for the investigation of structure and function of desert ecosystems. This started in 1971 when Otto Lange, Detlef Schulze, Ludger Kappen, and Uwe Buschbohm of the University of Wurzburg came to Avdat, together with their sophisticated equipment for measuring in situ photosynthesis and transpiration. They stayed with me in Avdat for eight months, undisturbed by any outside interference. During this time we worked as a happy team, living in our desert home, working in the field, and, during our meals and in the evenings, discussing our work, theories, and opinions about everything under the sun. It was like living in a scientific kibbutz. The results of our investigations were published in more than 30 scientific papers. This was only the beginning of the manifold multidisciplinary research projects at Avdat. We investigated the effect of salinity stress at different levels on photosynthesis, transpiration, ion uptake, water and osmotic potential, biomass and fruit set of the pistachio tree; also of various degrees of salinity on structure and composition of the soil. In parallel we investigated the effect of water stress on the almond tree. These two trees were chosen for these experiments because they are, together with the olive, of all the fruit trees tested the most promising for practical desert runoff agriculture. A third project deals with the decomposition process of organic material and the nutrient and especially the nitrogen cycle in three different desert ecosystems, with special emphasis on the question of who the main decomposers are.

As I arrive at the end of my story, I look at my scientific and personal life; it seems to consist of a series of lucky accidents that have enabled me to live a full and happy life and to fulfill most of my scientific ambitions in spite of World War I and II, the rise of Nazism, the Holocaust, and the continuous fighting in Palestine and Israel. But sometimes, lying awake at night, I ask myself if these "accidents" guiding me safely throughout my life were really only chance events. I will never know.

PHILOSOPHY OF NATURE

In telling the story of my scientific life I have here and there inserted some remarks of a general nature which I would now like to expand upon by summing up the lessons that my life and experiences have taught me concerning what was once called the "philosophy of nature." The motivation for giving my own view of the "philosophy of nature" here stems from my experiences with many students who harbor a naive trust in the absolute "truth" of science and a belief that, given time, science will enable us to understand the working of nature and of the whole universe. One of the reasons for our naive belief in the all-explaining nature of science is the equating of knowledge with understanding; another one is that we all grew up in an atmosphere of materialistic philosophy. I agree with this type of philosophy in one respect only: Scientific research is only possible if we act *as if* our mind would be able to unravel all the secrets of nature, but we should not fall into a trap of our own making in believing that this is really true. In carrying out our work we should always be conscious of the limitations of science. These limitations lie not in the fact well known to neurophysiologists that we use only 2–3% of the computing potential of our brain. In the future we may be able to use more of this innate potential. This will increase our knowledge of nature, leading to the invention of much more sophisticated tools of research and to the formulation of more advanced theories. It will not necessarily mean that we will apply this new rationally acquired knowledge in a rational way to our daily life, nor will it bring us nearer to any absolute truth about nature.

First of all, there is simply no "absolute truth," but only relative truths, theories which change with the accumulation of new knowledge. The absolute truth that we endeavor to chase is a phantom that recedes as we try to approach it. This statement can be based on various arguments. One is that each step taken on the so-called "frontier of science" into the yet unknown turns a seemingly simple problem into a more complex one that poses more questions than the original problem did. Du Bois-Reymond, a great physiologist of the past century (1818–1896), has illustrated this with a fitting simile: A man faces a closed door. He tries hard to open it. He succeeds—and finds himself in a passageway leading to many other closed doors.

I take photosynthesis as an example. In my early student days, the professor, when talking about photosynthesis, wrote the following formula on the black-board:

$$6CO_2 + 6H_2O \rightarrow C_6H_{12}O_6 + 6O_2$$

This was simplicity itself and seemed to explain the whole phenomenon. It was so simple that we were certain (including the professor) that soon man would be able to duplicate the process in factories. How naive we were! Since then thousands of scientists have worked on the problem, and the more we know the more we realize how complex the process is.

I take another example from physics. In my student days the Bohr model of the atom was the ultimate and seemingly final model explaining the structure of all matter. It was so beautiful and attractive because it seemed to mirror the solar system. It is superfluous to describe what has happened since then to the model. The innumerable new subatomic particles, some of them no more than mathematical symbols, detected since then have made the Bohr model unten-able today, and the structure of matter is still a riddle.

All this means that in basic research we continuously are faced with a multitude of new facts that we cannot put together into an understandable pattern. Then somebody with intuition will build a theory which puts all the pieces together into a unified model. When new facts are found, the theory has to be abandoned and so on. There is a continuous interplay and contradiction between accumulated knowledge and the meaning of it. We are dealing with the spiral of unification - discordance - synthesis. Interestingly enough, Hegel, the father of this idea of the triodic "progress" of science, human thought, and human society, has illustrated his system with an example taken from plant physiology: "The seed of a plant is an initial unit of life which, when placed in its proper soil, suffers disintegration into its constituents, and yet, in virtue of its vital unity keeps these divergent elements together and reappears as a plant with its members in organic union" until the whole process is repeated (my addition).

The quest for truth has therefore no end because the dialectic spiral points to infinity. Our motivation for questing is a combination of awe, curiosity, and doubt. It does not stem, as we sometimes pretend, from our noble wish to benefit mankind and to improve society.

There is also another reason to believe in the limitation of our cognitive ability. It concerns the nature of our brain, which, as we are told, works like one of the products of our brain, the computer. We improve almost daily on the quality of these man-made machines, increasing their storage and combining capacity. We teach them languages, word processing, drawing, music, etc, and in certain respects they are supposed to be more intelligent than we are. Can we in the same way improve our own brain computer? Certainly not, at least not above a certain innate potential. The limit to this potential lies in the physiological and material structures of our brain. We improve man-made computers by changing the material they are made of, putting in more and better computing units, changing circuits, etc. All this we cannot do with our brain because we have no control over its material structures as it emerged during evolution.

I can only allude to another possible limitation of our cognitive ability. We can only think in language, our thinking tool. The languages may be of different kinds, including the symbolic language of higher mathematics. Therefore our thinking activity is incarcerated in the cage of language. We can, to a certain extent, improve on language as we do in mathematics, but our thinking cannot jump out of language and be absolute, not tied to it.

In addition to these theoretical arguments concerning the limits of our thinking potential there is a more realistic argument regarding the physical existence of *Homo sapiens,* the only species that can think about the limits to thinking. I do not have to prove that there is a danger that our species may soon disappear. The question which I, and I suppose many others, have asked again and again is why and how mankind could have come to such a state?

I, for myself, can answer this question only by referring to what Arthur Koestler has called "the pitfalls of mental evolution" (9), i.e. the asymptotically increasing distance between the curves of what he calls "physical power of the race" and the curve of its "spiritual insight, moral awareness, charity and related values." The latter "curve" is nearly not curved.

It is more an overall straight horizontal line. Since the time our species appeared on the scene of evolutionary history, its structure and physiology have not changed and its emotional, spiritual, and moral qualities changed very little. Their time curve shows here and there some ups and downs but remains basically a nearly horizontal line. The species' physical power line, however, looks very different. For several hundred thousand years this time curve remained also more or less straight until at some point something happened that had never happened before in evolution. Man started to develop his inventive capacities on his own without being forced to do so by any evolutionary pressure. He invented tools, learned, and became a power never before seen on our globe, a power that could change the environment drastically and could even interfere in the order of the solar system.

In doing so the species jumped out of the normal evolutionary framework. It seems to me that the Bible has symbolized that event by telling us that the moment man ate from "the tree of knowledge of good and evil" (Genesis 2:17) he was chased out of paradise. He became "*Homo sapiens,*" the *knowing* man. From there on the physical power curve rises "in leaps and bounds; and in the last fifty years . . . the curve rises so steeply that it now points almost vertically upward" (9). The discrepancy between the two curves has a deep biological meaning. It points to a basic evolutionary disharmony.

The inventive capacity that our species freely evolved increasingly lost all ties with our biological and spiritual attributes. *Homo faber* (man the maker) came into being sometime in our past. In the beginning he created only simple stone tools. Then the pace increased dramatically until he invented atomic weapons, means to leave the globe, and so forth. But the spiritual, moral, and social means needed to control the use of these inventions remained at the stone age level, creating an evolutionary paradox. A creature of nature changed its environment without being able to adapt itself to it fast enough and in an adequate way. Theoretically, this creature should have been able to progress by consciously self-made adaptations parallel to its own inventional capability.

Evolution always shows that when a gap opens between environmental change and adaptability of a species the species disappears. When this happened in the evolutionary past it was a slow process. In man's case today it could be a very fast one leading to racial suicide, again an evolutionary first.

Now I would like to close with some more specific observations addressed to biologists and plant physiologists.

I belong to the fast disappearing race of "botanists," i.e. people who, in spite of the need to specialize in certain fields, still have a *working* knowledge of botany as a whole. I consider this to be so important because only thus can the plant as a complex unit be understood. Since the whole is more than the sum of its parts, it can be comprehended only by considering all the structures and functions in their totality and their mutual interplay.

The counterargument to such an attitude is that today no single individual is able to implement this, and anybody who tries to is a charlatan. I myself was a victim of such an argument. For many years I gave the course of general botany to our first-year students until one of my former students and younger colleagues asked me to discontinue the course claiming that it was too superficial. This may have been right but even a "superficial" survey of the whole field is better than no such survey at all, or a course on general biology given by a large number of specialists, each of whom naturally stresses their own field. In such a multi-man course the plant as a whole disappears under the hands (one should say: under the words) of the many lecturers.

I understand well that in order to survive as professionals we have to work and to train our students in very specialized fields. But if this is all we do, we fail to *educate,* which is very different from training. If we lose the general view of our field, we will know more and more about less and less and understand less and less about the working of the whole. We could counteract the dangerous trend of overspecialization by introducing *obligatory* courses on the philosophy of nature and of science history. I remember that once in

Germany a science student was obliged to read at least one course in philosophy.

Overspecialization is dangerous because it hinders us from realizing that "the organism in its totality is as essential to an explanation of its elements as its elements are to an explanation of the organism" (9). It leads also to what I have called the "forgotten problem syndrome." In two papers I have given many examples of this syndrome, which is simply the fact that in our ambition to be always at the "forefront of science," as the phrase goes, we leave behind a trail of unsolved problems, the existence of which we forget. It has, for example, been known for more than a hundred years that many plants have integumental stomata. Haberlandt, the first to report this, also asked if these stomata have a function. Even today we do not know because the problem was forgotten.

Another example concerns the process of double fertilization of angiosperms. Modern textbooks describe this process glibly as if we understood it perfectly. In reality the most important parts of this process are unknown to us and the problems involved forgotten, as I have tried to show elsewhere (1–3). Parenthetically I confess that I have a grudge against most textbooks because they do not stress enough the many things we do not know and pass over the unsolved problems giving the student a lopsided view of science.

I began my story telling how, at age 13, I was awed by the phenomenon of germination. Today, after having gained some scientific knowledge, I still harbor the same feeling when looking at the incredible orderliness of living beings, their high negative entropy, and their high degree of adaptability. What is the cause of this orderliness? The genes? What causes the orderly function of the genes exactly at the right time and place? Supergenes? And if so, what controls the supergenes? And so I conclude my story appropriately with a question mark.

Acknowledgments

I am deeply grateful to my colleague Nora Reinhold for revising my manuscript. She not only corrected my English but helped me improve the quality of the paper considerably with her good advice.

Literature Cited

1. Evenari, M. 1980/81. The history of germination research and the lesson it contains for us today. *Isr. J. Bot.* 29:4–21
2. Evenari, M. 1984. Seed physiology: Its history from antiquity to the beginning of the 20th century. *Bot. Rev.* 50:119–42
3. Evenari, M. 1984. Seed physiology: from ovule to maturing seed. *Bot. Rev.* 50:143–70
4. Evenari, M. (W. Schwarz), 1937. Physiological-ecological investigations in the wilderness of Judaea. *Linnean Soc. J. Bot.* 51:333–81
5. Evenari, M. (W. Schwarz), Richter, R. 1938. Root conditions of certain plants of the wilderness of Judaea. *Linnean Soc. J. Bot.* 51:383–88
6. Evenari, M. 1938. The physiological anatomy of the transpiring organs and the conducting system of certain plants typical of the wilderness of Judaea. *Linnean Soc. J. Bot.* 51:389–497
7. Evenari, M., Shanan, L., Tadmor, N. 1971, 1982. *The Negev: The challenge of a Desert.* Harvard Univ. Press. 1st and 2nd eds.

8. Francé, R. H. 1912. *Die Welt der Pflanze. Eine volkstuemliche Botanik.* Berlin-Wien:Ullstein. 455 pp.

8a. Haberlandt, G. 1918. *Physiologische Pflanzenanatomie.* Leipzig: Engelmann. 5th ed.

9. Koestler, A. 1959. *The Sleepwalkers.* London: Hutchinson. 624 pp.

10. Lang, A. 1980. Some recollections and reflections. *Ann. Rev. Plant Physiol.* 31:1–28

11. Maximov, N. A. 1928. *The Plant in Relation to Water.* London: Allen Unwin. 451 pp. (Original Russian title: *The Physiological Basis of Drought Resistance,* 1926)

12. Moebius, M. 1927. Die Farbstoffe der Pflanzen. *Linsbauer's Handb. Pflanzenanat.* 1. Abt., 1. Teil. Berlin: Gebr. Bontraeger. 200 pp.

13. Moebius, M. 1937. *Geschichte der Botanik von den ersten Anfaengen bis zur Gegenwart.* Jena: Fischer. 458 pp.

14. Molisch, H. 1937. *Der Einfluss einer Pflanze auf die andere. Allelopathie.* Jena: Fischer. 105 pp.

15. Muller, C. H. 1966. The role of chemical inhibition (allelopathy) in vegetational composition. *Bull. Torrey Bot. Club* 95:332–51

16. Oppenheimer, H. 1930. Reliquiae Aaronsohnianae I. Florula Transjordanica. *Bull. Soc. Bot. Genève* 22:126–409

17. Oppenheimer, H., Evenari, M. 1940. Reliquiae Aaronsohnianae II. Florula Cisjordanica. *Bull. Soc. Bot. Genève* 31:1–431

18. Schwarz, W. 1926. Die Wellung der Gefaessbuendel bei Heracleum. *Planta* 2:19–26

19. Schwarz, W. 1927. Die Entwicklung des Blattes bei Plectranthus fruticosus und Ligustrum vulgare und die Theorie der Periklinalchimaeren. *Planta* 3:302–8

20. Schwarz, W. 1928. Das Problem der mitogenetischen Strahlen. *Biol. Zentralblatt* 48:302–8

21. Schwarz, W. 1928. Zur Aetiologie der geaderten Panaschierung. *Planta* 5:660–80

22. Schwarz, W. 1931. Beitraege zur Entwicklungsgeschichte der Panaschierungen. 1. Entwicklungsgeschichte der Plastiden einiger gruener Pflanzen. *Zeitschr. f. Bot.* 25:1–57

23. Schwarz, W. 1933. Die Strukturaenderungen sprossloser Blattstecklinge und ihre Ursachen. (Habilitationsschrift). *Jahrb. wiss. Bot.* 78:92–155

24. Stocker, O. 1928. Der Wasserhaushalt aegyptischer Wuesten und Salzpflanzen vom Standpunkt einer experimentellen und vergleichenden Pflanzengeographie aus. *Bot. Abhandl.* 13:1–200

25. Volkens, G. 1887. Die Flora der aegyptische-arabischen Wueste auf Grundlage anatomisch-physiologischer Forschungen. Berlin: Gebr. Borntraeger. 156 pp.

26. Wagner, A. 1912. *Die Lebensgeheimnisse der Pflanze.* Leipzig: Thomas. 190 pp.

Norman Good

Ann. Rev. Plant Physiol. 1986. 37:1–22

CONFESSIONS OF A HABITUAL SKEPTIC

Norman E. Good

Department of Botany and Plant Pathology, Michigan State University, East Lansing, Michigan 48824

CONTENTS

When I was invited by the Editorial Committee of the *Annual Review of Plant Physiology* to prepare these memoirs, I was pleased and flattered. At last I had an opportunity to leave a legacy in the literature, limited only by the laws of libel. As it turned out, writing these memoirs has proved extraordinarily difficult. Heretofore my literary ramblings have been constrained by the subject matter and disciplined by editors. Now I must try to record the diffuse speculations of a lifetime without benefit of external restraint.

This article is only marginally autobiographical. It is primarily the history of my ideas on biological topics. It is only incidentally a brief account of a life that has been fairly uneventful except in terms of my exposure to the influence of outstanding intellects.

THE MAKING OF AN AGNOSTIC

I lived the first three decades of my life on a farm on the outskirts of the city of Brantford in the province of Ontario in Canada. Brantford is of some historical

2165

0066-4294/86/0601-0001$02.00

interest. Originally it was the new capitol of the Mohawk Nation of the Iroquois Federation when those allies of the British were invited to leave the United States after the American Revolution. The famous Joseph Brant, chief of the Mohawks, brought his people up the Grand River from Lake Erie to the ford at the headwaters of navigation, thereafter known as Brant's ford. In view of past and future events, there is some irony in the fact that the British granted to Brant and his people the lands along the river "from its mouth to its source, for them and their posterity to enjoy forever." These same lands had already been occupied by these same Iroquois for centuries. Their early occupation of the area blocked an otherwise easy spread of French settlements along the St. Lawrence river into the rich heartland of America. Indeed, the military prowess of the Iroquois and their occupation of my homeland are important reasons why this article is being written in English instead of French. It is also noteworthy that the "forever" mentioned in the treaty holds a record for brevity of eternity. Once the French political power in America was broken, the Iroquois ceased to be useful and were elbowed aside.

My paternal great-grandfather obtained a grant of the same lands from a forgetful British Crown in 1837. The framed original of this not so original deed still hangs in the stairwell of the Georgian house he built a few miles north of the main Mohawk settlement. One hundred and twenty-five years later, my wife and I added to our family of four children two Iroquois babies. This belated attempt to redress the wrong done their ancestors has been very satisfying, but I am not sure how much good it has done the remnant of the Six Nations.

More to the point of this narrative, there is in the old house an extensive library stocked with scientific and literary classics, written and printed over a period of 300 years. Among the books are two immense volumes of Joseph Black's lecture notes, published in the 18th century. It is probably not an exaggeration to attribute the birth of chemistry to Black. He developed ways of working with gases as chemical reactants, paving the way for Lavoisier; he discovered carbon dioxide; and he was the first to work with heat as a quantity (rather than with temperature which is an intensity). Not much of this leaps out from the books, however, because the language of chemistry that we use now has been invented since Black's time. There is also a treatise on electricity written and published about the time of Franklin's experiments by an author I do not now recall, a treatise remarkable for clarity and rigor of reasoning. Then there is a copy of Lyell's *Principles of Geology* published well before the appearance of Darwin's *Origin of Species*. When I looked into Lyell's book, I was astonished to find Darwin's principle already clearly and explicitly enunciated with respect to geology, namely that the things we see now can be explained in terms of processes now operating and need not be attributed to cataclysmic events. Only much later did I discover that Darwin made a point of acknowledging his debt to Lyell, the father of modern geology. In the library are many other texts of comparable importance to the history of science. Obviously, some of my ancestors had scientific interests in a day when science was mostly in the hands of laymen. If there are genes for an interest in things physical, chemical, and biological, I certainly have a right to a share.

I do not know why, but the area around Brantford attracted other immigrants with scientific or literary aspirations. The farm next to ours was and still is occupied by Carlyles, great-grandnephews and nieces and heirs of Thomas Carlyle. (Much as I love some of my Carlyle cousins, I still cannot read the outpourings of the senior Thomas). However, the best known among these immigrants was a certain Professor Bell from the University of London who sought the "wholesome" air of the New World to cure his tubercular sons. Unfortunately, in this he was only partly successful. His younger son shared his years of early manhood with my father's uncles and aunts. The senior Professor Bell was an "elocutionist," which in those days meant a specialist in speech. Among his many accomplishments was a truly phonetic alphabet, totally ignored by the public but much lauded by George Bernard Shaw. I well remember my father's old aunt telling a story that may have grown somewhat over the many intervening years. Apparently, the Professor had a dog that he had trained to emit a constant low growl, something not difficult to do. He then inserted his fingers into the dog's mouth and, by appropriate manipulations, made the dog utter comprehensible English words! Dr. Bell's surviving son followed in his father's footsteps; he also studied the physics of speech. But, in anticipation of modern techniques, he replaced the dog with instruments, converting the sounds of speech into electrical analogs and then converting these analogs back into sound. The trick was so successful that it became widely used for long-distance transmission of speech and for the recording of sounds. Unfortunately, none of my ancestors had the foresight to invest in the technique.

It is not altogether surprising that my father, steeped in such traditions, should have elected to study the then new discipline of physical chemistry at the University of Toronto. He was a natural scholar with a remarkable combination of manual, mathematical, and verbal skills. He could (and did) shoe horses with shoes he made in a forge he had built with his own hands. He laid bricks with as much skill as a professional. His intellectual accomplishments were formidable. He was a candidate for a Rhodes scholarship when these were new, but lost out because of his utter boredom with sports. I believe that he was at one time on the Board of Governors of the University of Toronto. Whenever I tested him with lists of little-used words, he was never at a loss. He looked up telephone numbers in the directory only once. He invariably solved the more difficult mathematical school problems of his children, more than 40 years after he had studied mathematics himself, and often with proofs that had teachers shaking their heads in amazement. He was physically very powerful, short and immensely broad, with muscular coordination that would have made him a good athlete if he had cared. He kept his great strength to old age. For instance, he was 79 years old before we began to relieve him of the most arduous farm work.

I suppose that my father's breadth of interests and his versatility were a kind of professional undoing. In any event, he early abandoned the study of the physical sciences (including aeronautics at a time when only birds could fly) for political science and economics. He was a minor authority on monetary theory and he was briefly in Canadian federal politics. When I was young the house was filled with economists and politicians but never scientists. However, a prophet is not without honor save in his own family, and my admiration for my father is moderated by the fact that he was not easy to live with. Although I did envy him his intellect and his skills and I admired his disinterested dedication to worthy social causes, my adulation is well within bounds. His understanding of people, including himself, was notoriously defective. He had a theoretical dedication to agriculture as a way of life, not as a business, and he did not hesitate to sacrifice his children on that altar. Although he was very hard working and energetic, he never took any responsibility for the boring routine work of raising livestock or the other repetitive aspects of farming that we dubbed "the chores." He was much too committed to his worthy causes to forego any of them in the interest of milking the cows twice daily. As a consequence, whichever of his sons must milk the cows had, perforce, a very erratic education.

The fact that I was often that son has long since ceased to rankle, if it ever did. I never was sure what I wanted to do with my life anyway. This uncertainty is still with me and will probably continue with me until time makes the question irrelevant. Besides, there were some benefits from the enforced absences from school. The intermittent nature of my formal education gave me long periods to consolidate my ideas and to clarify my understanding of basic concepts. (Nothing clarifies concepts like the dropping away of minutiae.) Another consequence of having my education interrupted all through the war years is that, when I did return to university, my colleagues and fellow graduate students were largely war veterans, an exceptionally mature and dedicated lot. Therefore, I have no fundamental professional regrets, although I did feel some nostalgia for the sparkle of a dark-browed Jeanie and a romance that never was.

There were three interruptions in my education totaling nine years. I did not graduate from the University of Toronto until 1948 when I was 31 years old. These interludes and the reason for them had profound and sometimes contradictory effects. I came to feel that school was a luxury pursuit and not a serious part of life. To this day, I cannot think of academic efforts as real work. (Some of my more candid colleagues might say that said frame of mind is obvious.) Deep inside me is a feeling that work is done with a plow and tractor, or a fork, or an axe. Consequently, I am always a little puzzled by the fact that I am being paid to play in the laboratory. Thus I am doomed forever to be, in my own mind, a dilettante, but being a dilettante has its advantages. I may be opinionated, dogmatic, and blunt to a point of incivility, but I am not too involved emotionally in my arguments. It is only a game.

Another consequence of my erratic schooling, intervals of reading being interrupted by long hours of repetitive physical work with no one to talk to, is that I early began to separate concepts from the words commonly used to express the concepts—just as an illiterate deaf-mute must. This mental trick puts a very different complexion on the contemplation of science. It can also make communication difficult when at last there is someone to communicate with. Thus having an idea and putting it into words can be quite separate processes in my mind. It may be that this mental trick is the most valuable residue from the breaks in my formal education. One of the pitfalls of sciences and indeed of all rational thought is the tendency to create words for concepts, words that then take on a "reality" in their own right, independent of the concept. Words separated from the parent concepts tend to usurp the role of concepts and to create "ideas" that are actually meaningless. Thus, if we are not careful, science can become a sort of litany, full of sound and fury, signifying nothing. This predilection for ideas over words was reinforced by my first chemistry teacher at the university, F. W. Kenrick, who was dedicated to but one cause. He spent several months lecturing us on the definition of a solution and on the definition of the terms used in the definition! In almost every lecture he implored us: "Know what you are talking about." In the idea section that follows, I am going to try to apply this criterion to a few of the dogmas of modern biology, but I am reconciled to the fact that there will be some annoyed readers.

After I graduated from the University of Toronto, I attended the California Institute of Technology. In those days G. W. Beadle of *Neurospora*-gene-enzyme fame, was head of the Biology Division, Linus Pauling was head of the Chemistry Division, and Max Delbrück was teaching us to use bacteriophages as biological tools. The whole Institute was in the birth pangs of so-called molecular (actually macromolecular) biology. Much of this passed me by because I was involved with other things, but the intellectual ferment could not fail to have an effect. My own research, under the direction of Herschell Mitchell, dealt with the chemistry and metabolism of small molecules, notably amino acids, but I did use *Neurospora* mutants as tools and some of my work did have genetic implications.

My first postdoctoral position involved a complete change of research area. I helped Allan Brown at the University of Minnesota in his task of differentiating respiration from concurrent photosynthesis, difficult because the two opposite processes both involve molecular oxygen and carbon dioxide. We succeeded by using isotopically labeled oxygen as the substrate for respiration, since the oxygen produced from water by photosynthesis is, of course, unlabeled. The problem was of considerable importance in those days because the great Otto Warburg had decreed that photosynthesis requires an immense uptake of oxygen, reconsuming three quarters of the photosynthetically produced oxygen. We succeeded in showing that Warburg was wrong, as we now know from many other kinds of experiments.

I have sometimes been asked how we came to miss the phenomenon of photorespiration since our instrument was ideally suited for detecting it. The simple answer is that we did not miss photorespiration at all. We observed it and adjusted conditions to eliminate it, using low concentrations of oxygen and high concentrations of carbon dioxide. We chose these conditions specifically to avoid a phenomenon that was already well known but irrelevant to our concerns. Photorespiration had been described many years before by Warburg, who called it "photocombustion."

My exposure to photosynthesis research with Brown introduced me to a new area in which I have dabbled in an intermittent and desultory way ever since. In late 1952, I joined Robin Hill in Cambridge, where Sanger was already determining the structure of insulin and Watson and Crick were theorizing on the structure of DNA. Again I let most of the excitement of the new molecular biology pass me by. Again my work was directed toward different goals. Robin and I developed a useful technique, since much employed, in which chloroplast lamellae reduce substances like flavins and viologens which then in turn reduce oxygen. Thus they are at the same time acceptors of electrons from chloroplast lamellae and catalysts of oxygen reduction by the lamellae. One of these substances, methylviologen, has since become popular as a herbicide under the alternative name, paraquat.

I cannot leave the account of this period of my life without acknowledging my appreciation of Robin Hill. He is an unusual person in many respects, among which is an unusual sensitivity and understanding coupled to an unusual lack of articulateness. The wisdom of his comments (which is often very great) only penetrates the average skull after a long delay, from hours to years. Withal, Robin is one of the most discerning and original scientists I have ever encountered. His contributions to our understanding of photosynthesis have been enormous. As early as 1953, he was speculating on the possibility of two sequential light reactions in photosynthesis. It is scandalous that the Nobel Prize Committee has continued to overlook him. Perhaps the oversight has occurred because he keeps a low profile, publishes rather infrequently, and abhors self-advertisement.

In 1954, I returned to Canada to take up a research position with the federal Department of Agriculture, in a new laboratory dedicated to fundamental research on biological problems. There, for about two years, I worked with the late Wolf Andreae on indoleacetic acid (auxin) metabolism in plants. We showed that a major product of the metabolism of exogenous indoleacetic acid is indoleacetylaspartic acid, an observation that may or may not have physiological importance. Then I went back to studying photosynthesis, especially photosynthetic electron transport coupled to ATP formation. I confirmed Jagendorf's observation that ammonium salts uncouple the two linked processes, and I extended the observation to a large number of aliphatic amines.

In the course of these studies on chloroplast photophosphorylation, I observed that many anions could also have uncoupling effects, especially at high concentrations, whereas "inner" salts such as glycine and other amino acids had no such effect. Therefore, it seemed to me that photophosphorylation in vitro might be laboring under a handicap because all of the buffers we used to control pH of necessity contained anions. So I made a series of amino acid buffers, a wide range of aminocarboxylic and aminosulfonic acids with pKa's appropriate for biological research. They have proved useful, not only in photosynthesis research but also in general biochemical research, in tissue culture, and in medicine.

When I came to Michigan State University in 1962, I had the good fortune to work with Seikichi Izawa. He is responsible for any reputation I may have for constructive and reliable research in photosynthesis. For 12 years he put up with me and my indolence while he trained some outstanding graduate students. For laboratory skills, hard work, scientific imagination, and ability to pick promising research topics, he is exceptional. The research done in the Izawa years in our laboratory is too diverse to catalog here. It was both in the mainstream of photosynthesis research and out of it; although we were looking at much discussed aspects of chloroplast electron transport and photophosphorylation, our conclusions were sometimes considered aberrant. Among our observations that have been much noted and well accepted are the in vitro unstacking and stacking of chloroplast lamellae and the role of lipophilic strong oxidants as acceptors of electrons directly from photosystem II. (Indophenol dyes which are reputed to accept electrons from photosystem II actually accept electrons primarily from photosystem I.)

Although I left a Department of Agriculture laboratory in Canada to come to an academic position at Michigan State University, it was only at Michigan State that I began to interact with agronomists. When the energy concerns of the 1970s dragged me kicking and screaming from my laboratory, I was forced to give some thought to where my skills and knowledge could be applied in the interest of growing better crops. In some ways this has been a recipe for schizophrenia. My analysis of the factors involved in crop yield (2) convinced me that my professional discipline had little to offer, but all of my instincts were crying out to me to involve myself with the cultivation of the soil or with subjects pertinent to the cultivation of the soil. Now that society has begun to return to its characteristic euphoria after its brief bout with hysteria in the 1970s, we scientists are being allowed to drift back to our various ivory towers. Meanwhile, I have made a personal compromise that has no rational justification but nevertheless soothes my soul. As I approach retirement, I find myself taking more and more satisfaction from growing things and working in the fields. No doubt this is a manifestation of a second childhood, of waning intellectual powers. Sometimes my more sedate colleagues raise their eyebrows when I arrive at the department in tattered and not too clean bib overalls—my badge of reverse snobbery. I play the lottery of growing thousands of seedling grapes and apples, not for hope of gain but for the joy of wielding a hoe in the still of the evening or the early morning. Besides, grapes taste much better than golf balls.

Now I close this brief biographical sketch and move on to more important things, a history of my ideas right and wrong in the context of evolving concepts in biology.

THE WATER RELATIONS OF PLANTS

When I returned to the University of Toronto after the war, I studied biology with an emphasis on botany. One of the subjects that intrigued me was the water economy of plants. It is obvious that terrestrial plants have a major problem in obtaining and conserving water; they must take up carbon dioxide out of an oftentimes very dry atmosphere without at the same time and by the same pathway losing an inordinate amount of water. Moreover, plants grow in large part by cellular expansion, that is by the uptake of water. For these reasons the water relations of plants have always played a major role in studies of plant physiology.

Unfortunately, the literature on plant reactions involving water has usually been couched in language that has done more to confuse than to clarify. Indeed, I found the terminology with its "osmosis" and "suction pressure deficits" so confusing that I had to stop, wipe my mental slate clean, and ask "but what is really going on?" When I did that, I found the deliberations of the experts simple and almost self-evident. Let me summarize without recourse to any of the conventional jargon:

1. Thermal agitation of the molecules of any fluid causes each molecule to move in a random way, that is to say equally in all directions. There is thus a net movement of molecules down gradients of concentration. This is obvious from experience and it is also conceptually obvious. If there are two molecules of water at point A and only one molecule of water at point B, the random motion of individual water molecules will cause more molecules of water to move from A toward B than from B toward A. There will be a net movement of water from A to B.

2. The effective "concentration" of water depends on the actual concentration but also on the presence of other substances that may bind water molecules or otherwise modify their ability to do things. It also depends on any differences in pressure on the water. This general ability to do things is referred to as the potential activity or sometimes simply as the "potential" or the "activity." I do not believe that any other terminology is necessary if we want to discuss the passive movement of water from regions of higher water potential to regions of lower water potential. This subject is practically important but conceptually trivial.

Botanists have tended to treat water relations in terms of the spontaneous water movements that I have been describing, but I doubt that the answers to many of the problems lie there. The really interesting thing about the water relations of living things is that not all is passive and the wedding of active and passive processes presents difficult problems. For instance, roots of many plants seem able to take up water *against* a potential gradient and to extrude almost pure water from cut stems in defiance of the forces for equilibration that we have been discussing. Clearly, such defiance of equilibration requires metabolic energy to do the necessary uphill work. On the other hand, the roots seem quite easily permeated by water (as they must be). Water moves out of them quite freely when the external water potential is lowered sufficiently and they take up water rapidly when the internal water potential is lowered sufficiently. There is a major problem in biological energetics here. Must we postulate the coexistence of a pump and rapid back-leak?

Many years ago Levitt (5) made a most important calculation. He showed that the entire respiration of roots could not produce enough energy to drive such a leaky system if some of the reported root pressures actually existed. Therefore, he concluded, quite logically and quite wrongly, that such great water disequilibria could not exist. The problem with his conclusion is that huge water disequilibria *do* exist in some biological systems and, indeed, ubiquitous disequilibria of all sorts of substances are characteristic of living things. Are we to postulate the maintenance of such disequilibria against back-leaks? For instance, considerable metabolic energy must be expended by the salt-glands of mangroves or the kidneys of desert animals, both of which excrete almost pure salt, just to separate the salts from the water. By Levitt's argument, these processes are made insupportable by the amounts of energy required to maintain the water concentration differences against the leaks implied by measured "permeability" to water. Since the costly processes do occur, I am forced to conclude that the pumps and back-leaks cannot be allowed to coexist. We must look very carefully at measurements of the conductivities of membranes to water and to all other substances. What are we really measuring? A "pump" can run in either direction depending on the demands on it, but in neither direction does it operate in a manner analogous to diffusion. The "pump" can operate as a "gate" or as a conduction channel without providing a pathway for diffusion. These considerations are necessarily vague but nevertheless extremely important.

A strange misconception that I have heard voiced over the years is that "pumps" concentrating solutes of various kinds do exist in plants but not "pumps" concentrating water. However, a massive solute pump is *ipso facto* a water pump, since separating solutes from water is the same thing as separating water from solutes. It all depends on which side of the membrane we place the observer.

I am forced to the somewhat revolutionary conclusion that the "permeabilities" of membranes as measured are totally misinterpreted. Movements in the direction of concentration (i.e. activity) differences must not be automatically equated with passive diffusion. Consider the unidirectional transport of auxins down their concentration gradients, which must owe nothing to ordinary diffusion. Perhaps if we study membrane function more carefully, we will find that all metabolites, even water, can be carried by "gated" processes, although I am not sure what "gated" means in mechanistic terms.

To reiterate: we must beware of arguments based on fancied "permeabilities" when there is as yet no evidence at all that the life processes studied are the typical first-order reactions of passive equilibration by diffusion. Diffusion and life are antithetical, even if life processes must be carried out in the context of diffusion-imposed stresses. Concentrating things against gradients, which is the essence of life, is inevitably energy-consuming anyway, but maintaining a vast multitude of concentration differences in the face of an equal multitude of rapid back-diffusion processes is unthinkable in view of the fact that living things are often fairly efficient engines. The downhill transport of substances we see may be pumps running backward and not diffusive leaks at all. Pumps running backward may be conserving energy, not wasting it. Understanding this aspect of membrane function is probably the most challenging and most important task in all of physiology.

PROTON PUMPS

After my undergraduate days I left the consideration of membrane function for many years. Only when it became apparent that electron transport in chloroplasts was linked to ATP formation through transmembrane phenomena did my attention return to the topic. It had long been known that concentration differences could be created during the hydrolysis of ATP and indeed this was not surprising. The creation of concentration differences requires an input of metabolic energy and ATP was known to be a ubiquitous reservoir of cellular energy. However, it was left to my colleague of Cambridge days, Peter Mitchell, to point out that the reverse process could be used in the generation of

ATP during electron transport. Thus he showed that electron transport in mitochondria creates ion activity differences across membranes, and André Jagendorf showed that electron transport in chloroplast lamellae does the same thing. It was also shown that these ion activity differences can drive ATP synthesis, as Mitchell had postulated. For this insight Peter Mitchell was awarded a well-deserved Nobel Prize.

Unfortunately, the language introduced to describe the phenomena and to explain the mechanisms has been murky at best and misleading at worst. The name of the process, "chemiosmosis," which has grasped the imagination of plant and animal physiologists alike, may have some historical justification in terms of a now long-dead version. It seems harmless. It is nevertheless nonsensical; "osmotic" refers to the effect of solutes on the potential activity of solvents, and no one now suggests that ATP is made by a local decrease in the potential of water.

Yet this is an unimportant matter compared to the very misleading term "proton pump," a term that conjures up a process having no meaning in the context of the chemistry of aqueous systems. Being meaningless, it blocks all rational analysis of events. Moving hydrogen ions from one place to another has no effect on the concentration of hydrogen ions! This fact is counterintuitive and nonchemists have difficulty in grasping it. However, I cannot understand why so many biochemists and biophysicists persist in the use of the misleading language (or at least tolerate it). Somehow they seem convinced that the term "proton pump" is a useful allegory, that we are, in any case, looking at a chicken-egg situation. It is not so. Electrical neutrality must be preserved (except at the surfaces of charged membranes) and therefore the sum of the cationic charges must equal the sum of the anionic charges at any place where we can make measurements. It follows that the concentration of hydrogen ions is prescribed absolutely by the difference between C^+ and A^-, where C^+ represents the sum of the cationic charges other than H^+ and A^- represents the sum of the anionic charges other than OH^-. Movements of H^+ are irrelevant, as we see when we note that concentration of hydrogen ions actually *decreases* toward the negative pole during electrophoresis in spite of the fact that hydrogen ions are moving *into* that region. This apparent conundrum can be resolved when we realize that the concentrations of H^+ and OH^- adjust, using the almost infinite reservoir of these ions in the slightly dissociated water, and that H^+ can be consumed or produced in reactions involving other weak acids. Thus, as Stewart has pointed out (7), the concentration of hydrogen ions is a dependent variable, a result not a cause. To assign phenomena to a "proton pump" is to center attention on a partial reaction, only one part of an equation and a dependent part at that. We must be prepared to write the entire reaction sequence if we want to gain an insight into what is actually occurring.

It is interesting to speculate on the origin of the unfortunate emphasis on "proton pumps." I suspect that a minor contribution may have come from phonetics, the tendency of the English language to alliteration, a holdover from medieval Germanic poetry where rhymes were at the beginning of words rather than at the end. The more important reason is recent history. The energy-conserving ion movements were first detected as pH changes simply because pH meters are available and cheap. Furthermore, very minute discrepancies in the movements of strong ions, cations vs anions, can be detected with a pH meter. For instance, if one has a 10 millimolar concentration of KCl and the concentration of K^+ is precisely equal to the concentration of Cl^-, the pH will be exactly 7.0. If one now removes 0.01% of the Cl^- or introduces 0.01% of additional K^+, there will be a *tenfold decrease* in the concentration of H^+ and the pH will be 8.0. The sensitivity of the measurement, however, does not lend any credibility to the movement of hydrogen ions as the cause of anything. Quite the contrary. I fear that further progress in the understanding of ATP synthesis (or indeed progress in the understanding of any transmembrane phenomenon attributed to "proton pumps") is mired in most unfortunate terminology.

The "electrogenic proton pumps" so dear to the hearts of the chemiosmoticists are not only nonsense, they are nonsense said backward. It would seem that the primary event that leads to the conservation of energy in ATP is a charge separation across a membrane. In photophosphorylation the reaction-center pigment seems to be oriented in such a way that the excitation transfers an electron across a dielectric layer. In oxidative phosphorylation, the same thing seems to be accomplished by placing catalysts of different partial reactions of the overall oxidation of substrates on different sides of a dielectric. Indeed, Mitchell has very wisely pointed out that the situation is analogous to the situation in a fuel cell (6). The charge separation is quickly followed by a migration of cations and anions in appropriate directions under the influence of the electric field, perhaps principally by Mg^{2+} migrations in vivo although ADP, P_i, or even malate, bicarbonate, and carbonate are candidates. In the suspensions of chloroplast lamellae usually used in phosphorylation studies, Mg^{2+} and Cl^- are the major ions migrating in opposite directions (4). Note again that the migrations of H^+ and OH^- are irrelevant. The consequent imbalance of cations and anions, which is far too large to be tolerated, is almost entirely obliterated by changes in the numbers of cationic and anionic charges on the membrane proteins. Thus the anionic charges left behind when Mg^{2+} migrates out of the membrane disappear when carboxylate anions are protonated to become carboxyl groups (anionic $-COO^-$ becomes neutral $-COOH$) or neutral amino groups ($-NH_2$) become cationic ($-NH_3^+$).

The change in pH, which originally alerted us to the phenomenon, is only a minor tertiary effect of a quite different initial cause, a minor process detected only because the concentration of hydrogen ions is such a sensitive indicator of strong ion imbalance. However, it is quite possible that the important end result for photophosphorylation, the "energization" of the membrane that makes ATP formation possible, represents a region of the membrane protonated in the manner described. If so, the protonated region must be oriented with respect to the sides of the membrane because the "sidedness" of the situation is preserved and acids stored on the inside of lamellar vesicles can be used to sustain phosphorylation.

It is time that we stop drawing cartoons with little H^+ arrows traversing membranes and think instead of what the measurements of pH changes really mean in terms of balanced equations. Also, it is time to stop thinking naively about the membrane "permeability" to hydrogen ions. In fact, as pointed out above, measurements of "permeability" to anything must be interpreted with great caution.

The above description of events is, of course, a very incomplete picture of how ATP is made, but it is not as incomplete as the conventional "chemiosmotic" picture. Moreover, the above description has the merit of being real and plausible chemistry without any appeal to magic proton pumps. It is not hypothetical at all. Rather, it is a description of the chemical stoichiometries that almost inevitably follow from the observations.

PHOTOSYNTHESIS AND CROP YIELD

When I moved to Michigan State University in 1962, I had over ten years of somewhat intermittent research in photosynthesis behind me. Therefore, it is not surprising that I was occasionally asked by production-minded agronomists for advice on methods for measuring photosynthesis. My response was to ask questions in return: What do you want to know and why do you want to know it? If their answer involved estimates of crop yield, I made myself unpopular by the only honest response possible, to wit: weigh the plant. As concern for energy sources mounted during the 1970s, this flip but appropriate answer became less and less appreciated, and the pressure on photosynthesis researchers to contribute to the practical problems of society mounted. I was even prevailed upon to coauthor a chapter of a book on the factors determining plant growth and crop yield (2). My conclusions were not altogether popular with my colleagues, who looked on the then current concerns of society as professionally providential.

It is a truism to say that crop yield is the result of photosynthesis and the translocation of the photosynthate into the economically important parts of the plant. It does not follow, however, that either the biochemistry of photosynthesis or the mechanism of translocation is an important limiting factor determining crop yield. The syllogism that, since plants grow entirely by photosynthesis, they must accumulate biomass to the extent that photosynthesis exceeds respiration is true but dreadfully misleading. One trouble lies in the distinction between actual photosynthesis and potential photosynthetic capacity. Moreover, the rate of growth of a plant depends almost equally on the rate of net photosynthesis and on the rate of reinvestment of the photosynthate in new photosynthetic machinery. In fact, the growth of a plant can be approximated by the following equation:

$$P_\tau = P_0 \epsilon^{\alpha R \tau}$$

where P_0 is the size at the beginning of the observation, P_τ is the size at time τ, ϵ is 2.71828---, α is the proportion of the photosynthate reinvested in more photosynthetic machinery, R is the rate of photosynthesis per unit of photosynthetic machinery, and τ is the time of measurement. Furthermore, R can be dependent on α because reinvestment of photosynthate provides a sink for the products, and this in itself often increases the proportion of the photosynthetic potential of the machinery actually used. But α is an expression of the *growth* potential of the organism, not an expression of the photosynthesis potential at all. Similarly, the relative growth rates of the various plant parts (and hence the harvest index) are expressions of differences in the growth potentials of these parts and almost certainly do not reflect limitations on the translocation pathway.

For these reasons, it is particularly important that photosynthesis research aimed at increasing crop yield be broadly interpreted in terms of constraints on photosynthesis and growth and not be over-concerned with the process of photosynthesis itself. For instance, it is obvious that the growth of plants, and therefore net photosynthesis, is greatly enhanced by available nitrogen. Yet nitrogen is not directly one of the reactants in photosynthesis. Nitrogen affects photosynthesis through the growth that builds new photosynthetic machinery. Too great a preoccupation with the details of photosynthesis, if it detracts from studies of stress, external deficiencies, and genetic determinants of growth, is a sure recipe for frustrating those geneticists and agronomists who would like guidance from plant physiologists.

This is not to denigrate photosynthesis research or any other search for knowledge for its own sake. We cannot afford to smother fundamental research, research not directed toward the solution of specific practical problems. It must never be forgotten that we run out of practical solutions very quickly when we run out of understanding. No matter how practical a problem, it must be approached from the standpoint of analysis and understanding. Unfortunately, the traditional category of research called "fundamental" is sometimes equated to no application (and by implication, useless) while "applied" is equated to description and testing, with no real need to understand. Either attitude is disastrous if we really are interested in solving problems.

GENES AND ENZYMES

When I went to Caltech in 1948, the dogma of one-gene-one-enzyme was at the height of its popularity, but I would have none of it. The concept seemed simplistic and the evidence minuscule. In hindsight, it would appear that I was both right and wrong; my objections were largely valid, but I underestimated the potential value of any theory, right or wrong, sophisticated or simplistic, if it directs research into useful channels. The alchemists discovered chemistry in their search for the philosopher's stone.

In order to appreciate my objections to the gene-enzyme dogma, one must realize that the gene as we knew it then was not a sequence of bases in DNA coding for a specific protein. Rather it was a lesion, something wrong with a region of a chromosome that interfered with some aspect of biological function or form. In fact, I used to bait my geneticist friends by pointing out that the gene was only a mathematical abstraction born of segregation ratios! Further, I used to say that enzymes, like everything else in the cell, must be susceptible to genetic modification. In that case, where was the evidence for any primacy of a gene-enzyme relationship? If one could modify a specific bristle on an insect by a gene, why should another gene not modify the conformation of an enzyme? The counterargument that, lacking a specific gene-enzyme relationship, it should be possible to modify an enzyme by several different unlinked genes, did not seem compelling. If one could find only one gene to modify a bristle, did that argue for a unique and direct gene-bristle relationship? Besides, the factual base for the gene-enzyme argument was woefully weak.

When I was a student, many genetic changes were attributed to the absence of particular enzymes in metabolic pathways, but as I remember it, an actual absence of the enzymes had only been demonstrated in two cases. In all other cases, the enzyme deficiencies were inferred from nutritional requirements of the mutants; inferences require less work than observations. My lab-mate Francis Haskins (3) worked out an elaborate and substantially correct scheme for the pathway of synthesis of tryptophan and niacin on the basis of a series of *Neurospora* mutants. Almost as an afterthought, and because he was thorough, he decided to do appropriate genetic studies, whereupon he discovered that the various mutants apparently blocked by deficiencies in quite different enzymes were, in fact, all alleles! Probably modern geneticists can interpret this observation to their satisfaction, but it was not comforting then in terms of the original gene-enzyme hypothesis.

The thing I failed to appreciate when I was exposed to the one-gene-one-enzyme dogma was the fact that, if it was true in even one instance, the phenomenon offered a wonderful way of looking at a gene-protein relationship. In other words, it offered an opportunity to follow the transfer of genetic information from DNA (which was even then accepted as the primary genetic material) to a protein, and proteins made up most of the structures and catalyzed most of the functions of cells.

Another factor contributing to my skepticism was the fact that I was then deeply immersed in studying the chemistry and metabolism of small molecules and I underestimated the potential of techniques for working with macromolecules. Consequently, my chance of studying the biology of macromolecules ("molecular biology") was missed.

I regret now that I did not keep up with the discipline of molecular biology as it was unfolding. Modern genetics is an incomparable tool for the study of a host of problems in biology. On the other hand, I worry lest the discipline of molecular biology become unduly parochial, lest molecular biologists approach the problems of plant physiology with the same naiveté that plant physiologists sometimes use in their approach to the problems of plant productivity. Unfortunately, the applications of genetics to plant physiology cannot be less naive than the physiological assumptions on which they are predicated. The nature of the lore of molecular biology as it stands today is also something of an impediment in making physiological or anatomical modifications of plants by the techniques of genetic engineering. It is not yet obvious how DNA-protein manipulations can answer some of the most important questions in biology, especially questions about the organization and regulation in eukaryote cells and in the tissues of multicellular organisms. How close are we to understanding the timing of the expression of genes, when they are expressed at all? Cut off my hand and I am forever handless but it is not so with my close cousin the salamander. Why? What tells a sieve tube cell to be a sieve tube cell? The list of problems of development yet to be solved is endless, yet such questions bear immediately on the genetic control of physiological functions.

In this case I am not pessimistic, however, and I hope that molecular biologist will continue to attack problems of regulation and differentiation. Already some of the changes in cells associated with specialized forms and functions have been correlated to DNA changes, sometimes reversible and sometimes irreversible. If differentiation and the multiplicity of functions of cell types is ever to be explained, it will be molecular biologists who do the explaining.

WHITHER BIOLOGY?

The impact of modern genetics on our perceptions of taxonomy and evolution is just beginning to be felt. Again I am at a loss for hard facts and again I regret my inattention to the science growing up around studies of nucleic acids and proteins. The little information I have acquired by diffusion has whetted my appetite for more, and I am excited by implications I can barely assess. Perhaps I will study these things in my next career.

Originally we knew genes primarily as malfunctions in development, by the consequent absence of some discernible structure or metabolic reaction. As more and more mutations blocking metabolic pathways were studied, a higher and higher proportion of mutant genes could be described in terms of inactive or less active enzymes. This state of affairs was desirable from the standpoint of the then current primitive analysis of gene action, but it posed something of a problem for those primarily interested in evolution and in the genetic variability that must provide the basis for natural selection. It was difficult to picture evolution in terms of losses of function when evolution is so obviously and so necessarily associated with changes improving functions. Of course, part of the dilemma lay in the nature of the mutations we felt obliged to study, the all-or-none effects of genetic disasters. It is not difficult to imagine small changes that would occasionally make small incremental improvements in gene function, but such small changes were too difficult to study when we were still incapable of refined genetic analysis.

Mutations, great or small, are presumably random in nature. However, their consequences in terms of survival of the organism must nearly always be bad, if only because the preexisting gene products had already been designed for particular functions by ages of natural selection. In other words, random changes, when the number of such possible changes is almost limitless and the changes are in every conceivable direction, should almost always be dis-advantageous over a wide range of conditions. One would expect that improvement by such a lottery would be slow indeed, so slow that we might not entertain the concept of evolution by mutagenesis at all if we were not impressed with the idea that evolutionary time is very long and that populations subject to natural selection are very large. But are the pertinent populations really large and do selection acts occur with great frequency? The answer is clearly yes when we consider many unicellular organisms, but it is certainly no when we consider many long-lived multicellular organisms.

Mutagenesis and selection must play by some remarkable rules in such multicellular organisms; consider as an example the elephant. Once the great majority of the cells has given up the ability to form gametes, mutations in them became irrelevant to evolution; the changes cannot be transmitted to progeny. If a mutation occurs that makes for better bones or brawn, it can only be selected when the individual elephant possessing it survives better. Thus selection in the germ line usually involves a characteristic not expressed in the germ line! Mutations that can be selected must occur very early in the development of the individual, before the cells are committed to their various specialized roles as sterile workers. Presumably the majority of such mutations in animals takes place in the germ lines of preceding generations. Furthermore, the world population of elephants has probably always been small and the same can be said for many organisms. The numbers are particularly unpromising when we consider that selection of improvements must be delayed for many years as the elephant matures; thus there can only be one selection act in 20 or 30 years, and the selection involves only one cell per generation, the fertilized ovum. If we accept the conventional picture of evolution by mutagenesis followed by natural selection, the problems are formidable. The entire world population of elephants would seem to be evolving with one cell selection each 20 years from a few thousand cells!

The numbers problem with higher plants is much less formidable. The selection unit is less well defined and the "germ line" is neither as clearly defined nor as protected from the environment as in animals. Indeed, each growing point of a plant can be looked upon as an "individual" producing either gametes or specialized cells for its own betterment. Moreover, the specialized cells behind the growing point (and genetically related to the growing point) almost immediately begin to contribute to the well-being of the growing point and thereby influence the production of and survival of gametes. Therefore, many mutations, though they affect only the specialized functions of nonreproducing cells, can nevertheless be promptly selected and transmitted to future generations through the better production of gametes and the better development of the propagules. Of course, some of the genetic changes in the "germ line" (in the special reproductive meristems) affect whole plant functions, and in these instances selection is again delayed for a whole plant generation. The evolutionary problems of elephants and oak trees are alike with regard to such changes.

Modern studies of molecular biology may provide a hint for a partial resolution of this numbers problem. It is now apparent that genetic information can sometimes be transferred among very different organisms by processes that have nothing to do with meiosis or the regular transfer of genes to progeny. Thus, in principle at least, all genes in all living things may be accessible to all other living things. Such lateral transfers of parts of genes, genes, or clusters of genes would tremendously expand the pool of diversity available to each organism. The evolution of elephants might be accelerated remarkably if preselection of some of the genes for elephants could be accomplished in rabbits! There are so many implications of lateral gene transfer that I cannot resist devoting a few paragraphs to a layman's speculations.

First, let us consider the frequency with which lateral gene transfer occurs. The transfer of genes among some prokaryotes is commonplace on the scale of evolutionary change, a regular laboratory exercise. The transfer of genes from prokaryotes to eukaryotes is less routine, but it is nevertheless well known. The best documented case is the crown gall disease of dicotyledonous plants, where bacteria transform the normal organized tissues of plants into tumors by transfer of genes from the bacterium to the host, genes that provide the plant with an unregulated supply of its own hormones. Other genes can be transferred along with the tumor-producing genes, and there is no reason why these other genes cannot also be transferred without the genes for tumor formation. The latter cryptic transformations could be happening with appreciable frequency all over the plant and animal kingdoms without our knowledge. None of the known mechanisms for lateral gene transfer can yet be assessed as potential factors in evolution. And who knows how many unknown mechanisms exist? The transfer of genetic information from eukaryotes to prokaryotes is less well documented, but it can be inferred in one or two instances (1).

However, I do not believe that the way to estimate the evolutionary importance of lateral gene transfer is by estimating the frequency of such transfers. Rather, we must ask ourselves how frequently such events are used by evolution when they have occurred and how effective they are if used. The importance of lateral gene transfer will only be assessed when the appropriate taxonomy of genes has been developed. Such studies are already in progress, but they have not been carried far enough to give us any general answers. For instance, the plant-derived globin in the leghemoglobin of nitrogen-fixing nodules (and the intron-interrupted DNA coding for the globin) are almost identical to the DNA and globin responsible for our hemoglobin. There are several ways in which this apparent homology could be explained, but in view of the known processes of lateral gene transfer, it seems to me that the most satisfactory explanation involves acceptance of the apparent homology as real, as a manifestation of the actual relationship of our oxygen-carrying system to the system in beans.

How is chaos to be avoided if genes migrate willy-nilly among organisms? Obviously chaos does occur occasionally; witness crown gall, some other cancers, and many other types of disorder. But disorders can be selected against and "foreign" intruders excluded by mechanisms known and unknown. I am reminded of the analogy of the introduction of foreign genes and foreign species into a plant community. No matter how abundant the propagules from species foreign to a forest community, the foreign genes and foreign species almost never gain a foothold unless there is a major disturbance in the community. I suggest that genes in a genome may share some of the characteristics of species in a biological community. The fact that ecologists cannot fully explain the stability of communities does not alter the fact that the stability is very real and very great. Perhaps we should spend more time training gene ecologists to study the ecology of the genome. If this concept of the genome as a relatively stable unit has validity, we may have to be less sanguine about attempts at genetic engineering.

It is even possible that the picture of lateral gene transfer and the concept of species as a community of genes may provide us with a clue to the reason for very different rates of evolution in different organisms and an explanation of "living fossils." Suppose a species develops some unusual activity, some critical metabolic function conducted in an unusual way. Then virtually any transfer of genetic material from another species might be disadvantageous or even lethal. I have no particular function in mind, but for the sake of argument let us consider a ridiculous extreme. Suppose that some species had discovered how to make and use D-amino acids instead of L-amino acids. Such an organism would not be able to use any genetic information not originating in itself and its evolution would, of necessity, be restricted to its private game of 20-sided dice.

The importance of horizontal gene transfers in evolution cannot long remain uncertain as more and more is learned of the relationships of genes among different organisms. It may be that all controversy will be taken out of the question before this article is published. Meanwhile, it seems to me that the topic is one of the most intriguing in biology. It is a topic that should bring together the interests of taxonomists, ecologists, and geneticists in a manner long overdue.

EPILOGUE

The period of my scientific training was so critical to the development of modern biology that it has already been described by many authors. Since I have no hope of rivaling Watson's *Double Helix*, I am tempted to refrain entirely from discussions of well-known personalities or of well-known events. On the other hand, such a course implies a graceless lack of appreciation of many colleagues, when there is no lack of appreciation in my heart. Forgive me one and all. At least you will be spared the odium of association with the more unorthodox of these discussions. However, I do want to acknowledge my appreciation of a thoroughly satisfactory family, biological and academic. It is a source of pride to me that two of my sons are already launched on successful scientific careers of their own and that several of my graduate students have achieved distinction. Strangely, I am also proud of the fact that I am known to my students and family alike as "The Old Goat." I am proud of the appellation because I feel that dignity implies a lack of self-confidence and reverence implies an element of fear. I hope that neither insecurity on my part nor fear on their part played any part in our relationships.

I also hope that I do not emerge from this account of my ideas as being only negative-minded, glorying in iconoclasm and the discomfiture of the doctrinaire. I am aware of the dangers of negativism for its own sake or even of justified skepticism if it stifles imagination; witness my objection to the one-gene-one-enzyme theory. Nevertheless, I do not apologize for negativism. I think that there is merit in some of the criticisms of dogma that I have presented here. Perhaps different ways of looking at things may get some mental flywheels off dead center, even if some of the different ways prove absurd. Moreover, no theory can be considered safe to use until it has been thoroughly tested by the attentions of professional doubters.

Many of the developments that have occurred during my career have pleased me and many have distressed me. Perhaps the most pleasing has been the changing role of women in science. When I was at Caltech there were no girls at all among the students, undergraduate or graduate. There were a very few female postdoctoral associates, but female students were specifically barred. An illustration of the change is ever before me. I have on my desk two photographs of international gatherings of photosynthesis researchers. The first was taken at Gatlinburg, Tennessee, in 1952. In that august assembly of 75 scientists there were only two women. The second was taken at Ventura, California, in 1983. In the latter assembly there were 23 women out of a total of about 140, quite an improvement in 31 years but not enough improvement; the disparity is still too large. However, the greatest cause for rejoicing is the change in attitude among the women themselves. In the early 1950s few of the male scientists I knew had any reservation about women in science, but many of the women seemed insecure. Now the level of confidence of women in science is justifiably high. It is good to have the missing half of the population accepting themselves in science at last. Welcome!

On the other hand, I worry about quite a number of educational problems. The university and high school educational systems are too chaotic and unstructured. Part of the trouble arises from a doctrinaire belief that early education should remain a regional or even a community affair, as though reading, writing, arithmetic, and quantum mechanics varied from state to state. Also, in the name of a "liberal" education quite nonsensical things are often taught, things that may be harmless in themselves but must not serve as alternatives to rigorous training. Languages, including the English language, are inadequately taught, if they are taught at all. Mathematics, the ultimate abstraction, is too often taught as an abstraction only, with the result that students rarely learn the mathematics of processes. For instance, few of my students realize that linear (arithmetic) processes do not occur since they imply reactions that do not change the conditions of the reaction, a contradiction. How many students appreciate that an exponential (logarithmic) plot implies that the process is proportional to the amount of something remaining?

The competitive grant system has been, on balance, a wonderful introduction. One need only compare institutionally supported science with grant supported science. The former is often riddled with local politics. Unfortunately, as the competitive grant system matures it is also subject to political pressures, the politics of sciencemanship. Scientists come to be evaluated by the amount of money they spend or by the number of papers they publish. Versatility on the part of scientists also suffers if money is too freely available.

It is more "economical" to buy things than to make them, which is all very well if one plods well-trodden paths, but where then are the new techniques? The economy argument can even become ridiculous: one of my associates took me to task for repairing a piece of glass equipment on the ground that my time was too valuable. While I enjoyed the mistaken estimate of my general usefulness, I found the argument unconvincing. I made the repair in about one-tenth of the time it would have taken to visit the glassblower. On the other hand, modern electronic equipment, sensors, separation devices, etc, are so accurate but so specialized that purchase of costly equipment is almost necessary. However, I hope that equipment will gradually become more and more modular, so that research versatility and inventiveness will not vanish from the face of the earth. Training students in the principles and practices of instrumentation must not be neglected simply because instrument companies are also in the business. I do not want instrument companies directing my research.

I cannot bear to close this account without recording an anecdote that should be part of the lore of plant physiology. This tale, which I know to be true, has to do with the origin of the term "phytotron," a term that is widely accepted and used although I suspect that it is bad Greek. Shortly before the newly built controlled climate greenhouse at Caltech was formally opened in 1949, James Bonner and Sam Wildman repaired to the local Greasy Spoon for morning coffee. Quoth James (whose aptness of phraseology is legendary): "They should call the place a phytotron, where they bombard spinach with high speed carrots." I am glad to have the opportunity to set the record straight on this important matter, which has been much and erroneously discussed. I fear that the facts may have been suppressed by influential biologists to whom science is a sacred cow and not a joyful lark.

Literature Cited

1. Bannister, J. V., Parker, M. W. 1975. The presence of copper/zinc superoxide dismutase in the bacterium *Photobacterium leiognathi*. A likely case of gene transfer from eukaryotes to prokaryotes. *Proc. Natl. Acad. Sci. USA* 82:149–52
2. Good, N. E., Bell, D. H. 1980. Photosynthesis, plant productivity and crop yield. In *The Biology of Crop Productivity*, ed. P. S. Carlson, pp. 3–51. New York: Academic
3. Haskins, F. A., Mitchell, H. K. 1952. An example of the influence of modifying genes in *Neurospora*. *Am. Nat.* 86:231–37
4. Hind, G., Nakatani, H. Y., Izawa, S. 1974. Light-dependent redistribution of ions in suspensions of chloroplast thylakoid membranes. *Proc. Natl. Acad. Sci. USA* 71:1484–88
5. Levitt, J. 1947. The thermodynamics of active (non-osmotic) water absorption. *Plant Physiol.* 22:514–25
6. Mitchell, P. 1968. *Chemiosmotic Coupling and Energy Transduction*. Printed for Glynn Research Ltd. by Hall Graphics, Plymouth, England
7. Stewart, P. A. 1983. Modern quantitative acid-base chemistry. *Can. J. Physiol. Pharmacol.* 61:1444–64

Beatrice M. Sweeney

Ann. Rev. Plant Physiol. 1987. 38:1–9

LIVING IN THE GOLDEN AGE OF BIOLOGY

Beatrice M. Sweeney

Department of Biological Sciences, University of California, Santa Barbara, California 93106

What great good luck for a biologist to live between 1914 and 1986, to come of age at the same time as the science of biology! Our predecessors were the herbalists, the catalogers, a few physiologists working with the most Stone-Age tools. Now instruments can measure biological parameters in seconds, automatically programmed by computers. With the help of isotopes of phosphorus, carbon, and sulfur, we can find a single gene in the tangle of chromosomes or a single protein with the help of antibodies specific for that protein. We can see the structure of a cell at the nanometer level. With these tools and more, much about how plants and animals, bacteria and protists manage to live and reproduce has been discovered during my lifetime.

My own life has reflected the development of biology in the last 70 years. My passion for botany began before I can really remember, when I was so small that my eyes were level with the celandine's translucent leaves on the stone walls along the road when we walked to Cabot Woods. A half mile then seemed a very great distance, so far that my brother, a year younger, rode there in a wicker baby carriage. In the wood at that time, a stream ran at the bottom of a rounded ridge covered with oaks and underbrush. By the brook we gathered dog tooth violets in the spring, feeling their smooth, mottled leaves and nodding lily flowers, pulling them up to show their pale underground stems. We discovered hepaticas and bloodroot under the dry oak leaves on the ridge. In the hot days of summer, we found forget-me-nots blossoming by the brook. Later Cabot Woods was given to the city to be preserved, but the city crews came to clean it up, burning the hill and bulldozing the stream until there was nothing left.

By the time I was six, I must have had a good idea of the flowers of New England, because I found the plants in Pasadena excitingly different from those at home when we spent a winter there in 1920. In the garden were Cecil

2189

0066-4294/87/0601-0001$02.00

Brunner doll roses and a thicket of pampas grass. My brother and I explored the Arroyo Seco, a wild tangle of rushes and chaparral with a small dirt path down the center. In the spring, we went for walks through the fields of lupines and poppies, fields long since solid with houses.

I lived my early scientific life entirely outside of school. Before I could read, I recorded the flowers I found in drawings, frustratingly primitive and inaccurate, later by dark and out-of-focus photographs made with my mother's ancient 4×5 view camera. Collecting and naming flowers was my greatest interest at that time. We had our first car then, driven by John, our Irish chauffeur. This dear man was very long-suffering with me and would stop the car as soon as I called out, so that I could look at a plant that I had spotted as I leaned out the car window. Cars then did not travel as fast as they do now. As I grew older, I learned the plants of the Adirondack forest where we spent the summers, the twin flowers, Indian pipe, and tawny hawkweed that I discovered while tramping in the mountains with my brother.

My mother took me to a biological supply house when I was probably no more than eight years old. This treasure house sold magnifying glasses with brass legs that came off, as well as "botany boxes," elongated lunch boxes with covers that folded back for collecting, and large sheets of grey blotting paper for drying plants. My taxonomic phase lasted until I was in my teens, but sooner than that I began to discover the microscopic world of cells and to wonder how cells worked.

My interest in physiology really began because of the rarity of yellow lady slippers and an advertisement in a plant catalog for roots of these plants. But we didn't have a bog in our garden. I described my rather unsuccessful efforts to create a bog in my back yard in an essay for school that was actually published, my first publication (7).

Perhaps today our children are so busy after school with music lessons, soccer, and gymnastics that there is no time to explore the world alone or to think. At school I had best friends and played on teams, but my time alone is what gave direction to my life.

Not until I was a freshman at Smith College did I take a formal course in biology: beginning botany. We looked in the microscope and I saw plants at another level, cells that did different things, cells with chloroplasts, xylem, epidermis with water-impermeable walls and stomata. At that time photosynthesis was taught as the transformation of CO_2 to carbohydrate by a single light step in which oxygen was evolved, perhaps with the formation of formaldehyde. It wasn't until I was in graduate school that I heard about Van Niel's deduction from observations of the sulfur bacteria that the oxygen evolved came from water!

Fortunately for me, I was admitted to the Special Honors Program as a junior at Smith. We were allowed to take fewer courses and to do small research projects in each, working independently with our professors. In botany, I did my first research under the direction of Miss Smith, a study of the effects of drugs on the rate of protoplasmic streaming in *Elodea* and *Nitella* (8). I worked in a laboratory in the greenhouse head-house, measuring rate of streaming with a stop watch. I had a key to the door, a microscope of my own, reagent bottles, cultures of *Nitella,* and an aquarium filled with *Elodea.* I was launched on my career.

My teachers in botany at college were fine women, but they were all spinsters almost without private lives. I resolved that I would be different, that I would marry and have children and be a scientist too. After all, men in science did not have to give up family life. After college I proceeded to marry and have children, and my life was pretty tempestuous for a number of years; but I didn't give up science. In this my graduate advisor, Kenneth Thimann, was a great help. He never in any way implied that I was inferior to his male graduate students. He quietly made it possible for me to continue my research during whatever time I could manage when I had my first and second children during graduate school. Once when he and I were working together on a paper and I hesitated to make suggestions, he told me "Two heads are better than one." Could he have considered our heads to be equal?

Folke Skoog was a postdoctoral fellow at Harvard at the time. He arranged double-blind tests for me for my research on the effects of auxin on protoplasmic streaming in *Avena* (18). We had many discussions about research while eating cherries and throwing the pits out the window at a chimney near by. At a AAAS meeting at Dartmouth, we listened together to the radio announcement of the beginning of the Second World War in Europe.

The state of the art in biology when I started graduate school in 1936 was as follows. Electron transport in photosynthesis and photophosphorylation had not yet been discovered. That chromosomes were equally divided among the progeny in cell division was understood; the genetic material was unknown, though Mendel's work had demonstrated its necessity. Microtubules were unknown—in fact, the electron microscope had not yet been invented. Mitochondria had been seen in the light microscope as barely visible elongated bodies. The citric acid cycle had just been discovered: I learned it for my orals. Plant hormones had been chemically characterized, and animal hormones were just being discovered. X-ray diffraction methods for determining molecular structure were not in general use, and there were of course no electronic calculators or computers to make analysis easy and quick. Cytoplasm was thought to be a gel-sol, essentially an aspic of enzymes and substrates.

Plant work was actually advanced compared to that in animals, plants being in some ways simpler to dissect at the physiological and biochemical level. For example photoperiodic control, discovered apropos flowering in plants in

1914 (19), was still very little known in animals. The circadian rhythms shown in the leaf sleep movements in leguminous plants were observed to continue in constant darkness in the 17th century (2), while animal circadian rhythms and the biological clock for which they are evidence began to be studied extensively only in the 1950s.

With a fresh PhD in my hand, I went with my physician husband to Rochester, Minnesota, so that he could be a Fellow there. We went in January. Snow was so deep in our backyard that the garbage pail was buried from sight; the temperature was $-20°F$, and the freshly washed diapers froze solid on the line. Our ears froze. There were no jobs available in botany at all. I worked half days at a farm that had been converted to a hormone-assay lab, learned to keep a rat colony and do animal surgery. I left by bus for the farm at 7:30 a.m., still night during winter, and returned at lunchtime to my two small kids. Finally I got a postdoctoral fellowship to study the effects of estrogens on respiration of rat uteri (10). The Warburg manometer was at first a mystery to me. Hardest was finding out the composition of Brodie's solution, but it finally turned up in one of the earliest of Warburg's papers.

Meanwhile the United States had entered the war. My husband left Rochester to be a flight surgeon in San Diego. We lived in La Jolla, near Scripps Institution of Oceanography, and I hurried there to find something to do in plant physiology. C. K. Tseng took me into his laboratory, where he was studying the agar-producing red alga, *Gelidium*. C. K. kept his plants in tanks of seawater bubbled with air, so that the lab resounded with this pleasant burbling. I measured photosynthetic gas exchange with an ancient Van Slyke apparatus (20) and for lunch ate Lipton's chicken soup with C. K., which he stirred with chopsticks.

During the 1950s, I shared a laboratory at Scripps Institution of Oceanography with Francis Haxo. He was interested in action spectra for photosynthesis in algae, including the red alga *Porphyridium*. The spectrum for this alga was peculiar in that those wavelengths absorbed by phycoerythrin were much more effective in photosynthesis than were the wavelengths absorbed by chlorophyll itself, yet it was known from fluorescence measurements that energy absorbed by the phycobilin pigments was transferred to chlorophyll. What was the matter with the light absorbed directly by chlorophyll *a*? Was it for some reason ineffective? Lawrence Blinks had the answer almost in his hand when he showed that when red wavelengths of light were exchanged for green light of the same effectiveness at steady state there was for a short time a peak of higher oxygen evolution. The explanation, however, did not come from a study of red algae but from Robert Emerson's careful measurements of photosynthesis of the green alga *Chlorella* at the red end of the spectrum. He noted that at wavelengths longer than 680 nm the efficiency of photosynthesis decreased a little faster than did the absorption of the chlorophyll. With brilliant intuition, Emerson irradiated *Chlorella* with two wavelengths at once. How he conceived this experiment is beyond understanding. The result, as you know, was his discovery that two photosystems with different pigment composition must be excited at the same time (3). Emerson immediately understood the explanation for the inefficiency of light absorbed by chlorophyll in red algae, where phycoerythrin is the light-harvesting pigment; this was in fact a much clearer case of the necessity for the "enhancement," as Emerson called it. How do I know he had understood? Because I went to see Emerson at Urbana just at this moment. He invited me to lunch with his family and after we had finished eating, he took me by the arm, led me into the living room, sat me down in a corner, drew up a chair, and started asking me questions about what we were doing at Scripps with the red algae—a very exciting experience for me, and a little scary.

It was by chance that I began to work with the dinoflagellates. It was difficult to get funds for equipment at Scripps in 1950. I wanted to work on photosynthesis in red and brown algae in different colors of light, for which I fancied I needed an integrating sphere, lights, and a monochromator. Marston Sargent overheard my complaints and said, "While you're waiting for money, Beazy, why don't you see if you can grow some of the dinoflagellates?" They had not been cultured at that time. It actually proved to be easy because fresh seawater and plankton samples were available at the end of the Scripps Pier. I could haul them up through a hatch with a bucket on a rope, as from a well. James Bonner wrote me a note praising the paper that described this research (11), such a splendid thing for a well-known scientist to do for a novice!

One of the dinoflagellates that I was able to culture was *Gonyaulax polyedra*. This dinoflagellate was bioluminescent in the laboratory! At that time the luminescence of *Renilla* had been shown to be inhibited by light, so I wanted to find out whether light might also inhibit the bioluminescence of *Gonyaulax*. A large number of tubes with samples from a culture were prepared and their light emission was measured at successive times with the photomultiplier photometer Jim Snodgrass had made for me. At first all went as expected: When samples from the constant light bank were darkened, the bioluminescence increased, reaching a maximum in 6 hr; but then it began to fall. Perhaps the cells were dying? Because I still had unused samples and I hate to waste anything, I continued to make measurements. I found that after about 24 hr in the dark, bioluminescence began again to increase, forming another peak and then decreasing once more. I presented this result at the conference on luminescence at Asilomar, California, that March, 1954 (6). There I met Woody Hastings, and we arranged to work together the following summer at my laboratory. He recognized right away that I had seen a circadian rhythm in the luminescence of *Gonyaulax,* because he was at the time an assistant professor at Northwestern where Frank Brown was experimenting with rhythms in crabs and potatoes. Well, I just never went back to the question of the spectral properties of algal photosynthesis.

Woody, Marlene Karakashian (an undergraduate at Northwestern), and I worked together at Scripps for three summers. During this time we characterized the circadian rhythm in stimulated bioluminescence in *Gonyaulax polyedra* (16) and the biochemistry of bioluminescence with respect to enzyme, substrate, oxygen requirement and lack of apparent cofactors (4). The structure of the substrate, a very unstable molecule, was only discovered recently.

This was a time of great excitement in our research. Data poured from our hands and interpretations were hammered out in late-night discussions while we waited for the time to make more readings of light emission. Perhaps the most exciting experiment was that determining the temperature effects on the period of the circadian rhythm in bioluminescence. Woody found out (I would never have dared to ask) that we could use all the small temperature-controlled rooms at the phytotron at California Institute of Technology in the late summer while the facility was being serviced. The three of us rented one bed at a graduate student house in Pasadena and occupied it in eight-hour shifts, each of us making measurements of the bioluminescence at the different temperatures for eight hours and then playing for eight hours. The first experiment failed because the lights in the temperature-controlled rooms, run at overvoltage, were bluer than our lights at Scripps. The extra blue light inhibited the mechanical stimulation of bioluminescence, as we learned later. The second experiment, using a lower irradiance, succeeded (5).

In the winters, beside doing experiments on bioluminescence and rhythms with *Gonyaulax*, I helped Francis Haxo with his measurements of the effect of different temperatures on photosynthesis of attached algae from arctic and temperate habitats. When Francis arranged to extend this study to the tropics with Pete Scholander's expedition in 1960, he asked me to come along. The Michael Pilson family, who were helping me look after my children, now four in number, agreed to take over for three months so that I could go. Twelve scientists traveled inside the Great Barrier Reef from Cairnes to Thursday Island in the 60-foot *Tropic Seas,* a crocodile-hunting boat skippered by Vince Vlasov. I occupied a bunk in the stern with the canned goods. When we reached Thursday Island, established ourselves in the CSIRO oyster laboratory, and waded out on the sand flat in front, we found *Acetabularia!*

For some time I had wanted to find out whether the nucleus was necessary for circadian timing. *Acetabularia* was the obvious organism to use because it can live and function normally for a long time after the nucleus is removed. However, I had no *Acetabularia* plants in La Jolla. Furthermore, no circadian rhythm was known in *Acetabularia* at that time. We had with us on Thursday Island only manometers to measure photosynthetic oxygen evolution, so I measured the oxygen evolution of *Acetabularia* at different times of day in natural light and in a makeshift artificial constant light. What luck! Rates of photosynthesis differed greatly between day and night, and this cycle continued in constant light. The nucleus can be removed by simply cutting off the basal part of an *Acetabularia* cell, so it was easy to test whether or not the rhythm continued in the absence of the nucleus. It did (17).

It's a wonderful thing to see something you've never seen before, the Great Barrier Reef or the ultrastructure of a cell you've been working with for a long time. In the light microscope, *Gonyaulax* is almost completely opaque. When I was at Yale I had the opportunity to see the details of the internal structure of *Gonyaulax* for the first time. Ben Bouck fixed and sectioned cells and we saw the chloroplasts, the mitochondria, and curious membrane-bound structures, square in cross sections, the trichocysts (1).

After this I learned to do electron microscopy myself, even freeze-fracture, a technique I found difficult. The limited number of cells one can examine in electron micrographs makes gathering quantitative data tiring and frustrating (13). However, some of the pictures, particularly of freeze-fractured cells, are truly beautiful.

It is wonderful, too, suddenly to understand something. The question of how a cell can keep time had long been in my mind. I was sitting at my typewriter composing a talk for an *Acetabularia* conference in Wilhelms-haven when all at once it came to me that membrane-bound organelles must be important in time-keeping, since not a single prokaryote is known to have a circadian rhythm. An exchange of ions or molecules must be taking place between cytoplasm and organelles in a controlled fashion, a feedback loop where the membrane must change the passage of ions and the passage of ions must change the membrane in a cycle (12). This might explain the peculiar observation in *Acetabularia* that, while the nucleus is clearly not necessary for timing, when it is present it can change timing. Of course this idea was not completely right, but it started a train of thought and a series of experiments both by myself and by others that confirmed the importance of membranes in generating time information in cells.

In 1967 I came to the University of California at Santa Barbara and the next year became an associate professor of biology, my first position as a real professor. I discovered the pleasures of working with students of my own— my graduate students and the undergraduates in the College of Creative Studies here. This institution is remarkable in being a small college embedded within a large university with all its resources. It was designed by Marvin Mudrick for the education of intellectually gifted students. In science, even the freshmen work in the faculty laboratories helping with research or even doing their own projects. I was very fortunate to have a number of these students in my laboratory, including several who are already well-known scientists. Their questions enlivened discussions and their ideas were an inspiration. To teach such students is pure fun.

During the days when the *R. V. Alpha Helix* was an active biological research ship, I was included in two expeditions to study bioluminescence, one to New Guinea and the other to South East Asia. These trips were productive for all of us. I studied the physiology of the *Noctiluca* with symbiotic green flagellates swimming in its vacuole (14) and brought back a culture of *Pyrocystis fusiformis,* a very large cell that proved useful for electrophysiology (21) and for a comparison of the cell cycle and the circadian cycle (15).

I have always enjoyed doing my own laboratory work, either alone or with one or two active collaborators—for example Francis Haxo and Woody Hastings, more recently Marie-Therese Nicolas and Goran Samuelsson (9). I find that I understand the advantages and flaws in an experimental technique when I do it myself, and most of my ideas have come when I am actually gathering data. Molecular biology, which has yielded so much information about the genes, has an inherent danger. It requires large research groups, as evidenced by the long list of authors on single papers, and thus requires large amounts of money for salaries and supplies. The senior scientist in such a group must thus act as administrator and fund raiser and will probably have difficulty finding time to work in the laboratory. We make a mistake when we convert our most able scientists into administrators and hence inactivate them.

People often ask me whether or not I have suffered from discrimination against women scientists. As a young girl, I really did not feel this. My family and my teachers were always encouraging, so I did not suffer any effects of discrimination until I had my PhD. The problems I met after that arose from attitudes that I had somehow acquired with respect to society's expectations of female behavior toward men. I instinctively felt that I should put men's interests ahead of my own. For example, I deferred to my husband in consenting to his taking a position at Mayo Clinic, in an isolated community where there was nothing for me to do scientifically. I deferred to my early collaborators in first authorship of the more important papers. I was satisfied with a research position when my children were small because I only wanted to be committed to work half-time. When they were grown and I could hold a full-time faculty position, it was difficult for me to get one. Universities considered that I had too many publications to be an assistant professor and too little teaching experience to be an associate professor. Finally I succeeded, but only barely—even in the 1960s when all universities were enlarging their faculties. The attitude of servility and guilt about pleasing ourselves that we learn as little girls is, in my experience, the greatest deterent to equal opportunity for women.

I advise young women in science to keep their maiden names, dare to be assertive, and refuse to get discouraged. They must keep learning and reading, especially if they have to take leave from science for family reasons. It may not be easy to combine a scientific career with a family but it can be done and it's worth all the effort. Research has sustained me through many otherwise unbearable times and given me much satisfaction, made my older years exciting. There is a temptation for women to settle for the family alternative as simpler. However, sooner or later the family will want a second income. A high percentage of all women do work at some point in their lives. When that point arrives it is crucial for a woman to be prepared to enter, or to resume, a challenging career.

As Goran Samuelsson said in introducing me recently at a seminar at the University of Umea, I have had a lifelong love affair with research.

Literature Cited

1. Bouck, G. B., Sweeney, B. M. 1966. The fine structure and ontogeny of trichocysts in marine dinoflagellates. *Protoplasma* 61:205–23
2. De Mairan, M. 1729. Observation botanique. *Hist. Acad. R. Sci., Paris,* p. 35
3. Emerson, R., Chalmers, R., Cederstrand, C. 1959. Some factors influencing the long-wave limit of photosynthesis. *Proc. Natl. Acad. Sci. USA* 43:133–43
4. Hastings, J. W., Sweeney, B. M. 1957. The luminescent reaction in extracts of the marine dinoflagellate, *Gonyaulax polyedra. J. Cell Comp. Physiol.* 49:209–26
5. Hastings, J. W., Sweeney, B. M. 1957. On the mechanism of temperature independence in a biological clock. *Proc. Natl. Acad. Sci. USA* 43:804–11
6. Haxo, F. T., Sweeney, B. M. 1955. Bioluminescence in *Gonyaulax polyedra.* In *The Luminescence of Biological Systems,* ed. F. H. Johnson, pp. 415-20. Washington, DC: AAAS
7. Marcy, B. 1931. In praise of bogs. In *Essays of Today,* ed. R. A. Witham, pp. 323–26. Cambridge: The Riverside Press
8. Marcy, B. 1937. Effect of ethylene chlorhydrin and thiourea on *Elodea* and *Nitella. Plant Physiol.* 21:207–12
9. Samuelsson, G., Sweeney, B. M., Matlick, H. A., Prezelin, B. B. 1983. Changes in photosystem II account for the circadian rhythm in photosynthesis in *Gonyaulax polyedra. Plant Physiol.* 73:329–31
10. Sweeney, B. M. 1944. The effect of estrone on anaerobic glycolysis of the uterus of the rat *in vitro. J. Lab. Clin. Med.* 29:957–62
11. Sweeney, B. M. 1954. *Gymnodinium splendens,* a marine dinoflagellate requiring vitamin B_{12}. *Am. J. Bot.* 41:821–24
12. Sweeney, B. M. 1974. A physiological model for circadian rhythms derived from the *Acetabularia* rhythm paradoxes. *Int. J. Chronobiol.* 2:95–110
13. Sweeney, B. M. 1976. Freeze-fracture studies of *Gonyaulax polyedra.* I. Membranes associated with the theca and circadian changes in the particles on one membrane face. *J. Cell Biol.* 68:451–61
14. Sweeney, B. M. 1976. *Pedinomonas noctilucae* (Prasinophyceae), the flagellate symbiotic in *Noctiluca* (Dinophyceae) in Southeast Asia. *J. Phycol.* 12:460–64
15. Sweeney, B. M. 1982. Interaction of the circadian cycle with the cell cycle in *Pyrocystis fusiformis. Plant Physiol.* 70:272–76
16. Sweeney, B. M., Hastings, J. W. 1957. Characteristics of the diurnal rhythm of luminescence in *Gonyaulax polyedra. J. Cell Comp. Physiol.* 49:115–28
17. Sweeney, B. M., Haxo, F. T. 1961. A persistence of a photosynthetic rhythm in enucleated *Acetabularia. Science* 134:1361–63
18. Sweeney, B. M., Thimann, K. V. 1938. The effect of auxin on protoplasmic streaming. II. *J. Gen. Physiol.* 21:439–161
19. Tournois, J. 1914. Études sur la sexualité du houblon. *Ann. Sci. Nat. Bot. Biol. Vég.* 19:49–191
20. Tseng, C. K., Sweeney, B. M. 1946. Physiological studies of *Gelidium cartilagineum.* I. Photosynthesis, with special reference to the carbon dioxide factor. *Am. J. Bot.* 33:706–15
21. Widder, E. A., Case, J. F. 1981. Bioluminescence excitation in a dinoflagellate. In *Bioluminescence, Current Perspectives,* ed. K. H. Nealson, pp. 125–32. Minneapolis: Burgess

SOCIOLOGY

Photograph by Christopher S. Johnson.

George C. Homans

Ann. Rev. Sociol. 1986. 12:xiii–xxx

FIFTY YEARS OF SOCIOLOGY

George C. Homans

Department of Sociology, Harvard University, Cambridge, Massachusetts 02138

Abstract

What follows is a personal appraisal of the development of sociology over the past 50 years. The 1930s were a time of high hopes, empirically (field studies, new statistical techniques) and theoretically (for instance, functionalism and operationalism). The great achievement of sociology has been its development of statistical techniques, but these have had the effect of inhibiting field studies. In theory, sociology has remained divided into a number of different schools, a condition maintained by the failure of most sociologists to consider what a theory *is*. The condition can be overcome only by sociologists' accepting the "covering law" view of theory, the covering laws referring to the behavior of individual persons as members of a single species, and the laws themselves being the laws of behavioral psychology. Recent favorable developments in sociology include network analysis, historical sociology, and sociobiology.

INTRODUCTION

In what follows I reflect, from my own point of view and from my experience with, and practice of, the discipline of sociology for 50 years, on what successes and failures the subject has met over this era, what new issues it faces, and what these have to teach us about the directions sociology should take in the future.

THE HOPES OF THE 1930S

I published my first book in sociology in 1934 (Homans & Curtis 1934), and I cannot be expected to publish many, if any, more. Age, together with many unpleasant things, brings a kind of freedom. In the societies of the East, we are told, an old man gives up his worldly responsibilities and goes off on pilgrimage, with nothing but a begging bowl in his hand. He feels free to act like an old

2201

0360-0572/86/0815-0000$02.00

rogue, exempt from the conventions and allowed to say anything he pleases, however outrageous, at the price, naturally, of being disregarded. I propose to take advantage myself of this privilege. I propose, indeed, to take advantage of it to the point of hurting some persons' feelings.

For American society at large, 1934 was a dreary year, because we were still in the grip of the depression. But in the social sciences it was a year of exhilaration. Many new statistical techniques were being invented, under the leadership of such men as Samuel Stouffer and Paul Lazarsfeld. At the other end of the technical spectrum, field work (the direct observation of small groups and the interviewing of their members) was being introduced into sociology from anthropology and was soon to show brilliant results in such studies as Arensberg & Kimball's *Family and Community in Ireland* (1941) and W. F. Whyte's *Street Corner Society* (1943). We were excited by the community studies of the Lynds (1928) and of W. L. Warner (1941–1963). Elton Mayo was beginning to introduce us to experimental and field studies of modern industry through the Western Electric researches (see Roethlisberger & Dickson 1939). I should even mention Moreno's invention of what was called sociometry in a book strangely entitled *Who Shall Survive?* (1934). I could mention others.

More important than the techniques were the ideas. The Russian Revolution had made Marxism almost unchallenged among some intellectual groups, especially Jews of poor parents who had yet managed to get a good education in New York City. This was also the time when Freud's ideas began to achieve general acceptance among intellectuals. These new ideas may have been the most prominent examples, though certainly not the most important in the long run. Keynsian economics and logical positivism also flourished, stimulated in the long run by the work of Whitehead and Russell, the popular philosophers of the age. Bridgman's operationalism (1936), Skinner's behaviorism (1938), at once broader and more rigorous than J. B. Watson's, and various forms of functionalism, derived again from anthropology, all attracted adherents. Closer to my own intellectual base, Talcott Parsons had begun his teaching at Harvard and was developing an extraordinary group of students which included Kingsley Davis and Robert Merton. Parsons introduced them to what are now considered the European classics of sociological thought: Durkheim, Weber, and Pareto (though I believe that Durkheim, at least, was already well known at Chicago). Why Tocqueville was not admitted to the canon, I have never discovered. Above all, Parsons and Merton made the idea of sociological theory respectable and even a means of gaining status, though they had no clear idea what a theory was.

More important still, what did we young social scientists believe we were going to be able to do? With our new techniques and ideas we were going to be able to solve, at least intellectually, most of the ills besetting the world. Certainly our first students, returning from World War II, were convinced of it and made our field among the most popular on our various campuses. Unfortunately we vastly and irresponsibly oversold ourselves, and disillusion rightly set in. (I myself was more skeptical, but my teacher Elton Mayo certainly was sure we could stop the rot, if not produce positive improvement.) We also thought that we were going to create an integrated and cumulative social science, cumulative because one piece of research would, with the help of a better theory, build on another.

STATISTICAL METHODS AND EMPIRICAL RESEARCH

Looking back over 50 years, how much have we done to realize the dreams of our youth? I have the impression of some advances, much wasted effort, some hopes disappointed, certainly much intellectual disarray, and even some losses. In one area at least we have made great progress. Our advances in statistical and mathematical techniques have been huge and consolidated. Multivariate- and path-analysis, coupled with high-speed computers, have made possible investigations that would have been impossible or impossibly costly even a few years ago. I am thinking especially of studies, notably the famous report by James Coleman (1966), on the effect of the character of schools on the intellectual attainments of their students. It was not Coleman's fault that the use of his findings by the Supreme Court and the federal hierarchy has, together with other factors, made conditions worse rather than better, at least in my own city of Boston.

I also have the impression that the new techniques are often used mechanically to produce some, indeed any, findings from data that just happen to be available. I am thinking here of one study of the wage differentials between supervisors and workers in a sample of factories. The data were readily available, but after the author had applied his statistical techniques to them, I still had the impression that he had never been inside a factory nor ever looked carefully at what a supervisor and his workers actually *did*.

Yet the general tendency has still been favorable. In the years immediately after World War II, if some group such as a government committee asked a sociologist what was known about a particular social problem, the best answer available was apt to be: "We know little about it, but give us enough money and we'll find out." Today in a number of fields a body of fairly solid knowledge is at once on hand. In this sense our subject has indeed become more cumulative. It has not become more cumulative in developing a powerful general theory, or at least in accepting such a theory.

The rise of the new statistics has actually encouraged us to abandon a technique—call it that—which was one of our glories in the period just before and after World War II. So powerful were the new statistics and so quickly could they generate findings that they made field studies of small groups seem almost a waste of time. Where are now such studies as that of the Bank Wiring Room in the Western Electric researches (Roethlisberger & Dickson 1939), of Whyte's *Street Corner Society* (1943), of Peter Blau's *The Dynamics of Bureaucracy* (1955)? I can think of some, but in no such quantity or quality.

The decline of these studies has had three unfortunate results. First, science begins, as L. J. Henderson used to say, with the scientist's acquiring an intuitive familiarity with the facts (see Barber 1970:67). In our field, this can only be acquired by the scientist's watching and talking to people at first hand, and field studies alone provide the opportunity. Science, of course, does not end there, but it certainly begins there.

Second, I believe that the fundamental principles of social behavior (an adequate microsociology, if you will) can best be worked out after direct observation and interviewing—and I do not mean with formal questionnaires— of a few people interacting with one another. At least since the time of Edgeworth (1881), economics has had an adequate microeconomics, on which it could fall back when its macrotheories yielded contradictory intellectual or practical results—as they do now, when Keynsianism has lost its dominance and macroeconomics is in disarray. We have no such microsociology. I tried to produce one, based on behaviorism, but I cannot say that it has been generally accepted (Homans 1961, 1974). Others have done microsociological work but have failed to produce a microsociology.

Third, we are unable to understand some (at least) of the results of survey research without the help of small-group research. Some years ago, Stanley Seashore (1954) made a survey of the workers in a large factory in the Middle West. Among other things he found that behavior interpretable as "restriction of output" was particularly likely to occur in departments that held high status. The survey itself provided no evidence by which one could explain this finding. I bet that, if the survey had been accompanied by detailed field studies of at least two departments (say, one of high status and one of low), the elements of an explanation would have been forthcoming. Why can we not combine our methods more often? Expense?

These considerations lead me to another technique, one which social psychology with its psychological tradition introduced into sociology: experimental methods. We used to argue about the degree to which these experiments corresponded to something we were pleased to call "real life." The underlying issue now seems to me a little different. It is whether in the course of establishing strong experimental controls the investigator may not destroy the very phenomena he is interested in. I have an example in mind, one unfortunately carried out by persons I otherwise admire and, what is more, published in a *festschrift* in my honor (Burgess & Nielsen 1977). The research was directed at what I call distributive justice, the relation between the rewards the members of a group receive in relation to what they contribute. I believe that some of the findings of the research were confused, but that is not the point I want to raise here. The subjects in the experiment used consoles through which they could deliver various sums of fictitious money to the parties with which they were in exchange, and to this extent each had some control over the others' behavior. With this exception, to ensure the purity of the findings, "every effort was made to prevent the subjects from meeting or identifying one another. During the course of the experimental sessions, no verbal communication between the subjects was possible." But it is only through social interaction, and repeated social interaction, that members of real groups discover what the others are gaining from or contributing to the exchange between them, and only thus do they manage to reach some degree of consensus as to the values of their rewards and contributions or the other resources by which they reckon distributive justice. Moreover, the rewards and contributions are not single but multiple. Here I think the experimenters, in a justifiable endeavor to establish strong experimental controls, managed to destroy the conditions under which problems of distributive justice manifest themselves in real life. So much for our obvious strengths and some of their ambiguities.

SOME DISAPPOINTING IDEAS

Let me turn now to the fate of some of the ideas of the 1930s; I will begin with Marxism. There are probably more American sociologists now who claim they are Marxists than there ever have been. But they are not the same young intellectuals who embraced Marx in one form or another in the New York universities before World War II. Glazer, Kristol, Bell, and the like are Marxists no longer. How then account for the recrudescence of Marxism? After all, it is now some 137 years since the publication of the Communist Manifesto. Through there are still great virtues in the ideas of historical materialism—after all, technology and the social organization the technology may favor are very good places from which to start the study of any society—none of Marx's specific predictions has been borne out, though they have had plenty of time. And Marxism has divided into a number of hairsplitting sects, which address themselves to what their rivals say rather than to how humans behave (the latter is, after all, what Marx himself tried to do).

The Bible has, in the Sermon on the Mount, a good adage: "By their fruits ye shall know them!" If ideas have consequences (and we are always being told they have), we should look at the consequences of trying to put into practice some Marxist ideas—if indeed they are Marxist and not Leninist. The results have been economic inefficiency and political repression, without even much progress towards a classless society. Under these conditions one might expect honest scientists to desert Marxism, and many of them have. But the recruits have more than replaced the deserters. Why? I think the reasons are what Pareto

(1917) would have called nonlogical (See Homans & Curtis 1934). Using Marxist language shows that you are on the side of the underdog or (which may not be quite the same thing) that you hate the bourgeoisie, though most Marxists are members of the bourgeoisie. Taking such impeccably moral positions gives many people a great deal of satisfaction. Indeed, for some there is no greater pleasure than embracing a good, simple, self-righteous moral position. Some few Marxists may believe that capitalism will collapse even in the Western democracies and that, by declaring themselves Marxists now, they will entitle themselves to positions of power in the ensuing Communist regime. Their hopes are almost wholly unfounded. As they ought to know by now, the revolution has a way of devouring its own.

At any rate, I think sociology would be much better off by abandoning Marxism and its controversies and, above all, by abandoning its language, for its substance being hollow, its language is about all Marxism has left. Abandoning Marxism would not mean that its adepts would have to change their concrete research interests. All they would have to abandon is a jargon that is increasingly hypocritical and out of touch with reality.

In a different way, the same is true of Freud, another of the thinkers who looked in the 1930s as if they would become intellectual liberators. Today what is left of Freud besides the concept of the unconscious, which even the behaviorists are eager to accept (and which does not always work in the ways Freud said it would), and some of his earlier clinical reports? The latter part of his work, if not merely the embodiment of old metaphors, is only speculation. Again, "By their fruits ye shall know them!" The application of Freud's ideas to psychotherapy has never been shown to have significant positive results. Statistically, that is, the Freudian cures have never been shown to be more numerous than those that would have occurred if the patients had just been left to the mercies of *vis medicatrix naturae* (see Brown & Herrnstein 1975, pp. 594–99.) Moreover, many psychic ills that once fell into the hands of Freudian and other psychotherapists whose techniques were limited to talking with the patient have now been shown to yield to treatment by drugs. And like Marxism, Freudianism has been riven by quibbling schools and by idiocies such as those of Jacques Lacan (see Turkle 1978). Happily, unlike Marxism, social scientists seem to me to use the ideas and language of Freudianism less and less often. It keeps its place as a convenient popular psychology for use at cocktail parties.

As for the other intellectual developments of the 1930s—functionalism in its old sense, as taken over by sociologists from anthropologists, is dead, at least in its collectivistic form, though Jeffrey Alexander (1982:55–63) promises to revive it. The individualistic form of functionalism can easily be identified with behavioral psychology (Homans 1984:343–45). Bridgman's operationalism certainly added rigor to our formulations of hypotheses, but the philosophers of science have shown that the doctrine cannot be generally applied. That is, all theories of any complexity must contain terms that cannot be defined operationally but only implicitly in the form of propositions. Thus, Newton's force law (f = ma) is not a proposition in which all the terms can be operationally defined (see Braithwaite 1953:50–87). Instead the proposition is an implicit definition of the concept *force* (f).

Strangely, the one intellectual enthusiasm of the 1930s that has shown it can maintain itself and develop scientifically is the one that has always been least popular. I mean behaviorism, a stripped-down form of which is called utilitarianism or rational-choice theory. I shall have much more to say about it later.

THE PROBLEM OF THEORY

If the chief strength of current sociology lies in its empirical studies controlled by new and powerful statistical and mathematical methods, its chief weakness lies in theory, which was once expected to redeem it from "mere fact-finding." Sociology is divided into a number of theoretical schools, each claiming to be different from the others: the Marxisms, symbolic interactionism derived from G. H. Mead (1934), exchange theory, dramaturgical theory (Goffman), ethnomethodology (Garfinkel), conflict theory (R. Collins), and others. Each has interesting empirical findings to its credit, but to my mind it is impossible to discover what the alleged theoretical differences are.

Among sociologists there is hardly anyone who does not sometimes set up to be a theorist. A theorist has high prestige. What really surprises me about the theorists is that, so far as I can tell, almost none of them, with the notable exception of myself, ever states what he or she believes a theory *is*. (See Homans 1967; another possible exception is Kaplan 1964.) True, there may be more than one definition of the word *theory,* but the theorists never specify which one is theirs. Perhaps they take the characteristics of a theory to be a matter of common knowledge. Or they may mean by *theory* simply *generalization.* If so, they are mistaken, for the issue is a tricky one, and the philosophers of science have been preoccupied with its trickiness. At any rate, the failure of sociologists even to discuss the nature of theory has meant that we cannot tell what it is they are trying to do. Or rather, we can see what in fact they do. They spend little time on developing a theory of actual human behavior and much time and enormous erudition in discussing the words of theorists like themselves. They produce words about words.

I myself as a confessed theorist have at least not neglected to consider what a theory *is*. Basing my description on the work of some, but not all, of the philosophers of science and a study of some, but not all, examples of what are accepted as good theories in the natural sciences, I have adopted and tried to use what is conventionally called the "covering law" view of theory.

I have said what follows many times, and I shall repeat it as long as I live or as long as I feel that repetition is necessary, whichever lasts longer. A theory of a phenomenon is also an explanation of the phenomenon. It consists of at least three propositions, as in a syllogism. A proposition is a statement of a relationship, often only approximately true, between properties of nature (variables). At least one of the propositions is more general than the others, in the sense that we cannot derive it logically from the conjuncture of the others, just as we cannot derive "All men are mortal" from the combination of "Socrates is a man" and "Socrates is mortal." Other propositions state the given conditions (parameters) to which the general propositions are to be applied. When the empirical proposition, the phenomenon, can be shown to follow deductively from the other propositions in the set (the deductive system), then the phenomenon is said to be explained. The deductive system is the theory of the phenomenon.

The "covering law" view takes its name from the general propositions. These can often in time be shown to follow from propositions still more general. Note that if the given conditions (parameters) change, different, though sometimes related, propositions will follow from the deductive system. And the nature of the given conditions can often themselves be explained. Indeed, what we usually call a theory is not a deductive system explaining a single phenomenon but a set of general propositions, a variety of given conditions, and a number of different empirical propositions which can be shown to follow from the general ones under different combinations of the given conditions.

The chief difference between real theories lies, it seems to me, in differences between their general propositions. Not only do sociological theorists never discuss what they mean by theory, a fortiori they never state what they consider their general propositions to be. In philosophical language, their logic is *enthymematic*. And since they do not state what they take their general propositions to be, we cannot tell whether or not, or in what ways, their theories are similar. They may well differ in their empirical content because they deal with different given conditions. That is not what counts. What does count is the nature of the general propositions.

I believe that, if theorists of sociology did make their general propositions explicit, they would find that they were all using some variety of the same set. The tragedy of sociology is that, since they do not do this (and perhaps are afraid of trying), the actual unity of our science and hence its possibilities for cumulative theoretical growth go unrecognized, and our intellectual chaos persists.

The theoretical unity of our science is well worth working toward. A good theory organizes a science, simplifies it by showing how its empirical propositions are related to one another, and often allows its adepts to predict new propositions that are worth testing.

UTILITARIANISM AND BEHAVIORISM

If it were made explicit, I think the set of general propositions in question would turn out, at the simplest level, to be something that is called rational-choice theory or utilitarianism. I do not like the word *rational* in rational-choice because *rational* is a normative term: it often refers to what someone believes to be a better way for a person to behave in order to achieve a particular result, rather than to how the person in question actually behaved. The latter is what sociologists, in the first instance, should be interested in. *Utilitarianism* does not bother me in the same way. No matter. The simplest way of putting the main proposition of rational-choice or utilitarian theory is this: In choosing between at least two alternative courses of action, a person is apt to choose the one for which the perceived value of the result multiplied by the perceived possibility that the action will achieve the result is the greater (Homans 1974:43–47).

Note that the two chief variables in the proposition are the value (degree of reward) of the result and the probability of attaining that result. The perceptions are the actor's own perceptions, which may be mistaken. They depend on his past experience of the degree of reward and success that have attended his actions and on the other circumstances (stimuli) accompanying his past actions in their bearing on value and success in the present case. They may depend on his observation of others performing similar actions and their success in obtaining valued results. (For model learning, see especially Bandura 1969.)

Rational-choice theory or utilitarianism is a first approximation to, or a simplification of, behavioral psychology. The two hold in common the main proposition cited above. Utilitarianism is a stripped-down version of behaviorism in a number of different ways. It often tends to take the perceived rewards and success of various actions as given—a statistically reasonable assumption for numerous actors pursuing widely shared values, when their success in obtaining them is virtually certain. But the more general science of behaviorism insists that individuals often learn different values, differ in their success in obtaining them, and acquire different perceptions. That is, utilitarianism has little to say about the effect of past history on present behavior. It also has little to say about emotional behavior and pays little attention to the niceties of behaviorism, such as the effect on behavior of different schedules of reward (Ferster & Skinner 1957). For instance, actions performed under intermittent schedules of reward (reinforcement) are particularly resistant to extinction (their elimination when reinforcement ceases). Utilitarianism will serve well enough as the general proposition in theories of many kinds of human behavior, but not for all. For the wider reach the full panoply of behavioral psychology is needed.

Behavioral psychology, together with cognitive psychology (which some take to be a separate science), is a set of propositions about human nature, or so much of it as we know now. It states the characteristics of behavior that human beings share as members of a single species. It obviously does not imply that all human beings behave exactly alike. The variables in the propositions remain; their coefficients, so to speak, differ. Individuals differ genetically in their behavior. They differ in what they have learned and therefore in their current behavior. Indeed, if we know enough about their pasts, including their genetic differences (which we rarely do), we can use the general propositions to explain why they differ in what they have learned. By the same token, human cultures differ because their past histories have differed, though we are often at a loss to explain why, because many differences were already present before the historical record begins. But the general propositions still hold, and human nature in this sense is indeed the same the world over—a view that would have been treated as a heresy in the heyday, again in the 1930s, of the "culture and personality" school of anthropology.

The general propositions of behaviorism are the general propositions not only of sociology but of all the social sciences. That does not mean that every social science uses them to explain the same phenomena, for the different disciplines apply them to different given conditions. The conditions of the classical market are not those of an enduring small group. What should serve as a warning to sociologists is that the other social sciences are increasingly using behaviorism, usually in its utilitarian form, to explain phenomena that used to lie within the field of sociology. Consider, just to take a few instances, the work of Olson (1965), Schelling (1984), and Barry (1970). Sociology is in danger of losing potentially important parts of its field to other disciplines because most sociologists (though, I am glad to say, not all) refuse to use this powerful theoretical tool.

I say that they *refuse* to use it because I observe that behaviorism and utilitarianism are sometimes mentioned by sociologists and then rejected, usually without familiarizing themselves with what they have rejected. The rejection also seems to vary by nationalities. Although most often rejected by British and American sociologists, behaviorism and utilitarianism are making their way even among them. For a few examples see Burgess & Bushell (1969), Hamblin (1971), Scott (1971), Kunkel (1975), and Hechter (1983). These theories seem to be most often accepted in West Germany, though the emphasis there is less often on behaviorism than on utilitarianism and methodological individualism. For a few examples, see Malewski (1967), Schwanenberg (1970), Opp (1972, 1985), Vanberg (1975), and Raub (1984). Since most American sociologists do not read German—another example of the weakness of language instruction in American schools—they do not know what they are missing. What are the reasons for this "great refusal"?

REASONS FOR THE REJECTION OF BEHAVIORISM

Six reasons have been put forward for the rejection of behaviorism as a theory.

1. By the very name of their subject, sociologists have to do with social behavior. But the propositions of behavioral psychology are individualistic. That is, the propositions apply to the behavior of individuals, not of groups. In the language of philosophy, such propositions should be formally stated, for instance, as "For all individuals, if a particular kind of action taken by an individual is rewarded, then . . . ." But this has nothing to do with individualism as a moral philosophy, which is repugnant to many persons. Behaviorism does not imply that the largest part of human behavior has ever been anything but social—that is, consisting of interactions between individuals. What it does imply is that no new proposition is needed to explain a person's behavior when his action is rewarded by another person from when it is rewarded by the nonhuman environment, as by his catching a fish. Of course the effects of social behavior may be much more complicated, because the given conditions to which the general propositions apply are more complicated too. Sociologists should heed statements like a recent one by Pierre Boudon (1984:39): "A fundamental principle of the sociologies of action is that social change ought to be analyzed as the resultant of an ensemble of individual actions" (my translation). The ensemble may be very complicated, requiring complicated mathematics for its analysis, and the resultant, one that no single individual intended.

   Behavioral psychology as a set of individualistic propositions implies that there can be no general laws of history as such, which so many historians have so long sought for in vain. The only general laws of history are the laws of human nature. There are many true historical, as there are sociological, propositions of less than full generality, because some social conditions may favor the appearance of similar "ensembles of individual actions;" but these conditions are never universal. Boudon's whole book *La Place du désordre* (1984) is devoted to this argument, though as an implication of individualism, not behaviorism.

2. Behaviorism, it has been said, robs human beings of purposes, which are one of their most precious perquisites (Coleman 1975:79). Behaviorism does no such thing. All it insists on is that purposes do not come to human beings out of the blue. Purpose does not imply teleology. Drives may be given genetically, but an individual must learn the particular actions that will satisfy these drives. He learns them from past experience, and he applies them under contingencies (stimuli) whose relevance he has also learned in the past. And the same is true of actions that receive secondary rather than primary rewards. A purpose is an action, or an approxima-

tion of that action, that has been rewarded in the past—that has been conditioned.

3. As a study of the effects of reward and punishment, behaviorism is hedonism in its crudest form, and good people—sociologists are of course good people—are never hedonists. Early behaviorist writings did often sound pretty crude. But modern behaviorism recognizes that altruism can be acquired through classical (respondent) conditioning which depends on genetically determined rewards. As H. J. Eysenck remarked, "Conscience is a conditioned reflex" (quoted in Wilson & Herrnstein 1985:48).

4. Behaviorism is a psychological doctrine. Indeed, once treated as a pariah, it has now become a part of mainstream psychology. What is worse, its propositions can be used to reduce sociology to psychology. What this means is that sociological propositions—to speak crudely, propositions of the sort usually put forward by sociologists—can be shown to follow under specified given conditions from psychological propositions. Many sociologists feel that this robs their subject of its status as an independent science; and its status is none too high as it is. But chemistry, for instance, does not seem to have lost its identity from being shown to be reducible to physics. And if sociology loses its sense of identity through psychological reductionism, so should all the social sciences. But they will not, because though they use the same psychological general propositions, they usually apply them to different given conditions. Perhaps the real source of the difficulty many sociologists (not the social psychologists) find with behaviorism derives from their limited knowledge of psychology. They prefer to read Habermas and forgo the greater advantage of really learning about social behavior in humans and animals.

5. A rather different kind of criticism is one that accuses some of the propositions of behaviorism of being tautological. For one example, consider the proposition: The more valuable a reward, the more likely a person is to take action that is followed by the reward. This is tautological, it is said, because there is no independent measure of value other than the frequency of action taken to get it. In this sense the proposition is indeed tautological. But we must always examine how statements like this are actually used in theories. The formula is a useful catchall. A large number of different kinds of result are known empirically to be reinforcing. Instead of enumerating them, which cannot be done, for new ones are being discovered all the time and learned, we substitute into our formula particular values that are relevant. After all, the statement that a person deprived of food is, until satiated, apt to take frequent action that is followed by food is true but *not* tautological. Braithwaite (1953:50–87) has shown that the kind of tautological statement which is an implicit definition of theoretical terms

must occur in many, if not all, scientific explanations. Incidentally, this is one of the most important arguments against strict operationalism.

6. I am not saying that the sociological theorists do not often have good ideas. They do. I am saying that their theories are inadequate because they leave their major premises unstated. Indeed, they seem to be unaware of what their major premises are. Let me take an example. One of the charges that some of the sociologists bring against their behaviorist colleagues is that they neglect the effects of social structures on the behavior of individuals. By social structures I shall mean—and I think the critics mean—more or less enduring practices followed by a number of persons, whether or not these practices are made explicit in norms or defended by sanctions. Structures include what we usually call institutions. To this controversy, Anthony Giddens introduces the idea of *structuration,* which he describes as follows: "The concept of structuration involves that of the *duality of structure,* which relates to the *fundamentally recursive character of social life, and expresses the mutual dependence of structure and agency.* By the duality of structure I mean that the structural properties of social systems are both the medium and the outcome of the practices that constitute those systems" (1979:69; Giddens' emphases).

If I understand it right, I think this is a good idea. I would translate it into what I think is clearer, less abstract, though more lengthy language: Individuals by their actions (agency), in concert or even at cross-purposes, intentionally or unintentionally, are always in the process of creating and maintaining social structures, at least for a time, for no structure endures forever. Indeed, the structures *are* the actions. But while structures are thus being created and maintained by the acts of individuals, these same structures are providing new contingencies (rewards, stimuli, possibilities of success or failure) for the maintenance, again by the actions of people, of old structures or the creation of new ones. This is the "fundamentally recursive character of social life." If I have translated *structuration* adequately, this is a sound view and one I have long held myself.

But the acute reader of Giddens will note that he provides no explanation how and why the actions of people create or maintain structures, nor how and why structures in turn maintain or change the actions of people by affecting the contingencies of their behavior. To do so, Giddens would have had to bring in behaviorism, under that name or another, and he does not even begin to do so. To that extent, his theory is a good description but an inadequate explanation.

The reasons for rejecting or ignoring behaviorism are often emotionally, but never intellectually, compelling.

RECENT DEVELOPMENTS

I have spoken of the failures and disappointments, except in the mathematical and technical fields, of the hopes entertained by sociologists in the 1930s. Especially disappointing have been the results of so-called theoretical work. They have not even produced conclusions on what a theory *is*. But at the end I want to mitigate the gloom and mention three recent developments I think are especially hopeful.

Social Networks

The first is the study of social networks, an application of mathematics to the analysis of small (and potentially to large) social structures. A social network is a pattern of positive or negative choices, on some criterion such as liking or disliking, made by individuals for or against other individuals in a small group or much larger assemblage of persons. The study of social networks grew out of such psychological researches as "balance theory" (Heider 1946) and the description of concrete networks in Homans (1950). The mathematical problem is that of devising equations from which the underlying and nonobvious characteristics of concrete networks can be generated. More generally, it is that of "generating quantified statements about the pattern of social relations in a community" (Barnes 1979:412). Network research has recently attracted an extremely able group of sociologists and mathematicians, such as Harrison White, James A. Davis, and their associates, Scott Boorman, Ronald Breiger, Mark Granovetter, Paul DiMaggio and many others. I am not enough of a mathematician myself to evaluate the work, but I can see that it can add an important dimension to our understanding of social structures. A good survey of recent research on social networks will be found in Holland & Lienhardt (1979).

But about one thing I believe I am clear: Social network theory cannot be a general sociological theory, for the existence of networks themselves still has to be explained. Social networks arise out of the choices individuals make for interacting with other individuals, from the development of cliques, and the development of hierarchies within cliques and among them. Social network research, with the exception of a part of "balance theory," cannot deal with these matters. They will have to be explained by a more general theory of social behavior, derived from an individualistic behavioral psychology. Indeed "balance theory" can itself be derived from such a psychology (see Homans 1974:59–65). But network analysis should help a behavioral psychology be much more clear about what social phenomena it has to explain.

Sociology and History

The second encouraging development is the increasing interest of sociologists in history. This too was, I believe, originally stimulated by work in the 1930s, especially by Marc Bloch (1931) and the *Annales* school in France, concentrating on the fields of economic and social history. But in those days American sociologists seemed to know no history except of the most contemporary American sort. Yet economic and social history obviously shares certain interests with sociology. No sharp line can be drawn between them, and history adds a dynamic dimension to sociology.

This condition has now changed, and American sociology in recent years has produced a number of interesting social and economic historians such as Barrington Moore, Jr. (1966), Charles Tilly (1975), Emmanuel Wallerstein (1974), Theda Skocpol (1979), and Arthur Stinchcombe (1978). The early members of the *Annales* school had usually limited themselves to close studies of particular countries or even districts. Perhaps under the influence of later *Annaliens* such as Fernand Braudel (1973), some of the Americans have taken on historical problems of much greater scope. Perhaps the most notable example is Wallerstein's attempt to delineate the development of a world economic system.

The problem faced by historians of this kind is that, with important exceptions such as Moore and Tilly, they must depend on secondary sources and on their own judgment of what sources are reliable. They lack the training in the use of primary sources which is one of the strengths of the professional historians. Their ambitions are so vast that they have not the time for this kind of training. The result is that professional historians often find gaps and weaknesses in particular steps in their arguments. My own suggestion is that sociologists should acquire training in the use of primary sources in at least some area of their interests. That experience would teach them to hold the other secondary sources in some abeyance of belief. With this reservation, I think many of the conceptions of the new sociological historians are undoubtedly important. This kind of work is promising; it enlarges our discipline and should be encouraged.

Charles Tilly is somewhat different from the others. The scope of the problems he tackles is a little smaller. He works systematically with certain kinds of primary sources and exploits them with modern statistical techniques. He can do this work only with the help of a team of well-integrated and well-trained colleagues. Indeed, I doubt if sound history on a grand scale can be carried out by individuals, in the manner of Spengler and Toynbee. But again the work is promising and should be encouraged.

Sociobiology

The third new, or revived, area of interest is that of genetic influences on behavior. Again, this interest goes back to the 1930s and to the controversy in psychology between the advocates of nature (genetics) or nurture (learning) as the main kind of influence on behavior. At that time the overwhelming weight of opinion and research came down on the side of nurture. Indeed, the question appeared to be settled, and nature was almost forgotten.

But later, stimulated by such pioneers in the study of the social behavior of animals as Lorenz and Tinbergen, the question was reopened, particularly in two books by Edward O. Wilson—*Sociobiology* (1975) and *On Human Nature* (1978). In one sense it had never been closed. Not just scholars but ordinary persons of sense had always known that animals, including humans, inherited their physical characteristics from their ancestors. And such persons of sense often shrewdly surmised that they had inherited some of their psychological characteristics too, and by physical genetics at that, not just because their parents had brought them up in the ways the parents themselves had learned.

The sociobiologists turned that shrewd surmise into a real scientific possibility. The new statistics then made it possible for social scientists to estimate *how much* of a given psychological trait (which of course would manifest itself in social behavior) could be attributed to genetic inheritance and *how much* to learning. The issue is highly controversial for three reasons. First, it raises the possibility of racial differences in such traits as intelligence—and I want to make it clear that for my part I am only concerned with individual differences. Second, the statistical techniques can be applied only under special and rather rare conditions, such as the existence of identical twins reared by different foster-parents. And third, genetic inheritance, which resists quick change, would inhibit the efforts of those high-minded persons who think human nature is a blank slate on which benign social arrangements can write through learning whatever behavior they wish. Will the Old Adam, for instance, abort the birth of the "new Communist man"?

Many sociologists are such high-minded persons, and others as usual are afraid that sociobiology will destroy the identity of sociology as a distinct science.

The most interesting question turns out not to be that of the place of genetics versus that of learning in human behavior, but rather of how the two interact. Thus, a person born genetically to be large and strong will learn to behave under contingencies different from those faced by one born to be small and puny, and therefore, the phenotypical behavior of the two is apt to be different for a combination of both kinds of reasons. Or again, some kinds of behavior that are certainly of genetic origin can be changed by learning much less easily than can some others.

But I shall say no more about sociobiology since both the research and the debate are young and still developing. The time is not ripe for firm conclusions. Yet the issues sociobiology raises for the understanding of human social behavior are momentous, and sociologists cannot afford to dismiss them, at least so long as they are more interested in the pursuit of truth than in the identity of their subject. Indeed if they ignore genetics as they have so long tried to ignore behavioral psychology, they will find their field contracting, not expanding.

Let me give an example of the possible shape of things to come. Sociologists have had for years a practical monopoly of the field of criminology; by and large they have adopted some form of the view that crime rates are determined by purely social conditions, and to some degree they are correct. But the best recent book on criminology, *Crime and Human Nature*, by James O. Wilson and Richard Herrnstein (1985), brings to the study of the subject both behaviorism and genetics. Yet neither of the authors are sociologists—one is a political scientist and the other a psychologist.

In the last 50 years, the great success of sociology has lain in its development of statistical and mathematical methods, which have added greatly to the soundness of our empirical knowledge in many fields. This success has been accomplished to some extent at the expense of field research. The weakness of sociology has continued to lie in theory. Not only have most sociologists failed to accept an adequate conception of what a theory *is,* but they have allowed themselves to fragment into a number of allegedly incompatible theoretical schools. These could be united by an individualistic behavioral psychology. Most sociologists have not been willing to write in this way, and their reluctance has allowed social scientists of other disciplines to make advances in areas that should be (and have been in the past) of great interest to sociology. The increasing interest of sociologists in historical problems is a favorable development, but their ambivalence towards sociobiology perpetuates their old and mistaken fears about the independent status of their discipline. Our field ought to be expanding, not contracting, and it should be open, as any great science must be, to all kinds of ideas and research that may have something to add to knowledge not only of our own field but of the behavior of all humankind.

Literature Cited

Alexander, J. 1982. *Theoretical Logic in Sociology,* Vol. 1. Berkeley: Univ. Calif. Press

Arensberg, C. M., Kimball, S. T. 1941. *Family and Community in Ireland.* Cambridge, Mass: Harvard Univ. Press

Bandura, A. 1969. *Principles of Behavior Modification.* New York: Holt, Rinehart & Winston

Barber, B. 1970. *L. J. Henderson on the Social System.* Chicago: Univ. Chicago Press

Barnes, J. A. 1979. Network analysis: Orienting notion, rigorous technique, or substantive field of study. See Holland & Lienhardt 1979, pp. 403–21.

Barry, B. 1970. *Sociologists, Economists, and Democracy.* London: Collier-Macmillan

Blau, P. M. 1955. *The Dynamics of Bureaucracy.* Chicago: Univ. Chicago Press

Bloch, M. 1931. *Les Caractères originaux de l'histoire rurale française.* Oslo: Inst. Sammenlignende Kulturforskning

Boudon, R. 1984. *La Place du désordre.* Paris: Presses Univ. France

Braithwaite, R. B. 1953. *Scientific Explanation.* Cambridge: Cambridge Univ. Press

Braudel, F. 1973. *The Mediterranean and the Mediterranean World in The Age of Philip II.* Vols. 1, 2. New York: Harper & Row. (Originally published as *La Méditerranée et le Monde Méditerranéen a l'Époque de Philippe II.* Paris: Armand Colin, 1949)

Bridgman, P. W. 1936. *The Nature of Physical Theory.* Princeton: Princeton Univ. Press

Brown, R., Herrnstein, R. J. 1975. *Psychology.* Boston: Little, Brown

Burgess, R. L., Bushell, D. Jr., eds. 1969.

Behavioral Sociology. New York: Columbia Univ. Press

Burgess, R. L., Nielsen, J. M. 1977. Distributive justice and the balance of power. In *Behavioral Theory in Sociology: Essays in Honor of George C. Homans*, ed. R. L. Hamblin, J. H. Kunkel, pp. 139–69. New Brunswick, NJ: Transaction

Coleman, J. S. 1975. Social structure and a theory of action. In *Approaches to the Study of Social Strucure*, ed. P. M. Blau, pp. 76–93. New York: Free Press

Coleman, J. S. et al 1966. *Equality of Educational Opportunity*, Vols. 1, 2. Washington, DC: Office Educ.

Edgeworth, F. Y. 1881. *Mathematical Psychics*. London: Kegan Paul

Ferster, C. B., Skinner, B. F. 1957. *Schedules of Reinforcement*. New York: Appleton-Century-Crofts

Giddens, A. 1979. *Central Problems in Social Theory*. Berkeley: Univ. Calif. Press

Hamblin, R. L. et al 1971. *The Humanization Processes*. New York: Wiley Intersci.

Hechter, M. ed. 1983. *The Microfoundations of Macrosociology*. Philadelphia: Temple Univ. Press

Heider, F. 1946. Attitudes and cognitive organization. *J. Psychol.* 21:107–12

Holland, P. W., Lienhardt, S., eds. 1979. *Perspectives on Social Network Research*. New York: Academic

Homans, G. C. 1950. *The Human Group*. New York: Harcourt Brace

Homans, G. C. 1961 (Rev. ed. 1974). *Social Behavior: Its Elementary Forms*. New York: Harcourt Brace Jovanovich

Homans, G. C. 1967. *The Nature of Social Science*. New York: Harcourt, Brace & World

Homans, G. C. 1984. *Coming to My Senses: The Autobiography of a Sociologist*. New Brunswick, NJ: Transaction

Homans, G. C., Curtis, C. P. Jr. 1934. *An Introduction to Pareto*. New York: Knopf

Kaplan, A. 1964. *The Conduct of Inquiry*. San Francisco: Chandler

Kunkel, J. H. 1975. *Behavior, Social Problems, and Change*. Englewood Cliffs, NJ: Prentice-Hall

Lynd, R. S., Lynd, H. M. 1929. *Middletown*. New York: Harcourt, Brace

Malewski, A. 1967. *Verhalten und Interaction*. Tübingen: Mohr. (Originally published 1964. *O zastosowaniach teorie zachowania*. Warsaw: Państwowe Wydawnictwo Naukowe)

Mead, G. H. 1934. *Mind, Self, & Society*. Chicago: Univ. Chicago Press

Moore, B. Jr. 1966. *Social Origins of Democracy and Dictatorship*. Boston: Beacon

Moreno, J. L. 1934. *Who Shall Survive?* Washington, DC: Nervous and Mental Disease Publ.

Olson, M. Jr. 1965. *The Logic of Collective Action*. Cambridge, Mass: Harvard Univ. Press

Opp, K-D. 1972. *Verhaltens-theoretische Soziologie*. Hamburg: Rohwolt

Opp, K-D. 1985. Sociology and Economic Man. *Zeitschrift für die gesamte Staatswissenschaft*. 141:213–43

Pareto, V. 1917. *Traité de sociologie générale*, Vols. 1, 2. Paris: Payot

Raub, W. 1984. *Rationale Akteure, institutionelle Regelungen und Interdependenzen*. Frankfurt am Main: Peter Lang

Roethlisberger, F. J., Dickson, W. J. 1939. *Management and the Worker*. Cambridge, Mass: Harvard Univ. Press.

Schelling, T. S. 1984. *Choice and Consequence*. Cambridge, Mass: Harvard Univ. Press

Schwanenberg, E. 1970. *Soziales Handel—Die Theorie und ihr Probleme*. Bern: Hans Huber

Scott, J. F. 1971. *The Internalization of Norms*. Englewood Cliffs, NJ: Prentice-Hall

Seashore, S. E. 1954. *Group Cohesiveness in the Industrial Work Group*. Ann Arbor, Mich: Univ. Mich. Inst. Soc. Res.

Skinner, B. F. 1938. *The Behavior of Organisms*. New York: Appleton-Century

Skocpol, T. 1979. *States and Social Revolutions*. Cambridge: Cambridge Univ. Press

Stinchcombe, A. L. 1978. *Theoretical Methods in Social History*. New York: Academic

Tilly, C., Tilly L., Tilly, R. 1975. *The Rebellious Century*. Cambridge, Mass: Harvard Univ. Press

Turkle, S. 1978. *Psychoanalytic Politics*. New York: Basic

Vanberg, V. 1975. *Die zwei Soziologien*. Tübingen: Mohr

Wallerstein, E. 1974. *The Modern World-System*, Vol. 1. New York: Academic Press

Warner, W. L. et al 1941–1963. *Yankee City Series*, Vols. 1–6. New Haven: Yale Univ. Press

Whyte, W. F. 1943. *Street Corner Society*. Chicago: Chicago Univ. Press

Wilson, E. O. 1975. *Sociobiology*. Cambridge, Mass: Harvard Univ. Press

Wilson, E. O. 1978. *On Human Nature*. Cambridge, Mass: Harvard Univ. Press

Wilson, J. Q., Herrnstein, R. J. 1985. *Crime and Human Nature*. New York: Simon & Schuster

Robert Merton

Ann. Rev. Sociol. 1987. 13:1–28

THREE FRAGMENTS FROM A SOCIOLOGIST'S NOTEBOOKS:
Establishing the Phenomenon, Specified Ignorance, and Strategic Research Materials

Robert K. Merton

University Professor Emeritus, Columbia University, New York, New York 10027, and Russell Sage Foundation, 112 East 64 Street, New York, New York 10021

Abstract

This occasionally biographical paper deals with three cognitive and social patterns in the practice of science (not '*the* scientific method'). The first, "establishing the phenomenon," involves the doctrine (universally accepted in the abstract) that phenomena should of course be shown to exist or to occur before one explains why they exist or how they come to be; sources of departure in practice from this seemingly self-evident principle are examined. One parochial case of such a departure is considered in detail. The second pattern is the particular form of ignorance described as "specified ignorance": the express recognition of what is not yet known but needs to be known in order to lay the foundation for still more knowledge. The substantial role of this practice in the sciences is identified and the case of successive specification of ignorance in the evolving sociological theory of deviant behavior by four thought-collectives is sketched out. Reference is made to the virtual institutionalization of specified ignorance in some sciences and the question is raised whether scientific disciplines differ in the extent of routinely specifying ignorance and how this affects the growth of knowledge. The two patterns of scientific practice are linked to a third: the use of "strategic research materials (SRMs)" i.e. strategic research sites, objects, or events that exhibit the phenomena to be explained or interpreted to such advantage and in such accessible form that they enable the fruitful investigation of previously stub-

2221

0360-0572/87/0815-0001$02.00

born problems and the discovery of new problems for further inquiry. The development of biology is taken as a self-exemplifying case since it provides innumerable SRMs for the sociological study of the selection and consequences of SRMs in science. The differing role of SRMs in the natural sciences and in the *Geisteswissenschaften* is identified and several cases of strategic research sites and events in sociology, explored.

INTRODUCTION

In his youthful journal, the exacting and agonistic literary scholar, C. S. Lewis (1975:76), makes benign reference to "the inexhaustible loquacity of educated age." Plainly alert to that capability, the Editors of *Annual Review of Sociology* wisely limit the space allotted prefatory chapters. In my own case, it was understood further that, unlike the prototypes that have long appeared in *Annual Reviews* of many other disciplines, this chapter would be neither a capsule intellectual autobiography nor an overview of the field. Instead, I tell only sporadically of biographical moments; for the rest, the asked-for personal aspect comes from my drawing upon fragments from notebooks assembled over the years and upon pieces published in obscure or improbable places. It was soon obvious that space would allow only for limited reflections on just 3 of the menu of 45 subjects I had itemized for the Editors—see Appendix, "The Menu"—the three being cognitive and social patterns in the practice of science that have long interested me. The patterns—"establishing the phenomenon," "specified ignorance" and "strategic research material"— have to do, not with scientific methods, let alone with "*the* scientific method," but with scientific practices (although there is, of course, much method in those practices).

ESTABLISHING THE PHENOMENON

In the abstract, it need hardly be said that before one proceeds to explain or to interpret a phenomenon, it is advisable to establish that the phenomenon actually exists, that it is enough of a regularity to require and to allow explanation. Yet, sometimes in science as often in everyday life, explanations are provided of matters that are not and never were. We need not reach back only to ancient days for such episodes as the younger Seneca explaining why some waters are so dense that no object, however heavy, will sink in them or explaining why lightning freezes wine. Our own century provides ample instances. There is René Blondlot's report of having discovered a "new species of invisible radiation," dubbed N rays. These were later "observed" by a dozen or so other investigators in France, but there were no comparable replications in England, Germany, or the United States. Intensive further inquiry, by French scientists as well as others, established the fact that the phenomenon was not one of N rays but rather of wishful perception and self-fulfilling prophecy. After that, N rays were no longer observed (Rostand 1960:12–29, Price 1961:85–90). Or again, there is Boris Deryagin's "discovery" of polywater in the 1960s, later found to be wholly artifactual (Franks 1981). Such episodes return us to Claude Bernard's observation that "if the facts used as a basis for reasoning are ill-established or erroneous, everything will crumble or be falsified; and it is thus that errors in scientific theories most often originate in errors of fact" (Bernard [1865] 1949:13).

No small part of sociological inquiry is given over to the establishing of social facts before proceeding to explain how they come to be. Often enough, the empirical data run contrary to widespread beliefs. Thus, it would seem premature to ask why "urbanization is accompanied by destruction of the social and moral order" inasmuch as evidence accumulates (Fischer 1977) to suggest that the connection is rather more an assumption than a repeatedly demonstrated fact.

In sociology as in other disciplines [see Leontief (1971) on economics], efforts to establish recurring social patterns are often described—sometimes of course with justice—as simply "fact-finding" or "fact-mongering" by those preferring swift explanation. Yet years ago, at the turn of the century, the exemplary scientist-philosopher C. S. Peirce was reminding us of the analytic function of fact-finding in what he described as the salient process of abduction (in turn related to processes of deduction and induction):

> Accepting the conclusion that an explanation is needed when facts contrary to what we should expect emerge, it follows that the explanation must be such a proposition as would lead to prediction of the observed facts, either as necessary consequences or at least as very probable under the circumstances. A hypothesis, then, has to be adopted, which is likely in itself, and renders the facts likely. This step of adopting a hypothesis *as being suggested by the facts,* is what I call *abduction* (Peirce 1958:VII 121–22; emphasis supplied).

The ex post facto phase of an empirical inquiry—or as some prefer; the post factum or post festem phase—has us introduce a hypothesis adopted, of course, only "on probation," while the ex ante phase draws out necessary and probable experiential consequences which can be put to falsifying or confirming test. Practiced investigators take it as a matter of course that, along with the free play of imagination drawing upon explicit and tacit knowledge, factual evidence often brings fruitful ideas to mind. To recognize this is not to engage in enumerative induction, pure and excessively simple. (I tried to elucidate these notions in "The bearing of empirical research upon the development of social theory," Merton 1948).

As I have noted, the basic role of empirical research designed to "establish the phenomenon" is at times downgraded as "mere empiricism." Yet we know that "pseudo-facts have a way of inducing pseudo-problems, which cannot be solved because matters are not as they purport to be" (Merton 1959:xv). Social scientists of diverse theoretical and value orientations have found it useful to address this matter of pseudo-facts; as examples, see Zeitlin (1974:1074–75) on the separation of ownership and control in large corporations, Gutman (1976:462–63) on the black family, and Sowell (1981:59) on ethnic education. To repeat myself: "only when tedious recitations of un-related fact [and fact-claims] are substituted for fact-related ideas does inquiry decline into 'mere fact-finding.' " Otherwise, of course, it is a crucial element in scientific inquiry.

As Neil Smelser reminded me upon reading this piece, establishing the phenomenon has its political dimension as well. In the cognitive domain as in others, there is competition among groups or collectivities "to capture what Heidegger ([1927]1962) called the 'public interpretation of reality.' With varying degrees of intent, groups in conflict want to make their interpretation the prevailing one of how things were and are and will be" (Merton 1973:110–11). In significant degree, that "interpretation of reality" involves establishing the phenomena that are an integral part of it.

The governing question in establishing the phenomenon—"Is it really so?"—holds as much for historical particularities as for sociological gener-alizations. Strongly held theoretical expectations or ideologically induced expectations can lead to perceptions of historical and social "facts" even when these are readily refutable by strong evidence close at hand. This is not so much wishful thinking as expectational thinking.

In the mode of collective biography called for by *Annual Review*, I turn for a parochial instance to the Department of Sociology at Columbia. In his widely read *The Coming Crisis of Western Sociology*, Alvin Gouldner (1970), himself a much-esteemed onetime student at Columbia (Merton 1982) observed that "C. Wright Mills never became a full professor" there. Having presented this as historical fact, he went on to draw its sociological and moral implications: Mills's " 'failure' may remind us that the serious players [in sociological criticism] are always those who have an ability to pay costs" (Gouldner 1970:15). Despite what I have reason to know was Alvin Gould-ner's commitment to scholarship, it appears that an overriding sense of the fitness of things and the expectation linked with it helped to create this pseudo-fact although evidence to the contrary was a matter of public record set down in easily accessible documents (such as the University bulletins with their rosters of faculty members). The evidentiary fact that Wright Mills, in a later academic cohort, had become a full professor at a younger age than the quintessential Establishment figure in sociology, Talcott Parsons, would scarcely have served to illustrate the premise or to reach the conclusion. In effect, Gouldner had tried to transform a historical event that never was into a social phenomenon that sometimes is.

The process of hagiographic creation of pseudo-facts did not stop there. Once set down in scholarly print as facts, pseudo-facts have a way of diffusing and becoming amplified (in the fashion long since established experimentally in the study of rumor). The same politically turbulent year of 1970 which saw the publication of Gouldner's book saw the translation into German of the book by Hans Gerth and C. Wright Mills, *Character and Social Structure* (for which I had happily written the foreword when it was first published in 1953). The German publishers went on to specify the pseudo-fact by declaring that Mills had "lost his professorship during the McCarthy period" ("verlor wahrend der McCarthy-Zeit seinen Lehrstuhl"). Not long afterward, an article in the Yugoslavian journal *Praxis,* also ignoring the biographical entry in the widely available *International Encyclopedia of the Social Sciences* which begins "C. Wright Mills (1916–62) was at his death professor of sociology at Columbia University" (Wallerstein 1968: Vol. 10, p. 362), explained the now elaborated nonfact to the contrary in these decisive terms: "C. Wright Mills was dismissed from Columbia University in USA because of his Marxist orientation" (Golubović 1973:363, noted by Oromaner 1974:7).

It is symbolically apt that a shared interest in the then nascent field of the sociology of knowledge had led Alvin Gouldner to adopt me as mentor when he arrived at Columbia in 1943—the story is told in Merton (1982)—just as a few years before, a similar interest had led Wright Mills, then still a graduate student at the University of Wisconsin, to have me vet his manuscripts in that field. So it was that, 30 years after our first meeting, when Alvin and I were reviewing this episode of the unwittingly fabricated 'fact,' he soon subordinated scholarly chagrin to shared intellectual pleasure in the episode as he agreed that it provided a sociological and methodological parable: Take care to establish a phenomenon (or a historical event) before proceeding to interpret or explain it.[1] As for the Wright Mills I knew, since the time in 1939 when he first sent me those manuscripts—he would probably have hooted at the ideological pieties that invited first the invention and then the successive explanations of these nonevents. Or perhaps Wright's ironic self might have argued for the symbolic if not the historical truth of that evolving myth of his never having become a full professor at Columbia, with all its seeming

---

[1]After our talk, Alvin Gouldner took quick action to erase that pseudo-fact, noting that "shortly after the publication of the *Crisis* I discovered (from Robert Merton) that my assertion that C. Wright Mills had never been made a Professor at Columbia was in error. Having discovered this, I immediately had this statement removed from the Avon paperback edition of the *Crisis,* which was then in production" (Gouldner 1973:130–131). The import of the episode apparently stayed with Alvin, for years later, and in quite a different context, he took care to observe: "Whether anything *might be* or even *should* have been is one thing; whether it was in fact, is quite another" (Gouldner 1980:281).

implications, and of his ideologically motivated dismissal from that professorship he had never held.[2]

The manifest advisability of establishing the phenomenon before undertaking to explain it has long been recognized in principle if not always observed in practice. They understood this abundantly well, for example, in seventeenth-century England (as I found during my years-long stay in that time and place). Consider only this reminder as set forth by the jurist and orientalist, John Selden, in his widely-read *Table Talk* ([1689]1890:139): "The Reason of a Thing is not to be enquired after, till you are sure the Thing itself be so. We commonly are at *What's the Reason of it?* before we are sure of the Thing." So, too, Bernard Fontenelle, the polymath destined to become a centenarian and the almost but not quite literally *secrétaire perpétuel* of the French Academy of Sciences—he served for only 42 years—was observing in his *Histoire des oracles* ([1686]1908:33): "I am convinced that our ignorance consists not so much in failing to explain what is as in explaining what is not. In other words, we not only lack principles that lead to the true but hold others that readily lead to the false." What Fontenelle did not take occasion to observe, however, is that a certain kind of ignorance advances scientific knowledge.

SPECIFIED IGNORANCE

It was Francis Bacon who made "the advancement of learning" a watchword in the culture of science emerging in the seventeenth century. From then till now, efforts to understand how science develops have largely centered on the modes of replacing ignorance by knowledge, with little attention to the formation of a useful kind of ignorance, as distinct from the manifestly

---

[2]That some (unknown number of) academic careers have been curbed or halted by political or ideological commitments is, of course, a matter of historical record (Lazarsfeld & Thielens 1958, Schrecker 1986). But it was Bernhard J. Stern, not Mills, who, an announced Marxist and cofounder of the Marxist journal *Science and Society*, never advanced beyond a lectureship in the Columbia Department of Sociology, despite departmental recommendations for promotion. Just as again, it was Bernhard J. Stern, not C. Wright Mills, who was attacked by Joe McCarthy as an alleged Communist only to have the University respond by continuing Stern in his marginal post as Lecturer. During the McCarthy period, actual events often transcended social categories such as Establishment and self-declared anti-Establishment figures. Thus, when McCarthy's associate, the then vice-presidential candidate, Richard Nixon, was charged by some Columbia professors with having "violated an elementary rule of public morality" by the way he had accumulated campaign funds, he responded by ransacking the files of the House Committee on UnAmerican Activities and then catapulting nine of those professors onto banner headlines in the New York Daily News and the Chicago Tribune as alleged subversives; the infamous nine included the literary critic Mark van Doren, the philosopher Irwin Edman, the historian Henry Steele Commager, and the sociologists, Robert M. MacIver, Paul F. Lazarsfeld, and Robert K. Merton—but not, as he himself ironically noted, C. Wright Mills.

dysfunctional kind. Karl Popper provides the monumental contemporary exception that illuminates the rule, most powerfully in his analytical essay "On the sources of knowledge and of ignorance" ([1960]1962). The general inattention to the formation of useful ignorance has long obtained as well in the sociology of scientific knowledge [but now see Smithson (1985) along with the early collateral paper on the functions of ignorance in social life by Moore & Tumin (1949)].

These retrospective notes focus on the dynamic cognitive role played by the particular form of ignorance I describe as "specified ignorance": "the express recognition of what is not yet known but needs to be known in order to lay the foundation for still more knowledge" (Merton 1971:191). "As the history of thought, both great and small, attests, *specified* ignorance is often a first step toward supplanting that ignorance with knowledge" (Merton 1957:417).

The concept of *specified* ignorance hints at various other kinds and shades of acknowledged ignorance in science. The familiar kind of a general, rote, and vague admission of ultimate ignorance serves little direct cognitive purpose though it may have symbolic significance in reminding us of our limitations. This kind, however, does not issue in definite questions. And vague questions evoke dusty answers. After all, it takes no great courage, or skill, in the domain of science to acknowledge a general want of knowledge. It is not merely that Socrates set an ancient pattern of announcing one's ignorance. Beyond that, the values of modern science have long put a premium on the public admission of one's limitations or the expression of humility in the face of the vast unknown. Scientists of epic stature have variously insisted on how little they have come to know and to understand in the course of their lives. We remember Galileo teaching himself and his pupils to reiterate: "I do not know." And then, inevitably, one recalls the "memorable sentiment" reportedly uttered by Newton "a short time before his death":

> I do not know what I may appear to the world, but to myself I seem to have been only like a boy playing on the seashore, and diverting myself in now and then finding a smoother pebble or prettier shell than ordinary, whilst the great ocean of truth lay all undiscovered before me (Brewster 1855: II, 407).

Or again, Laplace—the French Newton—is said to have put much the same sentiment in a typically Gallic epigram: "What we know is not much; what we do not know is immense" (Bell 1937:172). What the mathematician Bell (1931:204) describes elsewhere as "a common and engaging trait of the truly eminent scientist [found] in his frequent confession of how little he knows" can be identified sociologically as the living up to a normative expectation of ultimate humility in a community of sometimes egocentric scientists. It is not simply that a goodly number of scientists happen to express these self-belittling sentiments; they are applauded for doing so.

But of course these paradigmatic figures in science do not confine themselves to such generic confessions of ignorance as may reinforce the norm of a decent humility without directly shaping the growth of scientific knowledge. They repeatedly adopt the cognitively consequential practice of specifying this or that piece of ignorance derived from having acquired the added degree of knowledge that made it possible to identify definite portions of the still unknown. In workaday science, it is not enough to confess one's ignorance; the point is to specify it. That, of course, amounts to instituting, or finding, a new, worthy, and soluble scientific problem.

Thus, as I have had occasion to propose, the process of successive specification of our ignorance in light of newfound knowledge provides a recurrent sociocognitive pattern:

> As particular theoretical orientations come to be at the focus of a sufficient number of workers in the field to constitute a thought collective, interactively engaged in developing a distinctive thought style (Fleck [1935] 1979), they give rise to a variety of key questions requiring investigation. As the theoretical orientation is put to increasing use, further implications become identifiable. In anything but a paradoxical sense, newly acquired knowledge produces newly acquired ignorance. For the growth of knowledge and understanding within a field of inquiry brings with it the growth of *specifiable and specified ignorance:* a new awareness of what is not yet known or understood and a rationale for its being worth the knowing. To the extent that current theoretical frameworks prove unequal to the task of dealing with some of the newly emerging key questions, there develops a composite social-and-cognitive pressure within the discipline for new or revised frameworks. But typically, the new does not wholly crowd out the old, as [long as] earlier theoretical perspectives remain capable of dealing with problems distinctive to them (Merton 1981:v–vi).

It requires a newly informed theoretical eye to detect long obscured pockets of ignorance as a prelude to newly focussed inquiry. Each theoretical orientation or paradigm has its own problematics, its own sets of specified questions. As these questions about selected aspects of complex phenomena are provisionally answered, the new knowledge leads some scientists both within and without the given thought collective to become aware of other, newly identified aspects of the phenomena. There then develops a succession of specified ignorance.

As a case in point, consider the sociological theory of deviant behavior as it was developed in four thought collectives. (I draw upon the summary in Merton 1976.) Initiated in the 1920s, E. H. Sutherland's ([1925–1951] 1956) theory of differential association centered on the problem of the *social transmission* of deviant behavior. Its key question therefore inquired into the modes of socialization through which patterns of deviant behavior are learned from others. But as the brilliant philosopher of literature, Kenneth Burke, has reminded us: "A way of seeing is also a way of not seeing—a focus upon object A involves a neglect of object B" (Burke 1935:70). In this case, Sutherland's focus on the acquisition of these deviant patterns left largely untouched specifiable ignorance about the ways in which the patterns emerged in the first place.

Upon identifying that pocket of theoretical neglect, Merton (1938a) proposed the theory of anomie-and-opportunity-structures, that rates of various types of deviant behavior tend to be high among people so located in the social structure as to have little access to socially legitimate pathways for achieving culturally induced personal goals. The Sutherland and Merton theories were consolidated and extended by Cohen (1955) who proposed that delinquency subcultures arise as adaptations to this disjunction between culturally induced goals and the legitimate opportunity-structure and by Cloward & Ohlin (1960) who proposed that the social structure also provides differential access to *illegitimate* opportunities. Since that composite of theories centered on socially structured *sources* of deviant behavior, it had next to nothing to say about how these patterns of misbehavior are transmitted or about how these initial departures from the social rules sometimes crystallize into deviant careers, yet another sphere of specifiable ignorance.

That part of the evolving problematics was taken up in labeling (or societal reaction) theory as initiated by Lemert (1951) and Becker ([1963] 1973) and advanced by Erikson (1964), Cicourel (1968), and Kitsuse (1964). It centered on the processes through which some people are assigned a social identity by being labeled as "delinquents," "criminals," "psychotics," and the like and how, by responding to such stigmatization, they enter upon careers as deviants. In Becker's words: "Treating a person as though he were generally rather than specifically deviant produces a self-fulfilling prophecy. It sets in motion several mechanisms which conspire to shape the person in the image people have of him" (Becker [1963] 1973:34). With this problem as its focus, labeling theory has little to say about the sources of primary deviance or the making of societal rules defining deviance. As Lemert (1973:462) specified this ignorance: "When attention is turned to the rise and fall of moral ideas and the transformation of definitions of deviance, labeling theory and ethnomethodology do little to enlighten the process."

It is precisely this problem that the conflict theory of deviance took as central. Its main thrust, as variously set forth by Turk (1969) and Quinney (1970), for example, holds that a more or less homogeneous power elite incorporates its interests in the making and imposing of legal rules. It thus addresses questions neglected by the earlier theories: How do legal rules get formulated, how does this process affect their substance, and how are they differentially administered?

The case of deviance theory indicates how a dimly felt sense of sociological ignorance was successively specified for one class of social phenomena. But it is not yet known whether scientific disciplines differ in the practice of specifying ignorance—in the extent to which their practitioners state what it is about an established phenomenon that is not yet known and *why it matters* for generic knowledge that it become known.[3] Such specified ignorance is at a far remove from that familiar rote sentence which concludes not a few scientific papers to the effect that "more research is needed." Serendipity aside, questions not asked are questions seldom answered. The specification of ignorance amounts to problem-finding as a prelude to problem-solving.

It is being proposed that the socially defined role of the scientist calls for both the augmenting of knowledge and the specifying of ignorance. Just as yesterday's uncommon knowledge becomes today's common knowledge, so yesterday's unrecognized ignorance becomes today's specified ignorance (Merton 1957:417, Popper [1960] 1962, Sztompka 1986:97–98). As new contributions to knowledge bring about a new awareness of something else not yet known, the sum of manifest human ignorance increases along with the sum of manifest human knowledge.

STRATEGIC RESEARCH MATERIALS (SRMs)

Establishing the phenomenon and specifying ignorance link up with a third pattern of scientific practice that has long been of interest to me. This is the ongoing search, variously evident in the various sciences, for "strategic research material" (a cumbrous nine-syllable phrase better shortened to SRM). By SRM is meant the empirical material that exhibits the phenomena to be explained or interpreted to such advantage and in such accessible form that it enables the fruitful investigation of previously stubborn problems and

---

[3]Mathematics, of course, has a long tradition of publishing fundamental problems (long ago, in the form of challenges). Upon reading this portion of the chapter, my colleagues, Joshua Lederberg and Eugene Garfield, informed me of their episodic interest in institutionalizing what amounts to the specification of ignorance. For one expression of that interest in print, see Garfield's (1974) "The Unanswered Questions of Science." Lederberg has made me the beneficiary of his 1974 permuterm bibliography entitled "Unsolved Problems" in the various sciences and has referred me to a specimen volume entitled *100 Problems in Environmental Health* (McKee et al 1961). My attention was also redirected to that superb and lively anthology I had misplaced, *The Scientist Speculates: An Anthology of Partly Baked Ideas* (Good et al 1962), which is designed "to raise more questions than it answers." Of particular interest is the piece in the anthology happily entitled "Ignoratica" by one Félix Serratosa who ascribes the essential idea of a "science of unknowns" to the explosive imagination of that prolific and often paradoxical Florentine critic, novelist, poet, and journalist Giovanni Papini. However that may all be, it can be said in self-exemplifying style: That the specification of ignorance is indispensable to the advancement of knowledge, I do not doubt; whether disciplines do differ notably in the practice of such specification, I do not know. Since the phenomenon is not yet established, I do not undertake to explain such possible variation. But one can still speculate . . .

the discovery of new problems for further inquiry (Merton [1963a]1973:371–82). SRMs take differing forms in the various disciplines: among them, the (location) strategic research site (SRS) and the (temporal) strategic research event (SRE). Differing in operative detail, these forms have much the same functions. Just as the invention of new technologies for scientific investigation can facilitate the advance of scientific knowledge, so with the finding or creating of SRMs.

The concept of SRM provides a guide to the understanding of certain turning points in the sciences. Problems that have long remained intransigent become amenable as investigators identify new kinds of empirical materials that effectively exhibit the structure and workings of the phenomena to be understood. An inventory of SRMs in the history of the various sciences would, of course, run to unconscionable, not to say unmanageable, length, but even this capsule account has room for a conspicuous few, drawn from various times, places, and disciplines.

At times, scientists create an SRM or select one by design; at other times, they come upon such material serendipitously, recognizing its strategic character for the study of a particular problem only afterward. The seventeenth-century father of embryology, Marcello Malpighi, provides an SRS of the first kind: He elected to examine the lungs of frogs microscopically because of their great "simplicity and transparency" and thus observed for the first time so fine a feature as the capillary, not otherwise observable through microscopes of the time. It was this SRS—a "microscope of Nature" as the metaphor has it—that enabled Malpighi to see the blood move through the capillaries and thus helped him to round out Harvey's understanding of the greater circulation of the blood (Wilson 1960:165; Adelmann 1966).

A truly classic case of the second, serendipitous, kind of SRS was inadvertently provided by the Canadian trapper, Alexis St. Martin, when he suffered a gunshot wound that opened a large and permanent fistula into his stomach. This enabled his physician-and-friend, the early nineteenth-century physiologist, William Beaumont, to "look directly into the cavity of the Stomach, and almost see the process of digestion," as he put it in his notebook upon going on to his long series of pioneering experiments. The successful use of this serendipitous SRS in turn led the French chemist, Nicolas Blondlot (father of the hapless René), to create SRSs systematically by introducing similar fistulas in animals. But it was Beaumont's ingenious use of the singular fortituous SRS that deeply impressed the incomparable physician-humanist, William Osler, who held that it had led this "backwood physiologist" to the most consequential contributions to the physiology of digestion made in the nineteenth century. So impressed was Osler that, upon St. Martin's death—57 years after his scientifically fruitful accident (and his subsequent fathering of 20 children)—he wanted to conduct a postmortem examination and to deposit that strategic stomach, hole and all, in an appropriate museum. To round out the episode, I should report that intent was not translated into event: Osler refrained, upon receiving a warning telegram from St. Martin's French-Canadian community that read "Don't come for autopsy; will be killed" (Osler 1908:159–88, Cushing 1925: I, 177–79).

The early geneticists and especially the more recent molecular biologists hit upon a multitudinous variety of materials that strategically exhibit processes of reproduction and replication and lend themselves to the requisite research. In touching upon these, I surely indict myself as one of those benighted characters who insist on carrying coals to Newcastle, faggots into the wood, owls to Athens, and the concept of SRM to biology. I can only plead that biology is a self-exemplifying case: the history of biology itself provides strategic research materials for the study of the selection and consequences of strategic research materials.

Some time ago, there were, of course, Mendel's pea plants and then, de Vries' "pure species" of evening primrose with the ensuing complex story of his discovery of "mutation" (Mayr 1982:742–44). Harriet Zuckerman's unpublished inventory [1964] of research materials utilized in Nobel prize-winning work is fairly saturated with SRMs that gave rise to new lines of genetic inquiry and discovery. Among the many, I note only Morgan's choice of the fruit fly, so " 'easily and cheaply bred in the laboratory' " (Morgan in Allen 1975:331); Beadle & Tatum's "daring and astute selection of experimental material," the red bread mold *Neurospora crassa*, enabling them to advance biochemical genetics; Tatum & Lederberg's choice of *E. coli K-12* leading to the discovery of genetic recombination in bacteria and laying a "foundation of bacterial genetics and what has flowed from it" (Zuckerman & Lederberg 1986; Lederberg 1951, 1986); and to go no further, "that material of great convenience for studying many aspects of virus behavior," the filtrable virus bacteriophage (fondly shortened to phage) which, after their first collaboration in 1940, Delbrück and Luria converted into the SRM of that thought collective known as "the Phage group" that has contributed so much to the rise of molecular biology (Cairns et al 1966).

The recurrent pattern is one of identifying lineaments of materials that make them strategic for investigating a range of otherwise inaccessible scientific problems. Outside the sphere of genetics, Szent-Györgi's Hungarian paprika provided a rich source of ascorbic acid, enabling him to discover the role of Vitamin C in biological combustion, just as the newly available germanium and silicon crystals enabled Shockley, Bardeen, and Brattain to discover the transistor effect. Understandably, research workers become devoted to—not to say, captivated by—their fruitful SRMs. Hodgkin pays tribute to the nerve fiber of his giant squids as "an absolute gold mine," opening up all sorts of possibilities for study of physiological mechanisms in the transmission of messages.

My colleague, the neurobiologist, Eric Kandel, has also been known to wax eloquent about his prime SRM, the sea snail *Aplysia californica,* with its large and accessible nerve cells allowing investigation in molecular terms of such complex processes as learning and memory. Looking back on an earlier "encounter between neurobiology and molecular biology," he observes:

> These intellectual precursors shared an experimental approach that depended on model building and therefore on a willingness to study preparations that best exemplified the phenomena of interest. This led to a search for conveniently simple systems that provided abundant material. Thus, geneticists interested in inheritance in higher organisms first studied *Drosophila* and *Escherichia coli;* crystallographers first analyzed keratin and hemoglobin; and molecular biologists interested in replication of DNA studied bacterial viruses. Although the impetus was to understand complex phenomena, study was governed by optimization of simple experimental systems and by the presumed universality of the phenomena chosen for study (Kandel 1983:891).

Quite evidently, then, the biological sciences have long involved the search for SRMs and their sustained intensive investigation. That experimental tradition is at a considerable remove from the largely nonexperimental work in the social and behavioral sciences. Nevertheless, in those disciplines also we observe a hunt for empirical materials, research sites, and events that are judged strategic for investigating a generic scientific problem and for identifying new problems. Still, there is at times a profound difference in the orientations of biological scientists and social scientists toward the phenomena they establish and investigate. To a degree, that difference relates to the well-known distinction proposed by the philosopher Wilhelm Windelband (1884) and substantially developed by his student, Heinrich Rickert ([1902] 1921). That is the distinction between the *Naturwissenschaften* (readily translated as "the natural sciences") and the *Geisteswissenschaften* (not as readily and variously translated as "the human sciences," "the social sciences," or perhaps as "the sociocultural sciences"). Associated as he was with Rickert in several respects, Max Weber (1922) nevertheless transcended the Windelband-Rickert distinction between the natural sciences as adopting methods exclusively designed for nomothetic or generalizing objectives and the social sciences as exclusively adopting quite different methods for understanding the idiographic or individual character of a sociocultural reality.

For in place of this drastic, all-or-none choice between the two methodological orientations, one may choose the composite of intrinsic interest in understanding the particular "historical individuals"—for example, the capitalistic society of nineteenth-century England or the French Revolution or, for that matter, the Great Depression of the 1930s—*and* of an instrumental interest in those sociocultural phenomena as instructive specimens leading to discovery of general regularities which can then be drawn upon to understand other historical individuals. Thus, Sorokin (1925) examines a variety of revolutions over the centuries—from ancient Rome to our own time—to arrive at his nomothetic or generalizing work, *The Sociology of Revolution*, and to reach an understanding of the Russian Revolution he experienced at first hand. Or again, Thomas & Znaniecki (1918–1920) examine the historical case of *The Polish Peasant in Europe and America* both for its distinctive ("unique") characteristics and for its presumably generic patterns of social and personality change.

In short, it is being proposed that the history of sociological inquiry has its own complement of researches which relate variously to the use of strategic research sites and events: In one type, the empirical case is selected wholly because of intrinsic interest in it as a historical individual on grounds of its relevance to values *(Wertbeziehung)*, which Rickert held to be distinctive of the *Geisteswissenschaften*. In another type, the empirical case is regarded wholly as an SRS or SRE leading to provisional generalizations. And in what I take to be the most felicitous mode, the concrete materials hold both intrinsic interest as involving human values and instrumental interest as an SRS or SRE that may advance our general sociological knowledge.

Karl Marx provides us with a prime early instance of this last type. In the preface to the first German edition of *Capital*, he begins with a not uninteresting allusion to the logic of inquiry adopted by physicists and then goes on to the rationale for adopting a particular site for his own inquiry:

> The physicist either observes physical phenomena where they occur in their most typical form and most free from disturbing influence or, wherever possible, he makes experiments under conditions that assure the occurrence of the phenomenon in its normality. In this work I have to examine the capitalist mode of production, and the conditions of production and exchange corresponding to that mode. Up to the present time, their classic ground is England. That is the reason why England is used as the chief illustration in the development of my theoretical ideas (Marx [1867] 1906:12–13).

Marx goes on to elucidate the choice of England as an SRS(ite) by maintaining that the country which is "more developed industrially only shows, to the less developed, the image of its own future." And then, almost in the manner of an early biologist assaying a potential SRS, Marx assesses the research value of his elected case by noting that "The social statistics of Germany and the rest of Continental Western Europe are, in comparison of those of England, wretchedly compiled." Although this SRS scarcely has the same quality of exhibiting closely reproducible regularities on demand as SRMs in the physics to which Marx refers, it does not seem too much to suggest that Marx's choice of his SRS has had its own array of notable consequences, cognitive as well as social.

The sociological literature is chockfull of work that combines intrinsic interest in the particular sociocultural case with instrumental interest in it as leading to provisional general conclusions. Here, it is enough to instance Max Weber's monumental volumes (1910–1921) in the sociology of religion, with their intensive sociological analyses of Protestantism, Confucianism and Taoism, Hinduism and Buddhism, and ancient Judaism. The idiographic analyses of these historical materials which hold great intrinsic interest for many of us are powerfully joined with their instrumental use as SRSs leading to nomothetic hypotheses about such abstract sociological problems as the relations between institutionalized ideas and social organization as well as the modes and dynamics of structural interdependence of seemingly unconnected social institutions—all this best exemplified by the interplay between religious ideas and economic developments, not least in the prototypal case of ascetic Protestantism and the emergence of modern capitalism.

Other founders of modern sociology worked with a variety of strategic research sites and events. Durkheim, of course, notably so in his analyses of the division of labor, suicide, religious ceremony and ritual, and moral education, among others. As Hanan Selvin pointed out to me in correspondence (1976) on the evolving concept of SRS, Durkheim's first empirical study of suicide in 1888, antedating his famous monograph by a decade, rested on the strategic selection of "European nations as the units of recording and analysis. The availability of suicide rates as [assumed] indicators of national unhappiness was surely what led him to make this choice." Elsewhere, Selvin (1976) notes how Durkheim adopted a more fine-grained SRS to analyze—as it happens, erroneously—relationships between the proportions of German-speaking people and the suicide rate in 15 provinces of Austro-Hungary, this in an effort to identify the effects of German culture on the suicide rate while presumably neutralizing the effects of possible genetic dispositions to suicide. It is in this context, after the manner of one subjecting natural history to systematic analysis, that Durkheim (1888) emitted the metaphor: "Austria offers us the complete natural laboratory"—a kind of metaphor often echoed by Park, Burgess and others of that remarkable group of sociologists that made the city of Chicago a sociological "laboratory."

Familiar empirical materials were put to unfamiliar theoretical use. Thus, Durkheim (1899–1900) and, in kindred fashion, George H. Mead (1918) elected to tackle the problem of the social bases of moral indignation, integral to an understanding of mechanisms of social control, by turning to situations in which people react strongly to violations of social norms *even though they are not directly injured by them.* Systems of punishment and behavioral responses to violations of deep-seated rules provided an SRS not so much for the then-and-since traditional problem of the deterrent effects of punishment in curbing crime as for the problem of their other societal functions; in Mead's language, the "uniting all members of the community in the emotional solidarity of aggression." In an Excursus of the kind to which he was much given, to the lasting benefit of the rest of us, Simmel ([1908] 1950:402–408) focused on "the phenomenon of the stranger" in order to analyze how "the unity of nearness and remoteness in every human relation is organized" just as, in direct theoretical continuity, Park (1928) focused on the behavior of immigrants as providing strategic materials for coming to understand the structural bases of "the marginal man"—the men and women who, living in disparate social worlds, do not feel at home or fully accepted in any of them.

Following upon these and many another early prototype, the exponentially growing numbers of sociologists have adopted a numerous variety of strategic research sites and events. But of all these, nothing more can be said here. Instead, I obey the injunction of the Editors of *Annual Review* to make this prefatory piece as personal as I can bring myself to do and close out these capsule notes on the concept of SRM in two steps. First, I want to examine a turning point in the history of psychoanalysis that I have often singled out as a classic instance of an acute theoretical sensibility—to wit, Freud himself—transmuting seemingly trivial phenomena into strategic research material (however different present-day appraisals of that material may be). This invites attention to the apparently paradoxical theme of the occasional, perhaps frequent, importance in science and scholarship of what appear to be humanly insignificant phenomena. From that historic episode I move to three distinctly minor efforts on my own part to set forth an explicit rationale for adopting various kinds of sociological SRMs.

The "Trivial" as Strategic Research Material

It was back in the 1940s that I first found myself focussing on Freud's analytic decision to study seemingly trivial mistakes in everyday life as "strategic" in the sense being developed here:

> . . . in noting that the unexpected fact must be 'strategic,' *i.e.* that it must permit implications which bear upon generalized theory, we are, of course, referring rather to what the observer brings to the datum than to the datum itself. For it obviously requires a theoretically sensitive observer to detect the universal in the particular. After all, men had for centuries noticed such 'trivial' occurrences as slips of the tongue, slips of the pen, typographical errors, and lapses of memory, but it required the theoretic sensitivity of a Freud to see these as strategic data through which he could extend his theory of repression and symptomatic acts (Merton 1948:507).

Freud had signalled his intention of transmuting these seemingly trivial matters into basic theoretical matters by the emphasis given them in the title of the book where he first dealt systematically with them: *The Psychopathology of Everyday Life: Forgetting, Slips of the Tongue, Bungled Actions, Superstition and Errors* (Freud [1901] 1960). He proceeded to group these varied mishaps in the coined word-and-concept, *Fehlleistungen* (a psychological oxymoron translated in *The Standard Edition . . . of Freud* by the made-up Greek-like word, "parapraxes"[4] but as Bettleheim has tellingly noted, best rendered as "faulty achievements." Returning intensively to these same matters 15 years later in his *Introductory Lectures on Psycho-Analysis*, Freud forcefully states the case for his focussing on these "apparent trivialities":

> It is to these phenomena, then, that I now propose to draw your attention. But you will protest with some annoyance: 'There are so many vast problems in the [wide] universe, as well as within the narrower confines of our minds, . . . that it does really seem gratuitous to waste labour and interest on such trivialities. . . .'
>
> I should reply: Patience, Ladies and Gentlemen! I think your criticism has gone astray. It is true that psycho-analysis cannot boast that it has never concerned itself with trivialities. On the contrary, the material for its observations is usually provided by the inconsiderable events which have been put aside by the other sciences as being too unimportant—the dregs, one might say, of the world of phenomena. But are you not making a confusion in your criticism between the vastness of the problems and the consciousness of what points to them? Are there not very important things which can only reveal themselves, under certain conditions and at certain times, by quite feeble indications? (Freud [1916] 1961:26–27).

Freud is telling his audience that the seeming insignificance of these "phenomena" for everyday life says nothing about their significance for psychological science. That observation on the strategic theoretical value of such slips and errors holds quite apart from their evidentiary value for Freud's own theory that they result from repression [as is clear from the thoroughgoing analysis of Freud's "flawed reasoning" and from the review of alternative explanations of these phenomena by the philosopher of science, A. Grünbaum (1984:190–21)].

I cannot dwell on the enduring theme of the potential importance of the seemingly trivial in science and scholarship as it has appeared over the centuries—the seventeenth, for example, was chockfull of this theme, both as understood and as misunderstood. A few archetypal observations to this effect in our own century must serve. Having been taxed from time to time for

---

[4]As others have held and as Bettelheim has emphatically observed, this awkward term is a misleading translation of a central concept. Unable to improve upon Bettelheim's analysis, I do service by transmitting it here: "Freud coined *Fehlleistung* to signify a phenomenon that he had recognized—one that is common to the various ways in which our unconscious manages to prevail over our conscious intentions in everyday occurrences. The term combines two common, strangely opposite nouns, with which everybody has immediate and significant association. *Leistung* has the basic meaning of accomplishment, achievement, performance, which is qualified by the *Fehl* to indicate an achievement that somehow failed—was off the mark, in error. What happens in *Fehlleistung* is simultaneously—albeit on different levels of consciousness—a real achievement and a howling mistake. Normally, when we think of a mistake we feel that something has gone wrong, and when we refer to an accomplishment we approve of it. In *Fehlleistung*, the two responses become somehow merged: we both approve and disapprove, admire and disdain". (Bettelheim 1983:86–87).

attending to the apparently insignificant, Veblen (1932:42) took one occasion to observe that "All this may seem to be taking pains about trivialities. But the data with which any scientific inquiry has to do are trivialities in some other bearing than that one in which they are of account." And inevitably, in these reminiscent pages, I am put in mind of how this matter was being reiterated by teachers at Harvard during my time as a student and instructor there. Here is the biochemist and self-taught social scientist, L. J. Henderson, typically diluting his cogent observations by his passionate Paretan insistence that social scientists really must learn to quell their passions:

> This illustration has been chosen because, among other reasons, it is a simple case that is likely to seem trivial. Note well, however, that nothing is trivial, but thinking (or feeling) makes it so, and that we must ever guard against coloring facts with our prejudices. There was a time not so very long ago when electro-magnetic interactions, mosquitoes, and microorganisms seemed trivial. It is when we study the social sciences that the risk of mixing our prejudices and passions with the facts, and thus spoiling our analysis, is most likely to prevail (Henderson [1941] 1970:19; see also Bernard Barber's comment introducing this passage in Henderson's oral publication which Barber arranged to have put into print).

Whether Henderson alerted Talcott Parsons to this theme of the possible scientific importance of otherwise trivial phenomena, I cannot say. He may have done so during his close editing of Parsons' masterwork, *The Structure of Social Action* (1937), for, as the Preface gratefully states and as we young colleagues of them both knew, Henderson had "subjected the manuscript to important revision at many points, particularly in relation to general scientific methodology. . . ." In any case, Parsons picks up and develops the theme in the important section entitled "Theory and Empirical Fact," which virtually opens his immensely consequential treatise:

> A scientifically unimportant discovery is one which, however true and however interesting for other reasons, has no consequences for a system of theory with which scientists in that field are concerned. Conversely, even the most trivial observation from any other point of view—a very small deviation of the observed from the calculated position of a star, for instance—may be not only important but of revolutionary importance, if its logical consequences for the structure of theory are far-reaching (Parsons 1937:7–8).

In summary, then, it has long been recognized in a variety of disciplines that there is no necessary relation between the socially ascribed importance of the empirical materials under study and their importance for the better understanding of how nature or society works. The scientific and the human significance of those materials can be, although most emphatically they need not be, poles apart. This is often lost to view when a charge of triviality rests wholly on a commonsense appraisal of the subject-matter alone, as it often is, for example, by satirizing members of Congress. In gauging the human significance of the sociological *problem* rather than the empirical materials, many of us have argued, we sociologists have found no better general criterion than that advanced by Max Weber in the concept of *Wertbeziehung* (value relevance). Their values may lead scientists to refuse to work on certain scientific problems—for example, research that will lead to still more catastrophic weapon systems—or may lead them to focus on other scientific problems—for example, research on cancer or on the social mechanisms that perpetuate racial discrimination. There still remains the question of identifying the research materials that enable one to investigate these humanly important problems most effectively, the question of hitting upon strategic research sites or events.

In saying all this, I prefer not to be misunderstood. It is surely not the case that SRMs for investigating a particular scientific problem must be humanly trivial. Nor is it being said in the mood defensive that there is no authentically trivial work in today's sociology any more than it can be said that there was no trivial work in, for instance, the physical science of the seventeenth century. Our journals of sociology may have as impressive a complement of authentic trivia as the *Transactions* of the Royal Society had during its first century or so. But these are trivia in the strict rather than the unthinking rhetorical sense: They are inconsequential, both intellectually and humanly. The central point is only this: The social and the scientific significance of a concrete subject can be—although, of course, they need not be—of quite different magnitudes.

Some Personal Choices of SRMs

Responding again to the Editors' amiable reminder that these prefatory essays generally call for personal moments, I sketch out three abbreviated rationales, early and late, for my selecting or adopting certain empirical materials that seemed strategic for studying particular problems in sociology and social psychology.

I think back to the ancient days of 1943 and my interest, largely stimulated by my newfound collaborator, Paul F. Lazarsfeld, in understanding the workings and consequences of mass propaganda. "The radio marathon," then a wholly new historical phenomenon, promised to provide a strategic case for investigating the collective behavior of mass persuasion. In the course of 18 consecutive hours on the air, the pop singer Kate Smith, widely identified as the sincere patriot incarnate, spoke a series of prepared texts on 65 occasions, and elicited the then unprecedented sum of $39,000,000 in war bond pledges (Merton et al 1946). From the start, the concrete idiosyncratic and behavioral materials were delimited from their potential scientific interest: "Although her name inevitably recurs time and again throughout the book, this is *not* a study of Kate Smith." Rather, the collective bond-drive would "provide a peculiarly instructive case for research into the social psychology of mass persuasion."

Severely condensed, the stated attributes of this assumed strategic research

event were these: First, it was a "real-life" situation, not an isolated, recogniz-
ably contrived situation of the kind that limits the transferability of laboratory
findings in social psychology to the world outside. Second, the bond purchases provided a behavioral index of effective persuasion which, however crude,
was far better than the hypothetical pencil-and-paper responses common in
the laboratory research of the time. Third, there was reason to suppose that the
event would be emotionally freighted in varying degree for listeners, both
those who pledged bond purchases and those who did not. Fourth, unlike field
studies of other collective behavior, such as race riots, we would have full and
sustained access to parts of the developing collective situation in the form of
content analyses of the recorded broadcasts. Fifth, the self-selected individuals and groups engaging in this behavior would come from widely
differing social strata rather than being drawn, after the fashion of the time
(and, often enough, today) from the dependent, rather homogenous aggre-
gates of college students dragooned as "subjects" by their instructors. Finally,
it was assumed that this attempt at truly mass persuasion would link up with
identifiable sociocultural contexts.

In the course of the study, we did find such social phenomena, among
them, the operation of "pseudo-Gemeinschaft" (the feigning of common
values and primary concern with the other as a means of advancing one's own
interests); processes in the formation of what we described by the new concept
of "public image"; and a pervasive public distrust. Not least, in unanticipated
and self-exemplifying fashion, the study reactivated a sense of the moral
implications of the framing of scientific problems in one or another fashion,
leading to a specific elucidation of the Rickert-Weber idea of *Wertbeziehung*
that questioned a naive form of positivistic orientation common at the time:

> [The] social scientist investigating mass opinion may adopt the standpoint of the
> positivist, proclaim the ethical neutrality of science, insist upon his exclusive concern with
> the advancement of knowledge, explain that science deals only with the discovery of
> uniformities and not with ends and assert that in his role as a detached and dispassionate
> scientist, he has no traffic with values. He may, in short, affirm an occupational philosophy
> which appears to absolve him of any responsibility for the use to which his discoveries in
> methods of mass persuasion may be put. With its specious and delusory distinction [in this
> context] between 'ends' and 'means' and its insistence that the intrusion of social values
> into the work of scientists makes for special pleading, this philosophy fails to note that the
> investigator's social values do influence his choice and definition of problems. The
> investigator may naively suppose that he is engaged in the value-free activity of research,
> whereas in fact he may simply have so defined his research problems that the results will be
> of use to one group in the society, and not to others. His very choice and definition of a
> problem reflects his tacit values (Merton et al 1946:187–88).

And so on in further specifying detail drawn from this study of a public
event involving mass persuasion and the workings of what came to be
described as "technicians in sentiment." The point of dwelling on these

matters, I suppose, is simply to note, once again, that however focused an SRS is for the investigation of previously identified problems, it may lead to other, unanticipated, findings and problems.

By the way of necessarily quick conclusion, two contrasting episodes also involving my own work may illuminate a general point regarding SRMs in sociology: Studies of social institutions, social movements, and other macro-sociological inquiries require little explicit rationale, since their relation to values is taken as self-evident but the selection of seemingly peripheral, innocuous, or 'trivial' social data as strategic for investigating basic sociological problems, does, precisely because of the seeming distance of the data from prized values.

Back in the 1930s, when the sociology of science was far from having been legitimated as a scholarly field of inquiry, even the historians of science most critical of a study of the social and cultural contexts of the efflorescence of science in seventeenth-century England (Merton [1938] 1970) did not question its scholarly relevance. Some were even prepared to accept, on probation, its substantive hypotheses of linkages between Puritanism and the emergence of the new science as well as its hypotheses of the partial shaping of foci of scientific interest by economic and technological developments of the time. Some went on to take friendly note of the use in that study of the then newly developed procedures of prosopography (analysis of collective biography) (Stone 1971:50–51, Shapin & Thackray 1974:22) and the quantitative analysis of changing scientific foci through content-analysis of the new journal of the new science, *Philosophical Transactions. Wertbeziehung* gave immediate scholarly warrant to the "subjects" under study.

Not so, however, with another study of mine two decades later. In it I had elected to focus on a recurrent phenomenon in science over the centuries, though one which had been ignored for systematic study: priority-conflicts among scientists, including the greatest among them, who wanted to reap the glory of having been first to make a particular scientific discovery or scholarly contribution. This was paradoxically coupled with strong denials, by themselves and by disciples, of their ever having had such an "unworthy and puerile" motive for doing science.[5] The initial and subsequent response to that study is captured in a remarkably candid account by the historian of science and editor-in-chief of the 16-volume *Dictionary of Scientific Biography,*

---

[5]Those self-deprecatory words are Freud's. Still, his biographer and disciple, Ernest Jones (1957, III:105) writes that "Freud was never interested in questions of priority which he found merely boring," thus providing another case of fashioning a biographical pseudo-fact although abundant and accessible evidence testifies otherwise. Elinor Barber and I have identified some 150 occasions on which Freud exhibited an interest in priority. With typical self-awareness, he reports having even dreamt about priority and the credit normatively due scientists for their contributions (Merton [1963] 1973:385–91).

Charles C. Gillispie (1974:656–60). A colleague at a distance, he first responded with considerably less than enthusiasm. I can do no better than have him tell the telling story:

> Some years ago, probably in early 1958, Merton sent me an offprint of . . . his presidential address to the American Sociological Association on "Priorities in Scientific Discovery" [(1957) 1973:286–324]. It starts by noting (pp. 286–87) "the great frequency with which the history of science is punctuated by disputes, often by sordid disputes, over priority of discovery." As I read on, dismay overtook amusement at the parade of eminent scientists arguing and frequently quarreling with each other, not over what the truth was, but over who had it first, Newton or Leibniz, Newton or Hooke, Cavendish or Watt or Lavoisier, Adams or LeVerrier, Jenner or Pearson or Rabaut, Freud or Janet. Sometimes the great men themselves abstained from contending in the lists of professional recognition for title to their intellectual property only to have their claims championed by disciples or compatriots. All too clearly the particular instances that Merton adduced in a number of variations on the theme of intellectual possessiveness could have been multiplied almost indefinitely.
>
> In a note of acknowledgment to Merton, I wrote that, although it seemed surprising that the phenomenon was so nearly universal an accompaniment to scientific discovery, I did wonder whether the matter wasn't a bit trivial. I don't believe I also said "unworthy" but recollect that such a dark thought was in my mind (Gillispie 1974:656).

I do not recall Gillispie actually having said "unworthy." He did, however, signal his friendly concern over my having lavished so much attention on the distinctly minor "subject" of priority-conflicts. But a change in his own theoretical perspectives on the scope and character of the historiography of science evidently led to a changed perception. He no longer took the descriptive raw materials of priority-conflicts as the subject-matter in hand; rather, he came to see them for what they were being redesigned to be: as strategic research materials for identifying the reward system distinctive of the social institution of science, one in which peer recognition of original scientific work was the golden coinage of the scientific realm. Gillispie also came to see that the sociological analysis of priority-conflicts as SRMs led one to find a contradiction between that reward system and other parts of the social and normative structure of science, such as the system of free and open communication (at least, for scientists outside the world of industry).

Gillispie indicatively describes this shift in perception:

> Only a few years later, when I began to study and teach materials in the social and institutional as well as the more traditional internal and intellectual history of science, did I come to take the full thrust of what he had in fact said, and said clearly and convincingly. It was that such behavior occurs in service to social norms; that norms arise in the life of real communities governing the conduct of their members; that the phrase 'scientific community' is, therefore, no mere manner of speaking about some shared pleasure in the study of nature but refers to an effective social entity; and that, within its membership, which is bounded professionally and not geographically, two main sets of norms constrain behavior and do so in ways that conflict, the one enjoining selflessness in the advancement of knowledge, and the other ambition for professional reputation, which in science accrues from originality in discovery and from that alone. The analysis exhibits the scientific

community to be one wherein the dynamics derive from the competition for honor even as the dynamics of the classical economic community do from the competition for profit, and neither of those statements is in any way incompatible with agreeing that the competitors characteristically like their work and choose it for that reason (Gillispie 1974:656).

Gillispie goes on to report that the substance of James Watson's *The Double Helix* (1968) came as no surprise. After all, that confessional account of intense competition and marginal if not sharp practices in the author's quest for a Nobel variously exemplified what was set down in the sociological analysis of intellectual property and the race for priority in science which had appeared a decade before.

One further observation will round out this impromptu case study of a strategic research site in the sociology of science. Were Charles Gillispie reflecting on his shifting response to that early study of priority-conflicts today, after the quite recent spate of concern over the occurrence of fraud in science, he might have elucidated his account further. He might have gone on to observe that the study had proposed the strongly stated hypothesis that contradictions between the reward-system and the normative system of science made for such pathologies as the occasional felonies of plagiarism and the cooking of fraudulent data, the presumably more frequent misdemeanors of hoarding one's own data while making free use of others' data, and the breaching of the mores of science by failing to acknowledge the contributions of predecessors, the collective giant on whose shoulders one stood to see a bit or, rarely, a great deal farther. Gillispie might have noted that this 1957 paper was the first to set out a sociological analysis of fraud in science, a good many years before currently publicized cases of such scientific felonies had forced widespread attention, both scholarly and popular, to the phenomenon (Zuckerman 1977, Broad & Wade 1982). Now for me to visit this observation on Charles Gillispie might be taken as a self-exemplifying claim to priority (as no doubt it is). But the chief point is less a matter of priority than of attending to the sources of that early sociologically grounded focus on the phenomenon of fraud in science. That focus derived theoretically from anomie-and-opportunity-structure theory and empirically from the selection of priority-conflicts as a strategic research site. And this, in turn, suggests that once problems are theoretically identified, materials that were previously peripheral or of no interest at all become reassessed as, in effect, strategic research materials.

CODA

It is now plain why the preceding pages are described as fragments. Obviously, much more is needed to establish these three patterns of scientific practice as phenomena, to specify our current ignorance about each of them in the form of new feasible problems, and to propose a range of research materials strategic for their solution. To my way of thinking, that is work for the near future.

ACKNOWLEDGMENTS

I thank colleagues near and far for their thoughtful reading of early and late drafts of this paper; in the first instance, Joshua Lederberg, Robert C. Merton, and Harriet Zuckerman, and then, Orville G. Brim, Jr., Jonathan R. Cole, Cynthia F. Epstein, Jonathan Rieder, David L. Sills, D. K. Simonton, Neil J. Smelser, and Stephen M. Stigler. I acknowledge aid of another kind from the John D. & Catherine T. MacArthur Foundation and the Russell Sage Foundation.

APPENDIX: THE MENU

(On the suggestion of several readers by way of providing context, I append the list of subjects from which these three were drawn.)

I. Patterns of Scientific Practice

1. Establishing the Phenomenon
2. Specified Ignorance
3. Strategic Research Materials
4. Fact as Theory-Laden: A Periodic Rediscovery
5. Naive Falsificationism: When Trust Theory, When Trust Fact
6. Unanticipated Consequences of the Reward System in Science: A Model of the Sequencing of Problem-Choices (with R. C. Merton)
7. The Self-Fulfilling Prophecy in Scientific Work
8. Toward a Sociological Theory of Error:
   8a. Patterned Misunderstandings in Science and Learning
   8b. Fallacy of the Latest Word
   8c. The Phoenix Phenomenon
9. Disciplined Eclecticism
10. Confirmation and the Fallacy of Affirming the Consequent
11. A *Fortiori* Reasoning in the Design of Scientific Inquiry
12. Tacit Counterfactual History
13. The (William) James Distinction: Acquaintance With and Knowledge About
14. The (Kenneth) Burke Theorem: Seeing as a Way of Not-Seeing
15. The (L. J.) Henderson Maxim: It's a Good Thing to Know What You Are Doing

II. Patterns in Transmission, Change, and Growth of Scientific Knowledge

1. Selective Accumulation of Scientific Knowledge: Paradox of Progress
2. OBI: Obliteration (of Source of Ideas, Methods, or Findings) by Incorporation (in Canonical Knowledge)
3. "Trained Incapacity": A Case of OBI
4. Cognitive Conduits for the Blurred Central Message
5. The Retroactive Effect in the Transmission and Growth of Knowledge
6. The Matthew Effect II: Accumulation of Advantage and the Symbolism of Intellectual Property
7. Oral Publication and Publication in Print
8. The Scientific Paper as Tacit Reconstruction of Knowledge
9. Insiders and Outsiders: Privileged Access to Knowledge
10. The Adumbrationist Credo: What's New is Not True; What's True is not New
11. The Symbolism of Eponyms in Science
12. Fathers and Mothers of the Sciences
13. Fraud and Other Deviant Behaviors in Science: A Case of Goal Displacement
14. Taboo Knowledge
15. Givens: The 'Of-Course Mood' in Scientific Discourse
16. Francis Bacon as Sociologist of Knowledge
17. Organized Skepticism: The Social Organization and Functions of Criticism in Science and Scholarship

III. Neologisms as Sociological Concepts: History and Analysis

1. On The Origin and Character of the Word *Scientist*
2. Self-Exemplifying Ideas: in the Sociology of Science and Elsewhere
3. Influentials: Evolution of a Concept
4. Institutionalized Evasions and Other Patterned Evasions
5. SED: Socially Expected Durations as a Temporal Dimension of Social Structure
6. Homophily and Heterophily: Types of Friendship Patterns
7. "Whatever Is, Is Possible": A Brief Biography of the Theorem
8. Opportunity Structures: A Brief Biography of the Concept
9. "Haunting Presence of the Functionally Irrelevant Status": The Structural Analysis of Status-Sets
10. "Phatic Communion": Malinowski's Need of a Cognitive Conduit
11. Comte's "Cerebral Hygiene" and the Presumed Dangers of Erudition for Originality

12. *Veritas Filia Temporis:* Temporal Contexts of Scientific Knowledge
13. *Pseudo-Gemeinschaft* and Public Distrust
14. The Travels and Adventures of Serendipity: A Study in Historical
     Semantics and the Sociology of Science (with Elinor Barber)

Literature Cited

Adelmann, H. B. 1966. *Marcello Malpighi and the Evolution of Embryology.* Ithaca, NY: Cornell Univ. Press

Allen, G. E. 1975. The introduction of *Drosophila* into the study of heredity and evolution: 1900–1910. *Isis* 66:322–33

Becker, H. S. [1963] 1973. *Outsiders.* New York: Free Press

Becker, H. S., ed. 1964. *The Other Side: Perspectives on Deviance.* New York: Free Press

Bell, E. T. 1931. Mathematics and speculation. *Sci. Mon.* 32:193–209

Bell, E. T. 1937. *Men of Mathematics.* New York: Simon & Schuster

Bernard, C. [1865] 1949. *An Introduction to the Study of Experimental Medicine.* New York: Henry Schuman

Bettelheim, B. 1983. *Freud and Man's Soul.* New York: Knopf

Broad, W., Wade, N. 1982. *Betrayers of the Truth: Fraud and Deceit in the Halls of Science.* New York: Simon & Schuster

Brewster, D. 1855. *Memoirs of the Life, Writings, and Discoveries of Sir Isaac Newton.* 2 vols. Edinburgh: Thomas Constable

Burke, K. 1935. *Permanence and Change.* New York: New Republic

Cairns, J., Stent, G. S., Watson, J. D. eds. 1966. *Phage and the Origins of Molecular Biology.* Cold Spring Harbor, Me: Cold Spring Harbor Lab. Quant. Biol.

Cicourel, A. 1968. *The Social Organization of Juvenile Justice.* New York: Wiley

Cloward, R. A., Ohlin, L. E. 1960. *Delinquency and Opportunity.* New York: Free Press

Cohen, A. K. 1955. *Delinquent Boys.* New York: Free Press

Cushing, H. 1925. *The Life of Sir William Osler.* 2 vol. Oxford: Clarendon

Durkheim, E. 1899–1900. Deux lois de l'evolution pénale. *L'Année sociologique* 4:55–95

Durkheim, E. 1888. Suicide et natalité: études de statistique morale. *Revue philosophique* 26:444–63

Erikson, K. T. 1964. Notes on the sociology of deviance. See Becker 1964:9–21

Fischer, C. S. 1977. *Networks and Places: Social Relations in the Urban Setting.* New York: Free Press

Fleck, L. [1935] 1979. *Genesis and Development of a Scientific Fact,* ed. T. J. Trenn, R. K. Merton. Chicago: Univ. Chicago Press

Fontenelle, B. [1686] 1908. *Histoire des oracles.* Paris: Hachette

Franks, F. 1981. *Polywater.* Cambridge: MIT Press

Freud, S. [1901] 1960. *The Psychopathology of Everyday Life.* Vol. 6 In *The Standard Edition of the Complete Psychological Works of Sigmund Freud,* ed. J. Strachey. London: Hogarth

Freud, S. [1916] 1961. *Introductory Lectures on Psychoanalysis.* Vol. 15 in *The Standard Edition of the Complete Psychological Works of Sigmund Freud,* ed. J. Strachey. London: Hogarth

Garfield, E. The unanswered questions of science. 1974. *Curr. Contents* June 5:5–6

Gerth, H. H., Mills, C. W. 1953. *Character and Social Structure.* New York: Harcourt Brace Jovanovich

Gillispie, C. C. 1974. Mertonian theses. *Science* 184:656–60

Golubović, Z. 1973. Why is functionalism more desirable in present-day Yugoslavia than marxism? *Praxis* 4:357–68

Gouldner, A. W. 1970. *The Coming Crisis of Western Sociology.* New York: Basic Books

Gouldner, A. W. 1973. *For Sociology.* New York: Basic Books

Gouldner, A. W. 1980. *The Two Marxisms.* New York: Seabury

Grünbaum, A. 1984. *The Foundations of Psychoanalysis: A Philosophical Critique.* Berkeley: Univ. Calif. Press

Gutman, H. G. 1976. *The Black Family in Slavery and Freedom: 1750–1925.* New York: Pantheon

Henderson, L. J. [1941] 1970. *On the Social System,* ed. B. Barber. Chicago: Univ. Chicago Press

Heidegger, M. [1927] 1962. *Being and Time.* New York: Harper

Jones, E. 1957. *Sigmund Freud: Life and Work.* 3 vols. London: Hogarth

Kandel, E. R. 1983. Neurobiology and molecular biology. *Cold Spring Harbor Symposia on Quantitative Biol.* 48:891–908

Kitsuse, J. I. 1964. Societal reaction to deviant behavior. See Becker 1964, pp. 87–102

Lazarsfeld, P. F., Thielens, W. Jr. 1958. *The Academic Mind.* New York: Free Press

Lederberg, J. 1951. Genetic studies with bacteria. In *Genetics in the 20th Century,* ed. L. C. Dunn, pp. 263–89. New York: Macmillan

Lederberg, J. 1986. Forty years of genetic recombination in bacteria. *Nature* 324 (6098):627–28

Lemert, E. M. 1951. *Social Pathology.* New York: McGraw-Hill

Lemert, E. M. 1973. Beyond Mead: The societal reaction to deviance. *Soc. Probl.* 21:457–68

Leontief, W. 1971. Theoretical assumptions and nonobserved facts. *Am. Econ. Rev.* 21:457–68

Lewis, C. S. 1975. *Letters of C. S. Lewis,* ed. W. H. Lewis. New York: Harcourt Brace Jovanovich

Marx, K. [1867] 1906. *Capital: A Critique of Political Economy,* Vol. 1, (Ed. F. Engels). Chicago: C. H. Kerr

Mayr, E. 1982. *The Growth of Biological Thought.* Cambridge: Harvard Univ. Press

McKee, J. E. 1961. *100 Problems in Environmental Health.* Washington, D.C.: Jones Composition

Mead, G. H. 1918. The psychology of punitive justice. *Am. J. Sociol.* 23:577–602

Merton, R. K. 1938a. Social structure and anomie. *Am. Sociol. Rev.* 3:672–82

Merton, R. K. [1938b] 1970. *Science, Technology and Society in 17th-Century England.* New York: Howard Fertig

Merton, R. K. 1948. The bearing of empirical research upon the development of sociological theory. *Am. Sociol. Rev.* 13:505–15

Merton, R. K. 1957. *Social Theory and Social Structure.* New York: Free Press

Merton, R. K. 1959. Notes on problem-finding in sociology. In *Sociology Today: Problems and Prospects,* ed. R. K. Merton, L. Broom, L. S. Cottrell. New York: Basic Books

Merton, R. K. [1963a] 1973. Multiple discoveries as strategic research site. See Merton 1973, pp. 371–82

Merton, R. K. [1963b] 1973. The ambivalence of scientists. See Merton 1973, pp. 383–412

Merton, R. K. 1971. The precarious foundations of detachment in sociology. In *The Phenomenon of Sociology,* ed. E. A. Tiryakian, pp. 188–99. New York: Appleton-Century-Crofts

Merton, R. K. 1973. *The Sociology of Science,* ed. N. W. Storer. Chicago: Univ. Chicago Press

Merton, R. K. 1976. The sociology of social problems. In *Contemporary Social Problems,* ed. R. K. Merton, R. A. Nisbet, pp. 3–43. New York: Harcourt Brace Jovanovich

Merton, R. K. 1981. Remarks on theoretical pluralism. In *Continuities in Structural Inquiry,* ed. P. M. Blau, R. K. Merton, pp. i–vii. London: Sage

Merton, R. K. 1982. Alvin W. Gouldner: Genesis and growth of a friendship. *Theory and Society* 11:915–38

Merton, R. K., Fiske, M., Curtis, A. [1946] 1971. *Mass Persuasion.* Westport, Conn: Greenwood

Moore, W. E., Tumin, M. M. 1949. Some social functions of ignorance. *Am. Sociol. Rev.* 14:787–95

Oromaner, M. August, 1974. Critical function of errors. *Am. Sociol. Assn. Footnotes* 2:7

Osler, W. 1908. *An Alabama Student, and Other Biographical Essays,* pp. 159–88. New York: Oxford Univ. Press

Park, R. E. 1928. Human migration and the marginal man. *Am. J. Sociol.* 33:881–93

Parsons, T. 1937. *The Structure of Social Action.* New York: McGraw-Hill

Peirce, C. S. [c. 1903] 1958. *Collected Papers,* Vol. 7:121–144. Cambridge: Harvard Univ. Press

Popper, K. [1960] 1962. *Conjectures and Refutations: The Growth of Scientific Knowledge.* London: Routledge & Kegan Paul

Price, D. J. deS. 1961. *Science Since Babylon.* New Haven: Yale Univ. Press

Quinney, R. 1970. *The Social Reality of Crime.* Boston: Little, Brown

Rickert, H. [1902] 1921. *Die Grenzen der naturwissenschaftlichen Begriffsbildung.* Tübingen: J. C. Mohr. 4th ed.

Rostand, J. 1960. *Error and Deception in Science.* London: Hutchinson

Schrecker, E. 1986. *No Ivory Tower: McCarthyism in the Universities.* New York: Oxford Univ. Press

Selden, J. [1689] 1890. *Table Talk.* London: Reeves & Turner

Selvin, H. C. 1976. Durkheim, Booth and Yule: non-diffusion of an intellectual innovation. *Archives Européenes de sociol.* 17:39–51

Serratosa, F. 1962. Ignoratica. In *The Scientist Speculates,* ed. I. J. Good, pp. 4–9. New York: Basic Books

Shapin, S., Thackray, A. 1974. Prosopography as a research tool in history of science: The British scientific community 1700–1900. *Hist. Sci.* 12:1–28

Simmel, G. [1908] 1950. *The Sociology of Georg Simmel,* ed. K. H. Wolff, pp. 402–8. New York: Free Press

Smithson, M. 1985. Toward a social theory of ignorance. *J. Theory Soc. Behav.* 15:149–70

Sorokin, P. A. 1925. *The Sociology of Revolutions.* Philadelphia, Pa: Lippincott

Sowell, T. 1981. Assumptions versus history in ethnic education. *Teachers College Record* 83:37–69

Stone, L. 1971. Prosopography. *Daedalus.* Winter: 46–79

Sutherland, E. H. [1925–1951] 1956. *The Sutherland Papers,* ed. A. K. Cohen et al. Bloomington: Indiana Univ. Press

Sztompka, P. 1986. *Robert K. Merton: An Intellectual Profile.* New York: St. Martin's Press

Thomas, W. I., Znaniecki, F. [1918–20] 1927. *The Polish Peasant in Europe and America.* 2 vols. New York: Knopf

Turk, A. 1969. *Criminality and the Legal Order.* Chicago: Rand McNally

Veblen, T. 1932. *The Place of Science in Modern Civilization.* New York: Viking

Wallerstein, I. 1968. C. Wright Mills. *International Encyclopedia of the Social Sci-*ences, ed. D. L. Sills. New York: Macmillan & The Free Press

Watson, J. D. 1968. *The Double Helix.* New York: Atheneum

Weber, M. 1920–1921 *Gesammelte Aufsätze zur Religionssoziologie.* 3 vols. Tübingen: J. C. B. Mohr

Weber, M. 1922. *Gesammelte Aufsätze zur Wissenschaftslehre.* Tübingen: J. C. B. Mohr

Wilson, L. G. 1960. The transformation of ancient concepts of respiration in the 17th century. *Isis* 51:161–72

Windelband, W. 1884. *Präludien: Aufsätze und Reden zur Einleitung in die Philosophie.* Freiburg

Zeitlin, M. 1974. Corporate ownership and control. *Am. J. Sociol.* 79:1073–1119

Zuckerman, H., Lederberg, J. 1986. Postmature scientific discovery. *Nature* 324 (6098):629–31

Zuckerman, H. 1977. Deviant behavior and social control in science. In *Deviance and Social Change,* ed. E. Sagarin, pp. 87–138. Beverly Hills, Calif: Sage

NAME AND SUBJECT INDEXES

This volume includes indexes for Volumes 1 and 2 (not included when they were first published) and for Volume 3, Parts 1 and 2 combined. The name and subject index for Volume 3, Part 1 also appears at the end of Part 1 for convenience.

Ordering information for all three volumes appears in the back of this volume.

Name and Subject Indexes

NAME AND SUBJECT INDEX TO VOLUME 1

NAME AND SUBJECT INDEX TO VOLUME 2

2269

NAME AND SUBJECT INDEX TO VOLUME 3

APPENDIX: Historical and Prefatory Chapters Published in All *Annual Reviews* Series

Ackeret, J. See Rott, N.

Adolph, E.F. 1968. Research provides self-education. *Annu. Rev. Physiol.* 30:1-14 [Reprinted in *Exc. Fas. Sci.* Vol. 2]

Ainsworth, G.C. 1969. History of plant pathology in Great Britain. *Annu. Rev. Phytopathol.* 7:13-30

Akai, S. 1974. History of plant pathology in Japan. *Annu. Rev. Phytopathol.* 12:13-26

Alexander, C.P. 1969. Baron Osten Sacken and his influence on American dipterology. *Annu. Rev. Entomol.* 14:1-18

Ambartsumian, V.A. 1980. On some trends in the development of astrophysics. *Annu. Rev. Astron. Astrophys.* 18:1-13

Anastasi, A. 1986. Evolving concepts of test validation. *Annu. Rev. Psychol.* 37:1-15

Andrewartha, H.G., Birch, L.C. 1973. The history of insect ecology. *Hist. Entomol.*, pp. 229-66

Andrews, C.H. 1978. Fifty years with viruses. *Annu. Rev. Microbiol.* 27:19-32 [Reprinted in *Exc. Fas. Sci.* Vol. 2]

Anichkov, S.V. 1975. How I became a pharmacologist. *Annu. Rev. Pharmacol. Toxicol.* 15:1-10

Arensberg, C.M. 1972. Culture as behavior: structure and emergence. *Annu. Rev. Anthropol.* 1:1-26

Arthur, J.C. See Cummins, G.B.

Axelsson, J. 1971. Catecholamine functions. *Annu. Rev. Physiol.* 33:1-30

Ayala, F.J. 1976. Theodosius Dobzhansky: the man and the scientist. *Annu. Rev. Genet.* 10:1-6

Bailey, D.L. 1966. Whither pathology. *Annu. Rev. Phytopathol.* 4:1-8

Bak, T.A. 1974. The history of physical chemistry in Denmark. *Annu. Rev. Phys. Chem.* 25:1-10

Baker, K.F. 1982. Meditations on fifty years as an apolitical plant pathologist. *Annu. Rev. Phytopathol.* 20:1-25 [Reprinted in *Exc. Fas. Sci.* Vol. 3]

2319

Baker, K.F., Fischer, G.W. 1983. Pioneer leaders in plant pathology: F.D. Heald. *Annu. Rev. Phytopathol.* 21:13-20

Baker, W.O. 1976. Role of science and engineering in human use of materials. *Annu. Rev. Mater. Sci.* 6:35-52

Bard, P. 1973. The ontogenesis of one physiologist. *Annu. Rev. Physiol.* 35:1-16

Bardeen, J. 1980. Unity of concepts in the structure of matter. *Annu. Rev. Mater. Sci.* 10:1-18

Barker, H.A. 1978. Explorations of bacterial metabolism. *Annu. Rev. Biochem.* 47:1-33

Barker, K.R., Noffsinger, M., Griffin, G.D. 1981. Pioneer leaders in plant pathology: Gerald Thorne. *Annu. Rev. Phytopathol.* 19:21-28

Bawden, F.C. 1970. Musings of an erstwhile plant pathologist. *Annu. Rev. Phytopathol.* 8:1-12 [Reprinted in *Exc. Fas. Sci.* Vol. 2]

Beadle, G.W. 1974. Recollections. *Annu. Rev. Biochem.* 43:1-13 [Reprinted in *Exc. Fas. Sci.* Vol. 2]

Beals, R.L. 1982. Fifty years in anthropology. *Annu. Rev. Anthropol.* 11:1-23 [Reprinted in *Exc. Fas. Sci.* Vol. 3]

Bean, W.B. 1982. Personal reflections on clinical investigations. *Annu. Rev. Nutr.* 2:1-20 [Reprinted in *Exc. Fas. Sci.* Vol. 3]

Beier, M. 1973. The early naturalists and anatomists during the Renaissance and seventeenth century. *Hist. Entomol.*, pp. 81-94

Bennett, C.W. 1973. A consideration of some of the factors important in the growth of the science of plant pathology. *Annu. Rev. Phytopathol.* 11:1-10

Bennett, D. 1977. L.C. Dunn and his contribution to T-locus genetics. *Annu. Rev. Genet.* 11:1-12 [Reprinted in *Exc. Fas. Sci.* Vol. 3]

Beritashvili (Beritoff), J.S. 1966. From the spinal coordination of movements to the psychoneural integration of behaviour. *Annu. Rev. Physiol.* 28:1-16

Beyer, K.H. Jr. 1977. A career or two. *Annu. Rev. Pharmacol. Toxicol.* 17:1-10

Biale, J.B. 1978. On the interface of horticulture and plant physiology. *Annu. Rev. Plant Physiol.* 29:1-23 [Reprinted in *Exc. Fas. Sci.* Vol. 3]

Binnie, A.M. 1978. Some notes on the study of fluid mechanics in Cambridge, England. *Annu. Rev. Fluid Mech.* 10:1-10 [Reprinted in *Exc. Fas. Sci.* Vol. 3]

Birch, F. 1979. Reminiscences and digressions. *Annu. Rev. Earth Planet. Sci.* 7:1-9 [Reprinted in *Exc. Fas. Sci.* Vol. 3]

Birdsell, J.B. 1987. Some reflections on fifty years in biological anthropology. *Annu. Rev. Anthropol.* 16:1-12 [Reprinted in *Exc. Fas. Sci.* Vol. 3]

Bishop, G.H. 1965. My life among the axons. *Annu. Rev. Physiol.* 27:1-18 [Reprinted in *Exc. Fas. Sci.* Vol. 1]

Bitancourt, A.A. 1978. Phytopathology in a developing country. *Annu. Rev. Phytopathol.* 16:1-18 [Reprinted in *Exc. Fas. Sci.* Vol. 3]

Bjerknes, V. See Eliassen, A.

Black, L.M. 1981. Recollections and reflections. *Annu. Rev. Phytopathol.* 19:1-19 [Reprinted in *Exc. Fas. Sci.* Vol. 3]

Blaschko, H.K.F. 1980. My path to pharmacology. *Annu. Rev. Pharmacol. Toxicol.* 20:1-14 [Reprinted in *Exc. Fas. Sci.* Vol. 3]

Bloch, K. 1987. Summing up. *Annu. Rev. Biochem.* 56:1-19 [Reprinted in *Exc. Fas. Sci.* Vol. 3]

Bodenheimer, F.S. See Harpaz, I.

Boothroyd, C.W. 1982. Charles Chupp: extension plant pathologist. *Annu. Rev. Phytopathol.* 20:41-47

Brooks, H. 1978. Resources and the quality of life in 2000. *Annu. Rev. Mater. Sci.* 8:1-19

Brown, S.W. 1973. Genetics—the long story. *Hist. Entomol.*, pp. 407-32

Brown, W. 1965. Toxins and cell-wall dissolving enzymes in relation to plant disease. *Annu. Rev. Phytopathol.* 3:1-18

Brown, W. See Garrett, S.D.

Bruehl, G.W. 1980. James G. Dickson: the man and his work. *Annu. Rev. Phytopathol.* 18:11-18

Buchheim, R. See Habermann, E.R.

Bullard, E. 1975. The emergence of plate tectonics: a personal view. *Annu. Rev. Earth Planet. Sci.* 3:1-30

Buller, A.H.R. See Estey, R.H.

Bünning, E. 1977. Fifty years of research in the wake of Wilhelm Pfeffer. *Annu. Rev. Plant Physiol.* 28:1-22

Burgers, J.M. 1975. Some memories of early work in fluid mechanics at the Technical University of Delft. *Annu. Rev. Fluid Mech.* 7:1-11

Burmeister, H. See Ulrich, W.

Burton, A.C. 1978. Variety—the spice of science as well as of life: the disadvantages of specialization. *Annu. Rev. Physiol.* 37:1-12 [Reprinted in *Exc. Fas. Sci.* Vol. 2]

Busemann, A. 1971. Compressible flow in the thirties. *Annu. Rev. Fluid Mech.* 3:1-12

Cameron, J.W.M. 1973. Insect pathology. *Hist. Entomol.*, pp. 285-306

Campbell, C.L. 1983. Erwin Frink Smith—pioneer plant pathologist. *Annu. Rev. Phytopathol.* 21:21-27

Carlsson, A. 1987. Perspectives on the discovery of central monoaminergic neurotransmission. *Annu. Rev. Neurosci.* 10:19-40 [Reprinted in *Exc. Fas. Sci.* Vol. 3]

Chailakhyan, M. 1968. Internal factors of plant flowering. *Annu. Rev. Plant Physiol.* 19:1-36

Chandler, W.H. 1959. Plant physiology and horticulture. *Annu. Rev. Plant Physiol.* 10:1-12

Chargaff, E. 1975. A fever of reason: the early way. *Annu. Rev. Biochem.* 44:1-18

Chase, M.W. 1985. Immunology and experimental dermatology. *Annu. Rev. Immunol.* 3:1-29 [Reprinted in *Exc. Fas. Sci.* Vol. 3]

Chen, K.K. 1981. Two pharmacological traditions: notes from experience. *Annu. Rev. Pharmacol. Toxicol.* 21:1-6 [Reprinted in *Exc. Fas. Sci.* Vol. 3]

Chupp, C. See Boothroyd, C.W.

Chynoweth, A.G. 1975. Science-engineering coupling and some priorities in materials research. *Annu. Rev. Mater. Sci.* 5:27-42

Clark, G. 1979. Archaeology and human diversity. *Annu. Rev. Anthropol.* 8:1-20

Clark, W.M. 1962. Notes on a half-century of research, teaching, and administration. *Annu. Rev. Biochem.* 31:1-24 [Reprinted in *Exc. Fas. Sci.* Vol. 1]

Clarke, H.T. 1958. Impressions of an organic chemist in biochemistry. *Annu. Rev. Biochem.* 27:1-14 [Reprinted in *Exc. Fas. Sci.* Vol. 1]

Cole, K.S. 1979. Mostly membranes. *Annu. Rev. Physiol.* 41:1-24 [Reprinted in *Exc. Fas. Sci.* Vol. 3]

Coon, C.S. 1977. Overview. *Annu. Rev. Anthropol.* 6:1-10

Cosman, M.P. 1983. A feast for Aesculapius: historical diets for asthma and sexual pleasure. *Annu. Rev. Nutr.* 3:1-33

Cowling, T.G. 1985. Astronomer by accident. *Annu. Rev. Astron. Astrophys.* 23:1-18 [Reprinted in *Exc. Fas. Sci.* Vol. 3]

Craigie, J.H. See Green, G.J.

Crow, J.F. 1987. Population genetics history: a personal view. *Annu. Rev. Genet.* 21:1-22

Cummins, G.B. 1978. J.C. Arthur: the man and his work. *Annu. Rev. Phytopathol.* 15:19-30

Cushny, A.R. See MacGillivray, H.

Cutting, W.C. See Raffel, S.

Dagley, S. 1987. Lessons from biodegradation. *Annu. Rev. Microbiol.* 41:1-23 [Reprinted in *Exc. Fas. Sci.* Vol. 3]

Darby, W.J. 1985. Some personal reflections on a half century of nutrition science: 1930s-1980s. *Annu. Rev. Nutr.* 5:1-24 [Reprinted in *Exc. Fas. Sci.* Vol. 3]

Darwin, C. See Remington, J.E.

Davenport, H.W. 1985. The apology of a second-class man. *Annu. Rev. Physiol.* 47:1-14 [Reprinted in *Exc. Fas. Sci.* Vol. 3]

de Bary, H.A. See Horsfall, J.G. & Wilhelm, S.

Dickson, J.G. See Bruehl, G.W.

Dobzhansky, T. See Ayala, F.J.

Doisy, E.A. 1976. An autobiography. *Annu. Rev. Biochem.* 45:1-9

Dowlen, E. See Gardner, M.W.

Dowson, W.J. See Garrett, S.D.

Du Bois, C. 1980. Some anthropological hindsights. *Annu. Rev. Anthropol.* 9:1-13

Du Bois, E.F. 1950. Fifty years of physiology in America—a letter to the editor. *Annu. Rev. Physiol.* 12:1-12 [Reprinted in *Exc. Fas. Sci.* Vol. 1]

Duggar, B.M. See Walker, J.C.

Dunn, L.C. See Bennett, D.

Dupuis, C. 1974. Pierre Andre Latreille (1762-1833): the foremost entomologist of his time. *Annu. Rev. Entomol.* 19:1-13

Dupuis, C. 1984. Willi Hennig's impact on taxonomic thought. *Annu. Rev. Ecol. Syst.* 15:1-24

Eccles, J.C. 1978. My scientific odyssey. *Annu. Rev. Physiol.* 39:1-18 [Reprinted in *Exc. Fas. Sci.* Vol. 2]

Edsall, J.T. 1971. Some personal history and reflections from the life of a biochemist. *Annu. Rev. Biochem.* 40:1-28 [Reprinted in *Exc. Fas. Sci.* Vol. 2]

Eggan, F. 1974. Among the anthropologists. *Annu. Rev. Anthropol.* 3:1-19

Eliassen, A. 1982. Vilhelm Bjerknes and his students. *Annu. Rev. Fluid Mech.* 14:1-12

Emerson, S. 1971. Alfred Henry Sturtevant. *Annu. Rev. Genet.* 5:1-4

Engelhardt, W.A. 1982. Life and science. *Annu. Rev. Biochem.* 51:1-19 [Reprinted in *Exc. Fas. Sci.* Vol. 3]

Erlanger, J. 1964. A physiologist reminisces. *Annu. Rev. Physiol.* 26:1-14 [Reprinted in *Exc. Fas. Sci.* Vol. 1]

Estey, R.H. 1986. A.H.R. Buller: pioneer leader in plant pathology. *Annu. Rev. Phytopathol.* 24:17-25

Evenari, M. 1985. A cat has nine lives. *Annu. Rev. Plant Physiol.* 36:1-25 [Reprinted in *Exc. Fas. Sci.* Vol. 3]

Eyring, H. 1977. Men, mines, and molecules. *Annu. Rev. Phys. Chem.* 28:1-13 [Reprinted in *Exc. Fas. Sci.* Vol. 3]

Fabricius, J.C. See Tuxen, S.L.

Fenn, W.O. 1962. Born fifty years too soon. *Annu. Rev. Physiol.* 24:1-10 [Reprinted in *Exc. Fas. Sci.* Vol. 1]

Firth, R. 1975. An appraisal of modern social anthropology. *Annu. Rev. Anthropol.* 4:1-25

Fischer, H.O.L. 1960. Fifty years "synthetiker" in the service of biochemistry. *Annu. Rev. Biochem.* 29:1-14 [Reprinted in *Exc. Fas. Sci.* Vol. 1]

Fish, S. 1970. The history of plant pathology in Australia. *Annu. Rev. Phytopathol.* 8:13-36

Flügge-Lotz, I., Flügge, W. 1973. Ludwig Prandtl in the nineteen-thirties: reminiscences. *Annu. Rev. Fluid Mech.* 5:1-8

Fortes, M. 1978. An anthropologist's apprenticeship. *Annu. Rev. Anthropol.* 7:1-30

Fraisse, P. 1984. Perception and estimation of time. *Annu. Rev. Psychol.* 35:1-36

French, C.S. 1979. Fifty years of photosynthesis. *Annu. Rev. Plant Physiol.* 30:1-26 [Reprinted in *Exc. Fas. Sci.* Vol. 3]

Fretwell, S.D. 1975. The impact of Robert MacArthur on ecology. *Annu. Rev. Ecol. Syst.* 6:1-13

Fulton, R.W. 1984. Pioneer leaders in plant pathology: James Johnson. *Annu. Rev. Phytopathol.* 22:27-34

Gaffron, H. 1969. Resistance to knowledge. *Annu. Rev. Plant Physiol.* 20:1-40

Gardner, M.W. 1977. Little-known plant pathologists: Ethelbert Dowlen. *Annu. Rev. Phytopathol.* 15:13-15

Garrett, S.D. 1972. On learning to become a plant pathologist. *Annu. Rev. Phytopathol.* 10:1-8

Garrett, S.D. 1981. Pioneer leaders in plant pathology: W.J. Dowson. *Annu. Rev. Phytopathol.* 19:29-34

Garrett, S.D. 1985. William Brown: pioneer leader in plant pathology. *Annu. Rev. Phytopathol.* 23:13-18

Gaumann, E. See Kern, H.

Gerard, R.W. 1952. The organization of science. *Annu. Rev. Physiol.* 14:1-12 [Reprinted in *Exc. Fas. Sci.* Vol. 1]

Gilluly, J. 1977. American geology since 1910—a personal appraisal. *Annu. Rev. Earth Planet. Sci.* 5:1-12

Goldstein, S. 1969. Fluid mechanics in the first half of this century. *Annu. Rev. Fluid Mech.* 1:1-28

Good, N.E. 1986. Confessions of a habitual skeptic. *Annu. Rev. Plant Physiol.* 37:1-22 [Reprinted in *Exc. Fas. Sci.* Vol. 3]

Granit, R. 1978. Discovery and understanding. *Annu. Rev. Physiol.* 34:1-12 [Reprinted in *Exc. Fas. Sci.* Vol. 2]

Green, G.J., Johnson, T., Conners, I.L. 1980. Pioneer leaders in plant pathology: J.H. Craigie. *Annu. Rev. Phytopathol.* 18:19-25

Greenberg, J.H. 1986. On being a linguistic anthropologist. *Annu. Rev. Anthropol.* 15:1-24 [Reprinted in *Exc. Fas. Sci.* Vol. 3]

Greenstein, J.L. 1984. An astronomical life. *Annu. Rev. Astron. Astrophys.* 22:1-35 [Reprinted in *Exc. Fas. Sci.* Vol. 3]

Gregory, P.H. 1977. Spores in air. *Annu. Rev. Phytopathol.* 15:1-11 [Reprinted in *Exc. Fas. Sci.* Vol. 3]

Griffin, J.B. 1985. An individual's participation in American archaeology. *Annu. Rev. Anthropol.* 14:1-23 [Reprinted in *Exc. Fas. Sci.* Vol. 3]

Grogan, R.G. 1987. The relation of art and science of plant pathology for disease control. *Annu. Rev. Phytopathol.* 25:1-8

Gueron, J., Magat, M. 1971. A history of physical chemistry in France. *Annu. Rev. Phys. Chem.* 22:1-23

Gunsalus, I.C. 1984. Learning. *Annu. Rev. Microbiol.* 38:13-44 [Reprinted in *Exc. Fas. Sci.* Vol. 3]

Habermann, E.R. 1974. Rudolf Buchheim and the beginning of pharmacology as a science. *Annu. Rev. Pharmacol. Toxicol.* 14:1-8

Hadorn, E. See Mitchell, H.K.

Hagen, K.S., Franz, J.M. 1973. A history of biological control. *Hist. Entomol.*, pp. 433-76

Hales, A.L. 1986. Geophysics on three continents. *Annu. Rev. Earth Planet. Sci.* 14:1-20 [Reprinted in *Exc. Fas. Sci.* Vol. 3]

Hall, V.E., Sonnenschein, R.R. 1981. The *Annual Review of Physiology*: past and present. *Annu. Rev. Physiol.* 43:1-5

Harpaz, I. 1973. Early entomology in the Middle East. *Hist. Entomol.*, pp. 21-36

Harpaz, I. 1984. Frederick Simon Bodenheimer (1897-1959): idealist, scholar, scientist. *Annu. Rev. Entomol.* 29:1-23

Heald, F.D. See Baker, K.F. & Fischer, G.W.

Heidelberger, M. 1978. A "pure" organic chemist's downward path. *Annu. Rev. Microbiol.* 31:1-12 [Reprinted in *Exc. Fas. Sci.* Vol. 2]

Heidelberger, M. 1979. A "pure" organic chemist's downward path: chapter 2 – the years at P. and S. *Annu. Rev. Biochem.* 48:1-21 [Reprinted in *Exc. Fas. Sci.* Vol. 3]

Hendricks, S. B. 1970. The passing scene. *Annu. Rev. Plant Physiol.* 21:1-10

Hennig, W. See Dupuis, C.

Herzberg, G. 1985. Molecular spectroscopy: a personal history. *Annu. Rev. Phys. Chem.* 36:1-30 [Reprinted in *Exc. Fas. Sci.* Vol. 3]

Hewitt, W.B. 1979. Conceptualizing in plant pathology. *Annu. Rev. Phytopathol.* 17:1-12

Heymans, C. 1963. A look at an old but still current problem. *Annu. Rev. Physiol.* 25:1-14 [Reprinted in *Exc. Fas. Sci.* Vol. 1]

Heymans, C. 1967. Pharmacology in old and modern medicine. *Annu. Rev. Pharmacol. Toxicol.* 7:1-14

Hildebrand, J.H. 1981. A history of solution theory. *Annu. Rev. Phys. Chem.* 32:1-23

Hilgard, E.R. 1980. Consciousness in contemporary psychology. *Annu. Rev. Psychol.* 31:1-26

Hill, A.V. 1958. The heat production of muscle and nerve. *Annu. Rev. Physiol.* 21:1-18 [Reprinted in *Exc. Fas. Sci.* Vol. 1]

Hill, R. 1975. Days of visual spectroscopy. *Annu. Rev. Plant Physiol.* 26:1-11

Hirschfelder, J.O. 1983. My adventures in theoretical chemistry. *Annu. Rev. Phys. Chem.* 34:1-29 [Reprinted in *Exc. Fas. Sci.* Vol. 3]

Hodgkin, A.L. 1983. Beginning: some reminiscences of my early life (1914-1947). *Annu. Rev. Physiol.* 45:1-16 [Reprinted in *Exc. Fas. Sci.* Vol. 3]

Holmes, F.O. 1968. Trends in the development of plant virology. *Annu. Rev. Phytopathol.* 6:41-62

Homans, G.C. 1986. Fifty years of sociology. *Annu. Rev. Sociol.* 12:xiii-xxx [Reprinted in *Exc. Fas. Sci.* Vol. 3]

Horsfall, J.G. 1975. Fungi and fungicides, the story of a nonconformist. *Annu. Rev. Phytopathol.* 13:1-13

Horsfall, J.G. 1979. Roland Thaxter. *Annu. Rev. Phytopathol.* 17:29-35

Horsfall, J.G., Wilhelm, S. 1982. Heinrich Anton de Bary: nach einhundertfunfzig Jahren. *Annu. Rev. Phytopathol.* 20:27-32

Horst, R.K. 1984. Pioneer leaders in plant pathology: Cynthia Westcott, plant doctor. *Annu. Rev. Phytopathol.* 22:21-26

Houssay, B.A. 1956. Trends in physiology as seen from South America. *Annu. Rev. Physiol.* 18:1-12 [Reprinted in *Exc. Fas. Sci.* Vol. 1]

Howard, L.O. See Russell, L.M.

Hoyle, F. 1982. The universe: past and present reflections. *Annu. Rev. Astron. Astrophys.* 20:1-35

Hoytink, G.J. 1970. Physical chemistry in the Netherlands after Van't Hoff. *Annu. Rev. Phys. Chem.* 21:1-16

Humphrey, J.H. 1984. Serendipity in immunology. *Annu. Rev. Immunol.* 2:1-21 [Reprinted in *Exc. Fas. Sci.* Vol. 3]

Hungate, R.E. 1979. Evolution of a microbial ecologist. *Annu. Rev. Microbiol.* 33:1-20 [Reprinted in *Exc. Fas. Sci.* Vol. 3]

Hutt, P.B. 1984. Government regulation of the integrity of the food supply. *Annu. Rev. Nutr.* 4:1-20

Jasper, H., Sourkes, T.L. 1983. Nobel laureates in neuroscience: 1904-1981. *Annu. Rev. Neurosci.* 6:1-42

Jeffreys, H. 1973. Developments in geophysics. *Annu. Rev. Earth Planet. Sci.* 1:1-13

Johnson, J. See Fulton, R.W.

Jonas, J., Gutowsky, H.S. 1980. NMR in chemistry – an evergreen. *Annu. Rev. Phys. Chem.* 31:1-27

Jones, D.P. 1973. Agricultural entomology. *Hist. Entomol.*, pp. 307-32

Jones, L.R. See Walker, J.C.

Jones, R.T. 1977. Recollections from an earlier period in American aeronautics. *Annu. Rev. Fluid Mech.* 9:1-11 [Reprinted in *Exc. Fas. Sci.* Vol. 3]

Kabat, E.A. 1983. Getting started 50 years ago—experiences, perspectives, and problems of the first 21 years. *Annu. Rev. Immunol.* 1:1-32 [Reprinted in *Exc. Fas. Sci.* Vol. 3]

Kamen, M.D. 1986. A cupful of luck, a pinch of sagacity. *Annu. Rev. Biochem.* 55:1-34 [Reprinted in *Exc. Fas. Sci.* Vol. 3]

Kato, G. 1970. The road a scientist followed: notes of Japanese physiology as I myself experienced it. *Annu. Rev. Physiol.* 32:1-20

Keitt, G.W. See Leben, C.

Kelman, A. 1985. Plant pathology at the crossroads. *Annu. Rev. Phytopathol.* 23:1-11

Kent, G.C. 1979. Important little-known contributors to plant pathology: Mason Blanchard Thomas. *Annu. Rev. Phytopathol.* 17:21-28

Kerling, L.C.P., de Bruin-Brank, G., Ten Houten, J.G. 1986. Johanna Westerdijk: pioneer leader in plant pathology. *Annu. Rev. Phytopathol.* 24:33-41

Kern, H. 1985. Ernst Gaumann, 1893-1963: pioneer leader in plant pathology. *Annu. Rev. Phytopathol.* 23:19-22

Kety, S.S. 1979. The metamorphosis of a psychobiologist. *Annu. Rev. Neurosci.* 2:1-15 [Reprinted in *Exc. Fas. Sci.* Vol. 3]

Kiraly, Z. 1972. Main trends in the development of plant pathology in Hungary. *Annu. Rev. Phytopathol.* 10:9-20

Kleiber, M. 1967. Prefatory chapter: an old professor of animal husbandry ruminates. *Annu. Rev. Physiol.* 29:1-20 [Reprinted in *Exc. Fas. Sci.* Vol. 2]

Konishi, M., Ito, Y. 1973. Early entomology in east Asia. *Hist. Entomol.*, pp. 1-20

Kosterlitz, H.W. 1979. The best laid schemes o'mice an' men gang aft agley. *Annu. Rev. Pharmacol. Toxicol.* 19:1-12 [Reprinted in *Exc. Fas. Sci.* Vol. 3]

Kramer, P.J. 1973. Some reflections after 40 years in plant physiology. *Annu. Rev. Plant Physiol.* 24:1-24

Krogman, W.M. 1976. Fifty years of physical anthropology: the men, the material, the concepts, the methods. *Annu. Rev. Anthropol.* 5:1-14

Kuehn, J. See Wilhelm, S.

Kumagai, H. 1978. Pharmacology and medicine. *Annu. Rev. Pharmacol. Toxicol.* 10:277-84 [Reprinted in *Exc. Fas. Sci.* Vol. 2]

Lang, A. 1980. Some recollections and reflections. *Annu. Rev. Plant Physiol.* 31:1-28 [Reprinted in *Exc. Fas. Sci.* Vol. 3]

Latreille, P.A. See Dupuis, C.

Leach, E.R. 1984. Glimpses of the unmentionable in the history of British social anthropology. *Annu. Rev. Anthropol.* 13:1-23 [Reprinted in *Exc. Fas. Sci.* Vol. 3]

Leake, C.D. 1976. How I am. *Annu. Rev. Pharmacol. Toxicol.* 16:1-14

Leben, C. 1981. Pioneer leaders in plant pathology: G.W. Keitt. *Annu. Rev. Phytopathol.* 19:35-40

Lederberg, J. 1979. Edward Lawrie Tatum. *Annu. Rev. Genet.* 13:1-5 [Reprinted in *Exc. Fas. Sci.* Vol. 3]

Lederberg, J. 1987. Genetic recombination in bacteria – a discovery account. *Annu. Rev. Genet.* 21:23-46 [Reprinted in *Exc. Fas. Sci.* Vol. 3]

Leloir, L.F. 1983. Far away and long ago. *Annu. Rev. Biochem.* 52:1-15 [Reprinted in *Exc. Fas. Sci.* Vol. 3]

Lemberg, R. 1965. Chemist, biochemist, and seeker in three countries. *Annu. Rev. Biochem.* 34:1-20 [Reprinted in *Exc. Fas. Sci.* Vol. 1]

Libby, W.F. 1964. Thirty years of atomic chemistry. *Annu. Rev. Phys. Chem.* 15:241-54 [Reprinted in *Exc. Fas. Sci.* Vol. 1]

Liljestrand, G. 1957. The increasing responsibility of the physiological sciences. *Annu. Rev. Physiol.* 19:1-12 [Reprinted in *Exc. Fas. Sci.* Vol. 1]

Lindroth, C.H. 1973. Systematics specializes between Fabricus and Darwin: 1800-1859. *Hist. Entomol.*, pp. 119-54

Linnaeus, C. See Usinger, R.L.

Lipmann, F. 1984. A long life in times of great upheaval. *Annu. Rev. Biochem.* 53:1-33 [Reprinted in *Exc. Fas. Sci.* Vol. 3]

Loewi, O. 1954. Reflections on the study of physiology. *Annu. Rev. Physiol.* 16:1-10 [Reprinted in *Exc. Fas. Sci.* Vol. 1]

Loitsianskii, L.G. 1970. The development of boundary-layer theory in the USSR. *Annu. Rev. Fluid Mech.* 2:1-14

Luck, J.M. 1981. Confessions of a biochemist. *Annu. Rev. Biochem.* 50:1-22 [Reprinted in *Exc. Fas. Sci.* Vol. 3]

Lwoff, A. 1971. From protozoa to bacteria and viruses: fifty years with microbes. *Annu. Rev. Microbiol.* 25:1-26 [Reprinted in *Exc. Fas. Sci.* Vol. 2]

MacArthur, R. See Fretwell, S.D.

MacGillivray, H. 1968. A personal biography of Arthur Robertson Cushny, 1866-1926. *Annu. Rev. Pharmacol. Toxicol.* 8:1-24 [Reprinted in *Exc. Fas. Sci.* Vol. 2]

Mach, E., See Reichenbach, H.

MacLeod, R.A. 1985. Marine microbiology far from the sea. *Annu. Rev. Microbiol.* 39:1-20 [Reprinted in *Exc. Fas. Sci.* Vol. 3]

Mann, F.C. 1955. To the physiologically inclined. *Annu. Rev. Physiol.* 17:1-16 [Reprinted in *Exc. Fas. Sci.* Vol. 1]

Maren, T.H. 1982. Great expectations. *Annu. Rev. Pharmacol. Toxicol.* 22:1-18 [Reprinted in *Exc. Fas. Sci.* Vol. 3]

Mark, H.F., Atlas S.M. 1976. Polymers as building materials. *Annu. Rev. Mater. Sci.* 6:1-32

Matthews, R.E.F. 1987. The changing scene in plant physiology. *Annu. Rev. Phytopathol.* 25:11-23

Mayer, J.E. 1982. The way it was. *Annu. Rev. Phys. Chem.* 33:1-23 [Reprinted in *Exc. Fas. Sci.* Vol. 3]

McCallan, S.E.A. 1969. A perspective on plant pathology. *Annu. Rev. Phytopathol.* 7:1-12

McCarty, M. 1980. Reminiscences of the early days of transformation. *Annu. Rev. Genet.* 14:1-15 [Reprinted in *Exc. Fas. Sci.* Vol. 3]

McCollum, E.V. 1953. My early experiences in the study of foods and nutrition. *Annu. Rev. Biochem.* 22:1-16 [Reprinted in *Exc. Fas. Sci.* Vol. 1]

McCrea, W.H. 1987. Clustering of astronomers. *Annu. Rev. Astron. Astrophys.* 25:1-22 [Reprinted in *Exc. Fas. Sci.* Vol. 3]

McElroy, W.D. 1976. From the precise to the ambiguous: light, bonding, and administration. *Annu. Rev. Microbiol.* 30:1-20 [Reprinted in *Exc. Fas. Sci.* Vol. 2]

McLean, F.C. 1960. Prefatory chapter, physiology and medicine: a transition period. *Annu. Rev. Physiol.* 22:1-16 [Reprinted in *Exc. Fas. Sci.* Vol. 1]

Mead, M. 1973. Changing styles of anthropological work. *Annu. Rev. Anthropol.* 2:1-26

Mehl, R.F. 1975. A department and a research laboratory in a university. *Annu. Rev. Mater. Sci.* 5:1-26

Mellanby, E. See Platt, B.S.

Merton, R.K. 1987. Three fragments from a sociologist's notebooks: establishing the phenomenon, specified ignorance, and strategic research materials. *Annu. Rev. Sociol.* 13:1-28 [Reprinted in *Exc. Fas. Sci.* Vol. 3]

Mickel, C. 1973. John Ray: indefatigable student of nature. *Annu. Rev. Entomol.* 18:1-16

Miller, N.E. 1983. Behavioral medicine: symbiosis between laboratory and clinic. *Annu. Rev. Psychol.* 34:1-31

Millsaps, K. 1984. Karl Pohlhausen, as I remember him. *Annu. Rev. Fluid Mech.* 16:1-10

Mitchell, H.K. 1978. Ernst Hadorn. *Annu. Rev. Genet.* 12:1-3 [Reprinted in *Exc. Fas. Sci.* Vol. 3]

Mizushima, S. 1972. A history of physical chemistry in Japan. *Annu. Rev. Phys. Chem.* 23:1-14

Morge, G. 1973. Entomology in the western world in antiquity and in medieval times. *Hist. Entomol.*, pp. 37-80

Mudd, S. 1969. Sequences in medical microbiology: some observations over fifty years. *Annu. Rev. Microbiol.* 23:1-28 [Reprinted in *Exc. Fas. Sci.* Vol. 2]

Muller, H.J. See Pontecorvo, G.

Mulliken, R.S. 1978. Chemical bonding. *Annu. Rev. Phys. Chem.* 29:1-30

Munk, M.M. 1981. My early aerodynamic research—thoughts and memories. *Annu. Rev. Fluid Mech.* 13:1-7 [Reprinted in *Exc. Fas. Sci.* Vol. 3]

Munk, W.H. 1980. Affairs of the sea. *Annu. Rev. Earth Planet. Sci.* 8:1-16 [Reprinted in *Exc. Fas. Sci.* Vol. 3]

Munro, H.N. 1986. Back to basics: an evolutionary odyssey with reflections on the nutrition research of tomorrow. *Annu. Rev. Nutr.* 6:1-12

Muskett, A.E. 1967. Plant pathology and the plant pathologist. *Annu. Rev. Phytopathol.* 5:1-16

Nachmansohn, D. 1972. Biochemistry as part of my life. *Annu. Rev. Biochem.* 41:1-28 [Reprinted in *Exc. Fas. Sci.* Vol. 2]

Nanney, D.L. 1981. T.M. Sonneborn: an interpretation. *Annu. Rev. Genet.* 15:1-9 [Reprinted in *Exc. Fas. Sci.* Vol. 3]

Neel, J.V. 1983. Curt Stern, 1902-1981. *Annu. Rev. Genet.* 17:1-10 [Reprinted in *Exc. Fas. Sci.* Vol. 3]

Neergaard, P. 1986. Screening for plant health. *Annu. Rev. Phytopathol.* 24:1-16

Nelson, R.R. 1984. Pioneer leaders in plant pathology: E.C. Stakman. *Annu. Rev. Phytopathol.* 22:11-19

Newhall, A.G. 1980. Herbert Hice Whetzel: pioneer American plant pathologist. *Annu. Rev. Phytopathol.* 18:27-36

Nier, A.O. 1981. Some reminiscences of isotopes, geochronology, and mass spectrometry. *Annu. Rev. Earth Planet. Sci.* 9:1-17 [Reprinted in *Exc. Fas. Sci.* Vol. 3]

Nolla, J.A.B. 1976. Contributions to the history of plant pathology in South America, Central America, and Mexico. *Annu. Rev. Phytopathol.* 14:11-29

Norrish, R.G.W. 1969. Fifty years of physical chemistry in Great Britain. *Annu. Rev. Phys. Chem.* 20:1-24 [Reprinted in *Exc. Fas. Sci.* Vol. 2]

Northrop, J.H. 1961. Biochemists, biologists, and William of Occam. *Annu. Rev. Biochem.* 30:1-10 [Reprinted in *Exc. Fas. Sci.* Vol. 1]

Occam, William of, See Northrop, J.H.

Ochoa, S. 1980. The pursuit of a hobby. *Annu. Rev. Biochem.* 49:1-30 [Reprinted in *Exc. Fas. Sci.* Vol. 3]

Olson, G.B., Cohen, M. 1981. A perspective on martensitic nucleation. *Annu. Rev. Mater. Sci.* 11:1-30

Oort, J.H. 1981. Some notes on my life as an astronomer. *Annu. Rev. Astron. Astrophys.* 19:1-5

Öpik, E.J. 1977. About dogma in science, and other recollections of an astronomer. *Annu. Rev. Astron. Astrophys.* 15:1-17 [Reprinted in *Exc. Fas. Sci.* Vol. 2]

Orlob, G.B. 1971. History of plant pathology in the Middle Ages. *Annu. Rev. Phytopathol.* 9:7-20

Osterhout, W.J.V. 1957. The use of aquatic plants in the study of some fundamental problems. *Annu. Rev. Plant Physiol.* 8:1-10

Oswatitsch, K. 1987. Ludwig Prandtl and his Kaiser-Wilhelm-Institut. *Annu. Rev. Fluid Mech.* 19:1-25

Ou, S.H. 1984. Exploring tropical rice diseases: a reminiscence. *Annu. Rev. Phytopathol.* 22:1-10 [Reprinted in *Exc. Fas. Sci.* Vol. 3]

Pappenheimer, J.R. 1987. A silver spoon. *Annu. Rev. Physiol.* 49:1-15 [Reprinted in *Exc. Fas. Sci.* Vol. 3]

Paton, W.D.M. 1986. On becoming and being a pharmacologist. *Annu. Rev. Pharmacol. Toxicol.* 26:1-22 [Reprinted in *Exc. Fas. Sci.* Vol. 3]

Pauling, L. 1986. Early days of molecular biology in the California Institute of Technology. *Annu. Rev. Biophys. Biophys. Chem.* 15:1-9 [Reprinted in *Exc. Fas. Sci.* Vol. 3]

Pauling, L.C. 1965. Fifty years of physical chemistry in the California Institute of Technology. *Annu. Rev. Phys. Chem.* 16:1-14 [Reprinted in *Exc. Fas. Sci.* Vol. 1]

Payne-Gaposchkin, C. 1978. The development of our knowledge of variable stars. *Annu. Rev. Astron. Astrophys.* 16:1-13

Peters, R. 1957. Forty-five years of biochemistry. *Annu. Rev. Biochem.* 26:1-16 [Reprinted in *Exc. Fas. Sci.* Vol. 1]

Pfeffer, W. See Bunning, E.

Phaff, H.J. 1986. My life with yeasts. *Annu. Rev. Microbiol.* 40:1-28 [Reprinted in *Exc. Fas. Sci.* Vol. 3]

Philip, C.B., Rozeboom, L.E. 1973. Medico-veterinary entomology: a generation of progress. *Hist. Entomol.*, pp. 333-60

Piaget, J. 1979. Relations between psychology and other sciences. *Annu. Rev. Psychol.* 30:1-8

Pitts, R.F. 1976. Why a physiologist? *Annu. Rev. Physiol.* 38:1-6

Pitzer, K.S. 1987. Of physical chemistry and other activities. *Annu. Rev. Phys. Chem.* 38:1-25 [Reprinted in *Exc. Fas. Sci.* Vol. 3]

Platt, B.S. 1956. Sir Edward Mellanby, G.B.E., K.C.B., M.D., F.R.C.P., F.R.S. (1884-1955) the man, research worker, and statesman. *Annu. Rev. Biochem.* 25:1-28 [Reprinted in *Exc. Fas. Sci.* Vol. 1]

Pohlhausen, K. See Millsaps, K.

Pontecorvo, G. 1968. Hermann Joseph Muller. *Annu. Rev. Genet.* 2:1-10

Posnette, A.F. 1980. Recollections of a genetical plant pathologist. *Annu. Rev. Phytopathol.* 18:1-9 [Reprinted in *Exc. Fas. Sci.* Vol. 3]

Prandtl, L. See Oswatitsch, K.; see Flügge-Lotz, I.

Prosser, C.L. 1986. The making of a comparative physiologist. *Annu. Rev. Physiol.* 48:1-6 [Reprinted in *Exc. Fas. Sci.* Vol. 3]

Raffel, S. 1973. Windsor Cooper Cutting (1907-1972). *Annu. Rev. Pharmacol. Toxicol.* 13:1-4

Raffel, S. 1982. Fifty years of immunology. *Annu. Rev. Microbiol.* 36:1-26 [Reprinted in *Exc. Fas. Sci.* Vol. 3]

Ratner, S. 1977. A long view of nitrogen metabolism. *Annu. Rev. Biochem.* 46:1-24 [Reprinted in *Exc. Fas. Sci.* Vol. 3]

Ray, J. See Mickel, C.

Raychaudhuri, S.P., Verma, J.P., Nariani, T.K., Beneeta, S. 1972. The history of plant pathology in India. *Annu. Rev. Phytopathol.* 10:21-36

Reichenbach, H. 1983. Contributions of Ernst Mach to fluid mechanics. *Annu. Rev. Fluid Mech.* 15:1-28

Remington, J.E., Remington, C.L. 1961. Darwin's contributions to entomology. *Annu. Rev. Entomol.* 6:1-12

Revelle, R. 1987. How I became an oceanographer and other sea stories. *Annu. Rev. Earth Planet. Sci.* 15:1-23 [Reprinted in *Exc. Fas. Sci.* Vol. 3]

Rheingold, H.L. 1985. Development as the acquisition of familiarity. *Annu. Rev. Psychol.* 36:1-17

Rhoades, M.M. 1984. The early years of maize genetics. *Annu. Rev. Genet.* 18:1-29 [Reprinted in *Exc. Fas. Sci.* Vol. 3]

Richard, G. 1973. The historical development of nineteenth and twentieth century studies on the behavior of insects. *Hist. Entomol.*, pp. 477-502

Richards, A.G. 1973. Anatomy and morphology. *Hist. Entomol.*, pp. 185-202

Rodgers, J. 1985. Witnessing revolutions in the earth sciences. *Annu. Rev. Earth Planet. Sci.* 13:1-4 [Reprinted in *Exc. Fas. Sci.* Vol. 3]

Rohdendorf, B.B. 1973. The history of paleoentomology. *Hist. Entomol.*, pp. 155-70

Roman, H. 1986. The early days of yeast genetics: a personal narrative. *Annu. Rev. Genet.* 20:1-12 [Reprinted in *Exc. Fas. Sci.* Vol. 3]

Ross, H.H. 1973. Evolution and phylogeny. *Hist. Entomol.*, pp. 171-84

Rothlin, E. 1964. Outlines of a pharmacological career. *Annu. Rev. Pharmacol. Toxicol.* 4:9-32

Rott, N. 1985. Jakob Ackeret and the history of the Mach number. *Annu. Rev. Fluid Mech.* 17:1-9

Rouse, H. 1976. Hydraulics' latest golden age. *Annu. Rev. Fluid Mech.* 8:1-12

Rubey, W.W. 1974. Fifty years of the earth sciences—a renaissance. *Annu. Rev. Earth Planet. Sci.* 2:1-24

Russell, E.J. 1985. A history of mouse genetics. *Annu. Rev. Genet.* 19:1-28 [Reprinted in *Exc. Fas. Sci.* Vol. 3]

Russell, L.M. 1978. Leland Ossian Howard: a historical review. *Annu. Rev. Entomol.* 23:1-15

Sacken, Baron O. See Alexander, C.P.

Schafer, H. 1985. On the problem of polar intermetallic compounds: the stimulation of E. Zintl's work for the modern chemistry of intermetallics. *Annu. Rev. Mater. Sci.* 15:1-41

Scharrer, B. 1987. Neurosecretion: beginnings and new directions in neuropeptide research. *Annu. Rev. Neurosci.* 10:1-17 [Reprinted in *Exc. Fas. Sci.* Vol. 3]

Schmidt, C.F. 1965. Pharmacology in a changing world. *Annu. Rev. Physiol.* 23:1-14 [Reprinted in *Exc. Fas. Sci.* Vol. 1]

Schmitt, F.O. 1985. Adventures in molecular biology. *Annu. Rev. Biophys. Biophys. Chem.* 14:1-22 [Reprinted in *Exc. Fas. Sci.* Vol. 3]

Scholander, P.F. 1978. Rhapsody in science. *Annu. Rev. Physiol.* 40:1-17 [Reprinted in *Exc. Fas. Sci.* Vol. 2]

Schwerdtfeger, F. 1973. Forest entomology. *Hist. Entomol.*, pp. 361-86

Scrimshaw, N.S. 1987. The phenomenon of famine. *Annu. Rev. Nutr.* 7:1-21

Sears, W.R., Sears, M.R. 1979. The Kármán years at Galcit. *Annu. Rev. Fluid Mech.* 11:1-10 [Reprinted in *Exc. Fas. Sci.* Vol. 3]

Segrè, E. 1981. Fifty years up and down a strenuous and scenic trail. *Annu. Rev. Nucl. Part. Sci.* 31:1-18 [Reprinted in *Exc. Fas. Sci.* Vol. 1]

Sela, M. 1987. A peripatetic and personal view of molecular immunology for one third of a century. *Annu. Rev. Immunol.* 5:1-19 [Reprinted in *Exc. Fas. Sci.* Vol. 3]

Simpson, G.G. 1976. The compleat palaeontologist? *Annu. Rev. Earth Planet. Sci.* 4:1-13

Sloss, L.L. 1984. The greening of stratigraphy 1933-1983. *Annu. Rev. Earth Planet. Sci.* 12:1-10 [Reprinted in *Exc. Fas. Sci.* Vol. 3]

Smith, C.S. 1986. On material structure and human history. *Annu. Rev. Mater. Sci.* 16:1-11

Smith, E.F. See Campbell, C.L.

Smith, E.H. 1976. The Comstocks and Cornell: in the people's service. *Annu. Rev. Entomol.* 21:1-25

Snyder, W.C. 1971. Plant pathology today. *Annu. Rev. Phytopathol.* 9:1-6

Snyder, W.C. See Toussoun, T.A.

Sollman, T. 1965. Why an annual review of pharmacology?. *Annu. Rev. Pharmacol. Toxicol.* 1:453-60 [Reprinted in *Exc. Fas. Sci.* Vol. 1]

Sonneborn, T.M. See Nanney, D.L.

Sperry, R.W. 1981. Changing priorities. *Annu. Rev. Neurosci.* 4:1-15

Sproull, R.L. 1987. The early history of the materials research laboratories. *Annu. Rev. Mater. Sci.* 17:1-12

Stakman, E.C. 1964. Opportunity and obligation in plant pathology. *Annu. Rev. Phytopathol.* 2:1-12

Stakman, E.C. See Nelson, R.R.

Stanier, R.Y. 1980. The journey, not the arrival, matters. *Annu. Rev. Microbiol.* 34:1-48 [Reprinted in *Exc. Fas. Sci.* Vol. 3]

Stern, C. See Neel, J.V.

Steward, F.C. 1971. Plant physiology: the changing problems, the continuing quest. *Annu. Rev. Plant Physiol.* 22:1-22

Stewartson, K. See Stuart, J.T.

Stocking, C.R. 1984. Reminiscences and reflections. *Annu. Rev. Plant Physiol.* 35:1-14 [Reprinted in *Exc. Fas. Sci.* Vol. 3]

Stockmayer, W.H., Zimm, B.H. 1984. When polymer science looked easy. *Annu. Rev. Phys. Chem.* 35:1-21 [Reprinted in *Exc. Fas. Sci.* Vol. 3]

Stout, J.W. 1986. *The Journal of Chemical Physics*: the first 50 years. *Annu. Rev. Phys. Chem.* 37:1-23 [Reprinted in *Exc. Fas. Sci.* Vol. 3]

Strömgren, B. 1983. Scientists I have known and some astronomical problems I have met. *Annu. Rev. Astron. Astrophys.* 21:1-11

Stuart, J.T. 1986. Keith Stewartson: his life and work. *Annu. Rev. Fluid Mech.* 18:1-14

Stumpf, S.E. 1981. The moral dimensions of the world's food supply. *Annu. Rev. Nutr.* 1:1-25

Sturtevant, A.H. See Emerson, S.

Sweeney, B.M. 1987. Living in the golden age of biology. *Annu. Rev. Plant Physiol.* 38:1-9 [Reprinted in *Exc. Fas. Sci.* Vol. 3]

Swings, P. 1979. A few notes on my career as an astrophysicist. *Annu. Rev. Astron. Astrophys.* 17:1-7

Szentágothai, J. 1984. Downward causation? *Annu. Rev. Neurosci.* 7:1-11

Szent-Györgyi, A. 1959. Lost in the twentieth century. *Annu. Rev. Biochem.* 32:1-14 [Reprinted in *Exc. Fas. Sci.* Vol. 1]

Talmage, D.W. 1986. The acceptance and rejection of immunological concepts. *Annu. Rev. Immunol.* 4:1-11 [Reprinted in *Exc. Fas. Sci.* Vol. 3]

Tamiya, H. 1966. Synchronous cultures of algae. *Annu. Rev. Plant Physiol.* 17:1-26

Tang, P. 1983. Aspirations, reality, and circumstances: the devious trail of a roaming plant physiologist. *Annu. Rev. Plant Physiol.* 34:1-19 [Reprinted in *Exc. Fas. Sci.* Vol. 3]

Tani, I. 1977. History of boundary-layer theory. *Annu. Rev. Fluid Mech.* 9:87-111

Tatum, E.L. See Lederberg, J.

Taylor, G.I. 1974. The interaction between experiment and theory in fluid mechanics. *Annu. Rev. Fluid Mech.* 6:1-16

Taylor, H. 1962. Fifty years of chemical kineticists. *Annu. Rev. Phys. Chem.* 13:1-18 [Reprinted in *Exc. Fas. Sci.* Vol. 1]

Ten Houten, J.G. 1974. Plant pathology: changing agricultural methods and human society. *Annu. Rev. Phytopathol.* 12:1-11

Terroine, E.F. 1959. Fifty-five years of union between biochemistry and physiology. *Annu. Rev. Biochem.* 28:1-14 [Reprinted in *Exc. Fas. Sci.* Vol. 1]

Thaxter, R. See Horsfall, J.G.

Thimann, K.V. 1963. Plant growth substances: past, present and future. *Annu. Rev. Plant Physiol.* 14:1-18

Thomas, K. 1954. Fifty years of biochemistry in Germany. *Annu. Rev. Biochem.* 23:1-16 [Reprinted in *Exc. Fas. Sci.* Vol. 1]

Thomas, M.B. See Kent, G.C.

Thorne, G. See Barker, K.R. et al

Tiselius, A. 1968. Reflections from both sides of the counter. *Annu. Rev. Biochem.* 37:1-24 [Reprinted in *Exc. Fas. Sci.* Vol. 2]

Tomiyama, K. 1983. Research on the hypersensitive response. *Annu. Rev. Phytopathol.* 21:1-12 [Reprinted in *Exc. Fas. Sci.* Vol. 3]

Toussoun, T.A. 1986. William C. Snyder: pioneer leader in plant pathology. *Annu. Rev. Phytopathol.* 24:27-31

Townsend, G.F., Crane, E. 1973. History of agriculture. *Hist. Entomol.*, pp. 387-406

Truhlar, D.G., Wyatt, R.E. 1976. History of H_3 kinetics. *Annu. Rev. Phys. Chem.* 27:1-43

Turnbull, D. 1983. A commentary on the emergence and evolution of "materials science." *Annu. Rev. Mater. Sci.* 13:1-7

Tuxen, S.L. 1967. The entomologist, J.C. Fabricius. *Annu. Rev. Entomol.* 12:1-14

Tuxen, S.L. 1973. Entomology systematizes and describes: 1700-1815. *Hist. Entomol.*, pp. 95-118

Tyler, L.E. 1981. More stately mansions – psychology extends its boundaries. *Annu. Rev. Psychol.* 32:1-20

Uhlenbeck, G.E. 1979. Some notes on the relation between fluid mechanics and statistical physics. *Annu. Rev. Fluid Mech.* 12:1-9 [Reprinted in *Exc. Fas. Sci.* Vol. 3]

Ulrich, W. 1972. Hermann Burmeister, 1807 to 1892. *Annu. Rev. Entomol.* 17:1-20

Usinger, R.L. 1964. The role of Linnaeus in the advancement of entomology. *Annu. Rev. Entomol.* 9:1-16

Ussing, H.H. 1980. Life with tracers. *Annu. Rev. Physiol.* 42:1-16 [Reprinted in *Exc. Fas. Sci.* Vol. 3]

Uvnäs, B. 1984. From physiologist to pharmacologist—promotion or degradation? Fifty years in retrospect. *Annu. Rev. Pharmacol. Toxicol.* 24:1-18 [Reprinted in *Exc. Fas. Sci.* Vol. 3]

Vanderplank, J.E. 1976. Four essays. *Annu. Rev. Phytopathol.* 14:1-10

van Niel, C.B. 1962. The present status of the comparative study of photosynthesis. *Annu. Rev. Plant Physiol.* 13:1-26

van Niel, C.B. 1967. The education of a microbiologist: some reflections. *Annu. Rev. Microbiol.* 21:1-30 [Reprinted in *Exc. Fas. Sci.* Vol. 2]

van Overbeek, J. 1976. Plant physiology and the human ecosystem. *Annu. Rev. Plant Physiol.* 27:1-17

Vennesland, B. 1981. Recollections and small confessions. *Annu. Rev. Plant Physiol.* 32:1-20 [Reprinted in *Exc. Fas. Sci.* Vol. 3]

Verhoogen, J. 1983. Personal notes and sundry comments. *Annu. Rev. Earth Planet. Sci.* 11:1-9 [Reprinted in *Exc. Fas. Sci.* Vol. 3]

Vickery, H.B. 1972. A chemist among plants. *Annu. Rev. Plant Physiol.* 23:1-28

Villat, H. 1972. As luck would have it—a few mathematical reflections. *Annu. Rev. Fluid Mech.* 4:1-65

Virtanen, A.I. 1961. Some aspects of amino acid synthesis in plants and related subjects. *Annu. Rev. Plant Physiol.* 12:1-12

Visscher, M.B. 1978. A half century in science and society. *Annu. Rev. Physiol.* 31:1-18 [Reprinted in *Exc. Fas. Sci.* Vol. 2]

von Békésy, G. 1978. Some biographical experiments from fifty years ago. *Annu. Rev. Physiol.* 36:1-16 [Reprinted in *Exc. Fas. Sci.* Vol. 2]

von Euler, U.S. 1978. Pieces in the puzzle. *Annu. Rev. Pharmacol. Toxicol.* 11:675-86 [Reprinted in *Exc. Fas. Sci.* Vol. 2]

von Kármán, T. See Sears, W.R.

von Muralt, A. 1984. A life with several facets. *Annu. Rev. Physiol.* 46:1-13 [Reprinted in *Exc. Fas. Sci.* Vol. 3]

Wagner, C. 1977. Point defects and their interaction. *Annu. Rev. Mater. Sci.* 7:1-22

Walker, J.C. 1963. The future of plant pathology. *Annu. Rev. Phytopathol.* 1:1-4

Walker, J.C. 1975. Some highlights in plant pathology in the United States. *Annu. Rev. Phytopathol.* 13:15-29

Walker, J.C. 1979. Leaders in plant pathology: L.R. Jones. *Annu. Rev. Phytopathol.* 17:13-20

Walker, J.C. 1982. Pioneer leaders in plant pathology: Benjamin Minge Duggar. *Annu. Rev. Phytopathol.* 20:33-39

Wallach, H. 1987. Perceiving a stable environment when one moves. *Annu. Rev. Psychol.* 38:1-27

Warburg, O. 1964. Prefatory chapter. *Annu. Rev. Biochem.* 33:1-14 [Reprinted in *Exc. Fas. Sci.* Vol. 1]

Wareing, P.F. 1982. A plant physiological odyssey. *Annu. Rev. Plant Physiol.* 33:1-26 [Reprinted in *Exc. Fas. Sci.* Vol. 3]

Washburn, S.L. 1983. Evolution of a teacher. *Annu. Rev. Anthropol.* 12:1-24 [Reprinted in *Exc. Fas. Sci.* Vol. 3]

Welch, A.D. 1985. Reminiscences in pharmacology: auld acquaintance ne'er forgot. *Annu. Rev. Pharmacol. Toxicol.* 25:1-26 [Reprinted in *Exc. Fas. Sci.* Vol. 3]

Welker, H. 1979. From solid state research to semiconductor electronics. *Annu. Rev. Mater. Sci.* 9:1-21

Went, F.W. 1974. Reflections and speculations. *Annu. Rev. Plant Physiol.* 25:1-26

Westcott, C. See Horst, R.K.

Westerdijk, J. See Kerling, L.C.P.

Whetzel, H.H. See Newhall, A.G.

Whipple, F.L. 1978. The earth as part of the universe. *Annu. Rev. Earth Planet. Sci.* 6:1-8 [Reprinted in *Exc. Fas. Sci.* Vol. 3]

Whitford, A.E. 1986. A half-century of astronomy. *Annu. Rev. Astron. Astrophys.* 24:1-22 [Reprinted in *Exc. Fas. Sci.* Vol. 3]

Wiggers, C. 1951. Prefatory chapter: physiology from 1900-1920: incidents, accidents, and advances. *Annu. Rev. Physiol.* 13:1-20 [Reprinted in *Exc. Fas. Sci.* Vol. 1]

Wigglesworth, V.B. 1973. The history of insect physiology. *Hist. Entomol.*, pp. 203-28

Wilhelm, S., Tietz, H. 1978. Julius Kuehn—his concept of plant pathology. *Annu. Rev. Phytopathol.* 16:343-58

Williams, R.C. 1978. Spectroscopes, telescopes, microscopes. *Annu. Rev. Microbiol.* 32:1-18 [Reprinted in *Exc. Fas. Sci.* Vol. 3]

Wilson, E.B. 1979. Molecular spectroscopy. *Annu. Rev. Phys. Chem.* 30:1-27

Wilson, E.B., Ross, J. 1973. Physical chemistry in Cambridge, Massachusetts. *Annu. Rev. Phys. Chem.* 24:1-27

Wilson, J.T. 1982. Early days in university geophysics. *Annu. Rev. Earth Planet. Sci.* 10:1-14 [Reprinted in *Exc. Fas. Sci.* Vol. 3]

Wilson, M.K. 1975. The top twenty and the rest: big chemistry and little funding. *Annu. Rev. Phys. Chem.* 26:1-16

Wolman, A. 1986. Is there a public health function? *Annu. Rev. Publ. Health* 7:1-12

Wood, H. 1985. Then and now. *Annu. Rev. Biochem.* 54:1-41 [Reprinted in *Exc. Fas. Sci.* Vol. 3]

Wood, R.K.S. 1987. Physiological plant pathology comes of age. *Annu. Rev. Phytopathol.* 25:27-40

Woodruff, H.B. 1981. A soil microbiologist's odyssey. *Annu. Rev. Microbiol.* 35:1-28 [Reprinted in *Exc. Fas. Sci.* Vol. 3]

Wright, S. 1982. The shifting balance theory and macroevolution. *Annu. Rev. Genet.* 16:1-19

Yokoyama, T. 1973. The history of agricultural science in relation to industry. *Hist. Entomol.*, pp. 267-84

Zintl, E. See Schafer, H.

ANNUAL REVIEWS INC.

NONPROFIT SCIENTIFIC PUBLISHER

4139 El Camino Way
P.O. Box 10139
Palo Alto, CA 94303-0897 • USA

ORDER FORM

ORDER TOLL FREE
1-800-523-8635
(except California)

Annual Reviews Inc. publications may be ordered directly from our office; through **booksellers and subscription agents, worldwide; and through participating professional societies. Prices subject to change without notice.** ARI Federal I.D. #94-1156476

Individuals: Prepayment required on new accounts by check or money order (in U.S. dollars, check drawn on U.S. bank) or charge to credit card—American Express, VISA, MasterCard.
Institutional buyers: Please include purchase order.
Students: $10.00 discount from retail price, per volume. Prepayment required. Proof of student status must be provided (photocopy of student I.D. or signature of department secretary is acceptable). Students must send orders direct to Annual Reviews. Orders received through bookstores and institutions requesting student rates will be returned. You may order at the Student Rate for a maximum of 3 years.
Professional Society Members: Members of professional societies that have a contractual arrangement with Annual Reviews may order books through their society at a reduced rate. Check with your society for information.
Toll Free Telephone orders: Call 1-800-523-8635 (except from California) for orders paid by credit card or purchase order and customer service calls only. California customers and all other business calls use 415-493-4400 (not toll free). Hours: 8:00 AM to 4:00 PM, Monday-Friday, Pacific Time. **Written confirmation** is required on purchase orders from universities before shipment.
FAX: 415-855-9815 Telex: 910-290-0275

Regular orders: Please list below the volumes you wish to order by volume number.
Standing orders: New volume in the series will be sent to you automatically each year upon publication. Cancellation may be made at any time. Please indicate volume number to begin standing order.
Prepublication orders: Volumes not yet published will be shipped in month and year indicated.
California orders: Add applicable sales tax.
Postage paid (4th class bookrate/surface mail) **by Annual Reviews Inc.** Airmail postage or UPS, extra.

ANNUAL REVIEWS SERIES		Prices Postpaid per volume USA & Canada/elsewhere	Regular Order Please send:	Standing Order Begin with:
			Vol. number	Vol. number
Annual Review of **ANTHROPOLOGY**				
Vols. 1-16	(1972-1987)	$31.00/$35.00		
Vols. 17-18	(1988-1989)	$35.00/$39.00		
Vol. 19	(avail. Oct. 1990)	$39.00/$43.00	Vol(s). _____	Vol. _____
Annual Review of **ASTRONOMY AND ASTROPHYSICS**				
Vols. 1, 4-14, 16-20	(1963, 1966-1976, 1978-1982)	$31.00/$35.00		
Vols. 21-27	(1983-1989)	$47.00/$51.00		
Vol. 28	(avail. Sept. 1990)	$51.00/$55.00	Vol(s). _____	Vol. _____
Annual Review of **BIOCHEMISTRY**				
Vols. 30-34, 36-56	(1961-1965, 1967-1987)	$33.00/$37.00		
Vols. 57-58	(1988-1989)	$35.00/$39.00		
Vol. 59	(avail. July 1990)	$39.00/$44.00	Vol(s). _____	Vol. _____
Annual Review of **BIOPHYSICS AND BIOPHYSICAL CHEMISTRY**				
Vols. 1-11	(1972-1982)	$31.00/$35.00		
Vols. 12-18	(1983-1989)	$49.00/$53.00		
Vol. 19	(avail. June 1990)	$53.00/$57.00	Vol(s). _____	Vol. _____
Annual Review of **CELL BIOLOGY**				
Vols. 1-3	(1985-1987)	$31.00/$35.00		
Vols. 4-5	(1988-1989)	$35.00/$39.00		
Vol. 6	(avail. Nov. 1990)	$39.00/$43.00	Vol(s). _____	Vol. _____

ANNUAL REVIEWS SERIES	Prices Postpaid per volume USA & Canada/elsewhere	Regular Order Please send:	Standing Orde Begin with:
		Vol. number	Vol. number

Annual Review of COMPUTER SCIENCE

Vols. 1-2	(1986-1987)................$39.00/$43.00		
Vols. 3-4	(1988, 1989-1990)...........$45.00/$49.00	Vol(s). _____	Vol. _____

Annual Review of EARTH AND PLANETARY SCIENCES

Vols. 1-10	(1973-1982)................$31.00/$35.00		
Vols. 11-17	(1983-1989)................$49.00/$53.00		
Vol. 18	(avail. May 1990)...........$53.00/$57.00	Vol(s). _____	Vol. _____

Annual Review of ECOLOGY AND SYSTEMATICS

Vols. 2-18	(1971-1987)................$31.00/$35.00		
Vols. 19-20	(1988-1989)................$34.00/$38.00		
Vol. 21	(avail. Nov. 1990)..........$38.00/$42.00	Vol(s). _____	Vol. _____

Annual Review of ENERGY

Vols. 1-7	(1976-1982)................$31.00/$35.00		
Vols. 8-14	(1983-1989)................$58.00/$62.00		
Vol. 15	(avail. Oct. 1990)..........$62.00/$66.00	Vol(s). _____	Vol. _____

Annual Review of ENTOMOLOGY

Vols. 10-16, 18	(1965-1971, 1973)		
20-32	(1975-1987)................$31.00/$35.00		
Vols. 33-34	(1988-1989)................$34.00/$38.00		
Vol. 35	(avail. Jan. 1990)..........$38.00/$42.00	Vol(s). _____	Vol. _____

Annual Review of FLUID MECHANICS

Vols. 2-4, 7-19	(1970-1972, 1975-1987).......$32.00/$36.00		
Vols. 20-21	(1988-1989)................$34.00/$38.00		
Vol. 22	(avail. Jan. 1990)..........$38.00/$42.00	Vol(s). _____	Vol. _____

Annual Review of GENETICS

Vols. 1-21	(1967-1987)................$31.00/$35.00		
Vols. 22-23	(1988-1989)................$34.00/$38.00		
Vol. 24	(avail. Dec. 1990)..........$38.00/$42.00	Vol(s). _____	Vol. _____

Annual Review of IMMUNOLOGY

Vols. 1-5	(1983-1987)................$31.00/$35.00		
Vols. 6-7	(1988-1989)................$34.00/$38.00		
Vol. 8	(avail. April 1990).........$38.00/$42.00	Vol(s). _____	Vol. _____

Annual Review of MATERIALS SCIENCE

Vols. 1, 3-12	(1971, 1973-1982)...........$31.00/$35.00		
Vols. 13-19	(1983-1989)................$66.00/$70.00		
Vol. 20	(avail. Aug. 1990)..........$70.00/$74.00	Vol(s). _____	Vol. _____

Annual Review of MEDICINE

Vols. 9, 11-15	(1958, 1960-1964)		
17-38	(1966-1987)................$31.00/$35.00		
Vols. 39-40	(1988-1989)................$34.00/$38.00		
Vol. 41	(avail. April 1990).........$38.00/$42.00	Vol(s). _____	Vol. _____